Microbial Source Tracking: Methods, Applications, and Case Studies

Charles Hagedorn · Anicet R. Blanch
Valerie J. Harwood
Editors

Microbial Source Tracking: Methods, Applications, and Case Studies

Springer

Editors
Charles Hagedorn
Department of Crop and Soil
Environmental Sciences
Virginia Polytechnic Institute and State
University
Blacksburg, VA 24061, USA
chagedor@vt.edu

Valerie J. Harwood
Department of Integrative Biology
University of South Florida
Tampa, FL 33620, USA
vharwood@usf.edu

Anicet R. Blanch
Department of Microbiology
University of Barcelona
Barcelona, Spain
ablanch@ub.edu

ISBN 978-1-4419-9385-4 e-ISBN 978-1-4419-9386-1
DOI 10.1007/978-1-4419-9386-1
Springer New York Dordrecht Heidelberg London

Library of Congress Control Number: 2011928239

Printed on acid-free paper

Springer is part of Springer Science+Business Media (www.springer.com)

Contents

Contributors

Warish Ahmed CSIRO Land and Water, Ecosciences Precinct, 41 Boggo Road, Brisbane 4102, Australia
warish.ahmed@csiro.au

Abidemi O. Ajidahun Center for Environmental Diagnostics and Bioremediation, University of West Florida, 11000 University Parkway, Pensacola, FL 32514, USA

Susan Allender-Hagedorn Department of English, 207 Shanks Hall, Virginia Tech, Blacksburg, VA 24061-0112, USA
hagedors@vt.edu

Maria Alvarez Department of Biology, El Paso Community College, El Paso, TX, USA
mariaa@epcc.edu

Gary L. Andersen Lawrence Berkeley National Laboratory, One Cyclotron Road, Mail Stop 70A-3317, Berkeley, CA 94720, USA

Karina M. Barrella Texas AgriLife Research Center at El Paso, Texas A&M University System, El Paso, TX 79927, USA

Lluís A. Belanche Department of Software, School of Informatics, Technical University of Catalonia, Jordi Girona 1-3, Barcelona, Spain

Brian Benham Biological Systems Engineering, Virginia Tech, Blacksburg, VA, USA
benham@vt.edu

Anicet R. Blanch Department of Microbiology, University of Barcelona, Avda. Diagonal 645, Barcelona, Spain
ablanch@ub.edu

Günter Blöschl Centre for Water Resource Systems (CWRS), Vienna University of Technology, Karlsplatz 13/222, 1040 Vienna, Austria

Alexandria B. Boehm Department of Civil and Environmental Engineering, Stanford University, Stanford, CA, USA
and
University of Hawaii Center for Oceans and Human Health, Honolulu, HI, USA
aboehm@stanford.edu

Jane Caldwell USDA/ARS Food Science Research Unit, Department of Food, Bioprocessing, & Nutrition Sciences, NC State University, 323 Schaub Hall, Box 7624, Raleigh, NC 27695-7624, USA
Jane.caldwell@ars.usda.gov

Yiping Cao Southern California Coastal Water Research Project, 3535 Harbor Blvd, Suite 110, Costa Mesa, CA 92626, USA
yipingc@sccwrp.org

Elizabeth A. Casarez Texas AgriLife Research Center at El Paso, Texas A&M University System, El Paso, TX 79927, USA

J. Brooks Crozier Department of Biology, Roanoke College, Salem, VA, USA
crozier@roanoke.edu

George D. Di Giovanni Texas AgriLife Research Center at El Paso, Texas A&M University System, El Paso, TX 79927, USA
gdigiovanni@ag.tamu.edu

Theo Dillaha Biological Systems Engineering, Virginia Tech, Blacksburg, VA, USA

Joe Eugene Lepo Center for Environmental Diagnostics and Bioremediation, University of West Florida, 11000 University Parkway, Pensacola, FL 32514, USA
jlepo@uwf.edu

Andreas H. Farnleitner Institute of Chemical Engineering, Research Area Applied Biochemistry and Gene Technology, Research Group Environmental Microbiology and Molecular Ecology, Vienna University of Technology, Gumpendorferstraße 1a, 166/5-2, A-1060 Vienna Austria
and
InterUniversitary Cooperation Centre for Water and Health (ICC Water & Health), Vienna University of Technology, Gumpendorferstraße 1a, 166/5-2, A-1060 Vienna Austria
a.farnleitner@aon.at

Donna Ferguson Southern California Coastal Water Research Project, 3535 Harbor Blvd., Suite 110, Costa Mesa, CA 92626, USA
donnaf@sccwrp.org

Michele M. Garber Hazardous Substance & Waste Management Research, Tallahassee, FL 32309, USA
mgarber@hswmr.com

Brent Gilpin Environmental Health, Institute of Environmental Science & Research, PO Box 29-181, Christchurch 8041, New Zealand
brent.gilpin@esr.cri.nz

Willie Grabow Department of Microbiology, University of Pretoria, Pretoria, South Africa

Alexandria K. Graves Department of Soil Science, North Carolina State University, 3411E Williams Hall, Raleigh, NC 27695-7619, USA
alexandria_graves@ncsu.edu

Charles Hagedorn Department of Crop and Soil Environmental Sciences, 401 Price Hall, Virginia Tech, Blacksburg, VA 24061, USA
chagedor@vt.edu

Peter G. Hartel Department of Crop and Soil Sciences, University of Georgia, Athens, GA, USA
pghartel@uga.edu

Valerie J. Harwood Department of Integrative Biology, University of South Florida, 4202 E. Fowler Avenue, Tampa, FL 33620, USA
vharwood@usf.edu

Patricia A. Holden Donald Bren School of Environmental Science and Management, University of California, Santa Barbara, CA 93106, USA

Juan Jofre Department of Microbiology, School of Biology, University of Barcelona, Avinguda Diagonal 645, 00028 Barcelona, Spain
jjofre@ub.edu

David Kay Centre for Research into Environment and Health, Aberystwyth University, Ceredigion, Wales, UK SY23 3DB
dvk@aber.ac.uk

Elizabeth A. Kearns Department of Cell Biology, Microbiology and Molecular Biology, University of South Florida, Tampa, FL 33620-5150, USA

Julie Kinzelman City of Racine Health Department, Racine, WI 53403, USA
julie.kinzelman@cityofracine.org

Marek Kirs Aquatic Biotechnologies, Cawthron Institute, Private Bag 2, Nelson 7042, New Zealand
marek.kirs@cawthron.org.nz

Karen Kline Biological Systems Engineering, Virginia Tech, Blacksburg, VA, USA

Denny Kollanur Institute of Chemical Engineering, Research Area Applied Biochemistry and Gene Technology, Research Group Environmental Microbiology and Molecular Ecology, Vienna University of Technology, Getreidemarkt 9/166-5-2, 1060 Vienna, Austria

Leigh-Anne Krometis Biological Systems Engineering, Virginia Tech, Blacksburg, VA, USA

Stephaney D. Leskinen Department of Cell Biology, Microbiology and Molecular Biology, 4202 E. Fowler Avenue, BSF 218, University of South Florida, Tampa, FL 33620-5150, USA
Leskinen@usf.edu

Hua Li Environmental Resource and Soil Fertilizer, Zhejiang Academy of Agricultural Sciences, Hangzhou, China

S.M. McQuaig Natural Sciences, St. Petersburg College, 2465 Drew St., Clearwater 33765, FL, USA
mcquaig.shannon@spcollege.edu

Gerold Morrison BCI Engineers and Scientists, Inc., Lakeland, FL, USA
gmorrison@bcieng.com

Joseph A. Moss Center for Environmental Diagnostics and Bioremediation, University of West Florida, Pensacola, FL, USA

Joanna Mott Department of Life Sciences, Texas A&M University-Corpus Christi, 6300 Ocean Drive, Corpus Christi, TX 78412-5802, USA
Joanna.mott@tamucc.edu

Kristen Nicole Hellein Center for Environmental Diagnostics and Bioremediation, University of West Florida, 11000 University Parkway, Pensacola, FL 32514, USA

R.T. Noble Department of Cell Biology, Microbiology, and Molecular Biology, University of South Florida, Tampa, FL 33620, USA

Pierre Payment INRS-Institut Armand-Frappier, 531 Boulevard des Prairies, Laval, Québec, Canada, H7V 1B7

Kathy Pond Robens Centre for Public and Environmental Health, University of Surrey, Guilford, Surrey, GUZ 7XH, UK
K.Pond@surrey.ac.uk

Cheryl W. Propst PBS&J, Jacksonville, FL, USA
CMWapnick@pbsj.com

Georg H. Reischer Institute of Chemical Engineering, Research Area Applied Biochemistry and Gene Technology, Research Group Environmental Microbiology and Molecular Ecology, Vienna University of Technology, Getreidemarkt 9/166-5-2, A-1060 Vienna, Austria

Ana Maria de Roda Husman Laboratory for Zoonoses and Environmental Microbiology, National Institute of Public Health and the Environment, P.O. Box 1, 3720 BA Bilthoven, The Netherlands
ana.maria.de.roda.husman@rivm.nl

Jack F. Schijven Expert Centre for Methodology and Information Services, National Institute of Public Health and the Environment, P.O. Box 1, 3720 BA Bilthoven, The Netherlands
Jack.Schijven@rivm.nl

Troy M. Scott University of Miami Center for Oceans and Human Health, Key Biscayne, FL, USA
and
Source Molecular Corporation, Miami, FL, USA
dtrmscott@gmail.com

Caterina Signoretto Dipartimento di Patologia e Diagnostica, sezione di Microbiologia, Università di Verona, Strada Le Grazie, 8; 37134 Verona, Italy
caterina.signoretto@univr.it

Christopher D. Sinigalliano National Oceanic and Atmospheric Administration, Atlantic Oceanographic and Meteorological Laboratory, Miami, FL, USA
and
University of Miami Center for Oceans and Human Health, Key Biscayne, FL, USA
Christopher.Sinigalliano@noaa.gov

Amanda Smith Department of Life Sciences, Texas A&M University-Corpus Christi, 6300 Ocean Drive, Corpus Christi, TX 78412-5802, USA

Richard A. Snyder Center for Environmental Diagnostics and Bioremediation, University of West Florida, Pensacola, FL, USA
rsnyder@uwf.edu

Helena M. Solo-Gabriele Department of Civil, Architectural, and Environmental Engineering, University of Miami, Coral Gables, FL, USA
and
University of Miami, Center for Oceans and Human Health, Key Biscayne, FL, USA
hmsolo@miami.edu

Regina Sommer Institute of Hygiene and Applied Immunology, Medical University Vienna, Kinderspitalgasse 15, A-1090 Vienna, Austria

Hermann Stadler Institute of Water Resources Management, Hydrogeology and Geophysics, Joanneum Research, Elisabethstraße 16/II, A-8010 Graz, Austria

Jill R. Stewart Department of Environmental Sciences and Engineering, Gillings School of Global Public Health, University of North Carolina, Chapel Hill, NC, USA

Donald M. Stoeckel Battelle Memorial Institute, Columbus, OH, USA

Christopher M. Teaf Center for Biomedical & Toxicological Research and Waste Management, Florida State University, Tallahassee, FL 32310, USA
cteaf@fsu.edu

Joy A. Truesdale Texas AgriLife Research Center at El Paso, Texas A&M University System, El Paso, TX 79927, USA

Richard Villemur INRS-Institut Armand-Frappier, 531 Boulevard des Prairies, Laval, Québec, Canada, H7V 1B7

Dan Wang Department of Civil and Environmental Engineering, University of California, Davis, One Shields Avenue, Davis, CA 95616, USA

Stephen B. Weisberg Southern California Coastal Water Research Project, 3535 Harbor Blvd., Costa Mesa, CA 92626, USA

Cindy H. Wu Lawrence Berkeley National Laboratory, One Cyclotron Road, Mail Stop 70A-3317, Berkeley, CA 94720, USA

Stefan Wuertz Department of Civil and Environmental Engineering, University of California, Davis, One Shields Avenue, Davis, CA 95616, USA
swuertz@ucdavis.edu

Liang Xinqiang Institute of Environmental Science and Technology, College of Environmental and Resource Sciences, ZheJiang University, Hangzhou, China

Gene Yagow Biological Systems Engineering, Virginia Tech, Blacksburg, VA, USA

Wolfgang Zerobin Vienna Waterworks, Grabnergasse 4-6, A-1060 Vienna, Austria

Chapter 1
Overview

Charles Hagedorn, Valerie J. Harwood, and Anicet R. Blanch

Abstract Microbial source tracking (MST) is a still-new and emerging sub-discipline of Biology that allows practitioners to discriminate among the many possible sources of fecal pollution in environmental waters. MST's current and potential applications range from beach monitoring to total maximum daily load (TMDL) assessment of pollution sources, that in turn will mediate greater protection of public health and improvement of environmental water quality. This comprehensive book taps the expertise of many of the leading research scientists from an international assemblage, and contains chapters that range from China and developing nations (22) to New Zealand and Australia (21), plus the EU and USA. The book addresses subjects ranging from the fundamentals of performance criteria during method development (2), library-dependent (3) and library-independent (4) approaches with their pros and cons, and applications to case studies from agricultural (18), urban (19), and beach (20) watersheds. Separate chapters focus on viral (5), bacteriophage (6), protozoan (7), chemical (8), mitochondrial DNA (10), and community analysis (11) -based methods. Chapters that relate MST to the fecal indicator bacteria (15), determining when and where to use MST (16), and the environmental persistence of fecal bacteria (17) put MST in the context of environmental monitoring. Specialized topics include legal (13) and TMDL (14) -associated issues, public perceptions (12), statistical analysis (9), national security (23), risk assessment (24), food safety (25), and using MST in undergraduate education (26). We hope that this book will prove useful to new practitioners of MST as well as established researchers and scientists and that it will serve as a valuable reference for many years to come.

Keywords Source tracking methods • Case studies • Environmental persistence • Performance criteria • Monitoring and assessment • Water quality • Fecal indicator bacteria • Microbial tracers • Chemical tracers

---

C. Hagedorn (✉)
Department of Crop and Soil Environmental Sciences,
401 Price Hall, Virginia Tech, Blacksburg, VA 24061, USA
e-mail: chagedor@vt.edu

C. Hagedorn et al. (eds.), *Microbial Source Tracking: Methods, Applications, and Case Studies*, DOI 10.1007/978-1-4419-9386-1_1,

The progressive improvement of strategies for management of microbial quality of catchments during the last two centuries has played an essential role in the improvement of public health in human societies. The definition and implementation of microbial indicators to survey water quality and assess reductions in microbial pathogens of fecal origin have proven to be a practical and efficient measure for the protection and improvement of water resources. The citizens of developed countries are generally protected by legislation and regulations regarding water quality for many purposes, such as drinking, personal hygiene, recreational activities, agriculture watering, and food production. However, waterborne disease outbreaks remain an enormous burden in developing countries where management of water resources with the aim of reducing microbial contaminants is rare or nonexistent (Chap. 22).

It is important to understand that measurements of fecal indicator bacteria (FIB) for water quality do not provide information about the origin of fecal pollution, i.e., whether the host source is, for example, birds, dogs, cattle, or humans – or a combination of any of these. This limitation exists because the feces of most animals contain FIB concentrations that are great enough to affect water quality when many animals or their sewage impact a water body (Chap. 14). The detection of the origin of fecal pollution is assuming a prominent place in hazard identification related to host-specific pathogens (Chap. 24). Pathogens from infected animals or humans can be introduced into water resources through feces or sewage and can cause a human health risk. The identification of the fecal sources is important to protect the public from zoonotic pathogens that may be shed by animals such as wild birds, poultry, cattle, and pigs. The capability to detect human-source pollution is also crucial to management strategies, as sewage from human origin is generally expected to have a higher risk to public health than that of animal origin (Chap. 15). Consequently, understanding the origin of fecal pollution is essential in assessing potential human health risks as well as for determining the actions necessary to remediate the quality of waters contaminated by fecal matter.

The intensive research efforts directed at developing methods for detection of fecal pollution originated over the past few decades and have been grouped under the term microbial source tracking (MST). These studies began in the early 1980s (Geldreich 1976; Mara and Oragui 1981; Osawa et al. 1981; Mara and Oragui 1983), probably as a result of social and legal pressures. The term MST denotes procedures that use host-specific (found only in one host species or group) and host-associated (largely confined to one host species or group) microbial indicators to establish the origin of fecal pollution in water. From its inception, MST has experienced rapid growth in knowledge and technological capabilities, including PCR and quantitative PCR that have substantially augmented the established research field of water-quality microbiology.

The history of MST research could be divided into several phases. Phase 1 was the initial phase, when defining new indicators (Brown 1993; Awad-El-Kariem et al. 1995; Hsu et al. 1995; Tartera et al. 1989; Bernhard and Field 2000; Nebra et al. 2003) and appropriate methods for source discrimination (Hagedorn et al. 1999; Wiggins 1996; Parveen et al. 1997; Whitlock et al. 2002; Harwood et al. 2000; Manero et al. 2002;

Wallis and Taylor 2003) were emphasized. In response to the emergence of MST as a potential regulatory strategy, Phase 2 saw three large multilaboratory method comparison studies (two in USA and one in Europe) plus numerous workshops, book chapters, and review articles dedicated to synthesizing information on the topic (Field et al. 2003; Harwood et al. 2003; Griffith et al. 2003; Myoda et al. 2003; Noble et al. 2003; Ritter et al. 2003; Blanch et al. 2004; Blanch et al. 2006). Furthermore, a federal (US EPA) guide document that described the uses and limitations of MST methods was published in 2005 (US Environmental Protection Agency 2005), and a book dedicated to MST as an emerging issue in food safety was published in 2007 (SantoDomingo and Sadowsky 2007). Over the past ten years, library-dependent MST methods (Chap. 3), which require a large assemblage of typed organisms from various host sources, have been largely supplanted by library-independent methods (Chap. 4) that rely on detection of a particular host-specific organism or gene.

To date, there has been no widespread consensus among researchers or any regulatory agency regarding the best indicators for MST. Many studies still focus exclusively on the development of new MST indicators and the improvement of their methods of detection and quantification (Chaps. 3–8 and 10). These documents cited above provide a collective body of literature on MST that, although frequently complementary, is at times conflicting, repetitious, and difficult to condense and interpret. In addition, they do not reflect the current diversity of MST approaches with different organisms, newer methodologies such as quantitative PCR and anthropogenic chemicals, nor do they reflect the scope of MST research being conducted around the world (Chaps. 21 and 22).

The goal of this book is to serve as a valuable reference for all those who are involved with water quality, whether they are students, researchers, managers, or regulators. This book also aims to be the first comprehensive source to present the MST spectrum at the international level and to act as a future guide for researchers who need to use, apply, and interpret MST in all manner of watershed environments. For that reason, the editors have intentionally sought out authors who collectively represent a comprehensive expertise and whose work reflects the rich diversity and truly international scope of MST.

The unifying theme throughout the book is the design of more standardized approaches to MST that include performance criteria, regardless of method or organism (Chap. 2), plus recommendations for field study design and MST implementation (Chaps. 14 and 16). The content is structured in four sections to facilitate the search of topics and practical reading. The first is a "Method Development" section that includes a wide spectrum of different fecal source indicators that have been or are being developed. Here, readers can find not only the current state of the science for these indicators but also the historical track, present challenges, and future perspectives.

Microbial indicators based on the detection of bacteria or their components, e.g., genes, are described in two chapters that are delineated by the method's dependence (Chap. 3) or independence (Chap. 4) on reference libraries composed of typed organisms from various host sources (library-dependent and library-independent methods). Different approaches are also discussed and compared, including requirements for cultivation and the dependence on a priori developed reference libraries.

Other proposed MST indicators are also considered in detail within this section, i.e., viruses (Chap. 5), bacteriophages (Chap. 6), and protozoa (Chap. 7). The advantages and challenges for these microbial groups are analyzed, and the potential for practical applications is also explained. Moreover, chemical and eukaryotic (mitochondrial) indicators that have been developed and evaluated for MST uses also have their respective chapters (Chaps. 8 and 10), where advantages and drawbacks are also identified, and new perspectives are indicated. This section also includes three chapters for specific topics that are essential to implement of MST indicators and to evaluate their feasibility for routine analyses. To that end, performance criteria (Chap. 2), statistical approaches and modeling (Chap. 9), and the development of community-analysis-based methods (Chap. 11) each have a dedicated chapter.

Indicators, the methods used to detect and/or quantify them, and the appropriate performance characteristics need to be applied, understood, and properly interpreted by scientists, managers, and regulators who work on catchment management. The second section of the book covers "Use, Interpretation, and Application" and includes chapters on the public understanding of MST (Chap. 12), legal challenges (Chap. 13), and the use of MST indicators on the determination of the total load of fecal pollution that could support a catchment (i.e., TMDL) based primarily on the development of models for this purpose (Chap. 14). The relationship of MST indicators with respect to other standardized and routine microbiological parameters (i.e., microbial indicators and pathogens) that are used for water-quality management is also described in a specific chapter (Chap. 15). Designing representative sampling schemes and a decision-based matrix for when to use, or not use, MST are also included (Chap. 16). Lastly, this section includes a chapter on the persistence of indicator organisms in aquatic environments and sediments and sands, a very timely emerging issue (Chap. 17).

The third section is dedicated to "MST Case Studies." Field studies on agricultural and rural watersheds from different geographical areas are described, and implications for catchment management are discussed (Chap. 18). Many practical aspects of MST conducted in different geographic regions are described. Some are related to agricultural and rural watersheds (surface and karstic groundwater resources) but others to urban and suburban watersheds (Chap. 19). There is a chapter committed to the rationale for using microbial source tracking (MST) methods at beach sites and coastal water bodies (Chap. 20) and the use of MST methods for evaluating waters impacted by nonpoint sources of pollution. This chapter also describes the most common traditional and alternative MST markers used at beach sites. Lastly, this section contains two chapters outlining experiences and case studies on the application of MST methods in waterways in Australia and New Zealand (Chap. 21), and in China and developing countries (Chap. 22). The vast differences in the use of MST between developed and developing nations are readily apparent in these two final chapters of Sect. 3.

Finally, the fourth section is dedicated to the "Future Needs and Perspectives for MST Development." including more widespread application of MST on water management decisions. Issues and aspects of MST as related to national security (Chap. 23), quantitative risk assessment (Chap. 24), and food safety (Chap. 25) are all presented. Lastly, a chapter on education presents some available training resources for future scientists and technical staff and demonstrates how MST can be a component of undergraduate education in both the four-year and community college settings (Chap. 26).

We hope that this book will prove useful to new practitioners of MST as well as established researchers and scientists and that it will serve as a starting point into this fascinating area of MST that merges basic and applied science, field work and laboratory studies, theory and practicality, as well as any scientific endeavor in modern biology. We trust that this book will need substantial revision at some point as the field of MST continues to grow and that it will serve as a valuable reference for many years to come.

We are grateful to Andrea Macaluso (editor at Springer-US), who first proposed to us the idea of an interdisciplinary MST book. We especially acknowledge all the authors for their dedication and contribution and their efforts to relate the different chapters to each other. This greatly simplified the always-complex process of editing a book with many highly qualified authors who are experts in the wide range of topics covered in this book.

References

Awad-El-Kariem FM, Robinson HA, Dyson PA et al (1995) Differentiation between human and animal strains of *Cryptosporidium parvum* using isoenzyme typing. Parasitol 110:129–132.

Bernhard AE, Field KG (2000) Identification of nonpoint sources of faecal pollution in coastal waters by using host-specific 16S ribosomal DNA genetic markers from faecal anaerobes. Appl Environ Microbiol 66:1587–1594.

Blanch AR, Belanche-Munoz L, Bonjoch X et al (2004) Tracking the origin of faecal pollution in surface water: An ongoing project within the European Union research programme. J Wat Health 2:249–260.

Blanch AR, Belanche-Munoz L, Bonjoch X et al (2006) Integrated analysis of established and novel microbial and chemical methods for microbial source tracking. Appl Environ Microbiol 72:5915–5926.

Brown TJ (1993) *Giardia* and Giardiasis in New Zealand. Report to the Ministry of Health June 1991 – September 1993. Massey University/New Zealand Ministry of Health Giardia Unit. 37 pp.

Field KG, Chern EC, Dick LK et al (2003) A comparative study of culture-independent, library-independent genotypic methods of faecal source tracking. J Wat Health 1:181–194.

Geldreich EE (1976) Faecal coliforms and faecal streptococcus relationship in waste discharge and receiving waters. Crit Rev Environ Control 6:349–368.

Griffith JF, Weisbert SB, McGee CD (2003) Evaluation of microbial source tracking methods using mixed faecal sources in aqueous test samples. J Wat Health 1:141–151.

Hagedorn C, Robinson SL, Filtz JR et al (1999) Determining sources of faecal pollution in a rural Virginia watershed with antibiotic resistance patterns in faecal streptococci. Appl Environ Microbiol 65:5522–5531.

Harwood VJ, Whitlock J, Withington V (2000) Classification of antibiotic resistance patterns of indicator bacteria by discriminant analysis: use in predicting the source of faecal contamination in subtropical waters. Appl Environ Microbiol 66:3698–3704.

Harwood VJ, Wiggins B, Hagedorn C et al (2003) Phenotypic library-based microbial source tracking methods: Efficacy in the California collaborative study. J Wat Health 1:153–166.

Hsu FC, Shieh YSC, van Duin J et al (1995) Genotyping male-specific coliphages by hybridization with oligonucleotide probes. Appl Environ Microbiol 61:3960–3966.
Manero A, Vilanova X, Cerdà-Cuéllar M et al (2002) Characterization of sewage waters by biochemical fingerprinting of Enterococci. Wat Res 36:2831–2835.
Mara DD, Oragui JI (1981) Occurrence of *Rhodococcus coprophilus* and associated actinomycetes in feces, sewage and freshwater. Appl Environ Microbiol 51:85–93.
Mara DD, Oragui JI (1983) Sorbitol-fermenting bifidobacteria as specific indicators of human faecal pollution. J Appl Bacteriol 55:349–357.
Myoda SP, Carson CA, Fuhrmann JJ et al (2003) Comparison of genotypic-based microbial source tracking methods requiring a host origin database. J Wat Health 1:167–180.
Nebra Y, Bonjoch X, Blanch AR (2003) Use of *Bifidobacterium dentium* as an indicator of the origin of faecal water pollution. Appl Environ Microbiol 69:2651–2656.
Noble RT, Allen SM, Blackwood AD et al (2003) Use of viral pathogens and indicators to differentiate between human and non-human faecal contamination in a microbial source tracking comparison study. J Wat Health 1:195–209.
Osawa S, Furuse K, Watanabe I (1981) Distribution of ribonucleic acid coliphages in animals. Appl Environ Microbiol 41:164–168.
Parveen S, Murphree R, Edmiston L et al (1997) Association of multiple-antibiotic-resistance profiles with point and nonpoint sources of *Escherichia coli* in Apalachicola Bay. Appl Environ Microbiol 63:2607–2612.
Ritter KJ, Carruthers E, Carson CA et al (2003) Assessment of statistical methods used in library-based approaches to microbial source tracking. J Wat Health 1:209–223.
SantoDomingo JW, Sadowsky MJ (2007) Microbial source tracking. ASM Press, Washington, D.C.
Tartera C, Lucena F, Jofre J (1989) Human origin of *Bacteroides fragilis* bacteriophages present in the environment. Appl Environ Microbiol 55:2696–2701.
United States Environmental Protection Agency (2005) Microbial source tracking guide. U.S. Environmental Protection Agency, Washington, D.C. EPA/600/R-05/064.
Wallis JL, Taylor HD (2003) Phenotypic population characteristics of the enterococci in wastewater and animal faeces: implications for the new European directive on the quality of bathing waters. Wat Sc Technol 47:27–32.
Whitlock JE, Jones DT, Harwood VJ (2002) Identification of the sources of faecal coliforms in an urban watershed using antibiotic resistance analysis. Wat Res 36:4273–4282.
Wiggins BA (1996) Discriminant analysis of antibiotic resistance patterns in faecal streptococci, a method to differentiate human and animal sources of faecal pollution in natural waters. Appl Environ Microbiol 62:3997–4002.

Chapter 2
Performance Criteria

Valerie J. Harwood and Donald M. Stoeckel

Abstract The establishment of rigorous, consistent performance criteria for microbial source tracking (MST) methods is essential for their usefulness and widespread acceptance as research and regulatory tools. In this chapter, we focus on performance criteria for library-independent methods, although many aspects of the discussion are applicable to both library-independent and library-dependent methods. We separate these criteria into three levels for ease of discussion: (1) the intrinsic characteristics of the "marker" (target), (2) protocols for generating laboratory data, and (3) field applications. By ensuring that a consistent set of metrics for characteristics such as accuracy and precision be applied to field studies and published works, we can begin to circumscribe the set of MST tools that will be most useful for discriminating among fecal pollution sources in environmental waters.

Keywords qPCR • Performance • Efficiency • Accuracy • Precision • Error

2.1 Introduction

The nascent field of microbial source tracking has relied upon both library-dependent and library-independent approaches (see Chaps. 3 and 4, respectively) to detect fecal contamination from particular hosts. In particular, the library-dependent approach experienced a high level of application in first five or so years of the 21st century, which included the introduction of statistical methods such as discriminant analysis (Wiggins 1996), principle components analysis (Dombek et al. 2000), or nearest-neighbor analysis (Albert et al. 2003; Ritter et al. 2003;

V.J. Harwood (✉)
Department of Integrative Biology,
University of South Florida, 4202 E. Fowler Avenue, Tampa, FL 33620, USA
e-mail: vharwood@usf.edu

C. Hagedorn et al. (eds.), *Microbial Source Tracking: Methods, Applications, and Case Studies*, DOI 10.1007/978-1-4419-9386-1_2,

Robinson et al. 2007) to evaluate complex patterns generated by antibiotic resistance analysis (Hagedorn et al. 1999; Harwood et al. 2000; Wiggins 1996), rep-PCR (Carson et al. 2003; Dombek et al. 2000; McLellan et al. 2003), pulsed-field gel electrophoresis (Myoda et al. 2003), ribotyping(Parveen et al. 1999), and other methods. The validity of results from these library-dependent methods began to be questioned following proficiency testing with blind samples (Griffith et al. 2003; Harwood et al. 2003; Stoeckel et al. 2004). Other pressing concerns with library-dependent methods include the size and scope required for a "representative" library and concerns about broad geographic applicability and temporal stability (Stoeckel and Harwood 2007; US Environmental Protection Agency 2005; Wiggins et al. 2003).

As a result of these findings and concerns, library-independent methods, many of which showed better accuracy in limited proficiency testing compared with the library-dependent methods (Griffith et al. 2003; Harwood et al. 2003; Myoda et al. 2003), began to be more intensively developed and used in field studies. As was done with library-dependent methods, as these methods and markers emerge they should be routinely validated for provision of accurate results. The purpose of this chapter is to outline a strategy for method validation and proficiency testing that is applicable to library-independent MST methods, many of which utilize PCR and/or quantitative PCR (qPCR) to detect a host-associated target organism or gene. By establishment of rigorous performance criteria and application of proficiency tests, MST methods will be evaluated within a consistent framework, paving the way for more confident use in regulatory and legal contexts.

This chapter considers performance of MST methods separately at three levels – the genetic target or "marker," since interpretation of MST data for fecal source indication is dependent upon marker characteristics (sensitivity and specificity within the target population); the protocol for generating laboratory data, since without confidence in the data results cannot be interpreted; and field application, since interpretation of data collected from uncontrolled settings poses additional challenges beyond basic laboratory quality control. In this chapter, we use "performance" when referring to inherent characteristics of the method, e.g., sensitivity, specificity, evenness; and "proficiency" when referring to testing that is specifically designed to evaluate the quality and reliability of laboratory-generated data.

The use of common performance measures and validation strategies in the many studies that are expected over the next decade should facilitate rapid progress in this area, as we continue to work toward availability of reliable analyses, classification approaches, and interpretation strategies for tracking fecal contamination to its sources by use of MST tools. Although we focus here on methods that target specific genes via PCR, the general strategies and most of the considerations discussed here apply in some measure to all of the methodologies discussed in this book (see Chaps. 3 and 9 for criteria that are more appropriate for library- and chemical-based methods, respectively).

2.2 Evaluation of Target (MST Marker) Performance and Suitability

The various markers used for library-independent MST detect the presence of host-associated microbial populations. Sensitivity, or completeness of marker representation in the host population, along with specificity, or exclusivity of the host–microbe association, are critically important parameters (Table 2.1) (Stoeckel and Harwood 2007). Relatively poor sensitivity, which is associated with low-prevalence markers such as those that detect some pathogenic viruses (Noble et al. 2003; Stoeckel and Harwood 2007), frequently causes false-negative results. Incomplete specificity, which is associated with many existing genetic markers

Table 2.1 Characteristics of an ideal vs. a useful MST marker (Harwood 2007; US Environmental Protection Agency 2005)

Characteristic	Ideal marker	Useful marker
Specificity	Marker found only in target host species	Marker is differentially distributed among host species
Distribution in host population	Found in all members of all populations of target host species; contributes to sensitivity of method	Consistently found in host species whose feces could impact the target sites
Evenness	Quantity in the feces of individuals is similar	Quantity in aggregate sources, e.g., sewage influent; animal populations, is similar
Temporal stability in host	Frequency and concentration in host individuals and populations does not change over time	Despite variation in marker frequency and concentration in individuals, these characteristics are stable at the population level
Geographic range/stability	The frequency and concentration in geographically separated host populations are similar	The marker can consistently be detected and quantified across the geographic area to be studied
Environmental persistence	Consistent decay rate in various matrices and habitats; no increase under any conditions; response to treatment processes and environmental insults is similar to that of pathogens	Predictable decay rate in various matrices and habitats; no increase under ambient conditions; response to treatment processes and environmental insults is characterized
Quantitative assessment	Can be accurately quantified	Accurately indicates presence/absence of contamination source
Relevance to regulatory parameters	The marker is derived from an organism that is a regulatory tool	The marker is correlated with an organism that is a regulatory tool
Relevance to health risk	The marker is strongly correlated with risk of all types of waterborne disease, e.g., gastroenteritis, dermatitis, upper respiratory infections	The marker constitutes a health risk or is otherwise correlated with a subset of waterborne disease, e.g., viral gastroenteritis

(Harwood et al. 2009; Korajkic et al. 2009; Shanks et al. 2010), can cause false-positive results. The third major issue relevant to performance measurement for markers is evenness of marker distribution (in terms of prevalence and quantity), which applies both among populations and among individuals within a given host population. If the evenness of the marker is different from the evenness of fecal indicator bacteria or pathogens, then simple detection or even quantification of the marker may not be directly comparable to existing regulations or public health risk outcomes. These considerations are discussed in detail below.

2.2.1 Choosing the Tool(s) to Fit the Question

Potential applications of MST include (a) assessment of sources of fecal contamination in recreational or drinking source waters, (b) prioritization of impaired water bodies for total maximum daily load (TMDL) implementation or other interventions, (c) source apportionment for TMDL plans, and (d) forensic applications, i.e., assigning (or relieving) responsibility for pollution. The goals of a given study must be carefully considered when choosing or designing MST marker(s), and deciding whether conventional (presence/absence) PCR-based methods are sufficient or if quantitative PCR (qPCR) is required. For example, if one is most concerned about determining when and where contamination from human sources is present, a suite of human-specific markers may be chosen, and conventional PCR may be sufficient to achieve the study goals. If, however, one is attempting to apportion contributions from various fecal sources for TMDL applications, it would be necessary to use a suite of markers for the identified sources of fecal loading, and qPCR would be required.

Many authors have recommended toolbox or tiered approaches for MST study design, the first meaning that a group of MST methods is on hand and ready for deployment as the specific situation demands and the second meaning that lower cost methods that broadly measure contamination, such as conventional fecal indicator bacteria measurements, are used first, followed by more expensive, technically demanding methods such as PCR where they are needed to accomplish specific goals (Boehm et al. 2003; Lu et al. 2009; McQuaig et al. 2006; Noble et al. 2006; Vogel et al. 2007) (see also Chaps. 16 and 19). Another aspect of the toolbox approach is that multiple methods for detection of contamination from one source can be used to support one another (see below), alleviating the uncertainty that results from imperfections in all methods reported to date. On the contrary, the use of multiple tests increases the cost of a given study and can be unacceptably expensive for end users such as regulatory agencies. This situation can be a particular concern when multiple methods are used to identify one source.

One must also consider the performance characteristics of the methods and how they might affect interpretation of the results; for example, one could use a human-associated marker with high concentration in sewage but incomplete specificity to minimize the probability of false-negative results. Because use of such a marker could yield false-positive results, one might also use a highly human-specific marker that has the drawback of lower concentration in sewage to back up the indication of human fecal pollution. Other performance characteristics of MST methods, such as sensitivity to inhibition from interfering compounds (discussed below) should also be considered in the context of the characteristics of the water bodies that are sampled.

2.2.2 *Ideal Characteristics for MST Markers*

Each of the MST markers described in the literature to date has both positive and negative qualities. The same caveat applies to available chemical markers (Chap. 9). Some pathogen markers, such as enteroviruses and adenoviruses, tend to have relatively high false-negative rates in sewage, and particularly in individual human fecal samples (Griffith et al. 2003; Noble et al. 2003), which can lead to low sensitivity. Other markers, such as the human-associated *Bacteroidales* 16S rRNA sequence delineated by PCR primers HF183 and 708R (Bernhard and Field 2000), display incomplete specificity (a low but detectable rate of false positives against feces from nonhuman animals) (Balleste et al. 2010; Harwood et al. 2009). The ideal MST marker is described previously (Table 2.1), and since a marker that meets all these criteria has not been identified for any host, the characteristics of a useful marker are also described as adapted from (US Environmental Protection Agency 2005). The following sections further discuss key characteristics and how they are experimentally assessed.

2.2.2.1 Specificity

The central hypothesis of MST is that some microorganisms have an exclusive or preferential association with the gastrointestinal tract of a particular host species or group, and that these host-associated microorganisms are shed in feces and can be detected in water bodies. The detected markers may be extremely host-specific, such as human polyomaviruses (Ahmed et al. 2009a; Harwood et al. 2009; McQuaig et al. 2006; McQuaig et al. 2009) or they may have limited host specificity, such as some of the human-associated markers targeting *Bacteroidales* 16S rRNA genes (Ahmed et al. 2009b; Harwood et al. 2009; Layton et al. 2006; Shanks et al. 2007).

The specificity of a marker is generally assessed by analyzing fecal and/or sewage samples from animals other than the targeted host (nontarget hosts) (Harwood 2007; Shanks et al. 2010; Stoeckel and Harwood 2007; US Environmental Protection Agency 2005). Although DNA sequences that are candidates for MST markers can undergo a preliminary, *in silico* specificity assessment (i.e., a computer-generated BLAST search against the NCBI database of sequences), such an analysis should not be substituted for testing against nontarget fecal material. A quantitative expression of specificity is 1 minus the proportion of nontarget fecal samples in which the marker is detected, which is also 1 minus the false-positive rate (as described and compiled in (Stoeckel and Harwood 2007)). Specificity is generally expressed as a percentage; therefore, the calculations above would be multiplied by 100. Specificity testing can be accomplished with individual fecal samples or composite fecal samples (e.g., (Griffith et al. 2003)). Composite fecal samples can be further characterized as compilations of individual fecal samples made by the sampler, in which case one knows how many animals contributed to the sample, or large-scale composites such as sewage influent samples or slurries from dairy cattle operations. An obvious advantage of a composite sample is that, as long as tests negative for the marker, it can allow testing of more individuals with fewer negative-control analyses. Drawbacks to this approach are (a) if a false positive is obtained, one does not know how many individual scats contributed to that result, and (b) it is theoretically possible that the signal from one positive scat could be missed because it is diluted by the other scats in the composite. In the case of humans, it is highly recommended that sewage samples and, when applicable, onsite treatment and disposal systems (OSTDS, or septic systems) be tested because they are more likely to reach water bodies than waste from an individual, and because there is inherent selection for environmental survival or persistence within such systems (Gordon et al. 2002). A similar case can be made for some types of animal feces, e.g., slurries from cattle barns or egg layer poultry operations, or litter from broiler poultry production are very useful composite samples.

Box 2.2.2.1.1

The specificity of the marker commonly known as HF183 for human-source fecal contamination has been well documented both for conventional PCR (Bernhard and Field 2000; Harwood et al. 2009) and for the quantitative adaptations (Kildare et al. 2007; Seurinck et al. 2005; Shanks et al. 2009). Each of these reports is based on reference samples in North America or Europe. Specificity testing in New Zealand, however, indicated that the marker was commonly associated with a local species of opossum (Kirs et al. 2011). Furthermore, although the concentrations were not reported, extended sampling of nontarget sources not previously considered (e.g., fish) can identify additional sources of potential false-positive results (McLain et al. 2009).

Determination of the appropriate number of nontarget samples to include for specificity testing is not standardized, but should be based on the geographic area of the study, the intended use of the marker, and the distribution of host species in the study area that are reasonably expected to impact water quality. The USEPA MST Guide Document (2005) recommends that at least ten animals per host type are sampled for specificity. While it is not practical to sample the feces of more than a small subset of all individuals in a given area, a good faith effort should be made to capture the diversity among relevant host populations. For example, sampling the feces of five cattle from one farm in a study intended to characterize fecal sources in a watershed that is potentially impacted by ten cattle farms is clearly an inadequate effort. A more inclusive strategy in such a case would be to sample from three or more farms and to make five or more composite samples from 5 to 10 animals each for each farm. If one is testing the specificity of a marker with the goal of using it across a broad geographic region, the number and scope of samples tested should be broadened accordingly. For example, Shanks et al. (2009) tested the specificity of several qPCR methods for human markers using 265 fecal samples from 22 nontarget species collected across USA. The specificity of a qPCR assay for human polyomaviruses was tested with 127 fecal samples from 14 nontarget species (McQuaig et al. 2009).

The specificity parameters reported in the literature may not accurately reflect marker characteristics in a particular study due to factors such as geographic variability in marker distribution and/or the completeness of previous sampling effort(s). For these reasons, validation of the method protocol against reference fecal samples as part of the experimental start-up procedure is strongly recommended (see Sect. 2.3 for details on evaluation of data). Even relatively well-characterized MST markers such as the human-associated *Bacteroidales* 16S rRNA marker HF183 are subject to new findings when tested in a new geographic area or against previously untested host species (Box 2.2.2.1.1).

It is very important to characterize the error rate associated with specificity as thoroughly as is practically possible. There is no universally accepted criterion for the minimum specificity required of a useful MST marker. Of course, 100% specificity is ideal, but is rarely achieved. Even when observed in one study, this figure is rarely maintained over subsequent studies. It is generally agreed that methods with less than 80% specificity are not useful in most circumstances (US Environmental Protection Agency 2005), and the majority of recently published or frequently used methods have 90% or greater measured specificity, at least in the geographic area(s) for which they are characterized (Ahmed et al. 2009a; McQuaig et al. 2009; Shanks et al. 2009; Weidhaas et al. 2010).

When amplification of a particular marker from nontarget sources is noted (generally termed "false positive" in the literature) it may occur because the target sequence is present in the nontarget fecal or sewage sample, e.g., (Harwood et al. 2009). However, any number of other reasons may cause apparent false-positive results, including an uncalibrated thermocycler (annealing temperature too low), the existence of very similar, but demonstrably different, sequences in the sample, or contamination of the sample. These mistakes should be guarded against with adequate controls and method performance evaluations (see Sects. 2.3–2.6), and amplicons should be sequenced to determine whether they (a) are identical to the target, (b) are similar to the target, or (c) represent an unrelated PCR artifact. The latter concern is not as great for probe-based qPCR methods, as probe as well as primer must match the target sequence.

2.2.2.2 Distribution and Sensitivity

The distribution of a marker in the feces of individual members of a host species is a major contributor to method sensitivity. As discussed in Stoeckel and Harwood (2007), the sensitivity of a marker can be defined as the proportion of positive-control fecal or sewage samples (target samples) that yield positive results. Like specificity values, sensitivity is generally expressed as a percentage, so the figure obtained from the calculation above would be multiplied by 100. A more sensitive marker will be more frequently detectable in polluted water samples than a less sensitive marker, unless it is very widely distributed in the target host population although it is at low concentration.

Like specificity testing, sensitivity can be assessed in individual fecal samples or in composites. The small-scale composites described in specificity testing are not generally used for sensitivity assessment; however, large-scale composites can be particularly valuable when they represent the form of fecal material that is most likely to contaminate a water body. Sewage influent, septage in pump-out trucks or holding tanks, cattle or pig waste lagoons, and litter from poultry barns are some examples of useful, large-scale composite samples for assessment of method sensitivity. Individual fecal samples are useful for determining the evenness of marker distribution within a population (see below). Ideally, one would analyze both individual and composite fecal samples for better characterization of marker distribution, with the caveat that this practice makes specificity testing more costly. Furthermore, human fecal samples from healthy individuals (not clinical samples) can be very difficult to obtain, and at least in USA, permission to obtain such samples also can be a logistical challenge. Sewage samples, on the contrary, generally are very easy to obtain.

The number of samples needed to adequately assess sensitivity is another evaluator of method performance that has been approached in an *ad hoc* fashion. Many recent studies have included 20 or more sewage/septage samples when testing sensitivity of human markers (Harwood et al. 2009; McQuaig et al. 2009; Shanks et al. 2009). A study of the use of bovine polyomaviruses for detection of cattle waste tested 26 individual urine samples and ten individual fecal samples (Hundesa et al. 2010). Certainly, one must be cognizant of the geographic area represented by a given study and attempt to collect samples that adequately represent that area. For example, a study that examined the usefulness of MST markers for use across the US Gulf Coast states tested human sewage and septage from the Florida peninsula ($n=24$), the Florida panhandle several hundred miles away ($n=18$), and Mississippi ($n=11$) (Harwood et al. 2009). Data were also obtained from Louisiana and Texas. One hundred percent sensitivity was observed for the three human-associated MST markers (human polyomaviruses, HF183 *Bacteroidales*, and *M. smithii*), providing a strong indication that these markers are prevalent across the Gulf Coast of USA. In practice, the initial sensitivity testing for most new MST methods is more limited, but broadens as others use the methods and as more comprehensive studies are developed. It is highly recommended, however, that markers with limited or unknown specificity be fully vetted before publication of results or recommendation for wider usage.

The evenness of marker distribution among individuals within host populations can influence its usefulness in various locales or geographic ranges (among population distribution) and also becomes important when relatively small numbers of animals may impact a water body. Evenness within a host population is less important, however, in cases where homogenized waste is the source. For example, evenness is less important when leakage from a dairy lagoon is concerned compared with direct deposition of fecal material to a stream by individuals in a small herd of cattle, as the dairy lagoon waste is a composite from many individuals. A recent study (Shanks et al. 2010) of seven PCR methods that target bovine feces has found that marker prevalence and quantity varied widely among herds, even between herds housed at the same facility. These findings support the recommendation of Stoeckel and Harwood (2007) that preliminary testing of marker suitability in a given area is an extremely useful step for determining whether a given MST marker should be further vetted for a particular application and/or in a particular geographic region.

2.3 Evaluation of Data Quality

The previous section discusses considerations for selection of an appropriate analysis for MST based on method performance on laboratory samples. It is important to recognize that collection of reliable data about a given MST target in environmental waters is a daunting task that is quite a bit more complex than working with fecal and sewage samples. Regardless of whether the data are quantitative or qualitative, the researcher must start by evaluating the effectiveness of his or her analytical detection methods in the type(s) of environmental waters included in the study. Though it may be tempting to directly extrapolate bench-level results to environmental scenarios, potential errors introduced in the intermediate steps also must be considered. This section begins with a discussion of data quality assurance for laboratory results, i.e., the data delivered by the analytical protocol. In the next section, assurance of protocol consistency across extended time frames is evaluated. Various complexities introduced by the processes of sample concentration, purification, and storage are then discussed along with approaches to measure and correct for potential error added during these steps. Many of the problems and solutions presented in this section are couched in terms of marker detection by use of quantitative PCR; however, analogous situations and solutions should be apparent for other protocols.

2.3.1 Quality of Data Delivered by the Analytical Protocol and Other Preliminary Considerations

Before environmental samples are analyzed, it is critical to ensure that data of acceptable quality (that will provide meaningful results) can be generated by the analytical protocol. For example, qualitative presence/absence data are meaningless unless the laboratory has confidence in the consistency of detection on positive-control samples. Absence – more appropriately referred to as failure to detect – is much more meaningful when bounded by the limit of detection. Further, the integrity of the values provided by an analytical instrument must be supported by basic laboratory quality-control practices. The purpose of this section is to briefly present and describe the analytical quality-control measures that are necessary for provision of quality MST data. The "Quality Assurance/Quality Control Guidance for Laboratories Performing PCR Analyses on Environmental Samples" (US Environmental Protection Agency 2004) is a resource that might be consulted for related information.

2.3.1.1 Use and Composition of Negative and Positive Controls

Negative controls should be included in all protocols to guard against reagent contamination and (or) false-positive measurements by the instrument. The so-called "no-template control" for PCR is generally done by substituting reagent-grade water for test sample in the PCR reaction. In analyses other than the PCR, this type of negative control frequently is called the reagent blank. Positive results in the no-template control indicate contamination of reagents or equipment and the analysis should be repeated after identifying and correcting the source of contamination. A reagent blank is necessary with each batch of sample analyses to guard against false-positive results (US Environmental Protection Agency 2004). Recommendations for frequency of no-template controls range from inclusion in each standard curve run (Sigma qPCR Tech manual at http://www.sigmaaldrich.com/life-science/molecular-biology/pcr/quantitative-pcr/qpcr-technical-guide.html) to a rate representing 1 per 10 environmental samples analyzed (US Environmental Protection Agency 2004).

Inclusion of appropriate positive-control samples is also necessary during the initial evaluation of the PCR protocol, and, at minimum, each day the samples are run in the laboratory. Initial positive-control tests ideally should be done with well-defined material obtained from a colleague or culture collection (such as a pure culture of a target organism, if applicable), a plasmid containing the target, or a known-positive DNA extract or amplicon. In the event that such a control is not available, one could substitute sewage or feces from the target source; however, the resultant amplicon must be sequenced to determine that the correct product has been produced. The verified product from fecal material can then be cloned into a plasmid vector and subsequently used as positive control material. For quantitative methods, preliminary positive controls would include a dilution series on the positive-control material to evaluate amplification efficiency and the analytical limit of detection (described in more detail in Sect. 2.3.1.3).

After the protocol is demonstrated to consistently generate true-positive reactions, further tests must be done to characterize the method performance. It is essential that reaction positive controls be included with each set of test samples for analyses to guard against false-negative results. For quantitative methods, the standard curve (described in Sect. 2.3.1.2) may serve as the reaction positive control. For qualitative methods, a reaction positive control is necessary for each batch (as described in, for example, USEPA 2004). The concentration of analyte in the reaction positive control should be high enough to be consistently detected. On the contrary, the concentration in the reaction positive control must not be so high as to lack relevance to the environmental samples (and, as a practical note, excessive amounts of amplicon are more likely to produce laboratory contamination, which can be very difficult to eradicate). In our experience, use of a synthetic sample, such as a target-carrying plasmid, at 3–10× the protocol detection limit as the reaction positive control is useful for this purpose. Once the protocol is established, similar positive-control measures incorporated to establish ongoing method performance (Sect. 2.4) may be suitable or used in place of these initial positive-control samples.

2.3.1.2 Composition and Performance Characteristics of the Standard Curve

The primary purpose of the qPCR standard curve is to allow quantification of the target concentration in the purified DNA extract. To fulfill this purpose, the performance of the method (as indicated by the standard curve) should be evaluated. In cases where the model response is well understood – as, for example, in the qPCR – the slope of the standard curve can be used as a diagnostic test of protocol performance. In any case, the standard curve can be used to confirm that the dynamic range of the analytical protocol (the range of concentrations over which the target can be accurately quantified) is suitable for sample analysis. It is important to insure that plasmid control DNA is well purified and free of chromosomal DNA; otherwise, the concentration of target DNA will be overestimated and the standard curve will be erroneous (skewed high). The standard curve should include not fewer than three concentration levels (in addition to the blank) to ensure linearity of response. The concentration levels should be evenly distributed across the relevant range of concentrations that one wishes to detect, such as decimal (tenfold) dilutions for the qPCR.

Lack of linearity at the high or low end of the standard curve indicates that the curve extends beyond the dynamic range of either the measuring instrument or the protocol chemistry. When lack of linearity is noted, either by visual observation of the standard curve or by a coefficient of determination (R^2 value) less than 0.985 (Sigma qPCR tech manual), the detection method must be optimized or the standard curve must be truncated to the linear portion. Truncation of the standard curve may necessitate dilution of the sample extract to bring the sample concentration within the upper limit of detection. Observations higher and lower than the standard curve cannot be treated as reliable quantitative data.

The slope of the standard curve can be used as a performance criterion. In the PCR, for example, doubling of the target DNA is expected during each cycle. When the threshold cycle (Ct) is plotted against $\log_{10}$ (concentration) of target DNA, this leads to an idealized standard curve with a slope of −3.32. This slope generally is converted to amplification efficiency ($E = 10^{(-1/\mathrm{slope})} - 1$) when used as a performance criterion. Amplification efficiency reflects the relationship between the change in target concentration and the change in fluorescence measured; efficiencies between 0.8 and 1.1 often are considered to be acceptable (Sigma qPCR tech manual).

Replicate measurements for assessing the precision of the assay under ideal conditions (in buffer and water) are provided by the standard curve. Shanks et al. (2010) assessed the precision of standard curve values for several qPCR assays targeting human waste by calculating the mean percent coefficient of variation (CV) of the various data points included in the standard curve. Percent CV is the standard deviation divided by the mean and multiplied by 100 for expression as a

Box 2.3.1.2 The Importance of Accurate Dilutions

The slope of the standard curve should indicate the overall efficiency of the reaction. What if the standard curve reflects dilution error instead of nonideal reaction efficiency? The following table illustrates the error that would result from accepting and using a standard curve with a nonideal slope that indicates amplification efficiency of 0.8 (80%) or 1.1 (110%), if the true efficiency were 1.0 (100%). The results are presented in terms of log (concentration), the native output from the qPCR, as well as untransformed concentration to demonstrate the effect of slope on percent error of the measurement. Measurement error can be quite high, and because the standard curve is log normal, the measurement error increases as true concentration increases.

		Calculated concentration (per μL)				Calculated measurement error (%)	
		Log(conc)		Concentration			
True concentration	Measured C_T	$E=0.8$	$E=1.1$	$E=0.8$	$E=1.1$	$E=0.8$	$E=1.1$
10	36.667	0.85	1.08	7	12	40	16
100	33.333	1.71	2.15	51	141	95	29
1,000	30.000	2.56	3.23	367	1,682	173	41
10,000	26.667	3.42	4.30	2,623	20,002	281	50

percentage. Variability in the % CV for the various assays was observed, ranging from 1.03 to 3.00%.

Another important consideration related to standard curves is an understanding of what is being measured. A typical standard curve measures only the response of the instrument when exposed to a given concentration of analyte. For this purpose, qPCR standards can be prepared from extracted, transformed, or synthesized DNA for various targets. If more specific information on performance is desired then more complex approaches must be developed. For example, amplification efficiency within the specific sample matrix (i.e., DNA extracted from an environmental water sample) might be measured by spiking the sample with a known amount of target DNA (Sect. 2.6.1); this approach is helpful for determining whether the DNA extract contains substances that are inhibitory to the PCR. As another example, considering that the measurement typically is back-calculated to represent concentration in the original environmental matrix, some researchers choose to use standardized cell suspensions, rather than extracted DNA, for creation of the dilution series. Each diluted cell suspension must then be extracted independently prior to analysis. In this way, some of the uncertainties in sample processing are incorporated into the standard curve (see Sect. 2.4.2 for full discussion of the uncertainties in sample processing and ways to address those uncertainties).

For the purposes of this section, the standard curve is meant solely to represent the instrument response to a given concentration of analyte. Creation of a standardized solution of analyte can be challenging. For the example of qPCR, a suspension might be made of genomic DNA from the target organism, plasmid DNA with a target amplicon inserted, or synthesized oligomeric DNA representing the target. In any of these cases, determination of copy number concentration can be less than straightforward. If the total DNA is measured to standardize a genomic DNA extract, it is critical to know the molecular weight of the genome and confirm that the DNA extract is free from extrachromosomal DNA. Similarly, if the total DNA is measured for use to standardize a plasmid DNA extract, the molecular weight of the transformed plasmid must be known and the absence of unintended plasmids and of chromosomal DNA should be confirmed. If synthesized DNA is used as the standardized stock material, the lyophilized product must be rehydrated at a high enough concentration to allow quantification. Extreme care must be taken with this solution, as well as less-concentrated extracts of genomic or plasmid DNA, to avoid contamination of the working space with positive-control material. Care must also be taken with the standardized stock material to ensure that the concentration remains consistent throughout the experimental time frame, i.e., it should be aliquoted into volumes for one-time use and stored at −80°C if possible.

2.3.1.3 Estimation of the Protocol Limit of Detection (LOD)

Once a standardized solution of positive control material is created, the method protocol limit of detection can be measured by dilution to extinction. This protocol limit of detection can be used as a quantitative upper bound when reporting and analyzing no-detect data (e.g., <10 gene copies), as described by Helsel (Helsel 1990).

2.3.1.4 Estimation of the Protocol Limit of Quantification (LOQ)

The protocol limit of quantification cannot be lower than the lowest standard of the standard curve because, lacking evidence of linearity, one cannot say that the standard regression is applicable to C_T values above that representing the lowest standard. In many cases, the more dilute standard(s) of the standard curve generate variable responses. To evaluate whether a dilute standard generates less-precise estimates of original concentration relative to other standards, it is useful to test whether the standard error of the mean of replicate measurements is significantly higher than that for other samples. If so, the limit of quantification might more appropriately be set to the most dilute sample for which the standard error of the mean C_T among replicate measurements is consistent with the other standards. When presenting data, results that fall below the LOQ but above the LOD represent a quandary since the marker was detected; however, it was not quantifiable. It is good practice to censor these data and replace them with a notation to indicate detected, but not quantified (along with the LOQ), or to qualify the concentration calculated as "estimated."

2.3.2 *Evaluation of Recovery Methods*

The previous sections dealt with performance of the analytical protocol for detection of selected MST markers. In addition to these considerations, the processing of environmental samples (sampling, concentration, extraction, and purification) also must be considered. Decisions such as the use of centrifugation, filtration, or selective capture to concentrate a sample, and the protocol for extraction and purification of DNA-based or other markers, should be made based on both careful research and collection of performance data. The selected sample processing protocol should consider completeness of recovery, reproducibility of recovery, and purity of the final extract according to the data needs of the study. A test suspension of known-source fecal material, such as human wastewater for protocols designed for general or human-associated fecal contamination, can be useful in these tests.

2.3.2.1 Completeness of Recovery

Aliquots from a single sample should be processed by each candidate method in duplicate or triplicate. The analytical protocol developed using positive-control material (Sect. 2.4.1.2) should be used to detect the selected marker. Some processing methods do not result in full volumetric recovery of sample at each step as, for example, when the DNA extraction protocol requires leaving behind some portion of the total volume to prevent carryover of contaminants. To avoid bias in the final interpretation of recovery, the final results should be calculated in terms of copies or mass per volume of original sample. For example, the researcher may choose to test three protocols with the parameters against 100-mL aliquots of a dilute sewage sample, as shown in Table 2.2. If the researcher is content to evaluate the three protocols solely against the calculated target concentration per μL extract, he or she might choose to use Protocol 1, with nearly 5 times the apparent recovery (concentration in extract) relative to protocol 2 and 1.5 times the apparent recovery relative to protocol 3. If volumetric losses in the concentration, extraction, and purification steps are considered, however, it becomes apparent that protocol 3 outperformed the other two protocols.

If determination of absolute rather than relative recovery through the protocol is necessary, a standardized suspension of cells, i.e., (Stoeckel et al. 2009) or DNA i.e., (Siefring et al. 2008) must be used. In most cases, the target used to measure absolute recovery efficiency will not be the same as the MST marker. Rather, the target will be a surrogate measure and the degree to which it accurately reflects the recovery of MST marker must be tested.

2.3.2.2 Precision of Recovery

In cases where consistency is as or more important than completeness of recovery, the criterion described in the previous section changes from selection for highest

Table 2.2 The influence of recovery of extracted volume from environmental samples on the estimate of analyte concentration

Parameter	Protocol 1	Protocol 2	Protocol 3	Units
Initial volume	100	100	100	mL
Volume lysate	1.5	1	1.5	mL
Volume to extraction	1.3	1	1.3	mL
Volume to purification	1	1	1	mL
Volumetric percent recovery	67%	100%	67%	
Final extract	200	200	500	μL
C_T	27.254	29.547	26.548	Cycles
Standard curve	$C_T = -3.452 \times \log(\text{conc}) + 38.542$			
Log_{10} Conc	3.27	2.61	3.47	per rxn
Volume reaction	5	5	5	μL
Dilution factor	10	10	10	
Conc (in extract)	4.66E+02	1.01E+02	2.98E+02	per μL
Conc (in water)	1.40E+03	2.02E+02	2.24E+03	per 100 mL

recovery to selection for greatest precision (lowest variability among replicate measurements). Precision is frequently expressed as the relative standard deviation, which is the standard deviation of at least three replicated divided by the mean, then multiplied by 100 for expression as a percentage (US Environmental Protection Agency 2004).

2.3.2.3 Purity of DNA Extract

There is no clear-cut, simple test for the suitability of DNA extract for PCR amplification, i.e., freedom from inhibitory substances. If the researcher desires to evaluate extraction purity as a protocol selection criterion, then the candidate methods should be challenged by a series of samples that represents the range of contaminants that might be encountered in the experimental samples. Tests for matrix inhibition (described in Sect. 2.5.2) should be done to identify the methods that result in extracts that must be diluted or further purified to overcome inhibition.

2.4 Consistency of Results: Ongoing Proficiency Testing

Once the sample processing and analytical protocols are selected and characterized for use in the individual laboratory (as described in Sect. 2.2), they will be applied to environmental samples over the course of the study. Before discussing assurance of data quality for environmental samples (Sect. 2.5), this section examines ongoing quality-control tests that can be used to ensure that results across the entire study are comparable. Changes in equipment, staffing, or reagents may result in unexpected changes in method performance.

2.4.1 Use of Process Blanks to Establish a Demonstrated Limit of Detection

In every laboratory, no matter how careful the analyst is, there exists the possibility of false-positive results caused by cross-contamination by test samples or positive-control samples. Extra care must be taken to guard against contamination by the amplicons generated in the standard curve and in areas where concentrated positive-control materials are manipulated. Negative-control samples, generally called process blanks to distinguish them from analytical negative-control samples (frequently termed no-template controls), should be liberally included at all steps of the analysis. Field blanks test for contamination in collection and transport. These are samples in which uncontaminated laboratory water is collected, transported, and processed alongside the test sample. Extraction blanks test for contamination during concentration, extraction, and purification in the laboratory. For these samples, uncontaminated water is processed alongside test samples. Analytical blanks (as described in Sect. 2.3.1.1) test for contamination during the analytical protocol. In addition to detecting contamination problems at various stages of sample manipulation, results from these samples can be used to protect against false-positive results caused by laboratory contamination.

For qPCR-based analyses, the results of all blank analyses can be compiled over a reference time frame. Most of these analytical results likely will be nondetects; however, there likely will be several incidences of low-level detection. Depending on project needs, a level of acceptable error might be established. If, for instance, the researcher wishes to ensure that contamination is identified and rejected at least 95% of the time it causes detection, the fifth percentile C_T value among all blank analyses might be used in place of the analytical limit of detection. If the data pattern indicates that contamination was sporadic and concentrated in a particular time frame, and there are sufficient blank analyses in the time frame, then a rolling LOD or time-specific LOD might be applied to specific time frames. This strategy can be effective to discount low-level detections that have a high probability of being caused by contamination.

2.4.2 Amplification Efficiency

Amplification efficiency is an expression of the doubling rate of DNA during the exponential phase of the PCR. A perfectly efficient reaction precisely doubles the DNA at each cycle, resulting in 100% efficiency (Karlen et al. 2007). Amplification efficiency can be quite variable from one specific qPCR assay to the next, i.e., Shanks et al. (Shanks et al. 2010) observed values ranging from 89 to 99.5% for a variety of assays targeting human waste.

Variation in efficiency, and thus in the slope of the standard curve, can magnify into highly variable results (see box, Sect. 2.3.1.2). To guard against imprecision in slope measurement that is caused by inaccurate dilution, a control chart can be created (i.e., http://www.itl.nist.gov/div898/handbook/mpc/section3/mpc37.htm). In essence, the expected value (slope of the standard curve) and typical variation around the expected value are calculated. Runs in which the measured slope does not fall within the expected range can be subject to closer scrutiny, and may indeed be subject to reanalysis. In this way, false estimation of concentration can be avoided.

It is valuable to consider the possible causes of variation in amplification efficiency when determining good laboratory practices or troubleshooting a problem. The thermal cycler might be out of calibration, a situation that can be addressed by recalibration according to manufacturer's specifications. More difficult to address are problems with the standardized positive-control material and dilutions thereof that are used for the standard curve. Poorly calibrated pipettors and (or) poor technique can contribute to bias in the dilution series (dilution by a fixed factor other than 10; see Box 2.3.1.2). Stored standards may degrade, lose water to sublimation and (or) evaporation, or be sorbed out of solution.

To avoid these problems, proper calibration of thermal cycler and pipettors must be maintained. Technical staff should not only be properly trained, but they should be encouraged to demonstrate proficiency by pipetting control volumes. Concentrated positive control stocks should be stored frozen at −80°C in sealed cryogenic vials. Freeze–thaw cycles for stored stocks should be rigorously minimized – the practice of aliquoting storage material into multiple tubes with volumes suitable for one use is a good strategy for minimizing freeze–thaw cycles. DNA extracts should be stored in low-binding tubes and, particularly for dilute (low concentration) extracts, and incorporation of carrier DNA might prove to be a useful strategy.

2.4.3 Reproducibility (Precision) Over Time

Another way to evaluate data quality over time is to include repeated measures in each analytical batch. The repeated measure sample should be spiked with sewage or other target fecal source and could be DNA extracted from an environmental water sample, in other words, it should present the same challenges for analysis as typical test samples. The material chosen for the repeated measure should be aliqoted into multiple cryogenic vials and stored frozen at −80°C until use to prevent freeze–thaw degradation.

2.5 Performance in Environmental Samples

As described in the previous sections, reliable detection of MST marker presence and (or) concentration in a DNA extract using the PCR can be complex. The considerations outlined in Sects. 2.3 and 2.4 enhance our ability to accomplish this task in laboratory-generated positive-control samples. For test samples consisting of environmental waters or other matrices, additional effects should be considered. The efficiency of marker recovery through sample processing steps must be measured or assumed based on previous measurements. The potential for poor amplification efficiency in the individual sample, sometimes caused by inhibitors in the extract matrix, must be evaluated. Finally, variability must be measured to allow comparison of values or detection rates across samples.

2.5.1 *Recovery Efficiency*

In the rare instances that recovery efficiency through the processes of concentration, extraction, and purification has been measured, wide differences have been observed within sample sets. Several recent studies have used spike-and-recovery controls to estimate recovery efficiency. In one study (Stoeckel et al. 2009), recovery efficiency ranges were 1–5% in one sample set, 3–12% in another, and 0.2–30% in a third. These wide ranges in recovery efficiency are not unique. One study (Koike et al. 2007) reported a recovery range of 20–200% and another (Lebuhn et al. 2004) measured recoveries in the range of 10–70%. It is clear from these reports that variable recovery efficiencies can greatly affect the measured concentration of analyte, and lack of recovery efficiency data can impair cross-sample and cross-study comparisons by (1) masking true differences because of high variability and (2) showing false differences because of different recovery efficiencies in different studies. Use of measured recovery efficiency to adjust observed concentrations to account for analyte lost in processing can reduce variability in replicate measures (Stoeckel et al. 2009) and may be essential for accurate comparison of results between studies.

These considerations are not limited to detection using the qPCR. Divergent recovery efficiencies can complicate the interpretation of presence–absence analysis for endpoint PCR. For example, several samples that contain marker at approximately 5 times the limit of detection might be encountered. If the range of recovery efficiencies for a particular protocol is 10–60%, then the marker will be detected in some of the samples, but not all of the samples. Particularly in descriptive experiments with little replication, this scenario can lead to erroneous conclusions unless the limit of detection can be adjusted according to the recovery efficiency. With adjustment, it would be apparent to the researcher that the negative results were more common in samples with lower recovery efficiency, indicating that those samples were not necessarily less contaminated compared with the samples that tested positive.

2.5.2 *Matrix Inhibition*

An assumption of the quantitative PCR is that the efficiency of each reaction is the same as the efficiency measured in the standard curve. The matrix of the individual reaction tube may influence the efficiency of the PCR (Cankar et al. 2006), In the case of conventional PCR, differences in amplification efficiencies between tubes can alter the detection limit, making interpretation of nondetect results difficult. These considerations generally are discussed in terms of "matrix effect" (US Environmental Protection Agency 2004).

Four approaches are sometimes used to address the matrix effect on amplification efficiency. The first depends on generation of measureable product in the qPCR. The slope of the linear portion of the amplification curve is compared with the slope of the standard curve. If the amplification efficiency is consistent, the slopes will not be different (Ramakers et al. 2003).

The second approach is applicable only to the qPCR using TaqMan or related chemistry with a probe. A competitive internal positive control (CIPC) sequence, also known as internal amplification control (IAC) (Gregory et al. 2006; Sen et al. 2007), is developed with the same flanking primers as the marker but a different (artificial) internal sequence for probe hybridization. This positive control sequence is created and inserted into a carrier plasmid. A second probe, labeled with a different fluor and, therefore, detectable on a different channel from the marker detection probe, is added to the qPCR mix to create a multiplex qPCR using one primer pair and two probes. After carefully optimizing the multiplex reaction, the CIPC is added to every reaction mixture at low concentration (approximately 100 copies per reaction). Failure to detect the CIPC at the expected concentration indicates low amplification efficiency and inhibition in a reaction. In cases where the failure is due to matrix inhibition, dilution may result in a more accurate estimate of original marker concentration.

The third approach, also applicable to qPCR, is to use a dilution series to detect inhibition by amplification. Using the same premise as the standard curve, in which the slope of a series of dilutions measures amplification efficiency, a dilution series on a test sample also can be prepared. The difference between threshold cycles measured for decimal dilutions of a given sample should be approximately 3.3. Large differences in amplification efficiency are apparent when the difference is substantially less than 3.3. A benefit of this approach is that the appropriate dilution is known upon completion of the qPCR analysis. The assay does not have to be repeated, and the test simply indicates which dilution was appropriate. On the negative side, the dilution method is time-, labor-, and reagent-intensive; it cannot be used for samples in which the ambient concentration is near the detection limit, and it cannot detect slight deviations from the assumed amplification efficiency.

A traditional matrix spike is the fourth possible approach (USEPA 2004) and can serve a similar purpose for both conventional (end point) and quantitative PCR. The PCR master mix for a run is created and divided into two aliquots. One aliquot is used to generate the standard curves (if desired) and no-template control and the other is amended with a finite amount of the positive-control standard. A spiked no-template control must be run, and all test reactions also use the spiked master mix. Matrix inhibition is apparent in cases where the C_{T} of the test reaction is higher than the C_{T} of the spiked no-template control, or where a conventional PCR fails to amplify. This approach leaves one open to the possibility of partial inhibition (amplification present at low efficiency). It is theoretically possible to quantify the spiked PCR reaction and obtain an estimate of the spike concentration by subtracting the concentration in a parallel, unspiked test reaction. However, given the variability in concentration measurement by qPCR, it is virtually impossible to detect small differences, in the order of 100 copies, between two measurements that are substantially higher.

2.5.3 *Processing Precision*

Error measurement is critically important, particularly if the results might be challenged in a court of law (Chap. 13). One strategy by which processing and analytical error can be evaluated is incorporation of split replicate samples in the experimental design. A selected number of samples (in the order of 1 in 20, depending on project needs) is split into two samples, with unique identifying labels, and processed as separate test samples. The average standard error of the means can be used as a useful measure of precision in the overall process of sample processing and analysis.

2.6 Validation of Method Accuracy

Topics related to selection and (or) validation of an appropriate MST marker, initial evaluation of protocol performance against quality-control samples, ongoing tests of protocol performance, and performance measures to enhance the validity of data collected on test samples were discussed in detail in prior sections. None of those considerations is valuable if the data do not lead the researcher to correct information about environmental contaminants and processes. The purpose of this section is to describe broader quality-control measures to test the validity of interpretations.

2.6.1 *Confirmatory Reference Materials*

It is useful to check natural reference materials (i.e., feces or sewage) for sensitivity and specificity on a regular basis, particularly if a method is to be used over several years' period. It is essential to gather new reference materials for performance testing when a study is initiated in a new geographic area, as the distribution of the marker in hosts may vary. Note that quantitative data, e.g., target copies derived from qPCR, should be normalized to a common parameter such as grams of fecal material.

2.6.2 *Aqueous Suspensions*

The practice of suspending reference material in water and then testing the results provides a useful "reality check" for MST method performance (Field et al. 2003; Griffith et al. 2003; Harwood et al. 2003; 2009). Sensitivity, specificity, and a relative, semiquantitative assessment of the method limit of detection can be gained in this manner. For the latter, sewage (i.e., municipal sewage or water from animal-waste holding ponds) can be subjected to serial dilutions and added to buffer or an environmental water sample. These mixtures are then concentrated and DNA is extracted according to the laboratory's standard protocols. In this manner, the extent to which a pollution source can be diluted and still detected (Harwood et al. 2009) or quantified can be assessed. One can also add mixed sewage sources to aqueous samples and the detection and/or quantification of multiple targets can be assessed. An important consideration in interpreting results from multiple-fecal source samples is that the various targets may well be present at different concentrations in the reference material; therefore, if qPCR is used, the relative proportion of multiple targets quantified in the sample will not necessarily reflect the proportions of reference material added to the sample.

2.7 Conclusions

As the science of MST continues to grow and become more sophisticated, performance criteria will doubtlessly become more codified and consistent. All of the methods and applications discussed in this book depend upon the rigorous application of performance criteria and validation strategies to ensure the accuracy and usefulness of results. In a period of about two decades, we have made major strides toward legitimizing the use of MST for applications ranging from total maximum daily load assessment, to protection of the health of recreational water users, to pollution source identification in the legal area. We trust that this chapter will be obsolete within a decade, as further improvements to method performance criteria will doubtlessly be discovered and implemented by MST practitioners.

References

Ahmed, W., Goonetilleke, A., Powell, D., Chauhan, K., and Gardner, T. (2009a). Comparison of molecular markers to detect fresh sewage in environmental waters. *Water Res* **43**(19), 4908–17.

Ahmed, W., Goonetilleke, A., Powell, D., and Gardner, T. (2009b). Evaluation of multiple sewage-associated *Bacteroides* PCR markers for sewage pollution tracking. *Water Res* **43**(19), 4872–7.

Albert, J. M., Munakata-Marr, J., Tenorio, L., and Siegrist, R. L. (2003). Statistical evaluation of bacterial source tracking data obtained by rep-PCR DNA fingerprinting of *Escherichia coli*. *Environ Sci Technol* **37**(20), 4554–60.

Balleste, E., Bonjoch, X., Belanche, L. A., and Blanch, A. R. (2010). Molecular indicators used in the development of predictive models for microbial source tracking. *Appl Environ Microbiol* **76**(6), 1789–95.

Bernhard, A. E., and Field, K. G. (2000). A PCR assay To discriminate human and ruminant feces on the basis of host differences in *Bacteroides-Prevotella* genes encoding 16S rRNA. *Appl Environ Microbiol* **66**(10), 4571–4.

Boehm, A. B., Fuhrman, J. A., Mrse, R. D., and Grant, S. B. (2003). Tiered approach for identification of a human fecal pollution source at a recreational beach: case study at Avalon Bay, Catalina Island, California. *Environ Sci Technol* **37**(4), 673–80.

Cankar, K., Stebih, D., Dreo, T., Zel, J., and Gruden, K. (2006). Critical points of DNA quantification by real-time PCR--effects of DNA extraction method and sample matrix on quantification of genetically modified organisms. *BMC Biotechnol* **6**, 37.

Carson, C. A., Shear, B. L., Ellersieck, M. R., and Schnell, J. D. (2003). Comparison of ribotyping and repetitive extragenic palindromic-PCR for identification of fecal *Escherichia coli* from humans and animals. *Appl Environ Microbiol* **69**(3), 1836–9.

Dombek, P. E., Johnson, L. K., Zimmerley, S. T., and Sadowsky, M. J. (2000). Use of repetitive DNA sequences and the PCR To differentiate *Escherichia coli* isolates from human and animal sources. *Appl Environ Microbiol* **66**(6), 2572–7.

Field, K. G., Chern, E. C., Dick, L. K., Fuhrmann, J., Griffith, J., Holden, P., LaMontagne, M. G., Le, J., Olson, B., and Simonich, M. T. (2003). A comparative study of culture-independent, library-independent genotypic methods of fecal source tracking. *Journal of Water and Health* **01**(4), 181–94.

Gordon, D. M., Bauer, S., and Johnson, J. R. (2002). The genetic structure of *Escherichia coli* populations in primary and secondary habitats. *Microbiology* **148**(Pt 5), 1513–22.

Gregory, J. B., Litaker, R. W., and Noble, R. T. (2006). Rapid one-step quantitative reverse transcriptase PCR assay with competitive internal positive control for detection of enteroviruses in environmental samples. *Appl Environ Microbiol* **72**(6), 3960–7.

Griffith, J. F., Weisberg, S. B., and McGee, C. D. (2003). Evaluation of microbial source tracking methods using mixed fecal sources in aqueous test samples. *Journal of Water and Health* **01**, 141–151.

Hagedorn, C., Robinson, S. L., Filtz, J. R., Grubbs, S. M., Angier, T. A., and Reneau, R. B., Jr. (1999). Determining sources of fecal pollution in a rural Virginia watershed with antibiotic resistance patterns in fecal streptococci. *Appl Environ Microbiol* **65**(12), 5522–31.

Harwood, V. J. (2007). Assumptions and limitations of microbial source tracking methods. *In* "Microbial Source Tracking" (J. Santo-Domingo, Sadowsky, M., Ed.). ASM Press, Washington, D.C.

Harwood, V. J., Brownell, M., Wang, S., Lepo, J., Ellender, R. D., Ajidahun, A., Hellein, K. N., Kennedy, E., Ye, X., and Flood, C. (2009). Validation and field testing of library-independent microbial source tracking methods in the Gulf of Mexico. *Water Res* **43**(19), 4812–9.

Harwood, V. J., Whitlock, J., and Withington, V. (2000). Classification of antibiotic resistance patterns of indicator bacteria by discriminant analysis: use in predicting the source of fecal contamination in subtropical waters. *Appl Environ Microbiol* **66**(9), 3698–704.

Harwood, V. J., Wiggins, B., Hagedorn, C., Ellender, R. D., Gooch, J., Kern, J., Samadpour, M., Chapman, A. C. H., Robinson, B. J., and Thompson, B. C. (2003). Phenotypic library-based microbial source tracking methods: Efficacy in the California collaborative study. *J. Water Health* **01**, 153–66.

Helsel, D. R. (1990). Less than obvious: Statistical treatment of data below detection limit. *Environmental Science and Technology* (24), 1767–1744.

Hundesa, A., Bofill-Mas, S., Maluquer de Motes, C., Rodriguez-Manzano, J., Bach, A., Casas, M., and Girones, R. (2010). Development of a quantitative PCR assay for the quantitation of bovine polyomavirus as a microbial source-tracking tool. *J Virol Methods* **163**(2), 385–9.

Karlen, Y., McNair, A., Perseguers, S., Mazza, C., and Mermod, N. (2007). Statistical significance of quantitative PCR. *BMC Bioinformatics* **8**, 131.

Kildare, B. J., Leutenegger, C. M., McSwain, B. S., Bambic, D. G., Rajal, V. B., and Wuertz, S. (2007). 16S rRNA-based assays for quantitative detection of universal, human-, cow-, and dog-specific fecal Bacteroidales: a Bayesian approach. *Water Res* **41**(16), 3701–15.

Kirs, M., Harwood, V. J., Fidler, A. E., Gillespie, P. A., Fyfe, W. R., Blackwood, A. D., and Cornelisen, C. C. (2011). Source tracking faecal contamination in an urbanized and a rural waterway in the Nelson-Tasman region, New Zealand. *New Zealand J Freshwater Res* **45**:53–8.

Koike, S., Krapac, I. G., Oliver, H. D., Yannarell, A. C., Chee-Sanford, J. C., Aminov, R. I., and Mackie, R. I. (2007). Monitoring and source tracking of tetracycline resistance genes in lagoons and groundwater adjacent to swine production facilities over a 3-year period. *Appl Environ Microbiol* **73**(15), 4813–23.

Korajkic, A., Badgley, B. D., Brownell, M. J., and Harwood, V. J. (2009). Application of microbial source tracking methods in a Gulf of Mexico field setting. *J Appl Microbiol* **107**(5), 1518–27.

Layton, A., McKay, L., Williams, D., Garrett, V., Gentry, R., and Sayler, G. (2006). Development of Bacteroides 16S rRNA gene TaqMan-based real-time PCR assays for estimation of total, human, and bovine fecal pollution in water. *Appl Environ Microbiol* **72**(6), 4214–24.

Lebuhn, M., Effenberger, M., Garces, G., Gronauer, A., and Wilderer, P. A. (2004). Evaluating real-time PCR for the quantification of distinct pathogens and indicator organisms in environmental samples. *Water Sci Technol* **50**(1), 263–70.

Lu, J., Santo Domingo, J. W., Hill, S., and Edge, T. A. (2009). Microbial diversity and host-specific sequences of Canada goose feces. *Appl Environ Microbiol* **75**(18), 5919–26.

McLain, J. E., Ryu, H., Kabiri-Badr, L., Rock, C. M., and Abbaszadegan, M. (2009). Lack of specificity for PCR assays targeting human Bacteroides 16S rRNA gene: cross-amplification with fish feces. *FEMS Microbiol Lett* **299**:38–43.

McLellan, S. L., Daniels, A. D., and Salmore, A. K. (2003). Genetic characterization of *Escherichia coli* populations from host sources of fecal pollution by using DNA fingerprinting. *Appl Environ Microbiol* **69**(5), 2587–94.

McQuaig, S. M., Scott, T. M., Harwood, V. J., Farrah, S. R., and Lukasik, J. O. (2006). Detection of human-derived fecal pollution in environmental waters by use of a PCR-based human polyomavirus assay. *Appl Environ Microbiol* **72**(12), 7567–74.

McQuaig, S. M., Scott, T. M., Lukasik, J. O., Paul, J. H., and Harwood, V. J. (2009). Quantification of human polyomaviruses JC Virus and BK Virus by TaqMan quantitative PCR and comparison to other water quality indicators in water and fecal samples. *Appl Environ Microbiol* **75**(11), 3379–88.

Myoda, S. P., Carson, C. A., Fuhrmann, J. J., Hahm, B.-K., Hartel, P. G., Yampara-Iquise, H., Johnson, L., Kuntz, R. L., Nakatsu, C. H., Sadowsky, M. J., and Samadpour, M. (2003). Comparison of genotypic-based microbial source tracking methods requiring a host origin database. *Journal of Water and Health* **01**(4), 167–180.

Noble, R. T., Allen, S. M., Blackwood, A. D., Chu, W., Jiang, S. C., Lovelace, G. L., Sobsey, M. D., Stewart, J. R., and Wait, D. A. (2003). Use of viral pathogens and indicators to differentiate between human and non-human fecal contamination in a microbial source tracking comparison study. *J Water Health* **1**(4), 195–207.

Noble, R. T., Griffith, J. F., Blackwood, A. D., Fuhrman, J. A., Gregory, J. B., Hernandez, X., Liang, X., Bera, A. A., and Schiff, K. (2006). Multitiered approach using quantitative PCR to track sources of fecal pollution affecting Santa Monica Bay, California. *Appl Environ Microbiol* **72**(2), 1604–12.

Parveen, S., Portier, K. M., Robinson, K., Edmiston, L., and Tamplin, M. L. (1999). Discriminant analysis of ribotype profiles of *Escherichia coli* for differentiating human and nonhuman sources of fecal pollution. *Appl Environ Microbiol* **65**(7), 3142–7.

Ramakers, C., Ruijter, J. M., Deprez, R. H., and Moorman, A. F. (2003). Assumption-free analysis of quantitative real-time polymerase chain reaction (PCR) data. *Neurosci Lett* **339**(1), 62–6.

Ritter, K. J., Carruthers, E., Carson, C. A., Ellender, R. D., Harwood, V. J., Kingsley, K., Nakatsu, C., Sadowsky, M., Shear, B., West, B., Whitlock, J. E., Wiggins, B. A., and Wilbur, J. D. (2003). Assessment of statistical methods used in library-based approaches to microbial source tracking. *J Water Health* **1**(4), 209–23.

Robinson, B. J., Ritter, K. J., and Ellender, R. D. (2007). A statistical appraisal of disproportional versus proportional microbial source tracking libraries. *J Water Health* **5**(4), 503–9.

Sen, K., Schable, N. A., and Lye, D. J. (2007). Development of an internal control for evaluation and standardization of a quantitative PCR assay for detection of Helicobacter pylori in drinking water. *Appl Environ Microbiol* **73**(22), 7380–7.

Seurinck, S., Defoirdt, T., Verstraete, W., and Siciliano, S. D. (2005). Detection and quantification of the human-specific HF183 Bacteroides 16S rRNA genetic marker with real-time PCR for assessment of human faecal pollution in freshwater. *Environ Microbiol* **7**(2), 249–59.

Shanks, O. C., Domingo, J. W., Lu, J., Kelty, C. A., and Graham, J. E. (2007). Identification of bacterial DNA markers for the detection of human fecal pollution in water. *Appl Environ Microbiol* **73**(8), 2416–22.

Shanks, O. C., Kelty, C. A., Sivaganesan, M., Varma, M., and Haugland, R. A. (2009). Quantitative PCR for genetic markers of human fecal pollution. *Appl Environ Microbiol* **75**(17), 5507–13.

Shanks, O. C., White, K., Kelty, C. A., Sivaganesan, M., Blannon, J., Meckes, M., Varma, M., and Haugland, R. A. (2010). Performance of PCR-based assays targeting Bacteroidales genetic markers of human fecal pollution in sewage and fecal samples. *Environ Sci Technol* **44**(16), 6281–8.

Siefring, S., Varma, M., Atikovic, E., Wymer, L., and Haugland, R. A. (2008). Improved real-time PCR assays for the detection of fecal indicator bacteria in surface waters with different instrument and reagent systems. *J Water Health* **6**(2), 225–37.

Stoeckel, D. M., and Harwood, V. J. (2007). Performance, design, and analysis in microbial source tracking studies. *Appl Environ Microbiol* **73**(8), 2405–15.

Stoeckel, D. M., Mathes, M. V., Hyer, K. E., Hagedorn, C., Kator, H., Lukasik, J., O'Brien, T. L., Fenger, T. W., Samadpour, M., Strickler, K. M., and Wiggins, B. A. (2004). Comparison of seven protocols to identify fecal contamination sources using *Escherichia coli*. *Environ Sci Technol* **38**(22), 6109–17.

Stoeckel, D. M., Stelzer, E. A., and Dick, L. K. (2009). Evaluation of two spike-and-recovery controls for assessment of extraction efficiency in microbial source tracking studies. *Water Res* **43**(19), 4820–7.

U.S. Environmental Protection Agency (2004). Quality assurance/quality control guidance for laboratories performing PCR analyses on environmental samples. U.S. Environmental Protection Agency. EPA 815-B-04–001.

U.S. Environmental Protection Agency (2005). Microbial source tracking guide document. U.S. Environmental Protection Agency. EPA/600/R-05/064. June 2005.

Vogel, J. R., Stoeckel, D. M., Lamendella, R., Zelt, R. B., Santo Domingo, J. W., Walker, S. R., and Oerther, D. B. (2007). Identifying fecal sources in a selected catchment reach using multiple source-tracking tools. *J Environ Qual* **36**(3), 718–29.

Weidhaas, J. L., Macbeth, T. W., Olsen, R. L., Sadowsky, M. J., Norat, D., and Harwood, V. J. (2010). Identification of a Brevibacterium marker gene specific to poultry litter and development of a quantitative PCR assay. *J Appl Microbiol* **109**:334–47.

Wiggins, B. A. (1996). Discriminant analysis of antibiotic resistance patterns in fecal streptococci, a method to differentiate human and animal sources of fecal pollution in natural waters. *Appl Environ Microbiol* **62**(11), 3997–4002.

Wiggins, B. A., Cash, P. W., Creamer, W. S., Dart, S. E., Garcia, P. P., Gerecke, T. M., Han, J., Henry, B. L., Hoover, K. B., Johnson, E. L., Jones, K. C., McCarthy, J. G., McDonough, J. A., Mercer, S. A., Noto, M. J., Park, H., Phillips, M. S., Purner, S. M., Smith, B. M., Stevens, E. N., and Varner, A. K. (2003). Use of antibiotic resistance analysis for representativeness testing of multiwatershed libraries. *Appl Environ Microbiol* **69**(6), 3399–405.

Chapter 3
Library-Dependent Source Tracking Methods

Joanna Mott and Amanda Smith

Abstract A range of bacterial source tracking techniques is grouped under what is commonly referred to as library-dependent methods (LDM). The methods require the construction of a library of known source profiles that are used for comparison with environmental isolates to determine sources of contamination. Development of the library for a particular study requires consideration of many factors including the organism or group of organisms to be used, size of the library, proportionality, representativeness, and library stability. Appropriate performance testing and statistical analysis are critical for confidence in the results. Each method has advantages and disadvantages, and while many have been compared, there is no consensus on a "standard" method or group of methods due to the complexity of the field and range in the scope and goals of studies. Current recommendations include use of a toolbox approach (multiple methods) or limiting costs through use of a tiered, targeted design, initially with monitoring, followed by targeted source tracking.

Keywords Library-dependent • Geographic stability • Temporal stability • Representativeness • Statistics • Indicators • Method comparisons • Profiles

3.1 Introduction

This chapter focuses on a group of methods defined by their requirement for a dataset of characteristics (fingerprints, patterns, or profiles) of fecal isolates from different animal sources (host-origin) that can be used to compare with the same characteristics of isolates from the environment to determine source(s) of fecal contamination. The dataset is commonly called a library and the set of methods, library-dependent

---

J. Mott (✉)
Department of Life Sciences, Texas A&M University-Corpus Christi, 6300 Ocean Drive, Corpus Christi, TX 78412-5802, USA
e-mail: Joanna.mott@tamucc.edu

C. Hagedorn et al. (eds.), *Microbial Source Tracking: Methods, Applications, and Case Studies*, DOI 10.1007/978-1-4419-9386-1_3,

methods (LDM). The approach is based on the hypotheses that certain characteristics of fecal bacteria are associated with specific animals or groups of animals (Simpson et al. 2002; Ahmed 2007; Field and Samadpour 2007), that these characteristics in environmental strains are similar to those found in host groups (Ahmed 2007), and that the relative proportion of the identifying characteristic remains constant in the environment over time (i.e., survival of strains with each trait is the same). As pointed out by Field and Samadpour (2007), the latter assumption is usually untested.

The library can be composed of phenotypic (biochemical) or genotypic profiles of the bacteria. Early library-dependent methods were primarily based on phenotypic characters of isolates from known source fecal samples; for example, antibiotic resistance profiles of indicator bacteria such as enterococci or *E. coli* (Wiggins 1996; Parveen et al. 1997; Hagedorn et al. 1999; Wiggins et al. 1999; Harwood et al. 2000). As methods such as ribotyping and variations of PCR have been adopted for source tracking, libraries may comprise genotypic characteristics such as presence or absence of specific bands, or band patterns of isolates from different animals (Table 3.1).

3.2 General Library Characteristics

3.2.1 Organisms

The majority of library-dependent methods utilize characteristics of fecal indicator bacteria, mostly commonly *Escherichia coli* (Dombek et al. 2000; Buchan et al. 2001; McLellan et al. 2001; Guan et al. 2002; Leung et al. 2004; Lu et al. 2004; Webster et al. 2004; Venieri et al. 2004; Anderson et al. 2006; Casarez et al. 2007a; b; Vogel et al. 2007), or enterococci (Graves et al. 2002; Booth et al. 2003; Choi et al. 2003; Hagedorn et al. 2003; Genthner et al. 2005; Burtscher et al. 2006; Dickerson et al. 2007; Jiang et al. 2007; Giebel et al. 2008), but fecal coliforms (Harwood et al. 2000; Whitlock et al. 2002; Burnes 2003; Duran et al. 2006), fecal streptococci (Wiggins 1996; Hagedorn et al. 1999; Harwood et al. 2000; Geary and Davies 2003), and even total coliforms (Haznedaroğlu et al. 2005) also have been used. These bacteria have the advantage of linking source tracking results to the routinely monitored fecal indicator group(s) that provides regulatory agencies with a consistent parameter for evaluation. Coliphages have also been used in LDM; however, the "library" is comprised of only four subtypes that distinguish between human and animals, and the method is usually classified as a library-independent method (LIM) (USEPA 2005; Stoeckel and Harwood 2007).

The indicator bacteria group used for the method can affect the results, with various studies demonstrating that some can provide a higher level of accuracy than others, although this appears to be somewhat method dependent and other factors also play a role. For example, Yurtsever et al. (2007) showed FAME analyses were more successful using fecal coliforms or total coliforms rather than *E. coli* and suggested that the increasing accuracy with the increasing diversity of organisms (i.e., *E. coli* compared to total coliforms) may be due to differential distribution of

Table 3.1 Library-based phenotypic and genotypic methods

Method	Advantages	Disadvantages	References
Antibiotic resistance analysis (ARA)	Low cost Ease of use	Questionable stability of antibiotic resistance markers Geographical variability	Scott et al. (2002), Simpson et al. (2002), Meays et al. (2004), Seurinck et al. (2005), Ebdon and Taylor (2006)
Carbon source utilization (CSU)	Rapid Standardized	Unknown temporal Stability	Konopka et al. (1998), Hagedorn et al. (2003)
Fatty acid methyl ester (FAME) analysis	Potentially may use a small library	Still primarily in testing mode for MST	Haznedaroğlu et al. (2005), Field and Samadpour (2007)
Pulsed-field gel electrophoresis (PFGE)	Very sensitive Highly discriminatory Reproducible	Time-consuming Very extensive library needed due to sensitivity	Farber (1996), Olive and Bean (1999), Scott et al. (2002), Simpson et al. (2002), Lu et al. (2004)
Ribotyping	Excellent reproducibility High sensitivity	Costly Labor-intensive Diversity of ribotypes (geographical, temporal, and diet)	Van Belkum et al. (2001), Hartel et al. (2002), Scott et al. (2002), Jenkins et al. (2003), Hartel et al. (2003), Myoda et al. (2003)
Rep-PCR	Cost effective Less technically demanding than other genotypic methods	Temporal and geographical variability Moderate reproducibility	Carson et al. (2003), USEPA (2005), Seurinck et al. (2005), Hansen et al. (2009)
Amplified Fragment Length Polymorphism (AFLP) analysis	Reproducibility Robustness	Costly Limited information on method	Vos et al. (1995), Blears et al. (1998), Yan and Sadowsky (2007)
Random amplified polymorphic DNA (RAPD) analysis	Efficient Easy to use	Reproducibility problems Lack of standardization	Hopkins and Hilton (2000), Venieri et al. (2004), USEPA (2005)
Denaturing gradient gel electrophoresis (DGGE)	Community fingerprinting	Technically demanding Selection of proper gene target	Farnleitner et al. (2000), Simpson et al. (2002), Meays et al. (2004)
Matrix-assisted laser desorption/ ionization time of flight mass spectroscopy (MALDI-TOF-MS)	Rapid	Lack of information	Siegrist et al. (2007), Giebel et al. (2008)

species in host. In a performance comparison between carbon source utilization profiles of *Enterococcus* and *E. coli, Enterococcus* was a better predictor of source than *E. coli* (Harwood et al. 2003), and Griffith et al. (2003) also showed better results with *Enterococcus* antibiotic resistance profiles (ARA) than with that of *E. coli* . By contrast, Ahmed et al. (2006) used a relatively large existing (4,057 enterococci, 3,728 *E. coli*) library as well as a new watershed-specific library (~776 enterococci, 780 *E. coli*) of metabolic fingerprints and demonstrated that the *Enterococcus* and *E. coli* databases were equally capable of identifying sources.

3.2.2 Library Development

As the identification of fecal pollution sources(s) is based on the library, success of the method depends on development of an appropriate library with careful consideration of a number of criteria including distribution of fingerprint patterns among the potential sources, representation of each source, and method of statistical analysis. There has been considerable discussion on the requirements for a library, with some being dependent on the technique, while others are applicable to any library. A first step in determining the library composition should be a sanitary survey of the area with additional input from stakeholders, as appropriate, to determine the animals that are potential contributors to the fecal contamination of the watershed. One advantage of LDMs is that the study library can be tailored for a particular watershed. Libraries have been set up to distinguish human vs. nonhuman sources, to separate individual animal sources (e.g., cow, deer), to group animals into categories such as livestock, etc. The latter can then be assessed for regulatory processes, such as the US Total Maximum Daily Load (TMDL) programs, as "controllable" (human, livestock) vs. "uncontrollable," or at least more difficult to control (e.g., birds, wildlife), sources of contamination. Separate subcategories (e.g., dairy cow vs. range cattle) can provide an additional level of complexity to known source profiles.

3.2.3 Representativeness and Proportionality of a Library

In order for the source of environmental (water) isolates to be correctly identified, it is essential that the library be large enough and contain a sufficiently diverse set of profiles to be representative of all the potential animal sources in a particular watershed. The library should be tested before use for representativeness (Wiggins et al. 2003; USEPA 2005; Field and Samadpour 2007; Stoeckel and Harwood 2007) and isolates are added as needed.

While representativeness of phenotypic or genetic diversity in indicator organism populations is important, it is often not achieved due to the strain diversity of the organism, or sometimes due to difficulties in collection of certain source isolates. Additional issues arise as the sources are categorized into different numbers of categories i.e., for a six-way classification, there might be equal numbers from each source but when these are collapsed to examine fewer categories, the distribution becomes biased. This frequently occurs when human/sewage is compared with nonhuman in which the nonhuman contains data for multiple animals. This results in a disproportionate library (i.e., contains data from more of certain sources than others), which creates challenges in the statistical analysis as demonstrated by Robinson et al. (2007) who suggest using k-nearest neighbor analysis in these cases, rather than the more commonly used discriminant analysis.

Another debate for researchers using LDM has been the number of isolates that should be collected from a particular fecal sample for inclusion in the library. Depending on the method (and characteristic used), some have used as few as 2–5 for ribotyping (Hartel et al. 2002; Moore et al. 2005; Nelson et al. 2008), rep-PCR (Albert et al. 2003), or ARA (Whitlock et al. 2002; Moore et al. 2005), while others, particularly for antibiotic resistance libraries, have used ~10–12 (Wiggins et al. 2003), up to 20 from composite samples (Graves et al. 2002) or as many as ~48 (Hagedorn et al. 1999). Ebdon and Taylor (2006) in analyzing antibiotic resistance patterns of enterococci found 24 isolates per sample provided sufficient representation using discriminant function analysis. Many papers fail to report the number of isolates from a single fecal sample, providing only the number of isolates per each animal source. Decloning the library, basically removing those isolates from the same fecal sample that have identical profiles, is an additional factor for consideration and has been utilized in a number of studies (Wiggins et al. 2003; Ahmed et al. 2006; Casarez et al. 2007a, b; Graves et al. 2007) Researchers have suggested that removing clonal isolates improves prediction and library representativeness within a source tracking library (Wiggins et al. 2003; Hassan et al. 2005). Including clones can also bias the average rate of correct classification and overestimate the ability of the library to correctly classify the source of the isolate (Johnson et al. 2004). However, utilizing multiple isolates from one known source fecal sample does not necessarily imply that these isolates will be clonal. Results from Fogarty et al. (2003) showed that for gulls there were multiple *E. coli* rep-PCR genotypes in a sample and no identical genotypes were found except within a single sample.

The size of a library is also dependent on the method and organism used. The genetic diversity of *E. coli* has been demonstrated to result in the need for a large library with thousands of isolates to include the potential profiles of the organism. One strategy to minimize this problem, suggested by Johnson et al. (2004), is to use libraries generated for a specific watershed. Library size in studies using phenotypic characteristics have ranged from a few hundred (Geary and Davies 2003; Moussa and Massengale 2008) to one or several thousand isolates (Graves et al. 2002; Burnes 2003; Choi et al. 2003; Moore et al. 2005; Sayah et al. 2005). Libraries for genotypic methods tend to be smaller, typically less than 600 (Carson et al. 2001; Hartel et al. 2002; Albert et al. 2003; Scott et al. 2003) with a few over 800 (Casarez et al. 2007b). For enterococci antibiotic resistance profiles, Ebdon and Taylor (2006) demonstrated by using Shannon's diversity index that for a three source classification at least 400–500 isolates per source was needed, while Wiggins et al. (2003) showed that a minimum of 2,300 isolates were needed to produce a representative library. Olivas and Faulkner (2008) used 772 *Enterococcus* ARA profiles, but library representativeness testing suggested that this was not sufficient. For *E. coli*, the diversity of strains and temporal changes (discussed below) suggest that large numbers of isolates are needed to generate a representative library (Jenkins et al. 2003; McLellan 2004).

A complicating factor is the representativeness of host strains in the environment due to differences in survival and adaptation. Ahmed and Katouli (2008) found that >95% of the environmental strains were represented in their phenotypic enterococci library; however, for molecular methods with more discrimination, such as pulsed-field gel electrophoresis (PFGE), very large libraries must be created to represent the diversity found in the environment (Casarez et al. 2007b).

3.2.4 Stability of Libraries

3.2.4.1 Geographic Stability

Geographic stability is another factor that must be considered in developing and using a library, although relatively few studies have examined this issue. In Europe, patterns of antibiotic resistance of enterococci varied sufficiently between countries that a library developed in UK was not representative for the other locations (France, Sweden, and Spain), suggesting libraries may need to be developed more locally (Ebdon and Taylor 2006). Within smaller geographic areas such as within a US state, or between Australian catchments within a100-km radius, phenotypic libraries have been shown to be representative (Hagedorn et al. 1999; Ahmed and Katouli 2008), and small sublibraries can be merged, although some loss in accuracy may occur (Wiggins et al. 2003), but for larger geographic areas, such as between different states, separate libraries may be needed (Hartel et al. 2002; Wiggins et al. 2003).

Libraries of molecular profiles, which have been particularly well demonstrated for *E. coli*, may also have limited use across geographic regions due to the diversity within this species (Lasalde et al. 2005). For example, Hartel et al. (2002) found that the ribotypes of some *E. coli* strains were cosmopolitan (i.e., widespread), while others were endemic (found in only one geographic location) and that this varied with host animal species when examining ribotyping profiles. However, Scott et al. (2003) found that ribotypes of *E. coli* isolates from a wide area of Florida could be used to distinguish between human and nonhuman sources, although not between different animal sources. More recently, Hansen et al. (2009) have found that *E. coli* isolated from gull feces collected in some parts of the Great lakes region exhibited similar horizontal fluorophore-enhanced rep-PCR (HFERP) fingerprints, while those from other areas of the region were different.

3.2.4.2 Temporal Stability

Temporal stability of the library must be assessed in most cases, since the intention of creating a library is generally to use it over a period of months or years. Gordon (2001) raised major concerns over the use of *E. coli* as an indicator organism for source tracking, based on the evidence that there is little temporal stability in the clonal composition of populations in individual hosts, host populations, and locations over periods as short as weeks. Several other studies have illustrated changes in *E. coli* profiles over time (Aslam et al. 2003; Jenkins et al. 2003; Anderson et al. 2005; Hansen et al. 2009), and multiyear library development has been suggested based on changes in horizontal fluorophore-enhanced rep-PCR HFERP patterns of *E. coli* from waterfowl (Hansen et al. 2009). Based on ribotype data, some *E. coli* strains have been shown to be more persistent in the environment (Anderson et al. 2005), while many are transient (observed at only one sampling time) (Jenkins et al. 2003), and populations associated with a specific animal such as beef cattle have been shown to change over time (Aslam et al. 2003).

Phenotypic libraries of enterococci may be more stable; those composed of antibiotic resistance profiles have been reported as stable for 12 months (Wiggins et al. 2003), 36 months (Ebdon and Taylor 2006), or up to 5 years (Ahmed and Katouli 2008). Stability and representativeness of metabolic fingerprints of both *E. coli* and *Enterococcus,* tested following procedures described in Wiggins et al. (2003), were stable and representative over a 10-month period in separate and combined libraries (Ahmed et al. 2006). However, the distribution of individual *Enterococcus* species in fecal and environmental populations has been found to vary seasonally (Molina 2005). It has been suggested that both temporal and geographic stability may be related to the discriminatory ability of the methods used, with the most discriminatory methods being more affected, while those with less resolution, such as most phenotypic methods, being more stable (Ahmed and Katouli 2008).

Additional complexity in determining the performance of a library are other factors that can affect fingerprints such as animal diet (Hartel et al. 2003) and age of animals (Hansen et al. 2009).

3.2.5 Statistical Analysis

Once a library has been constructed, statistical analyses to validate and determine performance criteria are needed. As methods have been refined, so have statistical analyses. Recent reviews and papers have focused on aspects such as the validation of methods, performance criteria, and design and quantitative aspects (Ritter et al. 2003; Kaneene et al. 2007; Robinson et al. 2007; Stoeckel and Harwood 2007). The ultimate goal of LDM MST libraries is to successfully compare profiles of isolates from environmental waters (unknown sources) to those in the library for source identification.

LDM data analyses are generally either performed by use of statistical analyses to categorize isolates or, less commonly, by a direct comparison of unknown source isolates (matching) with the library. Statistical matching algorithms used to analyze LDM include discriminant analysis, principle components analysis, maximum similarity, average similarity, and k-means nearest neighbor. No single statistical approach has been found to be superior (Ritter et al. 2003). Discriminant analysis (DA) has probably been the most widely used method (Wiggins 1996; Parveen et al. 1999; Harwood et al. 2000; Moore et al. 2005; Moussa and Massengale 2008) and was shown to be a valid statistical analysis for MST studies (Kaneene et al. 2007). However, different analyses range in ability to correctly classify isolates in libraries with disproportionate representation. Robinson et al. (2007) found average similarity and discriminant analysis to be more robust than maximum similarity, but they found that these methods did not always provide the most accurate matching strategy; k-means nearest neighbor was suggested as a compromise for this purpose. Other methods of analysis include cluster analysis (Hagedorn et al. 1999; Kelsey et al. 2003; Webster et al. 2004; Lasalde et al. 2005; Molina 2005), and less commonly logistic regression (Dickerson et al. 2007), classification trees (Price et al. 2006; Seurinck et al. 2006; Price et al. 2007), and recently random forests (Smith 2009; Smith et al. 2010).

Different methods of analysis may affect the results of the source allocations. Lasalde et al. (2005) analyzed PFGE results using discriminant analysis and cluster analysis and found that discriminant analysis provided source identifications but that cluster analysis was unable to distinguish sources. DA is also more commonly used for phenotypic methods (e.g., for FAME (Yurtsever et al. 2007)) with Bionumerics® and other matching software utilized for genotypic profiles (bands, etc.) (Casarez et al. 2007a; b).

Statistical tests used to assess the classification accuracy of a library frequently provide an average rate of correct classification (ARCC) or for a specific group, rate of correct classification (RCC). Generally, smaller libraries provide higher ARCCs than larger ones; however, they are generally less representative and, thus, tend to have poor accuracy when classifying isolates that are not part of the library (i.e., challenge isolates or those from environmental samples) into source categories. Stoeckel and Harwood describe development of library validation techniques in their minireview (2007). Early studies used DA with only internal evaluation: "library self-cross" (Wiggins 1996; Parveen et al. 1999), which provided high assessments of within-library accuracy, but did not consider the ability of the library to accurately predict sources of isolates not in the library. This led to incorporation of cross-validation (also known as jackknife, leave one out, hold-out) where isolates are taken out of the library and then classified using the rest of the library. This is still considered an internal evaluation, unless all the isolates from a particular sample are removed from the library as a "pulled sample" (to avoid bias if an isolate is compared with a library still containing other isolates from the same sample) and compared with the library, rather than using a "pulled isolate" analysis (Wiggins et al. 2003; Moore et al. 2005) Another, more conservative (challenging) measure of library accuracy is to use challenge or blinded isolates collected independently of the library (Harwood et al. 2003; Wiggins et al. 2003; Hassan et al. 2005; Casarez et al. 2007a, b).

When assessing ARCCs, it is also important to understand the relationship between the number of source categories and library accuracy. For example, a three-way categorization would be expected to have ~33% ARCC by a random classification, whereas a six-way split would have a ~16.7%. Thus, an ARCC for three-way of 35% vs. an ARCC for a six-way of 35% indicates that the six-way split actually has a higher degree of accuracy. Stoeckel and Harwood (2007) proposed a Benefit over Random (BOR) calculation to address this issue. Misclassification levels have also been discussed, and ways to quantify these have been suggested (Harwood et al. 2000; Whitlock et al. 2002; Wiggins et al. 2003).

3.3 Methods in LDM

Selection of a particular method or methods for a study involves consideration of multiple factors including goals of the study (what data is needed and at what level of accuracy, sensitivity, specificity, and reproducibility), watershed characteristics, available resources including equipment and personnel, funding, time constraints, etc. The commonly used LDMs are described below, but even within these methods there are variations such as use of different antibiotics or restriction enzymes, method techniques such as replica plating vs. Kirby-Bauer for antibiotic resistance testing, automated ribotyping vs. manual, etc. While end users look for a standard or "best" method, neither the state of the science nor the range of applications make this assessment feasible. Each method has pros and cons, and each study has particular objectives to be met, all of which must be considered when choosing one or more methods. The following sections (and Table 3.1) provide more information on individual LDMs that have been evaluated in the peer-reviewed literature.

3.3.1 Phenotypic Typing Methods

3.3.1.1 Antibiotic Resistance Analysis (ARA)

Antibiotic resistance analysis (ARA), also known as antibiotic resistance profiling (ARP), relies on the phenotypic characteristic of bacterial resistance to antimicrobials to distinguish sources of fecal bacteria. The theory behind this method is that normal gut flora from different animal hosts are exposed to antibiotics in varying degrees and will develop resistance to antimicrobial agents over time due to selective pressure (Scott et al. 2002; Simpson et al. 2002). Patterns of resistance can be determined for isolates from different animal groups, which can then be used to identify sources of fecal pollution. Additionally, profiles from a mix of animals (e.g., rabbit, rodent, dog etc.) from a source such as "urban runoff" can be used to form more general source categories instead of using categories of individual animal species (Choi et al. 2003).

ARP as an MST method has its roots in earlier studies examining patterns of antibiotic resistance. Kibbey et al. (1978) examined antibiotic susceptibility of fecal streptococci in soil and suggested that multiple drug resistance could be a characteristic used in the future to ascertain the source of isolates obtained from environmental samples. Krumperman (1983) developed the multiple antibiotic resistance (MAR) index, which was originally used to determine human and animal sources of food contamination, but which has since been extended to several environmental studies. It has been used to examine resistance of bacteria in water samples from urban and rural areas (Kaspar and Burgess 1990) and to compare susceptibilities of environmental, food, and clinical isolates. (Knudtson and Hartman 1993).

As use of this method for MST became established, two methods for ARA testing emerged as the most commonly utilized – replica plating and Kirby-Bauer antibiotic disk diffusion. The replica plating method of Kelch and Lee (1978) was first adapted for MST by Wiggins (1996) and has since been used in a number of studies (Parveen et al. 1997; Hagedorn et al. 1999; Harwood et al. 2000; Graves et al. 2002; Stoeckel et al. 2004; Webster et al. 2004; Genthner et al. 2005; Anderson et al. 2006; Price et al. 2007). This method involves transferring isolates with a pronged replica plater to media plates containing an antibiotic. Multiple antibiotics are used at several concentrations, each on a separate plate, alongside control plates lacking antibiotics, with resistance to an antimicrobial agent defined as lack of growth on a plate (Wiggins 1996). The second common protocol, Kirby-Bauer antibiotic disk diffusion, is a standardized method commonly used in clinical laboratories. These guidelines, including measures for quality control, are published by the Clinical Laboratory Standards Institute (CLSI) (formerly the National Committee on Clinical Laboratory Standards) (CLSI 2009; 2010). The method involves the use of multiple disks, each impregnated with a single concentration of an antibiotic. Resistance is gauged by the size of the zone of growth inhibition around the disk, and susceptible, intermediate, and resistant values for individual organisms and antimicrobials are published and updated by CLSI. Although replica plating appears to be more widely used for MST studies, several groups have utilized the standardized Kirby-Bauer antibiotic disk diffusion approach (Dicuonzo et al. 2001; Mott and Lehman 2005; Samadpour et al. 2005; Sayah et al. 2005; Wilson 2005; Casarez et al. 2007a, b; Kaneene et al. 2007).

The ARP studies performed to date have shown reasonable success with respect with ARCCs. The ARA library constructed by Wiggins (1996) yielded an ARCC of 95.0% for discriminating between human and nonhuman sources ($n = 1{,}435$). When a new library was created to include six watersheds and 6,587 isolates, the ARCC for a three-way classification (human, domesticated, and wildlife) decreased to 57.0% (Wiggins et al. 2003). However, it has been demonstrated that larger libraries increase representativeness at the expense of the ARCC (Harwood et al. 2003; Wiggins et al. 2003). The results produced from a study by Hagedorn et al. (1999), with rates of correct classification for known sources greater than 95%, led to mitigation measures that were able to effectively reduce fecal contamination, demonstrating the applicability of the technique for implementation. ARP has been used in numerous investigations of fecal contamination (Booth et al. 2003; Burnes 2003; Choi et al. 2003; Geary and Davies 2003; Graves et al. 2007; Olivas and Faulkner 2008), in studies comparing MST methods (Farmer et al. 2003; Harwood et al. 2003; Stoeckel et al. 2004; Moore et al. 2005; Samadpour et al. 2005; Price et al. 2007), in composite data sets (Genthner et al. 2005; Casarez et al. 2007a, b; Moussa and Massengale 2008) and in studies using multiple methods (Dicuonzo et al. 2001; Ihrie et al. 2003; Anderson et al. 2006; Dickerson et al. 2007; Edge et al. 2007; Jiang et al. 2007).

ARA is commonly used for LDM as it is generally much less expensive and technically demanding than genotypic library-dependent methods. Price et al. (2007) found that ARA, on a per isolate basis, was 4–5 times less expensive than PFGE. The cost effectiveness of ARA allows for a larger number of samples to be analyzed, both for construction of the library, as well as for unknown source samples. ARP does not require extensive training of personnel or expensive equipment, which results in relatively easy implementation in a laboratory. Although specialized equipment is not required for this method, automated plate readers such as the BIOMIC system (Giles Scientific, Inc., Santa Barbara, California) may be used to read Kirby-Bauer plates to decrease processing, standardize measurements, and automate recording of results (Casarez et al. 2007a, b; Mott et al. 2008). One limitation of ARP as a MST tool is the questionable stability of antibiotic resistance of bacteria both in the laboratory and natural environments. As genetic elements containing multiple antibiotic resistance genes are mobile, gain or loss of resistance can change the profile of an isolate, which can complicate analysis (Simpson et al. 2002). The extent of this problem for MST studies has not been extensively investigated, although in one study, results implied that more than 50% of the antibiotic resistance markers in *E. coli* isolates were not stable throughout the various steps of analysis (Samadpour et al. 2005). ARA analysis of environmental isolates can be further complicated due to habitat sharing and diet overlap of wildlife and livestock (Meays et al. 2004) or the overlap in administration of antibiotics between classes of animals (Field and Samadpour 2007). High rates of false positives and problems with accuracy of identification of sources of blind samples have also been reported for this method (Harwood et al. 2003; Moore et al. 2005). Geographical stability is questionable for ARP profiles (Ebdon and Taylor 2006), but temporal stability has been demonstrated in libraries over a 1-year period (Wiggins et al. 2003) or even up to 5 years (Ahmed and Katouli 2008).

3.3.1.2 Biochemical Fingerprinting

Carbon source utilization methods of biochemical fingerprinting are based on the ability of bacteria to metabolize numerous carbon and nitrogen substrates. A high degree of phenotypic diversity can be resolved, even within a species such as *E. coli*. Commercial panels are available for this use, most notably the PhenePlate (PhP) System (Bio Sys inova, Stockholm, Sweden) and the Biolog MicroPlate™

System (Biolog, Inc., Hayward, California). PhenePlates use high-resolution plates that contain 24 or 48 reagents, or rapid screening plates with 11 reagents per isolate, while Biolog plates utilize a 96-well microplate with 95 substrates and one control well. The reagents in both systems vary depending on the type and class of organism to be analyzed (e.g., Gram-negative vs. Gram-positive). Both systems employ a plate reader connected to a computer with software that interprets intensity readings for each individual well of the microplate. The well intensities form a profile for each isolate that may then be analyzed by discriminant analysis or another statistical method.

The first published study using carbon source utilization (CSU) analysis for microbial source tracking purposes utilized the Biolog system for characterizing enterococci, and the analysis achieved a 92.7% ARCC for classifying human vs. nonhuman isolates (Hagedorn et al. 2003). Wallis and Taylor (2004) employed the use of the PhenePlate system to examine phenotypic diversity for distinguishing sources. Since these initial studies, several others have been published, using either Biolog or PhenePlate systems (Farmer et al. 2003; Harwood et al. 2003; Ihrie et al. 2003; Stoeckel et al. 2004; Ahmed et al. 2005, 2006; Moussa and Massengale 2008).

A comparison study by Harwood et al. (2003) reported an ARCC of 85% when distinguishing between human and nonhuman isolates for both *E. coli* and fecal streptococci CSU libraries. The fecal streptococci library was more accurate at predicting sources from water samples compared to the *E. coli* library. Stoeckel et al. (2004) examined the ability of CSU and other methods to classify blind replicate isolates and accuracy isolates into source categories. These isolates were not part of the initial library. The authors found that CSU was able to detect nonhuman sources at a high rate (98–100%). When compared head-to-head with ARA, CSU has provided higher ARCCs for enterococci (93% vs. 79% for ARA), *E. coli* (93.2% vs. 80.9% for ARA), and enteric bacteria (91.3% vs. 72.0% for ARA) (Farmer et al. 2003; Ihrie et al. 2003). Ahmed et al. (2005) utilized large carbon source utilization profile libraries ($n = 4{,}057$ (enterococci) and $n = 3{,}728$ (*E. coli*)) to successfully differentiate between human and nonhuman sources, as well as among animal sources. Moussa and Massengale (2008) utilized a smaller library of 596 carbon utilization profiles and generated ARCCs greater than 85% for two-way, three-way, and six-way classification to determine sources of *E. coli* from water samples.

Advantages to this approach include rapid analysis time and less personnel training than that required for molecular methods. Furthermore, this method utilizes plates and reagents that are commercially prepared; therefore, it is relatively standardized. The use of a plate reader and software to analyze plates removes the bias that may be introduced in manual readings. CSU profiles have also exhibited geographical stability for use in studies on watersheds with similar land use (Ahmed et al. 2006).

The paucity of literature available on this method has limited its establishment as a method in the MST community, despite some demonstrated potential, particularly for *Enterococcus*. Although this method has been used successfully, Stoeckel et al. (2004), in a comparison study, found CSU to be unable to accurately predict human sources, but it must be noted that the results were based on the utilization of only 22 wells of the possible 95 carbon sources. Data on the temporal stability and additional information on the geographical variability of profiles are needed to determine constraints on library validity.

3.3.1.3 Fatty Acid Methyl Esters (FAME)

Fatty acid compositions of bacterial cell walls exhibit species-specific characteristics in terms of types and chemical composition of fatty acids present (Haznedaroğlu et al. 2005), and this has been used to develop a MST method based on whole-cell fatty acid methyl ester (FAME) profiles. Fatty acid analysis has been used in clinical microbiology to identify organisms, in epidemiologic studies for typing (Mukwaya and Welch 1989; Kotilainen et al. 1991; Birnbaum et al. 1994; Kühn et al. 1997) as well as in environmental microbiology for community profiling, particularly in soil studies (Haack et al. 1994; Schutter and Dick 2000; Dunfield and Germida 2003).

The first published MST study involving FAME analysis examined 104 *E. coli* isolates from human and nonhuman sources, but the two groups could not be differentiated due to the lack of a signature FAME or significantly different FAMEs between sources (Parveen et al. 2001). FAME analysis of a limited library of *Enterococcus* spp. was performed by Genthner et al. (2005), as part of a study using several MST techniques. Dendrograms produced from the FAME analysis data did not show accurate clustering of sources and were unable to classify isolates beyond the species level (Genthner et al. 2005). Another FAME analysis study utilizing *E. coli* showed an effective differentiation between human and nonhuman sources but was unable to distinguish between different animal sources (Seurinck et al. 2006).

Parveen et al. (2001) suggested that the ability to differentiate between sources with FAME might be dependent on the bacterial species examined. Haznedaroğlu et al. (2005) examined FAME profiles of total coliforms and found statistically significant differences between known host profiles, with average rates of correct classification (ARCC) ranging from 81 to 84% for three-way classifications, depending on pooling strategies. ARCCs of 95 and 97% were produced for three-way classifications by FAME analysis of fecal coliforms, an improvement on the study using total coliforms (Duran et al. 2006). Yurtsever et al. (2007) took this one step further, by examining FAME profiles from four different indicator groups, enterococci, *E. coli*, fecal coliforms, and total coliforms. *E. coli* achieved the lowest accuracy of classification. The authors supported the hypothesis developed in a previous study (Duran et al. 2006) that the indicator groups composed of multiple species or genera may be better choices for FAME profiling, as differences in predominant organisms in the gut may differ between host groups (Yurtsever et al. 2007). However, enterococci has recently been put forth as a stronger candidate for FAME analysis, as indicator groups, such as fecal coliforms, include species from multiple genera that might have different die-off rates in the environment leading to possible errors in source predictions (Duran et al. 2009).

A key advantage of FAME analysis is the small library required for an MST study. FAME profiles rely on differences in signature fatty acids, forgoing the need for a large library (Haznedaroğlu et al. 2005). By the same token, the method is not as highly discriminatory among strains as other LDM. Although the studies suggest the potential of FAME analysis as a viable phenotypic MST method, its main limitation is that it is considered to be in testing mode, with further analysis required to determine its applicability for MST studies (Field and Samadpour 2007).

3.3.2 Genotypic Methods

3.3.2.1 Pulsed-Field Gel Electrophoresis (PFGE)

Pulsed-field gel electrophoresis (PFGE) is one of several genetic fingerprinting methods employed in MST. PFGE was developed by Schwartz and Cantor (1984) and involves the use of agarose gel electrophoresis with alternating pulsed electric fields. Prior to electrophoresis, the genomic DNA is digested with a rare-cutting restriction enzyme, which limits the number of fragments produced during digestion (Farber 1996).

This method is considered the "gold standard" of molecular typing methods (Olive and Bean 1999) and is widely used to fingerprint bacteria implicated in outbreaks, including those investigated by PulseNet, the national subtyping network for foodborne disease surveillance (Swaminathan et al. 2001). PFGE was also used as the primary method of analysis in the first published account of the isolation of an outbreak strain of *E. coli* O157:H7 from recreational waters (Samadpour et al. 2002).

Several early MST studies that used PFGE produced mixed results; however, the libraries were extremely small. Parveen et al. (2001) analyzed 32 *E. coli* isolates with PFGE and did not find an association between PFGE profile and host source of the isolate. However, Hahm et al. (2003) examined 54 environmental isolates and found that in some cases, the profiles clustered according to source, but the dataset was too small to be certain to draw conclusions.

Further studies have shown high accuracy for PFGE. A comparison study reported that PFGE produced comparable results, in terms of sensitivity and percentage of false positives, to ribotyping analysis of the same data (Myoda et al. 2003). The comparison study performed by Stoeckel et al. (2004) found PFGE to identify contributing sources in hypothetical water samples with accuracy, but the method was unable to identify any source for a large number of isolates. Similar results were found by Casarez et al. (a, b) with high confidence in the matches made by PFGE (>90% of blind QC challenge isolates correctly identified for precision, method accuracy, and source accuracy), but almost half of library isolates were left unidentified after jackknife analysis. Dickerson, Hagedorn and Hassall (2007) utilized multiple source tracking methods, including ARA and PFGE, as recommended by the Southern California Coastal Water Research Project (SCCWRP) methods comparison study (Stewart et al. 2003). PFGE results for the ARCCs of both the known source library isolates and the validation set were very similar, lending confidence to the analysis (Dickerson et al. 2007).

One advantage of PFGE is its extreme sensitivity to small genetic differences (Scott et al. 2002) such that it is highly discriminatory (Simpson, Santo Domingo and Reasoner 2002). The precise discrimination produced by this method has been suggested to be advantageous in situations where a limited number of sources are possible (Myoda et al. 2003). This method also has excellent reproducibility (Farber 1996), which is a highly desirable trait of an effective MST method.

Although discriminatory capability is an advantage of PFGE, this technique might actually be too sensitive to discriminate between host sources for MST purposes (Scott et al. 2002). McLellan et al. (2003) found PFGE to be the most discriminatory method used in a comparison study with two other DNA fingerprinting techniques, but profiles were so diverse that pattern comparisons for MST were difficult to perform. Lu et al. (2004) suggested that due to the extreme sensitivity of the method, a very extensive PFGE profile library is needed for source tracking in a complex watershed. Other limitations may constrain the number of isolates analyzed, i.e., the time-consuming nature of the analysis, (Olive and Bean 1999; Yan and Sadowsky 2007) and the problem that the genomic DNA of some bacterial strains cannot be effectively digested (Johnson et al. 1995; Casarez et al. 2007a, b). Complexity of band analysis, once a problem with this method (Tenover et al. 1995; Hopkins and Hilton 2000), has been resolved with the use of software.

3.3.2.2 Ribotyping

Ribotyping is a genotypic method that fingerprints bacteria based on sequence differences in genomic DNA, with steps including restriction enzyme digestion, electrophoresis, and probing via Southern blot (Farber 1996). MST studies have primarily used two ribotyping protocols, which differ in the type and number of restriction enzymes used. One protocol uses two restriction enzymes (*Eco*R1 and *Pvu*II) (Myoda et al. 2003; Samadpour et al. 2005) and the other uses a single enzyme, primarily *Hind*III (Parveen et al. 1999; Carson et al. 2001; Griffith et al. 2003; Scott et al. 2003; Moore et al. 2005; Casarez et al. 2007a; b).

The first published MST study using ribotyping (*Hind*III protocol) was able to distinguish between human source and nonhuman source *E. coli* isolates with an ARCC of 82% (with RCCs of 67 and 97%, respectively) (Parveen et al. 1999). Carson et al. (2001) also utilized the *Hind*III protocol to analyze a library of 287 *E. coli* isolates and obtained an even higher ARCC of 97.1% for the two-way classification between human and pooled nonhuman sources, although the library size must be considered in evaluating these results.

In addition to these pilot studies, ribotyping has been used successfully in several more recent MST studies. Ribotyping with the two-enzyme protocol was used to investigate the source of fecal contamination in Grand Teton National Park, resulting in *E. coli* isolates from unknown source water samples matching largely with wildlife sources (Farag et al. 2001). Scott et al. (2004) utilized the single-enzyme protocol to analyze *E. coli* in a South Carolina watershed to determine the source of fecal contamination and found 88% of environmental isolates to match animal profiles (ARCC=92.0%). Ribotyping data was also included in a four-method composite dataset used to analyze environmental samples from a Texas watershed (Casarez et al. 2007a, b).

Ribotyping offers the advantages of both excellent typeability and reproducibility (van Belkum et al. 2001). The section of DNA that encodes rRNA is highly conserved, and thus, ribotyping is considered to be one of the most reproducible of the molecular methods (Farber 1996; Hartel et al. 2003). However, Lefresne et al. (2004) found significant differences in between-laboratory variance, most likely due to differences in protocols.

Some limitations of the method are expense and the amount of labor involved (Scott et al. 2002). Ribotyping also requires a skilled technician (Carson et al. 2003). However, the amount of labor and skill level may be reduced with the use of an automated ribotyping system, but this in turn increases costs due to the initial investment of equipment and cost of consumables (Casarez et al. 2007a, b). Variations in methodology exist, making it difficult to compare results of studies (Meays et al. 2004). Accuracy of ribotyping has also been called into question with the study by Moore et al. (2005), where the ARCC of the *E. coli* ribotype profile library was 69%, but the ARCC for proficiency isolates (intended to assess predictive accuracy) was only 27%. Accuracy was also significantly lower for ribotyping analysis than for rep-PCR for the same isolates (Carson et al. 2003).

Ribotype profiles of *E. coli* exhibit geographic variability, with increased similarity of fingerprints shown with decreased distance (Hartel et al. 2002). This has been demonstrated with both the single-enzyme ribotyping protocol (Scott et al. 2003) and the two-enzyme protocol (Hartel et al. 2002). However, it must be noted that although geographical variability was seen in the study by Scott et al. (2003), profiles from animals still differed significantly from human profiles, suggesting that the distinction between human and nonhuman ribotype profiles may be more robust to geographical differences. Profiles are also affected temporally, so establishment of the library should be as close as possible in time to the collection of environmental samples (Jenkins et al. 2003; Anderson et al. 2006). Ribotype diversity in *E. coli* populations has also been shown to be affected by diet (Hartel et al. 2003; Nelson et al. 2008), which may impact study design if wild animals share a food source with another known host. Moore et al. (2005) also found ribotyping to lack sufficient accuracy to identify sources in a large, urban watershed, which may be linked to a variety of factors, including geographic variability, reproducibility, and library composition.

3.3.2.3 Rep-PCR

Repetitive element sequence-based PCR is a DNA fingerprinting technique that uses one primer targeting a repetitive, palindromic DNA sequence that is widespread in many bacterial genomes (Versalovic et al. 1991). The three most commonly studied prokaryotic repetitive sequences are the BOX element (composed of three subunits, Box A, Box B, and Box C) (Martin et al. 1992), enterobacterial repetitive intergenic consensus (ERIC) sequences (Hulton et al. 1991), and interspersed repetitive extragenic palindromes (REP) (Stern et al. 1984).

Several studies have compared the effectiveness of genotypic fingerprinting of *E. coli* with respect to the different primer methods. Lipman et al. (1995) differentiated strains of *E. coli* from cattle using both ERIC and REP-PCR methods and found similar results, as did McLellan et al. (2003) when analyzing isolates from several different host sources. ERIC-PCR profiles were deemed more reliable, in terms of reproducibility, than those of REP-PCR (Lipman et al. 1995), and BOX-PCR was more accurate than REP-PCR (Dombek et al. 2000). Baldy-Chudzik et al. (2003) found that REP-PCR produced more complex fingerprints, allowing a higher level of discrimination among strains but a more difficult interpretation of profiles when compared to ERIC-PCR. Five different primers, REP, ERIC, ERIC2, BOX, and $(GTG)_5$ were compared side-by-side, and ARCCs for a two-way classification ranged from 86.8 to 55.8%, with $(GTG)_5$ at the high end of the range and ERIC2 with the low end (Mohapatra and Mazumder 2008). $(GTG)_5$-PCR is being further explored for application to MST studies and has shown promise in terms of repeatability and accuracy for classification of *E. coli* population analysis in one investigation (Mohapatra et al. 2008).

The BOX primer has been the most widely used primer described in MST literature. BOX A1R-PCR analyses in MST studies in North American watersheds have yielded average rates of correct classification (ARCCs) ranging from 66.9 to 90.2% for libraries containing multiple sources (Carson et al. 2003; Seurinck et al. 2003; Somarelli et al. 2007; Mohapatra et al. 2007; Kon et al. 2009). BOX-PCR has also been used in several *E. coli* and *Enterococcus* studies using multiple methods to investigate a fecal contamination issue (Brownell et al. 2007; Edge et al. 2007; Vogel et al. 2007). REP and ERIC primers have been used in field investigations for *E. coli* and *Enterococcus* but less widely than BOX (Genthner et al. 2005; Casarez et al. 2007a, b).

Rep-PCR is advantageous for MST studies, as it is somewhat less technically demanding than PFGE and ribotyping (USEPA 2005). Additionally, Carson et al. (2003) found BOX PCR to be less expensive and more efficient than ribotyping. REP-PCR of *E. coli* may eliminate steps required in other genotypic methods by using a whole cell suspension, thus eliminating the need for DNA purification (Dombek et al. 2000).

Reproducibility of this method has been noted as moderate (Scott et al. 2002; Seurinck et al. 2005), which might be considered a limitation for MST purposes. However, factors potentially influencing reproducibility of BOX-PCR, such as changes in gel normalization, PCR reaction, DNA loading, and thermocycler settings were assessed, and fingerprints generated by the different procedures were generally shown to share a high degree of similarity (≥90%) (Albert et al. 2003). Libraries generated from REP-PCR profiles may produce clusters more closely related to the gels from which they were produced than to the source of the isolate (Johnson et al. 2004). To circumvent this problem, a computer-assisted analysis, known as horizontal fluorophore-enhanced rep-PCR (HFERP), may be used to reduce within-gel groupings of genotypic fingerprints, allowing the library to be more effective with classifications (Johnson et al. 2004). Temporal and spatial variability of BOX-PCR profiles have been demonstrated in bird populations (Hansen et al. 2009), but the stability of ERIC- and REP-PCR profiles have not been examined.

3.3.2.4 Amplified Fragment Length Polymorphism (AFLP)

Amplified fragment length polymorphism (AFLP) is another genotypic library-based method based on PCR detection of restriction fragments resulting from total genomic DNA digestion with restriction enzymes (Vos et al. 1995). AFLP has been used extensively for research with plant genomes and has recently been suggested for use in microbial source tracking.

The most common protocol for AFLP first digests the genomic DNA with two restriction enzymes, *Eco*RI and *Mse*I; the digest is then ligated to adapters that function as PCR primer binding sites that allow for a selective PCR amplification of a subset of genomic restriction fragments (Blears et al. 1998; Olive and Bean 1999). Several commercial kits, containing different combinations of primer pairs, are available for AFLP fingerprinting (USEPA 2005).

Few published studies have been conducted using AFLP as an MST method. Guan et al. (2002) applied AFLP to a very small library of 105 *E. coli* isolates from fecal sources. They were able to differentiate sources with rates of correct classification (RCC) of 94.4% for livestock animals, 97.1% for wildlife, and 97.1% for humans; however, the small library size suggests that they might not be able to accurately classify isolates that are not part of the library. Leung et al. (2004) found AFLP to be successful for discriminating between a geographically diverse collection of 110 pathogenic strains of *E. coli* from bovine, human, and pig sources, suggesting that AFLP profiles may be geographically stable, although the library size in the study constrains conclusions. Enterococci have also been analyzed with an AFLP procedure developed by Burtscher et al. (2006).

The users of AFLP cite its reproducibility and robustness, (Vos et al. 1995; Blears et al. 1998). This method is able to examine the whole genome for polymorphisms, which is an advantage over other fingerprinting methods (Simpson et al. 2002).

One major limitation of this method is cost. AFLP requires the use of a DNA sequencer, which can be a sizable initial investment for a laboratory. Another limitation is the lack of information on this method, such as the best primers to use for MST purposes. Hahm et al. (2003) found that AFLP profile clusters differed depending on the primer, and the primer sets used might have caused the inaccuracy of clustering of known source fecal isolates. Further studies are needed to assess this method using larger, more representative libraries, as previous studies (Guan et al. 2002; Leung et al. 2004) have utilized small known source libraries (Yan and Sadowsky 2007).

3.3.2.5 Random Amplified Polymorphic DNA (RAPD) Analysis

Random amplified polymorphic DNA (RAPD) analysis is also known as arbitrarily primed PCR (AP-PCR) (Welsh and McClellan 1990; Williams et al. 1990). Both methods produce similar results but have different reaction conditions. AP-PCR uses arbitrary primers at low stringency, while RAPD uses nonselective primers at high stringency (USEPA 2005).

A limited number of published MST studies have explored the use of RAPD. Venieri et al. (2004) utilized RAPD with seven different primer combinations and showed three sets able to successfully distinguish between human and animal sources of fecal *E. coli* isolates. RAPD profiles composed of composites from three DNA primer sets were utilized in a Great Lakes watershed to detect fecal contamination from multiple sources (Ting et al. 2003).

RAPD has demonstrated a higher discriminatory capacity and quicker analysis time than ribotyping for subtyping of *E. coli* (Vogel et al. 2000). Other advantages of RAPD include efficiency of analysis, ease of use, and cost-effectiveness (Hopkins and Hilton 2000; USEPA 2005). Questions about the reproducibility of RAPD analysis have arisen (Hopkins and Hilton 2000). Owing to the random relationship between the primer and target site, nonspecific hybridization of primer and template may occur under less than ideal conditions, making reaction conditions very sensitive for this method (Olive and Bean 1999). This technique is considered to be in test phase for MST studies (USEPA 2005); standardization of protocols and demonstration of reproducibility of patterns are necessary for this method to be further considered as a viable MST technique (Venieri et al. 2004).

3.3.2.6 Denaturing Gradient Gel Electrophoresis (DGGE)

Denaturing gradient gel electrophoresis (DGGE) is a genotypic technique that can separate closely related PCR products of the same length, based on their DNA sequence, which affects their melting properties and in turn influences their movement through a polyacrylamide gel (Muyzer et al. 1993). The gradient for separation in DGGE is generally composed of urea and formamide, and the fingerprints produced are for whole populations, not for individual isolates (Von Wintzingerode et al. 1997).

MST is a relatively new application for DGGE and most studies have focused on pinpointing a target area that will produce the most effective distinction between profiles. In one study, the intergenic spacer region (ISR) of 16S-23S rRNA *E. coli* isolates was found to produce highly diverse profiles, making it difficult to detect similarities between environmental water samples and potential host fecal samples. This failure could also be attributed to the small size of the library (Buchan et al. 2001). The *uidA* gene of *E. coli* has also been examined with DGGE in several studies. Farnleitner et al. (2000) examined DGGE profiles of the *uidA* gene to differentiate *E. coli* populations from environmental water samples, resulting in the successful generation of a community fingerprint. Studies have furthered this research by using *uidA* DGGE profiles to differentiate *E. coli* communities from different sample sites and sample types (water and sediment) (Sigler and Pasutti 2006). Other genes have been assessed as potential DGGE targets. Three genes (*mdh*, *phoE*, and *uidA*-4) produced DGGE profiles that were able to link known pollution sources to environmental water samples, while excluding noncontributing sources (Esseili et al. 2008).

Much is still unknown about the efficacy of DGGE in MST studies. The gene targeted by this analysis must have a variable sequence among strains to be a useful tool (Simpson et al. 2002). DGGE is a technically demanding method that requires long processing times (Meays et al. 2004). Fingerprinting of community populations may also have the added problem of bias arising from PCR analysis of mixed populations (Suzuki and Giovannoni 1996; von Wintzingerode et al.1997). Community fingerprints may also be misinterpreted due to heteroduplex formation and comigration of bands with similar melting behaviors but different sequences (Nübel et al. 1996; Casamayor et al. 2000; Esseili et al. 2008).

3.3.2.7 Matrix-Assisted Laser Desorption/Ionization Time-of-Flight Mass Spectroscopy (MALDI-TOF-MS)

MALDI-TOF-MS is among the most recently suggested MST methods. It has been used in bacteriology to investigate biopolymers, particularly in *E. coli* (Lay 2001), and has also been applied to food safety studies (Seiber 2007). Very recently this method has been extended to MST for both *E. coli* (Siegrist et al. 2007) and enterococci (Giebel et al. 2008). Siegrist et al. (2007) found that MALDI-TOF-MS produced results comparable to BOX-PCR. MALDI-TOF-MS had lower reproducibility but higher accuracy for source classification than BOX-PCR. Reproducibility studies were performed by Giebel et al. (2008) to further refine the method for MST purposes by improving preparation and data analysis.

The primary advantage of this method is speed of analysis. MALDI-TOF-MS can produce spectral fingerprints in as short as 2 h (Siegrist et al. 2007); however, the required earlier culturing steps add considerable time to the method. The technique still needs further development to reduce analysis time and investigate reproducibility factors.

3.3.3 Comparison of Methods

Numerous reviews have described or compared a variety of the source tracking methods described above (Scott et al. 2002; Simpson et al. 2002; Griffith et al. 2003; Stewart et al. 2003; Meays et al. 2004; Seurinck et al. 2005; Stoeckel and Harwood 2007; Field and Samadpour 2007; Yan and Sadowsky 2007). Multiple method studies provide important information on comparative performance under the conditions of the study; however, interpretation in terms of general evaluation of methods is limited because of biases toward certain methods due to design issues. Such issues include the number of source categories included, number of isolates analyzed, size of the library, and method-specific variables such as the number of wells used for carbon source utilization, antibiotics used for ARA, restriction enzyme used for PFGE (*Not*1, *Eco*RI, and *Pvu*II (RT-EcoR1)). Griffith et al. (2003) described results of a study of 12 LDM and LIM methods conducted by 22 researchers, showing genotypic methods overall performed better than phenotypic. However, they noted that overall, LDM were at a disadvantage due to the restrictions on library development. A study by Stoeckel et al. (2004) compared seven methods for analyzing *E. coli* isolates, including both phenotypic and genotypic methods. This study also demonstrated a range of results, including greatest accuracy for molecular typing methods, offset by identification of fewer isolates. A study comparing three genotypic and one phenotypic method (Casarez et al. 2007a; b) found that PFGE was the most accurate method, but patterns could not be generated for ~10% of the library and water isolates. Moore et al. (2005) evaluated performance of ARA and ribotyping of *E. coli* and ARA of enterococci and concluded that none of these were ready for applications in larger urban watersheds. Several other studies comparing only two or three LDM methods have been published, representing a variety of genotypic and phenotypic analyses (Carson et al. 2003; Harwood et al. 2003; Myoda et al. 2003; Leung et al. 2004; Genthner et al. 2005; Samadpour et al. 2005; Price et al. 2007).

As each method currently in use has advantages and disadvantages, recent recommendations have suggested a "toolbox" approach to include multiple methods (Stewart et al. 2003; McLellan 2004) with some studies including both LDM and LIM (for example McDonald et al. 2006; Vogel et al. 2007), while others have developed composite datasets and have shown enhanced performance of these libraries (Genthner et al. 2005; Casarez et al. 2007a; b; Edge et al. 2007; Moussa and Massengale 2008).

3.4 Summary

The majority of early microbial source tracking studies were conducted using library-dependent methods, and while library-independent methods are now being widely used, there are still a number of advantages to the LDM, particularly for use in TMDL studies where the relationship to fecal indicator bacteria and categorization of a number of sources is needed for development of loading models and strategies to reduce the impacts of contamination. LDMs can be tailored specifically for a watershed based on animal sources present in that watershed. However, LDMs can be costly and time consuming due to the library development involved, usually including isolating and culturing isolates from water sampling (Santo Domingo et al. 2007; Yurtsever et al. 2007). The construction of a representative library requires careful consideration of the types and diversity of animals in the watershed and appropriate method selection. Testing for performance of the library is essential.

Some factors affecting the representativeness of a library include geographic and temporal stability and diversity of the organism being used, and for many of the methods these have not been assessed.

A number of reviews are available, which describe general challenges for library-dependent methods (LDMs) in terms of library stability, representativeness, etc. (Gordon 2001; Ahmed 2007; Field and Samadpour 2007; Stoeckel and Harwood 2007) or compare performance criteria for specific LDM methods (Stewart et al. 2003; Stoeckel and Harwood 2007). The science of MST continues to evolve, and no one method or suite of methods has been identified as superior to others, partly due to the fact that a variety of factors must be taken into consideration when choosing an approach including the specific question to be answered, financial resources, complexity of the watershed, specificity required in terms of individual animal identification vs. category (e.g., wildlife), laboratory and personnel expertise available, time and turnaround constraints. A "toolbox" or group of methods is now frequently used to address some of these issues. As an alternative to this more costly approach, tiered or targeted strategies where the problem is addressed in a series of steps and MST is included only for an identified, "targeted" area after initial study of the watershed have also been recommended (McDonald et al. 2006; Stoeckel and Harwood 2007).

References

Ahmed W, Neller R, Katouli M (2005) Host species-specific metabolic fingerprint database for Enterococci and *Escherichia coli* and its application to identify sources of fecal contamination in surface waters. Appl Environ Microbiol 71:4461–4468

Ahmed W, Tucker J, Harper J et al (2006) Comparison of the efficacy of an existing versus a locally developed metabolic fingerprint database to identify non-point sources of faecal contamination in a coastal lake. Water Res 40:2339–2348

Ahmed W (2007) Limitations of library-dependent microbial source tracking methods. Water: J Aust Water Assoc 34:39–43

Ahmed W, Katouli M (2008) Phenotypic variations of enterococci in surface waters: analysis of biochemical fingerprinting data from multi-catchments. J Appl Microbiol 105:452–458

Albert JM, Munakata-Marr J, Tenorio L et al (2003) Statistical evaluation of bacterial source tracking data obtained by rep-PCR DNA fingerprinting of *Escherichia coli*. Environ Sci Tech 37:4554–4560

Anderson KL, Whitlock JE, Harwood VJ (2005) Persistence and differential survival of fecal indicator bacteria in subtropical waters and sediments. Appl Environ Microbiol 71:3041–3048

Anderson MA, Whitlock JE, Harwood VJ (2006) Diversity and distribution of *Escherichia coli* genotypes and antibiotic resistance phenotypes in feces of humans, cattle, and horses. Appl Environ Microbiol 72:6914–6922

Aslam M, Nattress F, Greer G et al (2003) Origin of contamination and genetic diversity of *Escherichia coli* in beef cattle. Appl Environ Microbiol 69:2794–2799

Baldy-Chudzik K, Niedback J, Stosik M (2003) Rep-PCR fingerprinting as a tool for the analysis of genomic diversity in *Escherichia coli* strains isolated from an aqueous/freshwater environment. Cell Mol Biol Lett 8:793–798

Birnbaum D, Herwaldt L, Low DE et al (1994) Efficacy of microbial identification system for epidemiologic typing of coagulase-negative staphylococci. J Clin Microbiol 32:2113–2119

Blears MJ, De Grandis SA, Lee H et al (1998) Amplified fragment length polymorphism (AFLP): a review of the procedure and its applications. J Ind Microbiol Biotechnol 21:99–114
Booth AM, Hagedorn C, Graves AK et al (2003) Sources of fecal pollution in Virginia's Blackwater River. J Environ Eng 129:547–552
Brownell MJ, Harwood VJ, Kurz RC et al (2007) Confirmation of putative stormwater impact on water quality at a Florida beach by microbial source tracking methods and structure of indicator organism populations. Water Res 41:3747–3757
Buchan A, Alber M, Hodson RE (2001) Strain-specific differentiation of environmental *Escherichia coli* isolates via denaturing gradient gel electrophoresis (DGGE) analysis of the 16S-23S intergenic spacer region. FEMS Microbiol Ecol 35:313–321
Burnes BS (2003) Antibiotic resistance analysis of fecal coliforms to determine fecal pollution sources in a mixed-use watershed. Environ Monit Assess 85:87–98
Burtscher MM, Köllner KE, Sommer R et al (2006) Development of a novel amplified fragment length polymorphism (AFLP) typing method for enterococci isolates from cattle faeces and evaluation of the single versus pooled faecal sampling approach. J Microbiol Methods 67:281–293
Carson CA, Shear BL, Ellersieck MR et al (2001) Identification of fecal *Escherichia coli* from humans and animals by ribotyping. Appl Environ Microbiol 67:1503–1507
Carson CA, Shear BL, Ellersieck MR et al (2003) Comparison of ribotyping and repetitive extragenic palindromic-PCR for identification of fecal *Escherichia coli* from humans and animals. Appl Environ Microbiol 69:1836–1839
Casamayor EO, Schafer H, Bañeras L et al (2000) Identification of and spatio-temporal differences between microbial assemblages from two neighboring sulfurous lakes: comparison by microscopy and denaturing gradient gel electrophoresis. Appl Environ Microbiol 66:499–508
Casarez EA, Pillai SD, Mott JB et al (2007) Direct comparison of four bacterial source tracking methods and use of composite data sets. J Appl Microbiol 103:350–364
Casarez EA, Pillai DS, Di Giovanni GD (2007) Genotype diversity of *Escherichia coli* isolates in natural waters determined by PFGE and ERIC-PCR. Water Res 41:3643–3648
Choi S, Chu W, Brown J et al (2003) Application of enterococci antibiotic resistance patterns for contamination source identification at Huntington Beach, California. Mar Pollut Bull 46:748–755
Clinical Laboratory Standards Institute (2009) Performance standards for antimicrobial disc susceptibility tests. 10th edn, Clinical Laboratory Standards Institute, Wayne PA, USA
Clinical Laboratory Standards Institute (2010) Performance standards for antimicrobial testing; sixteenth information supplement. 20th edn, Clinical Laboratory Standards Institute, Wayne PA, USA
Dickerson JW Jr, Hagedorn C, Hassall A (2007) Detection and remediation of human-origin pollution at two public beaches in Virginia using multiple source tracking methods. Water Res 41:3758–3770
Dicuonzo G, Gherardi G, Lorino G et al (2001) Antibiotic resistance and genotypic characterization by PFGE of clinical and environmental isolates of enterococci. FEMS Microbiol Lett 201:205–211
Dombek PE, Johnson LK, Zimmerly ST et al (2000) Use of repetitive DNA sequences and the PCR to differentiate *Escherichia coli* isolates from human and animal sources. Appl Environ Microbiol 66:2572–2577
Dunfield KE, Germida JJ (2003) Seasonal changes in the rhizosphere microbial communities associated with field-grown genetically modified Canola (*Brassica napus*). Appl Environ Microbiol 69:7310–7318
Duran M, Haznedaroğlu BZ, Zitomer DH (2006) Microbial source tracking using host specific FAME profiles of fecal coliforms. Water Res 40:67–74
Duran M, Yurtsever D, Dunaev T (2009) Choice of indicator organism and library size considerations for phenotypic microbial source tracking by FAME profiling. Water Sci Tech 60: 2659–2668
Ebdon JE, Taylor HD (2006) Geographical stability of enterococcal antibiotic resistance profiles in Europe and its implications for the identification of fecal sources. Environ Sci Tech 40:5327–5332

Edge TA, Hill S, Stinson G et al. (2007) Experience with the antibiotic resistance analysis and DNA fingerprinting in tracking faecal pollution at two lake beaches. Water Sci Tech 56:51–58

Esseili MA, Kassem II, Sigler V et al (2008) Optimization of DGGE community fingerprinting for characterizing *Escherichia coli* communities associated with fecal pollution. Water Res 42:4467–4476

Farag AM, Goldstein JN, Woodward DF et al (2001) Water quality in three creeks in the backcountry of Grand Teton National Park, USA. J Freshwat Ecol 16:135–143

Farber JM (1996) An introduction to the hows and whys of molecular typing. J Food Prot 59:1091–1101

Farmer JJ, Ejiofor AO, Johnson TL (2003) Discriminant analysis of multiple antibiotic resistance patterns contrasted to carbon utilization profiles in bacterial source tracking in the karstic Duck River watershed of middle Tennessee. In: Abstracts of the 103rd American Society for Microbiology General Meeting Washington, D.C, 2003

Farnleitner AH, Kreuzinger N, Kavka GG et al (2000) Simultaneous detection and differentiation of *Escherichia coli* populations from environmental freshwaters by means of sequence variations in a fragment of the beta-D-glucuronidase gene. Appl Environ Microbiol 66:1340–1346

Field KG, Samadpour M (2007) Fecal source tracking, the indicator paradigm, and managing water quality. Water Res 41:3517–3538

Fogarty LR, Haack SK, Wolcott MJ et al. (2003) Abundance and characteristics of the recreational water quality indicator bacteria *Escherichia coli* and enterococci in gull faeces. J Appl Microbiol 94:865–878

Geary PM, Davies CM (2003) Bacterial source tracking and shellfish contamination in a coastal catchment. Water Sci Tech 47:95–100

Genthner FJ, James JB, Yates DF et al (2005) Use of composite data sets for source-tracking enterococci in the water column and shoreline interstitial waters on Pensacola Beach, Florida. Mar Pollut Bull 50:724-732

Giebel RA, Frendenberg W, Sandrin TR (2008) Characterization of environmental isolates of Enterococcus spp. by matrix-assisted laser desorption/ionization time-of-flight mass spectrometry. Water Res 42:931–940

Gordon DM (2001) Geographical structure and host specificity in bacteria and the implications for tracing the source of coliform contamination. Microbiology 147:1079–1085

Graves AK, Hagedorn C, Teetor A et al (2002) Antibiotic resistance profiles to determine sources of fecal contamination in a rural Virginia watershed. J Environ Qual 31:1300–1308

Graves AK, Hagedorn C, Brooks A et al (2007) Microbial source tracking in a rural watershed dominated by cattle. Water Res 41:3729–3739

Griffith JF, Weisberg SB, McGee CD et al (2003) Evaluation of microbial source tracking methods using mixed fecal sources in aqueous test sample. J Water Health 1:141–151

Guan S, Xu R, Chen S et al (2002) Development of a procedure for discriminating among *Escherichia coli* isolates from animal and human sources. Appl Environ Microbiol 68: 2690–2698

Haack SK, Garchow H, Odelson DA et al (1994) Accuracy, reproducibility, and interpretation of fatty acid methyl ester profiles of model bacterial communities. Appl Environ Microbiol 60:2483–2493

Hagedorn C, Robinson SL, Filtz JR et al (1999) Determining sources of fecal pollution in a rural Virginia watershed with antibiotic resistance patterns in fecal streptococci. Appl Environ Microbiol 65:5522–5531

Hagedorn C, Crozier JB, Mentz KA et al (2003) Carbon source utilization profiles as a method to identify sources of faecal pollution in water. J Appl Microbiol 94:792–799

Hahm B-K, Maldonado Y, Schreiber E et al (2003) Subtyping of foodborne and environmental isolates of *Escherichia coli* by multiplex-PCR, rep-PCR, PFGE, ribotyping and AFLP. J Microbiol Meth 53:387–399

Hansen DL, Ishii S, Sadowsky MJ et al (2009) *Escherichia coli* populations in Great Lakes waterfowl exhibit spatial stability and temporal shifting. Appl Environ Microbiol 75:1546–1551

Hartel PG, Summer JD, Hill JL et al (2002) Geographic variability of *Escherichia coli* ribotypes from animals in Idaho and Georgia. J Environ Qual 31: 1273–1278

Hartel PG, Summer JD, Segars WI (2003) Deer diet affects ribotype diversity of *Escherichia coli* for bacterial source tracking. Water Res 37:3263–3268

Harwood VJ, Whitlock J, Withington V (2000) Classification of antibiotic resistance patterns of indicator bacteria by discriminant analysis: use in predicting the source of fecal contamination in subtropical waters. Appl Environ Microbiol 66:3698–3704

Harwood VJ, Wiggins BA, Hagedorn C et al (2003) Phenotypic library-based microbial source tracking methods: efficacy in the California collaborative study. J Water Health 1:153–166

Hassan WM, Wang SY, Ellender RD (2005) Methods to increase fidelity of repetitive extragenic palindromic PCR fingerprint-based bacterial source tracking efforts. Appl Environ Microbiol 71: 512–518

Haznedaroğlu BZ, Zitomer DH, Hughes-Strange GB et al (2005) Whole-cell fatty acid composition of total coliforms to predict sources of fecal contamination. J Environ Eng 131:1426–1432

Hopkins KL, Hilton AC (2000) Methods available for the sub-typing of *Escherichia coli* O157. World J Microbiol Biotechnol 16:741–748

Hulton CS, Higgins CF, Sharp PM (1991) ERIC sequences: a novel family of repetitive elements in the genome of *Escherichia coli*, *Salmonella typhimurium* and other enteric bacteria. Mol Microbiol 5:825–834

Ihrie PD, Farmer JJ, Bailey EC (2003) Use of bacterial antibiotic resistance patterns and carbon source utilization fingerprints to determine sources of enterococcal fecal pollution. In: Abstracts of the 103rd American Society for Microbiology General Meeting Washington, D.C, 2003

Jenkins MB, Hartel PG, Olexa TJ et al (2003) Putative temporal variability of *Escherichia coli* ribotypes from yearling steers. J Environ Qual 32:305–309

Jiang SC, Chu W, Olson BH et al (2007) Microbial source tracking in a small southern California urban watershed indicates wild animals and growth as the source of fecal bacteria. Appl Microbiol Biotechnol 76:927–934

Johnson JM, Weagant SD, Jinneman KC et al (1995) Use of pulsed-field gel electrophoresis for epidemiological study of *Escherichia coli* O157:H7 during a food-borne outbreak. Appl Environ Microbiol 61:2806–2808

Johnson LK, Brown MB, Carruthers EA et al (2004) Sample size, library composition, and genotypic diversity among natural populations of *Escherichia coli* from different animals influence accuracy of determining sources of fecal pollution. Appl Environ Microbiol 70:4478–4485

Kaneene JB, Miller R, Sayah R et al (2007) Considerations when using discriminant function analysis of antimicrobial resistance profiles to identify sources of fecal contamination of surface water in Michigan. Appl Environ Microbiol 73:2878–2890

Kaspar CW, Burgess JL (1990) Antibiotic resistance indexing of *Escherichia coli* to identify sources of fecal contamination in water. Can J Microbiol 36:891–894

Kelch WJ, Lee JS (1978) Antibiotic resistance patterns of gram-negative bacteria isolated from envirornmental sources. Appl Environ Microbiol 36:450–456

Kelsey RH, Scott GI, Porter DE et al (2003) Using multiple antibiotic resistance and land use characteristics to determine sources of fecal coliform bacterial pollution. Environ Monit Assess 81:337–348

Kibbey HJ, Hagedorn C, McCoy EL (1978) Use of fecal streptococci as indicators of pollution in soil. Appl Environ Microbiol 35:711–717

Knudtson LM, Hartman PA (1993) Antibiotic resistance among enterococcal isolates from environmental and clinical sources. J Food Protect 56:489–492

Kon T, Weir SC, Howell ET et al (2009) Repetitive element (REP)-polymerase chain reaction (PCR) analysis of *Escherichia coli* isolates from recreational waters of southeastern Lake Huron. Can J Microbiol 55:269–276

Konopka A, Oliver L, Turco RF Jr (1998) The use of carbon substrate utilization patterns in environmental and ecological microbiology. Microb Ecol 35:103–115

Kotilainen P, Huovinen P, Eerola E (1991) Application of gas-liquid chromatographic analysis of cellular fatty acids for species identification and typing of coagulase-negative Staphylococci. J Clin Microbiol 29:315–322

Krumperman PH (1983) Multiple antibiotic resistance indexing of *Escherichia coli* to identify high-risk sources of fecal contamination of foods. Appl Environ Microbiol 46:165–170

Kühn I, Albert MJ, Ansaruzzaman M et al (1997) Characterization of *Aeromonas* spp. isolated from humans with diarrhea, from healthy controls, and from surface water in Bangladesh. J Clin Microbiol 35:369–373

Lasalde C, Rodriguez R, Toranzos GA (2005) Statistical analyses: possible reasons for unreliability of source tracking efforts. Appl Environ Microbiol 71:4690–4695

Lay JO (2001) MALDI-TOF mass spectrometry of bacteria. Mass Spectrom Rev 20:172–194

Lefresne G, Latrille E, Irlinger F et al (2004) Repeatability and reproducibility of ribotyping and its computer interpretation. Res Microbiol 155:154–161

Leung KT, Mackereth R, Tien Y-C et al (2004) A comparison of AFLP and ERIC-PCR analyses for discriminating *Escherichia coli* from cattle, pig and human sources. FEMS Microbiol Ecol 47:111–119

Lipman L, deNijs A, Lam T et al (1995) Identification of *Escherichia coli* strains from cows with clinical mastitis by serotyping and DNA polymorphism patterns with REP and ERIC primers. Vet Microbiol 43:13–19

Lu L, Hume ME, Sternes KL et al (2004) Genetic diversity of *Escherichia coli* isolates in irrigation water and associated sediments: implications for source tracking. Water Res 38:3899–3908

Martin B, Humbert O, Camara A et al (1992) A highly conserved repeated DNA element in the chromosome of *Streptococcus pneumoniae*. Nucleic Acids Res 20:3479–3483

McDonald JL, Hartel PG, Gentit LC et al (2006) Identifying sources of fecal contamination inexpensively with targeted sampling and bacterial source tracking. J Environ Qual 35:889–897

McLellan SL (2004) Genetic diversity of *Escherichia coli* isolated from urban rivers and beach water. Appl Environ Microbiol 70:4658–4665

McLellan SL, Daniels AD, Salmore AK (2001) Clonal Populations of thermotolerant *Enterobacteriaceae* in recreational water and their potential interference with fecal *Escherichia coli* counts. Appl Environ Microbiol 67:4934–4938

McLellan SL, Daniels AD, Salmore AK (2003) Genetic characterization of *Escherichia coli* populations from host sources of fecal pollution by using DNA fingerprinting. Appl Environ Microbiol 69:2587–2594

Meays CL, Broersma K, Nordin R, Mazumder A (2004) Source tracking fecal bacteria in water: a critical review of current methods. J Environ Manag 73:71–79

Mohapatra BR, Broersma K, Nordin R et al (2007) Wildlife identified as major source of *Escherichia coli* in agriculturally dominated watersheds by BOX A1R-derived genetic fingerprints. Microbiol Immunol 51:733–740

Mohapatra BR, Mazumder A (2008) Comparative efficacy of five different rep-PCR methods to discriminate *Escherichia coli* populations in aquatic environments. Water Sci Tech 58:537–547

Mohapatra BR, Broersma K, Mazumder A (2008) Differentiation of fecal *Escherichia coli* from poultry and free-living birds by (GTG)5-PCR genomic fingerprinting. Int J Med Microbiol 298:245–252

Molina M (2005) Temporal and spatial variability of fecal indicator bacteria: implications for the application of MST methodologies to differentiate sources of fecal contamination. USEPA, Washington D.C.

Moore DF, Harwood VJ, Ferguson DM et al (2005) Evaluation of antibiotic resistance analysis and ribotyping for identification of faecal pollution sources in an urban watershed. J Appl Microbiol 99:618–628

Mott J, Lehman R (2005) Bacteria source tracking in Copano Bay. Texas General Land Office, Austin TX, USA

Mott JB, Lehman R, Smith A (2008) Bacteria source tracking on the Mission and Aransas Rivers. Publication Number CBBEP-56. Coastal Bend Bays and Estuaries Program, Corpus Christi TX, USA. http://www.cbbep.org/publications/virtuallibrary/0630final.pdf. Accessed May 17, 2010

Moussa SH, Massengale RD (2008) Identification of the sources of Escherichia coli in a watershed using carbon-utilization patterns and composite data sets. J Water Health 6:197–207

Mukwaya GM, Welch DF (1989) Subgrouping of *Pseudomonas cepacia* by cellular fatty acid composition. J Clin Microbiol 27:2640–2646

Muyzer G, De Waal EC, Uitterlinder AG (1993) Profiling of complex microbial populations by denaturing gradient gel electrophoresis analysis of polymerase chain reaction-amplified genes coding for 16S rRNA. Appl Environ Microbiol 59:695–700

Myoda SP, Carson CA, Fuhrmann JJ et al (2003) Comparison of Genotypic Based Microbial Source Tracking Methods Requiring a Host Origin Database. J Water Health 1:167–171

Nelson M, Jones SH, Edwards C et al (2008) Characterization of *Escherichia coli* populations from gulls, landfill trash, and wastewater using ribotyping. Dis Aquat Org 81:53–63

Nübel U, Engelen B, Felske A et al (1996) Sequence heterogeneities of genes encoding 16S rRNAs in *Paenibacillus polymyxa* detected by temperature gradient gel electrophoresis. J Bacteriol 178:5636–5643

Olivas Y, Faulkner BR (2008) Fecal source tracking by antibiotic resistance analysis on a watershed exhibiting low resistance. Environ Monit Assess 139:15–25

Olive DM, Bean P (1999) Principles and applications of methods for DNA-based typing of microbial organisms. J Clin Microbiol 37:1661–1669

Parveen S, Murphree RL, Edmiston L et al (1997) Association of multiple-antibiotic-resistance profiles with point and nonpoint sources of Escherichia coli in Apalachicola Bay. Appl Environ Microbiol 63:2607–2612

Parveen S, Portier KM, Robinson K et al (1999) Discriminant analysis of ribotype profiles of *Escherichia coli* for differentiating human and nonhuman sources of fecal pollution. Appl Environ Microbiol 65:3142–3147

Parveen S, Hodge NC, Stall RE et al (2001) Phenotypic and genotypic characterization of human and nonhuman *Escherichia coli*. Water Res 35:379–386

Price B, Venso EA, Frana MF et al (2006) Classification tree method for bacterial source tracking with antibiotic resistance analysis data. Appl Environ Microbiol 72:3468–3475

Price B, Venso E, Frana M et al (2007) A comparison of ARA and DNA data for microbial source tracking based on source-classification models developed using classification trees. Water Res 41:3575–3584

Ritter KJ, Carruthers E, Carson CA et al (2003) Assessment of statistical methods used in library-based approaches to microbial source tracking. J Water Health 1:209–223

Robinson BJ, Ritter KJ, Ellender RD (2007) A statistical appraisal of disproportional versus proportional microbial source tracking libraries. J Water Health 5:503–509

Samadpour M, Stewart J, Steingart K et al (2002) Laboratory investigation of an *E. coli* O157:H7 outbreak associated with swimming in Battle Ground Lake, Vancouver, Washington. J Environ Health 64:16–20

Samadpour M, Roberts MC, Kitts C et al (2005) The use of ribotyping and antibiotic resistance patterns for identification of host sources of *Escherichia coli* strains. Lett Appl Microbiol 40:63–68

Santo Domingo JW, Bambic DG, Edge TA et al (2007) Quo vadis source tracking? Towards a strategic framework for environmental monitoring of fecal pollution. Water Res 41:3539–3552

Sayah RS, Kaneene JB, Johnson Y et al (2005) Patterns of antimicrobial resistance observed in *Escherichia coli* isolates obtained from domestic- and wild-animal fecal samples, human septage, and surface water. Appl Environ Microbiol 71:1394–1404

Schutter ME, Dick RP (2000) Comparison of fatty acid methyl ester (FAME) methods for characterizing microbial communities Soil Sci. Soc. Am. J. 64:1659–1668

Schwartz DC, Cantor CR (1984) Separation of yeast chromosome-sized DNAs by pulsed field gradient gel electrophoresis. Cell 37:67–75

Scott TM, Rose JB, Jenkins TM et al (2002) Microbial source tracking: current methodology and future directions. Appl Environ Microbiol 68:5796–5803

Scott TM, Parveen S, Portier KM et al (2003) Geographical variation in ribotype profiles of *Escherichia coli* isolates from humans, swine, poultry, beef, and dairy cattle in Florida. Appl Environ Microbiol 69:1089–1092

Scott TM, Caren J, Nelson GR et al (2004) Tracking sources of fecal pollution in a South Carolina watershed by ribotyping *Escherichia coli*: a case study. Environ Forensics 5:15–19

Seiber JN (2007) New dimensions of food safety and food quality research In: Ohkawa H, Miyagawa H, Lee PW (ed) Pesticide chemistry, crop protection, public health, environmental safety, Wiley-VCH, Germany

Seurinck S, Verstraete W, Siciliano SD (2003) Use of 16S-23S rRNA intergenic spacer region PCR and repetitive extragenic palindromic PCR analyses of *Escherichia coli* isolates to identify nonpoint fecal sources. Appl Environ Microbiol 69:4942–4950

Seurinck S, Verstraete W, Siciliano S (2005) Microbial source tracking for identification of fecal pollution. Rev Environ Sci Biotechnol 4:19–37

Seurinck S, Verdievel M, Verstraete W et al (2006) Identification of human fecal pollution sources in a coastal area: a case study at Oostende (Belgium). J Water Health 4:167–175

Siegrist TJ, Anderson PD, Huen WH et al (2007) Discrimination and characterization of environmental strains of *Escherichia coli* by matrix-assisted laser desorption/ionization time-of-flight mass spectrometry (MALDI-TOF-MS). J Microbiol Meth 68:554–562

Sigler V, Pasutti L (2006) Evaluation of denaturing gradient gel electrophoresis to differentiate *Escherichia coli* populations in secondary environments. Environ Microbiol 8:1703–1711

Simpson JM, Santo Domingo JW, Reasoner DJ (2002) Microbial source tracking: state of the science. Environ Sci Tech 36:5279–5288

Smith AK (2009) A comparison between discriminant analysis and Random Forests as techniques to determine sources of fecal contamination in Cedar Lakes, Texas. Thesis, Texas A&M University-Corpus Christi

Smith AK, Sterba Boatwright B, Mott, JB (2010) Novel application of a statistical technique, Random Forests, in a bacterial source tracking study. Water Res. doi:10.1016/j.watres.2010.05.019

Somarelli JA, Makarewicz JC, Sia R et al (2007) Wildlife identified as major source of *Escherichia coli* in agriculturally dominated watersheds by BOX A1R-derived genetic fingerprints. J Environ Manag 82:60–65

Stern MJ, Ames GF-L, Smith NH et al (1984) Repetitive extragenic palindromic sequences: a major component of the bacterial genome. Cell 37:1015–1026

Stewart J R, Ellender RD, Gooch JA et al (2003) Recommendations for Microbial Source Tracking: Lessons from a Methods Comparison Study. J Water Health 1:225–231

Stoeckel DM, Mathes MV, Hyer KE et al (2004) Comparison of seven protocols to identify fecal contamination sources using *Escherichia coli*. Environ Sci Tech 38:6109–6117

Stoeckel DM, Harwood VJ (2007) Performance, design, and analysis in microbial source tracking studies. Appl Environ Microbiol 73:2405–2415

Suzuki MT, Giovannoni SJ (1996) Bias caused by template annealing in the amplification of mixtures of 16S rRNA genes by PCR. Appl Environ Microbiol 62:625–630

Swaminathan B, Barrett TJ, Hunter SB et al (2001) PulseNet: The molecular subtyping network for foodborne bacterial disease surveillance, United States. Emerg Infect Dis 7:382–389

Tenover FC, Arbeit RD, Goering RV et al (1995) Interpreting chromosomal DNA restriction patterns produced by pulsed-field gel electrophoresis: criteria for bacterial strain typing. J Clin Microbiol 33:2233–2239

Ting WTE, Johnson DS, Holler AM et al (2003) A study of the sources of *E. coli* contamination oat Marquette Park Beach by random amplified polymorphic DNA typing. In: Abstracts of the 103rd American Society for Microbiology General Meeting Washington, D.C, 2003

USEPA (2005) Microbial source tracking guide document. EPA Publication No. EPA/600-R-05-064. USEPA: Cincinnati, OH

Van Belkum A, Struelens M, de Visser A et al (2001) Role of genomic typing in taxonomy, evolutionary genetics, and microbial epidemiology. Clin Microbiol Rev 14:547–560

Venieri D, Vantarakis A, Konminou G et al (2004) Differentiation of faecal *Escherichia coli* from human and animal sources by random amplified polymorphic DNA-PCR (RAPD-PCR). Water Sci Tech 50:193–198

Versalovic J, Koeuth T, Lupski JR (1991) Distribution of repetitive DNA sequences in eubacteria and application to fingerprinting of bacterial genomes. Nucleic Acids Res 19:6823–6831

Vogel L, van Oorschot E, Maas HME et al (2000) Epidemiologic typing of *Escherichia coli* using RAPD analysis, ribotyping and serotyping. Clin Microbiol Infect Dis 6:82–87

Vogel JR, Stoeckel DM, Lamendella R et al (2007) Identifying fecal sources in a selected catchment reach using multiple source-tracking tools. J Environ Qual 36:718–729

Von Wintzingerode F, Göbel UB, Stackebrandt E (1997) Determination of microbial diversity in environmental samples: pitfalls of PCR-based rRNA analysis. FEMS Microbiol Rev 21:213–229

Vos P, Hogers R, Bleeker M et al (1995) AFLP: a new technique for DNA fingerprinting. Nucleic Acids Res 23:4407–4414

Wallis JL, Taylor HD (2004) Phenotypic population characteristics of the enterococci in wastewater and animal faeces: implications for the new European directive on the quality of bathing waters. Water Sci Tech 47:27–32

Webster LF, Thompson BC, Fulton MH et al (2004) Identification of sources of *Escherichia coli* in South Carolina estuaries using antibiotic resistance analysis. J Exp Mar Biol Ecol 298:179–195

Welsh J, McClellan M (1990) Fingerprinting genomes using PCR with arbitrary primers. Nucleic Acids Res 18:7213–7218

Whitlock JE, Jones DT, Harwood VJ (2002) Identification of the sources of fecal coliforms in an urban watershed using antibiotic resistance analysis. Water Res 36:4273–4282

Wiggins BA (1996) Discriminant analysis of antibiotic resistance patterns in fecal streptococci, a method to differentiate human and animal sources of fecal pollution in natural waters. Appl Environ Microbiol 62:3997–4002

Wiggins BA, Andrews RW, Conway RA et al (1999) Use of antibiotic resistance analysis to identify nonpoint sources of fecal pollution. Appl Environ Microbiol 65:3483–3486

Wiggins BA, Cash PW, Creamer WS et al (2003) Use of antibiotic resistance analysis for representativeness testing of multiwatershed libraries. Appl Environ Microbiol 69:3399–3405

Williams JGK, Kubelik AR, Livak KJ et al (1990) DNA polymorphisms amplified by arbitrary primers are useful as genetic markers. Nucleic Acids Res 18:6531–6535

Wilson JE (2005) The application of antibiotic resistance analysis using Kirby Bauer disk diffusion to determine sources of fecal contamination in Copano Bay, Texas. Thesis, Texas A&M University-Corpus Christi

Yan T, Sadowsky M (2007) Determining sources of fecal bacteria in waterways. Environ Monit Assess 129:97–106

Yurtsever D, Haznedaroglu BZ, Dunaev T et al (2007) The effects of indicator organism type on phenotypic characterization of host-specificity and the implications for microbial source tracking. Proceedings of the Water Environment Federation, 7063–7071

Chapter 4
Library-Independent Bacterial Source Tracking Methods

Stefan Wuertz, Dan Wang, Georg H. Reischer, and Andreas H. Farnleitner

Abstract In recent years numerous library-independent methods for microbial source tracking have become available either relying on selective cultivation of source-specific bacteria or, increasingly, on direct detection of source-specific genetic markers. The scientific foundation for the detection of source-specific bacterial populations is discussed and an overview is provided of the methods developed in this field in the last 30 years. Another focus is on potential advantages and drawbacks as well as method performance characteristics in method development, evaluation and application. Unfortunately, few methods have been evaluated and applied beyond the regional geographical scale, making it clear that the global toolbox for bacterial MST is still in the development and evaluation stage. However, recent advances in statistical methods for interpretation of MST results will help account for less than perfect diagnostic sensitivities and specificities, while integrated study design must consider pollution source complexity and dynamics. Numerous successful MST applications have proven the practicality and potential of library-independent bacterial MST methods for the characterization and identification of fecal pollution sources.

Keywords Bacterial fecal sourc tracking • Cultivation • Direct detection • Molecular analysis • PCR • qPCR • Bacteroidales

4.1 Introduction

Routine detection of fecal pollution is still based on the selective growth of standard fecal indicator bacteria (FIB) including *Escherichia coli* (the most abundant representative of thermotolerant coliforms) and intestinal enterococci. Without

S. Wuertz (✉)
Department of Civil and Environmental Engineering, University of California, Davis, One Shields Avenue, Davis, CA 95616, USA
e-mail: swuertz@ucdavis.edu

C. Hagedorn et al. (eds.), *Microbial Source Tracking: Methods, Applications, and Case Studies*, DOI 10.1007/978-1-4419-9386-1_4,

doubt, water-quality testing based on the application of standard FIB has contributed to a fundamental improvement in water safety management during the last century (Tallon et al. 2005). There are several convincing arguments for using bacterial cells in the detection and characterization of fecal pollution: (1) bacteria are highly abundant and prevalent in fecal material (Suau et al. 1999); (2) sampling and sample concentration are relatively easy; (3) detection methods are well established; and (4) no particular health risks for the laboratory personal during analysis exist. Since *E. coli* and enterococci occur in human, livestock, and wildlife fecal pollution sources, they are considered as indicators of general fecal pollution (Klein and Houston 1898; Houston 1902; Geldreich 1976; Farnleitner et al. 2010). Fecal source identification based on these indicators requires library-dependent microbial source tracking (MST) methods (see Chap. 3). By contrast, bacterial targets for library-independent MST have to be source-specific.[1]

The aim of this chapter is to describe and discuss MST methods based on the selective detection of source-specific bacterial populations (Fig. 4.1). Alternative MST methods including bacteriophages (see Chap. 6), viruses (see Chap. 5), and mitochondrial host DNA (see Chap. 10) are covered in separate chapters of this

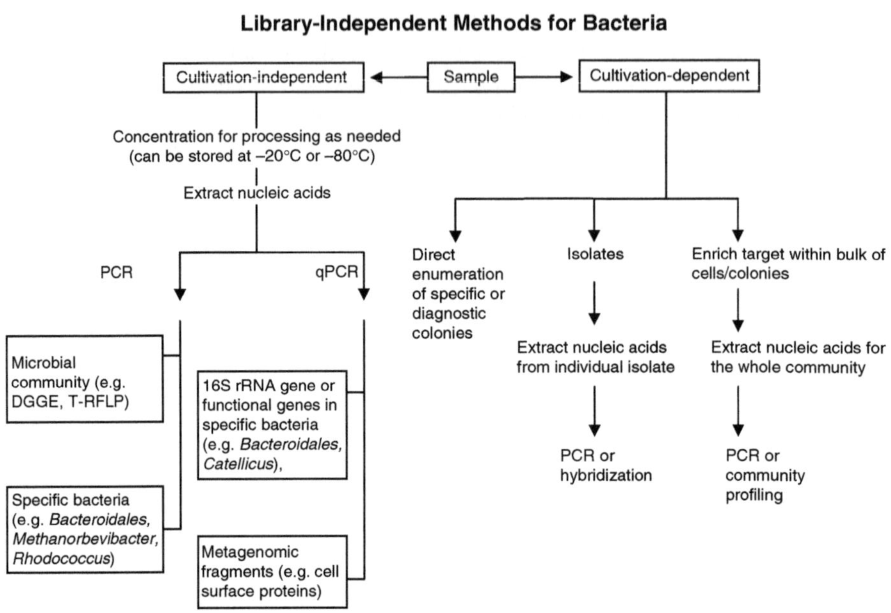

Fig. 4.1 Library-independent bacterial methods based on the detection of nucleic acids in prokaryotes

[1] The terms "bacteria" and "bacterial" are used synonymous to prokaryotic organisms encompassing the superkingdoms Bacteria and Archaea.

book. Microbial community analysis is briefly introduced (see Sect. 4.3.1), and the reader is referred to Chap. 11 for a more detailed discussion of this topic.

4.1.1 Basic Characteristics of Intestinal Bacterial Communities

Understanding the composition and structure of intestinal bacterial communities forms the scientific basis for the application of library-independent MST. Prokaryotic abundance in the intestine of animals and humans can reach cell numbers of more than 10^{11} cells/g (Whitman et al. 1998). By contrast, nonintestinal systems such as natural aquatic habitats or soil environments exhibit cell numbers between 10^4 and 10^8 cells per mL (Kirschner et al. 2002; Whitman et al. 1998; Wilhartitz et al. 2007) and 10^6 and 10^9 cells per g (Richter and Markewitz 1995), respectively. Despite this exceptional abundance, intestinal bacteria show surprisingly low diversity on higher taxonomic levels when compared to other ecosystems. For example, the intestinal microbiota of humans have been found to contain populations from only nine different phyla and the dominating phyla *Firmicutes* and *Bacteroidetes* accounted for more than 98% of the bacterial abundance (Fig. 4.2) (Backhed et al. 2005; Eckburg et al. 2005; Ley et al. 2008a; Suau et al. 1999). Other habitats such as soil harbor at least 20 different phyla with a more balanced distribution among lineages (Dunbar et al. 2002). On the contrary, diversity on lower taxonomic levels of the dominating intestinal phyla (i.e., species to subspecies range) proved to be enormous. In one study, 7,555 of the 13,335 sequences collected from colonic community samples were recovered only once (Eckburg et al. 2005). Notably, members of the phylum Proteobacteria (including coliforms and *E. coli*) that play an important role in many other microbial habitats represent only a minority of the intestinal community. The high phylogenetic diversity of the abundant phyla goes hand in hand with a pronounced metabolic diversity of the gut populations (Turnbaugh et al. 2006). With respect to fecal pollution analysis, physiological properties regarding the mode of electron acceptor usage (anaerobes vs. facultative aerobes) and the possibility to form resting stages (vegetative cells vs. spores) have major implications on their fate in nonintestinal environments after release from the intestinal tract.

4.1.2 Ability to Detect Intestinal Populations in the Aquatic Environment

Analysis of fecal pollution based on standard indicators predominantly relies on facultative aerobic or oxygen tolerant intestinal populations. Intestinal organisms such as *E. coli* or enterococci are not harmed by the presence of oxygen in aerobic aquatic habitats. They constitute appropriate targets for cultivation-based methods using simple aerated incubation conditions.

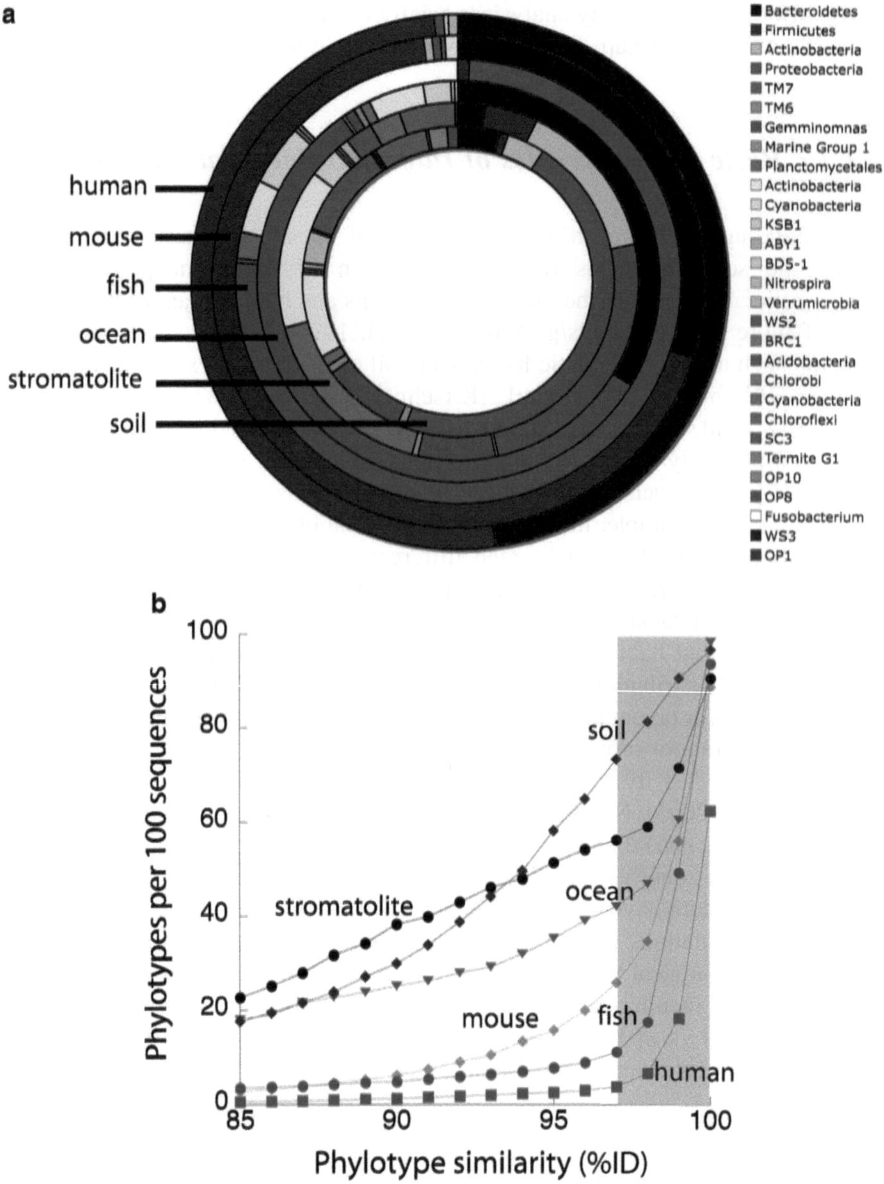

Fig. 4.2 Comparison of microbial diversity in the human colon, mouse cecum, ocean, and soil (reproduced by permission from Ley et al. 2006). (**a**) Percent representation of divisions in each environment; (**b**) phylogenetic architecture of the microbial communities shown in (**a**). For each habitat, the number of phylotypes per 100 16S rRNA gene sequences is shown for differing thresholds of 16S rRNA gene pairwise sequence identity (%ID). The *grey bar* highlights the phylotypes with ≥97%ID, the cutoff used to designate species and subspecies-level taxa. Note that compared to the soil and ocean, the gut shows the steepest decline in phylotype abundance at %IDs ≤ 97%. The shape of the curve reflects the structure of diversity. For details and methods see Ley et al. (2006)

For obligate anaerobic populations, oxygen presents a very toxic chemical species. Depending on the actual oxygen sensitivity of the considered population and the type of aquatic habitat (e.g., oxygen saturation,) abundant intestinal anaerobes will, thus, only be cultivable for short periods of time (diagnostic time frame: hours to days) after their introduction into extraintestinal environments (Bonjoch et al. 2009). For their cultivation, anoxic incubation techniques have to be applied. In contrast to cultivation, detection of intestinal anaerobic bacterial populations by cultivation-independent DNA-based molecular-biological methods is possible irrespective of cell viability (Bonjoch et al. 2009; Okabe and Shimazu 2007). Maintenance of cell integrity seems to be the main criterion for successful cell enrichment using filtration-based analysis procedures (Okabe and Shimazu 2007). Depending on biotic (e.g., trophic status and grazing pressure) and abiotic (e.g., temperature and sunlight) environmental factors, such molecular markers are detectable for time periods ranging from days to months (Bae and Wuertz 2009b; Bonjoch et al. 2009; Okabe and Shimazu 2007). Direct detection by molecular-biological techniques is the method of choice if fecal pollution signatures from abundant anaerobic populations are to be detected in aquatic environments in regard to such time frames.

On the contrary, spores of spore-forming intestinal populations have been shown to persist for extremely long periods in the environment (diagnostic time frame: years to decades) and still being detectable by cultivation-based techniques (Skanavis and Yanko 2001). Intestinal spores can, thus, be useful as conservative markers of fecal pollution in aquatic environments.

4.1.3 Basic Requirements for Library-Independent Bacterial MST

The ideal MST methods/targets should meet the following basic requirements (see also Sect. 4.3.3.3 for numerical indices and statistical considerations):

1. Bacterial MST targets should only be present in the fecal material of the respective source group considered. Consequently, the target should be absent in the fecal material of all other source groups, even in those that are closely related to the specific host (i.e., source-specificity criterion).
2. MST targets should be present in comparable numbers in the feces of all subgroups of the targeted sources (i.e., source-sensitivity criterion). In addition, markers should be highly abundant in source feces with concentrations comparable to or exceeding the concentrations of traditional fecal indicators.
3. If the significance of fecal pollution from different source groups is to be compared, knowledge on quantitative target occurrence in the respective source group is required. Furthermore, information on the environmental persistence and potential proliferation of the target is essential. Some targets might cease to be detectable very quickly under certain circumstances, while others tend to persist for prolonged periods of time (see above). Differential persistence of targets

can lead to noncomparable results even if correction measures are applied. In this context, it is necessary to experimentally demonstrate the comparability of different approaches (i.e., source-group comparability criterion).

At this point, it is necessary to clarify the term "source group" as it is used in this chapter. A fecal source group can be defined on different levels of resolution depending on the actual environmental MST problem. A source group might be an animal species (e.g., cattle), a larger group of species sharing specific traits (e.g., ruminant animals), or a systematically broad group (e.g., the class of *Mammalia*). Therefore, it is essential to clearly define the boundaries of a targeted source group to make proper assessment of source-group sensitivity and specificity possible.

The actual "target" of a library-independent MST method can be situated on various levels of interaction between microbiota and host. First, a method might detect a source-specific bacterial gene (or its product) that has been adapted during its evolution by interaction on the molecular level with the host. Horizontal gene transfer might contribute to the distribution of such a source-specific-gene among members of a bacterial community (e.g., antibiotic resistance genes). Second, the target might be a host-associated bacterial population (e.g., a lineage in the phylum *Bacteroidetes*) that has formed a stable symbiosis with a host group. And third, a bacterial community or consortium might also be indicative for its host group (community-based analysis). Most of the currently used MST methods fall in the second group and target host-associated populations.

4.1.4 Do the Ideal Target Populations for MST Exist in Reality?

Recent years have brought mounting scientific evidence that there are in fact intestinal bacterial populations that might meet the set requirements. In general terms, it has been demonstrated that abundant intestinal bacterial communities seem to be different from those found in extraintestinal habitats (Ley et al. 2008b). These differences are evident in qualitative (specific phylogenetic lineages) as well as quantitative community composition (differing abundances of distinct and/or common lineages). The difference might be explained by the evolving adaptive immune system of the host as well as evolutionary selection pressure affecting the host ("selection pressure on the habitat") and the microbiota, a factor absent in most other microbial habitats (Ley et al. 2008b).

Another study showed that the composition of the intestinal microbiota of mammals exhibit obvious signs of coevolution and codiversification (Ley et al. 2008a). It has become evident that host phylogeny is reflected in the diversification of gut microbiota. This evolutionary association of hosts and intestinal communities strongly supports the hypothesis on the existence of source-specific bacterial lineages (Ley et al. 2006). However, results also emphasize the importance of additional influencing factors such as diet, host morphology, and to a smaller extent geographical provenance of the host on intestinal bacterial communities (Ley et al. 2008a, b).

These recent findings support the suggestion that there are promising target populations for MST, even if it is unlikely that they will be perfect in every respect.

At the moment, our search for the most appropriate targets is still hampered by the lack of comprehensive knowledge about intestinal and extraintestinal bacterial habitats.

4.1.5 Library-Independent Bacterial MST Methods

In the context of MST this portrayal of the intestinal microbiota as a complex ecosystem underscores the importance of carefully choosing and evaluating the MST method to be applied to a specific setting or problem. Some of the issues that have to be considered are: How many possible source groups are there? What level of source-group resolution has to be achieved to differentiate those sources? Are MST methods available, which support the required source specificity and sensitivity in regard to the question being asked? This latter question has been addressed through the development of qualitative and quantitative models for genetic assays targeting the *Bacteroidales* and is discussed in Sects. 4.3.3.3 and 4.3.3.4, respectively.

When investigating the wealth of available literature on MST methods, it is often hard to compare the performance characteristics of the proposed methods. Methods might be based on very different detection techniques and on their mode of application. Furthermore, studies often lack comprehensive information on method source specificity and sensitivity or abundance of the MST target in feces of the respective source group. This chapter attempts to give a balanced indication of the advantages and disadvantages of the presented approaches. A guide to available library-independent methods based on the detection of bacterial targets and their applications in the field is presented in Tables 4.1 and 4.2.

4.2 Cultivation-Dependent Methods

As with all other methodical approaches there are some fundamental characteristics that have to be kept in mind when applying culture-based methods to MST. Formulation of media and cultivation conditions used in these approaches will critically influence the outcomes. Efforts to standardize used media and techniques are a prerequisite for reproducible results. In addition, taxon-targeted microbiological detection methods are hampered by the constantly evolving taxonomy and systematics of prokaryotes (Leclerc et al. 1996).

4.2.1 Supporting Methods Based on Standard Fecal Indicator Bacteria (FIB)

One of the first attempts to differentiate human from nonhuman fecal pollution was based on the ratio of fecal coliforms to fecal streptococci (Geldreich 1976; Geldreich and Kenner 1969). The authors concluded that a ratio greater than 4 (FC/FS > 4) is indicative for pollution from humans, whereas a ratio smaller than 0.7 (FC/FS < 0.7)

Table 4.1 Library-independent methods and applications for microbial source tracking (non-*Bacteroidales* targets)

Targeted animal host	Target organism	Target	Procedure	Method development	Representative application	Characteristics
Human	Sorbitol-fermenting *Bifidobacteria*	Cultivable anaerobic cells	Membrane filtration and selective cultivation	Mara and Oragui (1983)	Mara and Oragui (1985), Jagals et al. (1995), Jagals and Grabow (1996), Rhodes and Kantor (1999), Long et al. (2003), Blanch et al. (2004), Bonjoch et al. (2005), Climenti et al. (2005), and Mushi et al. (2010)	Gram-positive, anaerobic organism, fast dieoff in the aquatic environment; "low-tech" alternative for the detection of very recent fecal pollution; fast dieoff needs to be considered at applied sampling strategy
	Bifidobacterium adolescentis *Bifidobacterium dentium*	16S rRNA	DNA extraction Two-step PCR	Bonjoch et al. (2004) and King et al. (2007)	Blanch et al. (2006), King et al. (2007), Amador et al. (2008), Morrison et al. (2008), Bachoon et al. (2009), and Bonjoch et al. (2009)	Gram-positive, anaerobic organism, survival highly variable signal decay fast in environment some false-positive/negative signal
	Enterococcus faecium	*esp* gene	Enrichment (membrane filtration and incubation for 48 h) DNA extraction PCR	Hammerum and Jensen (2002), Scott et al. (2005), McQuaig et al. (2006), and Ahmed et al. (2008b)	McDonald et al. (2006), McQuaig et al. (2006), Ahmed et al. (2008a, b), Knee et al. (2008), Hartel et al. (2008), Roslev et al. (2008), Wong et al. (2009), and Balleste et al. (2010)	Unique to human waste; relatively low prevalence of this marker in enterococci population
	Faecalibacterium	16S rRNA	DNA extraction PCR	Zheng et al. (2009)		Gram-negative; anaerobe; no false-positive signal for 10 ng per reaction; BLAST result indicate this marker distributed globally; field application not done yet

	Methanobrevibacter smithii	*nifH*	DNA extraction PCR	Ufnar et al. (2006)	McQuaig et al. (2009)	Highly specific to and abundant in human fecal material; does not frequently inhabit marine/fluvial environment or sediment; signal indicates recent contamination
	Diarrheic *E. coli*	STIb gene	DNA extraction PCR	Olson and Oshiro (1997) and Field et al. (2003)		Can detect all human sewage samples; 50% false-positive rate for nonhuman fecal pollution
	Diarrheic *E. coli*	STh gene	Membrane filtration Growth of colonies DNA extraction PCR	Olson and Oshiro (1997)	Jiang et al. (2007)	Negative in many environmental water samples
Anthropogenic (human and livestock)	*Clostridium perfringens*	Cultivable spores	Membrane filtration, selective anaerobic growth	ISO (2002)	Hill et al. (1996), Skanavis and Yanko (2001), Sorensen (1989), and Farnleitner et al. (2010)	Anaerobic spore former; spores extremely persistent; applied as alternative indicator of fecal pollution; also used as conservative tracer of sewage and sewage sludge; recently suggested as conservative indicator for anthropogenic fecal pollution; relative low abundance

(continued)

Table 4.1 (continued)

Targeted animal host	Target organism	Target	Procedure	Method development	Representative application	Characteristics
Herbivore	*Rhodococcus coprophilus*	Cultivable cells	Membrane filtration and selective growth	Rowbotham and Cross (1977)	Mara and Oragui (1981) and Jagals et al. (1995)	Actinomycete; aerobe; naturally growing on herbivorous dung and feces and persists in water and sediment; signal may not indicate fresh fecal pollution; detect total herbivore waste; relative low concentration in feces; long incubation time
		16S rRNA	DNA extraction qPCR	Savill et al. (2001)	Gilpin et al. (2002, 2003)	
Ruminants	*Methanobrevibacter ruminantium*	*nifH*	DNA extraction PCR	Ufnar et al. (2007b)		Positive for domesticated bovine; sheep, and goat feces; abundant in rumen fluid
Cattle	Diarrheic *E. coli*	IIA (LTIIa) toxin gene	Centrifugation or membrane filtration DNA extraction PCR or nested qPCR	Khatib et al. (2002) and Chern et al. (2004)	Tsai et al. (2003) and Jiang et al. (2007)	Marker was stable for 2–4 weeks in the environment; low prevalence of the toxin gene in *E. coli* population led to low sensitivity; no cross-reactivity observed in USA; nested qPCR increased sensitivity of the assay
Swine	Diarrheic *E. coli*	STII toxin gene	Centrifugation or membrane filtration DNA extraction PCR or nested qPCR	Khatib et al. (2003) and Chern et al. (2004)	Chern et al. (2004)	Highly specific to swine waste; 100% positive when more than 35 *E. coli* were screened; nested qPCR increased sensitivity of the assay
	Uncultured methanogen	*mcrA*	DNA extraction qPCR	Ufnar et al. (2007a)	Mieszkin et al. (2009) and Lamendella et al. (2009)	Specific to swine fecal material; signal indicates recent pollution; low signal from environmental samples

Dog	Diarrheic *E. coli*	*papG* III gene	Membrane filtration Growth of colonies DNA extraction PCR	Johnson et al. (2000) and Jiang et al. (2007)		
Rabbit	Diarrheic *E. coli*	*ralG* gene	Membrane filtration Growth of colonies DNA extraction PCR	Adams et al. (1997) and Jiang et al. (2007)		
Bird	Diarrheic *E. coli*	*tsh* gene	Membrane filtration Growth of colonies DNA extraction PCR	Provence and Curtiss (1994) and Jiang et al. (2007)		
	Enterohemorrhagic *E. coli*	*hlyE* gene	DNA extraction PCR	Chern and Olson (2004)		
Gull	*Catellicoccus marimammalium*	16S rRNA	DNA extraction qPCR	Lu et al. (2008)	Sirikanchana et al. (submitted)	Low G+C, gram-positive, catalase-negative; positive for other seabirds; minor detection of marine mammals

Table 4.2 Library-independent methods and applications for microbial source tracking (*Bacteriodales* targets)

Targeted animal host	Target gene	Assay name	Procedure	Method development	Representative application	Characteristics and PCR platform
Bacteroides-Prevotella	16S rRNA	Bac32F/708R	DNA extraction PCR	Bernhard and Field (2000a)	Bower et al. (2005), Shanks et al. (2006b), Walters and Field (2006), Gourmelon et al. (2007), Ahmed et al. (2008c), Lamendella et al. (2009), McQuaig et al. (2009), and Saunders et al. (2009)	Thermal minicycler (MJ Research)
Universal	16S rRNA	GenBac2	DNA extraction qPCR	Dick and Field (2004)	Wade et al. (2006) and Siefring et al. (2008)	ABI prism 7700; probe-based assay; TaqMan Buffer A
	16S rRNA	GenBac3	DNA extraction qPCR	Dick and Field (2004) and Siefring et al. (2008)	Chern et al. (2009) and Shanks et al. (2009, 2010)	Compared three qPCR platforms: ABI model 7700/7900 with TaqMan Mix, ABI model 7900 with Fast Mix, and Cepheid SmartCycler with OmniMix
	16S rRNA	AllBac	DNA extraction qPCR	Layton et al. (2006)	Bell et al. (2009) Mieszkin et al. (2009a), and Dick et al. (2010)	DNA Engine Opticon continuous fluorescence detection system (MJ research); FAM, BHQ1 labeled probe-based assay; QuantiTect PCR mix (QIAGEN)
	16S rRNA	BacUni	DNA extraction qPCR	Kildare et al. (2007)	Kildare et al. (2006), Bambicet al. (2007), Bae and Wuertz (2009a, b), Miura et al. (2009), Silkie and Nelson (2009), Schriewer et al. (2010), and Wang et al. (2010)	TaqMan 7000 system (ABI); FAM, MGB labeled probe-based assay; TaqMan Universal PCR Master Mix (ABI)
Bacteroides-Prevotella	16S rRNA	BacPre1	DNA extraction qPCR	Okabe et al. (2007)		TaqMan 7000 system (ABI); SYBR Green assay; TaqMan Universal PCR Master Mix (ABI)

Human	16S rRNA	HF 134F/654R; HF 183F/ Bac708R; HF 183F/264R	DNA extraction PCR qPCR	Bernhard and Field (2000b) and Seurinck et al. (2005)	Bower et al. (2005), Shanks et al. (2006b), Walters and Field (2006), Ahmed et al. (2008b, d, 2009a, b), Gourmelon et al. (2007), Santoro and Boehm (2007), Mieszkin et al. (2009a), McLain et al. (2009), McQuaig et al. (2009), Rosario et al. (2009), Saunders et al. (2009), Balleste et al. (2010), and Dick et al. (2010)	Sequences identified by LH-PCR and T-RFLP; original PCR assay was modified to be a SYBR Green assay
	16S rRNA	HuBac	DNA extraction qPCR	Layton et al. (2006)	Ahmed et al. (2009a, b), and McLain et al. (2009)	DNA Engine Opticon continuous fluorescence detection system (MJ research); FAM, BHQ1 labeled probe-based assay; QuantiTect PCR mix (QIAGEN)
	16S rRNA	BacHum	DNA extraction qPCR	Kildare et al. (2007)	Kildare et al. (2006), Bambic et al. (2007), Ahmed et al. (2009a, b), Bae and Wuertz (2009a, b), McLain et al. (2009), Silkie and Nelson (2009), Walters et al. (2009), Dick et al. (2010), Reischer et al. (2010), Schriewer et al. (2010), and Wang et al. (2010)	TaqMan 7000 system (ABI); FAM, TAMRA labeled probe-based assay; TaqMan Universal PCR Master Mix (ABI)
	16S rRNA	BacH	DNA extraction qPCR	Reischer et al. (2007)	Reischer et al. (2008), Ahmed et al. (2009a, b), McLain et al. (2009), and Reischer et al. (2011)	iQ Real-Time system (Biorad); FAM, MGB labeled probe-based assay; iQ Supermix

(continued)

Table 4.2 (continued)

Targeted animal host	Target gene	Assay name	Procedure	Method development	Representative application	Characteristics and PCR platform
	16S rRNA	Human-Bac1	DNA extraction qPCR	Okabe et al. (2007)	Ahmed et al. (2009a, b)	TaqMan 7000 system (ABI); FAM, MGB labeled probe-based assay; TaqMan Universal PCR Master Mix (ABI)
	16S rRNA	HF134f, Bac654r, HI408r	DNA extraction qPCR	Stricker et al. (2008)		iCycler iQ, Real-Time Detection System (Biorad); Duplex Scorpion probe-based qPCR; iQ Supermix (Biorad)
	α -1-6 mannanase gene	BtH	DNA extraction qPCR	Yampara-Iquise et al. (2008)		Targets α-1-6 mannanase gene of *B. thetaiotaomicron*, a species only found in human guts; this gene is associated with symbiont host-microbe interaction; FAM, NFQ labeled probe-based assay; StepOne Real-Time PCR system (ABI); TaqMan Universal PCR Master Mix (ABI)
	Cell surface-associated genes	HumM2 HumM3	DNA extraction qPCR	Shanks et al. (2007, 2009)		*Bacteroidales*-like genes markers identified by comparing of human feces against pig feces, prevalent in ambient environment; FAM, TAMRA labeled probe-based assay; 7900HT fast real-time sequence detector (ABI); TaqMan Universal Master Mix
	16s rRNA	BFD	DNA extraction qPCR	Converse et al. (2009)		FAM labeled probe-based assay; SmartCycler II (Cepheid); OmniMix HS-50 reagents
	16S rRNA	YHF	DNA extraction qPCR	Jeong et al. (2010)		SYBR Green Assay; Lightcycler 480 (Roche); Roche SYBR Green master mix

Ruminant	16S rRNA	BacR	DNA extraction qPCR	Reischer et al. (2006)	Reischer et al. (2008), Mieszkin et al. (2009a), McLain et al. (2009), Stapleton et al. (2009), Gourmelon et al. (2010), and Reischer et al. (2011)	iQ Real-Time system (Biorad); FAM, MGB labeled probe-based assay; iQ Supermix
	16S rRNA	Rum-2-BacqPCR	DNA extraction qPCR	Mieszkin et al. (2009b)		FAM, BHQ1 labeled probe-based assay; TaqMan Brilliant II QPCR Master Mix kit (Stratagene); Stratagene MX 3000 P
Bovine[a]	16S rRNA	BoBac	DNA extraction qPCR	Layton et al. (2006)	Shanks et al. (2010)	DNA Engine Opticon continuous fluorescence detection system (MJ research); FAM, BHQ1 labeled probe-based assay; QuantiTect PCR mix (QIAGEN)
Cow[a]	16S rRNA	CF128F/Bac708R CF193F/Bac708R	DNA extraction PCR	Bernhard and Field (2000b)	Bower et al. (2005), Shanks et al. (2006b), Gourmelon et al. (2007), Ahmed et al. (2008a–d), Balleste et al. (2010), and Shanks et al. (2010)	Sequences identified by LH-PCR and T-RFLP
	16S rRNA	BacCow	DNA extraction qPCR	Bernhard and Field (2000b) and Kildare et al. (2007)	Kildare et al. (2006), Bambic et al. (2007), Silkie and Nelson (2009), Schriewer et al. (2010), and Wang et al. (2010)	TaqMan 7000 system (ABI); FAM, TAMRA labeled probe-based assay; TaqMan Universal PCR Master Mix (ABI)
	16S rRNA	Cow-Bac 1, 2, 3	DNA extraction qPCR	Okabe et al. (2007)		TaqMan 7000 system (ABI); SYBR Green assay; TaqMan Universal PCR Master Mix (ABI)
	16S rRNA	CI125f, Bac708r, I408r	DNA extraction qPCR	Stricker et al. (2008)		iCycler iQ, Real-Time Detection System (Biorad); Duplex Scorpion probe-based qPCR; iQ Supermix (Biorad)

(continued)

Table 4.2 (continued)

Targeted animal host	Target gene	Assay name	Procedure	Method development	Representative application	Characteristics and PCR platform
	16S rRNA	YCF	DNA extraction qPCR	Jeong et al. (2010)		SYBR Green Assay; Lightcycler 480; Roche SYBR Green master mix
	Genes encode membrane-associated and secreted proteins	Bac2, Bac3 CowM2, CowM3	DNA extraction qPCR	Shanks et al. (2006a, b, 2008)	Shanks et al. (2010)	*Bacteroidales*-like gene markers identified by comparing bovine feces against porcine feces, distinguish bovine from other ruminant sources; ABI 7000 or 7900 HT fast real-time system (ABI); FAM, TAMRA labeled probe-based assay; TaqMan Universal PCR Master Mix (ABI)
Elk	16S rRNA	EF447F/990R	DNA extraction PCR	Dick et al. (2005a)	Shanks et al. (2006b)	Sequence identified by microplate subtractive hybridization
Horse	16S rRNA	HoF597F/ Bac708R; HorseBact	DNA extraction PCR qPCR	Simpson et al. (2004), Dick et al. (2005b), Layton et al. (2006), and Silkie and Nelson (2009)		Sequence originally identified by comparing the feces of eight hosts: human, bovine, pig, horse, dog, cat, gull, and elk; TaqMan 7000, 7300 or StepOne system (ABI); FAM, MGB labeled probe-based assay; TaqMan Universal PCR Master Mix (ABI)
Dog	16S rRNA	DF475F/ Bac708R	DNA extraction PCR	Dick et al. (2005a)		Sequence identified by microplate subtractive hybridization
	16S rRNA	BacCan	DNA extraction qPCR	Kildare et al. (2007)	Kildare et al. (2006), Bambic et al. (2007), Ahmed et al. (2008a–d), Silkie and Nelson (2009), Schriewer et al. (2010), and Wang et al. (2010)	TaqMan 7000 system (ABI); FAM, MGB labeled probe-based assay; TaqMan Universal PCR Master Mix (ABI)

Pig	16S rRNA	PF163F/ Bac708R	DNA extraction PCR	Dick et al. (2005b)	Gourmelon et al. (2007) and Lamendella et al. (2009)	Sequence identified by comparing the feces of eight hosts: human, bovine, pig, horse, dog, cat, gull, and elk
	16S rRNA	Pig-Bac 1, 2	DNA extraction qPCR	Okabe et al. (2007)	Lamendella et al. (2009)	TaqMan 7000 system (ABI); SYBR Green assay; TaqMan Universal PCR Master Mix (ABI)
	16S rRNA	Pig-1-Bac, Pig-2-Bac	DNA extraction qPCR	Mieszkin et al. (2009a)		FAM, BHQ1 labeled probe-based assay; TaqMan Brilliant II QPCR Master Mix kit (Stratagene); Stratagene MX 3000 P

[a] Bovine and cow assays will likely also detect *Bacteroidales* populations in other ruminants

points to animal fecal sources (Geldreich 1976). It should be highlighted that this ratio is no longer considered to be significantly different between fecal samples from humans and animals (Farnleitner et al. 2010; Pourcher et al. 1991). The different ratios observed were rather a consequence of differential dieoff of the respective populations during wastewater treatment or storage in receiving waters (Feachem 1975; Geldreich et al. 1968). Later studies have concluded that different survival rates of fecal coliforms and fecal streptococci strains make the FC/FS ratio unreliable for source identification (Blanch et al. 2006; Jagals and Grabow 1996; McFeters et al. 1974; Pourcher et al. 1991). Nevertheless, monitoring differential dieoff kinetics of fecal coliforms and fecal enterococci in microcosms (e.g., in stored water samples from routine investigations) can provide ancillary information in fecal pollution characterization (Farnleitner et al. 2010; Feachem 1975). Yet, FC/FS ratios do not provide a means for source identification.

Another approach employing microbiological parameters from routine monitoring is the ratio of atypical to typical colonies occurring during total coliform analysis (Nieman and Brion 2003) according to standard methods (APHA 1992). This method was originally proposed as a method for the determination of the age of fecal contamination but has also found use in source identification (Booth and Brion 2004; Coakley et al. 2009; Ward et al. 2009). Again, this approach might support fecal pollution characterization but was not suggested as a stand-alone determinant parameter for MST (Coakley et al. 2009).

Several studies have proposed that some *Enterococcus* species might have a limited host range (Geldreich and Kenner 1969; Leclerc et al. 1996). For example, *Enterococcus faecalis* has been found to be restricted to human and poultry sources in some studies (Kuntz et al. 2004; Pourcher et al. 1991; Wheeler et al. 2002). Those studies were mainly based on selective cultivation of enterococci and characterization by biochemical or molecular typing and are therefore considered library-based approaches. However, several attempts have been undertaken to allow the species identification of isolates by species-specific PCR assays (Harwood et al. 2004; Layton et al. 2010). The most recent study showed that no single species of *Enterococcus* seems to be indicative for a specific source, but assemblages as determined by multiplex PCR applied on *Enterococcus* enrichment cultures may support source identification to some extent (Layton et al. 2010).

A recent paper suggests that spores of the anaerobe *Clostridium perfringens* might be a good conservative indicator for anthropogenic fecal influence (humans and livestock). *C. perfringens* was found in human, livestock, and carnivore animal feces in high numbers but rarely in herbivore wildlife fecal material (Farnleitner et al. 2010). These findings are in accordance with studies using *C. perfringens* spores successfully as conservative tracer of wastewater sludge pollution in marine systems (Hill et al. 1996; Skanavis and Yanko 2001; Sorensen et al. 1989). It has also been suggested as an alternate fecal indicator in tropical waters (Roll and Fujioka 1997; Byamukama et al. 2005) because, unlike FIB, *C. perfringens* does not multiply in the tropical environment. One potential drawback of this approach is that the indicator is about 100-fold less abundant in many anthropogenic fecal sources than *E. coli*, which can limit detectability and calls for larger sampling volumes (Farnleitner et al. 2010).

4.2.2 Combining Cultivation with Genetic Marker Detection

A combination of cultivation and genetic marker detection is used in the detection of some toxin and virulence genes for MST. Several methods have been proposed for the identification of human-specific (Oshiro and Olson 1997) and pig-specific (Khatib et al. 2003) variants of heat-stable STII toxin and a cattle-specific variant of the heat-labile protein IIA in enterotoxigenic *E. coli* (ETEC) (Khatib et al. 2002; Tsai et al. 2003). The major drawback of this approach is that those toxin genes are shed only by infected hosts. As with the detection of many other pathogens, the detection is dependent on the prevalence of the disease and the concentrations in the feces of infected animals. Therefore, those methods seem unlikely to meet the requirements for source sensitivity, potentially restricting their general applicability.

The enterococcal surface protein (*esp*) marker found in *Enterococcus faecium* has been proposed as a marker for human fecal pollution (Scott et al. 2005). The method is based on the enrichment of membrane-filtered enterococci on mEnterococcus agar and subsequent detection of the marker by PCR. Studies evaluating this marker reached contrasting conclusions. Several studies attested a good source specificity and sensitivity (Ahmed et al. 2008a–d; Scott et al. 2005), while others did not recommend the marker due to insufficient prevalence and specificity (Byappanahalli et al. 2008; Layton et al. 2009; Whitman et al. 2007). Regardless of these not yet resolved issues, the applications of this approach seem to be restrained by two drawbacks: (1) the enrichment step before PCR complicates the ability to gain at least semiquantitative results on the abundance of this marker in a sample, and (2) the marker is only found in about 1% of the enterococci populations growing on mEI agar (Scott et al. 2009), which might limit the ability to detect moderate or low levels of human fecal pollution.

4.2.3 Sorbitol-Fermenting Bifidobacteria

Bifidobacteria have been proposed as possible indicators of fecal pollution for 50 years (Evison and James 1975; Gyllenberg et al. 1960). They are gram-positive anaerobes with high abundance in fecal material of humans and animals (Bahaka et al. 1993; Suau et al. 1999). Since the 1980s, subgroups of bifidobacteria have been suggested as specific indicators for human fecal pollution. Resnick and Levin (1981a) used YN-6 medium (Resnick and Levin 1981b) to enumerate *Bifidobacterium* sp. in fecal samples and raw wastewater. They found bifidobacteria only in human and swine fecal samples (Resnick and Levin 1981a). Mara and Oragui (1983) extended this work and formulated a medium for the selective cultivation of sorbitol-fermenting bifidobacteria (SFB). Using the developed human bifid sorbitol agar (HBSA), they found high numbers of bifidobacteria in human feces and wastewater but not in animal feces, making it the most promising candidate human-specific MST method at that time. Most of the strains from human feces growing on this medium were identified as *Bifidobacterium adolescentis* and *B. breve* (Mara and Oragui 1983).

Good source specificity of this approach was subsequently found in several studies in the USA (Resnick and Levin 1981a), Spain (Blanch et al. 2004, 2006; Bonjoch et al. 2005), Nigeria and Zimbabwe (Mara and Oragui 1985), and in Tanzania (Mushi et al. 2010). Mean concentrations of SFB enumerated on the HBSA medium in those studies averaged at 9.3 $\log_{10}$ CFU/g feces and 6.6 $\log_{10}$ CFU/100 mL raw wastewater, respectively. Several studies investigated the persistence of cultivable SFB in water (Bonjoch et al. 2009; Cimenti et al. 2005; Mushi et al. 2010; Resnick and Levin 1981a; Rhodes and Kator 1999; Ottoson 2009). The results demonstrated SFB stay cultivable only for a relatively short time in environmental waters (T_{90} values of 5 h up to 25 days), especially at higher temperatures and lower salinities (Bonjoch et al. 2009). Nevertheless, the numerous published applications of cultivation-based SFB detection (Cimenti et al. 2005; Jagals and Grabow 1996; Jagals et al. 1995; Long et al. 2003; Mara and Oragui 1985; Mushi et al. 2010; Rhodes and Kator 1999) indicate it may be a relatively low-tech alternative to molecular methods for the specific detection of very recent human fecal contamination in water. The fast dieoff of SFB might necessitate the adaption of sampling strategies with shorter intervals (temporal) or distances (spatial) between periods of sample collection or different sites, respectively, to keep comparable sensitivity in regard to standard approaches (Mushi et al. 2010).

Several studies have proposed cultivation-independent molecular marker detection for source-specific bifidobacteria populations, which are discussed in detail in Sect. 4.3.2.1.

4.2.4 *Rhodococcus coprophilus*

Rhodococcus coprophilus is an anaerobic actinomycete of the family Nocardiaceae and has been proposed as a specific indicator for farm animal fecal contamination. Its natural habitat is herbivore dung on which the bacterium proliferates and can be found in high numbers (Rowbotham and Cross 1977). It is washed into waterways with the fecal material and can be selectively cultivated (Mara and Oragui 1981). *R. coprophilus* has been evaluated repeatedly as a MST target and has shown high source specificity for animal feces and high environmental persistence (Jagals et al. 1995). However, the relatively low concentrations in fecal samples (3–4 $\log_{10}$ CFU/g feces) (Mara and Oragui 1981) and the time-consuming cultivation technique (18 days) may limit its practical applicability (Jagals et al. 1995).

4.3 Cultivation-Independent Methods

Methods that do not rely on detection based on growth of bacterial populations or isolated bacteria present in a sample have become increasingly widespread in recent years and mostly target either the 16S rRNA gene or sequences obtained from metagenomic fragments (Fig. 4.1). They usually involve target enrichment by filtration followed by extraction of nucleic acids and storage at low temperature prior to amplification of target genes by PCR or qPCR. So far, members of the order *Bacteroidales*, the genera *Methanobrevibacter*, *Rhodococcus*, *Faecalibacterium*, *Catellicoccus*, pathotypes of *Enterococcus* and *Escherichia*, and the bifidobacteria have been reported as being associated with specific animal hosts or pollution sources (see Tables 4.1 and 4.2) and are discussed below.

4.3.1 Microbial Community Analysis

Tools for the investigation of the total bacterial community composition (e.g., terminal restriction fragment length polymorphism [T-RFLP] or denaturing gradient gel electrophoresis [DGGE]) have not been widely used in MST. The application is often hampered due to the fact that bacteria of fecal origin usually constitute a relatively small part of the total bacterial community in an impacted water resource (Farnleitner et al. 2005). Only in heavily polluted water can the contribution of microbial fecal pollution reach levels that make it detectable above the "background" of the natural aquatic microbial community by most of the available methods. For example, the limit of detection for a population by PCR-based DGGE community profiling was considered to be in the range of ≥1% target contribution to the total microbial community (Muyzer 1999).

Second, it is very difficult to draw a line between library-based and library-independent total community analysis. Many of these methods can be used for typing of isolated microbes in MST strain libraries (e.g., DGGE) (Buchan et al. 2001). In addition, total community profiling often yields fingerprints that have to be assembled into libraries to be useful for source tracking applications (Burtscher et al. 2009; Farnleitner et al. 2004), and the focus of this chapter is on library-independent methods. The reader is referred to Chap. 11 for a detailed analysis of microbial community-based MST methods.

4.3.2 Genetic Markers for Host-Associated Bacterial Populations

Despite the advent of whole genome sequencing, there is still no comprehensive sequence database comprising all known fecal bacteria that would allow one to identify host-associated genetic markers directly based on in silico tools. To date, researchers have used two basic approaches to search for such sequences. First, if a given gene or intragenic region from a taxon of bacteria is suspected to exhibit some degree of host specificity, a universal PCR primer set can amplify the respective gene sequence from feces of the target host group as well as from nontarget hosts, therefore, permitting sequence alignment and identification of host-associated markers. In this scenario, the researchers already know the target microorganism and gene, and typically the function of that gene. The most popular target is the 16S rRNA gene, and both length heterogeneity PCR (LH-PCR) and T-RFLP analysis have been used to identify suitably conserved DNA regions (Bernhard and Field 2000b). In the second approach, there is no prior knowledge about a given gene, and all of the DNA extracted from the target fecal material must be compared with that from other hosts. With the appropriate technique, a wide range of possible gene candidates can be found, the function of some of which may not have been fully characterized. Subtractive hybridization is typically used to enrich the host group-specific DNA fragments by selectively removing the DNA from other related hosts (Dick et al. 2005a).

4.3.2.1 16S rRNA Gene

The 16S rRNA gene is widely used for phylogenetic studies (Weisburg et al. 1991). It is highly conserved, meaning universal or quasi-universal primers are available for PCR. On the contrary, this gene also contains variable regions which, depending on the considered phylogenetic groups, can allow discrimination of bacteria down to the subspecies level (although the 16S rRNA gene shows highest phylogenetic resolution on levels higher than species). It also has multiple copies in each cell, making it easily detectable even when diluted in the environment (Shanks et al. 2008). It is the most common bacterial target gene in MST at this time. The order *Bacteroidales* is the most widely used taxon targeted for source identification for livestock (pigs, cattle, sheep, horses, and chicken) and domestic pets (dogs and cats), and *Bacteroides* is considered the predominant genus of human fecal bacteria (Holdeman et al. 1976). The reader is referred to Sect. 4.3.3 for an extensive discussion of *Bacteroidales* assays.

Several other bacterial MST methods are also based on the 16S rRNA gene, targeting *Bifidobacterium* for human waste (Bonjoch et al. 2004; Matsuki et al. 2004; Lamendella et al. 2008), *Faecalibacterium* for human waste (Dick et al. 2005a, b; Zheng et al. 2009), *Rhodococcus* for farmed herbivores (Savill et al. 2001; Gilpin et al. 2002; Cimenti et al. 2005), and *Catellicoccus* for gull feces (Lu et al. 2008).

Bifidobacteria spp. are gram-positive rods and strict anaerobes commonly associated with human and animal intestines. A 16S rRNA gene fragment from *Bifidobacteria* species was used in the first library-independent molecular/genetic MST method (Bernhard et al. 2000a, b). Later, nine species were identified in human fecal material of which *B. adolescentis*, *B. dentium*, and *B. longum* are thought to exist exclusively in human sewage (Lynch et al. 2002; Nebra et al. 2003; Bonjoch et al. 2004). *B. adolescentis* was found to be the most dominant *Bifidobacterium* species in the human intestinal flora (Matsuki et al. 2004), and thus a very promising candidate to identify human fecal sources. A nested PCR approach was designed to detect human fecal pollution, involving an initial amplification step with universal *Bifidobacterium* genus primers followed by a second amplification step with human strain-specific primers for *B. adolescentis* and *B. dentium* (Bonjoch et al. 2004). Both false-positive (one poultry and two cattle fecal samples were positive) and false-negative (one sewage sample was negative) results were observed. The nested PCR assay for *B. adolescentis* was later modified using a different set of genus primers, and the assay's sensitivity and stability were improved (King et al. 2007). However, *Bifidobacterium* cells apparently have low persistence in the environment; therefore, the sensitivity in terms of detection limits of these species-specific assays is not high when applied to moderately or lowly polluted waters. In a comparative test, the presence of bifidobacteria among 269 fecal DNA extracts from 32 different animals was studied using genus- and species-specific 16S rRNA gene PCR assays. *B. adolescentis*, *B. bifidum*, *B. dentium*, and *B. catenulatum* had the broadest host distribution (11.9–17.4%) using the species-specific assays, whereas *B. breve*, *B. infantis*, and *B. longum* were detected in fewer than 3% of all fecal samples. Genus-specific assays were positive in five of nineteen human fecal samples and in three of nine septic tank samples. The latter were also positive for about 1/5 of the chicken, dairy cattle, cat, goat, pig and sheep fecal samples, but negative for several other animals (armadillo, dove, fox, guinea pig, hedgehog, raccoon, squirrel and vulture) (Lamendella et al. 2008). Until now, no host-associated *Bifidobacterium* assays have been designed for animals.

Faecalibacterium in the phylum *Firmicutes* is a gram-negative, obligate anaerobe commonly found in human feces (Dick et al. 2005a, b). *Faecalibacterium*, followed by *Bacteroides–Prevotella* in the *Bacteroidetes*, was found to be most abundant in the 16S rRNA gene clone library obtained by suppression subtractive hybridization of human fecal DNA extract against dog fecal DNA extract. A qPCR assay designed to target *Faecalibacterium* was only positive for human fecal material, sewage, and human-impacted natural waters and not for other animal-derived materials. Further studies on environmental samples need to be performed to fully evaluate this assay (Zheng et al. 2009).

R. coprophilus is an actinomycete found in herbivore dung (about 10^5 CFU/g of feces), but it is absent in human feces. These bacteria naturally exist in grass and hay and can survive passage through herbivore's digestive systems (Al-Diwany and Cross 1978). PCR and qPCR assays were designed to target *R. coprophilus* that were positive for herbivores such as cow, sheep, horse, donkeys, goats, and deer feces, but negative for human, pig, possum, duck, and rabbit. The target sequence was also found in poultry reared in proximity to farm animals (Savill et al. 2001; Gilpin et al. 2002). This approach was proposed by the authors for the detection of herbivore waste, but it cannot distinguish between animal species in herbivores. To date, the assay has been used only in New Zealand (Savill et al. 2001).

Catellicoccus marimammalium is a gram-positive, catalase-negative bacterium that represented 26% of the sequences in a 16S rRNA gene clone library from gull feces. Its widespread presence in gulls and gull-impacted waters make it a potential indicator for gull fecal contamination. A *C. marimammalium* species-specific qPCR assay was designed to detect gull fecal pollution (Lu et al. 2008). Notably, this species, a low G+C gram-positive member of the Enterococcaceae family, was originally isolated from a porpoise and a gray seal (Lawson et al. 2006). Subsequently, this assay was found to be positive not only for sea gulls but also for other seabirds, with minor detection of marine mammal and some warm-blooded animals at levels four orders of magnitude lower than in Western seagulls. The assay can detect fecal contamination originating from the source group of seabirds along the California coast (Sirikanchana et al. submitted).

4.3.2.2 Bacteria–Host Interaction Genes

In animal guts, genes that are directly involved in bacterium–host interactions may display increased levels of host-associated genetic variation, making them good candidates for fecal source tracking. Most of the methods developed fall into the category of cultivation-dependent methods and hence are not included here (see Sect. 4.2); for example, the goose- and duck-specific DNA markers identified by Hamilton et al. (2006) were used in colony hybridization against *E. coli* isolates. The methods for *Bacteroidales* are described in Sect. 4.3.3.1.

4.3.2.3 Toxin Genes

Traditional fecal indicators such as *E. coli* and *Enterococcus* are considered indicators of general fecal pollution, and library-dependent methods are used for source tracking applications (see Chap. 3). Instead, some of the toxin genes of pathogenic fecal indicator strains were found to be host-associated. However, horizontal gene transfer events may shift the host specificity of these toxin genes (Ochman et al. 2000).

There are several host-specific PCR assays targeting the toxin genes of diarrheic *E. coli* to identify cattle and swine fecal pollution. A cattle-specific *E. coli* assay targeting a portion of the heat-labile toxin IIA (LTIIa) gene showed no cross-reactivity against human sewage and swine, sheep, goat, horse, cat, dog, bison, or bird fecal samples. The genetic marker was stable in natural waters for 2–4 weeks. Prevalence of the toxin gene in the *E. coli* population varied from farm to farm but was relatively low (estimated to be 1:1,000) resulting in a high detection limit (Khatib et al. 2002). This research group also developed a swine-specific *E. coli* assay targeting the heat stable toxin II (STII) gene. The latter assay showed no cross-reactivity against human sewage and cow/steer, sheep, goat, horse, cat, dog, bison, agouti, or bird fecal samples. The genetic markers remained stable for approximately 2 weeks when seeded into environmental waters. The assay was 100% positive when there were more than 35 *E. coli* cells in swine waste samples (Khatib et al. 2003). Two nested-qPCR assays have since been designed based on Khatib's assays. The sensitivity of the cattle- and swine-associated assays was improved 32 and 10 times, respectively, and the prevalence of the LTIIa gene and STII gene was proven to be much greater than originally reported (Chern et al. 2004).

The heat-stable enterotoxin gene, STIb, of pathogenic *E. coli* was found to be associated with human fecal pollution, and both PCR and nested-PCR assays have been developed. One assay could detect all human sewage samples but also had a high false-positive rate (Oshiro and Olson 1997; Field et al. 2003). More nested-PCR assays were designed to target rabbit- (*ralG* gene), bird- (*tsh* gene), and dog- (*papG* III gene) specific toxin genes of diarrheic *E. coli* (Jiang et al. 2007). However, the specificity and sensitivity of these assays need to be further evaluated. Another study reported that the *tsh* gene associated with birds was also found in 46% of dog fecal samples, leading the authors to propose that the hemolysin E gene (*hlyE*) of the enterohemorrhagic *E. coli* would be a better source tracking marker for bird feces (Chern and Olson 2004).

A PCR assay targeting a putative virulence factor, the enterococcal surface protein (*esp*) in *E. faecium*, was designed to identify human fecal pollution (see also Sect. 4.2.2 for cultivation-dependent studies). This assay was positive in 97% of sewage and septic samples, but negative for livestock and bird waste. Water samples needed to be concentrated by membrane filtration and were incubated for 48 h before PCR detection (Scott et al. 2005). When the analysis was carried out without the initial enrichment step, its sensitivity and percentage of correct classification dropped to 4.4 and 5.9%, respectively (Balleste et al. 2010). The reported detection limit for this assay was about 100 CFU of human-derived enterococci (Scott et al. 2005). This marker was found in some near-shore water samples from Hawaii (Knee et al. 2008), but not in sediments or recreational waters in a Danish Estuary that received occasional input of human waste from storm-water runoff (Roslev et al. 2008). Detection did not correlate significantly with the presence of Enterovirus, Adenovirus, or Rotavirus (Wong et al. 2009). Another study used the forward primer from Scott et al. (2009) and a conservative reverse primer from Hammerum and Jensen (2002). This modified assay showed a correlation of 0.371 with the human-specific *Bacteroides* PCR assay (Bernhard and Field 2000a, b) and a correlation of 0.558 with a human polyomavirus assay (McQuaig et al. 2006).

The modified *esp* PCR assay was positive for 90.5% of sewage and septic system samples and negative for fecal samples from cow, chicken, deer, dog, duck, goat, horse, kangaroo, pelican, pig, sheep, and wild birds in Australia. The *esp* marker was also detected in some water samples from nonsewered catchments (Ahmed et al. 2008a). Using a SYBR Green qPCR assay, researchers found about seven gene copies of the *esp* marker for every 100 CFU of enterococci in sewage in Australia (Ahmed et al. 2008b). Several recent studies have suggested that the *esp* gene may not be exclusive to human fecal pollution (Byappanahalli et al. 2008; Layton et al. 2009; Balleste et al. 2010).

4.3.2.4 Other Metabolic Genes

Many other metabolic genes may show some level of host specificity due to the difference in a host's digestive system, genotype and diet. Yet not much research has been done to specifically focus on these genes due to technical difficulties when sorting through such a large pool of potential candidates.

The β-glucuronidase of *E. coli* is frequently used in detecting *E. coli* and it forms one of the basic diagnostic biochemical features of the U.S. EPA approved Colilert (IDEXX Laboratories, Inc.) method of enumerating *E. coli*. Sequencing analysis of a fragment of the β-glucuronidase gene, *uidA*, exhibited some level of host specificity. Many alleles occurred in a subgroup of host species, while 3 and 2 out of 81 alleles were found to be exclusively human- and bird-specific, respectively (Ram et al. 2004). These host-specific alleles might be used to develop human- and bird-specific source tracking assays. Further studies on the prevalence and distribution of these alleles in animal feces and in environmental water samples are needed (see also Sect. 4.1.2).

Methanogens, a group of archaea rather than bacteria, represent good candidate indicators for human, ruminant, and swine fecal pollution. The genus *Methanobrevibacter* is in the order of Methanobacteriales and includes human-specific *Methanobrevibacter smithii* and ruminant-specific *Methanobrevibacter ruminantium*. Neither *M. smithii* nor *M. ruminantium* occurs frequently in the marine/fluvial environment or sediment; therefore, they indicate recent human or ruminant fecal contamination once found in these environments. A PCR assay was designed to identify human fecal pollution by targeting the *nifH* gene of *M. smithii*. This assay was only positive for human fecal material, sewage, and sewage-impacted waters, and negative for other animal fecal samples (cow, sheep, swine, horse, deer, goat, turkey, goose, chicken, and dog) and environmental waters impacted by bovine waste (Ufnar et al. 2006). *M. smithii* is abundant in the human gut, at a concentration of 10^7–10^{10} cells/g of dry feces (Lin and Miller 1998). The same group of researchers also designed a PCR assay to identify pollution from domesticated ruminants by targeting the *nifH* gene of *M. ruminantium*, which is known to inhabit only the rumen at a concentration of 10^6–10^8 per mL of rumen fluid. The assay was positive for domesticated bovine, sheep, and goat fecal materials and impacted waters, but negative for human and other animal fecal material (Ufnar et al. 2007b). A PCR assay targeting *mcrA* genes of an uncultured swine-specific methanogen was designed to identify swine waste. It was positive for swine waste, but negative for human and non-swine animals (cow, dog, sheep, goat, deer, rat, horse, and chicken) feces. This assay also did not amplify 138 nonpolluted environmental samples; therefore, it was suggested as a good indicator of fresh swine waste pollution once found in these environments (Ufnar et al. 2007a).

4.3.2.5 Extragenic Regions, Multiple Genes, and Metagenomic Approaches

When a single gene of a single bacterium target cannot identify the fecal source unambiguously, extragenic sequences or multiple genes of a bacterial species may be used. For example, *E. coli* does not reveal much host specificity based on its 16S rRNA gene sequence; however, its DGGE fingerprinting pattern of the *rrnB* ribosomal operon in the 16S–23S intergenic spacer region was used to differentiate human and Canadian goose fecal contamination from other animal sources (D'Elia et al. 2007).

Some assays that target multiple genes of a bacteria species were originally designed to be cultivation-dependent but can be adapted as cultivation-independent method. Fifteen host-associated genetic markers of *Enterococcus* genomic DNA were identified by microarray hybridization; two were exclusively specific for cattle, two for elk/deer, and four for human feces. The markers were used in colony hybridization against *Enterococcus* isolates (Soule et al. 2006). Others identified seven goose- and duck-specific markers from *E. coli* by suppressive subtractive hybridization and used them together in colony hybridizations to identify the sources of *E. coli* isolates (Hamilton et al. 2006). Further studies are needed on how to incorporate these markers into microarray analysis against genomic DNA extracted from environment water sample or use them to build host-specific PCR assays.

The metagenomic approach has been the most recent methodological advance in MST (Sadowsky et al. 2007). Researchers can use it to analyze genomic information extracted directly from environmental communities without prior cultivation. For example, Lu et al. (2007) applied a genome fragment enrichment method to fecal metagenomes and screened for chicken-specific DNA fragments. These identified fragments encode genes functioning for information storage and processing, cellular processes and signaling, as well as metabolism. Twenty-one chicken-specific fecal PCR assays were designed with various levels of specificity and sensitivity. The authors suggested multiple assays should be used simultaneously to detect poultry fecal sources of pollution (Lu et al. 2007).

4.3.3 *Bacteroidales*

Of the bacteria that demonstrate some kind of host specificity, strictly anaerobic members of the order *Bacteroidales* are currently the most widely used fecal source identifiers in water. *Bacteroides* are believed to be found exclusively in feces, animal rumen, and other cavities of humans and animals (Paster et al. 1994). They constitute about one third of the human fecal bacterial population (Holdeman et al. 1976) and outnumber coliforms by two to three orders of magnitude in the human and animal intestine. *Bacteroidales* are also numerically abundant in the digestive tract of other warm-blooded animals (Meays et al. 2004), but are not as dominant as in humans (Kreader 1995). For example, their abundance in avian (chicken, gull, turkey, and goose) species has been consistently reported as low (Lu et al. 2007, 2008, 2009; Lu and Santo Domingo 2008; Jeter et al. 2009; Sirikanchana et al. submitted). *Bacteroides* can be indicative of pollution when found in high concentrations in natural waters.

Almost all assays in Table 4.2 are based on the 16S rRNA gene with the exception of some human- and bovine-specific assays from Shanks' group (Shanks et al. 2008, 2009). The prevalence of *Bacteroidales* in the human gut makes human-associated *Bacteroidales* assays highly sensitive. The marker BacHum (Kildare et al. 2007) occupies on average 82% of total *Bacteroidales* in human guts detected by the general BacUni assay (Silkie and Nelson 2009). Thus, it is easily detectable if there is human fecal contamination present in water samples. However, a set of cow-, dog-, and horse-associated *Bacteroidales* markers was much less sensitive, occupying on average 4, 6, and 2% of the total *Bacteroidales* populations in their corresponding hosts (Silkie and Nelson 2009). There is some cross-reactivity of host-associated assays, namely, these assays are not 100% specific for their target hosts (Silkie and Nelson 2009; McLain et al. 2009). A statistical model has been developed to account for any false-positive and false-negative information arising from the less-than-perfect sensitivity and specificity of genetic assays (Wang et al. 2010, see Sects. 4.3.3.3 and 4.3.3.4).

4.3.3.1 Method Development for *Bacteroidales*

16S rRNA Gene

Kreader first developed PCR-based assays to amplify genes from three cultivated strains of *Bacteroides* to monitor human fecal pollution in water (Kreader 1995) and evaluated the persistence of *Bacteroides distasonis* in the environment (Kreader 1998). Bernhard and Field (2000a) identified host-associated *Bacteroides–Prevotella* genetic markers for humans and cows based on the 16S rRNA gene using length heterogeneity-PCR (LH-PCR) and terminal restriction length polymorphism (T-RFLP) analysis to screen fecal bacterial DNA extracts. With information obtained from this study, they identified cluster-specific primer sets that are useful in discriminating human and ruminant feces (Bernhard and Field 2000b). The system of primers was further tested and shown to reliably and specifically detect host-associated markers from feces and polluted water samples (Field et al. 2003).

To date, four general (Dick et al. 2005b; Layton et al. 2006; Kildare et al. 2007; Okabe et al. 2007) and a panel of host-associated assays have been designed targeting *Bacteroidales*, including human, dog, ruminant (sometimes referred to as bovine or cow but usually not capable of distinguishing between different ruminant animal species), swine (pig), equine (horse), elk, and gull (Table 4.2) (Bernhard and Field 2000a, b; Dick et al. 2005a, b; Layton et al. 2006; Reischer et al. 2006, 2007; Kildare et al. 2007; Okabe et al. 2007; Shanks et al. 2008, 2009, 2010; Jeter et al. 2009; Dorai-Raj et al. 2009; Lamendella et al. 2009; Silkie and Nelson 2009; Converse et al. 2009; Mieszkin et al. 2010; Jeong et al. 2010). The existence of a wide spectrum of host-associated assays is advantageous when there is more than one dominant fecal source in environmental samples because it allows the researcher to detect multiple sources and compare them.

Nearly all of the early species-associated *Bacteroidales* assays were intended for end-point PCR applications, which provide only information about presence or absence of a target sequence (Bernhard and Field 2000a, b; Dick et al. 2005a, b). This information is still helpful to watershed managers and regulators provided the probability that bacteria from nontarget hosts have been detected is known and reasonably low. For any statistical model, it is true that the higher the diagnostic sensitivity and specificity of the assay, the lower is the likelihood of false-positive detection in a specific watershed (see Sect. 4.3.3.3).

This advance was followed by real-time quantitative PCR (qPCR) assays for the detection of general and host-associated fecal pollution using 16S rRNA markers for *Bacteroidales*. Three of the assays designed to detect total fecal pollution – the total (Dick et al. 2004), Allbac (Layton et al. 2006) and BacUni (Kildare et al. 2007) – are based on TaqMan chemistry. This type of assay requires the use of a TaqMan probe that is labeled with a fluorescent reporter dye on its 5′ end and a quencher dye on its 3′ end. The probe attaches only to the desired target sequence during PCR amplification of the target sequence. As the PCR product is synthesized, the probe is cleaved, releasing the reporter dye from its quencher, and thus emitting fluorescence. The fluorescence signal increases in intensity as a function of the amount of PCR product (double-stranded DNA) in a reaction.

Another type of quantitative *Bacteroidales* assay is a real-time PCR assay that utilizes SYBR Green 1 Dye detection. These assays target, for example, a total *Bacteroidales* (Okabe et al. 2007a) and targets a human-specific 16S rRNA genetic marker (Seurinck et al. 2005). The SYBR Green dye is a DNA-intercalating dye, and is used in this assay as the reporter fluorophore. SYBR Green dye fluoresces when it binds to double-stranded DNA. There has been some discussion concerning the benefits in terms of sensitivity and specificity of TaqMan over SYBR Green assays, and the relative sensitivities of the two technologies have been debated (Wittwer et al. 1997; Morrison et al. 1998). With regard to specificity, TaqMan assays are undoubtedly the better choice for differentiation of host-associated DNA sequences when environmental samples are considered. This is because nonspecific products, which are especially formed during the later cycles, may be detected with the SYBR Green technology. Melting curve analysis can assist in proper identification of these false-positive sequences, but absolute quantification may be compromised. Comparisons of numerical data generated by SYBR Green technology with quantitative results obtained by TaqMan analysis has not been performed for *Bacteroidales*. By contrast, probe-based instrument and reagent systems from Applied Biosystems, Inc. and Cepheid have been compared for *Bacteroidales* and *Enterococcus* using previously reported (Dick and Field 2004; Ludwig and Schleifer 2000) and redesigned primer and probe assays (Siefring et al. 2008). The consistency of results varied between platforms and reproducibility improved when the original assays were modified.

The number of host-associated assays has been steadily increasing. Genetic markers were broadly distributed among individual bovine samples in a performance evaluation of seven end-point PCR and real-time quantitative PCR (qPCR) assays reported to be associated with either ruminant or bovine feces on 247 individual bovine fecal samples representing 11 different populations and 175 fecal DNA extracts from 24 different animal species (Shanks et al. 2010). Specificity levels of the assays ranged from 47.4 to 100%. Overall, there were large discrepancies in the performance of bovine-associated assays across different bovine populations, and a major recommendation of that study was to perform a priori characterization at each watershed of interest before applying any bovine genetic assay targeting *Bacteroidales* in MST monitoring studies. This approach was followed in the Njoro watershed in Kenya, where the BacCow assay showed 100% specificity and 94% sensitivity for bovine feces (Jenkins et al. 2009). Other assays such as the BacHum marker targeting human-associated fecal pollution did not perform nearly as well in Kenya at 100% specificity and 25% sensitivity. It follows that method validation must continue for library-independent methods as it does for library-dependent methods when different geographic regions are to be monitored (see Sect. 4.3.3.2).

Bacteroidales–Host Interaction Genes

qPCR assays amplifying DNA sequences of *Bacteroidales* predicted to encode membrane-associated and secreted proteins (Shanks et al. 2006a, b) can differentiate bovine from other ruminant species, which to date has not been achieved by targeting the 16S rRNA gene (Shanks et al. 2008). Other qPCR assays amplify two human-specific genetic markers for *Bacteroidales*-like cell surface-associated genes (Shanks et al. 2007). These genetic markers are highly specific for human fecal extract and prevalent in ambient environmental waters, making them good candidates for environmental MST (Shanks et al. 2009).

A qPCR assay targeting the $\alpha-1-6$ mannanase gene of *Bacteroides thetaiotaomicron* was designed to identify human fecal pollution. *B. thetaiotaomicron* is considered to be a symbiont in the human gut, i.e., the target gene is associated with bacteria–host interaction (Xu et al. 2003). *B. thetaiotaomicron* is abundant in the human gut, with an average concentration of 10^8 cells per gram of feces. This assay was reported to be positive for human fecal and sewage samples, but negative for nonhuman samples (Yampara-Iquise et al. 2008).

4.3.3.2 Geographic Range

Continental *Bacteroidales* sequences from individual mammalian species have been shown to be geographically similar in USA, thus providing a universal standard for their use (Field et al. 2003). Assays developed in the US states of Oregon (Bernhard and Field 2000a, b; Dick et al. 2004) and California (Kildare et al. 2007) have been shown to work well in several European countries (Gawler et al. 2007) but had relatively low sensitivity in Spain (Balleste et al. 2010) in Australia (Ahmed et al. 2009a, b), and in Kenya (Jenkins et al. 2009). A set of host-associated markers developed in Japan (Okabe et al. 2007) did not show high specificity in Korea (Jeong et al. 2010) and other world regions such as Australia (Ahmed et al. 2009a, b).

The human-associated assay BacH (Reischer et al. 2007) and bovine-associated assay BacR (Reischer et al. 2006) were developed for Austrian alpine karst spring catchments and subsequently tested on a global scale by the authors (Reischer et al. 2009) together with four other markers using probe-based 5′-nuclease assays to ensure high marker sensitivity and specificity: BacHum (Kildare et al. 2007) and HuBac (Layton et al. 2006) targeting human feces, and BacCow (Kildare et al. 2007) and BoBac (Layton et al. 2006) targeting bovine feces. In addition, one SYBR Green qPCR assay based on the human marker HF183 (Seurinck et al. 2005) was included. In total, 320 individual fecal samples of known origin were collected from 15 countries (Argentina, Australia, Austria, Germany, Hungary, Korea, Nepal, Netherlands, Romania, Spain, Sweden, Tanzania, Uganda, United Kingdom, and USA). Of these samples, 25% were from humans, 25% from ruminants, and 50% from "other" animal sources (e.g., wild birds, pigs, horses, companion animals).

The number of false positives with the human-specific SYBR Green assay was extremely high, mostly because qPCR measurements were confounded by formation of primer dimers and other unspecific PCR products. Among the other probe-based 5′-nuclease qPCR assays, the diagnostic sensitivity of the tested probe-based assays proved to be fairly high (>80%). On the contrary, the source specificity was often below 80% with strong differences in results between the various countries. However, the average marker abundance in false positives was usually two to three orders of magnitude lower than in true positive samples. On a global scale, the BacHum assay performed best among the human-specific assays (63% specificity), and the BacR assay had the highest specificity (83%) among the ruminant-specific assays (Reischer et al. 2009).

4.3.3.3 Interpreting Environmental Monitoring Data: Qualitative Analysis of Conditional Probabilities

The ultimate goal of MST is to determine the relative amounts of host-associated fecal pollution in a water sample. As a first step toward this goal, there is a need for *Bacteroidales* monitoring data to be analyzed statistically to calculate conditional probabilities of correctly identifying sources of fecal pollution in a watershed, given that a particular host-associated assay tests positive in environmental samples. Such a qualitative approach based on Bayes' theorem was developed for *Bacteroidales* (Kildare et al. 2007) and then applied in a year-long MST study in Calleguas Creek watershed in southern California (http://www.calleguas.com/ccwmp/DRAFT_CCW_MST_061406.pdf) and in the River Njoro watershed in Kenya (Jenkins et al. 2009).

Briefly, assays are assumed to be independent discrete random variables. The method uses presence/absence data to calculate conditional probabilities based on information gained from source-specificity tests and the frequency of occurrence of a genetic marker in the monitored watershed. For example, when using the BacHum assay to detect human-associated fecal pollution in a water sample from Calleguas Creek watershed, one can calculate the conditional probability of the BacHum assay to detect fecal pollution originating from humans in a water sample (true positive), and not fecal *Bacteroidales* sequences originating from dog (false positive). Potential errors due to other false positives from animal hosts for which the assay has not been validated are assumed to be negligible. Equation (4.1) estimates $P(H/T)$, the probability of a human source of contamination (H) in an analyzed water sample given a positive test result (T) with BacHum (Kildare et al. 2007):

$$P(H/T) = \frac{P(T/H)\cdot P(H)}{P(T/H)\cdot P(H) + P(T/H')\cdot P(H')}, \tag{4.1}$$

where

$P(T/H)$ is the probability of a positive signal with the BacHum assay in a fecal sample that is human-derived. This value was obtained from a laboratory validation study (Kildare et al. 2007) as 1.00 due to the 100% detection of mixed human samples screened with this assay.

$P(T/H')$ is the probability of positive signal with the BacHum assay in a fecal sample that is not human-derived. This value was obtained from the laboratory validation study as 0.13 due to the 13% detection of dog-derived fecal sources by this assay.

$P(H)$ is the background probability of detecting the BacHum marker in the Calleguas Creek watershed. This value is 0.89, since the marker was detected in 65 of 73 samples.

$P(H')$ is the background probability that the H160F marker is absent in CCW. This value is $1-P(H)$, or 0.11.

Substituting these values into (4.1) gives

$$P(H/T)=\frac{(1.00)\cdot(0.89)}{(1.00)\cdot(0.89)+(0.13)\cdot(0.11)}=0.98.$$

In other words, based on water and fecal samples analyzed with the methods employed during this study, there is a 98% probability that a detection of the BacHum marker in a water sample from Calleguas Creek watershed is due to mixed human contamination (not fecal *Bacteroidales* sequences originating from dogs).

In table form, the following probabilities are determined:

Test	Tested positive	Tested negative	Total probability
Samples containing mixed-human feces	0.89 * 1 = 0.89	0.89 * 0 = 0	0.89
Samples without mixed-human feces	0.11 * 0.13 = 0.0143	0.11 * 0.87 = 0.0957	0.11
Total probability	0.90	0.10	1.00

Then one can let:

TPC = tested positive correctly
TPI = tested positive incorrectly
TNC = tested negative correctly
TNI = tested negative incorrectly.

And the above table can be categorized as follows:

Test	Tested positive	Tested negative	Total probability
Samples with mixed-human feces	TPC	TNI	TPC + TNI
Samples without mixed-human feces	TPI	TNC	TPI + TNC
Total probability	TPC + TPI	TNI + TNC	TNI + TNC + TPC + TPI

Based on these categories, one can define the test sensitivity, test specificity, and specify predictive values and the prevailing rates (Deep 2006).

Sensitivity is the ratio of those samples that correctly tested positive to all those samples that actually experienced fecal contamination of mixed human origin.

$$\text{Sensitivity}=\frac{\text{TPC}}{(\text{TPC}+\text{TNI})}=\frac{0.89}{(0.89+0)}=1.00.$$

Specificity is the ratio of those samples that correctly tested negative to all those samples that actually did not experience fecal contamination of mixed-human origin.

$$\text{Specificity} = \frac{\text{TNC}}{(\text{TNC}+\text{TPI})} = \frac{0.0957}{(0.0143+0.0957)} = 0.87.$$

The positive predictive value of the test is the ratio of the number of samples that correctly tested positive to the total number of samples that tested positive.

$$\text{Positive predictive value} = \frac{\text{TPC}}{(\text{TPC}+\text{TPI})} = \frac{0.89}{(0.89+0.0143)} = 0.98.$$

The negative predictive value of the test is the ratio of the number of samples that correctly tested negative to the total number of samples that tested negative.

$$\text{Negative predictive value} = \frac{\text{TNC}}{(\text{TNC}+\text{TNI})} = \frac{0.0957}{(0.0957+0)} = 1.00.$$

The prevailing rate is the proportion of the total number of samples that actually experienced mixed-human fecal contamination.

$$\text{Prevailing rate} = \frac{(\text{TPC}+\text{TNI})}{(\text{TPC}+\text{TNI}+\text{TPI}+\text{TNC})} = \frac{(0.89+0)}{(0.89+0+0.0143+0.0957)} = 0.89.$$

This approach simplifies the interpretation of MST monitoring data and takes into account the applicability of a particular genetic assay in a given watershed.

4.3.3.4 Interpreting qPCR Monitoring Data: Probabilistic Model for Quantitative MST

The development of molecular target detection assays for *Bacteroidales* has provided a fast, reliable, and relatively inexpensive means by which to diagnose the source of fecal contamination to natural waters. While the conditional probability of correctly identifying feces from a specific host can be calculated for any water sample (see Sect. 4.3.3.3), this approach does not harness the potential of quantitative measurements. Quantitative PCR provides numerical data in terms of gene copies of a particular host-associated genetic marker, yet it would be inappropriate to conclude based on these numbers what the relative fecal contributions of the measured host species are in a water sample. In other words, qPCR does not by necessity imply quantitative microbial source tracking (QMST).

Recently, a probabilistic model has been developed that aims to account for uncertainties in qPCR measurements in addition to the less than perfect diagnostic sensitivity and specificity of *Bacteroidales* assays (Wang et al. 2010). The main features of the model are listed here together with an example for its application.

The factors that can lead to false-positive and false-negative information in qPCR results are known and well defined. For *Bacteroidales* assays, false or variable information arises from the following situations:

- A primer set for a specific host might amplify *Bacteroidales* DNA from other hosts (false-positive information for that specific host).
- *Bacteroidales* DNA from a specific host might not be amplified by a primer set designed for that host (false-negative information for that specific host).
- The distribution (number of gene copies) of target genetic markers varies among individuals leading to nonconstant ratios of specific marker to universal marker (variable ratios affect any prediction of relative amounts of fecal host pollution as needed for quantitative MST).
- There can be measurement errors associated with the instrument used.

It is highly desirable to have a way of removing such false information, and to estimate the true concentration of host-associated genetic markers and help guide the interpretation of environmental monitoring studies. The statistical model is based on the Law of Total Probability to predict the true concentration of these markers. For example, the DNA sequences in a water sample that can be amplified by the human specific primer set may be derived from human, cow, dog, or other minor sources of fecal contamination:

Measured *Bacteroidales* DNA concentration by human - specific assay:
= DNA targets originating from human source and amplified
by human - specific assay
+ DNA targets originating from cow source and amplified
by human - specific assay
+ DNA targets originating from dog source and amplified
by human - specific assay
+ DNA targets originating from other sources and amplified
by human-specific assay
+ measurement error

= (DNA targets from human source) × probability that amplified
by human -specific assay
+ (DNA targets from cow source) × probability that amplified
by human - specific assay
+ (DNA targets from dog source) × probability that amplified
by human - specific assay
+ (DNA targets from other sources) × probability that amplified
by human - specific assay
+ measurement error.

Mathematically, this leads to one equation:

$$C(h) = C(H) \times p(h / H) + C(D) \times p(h / D) + C(C) \times p(h / C) + C(O) \times p(h / O) + e,$$

where

$C(h)$ is the measured concentration using the human-specific assay
$C(H)$ is the true concentration of DNA originating from human sources

p(*h*/*H*), *p*(*h*/*D*), *p*(*h*/*C*) and *p*(*h*/*O*) are the conditional probabilities that DNA originating from human, dog, cow, and other minor sources is amplified by the human-specific assay
e is the measurement error.

Analogous to the human assay described above, three more equations can be formulated for the dog-specific assay, the cow-specific assay, and an imaginary assay that targets other sources.

$$C(d) = C(H) \times p(d / H) + C(D) \times p(d / D) + C(C) \times p(d / C) + C(O) \times p(d / O) + e,$$

$$C(c) = C(H) \times p(c / H) + C(D) \times p(c / D) + C(C) \times p(c / C) + C(O) \times p(c / O) + e,$$

$$C(o) = C(H) \times p(o / H) + C(D) \times p(o / D) + C(C) \times p(o / C) + C(O) \times p(o / O) + e.$$

$C(h)$, $C(d)$, and $C(c)$ represent measured concentrations in terms of gene copies per unit volume when applying human-, dog- and cow-specific assays, respectively, and $C(u)$ represents the measured concentration obtained after running the universal assay. $C(o)$ cannot be measured but is estimated by the equation $C(o) = C(u) - C(h) - C(d) - C(c)$.

Once the conditional probabilities like *p*(*h*/*H*) and the measurement error are estimated, the true concentrations *C*(*H*), *C*(*D*), *C*(*C*), and *C*(*O*) can be calculated using the above four equations. The distributions of these conditional probabilities were estimated using representative fecal samples of known origins provided in Silkie and Nelson (2009). For example, 12 pooled human fecal samples were tested with the universal, human-, cow- and dog-specific *Bacteroidales* assays (Kildare et al. 2007). Then, the set of probabilities for human fecal material, *p*(*/*H*), was calculated for each pooled sample in the following way:

$$p(h / H) = C(h) / C(u),$$

$$p(c / H) = C(c) / C(u),$$

$$p(d / H) = C(d) / C(u),$$

$$p(o / H) = C(o) / C(u) = (C(u) - C(h) - C(d) - C(c)) - C(u).$$

The model was validated using DNA from the 12 pooled human fecal samples, resulting in 12 sets of values for *p*(*/*H*) (Fig. 4.3). Random weights were assigned to the four probabilities, *p*(*/*H*), associated with each of the 12 pooled samples; their weighted sum was assumed to be the set of probabilities for a environmental sample. The distributions of these probabilities were obtained by repeating this weighted-sum process 10,000 times. Percentage measurement error, i.e., the precision error associated with qPCR, was assumed to be normally distributed $(0,\sigma^2)$ since many factors contribute to it. σ^2 was estimated from the sample standard errors of replicated qPCR reactions on some samples.

Then, the Monte-Carlo method was used to sample from these distributions of probabilities and measurement errors. The set of equations given by the Law of Total Probability allows for the calculation of the distribution of true concentrations,

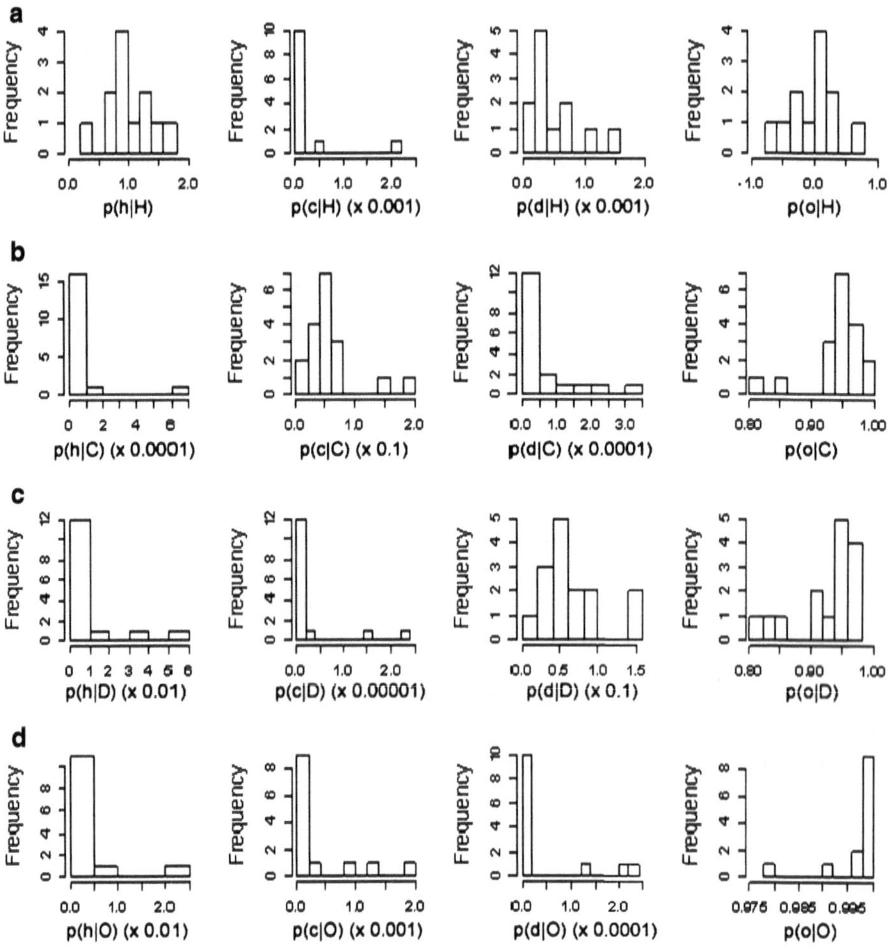

Fig. 4.3 Histograms of the conditional probabilities of representative fecal extracts. The four subpanels in (**a**) are the histograms for $p(h|H)$, $p(c|H)$, $p(d|H)$, and $p(o|H)$ of 12 human-derived fecal extracts; the four subpanels in (**b**) are the histograms for $p(h|C)$, $p(c|C)$, $p(d|C)$, and $p(o|C)$ of 18 cow related fecal extracts; the four subpanels in (**c**) are the histograms for $p(h|D)$, $p(c|D)$, $p(d|D)$, and $p(o|D)$ of 15 dog related fecal extracts; and the four subpanels in (**d**) are the histograms for $p(h|O)$, $p(c|O)$, $p(d|O)$, and $p(o|O)$ of 13 other animal fecal extracts (reproduced by permission from Wang et al. 2010)

from which their expected value, confidence interval, and other statistical characteristics can be easily evaluated.

The model was tested using the qPCR *Bacteroidales* assays BacUni, BacHum, BacCow, and BacCan (Kildare et al. 2007); validation was done via statistical simulations and experimental analysis of real samples. The model performed well when human, dog, and cow feces were the dominant sources of fecal pollution. For example, we tested a known mixed fecal sample with equal amounts of feces, in terms of total *Bacteroidales* gene copies measured by the BacUni assay (Kildare et al. 2007), from human, dog, cow, and other sources and obtained the vector of measured concentrations ($C(u)$, $C(h)$, $C(c)$, $C(d)$) in gene copies/mL as (3.61×10^7, 8.83×10^6, 4.41×10^5, 6.23×10^5). From the measured qPCR results alone, it seemed that the relative contribution to the pollution from human, dog, and cow would be ($8.83\times10^6/3.61\times10^7$, $4.41\times10^5/3.61\times10^7$, $6.23\times10^5/3.61\times10^7$) = (0.2445, 0.0122, 0.0172), which clearly underestimated the contribution of feces from cow and dog. The probabilistic model allows for correction of the measured qPCR results to estimate true concentrations of individual fecal host contributions. Assuming a 5% precision error for qPCR measurements, the expected values for *Bacteroidales* concentrations from human, cow, and dog hosts are (9.57×10^6, 8.34×10^6, 9.42×10^6) with corresponding relative contributions (0.2650, 0.2311, 0.2611), which is a good estimate of their actual contributions (0.25, 0.25, 0.25) (Fig. 4.4).

Real-world samples from Calleguas Creek Watershed (see also Sect. 4.3.3.3) were measured by qPCR for the presence of *Bacteroidales* genetic markers and then reanalyzed with the model. Estimated "true" concentrations of host-associated markers and relative contributions from different fecal sources were either close to those measured by qPCR (Fig. 4.5a) or underestimated the contribution of host-associated markers (Fig. 4.5b).

When feces from other animal hosts constituted the primary source, the model tended to underestimate the contribution of these other sources. This result is to be expected if one does not fully characterize the contribution from other sources when estimating the distribution of conditional probabilities. Once other relevant host-associated *Bacteroidales* qPCR assays are included in the model, for example, a horse-specific assay, its predictive power will improve. Even if a host-associated assay is not available for a dominant fecal source, the model will still point out that the sources accounted for (here, feces from human, cow, or dog) were not dominant

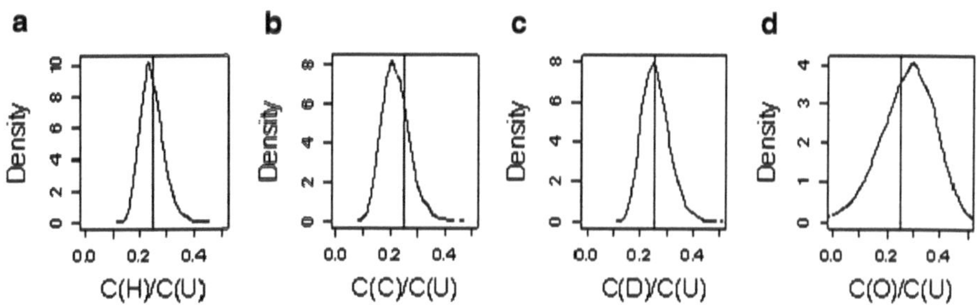

Fig. 4.4 Model output distributions of fecal source composition for a mixed fecal sample containing equal amounts of total *Bacteroidales* from human, dog, cow, and other sources based on the BacUni assay. (**a**) Represents human fecal source contribution to the total *Bacteroidales* signal, $C(H)/C(U)$; (**b**) cow fecal source contribution to the total *Bacteroidales* signal, $C(C)/C(U)$; (**c**) dog fecal source contribution to the total *Bacteroidales* signal, $C(D)/C(U)$; and (**d**) other fecal sources contribution to the total *Bacteroidales* signal, $C(O)/C(U)$. Curves represent the predicted distribution of source compositions (the density is fitted by kernel density estimates in the software R) and vertical lines the true compositions. The true composition is located within the 95% confidence interval of its predicted distribution for all four panels (reproduced by permission from Wang et al. 2010)

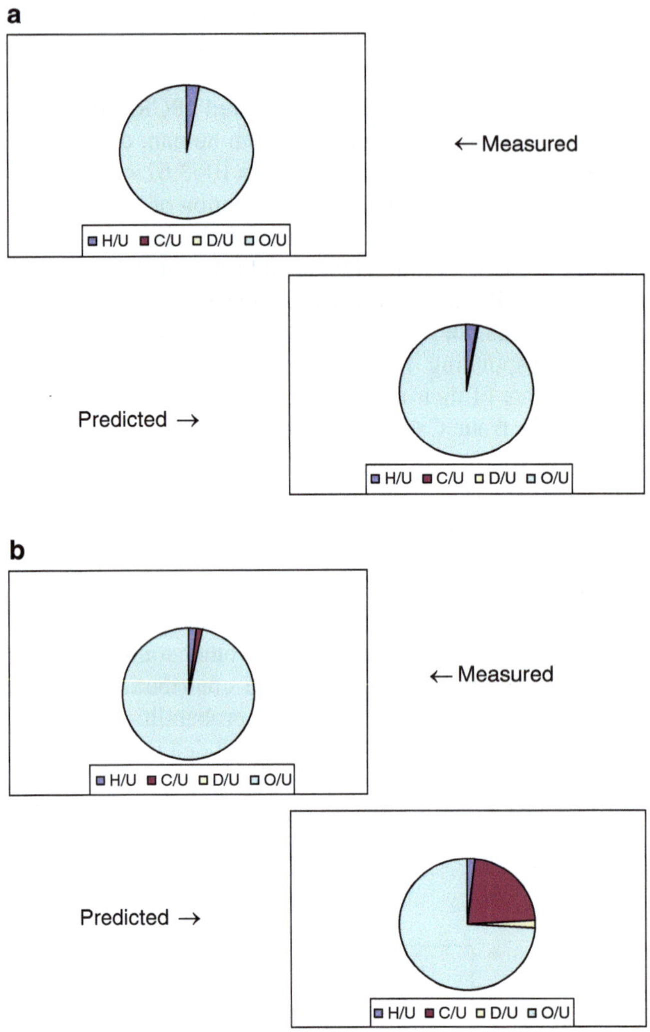

Fig. 4.5 Model output for two unknown water samples. Measured and predicted *Bacteroidales* concentrations in Calleguas Creek Watershed in samples taken in (**a**) June and (**b**) October. $H/U = C(H)/C(U)$, $C/U = C(C)/C(U)$, $D/U = C(D)/C(U)$, $O/U = C(O)/C(U)$. Measured concentrations in the October sample underestimate the contributions of cow and dog feces

and hence lead the researcher to look for other sources. Simulation results also indicate that the percentage measurement error contributes greatly to the stability of the model. A large measurement error could alter the output results dramatically. It is recommended that researchers continue to work toward minimization of the measurement error associated with qPCR.

This methodology is not limited to qPCR measurements of *Bacteroidales* and is readily transferable to other indicators where a universal assay for that indicator exists. The output distributions of predicted true concentrations can then be used as input to total maximum daily load (TMDL) studies, for quantitative microbial risk assessment, and for environmental models.

4.3.3.5 Detecting Genetic Markers in Intact Cells by PCR

PCR or qPCR based on genomic DNA detects both viable cells and dead cells; even particle-attached DNA might be targeted (see introduction – ability to detect intestinal population in the aquatic environment). For those cases where PCR/qPCR results have to be compared with results from cultivation-based approaches, they tend to overestimate the number of viable bacterial cells. Propidium monoazide (PMA) and ethidium monoazide (EMA) have thus been used to distinguish "live" cells from "dead" cells, or more precisely, from DNA in both intact cells and membrane-impaired cells. They are both intercalating DNA-binding chemicals that prevent PCR amplification once they bind to DNA. At the proper concentration, which must be optimized empirically for each target organism and matrix studied, they can selectively penetrate only membrane-impaired cells, thus preventing the amplification of DNA from these cells (Nogva et al. 2003; Rudi et al. 2005; Nocker and Camper 2006; Nocker et al. 2007a; Pan and Breidt 2007; Wagner et al. 2008; Bae and Wuertz 2009a). Consequently, only physically intact cells are subject to detection by qPCR under these conditions. Both agents have been used to detect viable cells of bacteria, spores, and fungi (Rudi et al. 2005; Soejima et al. 2007; Cawthorn and Witthuhn 2008; Vesper et al. 2008; Bae and Wuertz 2009b; Rawsthorne et al. 2009). Several studies reported that EMA can penetrate some intact cells, resulting in a significant loss of measurable genomic DNA in viable cells as well (Nocker and Camper 2006; Flekna et al. 2007; Cawthorn and Witthuhn 2008). The extent of EMA uptake by intact cells depends on the bacterial species (Nocker et al. 2006).

By comparison, PMA more specifically penetrates dead (membrane-impaired) cells (Nocker and Camper 2006; Nocker et al. 2006; Flekna et al. 2007). For example, EMA was not nearly as effective in differentiating *Bacteroidales* markers found in viable and dead cells in feces as PMA (Bae and Wuertz 2009a). PMA seems to be effective in Gram-negative bacteria such as *Bacteroidales* and *Salmonella* and much less so in Gram-positive bacteria such as enterococci. Neither PMA nor EMA method allows for the monitoring of inactivation mechanisms that do not directly target cell membranes (Kim et al. 2008) such as UV treatment (Nocker et al. 2007b) and chloramine treatment (Stewart and Olson 1996; Gedalanga and Olson 2009).

To conclude, using PMA/EMA to detect DNA in intact cells can provide some insights into how recent the fecal contamination is. However, one needs to be extremely cautious when interpreting the results for three reasons: (1) membrane-impaired cells are not necessarily dead cells; (2) the effectiveness of the method may vary from one microorganism to the next and from one sample to another; and (3) DNA from inactivated cells with intact membranes will still be amplified. The use of activity-labile nucleic acid-modifying compounds (ALCs) has been suggested in place of PMA (Nocker and Camper 2009) to detect nucleic acids in cells with an active metabolism. There are very few practical applications to date, and proteins rather than membranes or nucleic acids could be early targets of inactivation caused by sunlight on bacteria (Bosshard et al. 2010).

4.3.3.6 Fate and Transport of *Bacteroidales* in Surface Waters

In most MST studies, samples are taken from large water bodies in specific (few) locations and at certain (discontinuous) times. However, only investigation procedures taking the spatial and temporal variability characteristics of the respective watershed into account support reliable source tracking and identification (e.g., Reischer et al. 2010; see also Chap. 18). Furthermore, well-formulated and developed mathematical and numerical transport models can predict continuous concentrations of pathogens under diverse scenarios of interest, and can significantly facilitate source identification (e.g., Wuertz et al. 2009, see also Chaps. 9, 15, and 19).

Several blocks of information on the genetic markers are required to build a fate and transport model: death/growth rate in the water body and in sediments, sorption/desorption to particles and flocculation, resuspension and deposition of sediments. The latter two only involve physical processes that have been extensively modeled in other fields. The former one involves both biological and physical processes and needs to be experimentally studied. To date, the decay rate of these genetic markers have been evaluated in the lab (Seurinck et al. 2005; Okabe and Shimazu 2007; Bell et al. 2009) and in microcosms for fresh and saline water (Anderson et al. 2005; Bae and Wuertz 2009b; Walters et al. 2009; Dick et al. 2010). Information on the decay/growth rate in the sediment as well as the impact of predation is still missing. Further research to build a more comprehensive and efficient model incorporating the sediment/water interface and predation is also needed.

4.4 Conclusions and Perspectives

There are many library-independent MST methods that have been proposed for the identification of sources of fecal pollution. With so many available assays, it is difficult to decide which ones to use. Currently, there is no "universal and easy to use system" available supporting fecal source tracking in water resources without sound definition of background conditions, study design, method selection, and performance evaluation. Most methods developed are based on their regional boundary conditions and have not yet been evaluated at broader geographic scales. Although several studies have compared the performance of some assays (Griffith et al. 2009; Harwood et al. 2009; Jenkins et al. 2009; Lamendella et al. 2009; Reischer et al. 2009;

Shanks et al. 2010), more comprehensive comparative studies are needed. However, numerous successful MST applications have proven the practicability and potential of library-independent bacterial MST methods for the characterization of fecal pollution and their identification of respective sources (see case studies described in this book and references). Nonetheless, MST is a very young discipline, and many methods are still in the development phase. Proper selection of currently available tools and their verification on the watershed level is an absolute key requirement defining respective capacities and limitations (Chap. 2).

Depending on the methods applied (e.g., cultivation and enumeration of diagnostic populations, cultivation-dependent PCR, cultivation-independent PCR or qPCR) specific microbial populations or gene targets methods can be determined in a qualitative, semiquantitative, or quantitative way. It is important to note that even when quantitative analytical results are determined, concentrations or ratios between MST targets do not necessarily reflect actual source contributions. Besides basic analytical performance characteristics (e.g., sensitivity, specificity, detection limit; Chap. 2), QMST has to consider target distribution of different host groups with respect to abundance in excretion sources and their persistence and mobility in the investigated watershed. Furthermore, temporal and spatial variability of the water body and the pollution sources have to be described (e.g., Reischer et al. 2008 and Chap. 18).

Many methodical issues need further consideration or improvement regarding the analytical performance characteristics of library-independent bacterial MST. For example, it is well known that no single genetic marker can be 100% specific and sensitive at the same time and each assay has its own bias. A statistical model based on the law of total probability has, thus, been recently suggested to address this problem (Wang et al. 2010). Another approach is to compensate the bias of individual assays by combining the results of several assays. Some studies have tried this approach using their newly found sequences (Soule et al. 2006; Hamilton et al. 2006; Lu et al. 2007). Efforts are needed to combine some of the existing, individually designed assays. Bioinformatics, especially skills learnt from the design of microarray chips, can help in the design of new platforms and in the analysis of results.

The major challenges for accurate quantification of genetic markers from natural water samples are twofold. The first challenge is to assess the recovery efficiency in sample process steps, namely, concentration of targets and extraction and purification of nucleic acids. Several studies have utilized some kind of spike to assess the recovery efficiency (Rajal et al. 2007; Silkie and Nelson 2009; Stoeckel et al. 2009). The second challenge is to assess the impact of PCR inhibitors. General or assay-specific internal amplification control (IAC) has been utilized to assess the inhibition (Haugland et al. 2005; Shanks et al. 2008, 2009). However, the use of any type of general IAC is of limited value since different assays suffer different levels of inhibition in the same water matrix (Boehm et al. 2009). Another problem with IAC is the competition and interference of IAC with the detection of targeted genetic markers. Efforts are needed to identify the exact working mechanisms of various inhibitors and the competition/interference effect of IAC (Opel et al. 2010), based on which more resilient qPCR master mixes can be prepared and proper IACs can be designed.

Acknowledgments This study was sponsored in part by California Department of Transportation task order 23 of 43A01684 and Water Environment Research Foundation grant PATH2R08 to SW. The Austrian part of the work was supported by the Austrian Science Fund (FWF) projects #P22309-B20 and DK plus #W1219-N22 (Vienna Doctoral Programme on Water Resource Systems) granted to A.H.F.

References

Adams LM, Simmons CP, Robins-Browne RM et al (1997) Identification and characterization of a K88- and CS31A-like operon of a rabbit enteropathogenic *Escherichia coli* strain which encodes fimbriae involved in the colonization of rabbit intestine. Infect Immun 65:5222–5230

Ahmed W, Stewart J, Powell D et al (2008a) A real-time polymerase chain reaction assay for quantitative detection of the human-specific Enterococci surface protein marker in sewage and environmental waters. Environ Microbiol 10(12):3255–3264

Ahmed W, Stewart J, Powell D et al (2008b) Evaluation of the host-specificity and prevalence of Enterococci surface protein (*esp*) marker in sewage and its application for sourcing human fecal pollution. J Environ Qual 37:1583–1588

Ahmed W, Powell D, Gardner T et al (2008c) Detection and source identification of faecal pollution in non-sewered catchment by means of host-specific molecular markers. Water Sci Technol 58:579–586

Ahmed W, Stewart J, Gardner T et al (2008d) Evaluation of *Bacteroides* markers for the detection of human faecal pollution. Lett Appl Microbiol 46:237–242

Ahmed W, Goonetilleke A, Gardner T et al (2009a) Evaluation of multiple sewage-associated *Bacteroides* PCR markers for sewage pollution tracking. Water Res 43:4872–4877

Ahmed W, Goonetilleke A, Gardner T et al (2009b) Comparison of molecular markers to detect fresh sewage in environmental waters. Water Res 43(9):4908–4917

Al-Diwany LJ, Cross T (1978) Ecological studies on Nocardioforms and other *Actinomycetes* in aquatic habitats. In: Mordarski, M., Kurylowicz, W. and Jeljaszewicz, J. (ed) *Nocardia* and *Streptomyces*, Proceedings of the international symposium on *Nocardia* and *Streptomyces*, 153–160. Gustav Fischer Verlag, New York, NY

Amador JA, Sotomayor-Ramírez D, Bachoon D et al (2008) Tracking human faecal contamination in tropical reservoirs in Puerto Rico. Lakes & Reservoirs: Research and Management 13(4):301–317

Anderson KL, Whitlock JE, Harwood VJ (2005) Persistence and differential survival of fecal indicator bacteria in subtropical waters and sediments. Appl Environ Microbiol 71(6):3041–3048

APHA (1992) *Standard methods for the examination of water and wastewater, 16th edn*. American Public Health Association, Washington D.C.

Bachoon DS, Nichols TW, Oetter DR et al (2009) Assessment of faecal pollution and relative algal abundances in Lakes Oconee and Sinclair, Georgia, USA. Lakes & Reservoirs: Research and Management 14:139–149

Backhed F, Ley RE, Sonnenburg JL et al (2005) Host-bacterial mutualism in the human intestine. Science 307:1915–1920

Bae S, Wuertz S (2009a) Discrimination of viable and dead fecal *Bacteroidales* bacteria by quantitative PCR with propidium monoazide. Appl Environ Microbiol 75(9):2940–2944

Bae S, Wuertz S (2009b) Rapid decay of host-specific fecal *Bacteroidales* cells in seawater as measured by quantitative PCR with propidium monoazide. Water Res 43(19):4850–4859

Bahaka D, Neut C, Khattabi A et al (1993) Phenotypic and genomic analyses of human strains belonging or related to *Bifidobacterium-longum, Bifidobacterium-infantis, and Bifidobacterium-breve.* Int J Syst Bacteriol 43:565–573
Balleste E, Bonjoch X, Belanche LA et al (2010) Molecular indicators used in the development of predictive models for microbial source tracking. Appl Environ Microbiol 76(6):1789–1795
Bambic D, Wuertz S, Ban M (2007) Fecal source tracking using human toolkits based on library-independent chemical and microbial markers. P Water Environ F 15:931–945
Bell A, Layton AC, Sayler GS et al (2009) Development of *Bacteroides* 16S rRNA gene TaqMan-based real-time PCR assays for estimation of total, human, and bovine fecal pollution in water. J Environ Qual 38:1224–1232
Bernhard AE, Field KG (2000a) Identification of nonpoint sources of fecal pollution in coastal waters by using host-specific 16S ribosomal DNA genetic markers from fecal anaerobes. Appl Environ Microbiol 66:1587–1594
Bernhard AE and Field KG (2000b) A PCR assay to discriminate human and ruminant feces on the basis of host differences in *Bacteriodales-Prevotella* genes encoding 16S rRNA. Appl Environ Microbiol 66(10):4571–4574
Blanch AR, Belanche-Munoz L, Bonjoch X et al (2004) Tracking the origin of faecal pollution in surface water: an ongoing project within the European Union research programme. J Water Health 2:249–260
Blanch AR, Belanche-Munoz L, Bonjoch X et al (2006) Integrated analysis of established and novel microbial and chemical methods for microbial source tracking. Appl Environ Microbiol 72:5915–5926
Boehm AB, Yamahara KM, Nelson KL (2009) Covariation and photoinactivation of traditional and novel indicator organisms and human viruses at a sewage-impacted marine beach. Environ Sci Technol, 43:8046–8052
Bonjoch X, Balleste E, Blanch AR (2004) Multiplex PCR with 16S rRNA gene-targeted primers of *Bifidobacterium* spp. to identify sources of fecal pollution. Appl Environ Microbiol 70:3171–3175
Bonjoch X, Balleste E, Blanch AR (2005) Enumeration of bifidobacterial populations with selective media to determine the source of waterborne fecal pollution. Water Res 39:1621–1627
Bonjoch X, Lucena F, Blanch AR (2009) The persistence of bifidobacteria populations in a river measured by molecular and culture techniques. J Appl Microbiol 107:1178–1185
Booth J, Brion GM (2004) The utility of the AC/TC ratio for watershed management: a case study. Water Sci Technol 50:199–203
Bosshard F, Riedel K, Schneider T et al (2010) Protein oxidation and aggregation in UVA-irradiated *Escherichia coli* cells as signs of accelerated cellular senescence. Environ Microbiol 12(11):2931–45
Bower PA, Scopel CO, McLellan SL (2005) Detection of genetic markers of fecal indicator bacteria in Lake Michigan and determination of their relationship to Escherichia coli densities using standard microbiological methods. Appl Environ Microbiol 71(12):8305–8313
Buchan A, Alber M, Hodson RE (2001) Strain-specific differentiation of environmental *Escherichia coli* isolates via denaturing gradient gel electrophoresis (DGGE) analysis of the 16S-23S intergenic spacer region. FEMS Microbiol Ecol 35:313–321
Burtscher MM, Zibuschka F, Mach RL et al (2009) Heterotrophic plate count vs. in situ bacterial 16S rRNA gene amplicon profiles from drinking water reveal completely different communities with distinct spatial and temporal allocations in a distribution net. Water SA 35:495–504
Byamukama D, Mach RL, Farnleitner AH (2005) Discrimination efficacy of fecal pollution detection in different aquatic habitats of a high altitude tropical country using presumptive coliform, *Escherichia coli* and *Clostridium perfringens* spores. Appl Environ Microbiol 71(1):65–71
Byappanahalli MN, Przybyla-Kelly K, Shively DA et al (2008) Environmental occurrence of the enterococcal surface protein (esp) gene is an unreliable indicator of human fecal contamination. Environ Sci Technol 42:8014–8020
Cawthorn DM, Witthuhn RC (2008) Selective PCR detection of viable *Enterobacter sakazakii* cells utilizing propidium monoazide or ethidium bromide monoazide. J Appl Microbiol 105:1178–1185

Chern EC, Tsai Y, Olson BH (2004a) Occurrence of genes associated with enterotoxigenic and enterohemorrhagic *Escherichia coli* in agricultural waste lagoons. Appl Environ Microbiol 70(1):356–362

Chern EC, Olson BH (2004b) Development of a biomarker to detect bird fecal waste in environmental waters. In: Perez JMS, Andion LG, Brebbia CA (ed) Coastal environment V: Incorporating oil spill studies. 95-102, WIT Press, Southampton, UK.

Chern EC, Brenner KP, Haugland RA et al (2009) Comparison of fecal indicator bacteria densities in marine recreational waters by QPCR. Water Qual Expo Health 1:203–214

Cimenti M, Biswas N, Bewtra JK et al (2005) Evaluation of microbial indicators for the determination of bacterial groundwater contamination sources. Water Air Soil Poll 168:157–169

Coakley T, Brion GM, Fryar A (2009) Relationships between indicators of faecal load, source, and age: developing a multi-indicator approach for risk characterization. In: Rose JB (ed) 15th Int Symp Health Related Microbiol/Internat Water Association, Naxos, Greece

Converse RR, Blackwood AD, Noble RT et al (2009) Rapid QPCR-based assay for fecal *Bacteroides* spp. as a tool for assessing fecal contamination in recreational waters. Water Res 43(19):4828–4837

Deep R (2006) Probability and statistics: with integrated software routines. Elsevier: Amsterdam.

D'Elia TV, Cooper CR, Johnston CG (2007) Source tracking of *Escherichia coli* by 16S-23S intergenic spacer region denaturing gradient gel elctrophoresis (DGGE) of the rrnB ribosomal operon. Can J Microbiol 53(10):1174–1184

Dick LK, Field KG (2004) Rapid estimation of numbers of fecal *Bacteroidetes* by use of a quantitative PCR assay for 16S rRNA genes. Appl Environ Microbiol 70(9):5695–5697

Dick LK, Simonich MT, Field KG (2005a) Microplate subtractive hybridization to enrich for *Bacteroidales* genetic markers for fecal source identification. Appl Environ Microbiol 71(6):3179–3183

Dick LK, Bernhard AE, Field KG et al (2005b) Host distribution of uncultivated fecal *Bacteroidales* bacteria reveal genetic markers for fecal source identification. Appl Environ Microbiol 71(6):3184–3191

Dick LK, Stelzer EA, Stoeckel DM et al (2010) Relative decay of *Bacteroidales* microbial source tracking markers and cultivated *Escherichia coli* in freshwater microcosms. Appl Environ Microbiol 76(10):3255–3262

Dorai-Raj S, O'Grady J, Colleran E (2009) Specificity and sensitivity evaluation of novel and existing *Bacteroidales* and *Bifidobacteria*-specific PCR assays on feces and sewage samples and their application for microbial source tracking in Ireland. Water Res 43(19):4980–4988

Dunbar J, Barns SM, Ticknor LO et al (2002) Empirical and theoretical bacterial diversity in four Arizona soils. Appl Environ Microbiol 68:3035–3045

Eckburg PB, Bik EM, Bernstein CN et al (2005) Diversity of the human intestinal microbial flora. Science 308:1635–1638

Evison LM, James A (1975) *Bifidobacterium* as an indicator of faecal pollution in water. Prog Water Technol 7:57–66

Farnleitner AH, Zibuschka F, Burtscher MM et al (2004) Eubacterial 16S-rDNA amplicon profiling: a rapid technique for comparison and differentiation of heterotrophic plate count communities from drinking water. Int J Food Microbiol 92:333–345

Farnleitner AH, Wilhartitz I, Ryzinska G et al (2005) Bacterial dynamics in spring water of alpine karst aquifers indicates the presence of stable autochthonous microbial endokarst communities. Environ Microbiol 7:1248–1259

Farnleitner AH, Ryzinska-Paier G, Reischer GH et al (2010) Escherichia coli and enterococci are sensitive and reliable indicators for human, livestock and wildlife faecal pollution in alpine mountainous water resources. J Appl Microbiol 109:1599–1608

Feachem R (1975) An improved role for faecal coliform to faecal streptococci ratios in the differentiation between human and non-human pollution sources. Water Res 9:689–690

Field KG, Chern EC, Dick LK et al (2003) A comparative study of culture-independent, library-independent genotypic methods of fecal source tracking. J Water Health 1:181–194

Flekna G, Stefanic P, Hein I et al (2007) Insufficient differentiation of live and dead *Campylobacter jejuni* and *Listeria monocytogenes* cells by ethidium monoazide (EMA) compromises EMA/real-time PCR. Res Microbiol 158:405–412

Gawler AH, Beechere JE, Brandãoc J, Carroll NM, Falcãoc L, Gourmelon M, Masterson B, Nunes B, Portera J, Rincé A, Rodrigues R, Thorp M, Walters JM and Meijer WG. Validation of host-specific Bacteriodales 16S rRNA genes as markers to determine the origin of faecal pollution in Atlantic Rim countries of the European Union. Water Res 41(16):378–384

Gedalanga PB, Olson BH (2009) Development of a quantitative PCR method to differentiate between viable and nonviable bacteria in environmental water samples. Appl Microbiol Biotechnol 82:587–596

Geldreich EE, Best LC, Kenner BA et al (1968) Bacteriological aspects of stormwater pollution. J Water Pollut Con F 40:1861

Geldreich EE, Kenner BA (1969) Concepts of fecal streptococci in stream pollution. J Water Pollut Con F 41:R336

Geldreich EE (1976) Fecal coliform and fecal *Streptococcus* density relationships in waste discharges and receiving waters. Crit Rev Env Contr 6:349–369

Gilpin BJ, Gregor JE, Savill MG (2002) Identification of the source of faecal pollution in contaminated rivers. Water Sci Technol 46(3):9–15

Gilpin BJ, James T, Savil MG et al (2003) The use of chemical and molecular microbial indicators for faecal source identification. Water Sci Technol 47(3):39–43

Gourmelon M, Caprais MP, Rince A et al (2007) Evaluation of two library-independent microbial source tracking methods to identify sources of fecal contamination in French estuaries. Appl Environ Microbiol 73(15):4857–4866

Gourmelon M, Caprais MP, Le Mennec C, et al. (2010) Application of library-independent microbial source tracking methods for identifying the sources of faecal contamination in coastal areas. Water Sci Technol 61:1401–1409

Griffith JF, Cao Y, Weisberg SB et al (2009) Evaluation of rapid methods and novel indicators for assessing microbiological beach water quality. Water Res 43(19):4900–4907

Gyllenberg H, Niemela S, Sormunen T (1960) Survival of bifidobacteria in water as compared with that of coliform bacteria and *Enterococci*. Appl Microbiol 8:20–22

Hamilton MJ, Yan T, Sadowsky MJ (2006) Development of goose- and duck-specific DNA markers to determine sources of *Escherichia coli* in waterways. Appl Environ Microbiol 72(6): 4012–4019

Hammerum AM, Jensen LB (2002) Prevalence of *esp*, encoding the enterococcal surface protein, in *Enterococcus faecalis* and *Enterococcus faecium* isolates from hospital patients, poultry, and pigs in Denmark. *J Clin Microbiol 40*(11):4396

Hartel PG, Rodgers K, McDonald JL et al (2008) Combining targeted sampling and fluorometry to identify human fecal contamination in a freshwater creek. J Water Health 6(1):105

Harwood VJ, Delahoya NC, Ulrich RM et al (2004) Molecular confirmation of *Enterococcus faecalis* and *E-faecium* from clinical, faecal and environmental sources. Lett Appl Microbiol 38:476–482

Harwood VJ, Brownell M, Wang S, Lepo J, Ellender RD, Ajidahun A, Hellein KN, Kennedy E, Ye X, Flood C (2009) Validation and field testing of library-independent microbial source tracking methods in the Gulf of Mexico. Water Res 43:4812–4819

Haugland RA, Siefring SC, Dufour AP et al (2005) Comparison of *Enterococcus* measurements in freshwater at two recreational beaches by quantitative polymerase chain reaction and membrane filter culture analysis. Water Res 39:559–568

Hill RT, Straube WL, Palmisano AC et al (1996) Distribution of sewage indicated by *Clostridium perfringens* at a deep-water disposal site after cessation of sewage disposal. Appl Environ Microbiol 62:1741–1746

Holdeman V, Cato ET, Moore WEC (1976) Human fecal flora: variation in bacterial composition within individuals and a possible effect of emotional stress. Appl Environ Microbiol 31:359–375

Houston, A.C. (1902) Second report of the Royal Commission on treating and disposing of sewage, London

ISO (2002) Water Quality. Detection and enumeration of *Clostridium perfringens* – Part 2: Method by membrane filtration (ISO/CD 6461-2). Geneva, Switzerland, International Organization of Standardization

Jagals P, Grabow WOK, de Villiers JC (1995) Evaluation of indicators for assessment of human and animal faecal pollution of surface run-off. Water Sci Technol 31:235–241

Jagals P, Grabow WOK (1996) An evaluation of sorbitol-fermenting bifidobacteria as specific indicators of human faecal pollution of environmental water. Water SA 22:235–238

Jenkins MW, Tiwari S, Wuertz S et al (2009) Identifying human and livestock sources of fecal contamination in Kenya with host-specific *Bacteroidales* assays. Water Res 43(19):4956–4966

Jeong JY, Park HD, Lee KH et al (2010) Quantitative analysis of human- and cow-specific 16S rRNA gene markers for assessment of fecal pollution in river waters by real-time PCR. J Microbiol Biotechn 20:245–253

Jeter SN, McDermott CM, McLellan SL et al (2009) *Bacteroidales* diversity in ring-billed gulls (*Laurus delawarensis*) residing at lake Michigan beaches. Appl Environ Microbiol 75(6):1525–1533

Jiang SC, Chu W, Gedalanga PB et al (2007) Microbial source tracking in a small southern California urban watershed indicates wild animals and growth as the source of fecal bacteria. Appl Environ Microbiol 76:927–934

Johnson JR, O'Bryan TT, Stell AL et al (2000) Evidence of commonality between canine and human extraintestinal pathogenic *Escherichia coli* strains that express *papG* allele III. Infect Immun 68:3327–3336

Kreader CA (1995) Design and evaluation of *Bacteroides* DNA probes for the specific detection of human fecal pollution. Appl Environ Microbiol 61:1171–1179

Khatib LA, Tsai YL, Olson BH (2002) A biomarker for the identification of cattle fecal pollution in water using the *LTIIa* toxin gene from enterotoxigenic *Escherichia coli.* Appl Microbiol Biotechnol 59:97–104

Khatib LA, Tsai YL and Olson BH (2003) A biomarker for the identification of swine fecal pollution in water, using the *STII* toxin gene from enterotoxigenic *Escherichia coli.* Appl Microbiol Biotechnol 63:231–238

Kildare BJ, Rajal V, Wuertz S et al (2006) Calleguas creek watershed quantitative microbial source tracking study. http://www.calleguas.com/ccwmp/DRAFT_CCW_MST_061406.pdf

Kildare BJ, Leutenegger CM, Wuertz S et al (2007) 16S rRNA-based assays for quantitative detection of universal, human-, cow-, and dog-specific fecal *Bacteroidales*: a Bayesian approach. Water Res 41:3701–3715

Kim J, Pitts B, Yoon J et al (2008) Comparison of the antimicrobial effects of chlorine, silver ion, and tobramycin on biofilm. Antimicrob Agents Ch 52:1446–1453

King EL, Bachoon DS, Gates KW (2007) Rapid detection of human fecal contamination in estuarine environments by PCR targeting of *Bifidobacterium adolescentis.* J Microbiol Meth 68(1):76-81

Kirschner AK, Eiler A, Zechmeister TC et al (2002) Extremely productive microbial communities in shallow saline pools respond immediately to changing meteorological conditions. Environ Microbiol 4:546–555

Klein and Houston (1898) Report on bacteriological evidence of presumably recent, and therefore dangerous, sewage pollution of elsewise potable waters. 27th Report Annual Report of the Local Government Board 1897-98, Supplement Report of Local Officer. 4:318–325

Knee KL, Layton BA, Paytan A (2008) Sources of nutrients and fecal indicator bacteria to nearshore waters on the north shore of Kauai (Hawaii, USA). Estuar Coast 31:607–622

Kreader CA (1998) Persistence of PCR-detectable *Bacteroides distasonis* from human feces in river water. Appl Environ Microbiol 64(10):4103–4105

Kuntz RL, Hartel PG, Rodgers K et al (2004) Presence of *Enterococcus faecalis* in broiler litter and wild bird feces for bacterial source tracking. Water Res 38:3551–3557

Lamendella R, Santo Domingo JW, Oerther DB et al (2008) Bifidobacteria in feces and environmental waters. Appl Environ Microbiol 74(3):575–584

Lamendella R, Santo Domingo JW, Oerther DB et al (2009) Evaluation of swine-specific PCR assays used for fecal source tracking and analysis of molecular diversity of swine-specific "*Bacteroidales*" populations. Appl Environ Microbiol 75(18):5787–5796

Lawson PA, Collins MD, Foster G et al (2006) *Catellicoccus marimammalium* gen. nov., sp. nov., a novel gram-positive, catalase-negative, coccus-shaped bacterium from porpoise and grey seal. Int J Syst Evol Micr 56:429–432

Layton BA, McKay L, Sayler G et al (2006) Development of Bacteroides 16S rRNA gene TaqMan-based real-time PCR assays for estimation of total, human, and bovine fecal pollution in water. Appl Environ Microbiol 72(6):4214–4224

Layton BA, Walters SP, Boehm AB (2009) Distribution and diversity of the enterococcal surface protein (*esp*) gene in animal hosts and the Pacific coast environment. J Appl Microbiol 106:1521–1531

Layton BA, Walters SP, Lam LH et al (2010) *Enterococcus* species distribution among human and animal hosts using multiplex PCR. J Appl Microbiol 109(2):539–547

Leclerc H, Devriese LA, Mossel DAA (1996) Taxonomical changes in intestinal (faecal) *Enterococci* and *Streptococci*: consequences on their use as indicators of faecal contamination in drinking water. J Appl Bacteriol 81:459–466

Ley RE, Peterson DA, Gordon JI (2006) Ecological and evolutionary forces shaping microbial diversity in the human intestine. Cell 124:837–848

Ley RE, Hamady M, Lozupone C et al (2008a) Evolution of mammals and their gut microbes. Science 320:1647–1651

Ley RE, Lozupone CA, Hamady M et al (2008b) Worlds within worlds: evolution of the vertebrate gut microbiota. Nat Rev Microbiol 6:776–788

Lin C, Miller TL (1998) Phylogenetic analysis of *Methanobrevibacter* isolated from feces of humans and other animals. Archeol Microbiol 169:397–403

Long SC, Shafer E, Arango C et al (2003) Evaluation of three source tracking indicator organisms for watershed management. Water Supply 52:565–575

Lu J, Santo Domingo JW, Shanks OC (2007) Identification of chicken-specific fecal microbial sequences using a metagenomic approach. Water Res 41:3561–3574

Lu J, Santo Domingo JW, Hill S et al (2008a) Phylogenetic diversity and molecular detection of bacteria in gull feces. Appl Environ Microbiol 74(13):3969–3976

Lu J, Santo Domingo JW (2008b) Turkey fecal microbial community structure and functional gene diversity revealed by 16S rRNA gene and metagenomic sequences. J Microbiol 46(5):469–477

Lu J, Santo Domingo JW, Edge TA et al (2009) Microbial diversity and host-specific sequences of Canada goose feces. Appl Environ Microbiol 75(18):5919–5926

Ludwig W, Schleifer KH (2000) How quantitative is quantitative PCR with respect to cell counts? Syst Appl Microbiol 23(4):556–562

Lynch PA, Gilpin BJ, Savill MG et al (2002) The detection of *Bifidobacterium adolescentis* by colony hybridization as an indicator of human fecal pollution. J Appl Microbiol 92:526–533

Mara DD, Oragui JI (1981) Occurrence of *Rhodococcus coprophilus* and associated Actinomycetes in feces, sewage, and freshwater. Appl Environ Microbiol 42:1037–1042

Mara DD, Oragui JI (1983) Sorbitol-fermenting bifidobacteria as specific indicators of human faecal pollution. J Appl Bacteriol 55:349–357

Mara DD, Oragui J (1985) Bacteriological methods for distinguishing between human and animal faecal pollution of water: results of fieldwork in Nigeria and Zimbabwe. B World Health Organ 63:773–783

Matsuki T, Watanabe K, Tanaka R et al (2004) Quantitative PCR with 16S rRNA-gene-targeted species-specific primers for analysis of human intestinal bifidobacteria. Appl Environ Microbiol 70:167–173

McDonald JL, Hartel PG, Payne KA et al (2006) Identifying sources of fecal contamination inexpensively with targeted sampling and bacterial source tracking. J Environ Qual 35:889-897

McFeters GA, Bissonne GK, Jezeski JJ et al (1974) Comparative survival of indicator bacteria and enteric pathogens in well water. Appl Microbiol 27:823–829

McLain JET, Ryu H, Abbaszadegan M et al (2009) Lack of specificity for PCR assays targeting human *Bacteroides* 16S rRNA gene: cross-amplification with fish feces. FEMS Microbiol Lett 299(1):38–43

McQuaig SM, Scott TM, Lukasik JO et al (2006) Detection of human-derived fecal pollution in environmental waters by use of a PCR-based human polyomavirus assay. Appl Environ Microbiol 72:7567–7574

McQuaig SM, Scott TM, Harwood VJ et al (2009) Quantification of human polyomaviruses JC virus and BK virus by TaqMan quantitative PCR and comparison to other water quality indicators in water and fecal samples. Appl Environ Microbiol 75(11):3379–3388

Meays CL, Broersma K, Mazumder A et al (2004) Source tracking fecal bacteria in water: a critical review of current methods. J Environ Manage 73:71–79

Mieszkin S, Yala JF, Gourmelon M et al (2009) Phylogenetic analysis of *Bacteroidales* 16S rRNA gene sequences from human and animal effluents and assessment of ruminant faecal pollution by real-time PCR. J Appl Microbiol 108(3):974–984

Mieszkin S, Furet JP, Gourmelon M et al (2010) Estimation of pig fecal contamination in a river catchment by real-time PCR using two pig-specific *Bacteroidales* 16S rRNA genetic markers. Appl Environ Microbiol 75(10):3045–3054

Miura T, Masago Y, Omura T et al (2009) Detection of bacteria and enteric viruses from river and estuarine sediment. J Water Environ Technol. 7(4):307–316

Morrison TB, Weis JJ, Wittwer CT (1998) Quantification of low-copy transcripts by continuous SYBR Green I monitoring during amplification. BioTechniques 24(6):954–958

Morrison CR, Bachoon DS, Gates KW (2008) Quantification of enterococci and bifidobacteria in Georgia estuaries using conventional and molecular methods. Water Res 42(14): 4001–4009

Mushi DW, Byamukama D, Kivaisi KA et al (2010) Sorbitol-fermenting Bifidobacteria are indicators of very recent human faecal pollution in streams and groundwater habitats in urban tropical lowlands. J Water Health 8(3):466–478

Muyzer G (1999) DGGE/TGGE a method for identifying genes from natural ecosystems. Curr Opin Microbiol 2:317–322

Nebra Y, Bonjoch X, Blanch AR (2003) Use of *Bifidobacterium dentium* as an indicator of the origin of fecal water pollution. Appl Environ Microbiol 69:2651–2656

Nieman J, Brion GM (2003) Novel bacterial ratio for predicting faecal age. Water Sci Technol 47:45–49

Nocker A, Camper A K (2006) Selective removal of DNA from dead cells of mixed bacterial communities by use of ethidium monoazide. Appl Environ Microbiol 72:1997–2004

Nocker A, Cheung C, Camper AK (2006) Comparison of propidium monozide with ethidium monoazide for differentiation of live vs. dead bacteria by selective removal of DNA from dead cells. J Microbiol Meth 67:310–320

Nocker A, Sossa-Fernandez P, Camper A K et al (2007a) Use of propidium monoazide for live/dead distinction in microbial ecology. Appl Environ Microbiol 73:5111–5117

Nocker A, Sossa KE, Camper AK (2007b) Molecular monitoring of disinfection efficacy using propidium monoazide in combination with quantitative PCR. J Microbiol Meth 70: 252–260

Nocker A, Camper AK (2009) Novel approaches toward preferential detection of viable cells using nucleic acid amplification techniques. FEMS Microbiol Lett 291:137–142

Nogva H, Dromtorp S, Rudi K et al (2003) Ethidium monoazide for DNA-based differentiation of viable and dead bacteria by 5′-nuclease PCR. BioTechniques 34(4):804–813

Ochman H, Lawrence JG, Groisman EA (2000) Lateral gene transfer and the nature of bacterial innovation. Nature 405:299–304

Okabe S, Okayama N, Ito T et al (2007a) Quantification of host-specific *Bacteroides-Prevotella* 16S rRNA genetic markers for assessment of fecal pollution in freshwater. Appl Environ Microbiol 74:890–901

Okabe S, Shimazu Y (2007b) Persistence of host-specific *Bacteroides-Prevotella* 16S rRNA genetic markers in environmental waters: effects of temperature and salinity. Appl Microbiol Biot 76:935–944

Opel KL, Chung D, McCord BR (2010) A study of PCR inhibition mechanisms using real time PCR. J Forensic Sci 55(1):25–33

Oshiro RK, Olson BH (1997) Occurrence of *STh* toxin gene in wastewater. In: Kay D, Fricher C (ed), Coliforms and *E coli*: problem or solution? The Royal Society of Chemistry, Cambridge. pp 255–259

Ottoson JR (2009) Bifidobacterial survival in surface water and implications for microbial source tracking. Can J Microbiol 55(6):642–647

Pan, Y, Breidt F (2007) Enumeration of viable *Listeria monocytogenes* cells by real-time PCR with propidium monoazide and ethidium monoazide in the presence of dead cells. Appl Environ Microbiol 73:8028–8031

Paster BJ, Dewhirst FE, Olsen I et al (1994) Phylogeny of *Bacteroides, Prevotella* and *Porphyromonas* spp. and related bacteria. J Bacteriol 176(3):725–732

Pourcher AM, Devriese LA, Hernandez JF et al (1991) Enumeration by a miniaturized method of *Escherichia-Coli, Streptococcus bovis* and enterococci as indicators of the origin of fecal pollution of waters. J Appl Bacteriol 70:525–530

Provence DL, Curtiss R (1994) Isolation and characterization of a gene involved in hemagglutination by an avian pathogenic *Escherichia coli* strain. Infect Immun 62:1369–1380

Rajal VB, McSwain BS, Wuertz S et al (2007) Validation of hollow fiber ultrafiltration and real time PCR using bacteriophage PP7 as surrogate for the quantification of viruses from water samples. Water Res 41(7):1411–1422

Ram JL, Ritchie RP, Selegean JP et al (2004) Sequence-based source tracking of *Escherichia coli* based on genetic diversity of beta-glucuronidase. J Environ Qual 33:1024–1032

Rawsthorne H, Dock CN, Jaykus LA (2009) PCR-based method using propidium monoazide to distinguish viable from nonviable *Bacillus subtilis* spores. Appl Environ Microbiol 75: 2936–2939

Reischer GH, Kasper DC, Farnleitner AD et al (2006) Quantitative PCR method for sensitive detection of ruminant fecal pollution in freshwater and evaluation of this method in alpine karstic regions. Appl Environ Microbiol 72:5610–5614

Reischer GH, Kasper DC, Mach RL et al (2007) A quantitative real-time PCR assay for the highly sensitive and specific detection of human faecal influence in spring water from a large alpine catchment area. Lett Appl Microbiol 44(4):351–356

Reischer GH, Haider JM, Farnleitner AH et al (2008) Quantitative microbial faecal source tracking with sampling guided by hydrological catchment dynamics. Environ Microbiol 10(10): 2598–2608

Reischer GH, Haider JM, Farnleitner AH et al (2009) A global assessment of the source specificity, sensitivity and geographical stability of *Bacteroides* qPCR assays for microbial source tracking. Proceedings of the 15th International Symposium on Health-Related Water Microbiology, 31 May–5 June 2009 Naxos, Greece

Reischer GH, Kollanur D, Vierheilig J et al (2011) Hypothesis-driven approach for the identification of fecal pollution sources in water resources. Environ Sci Technol 45(9):4038–4045

Resnick IG, Levin MA (1981a) Assessment of bifidobacteria as indicators of human fecal pollution. Appl Environ Microbiol 42:433–438

Resnick IG, Levin MA (1981b) Quantitative procedure for enumeration of bifidobacteria. Appl Environ Microbiol 42:427–432

Rhodes MW, Kator H (1999) Sorbitol-fermenting bifidobacteria as indicators of diffuse human faecal pollution in estuarine watersheds. J Appl Microbiol 87:528–535

Richter DD, Markewitz D (1995) How deep is soil – soil, the zone of the earths crust that is biologically-active, is much deeper than has been thought by many ecologists. Bioscience 45:600–609

Roll BM, Fujioka RS (1997) Sources of faecal indicator bacteria in a brackish, tropical stream and their impact on recreational water quality. Water Sci Technol 35(11):179–186

Rosario K, Symonds EM, Breitbart M et al (2009) Pepper mild mottle virus as an indicator of fecal pollution. Appl Environ Microbiol 75(22):7261–7267

Roslev P, Bastholm S, Iversen N (2008) Relationship between fecal indicators in sediment and recreational waters in a Danish estuary. Water Air Soil Poll 194:13–21

Rowbotham TJ, Cross T (1977) Ecology of *Rhodococcus coprophilus* and associated Actinomycetes in fresh water and agricultural habitats. J Gen Microbiol 100:231–240

Rudi K, Moen B, Holck AL et al (2005) Use of ethidium monoazide and PCR in combination for quantification of viable and dead cells in complex samples. Appl Environ Microbiol 71:1018–1024

Sadowsky MJ, Call DR, Santo Domingo JW (2007) The future of microbial source tracking studies. In: Santo Domingo JW, Sadowsky MJ (eds) Microbial source tracking. Washington, DC: ASM Press, pp. 235–277

Santoro AE, Boehm AB (2007) Frequent occurrence of the human-specific *Bacteroides* fecal marker at an open coast marine beach: relationship to waves, tides and traditional indicators. Environ Microbiol. doi:10.1111/j.1462-2920.2007.01319.x

Saunders AM, Kristiansen A, Schramm A et al (2009) Detection and persistence of fecal *Bacteroidales* as water quality indicators in unchlorinated drinking water. Syst Appl Microbiol 32(5):362–370

Savill MG, Murray SR, Gilpin BJ et al (2001) Application of polymerase chain reaction (PCR) and TaqMan™ PCR techniques to the detection and identification of *Rhodococcus coprophilus* in faecal samples. J Microbiol Meth 47:355–368

Schriewer A, Miller WA, Byrne BA et al (2010) *Bacteroidales* as a predictor of pathogens in surface waters of the central California coast. Appl Environ Microbiol 76:5802–5814

Scott TM, Jenkins TM, Rose JB et al (2005) Potential use of a host associated molecular marker in *Enterococcus faecium* as an index of human fecal pollution. Environ Sci Technol 39:283–287

Scott TM, Harwood VJ, Ahmed W et al (2009) Comment on "environmental occurrence of the enterococcal surface protein (esp) gene is an unreliable indicator of human fecal contamination". Environ Sci Technol 43:6434–6435

Seurinck S, Defoirdt T, Sciliano S et al (2005) Detection and quantification of the human-specific HF183 *Bacteroides* 16S rRNA genetic marker with real-time PCR for assessment of human faecal pollution in freshwater. Environ Microbiol 7(2):249–259

Shanks OC, Santo Domingo JW, Graham JE et al (2006a) Competitive metagenomic DNA hybridization identifies host-specific microbial genetic markers in cow fecal samples. Appl Environ Microbiol 72:4054-4060

Shanks OC, Nietch C, Field KG et al (2006b) Basin-wide analysis of the dynamics of fecal contamination and fecal source identification in Tillamook Bay, Oregon. Appl Environ Microbiol 72(8):5537–5546

Shanks OC, Santo Domingo JW, Graham JE et al (2007) Identification of bacterial DNA markers for the detection of human fecal pollution in water. Appl Environ Microbiol 73:2416–2422

Shanks OC, Atikovic E, Haugland RA et al (2008) Quantitative PCR for detection and enumeration of genetic markers of bovine fecal pollution. Appl Environ Microbiol 74:745–752

Shanks OC, Kelty CA, Haugland RA et al (2009) Quantitative PCR for genetic markers of human fecal pollution. Appl Environ Microbiol 75(17):5507–5513

Shanks OC, White K, Haugland RA (2010) Performance assessment PCR-based assays targeting Bacteroidales genetic markers of bovine fecal pollution. Appl Environ Microbiol 76(5):1359–1366

Siefring S, Varma M, Haugland RA (2008) Improved real-time PCR assays for the detection of fecal indicator bacteria in surface waters with different instrument and reagent systems. J Water Health 6(2):225

Silkie SS, Nelson KL (2009) Concentrations of host-specific and generic fecal markers measured by quantitative PCR in raw sewage and fresh animal feces. Water Res 43(19):4860–4871

Simpson JM, Santo Domingo JW, Reasoner DJ (2004) Assessment of equine fecal contamination: the search for alternative bacterial source-tracking targets. FEMS Microbiol Ecol 47(1):65–75

Sirikanchana K, Wang D, Bombardelli FA et al. Microbial source tracking using *Bacteroidales* and quantification of bacterial indicators and human pathogens in a major estuary. *Submitted*

Skanavis C, Yanko WA (2001) *Clostridium perfringens* as a potential indicator for the presence of sewage solids in marine sediments. Mar Pollut Bull 42:31–35

Soejima T, Iida K, Yoshida S et al (2007) Photoactivated ethidium monoazide directly cleaves bacterial DNA and is applied to PCR for discrimination of live and dead bacteria. Microbiol Immunol 51:763–775

Sorensen DL, Eberl SG, Dicksa RA (1989) *Clostridium perfringens* as a point-source indicator in non-point polluted streams. Water Res 23:191–197

Soule M, Kuhn E, Call DR et al (2006) Using DNA microarrays to identify library-independent markers for bacterial source tracking. Appl Environ Microbiol 72(3):1843–1851

Stapleton CM, Kay D, Wyer MD, et al. (2009) Evaluating the operational utility of a *Bacteroidales* quantitative PCR-based MST approach in determining the source of faecal indicator organisms at a UK bathing water. Water Res 43:4888–4899

Stewart MH, Olson B (1996) Bacterial resistance to potable water disinfectants. In: Hurst CJ (ed) Modeling disease transmission and its prevention by disinfection. Cambridge: Cambridge University Press, pp 140–192

Stoeckel DM, Stelzer EA, Dick LK (2009) Evaluation of two spike-and-recovery controls for assessment of extraction efficiency in microbial source tracking studies. Water Res 43(19): 4820–4827

Stricker AR, Wilhartitz I, Mach RL et al (2008) Development of a scorpion probe-based real-time PCR for the sensitive quantification of *Bacteroides* sp. ribosomal DNA from human and cattle origin and evaluation in spring water matrices. Microbiol Res 163(2):140–147

Suau A, Bonnet R, Sutren M et al (1999) Direct analysis of genes encoding 16S rRNA from complex communities reveals many novel molecular species within the human gut. Appl Environ Microbiol 65:4799–4807

Tallon P, Magajna B, Lofranco C et al (2005) Microbial indicators of faecal contamination in water: A current perspective. Water Air Soil Poll 166:139–166

Tsai YL, Le JY, Olson BH (2003) Magnetic bead hybridization to detect enterotoxigenic *Escherichia coli* strains associated with cattle in environmental water sources. Can J Microbiol 49:391–398

Turnbaugh PJ, Ley RE, Mahowald MA et al (2006) An obesity-associated gut microbiome with increased capacity for energy harvest. Nature 444:1027–1031

Ufnar JA, Wang SY, Ellender RD et al (2006) Detection of the *nifH* gene of *Methanobrevibacter smithii*: a potential tool to identify sewage pollution in recreational waters. J Appl Microbiol 101:44–52

Ufnar JA, Ufnar DF, Ellender RD et al (2007a) Development of a swine-specific fecal pollution marker based on host differences in Methanogen *mcrA* genes. Appl Environ Microbiol 73(16): 5209–5217

Ufnar JA, Wang SY, Ellender RD et al (2007b) *Methanobrevibacter ruminantium* as an indicator of domesticated-ruminant fecal pollution in surface waters. Appl Environ Microbiol, 73(21): 7118–7121

Vesper S, McKinstry C, Vesper A et at (2008) Quantifying fungal viability in air and water samples using quantitative PCR after treatment with propidium monoazide (PMA). J Microbiol Meth 72:180–184

Wade T, Calderon RL, Dufour AP (2006) Rapidly measured indicators of recreational water quality are predictive of swimming-associated gastrointestinal illness. Environ Health Persp 114(1):24–28

Wagner AO, Malin C, Illmer P et al (2008) Removal of free extracellular DNA from environmental samples by ethidium monoazide and propidium monoazide. Appl Environ Microbiol 74:2537–2539

Walters SP, Field KG (2006) Persistence and growth of fecal *Bacteroidales* assessed by bromodeoxyuridine immunocapture. Appl Environ Microbiol 72(7):4532–4539

Walters SP, Yamahara KM, Boehm AB (2009) Persistence of nucleic acid markers of health-relevant organisms in seawater microcosms: implications for their use in assessing risk in recreational waters. Water Res 43(19):4929–4939

Wang D, Silkie SS, Nelson K et al (2010) Estimating true human and animal host source contribution in quantitative microbial source tracking using the Monte Carlo method. Water Res 44(16):4760–4775

Ward JW, Reed TM, Fryar AE et al (2009) Using the AC/TC ratio to evaluate fecal inputs in a karst groundwater basin. Environ Eng Geosci 15:57–65

Weisburg WG, Barns SM, Lane DJ et al (1991) 16S ribosomal DNA amplification for phylogenetic study. J Bacteriol 173(2):697–703

Wheeler AL, Hartel PG, Godfrey DG et al (2002) Potential of *Enterococcus faecalis* as a human fecal indicator for microbial source tracking. J Environ Qual 31:1286–1293

Whitman WB, Coleman DC, Wiebe WJ (1998) Prokaryotes: the unseen majority. P Natl Acad Sci USA 95:6578–6583

Whitman RL, Przybyla-Kelly K, Shively DA et al (2007) Incidence of the enterococcal surface protein (esp) gene in human and animal fecal sources. Environ Sci Technol 41:6090–6095

Wilhartitz I, Mach RL, Teira E et al (2007) Prokaryotic community analysis with CARD-FISH in comparison with FISH in ultra-oligotrophic ground- and drinking water. J Appl Microbiol 103:871–881

Wittwer CT, Herrman MG, Moss AA (1997) Continuous fluorescence monitoring of rapid cycle DNA amplification. BioTechniques 22(1):130

Wong M, Kumar L, Rose JB et al (2009) Evaluation of public health risks at recreational beaches in Lake Michigan via detection of enteric viruses and a human-specific bacteriological marker. Water Res 43 (4):1137–1149

Wuertz S, Bombardelli F, Sirikanchana K et al (2009) Quantitative pathogen detection & microbial source tracking combined with modeling the fate and transport of *Bacteroidales* in San Pablo Bay. CICEET report: http://ciceet.unh.edu/news/releases/fall09_reports/pdf/wuertz06_fr_fall06.pdf

Xu J, Bjursell MK, Gordon GI et al (2003) A genomic view of the human-*Bacteroides thetaiotaomicron* symbiosis. Science 299:2074–2076

Yampara-Iquise H, Zheng G, Carson CA et al (2008) Use of a *Bacteroides thetaiotaomicron*-specific alfa-1-6, mannanase quantitative PCR to detect human faecal pollution in water. J Appl Microbiol 105:1686–1693

Zheng G, Yampara-Iquise H, Carson CA et al (2009) Development of *Faecalibacterium* 16S rRNA gene marker for identification of human faeces. J Appl Microbiol 106:634–641

Chapter 5
Viruses as Tracers of Fecal Contamination

S.M. McQuaig and R.T. Noble

Abstract In assessments of water quality, determining the source of fecal contamination is of paramount importance for mitigating contamination, maintaining and restoring healthy ecosystems, and protecting public health. Historically, attention has focused on the use of fecal indicator bacteria (FIB) to indicate the level of fecal contamination, but over the past decade some attention has shifted to the use of pathogenic viruses to determine source(s) of fecal contamination and to assess risks to human health. Initially, viral detection methodologies were used to examine drinking water quality; however, the power of those methodologies has been recognized by those concerned with environmental water quality. Efforts have been made to apply these methods to natural water systems, including recreational waters, stormwater, and shellfishing areas. This chapter addresses the use of viruses in microbial source tracking (MST), specifically the application of quantitative tests for specific types of human pathogenic viruses, and how to choose the most appropriate assay for a particular study. Examples of studies utilizing a range of human pathogenic viruses, concentration and quantification approaches, successful case studies, and challenges associated with the use of viral MST methods are discussed.

Keywords Virus • Enterovirus • Adenovirus • Human polyomavirus • Hepatitis A virus • Water quality • Public health • Concentration

5.1 Enteric Viruses as Water Quality Indicators

The use of fecal indicator bacteria (FIB) as proxies for fecal contamination in water has long been a common practice; however, the lack of correlation between FIB concentrations and either pathogen density or risk of gastrointestinal illness in

S.M. McQuaig (✉)
Natural Sciences, St. Petersburg College,
2465 Drew St., Clearwater, FL 33765, USA
e-mail: mcquaig.shannon@spcollege.edu

C. Hagedorn et al. (eds.), *Microbial Source Tracking: Methods, Applications, and Case Studies*, DOI 10.1007/978-1-4419-9386-1_5,

recreational waters has encouraged some researchers to consider other microbial targets to assess water quality (Jiang et al. 2001; Colford et al. 2007). The direct monitoring of all pathogenic organisms would provide the most ideal and accurate assessment of water quality and risk potential, yet this monitoring program would be tedious, cost-ineffective, labor-intensive, and impractical because of the wide array of relevant pathogens. Therefore, many studies have focused on the use of one or more relatively prevalent pathogen(s) as an indicator of water quality. The use of enteric viruses as markers of fecal contamination in water has been proposed because it is thought that viruses are the causative agent of a large proportion of waterborne disease cases (Moore et al. 1993; Kramer et al. 1996; Levy et al. 1998; Barwick et al. 2000; Lee et al. 2002; Yoder et al. 2004, 2008; Dziuban et al. 2006). Furthermore, enteric viruses tend to be more resilient to environmental stresses (e.g., temperature and salinity) (Neefe et al. 1947; Havelaar et al. 1990; Jiang 2006) and water treatment practices (e.g., ultraviolet radiation and chlorination) as compared to FIB (Thurston-Enriquez et al. 2003a, b; Hijnen et al. 2006; Blatchley et al. 2007). Finally, the presence of even a small concentration of viruses (as low as 1–10 infectious units, (Centers for Disease Control and Prevention 2002)) may represent a significant health risk. This is because many enteric viruses have a relatively small infectious dose, particularly compared to bacteria (e.g., infectious dose of *Salmonella typhi* is estimated at between 100 and 1×10^5 cells) (Bitton 2005). The most prevalent viruses in human-derived sewage include adenoviruses, enteroviruses, and human polyomaviruses (Table 5.1). While researchers have suggested the use of animal-specific viruses as well as human-specific viruses as MST markers, the use of human-specific viruses as indicators of environmental water quality and direct health risks has grown at a rapid pace over the past two decades (Fong and Lipp 2005). This chapter focuses primarily on studies utilizing human-specific viruses.

Table 5.1 Concentrations of human enteric viruses in untreated sewage in gene copies/liter (GC/L)

	Concentration (GC/L)			
Virus type	Low	High	Countries	Reference(s)
Adenovirus	36	4.63×10^5	Japan, New Zealand, United States	Wolf et al. (2010); Katayama et al. (2008); Fong et al. (2010)
Enterovirus	9	1.8×10^8	Japan, New Zealand	Wolf et al. (2010); Katayama et al. (2008)
Norovirus	4.9×10^3	1.0×10^9	France, Japan	da Silva et al. (2007); Katayama et al. (2008)
Human polyomaviruses[a]	1.3×10^7	4.7×10^7	United States	McQuaig et al. (2009)

[a]Assay targets both BKV and JCV

5.2 Introduction to the Enteric Viruses

5.2.1 Traditional Enteric Viruses

Viruses were first discovered in 1892 by Dmitri Iwanowski, who found evidence of filterable, disease-causing agents from tobacco plants (tobacco mosaic virus) (Zaitlin 1999). Since then, thousands of viruses have been identified. To date, there are over 100 recognized enteric virus species that infect humans (Murray et al. 2002; Bitton 2005). Enteric viruses are defined as viruses that enter the gastrointestinal tract and subsequently establish infection. In general, enteric virus infections are typically the result of poor sanitary conditions or consumption of contaminated food or water. Clinical symptoms associated with these infections can be intestinal (e.g., abdominal cramps, gastroenteritis) or extraintestinal (e.g., fever, headache, jaundice) (Murray et al. 2002). Human infections caused by enteric viruses can cause asymptomatic or mild to serious illnesses, and in some cases may even be fatal. The most common human enteric viruses include: adenoviruses, astroviruses, enteroviruses, hepatitis A viruses, hepatitis E viruses, noroviruses, and rotaviruses (Murray et al. 2002; Bitton 2005). Table 5.2 provides a brief overview of the characteristics of common animal and human enteric viruses.

5.2.2 A Nontraditional "Enteric" Virus

In the late 1990s, researchers began to document the presence of the human polyomavirus (HPyV) species BKV and JCV in urban raw sewage (Bofill-Mas et al. 2000). These viruses are double-stranded DNA viruses frequently isolated from the urine, and in some cases feces, of both healthy and immunocompromised individuals (Zhong et al. 2007; Bialasiewicz et al. 2009). It has been suggested that JCV and BKV are spread via the urine–oral route (Kunitake et al. 1995; Bofill-Mas et al. 2001), and therefore, these two HPyVs are considered "nontraditional enteric viruses." An asymptomatic primary infection typically occurs during childhood, followed by latent infections in the renal tissue, which can persist indefinitely (Shah 1996; Dorries 1998). Asymptomatic viruria can occur occasionally or continuously in infected individuals (Hogan et al. 1980; Arthur et al. 1989; Markowitz et al. 1993; Polo et al. 2004; Vanchiere et al. 2005). Serological studies have reported that 60–90% of the adult population harbor antibodies against JCV and BKV (from this point on JCV and BKV will be referred to as HPyVs) (Shah et al. 1973; Hirsch and Steiger 2003; Polo et al. 2004). Disease generally occurs only when the host's immune system becomes suppressed by conditions such as AIDS (Shah 1996; White et al. 2005).

Table 5.2 Common enteric viruses of humans

Genus	Common virus name(s)	Diameter (nm)	Capsid	Nucleic acid	Genome size (kbp)	Current types associated with human disease	Symptoms	References
Mastadenovirus	Adenovirus	80–100	Icosahedral, non-enveloped	Double-stranded DNA	35	51 Serotypes of the species: human adenovirus A, B, C, D, E, and F	Respiratory illness, conjunctivitis, and gastroenteritis	Murray et al. (2002); Büchen-Osmond (2006e); Jiang (2006)
Mamastrovirus	Astrovirus	28–33	Icosahedral, non-enveloped	Single-stranded, positive-sense RNA	6–8	8 Serotypes: human astrovirus 1, 2, 3, 4, 5, 6, 7, 8	Diarrhea, vomiting, and fever	Lee and Kurtz (1982); Walter and Mitchell (2003); Büchen-Osmond (2006f); Finkbeiner et al. (2008)
Enterovirus	Enterovirus, poliovirus, coxsackie-virus, echovirus	30	Icosahedral, non-enveloped	Single-stranded, positive-sense RNA	7–8.5	62 Serotypes: 3 polioviruses, 29 coxsackieviruses, 28 echoviruses, and 5 enteroviruses	May be asymptomatic or cause gastroenteritis, hand foot and mouth disease, respiratory illness, and conjunctivitis to meningitis, myocarditis, poliomyelitis, and paralysis	Oberste et al. (1999); Büchen-Osmond (2006b); Khetsuriani et al. (2006)
Hepatovirus	Hepatitis A virus	27	Icosahedral, non-enveloped	Single-stranded, positive-sense RNA	7.5	1 Serotype: hepatitis A virus	Acute hepatitis with fever, fatigue, nausea, abdominal pain, and jaundice	Adkins (1997); Murray et al. (2002); Büchen-Osmond (2006c)

Hepevirus	Hepatitis E virus	27–34	Icosahedral, non-enveloped	Single-stranded, positive-sense RNA	7.2	1 Serotype: hepatitis E virus	Similar to Hepatitis A, can lead to fulminant hepatic failure in pregnant women with ~20% mortality rate during the third trimester	Büchen-Osmond (2006g); Kuniholm et al. (2009)
Norovirus	Norovirus	35–39	Icosahedral, non-enveloped	Single-stranded, positive-sense RNA	7.3–7.7	3 Genogroups of 1 serotype: GI, GII, and GIV	Vomiting, watery nonbloody diarrhea, abdominal cramps and nausea	Büchen-Osmond (2006a); Patel et al. (2009)
Rotavirus	Rotavirus	60–80	Nonenveloped, triple-layer of protein	11 Segments of double-stranded RNA	11–25	3 Genotypes: RV-A, RV-B, and RV-C	Diarrhea, fever, vomiting, and in some cases severe dehydration	Murray et al. (2002); Büchen-Osmond (2006d); Phan et al. (2007); Greenberg and Estes (2009)
Polyomavirus	Polyomavirus	40–55	Icosahedral, non-enveloped	Double-stranded DNA	5	2 Species: BKV and JCV	Asymptomatic primary infection, latent renal infections; kidney nephratitis or progressive multifocal leukoencephalopathy in immunocompromised individuals	Khalil and Stoner (2001)

5.2.3 Enteric Viruses in Sewage

In general, enteric viruses establish infection in an individual, undergo replication, and are then released into the environment through the gastrointestinal (or urogenital) tract. The excretion of these viruses in feces or urine leads to the presence of various enteric viruses in communal waste systems such as sewage, and the potential presence of these viruses in smaller-scale waste handling systems (e.g., septic tanks, also called onsite wastewater treatment systems, and their drainfields). Recent studies using quantitative methods have found a range of enteric and nontraditional enteric viruses in raw sewage at varying concentrations (Table 5.1).

5.3 Viral Microbial Source Tracking Methods

In general, enteric viruses exhibit a high degree of host specificity, which has lead to the increased use of viruses as species-specific water quality indicators. In addition, discrepancies between viral presence and exceedance of bacterial regulatory standards in many studies have lead to the incorporation of virus-based methods for water quality studies by many MST researchers. The following is an overview of selected studies examining viral detection, concentration, method evaluation, and proposed methodologies for detecting human or animal fecal contamination in water.

5.3.1 Choosing the Right Virus

Choosing the right virus and the appropriate concentration method for quantification of enteric viruses from water has proven to be a challenge for water quality scientists. A consistent finding after decades of human and animal virus studies in aquatic systems is that not all viral markers are useful in all areas, and in some cases certain viruses exhibit distinct geographical and seasonal distributions. For example, Fong et al. (2005) consistently detected human adenoviruses in raw river water samples during the summer and fall months but did not detect the virus during the winter months. Moreover, Haramoto et al. (2006) conducted a study in Japan in which the group monitored the number of noroviruses in raw sewage throughout a year-long period. The group found norovirus concentrations ranging from 0.17 to 1,900 copies per mL, with the highest concentrations detected during the winter months (Haramoto et al. 2006). This study also enumerated FIB and found relatively consistent concentrations and the absence of seasonal distributions (Haramoto et al. 2006). While supporting the possibility of viral seasonal distribution, this finding further exemplified the disconnect between FIB and pathogenic viruses. A study conducted in Greece reported the detection of both enteroviruses and adenoviruses in raw sewage during the spring and fall months. The same study reported the absence of the hepatitis A virus (HAV) in all raw sewage samples taken over a 2-year period (Vantarakis and Papapetropoulou 1999). While a complete summary of the literature supporting the variance of viral distribution across temporal periods and geographical areas is outside the scope of this chapter, the cited works provide some pertinent examples.

In 2005, Fong and Lipp wrote a review in which they recounted major developments in molecular detection methods over the past decades, including applications of human and animal viruses to MST approaches (Fong and Lipp 2005). In particular, they present a summary of information of the myriad of animal viruses recently being applied to source tracking studies of animal fecal contamination, including bovine and porcine enteroviruses and adenoviruses (see review for specific references). Inclusion of specific animal virus methods and their application are important for advancement of the field, as it was noted that the hurdles to successful detection and/or quantification of animal viruses are similar to those encountered when quantifying human pathogenic viruses.

The first step in a prospective MST study is to determine which target viruses (whether human or animal) might be most useful for study in certain regions, based upon parameters such as viral incidence, seasonal distribution, and concentration in the local population and urban or agricultural wastewater. Fong et al. (2005) conducted a study of the presence of human adenoviruses, enteroviruses, and HAVs in sewage, environmental, and shellfish samples using PCR. The samples that they collected ranged from urban locations (raw sewage), slaughterhouse sewage, river, seawater, and shellfish tissue. Interestingly, they found that every sample that was positive for human enteroviruses or HAV also contained human adenoviruses. This study showed some level of agreement among the viral methods being applied. However, for most studies, it is necessary to do an initial screen of an array of waterborne enteric viruses, then base decisions for selection of the right virus on the seasonality of the viral patterns observed, the viral concentrations, the prevalence over a range of samples, and the goals of the project. Finally, it is important to consider the composition of the viruses. DNA viruses are relatively easy to work with in the laboratory compared to RNA-based viruses such as enteroviruses and noroviruses. If novice students and technicians are to conduct the sample processing, it is sometimes beneficial to start by working with DNA viruses, then move to RNA viruses as proficiency increases.

Once a suitable virus (or viruses) is identified the next vital step is to determine the appropriate concentration method. In the past, water quality researchers have thought that concentration of very large volumes of water (20–100 L) would be the most effective means for successful detection and quantification. An elegant study conducted by Rajal et al. (2007) showed that the use of large volumes of stormwater (70–881 L) and a two-step, hollow-fiber ultrafiltration procedure yielded only one sample out of 61 with measurable viruses. Filtration recovery efficiency ranged widely, from 9.7 to 97.9%. They also observed inhibition of the QPCR reactions in almost all samples that required dilutions from 10 to 500-fold to relieve. The authors developed an equation for the sample limit of detection that accounted for variables such as sample volume, filter recovery efficiency and inhibition.

Other scientists have suggested that using smaller volumes, with excellent recovery tactics, and strong use of controls are more beneficial to sensitive detection. Work conducted by Gregory et al. (2006) and He and Jiang (2005) demonstrated the successful use of a range of small-volume sample processing (<10 L), with detection limits for both studies reported at under 100 virus particles/100 mL. The type of water being analyzed is an important consideration for volume studied; drinking water (absent of materials that inhibit molecular analyses) volumes can be large (100–1,000 L), while for stormwater and eutrophic recreational waters, analysis of 250 mL to 5 L is recommended. For example, Fuhrman et al. (2005) assessed the recovery efficiencies of enteroviruses in natural water samples. After testing efficiencies of a range of methods for enterovirus quantification using either glass fiber filters (GF/F) or cellulose acetate/nitrate (HA) filters, they found HA filters to be more efficient. Those filters were used to concentrate 1-L volumes of urban creek samples. Using this relatively small volume, they reported acceptable enteroviral detection limits in complex environmental samples at concentrations of 10^2–10^3 virus particles per mL.

In any study, virus selection and determination of ideal sample volumes, along with detection limit assessment and assessment of potential inhibition problems (if using molecular techniques), should be conducted before sample processing is initiated. The following sections describe studies conducted with a range of different virus types and can be used to guide the virus selection for a particular study.

5.3.1.1 Enterovirus as a Water Quality Indicator

During the 1970s and up until 2000, the poliovirus vaccine was routinely administered entirely or in part as a live attenuated virus (http://www.cdc.gov/vaccines/vpd-vac/polio/vac-faqs.htm). Research conducted at that time found the circulation of this attenuated virus in raw sewage from various communities, and based on these findings Payment et al. (1979) suggested the use of these viruses as indicators of virological quality of water. Poliovirus has since been eradicated in USA, and only inactivated poliovirus vaccines are currently administered to the general public. However, the initial identification of these enteroviruses in communal waste samples paved the way for the use of other *Enterovirus* species as human-specific water quality indicators. Various studies have been conducted throughout the USA (Griffin et al. 1999; Jiang and Chu 2004; Wetz et al. 2004; Betancourt and Rose 2005; Fong et al. 2005; Gersberg et al. 2006) and across the world utilizing enteroviruses as water quality indicators (Guttman-Bass and Nasser 1984; Schwartzbrod et al. 1991; Gilgen et al. 1995; Pianetti et al. 2000; Pusch et al. 2005; Rose et al. 2006; Hsu et al. 2008; Lee and Lee 2008). Collectively, these studies reported the presence of human-specific enterovirus species in areas suspected of human fecal contamination.

In 2002, bovine enteroviruses (BEV) were proposed as an indicator of cattle and/or deer fecal contamination (Ley et al. 2002). Ley et al. (2002) used RT-PCR to detect BEV in fecal samples of cattle, animals surrounding the farm, and various water systems adjacent to the farm. They reported 76% of cattle fecal samples and 38% of white-tailed deer fecal samples analyzed were positive for BEV, indicating this virus may be a potential ruminant marker. In addition, various natural water systems adjacent to the cattle farm, as well as oysters downstream from the farm, were positive for BEV, illustrating the power of the virus to be used as a water quality indicator. The researchers also noted BEV was amplified from one of three Canadian geese feces. In 2005, Jimenez-Clavero et al. published similar findings with 78% of all cattle fecal samples positive for BEV (Jimenez-Clavero et al. 2005). In addition, the same group reported the detection of BEV in feces collected from sheep and goats, again illustrating the potential of BEV to predict ruminant fecal contamination.

5.3.1.2 Adenovirus as a Water Quality Indicator

In 1998, Pina et al. tested wastewater treatment plant influent samples for human adenoviruses, hepatitis A viruses, and enteroviruses. Adenoviruses were the only viruses consistently detected throughout the year (Pina et al. 1998). In addition, adenoviruses were routinely detected in environmental samples when fecal coliform concentrations were below regulatory standards (400 CFU/100 mL). Based on these results, Pina et al. suggested the use of adenoviruses to monitor viral contaminants in natural waters (1998). Since then, the presence of enteric adenovirus type 40 and 41 (Ad 40 and Ad 41) has been utilized as an indication of human fecal pollution in various studies across the United States. Jiang et al. (2001) examined the water quality at several California beaches receiving urban runoff. Four of the 12 sites sampled were positive for adenoviruses, with concentrations ranging from 0.9 to 7.5×10^3 genomes/L (Jiang et al. 2001). Of those four sites, only one site exceeded bacterial regulatory standards. In 2004, Jiang et al. found a regular occurrence of adenoviruses in rivers and creeks exposed to urban runoff in the California area (Jiang and Chu 2004). In this study, more than 60% of the samples positive for adenovirus did not meet the regulatory standards for FIB. The use of adenoviruses to assess water quality has not been limited to California; Fong et al. targeted human adenoviruses to determine sources of contamination in a Georgia river (Fong et al. 2005) as well as the Grand River in Michigan (Fong et al. 2010). Lipp et al. (2007) used the detection of adenovirus to assess the effects of contaminated groundwater transport on aquatic environment of coral reefs. Adenoviruses have also been successfully used to assess water quality in New Hampshire (Chapron et al. 2000). The use of adenoviruses to assess water quality spans the globe with successful utilization of the target in Germany (Pusch et al. 2005), South Africa (van Heerden et al. 2005), Japan (Haramoto et al. 2010a), and Korea (Lee et al. 2005).

In 2004, Maluquer de Motes et al. suggested the potential of porcine adenovirus (PAV) and bovine adenovirus (BAV) to indicate swine and cattle fecal contamination, respectively (Maluquer de Motes et al. 2004). The PCR methods detected PAV in 70% of swine samples and BAV in 75% of cattle samples analyzed. In 2006, Hundesa et al. reported 100% frequency of detection of PAV in swine samples ($n = 10$); however, only 10% of cattle fecal samples were positive by PCR for BAV (Hundesa et al. 2006). Hundesa et al. (2009) continued working with the PAV methodology and in 2009 published a real-time PCR assay to quantify PAV in animal and environmental samples. In this study, PAV was detected in 76–100% of swine feces from various areas and in 100% of slaughterhouse effluent. Moreover, the concentrations of PAV in feces and slaughterhouse effluent were relatively high (~6×10^5 GC/g and ~1.6×10^3 GC/mL, respectively). These studies illustrate the need to determine geographical distribution of viruses and their relative abundance in target species.

5.3.1.3 Human Polyomaviruses as Water Quality Indicators

In 2000, Bofill-Mas et al. (2000) designed two nested-PCR assays specific for JCV or BKV. This group then applied the newly developed assays to raw sewage samples from Europe and Africa and reported that JCV was detected in a high proportion of samples, prompting the proposal of JCV as a human-specific water quality indicator. McQuaig et al. (2006) next proposed the use a nested PCR assay to simultaneously detect both JCV and BKV (HPyVs) to increase the sensitivity of the assay. HPyVs were consistently detected in samples suspected of human contamination, and also their presence was also highly correlated with other human-specific molecular indicators (i.e., human-associated *Bacteroidetes* spp. and the *esp* gene of *Ent. faecium*) (McQuaig et al. 2006). Quantitative PCR (QPCR) assays have now been developed to both detect and quantify HPyVs (McQuaig et al. 2009; Fumian et al. 2010). The use of HPyVs as a human-specific viral marker is increasing with studies reporting the presence of these viruses in raw sewage samples and/or environmental waters from Florida (McQuaig et al. 2009; Leskinen et al. 2010), California (Rafique and Jiang 2008; McQuaig et al. 2009), Australia (Ahmed et al. 2009, 2010), Spain (Albinana-Gimenez et al. 2009), Germany (Hamza et al. 2009) and Japan (Haramoto et al. 2010b).

5.3.2 Field Studies Employing Viral Markers

PCR is becoming an increasingly accessible research tool for those investigating viral contaminants in environmental waters. For example, Fong et al. (2005) published a study using human and bovine enteroviruses, and human adenoviruses in the urbanized areas of the Altamaha River, Georgia. They used end-point (presence–absence) approaches, including PCR (RT-PCR), and nested-PCR, and determined that the human viruses co-occurred in 26% of samples and that human enteroviruses were present in over half of the samples collected (Fong and Lipp 2005).

In 2006, Noble et al. published a study conducted in Ballona Creek, California. This area is a highly urbanized watershed comprising most of the populated area that is west of downtown Los Angeles and has about 85% impervious surface coverage. The researchers used a tiered approach combining a mass-based design at six mainstem sites and four major tributaries. The mass-based design utilized the hydrologic budget to determine if most of the flow was sampled, and was conducted by comparing the volume from each of the tributaries to the volumetric discharges along the mainstem of Ballona Creek. Next, trends in the flux of FIB cells per hour were determined by multiplying the FIB concentration by the flow rate. The mean hourly flux was calculated by averaging the flux of FIB at the locations in the mainstem each hour. The flux at each location was then calculated by averaging the flux at each main-stem or tributary site for hourly. Culture-based FIB quantification and simultaneous assessment of flow were used to quantify the flux of enterococci and *E. coli*. Human-associated *Bacteroidetes* spp. and human pathogenic enteroviruses were measured by QPCR. Ninety-seven percent of the flow was attributed to monitored inputs from several tributaries, which were all far upstream from the Pacific Ocean outlet. FIB concentrations were ubiquitously high throughout Ballona Creek, and no single tributary appeared to dominate the fecal bacteria inputs. An important component of this study was the ability to fully quantify human pathogenic enterovirus, using appropriate controls (Gregory et al. 2006). The study showed that enterovirus and the human-associated *Bacteroidetes* marker presence agreed on the presence of human sewage in 92% of samples, confirming the presence of human fecal contamination.

Another case study conducted in southern California using multiple approaches and including a viral marker was presented by Jiang et al. (2007a, b). Three completely different microbial source tracking approaches were used: library-dependent antibiotic resistance analysis (ARA), PCR-based detection of human pathogenic enteroviruses and adenoviruses, and *E. coli* toxin gene assessment for both humans and animals (Jiang et al. 2007b). The approach was conducted in a small watershed in southern California. ARA of *Enterococcus* spp. and *E. coli* toxin gene results determined that the primary sources of fecal contamination were from animals. Only two samples from the entire study were positive for human adenoviruses; however, when adenoviruses were detected, the concentration of FIB in the sample was negligible. This approach, using multiple lines of evidence, led the researchers to be able to use the measurements of the human viruses in a converse sense, to determine that the dominant source of the FIB contamination in the area was not related to human fecal contamination.

Ahmed et al. (2010) incorporated the use of both human and animal adenoviruses to assess the source of contamination in river water samples. In addition, the study enumerated the following zoonotic pathogens: *C. jejuni*, *Salmonella* spp., and *G. intestinalis*. This study also assessed presence and quantity of microbial agents during wet and dry conditions (Ahmed et al. 2010). The results included the observation that human-specific adenoviruses (HAV) and bovine-specific adenoviruses (BAV) were not codetected in any samples. Samples positive for HAV by PCR did not exceed *E. coli* and enterococci levels of 160 CFU/100 mL or 390 CFU/100 mL, respectively. In contrast, samples positive by PCR for BAV had an average fecal bacteria concentration of 1,596 CFU/100 mL. In addition, samples positive by PCR for BAV were also positive for at least one other zoonotic pathogen. This study illustrates the possible potency of viral markers to detect and accurately predict sources of fecal contamination. This study also shows the analytical power of viral indicators to predict the presence of other pathogens.

In 2007, a study was conducted at Doheny and Avalon beaches in California (McQuaig 2009). The study assessed water quality every Friday, Saturday, and

Sunday (high beach-use days) over a 4-month period in the summer. Water samples were analyzed for FIB concentrations (total coliforms, fecal coliforms, and enterococci) and for the presence of four human-associated markers (*Methanobrevibacter smithii*, human-associated *Bacteroidetes*, human polyomaviruses, and adenoviruses). Adenovirus and human polyomaviruses were highly correlated at sites impacted by human input. In addition, in areas where failing septic systems were suspected, human polyomaviruses were the most prevalent, suggesting that these viruses may be very useful indicators in areas suspected of contamination due to faulty septic tanks.

A field study has been recently conducted at a Florida beach impacted by non-point source runoff (Abdelzaher et al. 2010). Water and sand samples were collected during high- and low-tide conditions over four sampling periods. Included in that study were measurements of enterococci, *E. coli*, *C. perfringens*, and *S. aureus*, norovirus, HAV, enterovirus, and HPyVs. During that study of a wide array of measured viruses, very few virus targets were quantifiable. Only HPyVs were quantified, and only on one particular day of the entire study. This study points out the difficulty with conducting large-scale studies of multiple indicators with infrequent detection and lack of agreement among markers, and the potential problems still associated with specific methods used to concentrate and quantify human viruses. In particular, it demonstrates the importance of quantitatively assessing recovery efficiency of specific virus concentration protocols and that there is no universally applicable, successful viral concentration approach for aquatic samples.

5.3.3 Detection and Quantification Methods

In the past decades, the viral detection and quantification assays recognized by regulatory agencies for drinking water analysis have been based upon cell-culture of pathogenic viruses. The total cultivatable virus assay (TCVA) has been the industry standard, but it is extremely time-consuming (>14 d), costly, insensitive, and at best only semiquantitative (US Environmental Protection Agency 1995). There is no regulatory standard in the USA for viral parameters in recreational water and stormwater, and therefore, virus quantification methods have largely evolved on an ad hoc, rather than agency-guided basis. Most approaches have emphasized human pathogenic viruses as the intended target. In 2003, Griffin et al. reviewed the literature regarding the use of human pathogenic viruses as MST tools (Griffin et al. 2003). Early in the decade, PCR-based methods were used in water quality studies to confirm the presence of human pathogenic viruses stemming from human fecal contamination. The methods were largely presence–absence or semiquantitative and were used as indicators of the presence of fecal contamination. During this decade, end-point (presence–absence) PCR detection of human viruses, including enteroviruses, adenoviruses, HAV, and others, gave way to quantification utilizing QPCR (Gregory et al. 2006; Jiang 2006; Villar et al. 2006; McQuaig et al. 2009).

More recently, additional methods have been developed that are fully quantitative and can be used in conjunction with other approaches to confirm the presence of human fecal contamination, to identify the specific viral types in a water sample (via sequence-based analysis), and even be combined with flow information to develop schemes for loading of microbial contaminants to receiving waters. Microarrays have also been suggested to target various water quality indicators simultaneously (Lemarchand et al. 2004).

While nucleic-acid based techniques have shown great promise, many water quality managers have speculated that many of the human viruses found to be present in natural waters were not infective and, therefore, not a public health risk concern. In the past decade, QPCR and quantitative reverse-transcriptase PCR (QRTPCR)-based methods have been applied successfully to human viral pathogens and have included integrated cell culture (ICC)-based methods to determine infectivity of the detected virus particles. Based on these types of infectivity studies, there has recently been a shift in the mindset of water quality managers with the acceptance of nucleic-acid-based methods as excellent indicators of specific types of fecal contamination in aquatic systems (Fong and Lipp 2005).

5.3.4 Concentration

While virus detection and quantification methods in water quality assessments have evolved and grown in number in the past decades (Sobsey and Glass 1984; Fong and Lipp 2005), only incremental advancements have been made as related to novel approaches for sample concentration, detection of low target numbers of viruses, and applicability of methods across water matrices. Human pathogenic viruses in recreational and bathing waters differ from their bacterial pathogen counterparts in that their infectious doses are orders of magnitude less (i.e., approximately 1–10 viruses vs. >1,000 *Salmonella typhi* upon ingestion) (Murray et al. 2002; Bitton 2005); therefore, dilute concentrations of pathogenic viruses are more likely to cause infection than similar concentrations of pathogens with higher infectious doses. Furthermore, a small percentage of any given human population will be shedding pathogenic viruses, so they are at relatively dilute concentrations in sewage. Consequently, one of the largest hurdles to accurate quantification of pathogenic viruses has been sample concentration and processing.

In past decades, it was common for water quality scientists and researchers alike to attempt to concentrate large volumes of sample water (10–40 L) using a range of approaches such as tangential flow filtration (Alonso et al. 1999; Jiang et al. 2001), vortex flow filtration (Paul et al. 1991; Tsai et al. 1993, 1994) and even large-scale precipitation (Dahling and Wright 1986; Shields and Farrah 1986; Payment et al. 1989). Dong et al. (2009) used a dialysis filtration approach combined with a range of PCR-based approaches to detect and quantify human adenovirus in an array of environmental water types (Dong et al. 2009). In this study, volumes ranging from 50 to 200 L were concentrated using hollow-fiber microfiltration and variable rates of recovery were reported in natural waters (sea and stream water) after seeding the samples with MS2 coliphage (recovery ranged from 23 to 98%). The method used did not perform as well for stormwater samples, showing much lower recovery rates. The researchers relied on the use of a range of only semiquantitative PCR-based methods, which worked with varying levels of success. It was noted that only periodic detection of adenoviruses occurred in the environmental waters studied, which presented a problem in the evaluation of the methods in natural environmental waters.

Jiang et al. (2001) compared the efficiency of tangential flow filtration (TFF) and vortex flow filtration (VFF) for viral recovery using PCR. The two ultrafiltration concentration methods were compared using six environmental samples. Human adenoviruses were detected in 0% and 17% in the VFF and TFF samples, respectively. In this study, it was reported adenoviruses were concentrated by 10^5-fold using TFF in combination with Centriprep-30; however, PCR inhibition was observed in both TFF and VFF samples seeded with adenoviruses, while a 1:5 dilution in DI water decreased PCR inhibition. To remove PCR-interfering substances, GuSCN-silica bead extraction was used to purify viral nucleic-acid extracts (Jiang et al. 2001). After extraction, three of the six VFF samples that were initially negative tested positive for adenovirus. Olszewski st al. (2005) seeded 2-L water samples with human enteroviruses and concentrated the samples using either hollow-fiber filtration (HFF) or TFF. Enteroviruses were detected using cell culture plaque assay. In this study, enteroviruses recoveries were 82% and 95% for HFF and TFF, respectively (Olszewski et al. 2005). Despite the relatively high recovery rates, the researchers noted persistent clogging of filters with turbid water samples and suggested prefiltering water samples. It is important to note that prefiltration may also lead to a decrease of viral recovery, since viruses tend to be particle associated in the environment (Bitton 2005).

During the early 2000s, QPCR was not in widespread use for water quality assessments, so most scientists used PCR for detection, either coupled or uncoupled with cell culture. It was widely found that PCR and cell culture analyses alike were inhibited in natural water samples due to the concentration of organic matter, phytoplankton exudates, sand, total suspended solids, and similar materials (Jiang et al. 2001). Furthermore, decades old methods using 1MDS electropositive filters (Sobsey and Glass 1984) were known to suffer from clogging and recovery problems. In the last decade, a plethora of combined sample concentration approaches have evolved; however, issues remain in that some procedures work better on certain types of samples than others, and no single concentration method has been found to work equally across marine, fresh, storm, waste, or brackish waters. In addition, methods for viral quantification in alternative matrices, such as oyster tissue or sediment, have been even more problematic. The following is a review of some of the most recent advancements and modifications to sample processing. The main take-home message for those designing a new viral quantification assay in natural waters is to learn from the research conducted on similar water matrices, virus types, and quantification approaches, as a wealth of literature exists on the subject.

In 2002, Katayama et al. reported a viral elution method that has been successfully adapted and used by an array of researchers (Gregory et al. 2006; Jiang et al. 2007a). They used a negatively charged membrane filter and rinsed the membrane with 0.5 mM H_2SO_4 (pH 3.0) to remove cations before eluting the viruses with 1 mM NaOH (pH 10.5). They reported recoveries of 33–95% for poliovirus when they applied this approach to purified water and recoveries of 38–89% when the method was utilized on natural seawater. They also conducted a comparison with positively charged membranes and found that their approach strongly outperformed the positively charged membrane. An important facet of this methodological improvement was the lack of beef extract to elute the viruses from the membrane. Beef extract elution had been previously a commonly used strategy, but upon the widespread application of PCR, RT-PCR, and QPCR for amplification of virus signals, inhibition of enzyme activity was frequently noted (Tsai et al. 1993; Burgener et al. 2003). Their approach was clearly superior for amplification of RNA templates (RT PCR) of human pathogenic viruses such as enterovirus, norovirus, and HAV. This group also focused on the filtration of smaller volumes of seawater and other source waters (e.g., 50–1,000 mL) than had previously been used. Instead, they relied on high recovery rates of the viruses and sensitive detection methods. More recently, Rigotto et al. (2009) have conducted an additional study of the same negatively charged "HA" filters used by Katayama et al. (2002). They conducted their study on human adenoviruses and HAV in a range of waters, from distilled water, treated wastewater, seawater and recreational lagoon water (Rigotto et al. 2009). They found nearly 100% recovery for distilled water and treated wastewater but very different results depending on virus type and water matrix for the seawater and recreational waters. This very recent study highlights the variable performance of specific approaches on different sample types and the need for optimization of approaches for the water type in question.

In 2010, Haramoto et al. reported on the recovery of human norovirus from water samples by a variety of virus concentration methods (Haramoto et al. 2010b). The aim of the study was to compare the use of the 1MDS electropositive filter (presented by Sobsey and Glass 1984, and utilized in the regulatory-approved EPA TCVA-based method (American Public Health Association 1998)) and two filtration methods using electronegative filters. The effect of added magnesium and aluminum on recovery was also assessed. They conducted recovery tests of human norovirus and poliovirus in an array of different water sample types, ranging from nanopure water to river water. Norovirus recovery ranged from 100% in nanopure water to 15% in river water when magnesium was added. Results were similar for poliovirus. They had little to no success with the electropositive 1 MDS filter for either norovirus or poliovirus, showing lower rates of recovery, which is surprising, as it has been held as one of the standard virus concentration methods for water assessment for decades (US Environmental Protection Agency 1995).

In summary, many methods exist for the concentration of viruses from environmental waters. Careful consideration must be exercised when choosing a concentration method. The methods must be chosen based on type of water sampled, target virus concentration, particle concentration and type of detection/quantification assay.

5.3.5 *Method Evaluation*

A publication by Rajal et al. (2007) details simultaneously the challenges of quantifying human and animal pathogenic viruses in water samples, and the relative merits of specific approaches. In a study of 61 stormwater samples collected from an array of storm drains in California, viruses were concentrated using a hollow-fiber ultrafiltration protocol (Rajal et al. 2007). Both human enteroviruses and adenoviruses were quantified using Taqman™-based QPCR methods. Only one sample was positive for Adenovirus 40/41. The authors calculated the sample specific limit of detection based upon the starting volume, a tracer virus added to the concentration process to specifically follow recovery, and also by understanding the efficiency of the QPCR reactions. This publication was the first attempt to assess the effectiveness of each step of virus quantification from sample collection to nucleic acid amplification. In 2006, Gregory et al. reported the first use of a competitive internal positive control (CIPC) to aid in quantification of RNA viruses for water quality applications. The purpose of this internal control was to assess the efficiency of the combined reverse-transcriptase and PCR steps for enterovirus quantification and to provide a scenario for CIPC design approaches for other virus types. Incorporating the use of an internal positive control may allow for accurate assessments of assay efficacy and possible inhibition from environmental samples, thereby increasing the accuracy and predictive power of any assay. The approach outlined in Gregory et al. (2006) was also recently applied to norovirus quantification in aquatic samples using a similarly designed CIPC, and represents an significant advancement in quantification of viral targets (Gregory et al. 2011).

While not all viruses in a water sample might be infectious, accurate quantification of human pathogenic viruses is vital as this information can be coupled with water flow measurements thereby permiting loading estimates of potentially pathogenic material, specific to source. This information can be used to make informed decisions for contamination control, and can permit water quality managers to prioritize impaired systems. The accurate quantification of human pathogenic viruses is also an important parameter in accurate estimation of public health risk and can be a key component to Quantitative Microbial Risk Assessment (QMRA). This type of approach was applied by Wong et al. (2009) in the Great Lakes area, using a most probable number approach system for semiquantification of human adenoviruses. A fully quantitative approach to this study would have been even more powerful. Certainly, that capability is coming in the near future for human virus quantification research in recreational waters.

5.4 Conclusions and Recommendations

The advantages of utilizing pathogenic viruses for water quality management are significant. Viruses offer a high degree of host specificity, and detection of various source-specific viral targets can solidify the determination of the source of contamination. Human pathogenic virus quantification in stormwater samples and contaminated recreational waters can be an important tool for quantifying contaminant load and potential human health risk, respectively. Recent methodological developments suggest that smaller volumes of environmental samples can be analyzed with confidence, making these methods more practical than previously thought. The recent incorporation of better internal and external controls in quantitative methods such as QPCR improves the accuracy of the methods. Moreover, the incorporation of quantitative viral methods into QMRA approaches improves our understanding of the relative public health risk associated with sewage inputs vs. stormwater inputs. As with any MST methodology, careful consideration of the water type, potential for inhibitors of molecular methods, expected concentration in source water samples, and detection and quantification approaches combined with appropriate controls is vital to success. The need for statistically rigorous study designs with appropriate numbers of samples collected to give statistical power, and replication of sample analysis, along with collection of other relevant environmental data, extend to all viral quantification and detection approaches. Increased research addressing the prevalence, distribution, and stability of these viruses in environmental waters will lead to more understanding and confidence in applying of viral pathogen-based approaches. Viral markers are invaluable for assessment of potential public health risk, prioritization for water quality structural improvements, and development of mitigation scenarios.

References

Abdelzaher, A.M., Wright, M.E., Ortega, C., Solo-Gabriele, H.M., Miller, G., Elmir, S., Newman, X., Shih, P., Bonilla, J.A., Bonilla, T.D., Palmer, C.J., Scott, T., Lukasik, J., Harwood, V.J., McQuaig, S., Sinigalliano, C., Gidley, M., Plano, L.R., Zhu, X., Wang, J.D. and Fleming, L.E. (2010) Presence of pathogens and indicator microbes at a non-point source subtropical recreational marine beach. *Appl Environ Microbiol* **76**, 724–732.

Adkins, S.B. (1997) Immunizations: current recommendations. *Am Fam Physician* **56**, 865–874.

Ahmed, W., A.Goonetilleke and T.Gardner (2010) Human and bovine adenoviruses for the detection of source-specific fecal pollution in coastal waters in Australia. *Water Res* **44**, 4662–4673.

Ahmed, W., Goonetilleke, A., Powell, D., Chauhan, K. and Gardner, T. (2009) Comparison of molecular markers to detect fresh sewage in environmental waters. *Water Res* **43**, 4908–4917.

Albinana-Gimenez, N., Clemente-Casares, P., Calgua, B., Huguet, J.M., Courtois, S. and Girones, R. (2009) Comparison of methods for concentrating human adenoviruses, polyomavirus JC and noroviruses in source waters and drinking water using quantitative PCR. *J Virol Methods* **158**, 104–109.

Alonso, M.C., Rodriguez, J. and Borrego, J.J. (1999) Enumeration and isolation of viral particles from oligotrophic marine environments by tangential flow filtration. *Int Microbiol* **2**, 227–232.

American Public Health Association (1998) *Standard Methods for the Examination of Water and Wastewater*. Maryland. pp. 9.137–9.141: American Public Health Association.

Arthur, R.R., Dagostin, S. and Shah, K.V. (1989) Detection of BK virus and JC virus in urine and brain tissue by the polymerase chain reaction. *J Clin Microbiol* **27**, 1174–1179.

Barwick, R.S., Levy, D.A., Craun, G.F., Beach, M.J. and Calderon, R.L. (2000) Surveillance for waterborne-disease outbreaks--United States, 1997-1998. *MMWR CDC Surveill Summ* **49**, 1–21.

Betancourt, W.Q. and Rose, J.B. (2005) Microbiological assessment of ambient waters and proposed water sources for restoration of a Florida wetland. *J Water Health* **3**, 89–100.

Bialasiewicz, S., Whiley, D.M., Lambert, S.B., Nissen, M.D. and Sloots, T.P. (2009) Detection of BK, JC, WU, or KI polyomaviruses in faecal, urine, blood, cerebrospinal fluid and respiratory samples. *J Clin Virol* **45**, 249–254.

Bitton, G. (2005) *Wastewater Microbiology*. Hoboken, N.J.: Wiley, Johns & Sons, Incorporated.

Blatchley, E.R., 3rd, Gong, W.L., Alleman, J.E., Rose, J.B., Huffman, D.E., Otaki, M. and Lisle, J.T. (2007) Effects of wastewater disinfection on waterborne bacteria and viruses. *Water Environ Res* **79**, 81–92.

Bofill-Mas, S., Formiga-Cruz, M., Clemente-Casares, P., Calafell, F. and Girones, R. (2001) Potential transmission of human polyomaviruses through the gastrointestinal tract after exposure to virions or viral DNA. *J Virol* **75**, 10290–10299.

Bofill-Mas, S., Pina, S. and Girones, R. (2000) Documenting the epidemiologic patterns of polyomaviruses in human populations by studying their presence in urban sewage. *Appl Environ Microbiol* **66**, 238–245.

Büchen-Osmond, C. (2006a) ICTVdB Management (2006) 00.012.0.03. *Norovirus. In: ICTVdB – The Universal Virus Database, version 4 Columbia University, New York, USA* http://wwwncbinlmnihgov/ICTVdb/Ictv/fs_indexhtm.

Büchen-Osmond, C. (2006b) ICTVdB Management (2006) 00.052.0.01. *Enterovirus. In: ICTVdB – The Universal Virus Database, version 4 Columbia University, New York, USA* http://wwwncbinlmnihgov/ICTVdb/Ictv/fs_indexhtm.

Büchen-Osmond, C. (2006c) ICTVdB Management (2006) 00.052.0.03. *Hepatovirus. In: ICTVdB – The Universal Virus Database, version 4 Columbia University, New York, USA* http://wwwncbinlmnihgov/ICTVdb/Ictv/fs_indexhtm.

Büchen-Osmond, C. (2006d) ICTVdB Management (2006) 00.060.0.03. *Rotavirus. In: ICTVdB – The Universal Virus Database, version 4 Columbia University, New York, USA* http://wwwncbinlmnihgov/ICTVdb/Ictv/fs_indexhtm.

Büchen-Osmond, C. (2006e) Index of Viruses – *Adenoviridae* (2006). *In: ICTVdB – The Universal Virus Database, version 4 Columbia University, New York, USA* http://wwwncbinlmnihgov/ICTVdb/Ictv/fs_indexhtm.

Büchen-Osmond, C. (2006f) Index of Viruses – *Astroviridae* (2006). *In: ICTVdB – The Universal Virus Database, version 4 Columbia University, New York, USA* http://wwwncbinlmnihgov/ICTVdb/Ictv/fs_indexhtm.

Büchen-Osmond, C. (2006g) Index of Viruses – *Hepevirus* (2006). *In: ICTVdB - The Universal Virus Database, version 4 Columbia University, New York, USA* http://wwwncbinlmnihgov/ICTVdb/Ictv/fs_indexhtm.

Burgener, M., Candrian, U. and Gilgen, M. (2003) Comparative evaluation of four large-volume RNA extraction kits in the isolation of viral RNA from water samples. *J Virol Methods* **108**, 165–170.

Centers for Disease Control and Prevention (2003) Norovirus Activity-United States, 2002. *MMWR* **52,** 41–45.

Chapron, C.D., Ballester, N.A., Fontaine, J.H., Frades, C.N. and Margolin, A.B. (2000) Detection of astroviruses, enteroviruses, and adenovirus types 40 and 41 in surface waters collected and evaluated by the information collection rule and an integrated cell culture-nested PCR procedure. *Appl Environ Microbiol* **66**, 2520–2525.

Colford, J.M., Wade, T.J., Schiff, K.C., Wright, C.C., Griffith, J., Sandhu, S.K., Burns, S., Sobsey, M.D., Lovelace, G.L. and Weisberg, S.B. (2007) Water quality indicators and the risk of illness at beaches with nonpoint sources of fecal contamination. *Epidemiology* **18**, 27–35.

da Silva, A.K., Le Saux, J.C., Parnaudeau, S., Pommepuy, M., Elimelech, M. and Le Guyader, F.S. (2007) Evaluation of removal of noroviruses during wastewater treatment, using real-time reverse transcription-PCR: different behaviors of genogroups I and II. *Appl Environ Microbiol* **73**, 7891–7897.

Dahling, D.R. and Wright, B.A. (1986) Recovery of viruses from water by a modified flocculation procedure for second-step concentration. *Appl Environ Microbiol* **51**, 1326–1331.

Dong, Y., Kim, J. and Lewis, G.D. (2009) Evaluation of methodology for detection of human adenoviruses in wastewater, drinking water, stream water and recreational waters. *J Appl Microbiol* **108**, 800–809.

Dorries, K. (1998) Molecular biology and pathogenesis of human polyomavirus infections. *Dev Biol Stand* **94**, 71–79.

Dziuban, E.J., Liang, J.L., Craun, G.F., Hill, V., Yu, P.A., Painter, J., Moore, M.R., Calderon, R.L., Roy, S.L. and Beach, M.J. (2006) Surveillance for waterborne disease and outbreaks associated with recreational water--United States, 2003-2004. *MMWR Surveill Summ* **55**, 1–30.

Finkbeiner, S.R., Kirkwood, C.D. and Wang, D. (2008) Complete genome sequence of a highly divergent astrovirus isolated from a child with acute diarrhea. *Virol J* **5**, 117.

Fong, T., Phanikumar, M., Xagoraraki, I. and Rose, J.B. (2010) Quantitative detection of human adenoviruses 1 in waste water and two combined sewer overflows influencing a Michigan river. *Appl Environ Microbiol* **76**(3): 715–723.

Fong, T.T., Griffin, D.W. and Lipp, E.K. (2005) Molecular assays for targeting human and bovine enteric viruses in coastal waters and their application for library-independent source tracking. *Appl Environ Microbiol* **71**, 2070–2078.

Fong, T.T. and Lipp, E.K. (2005) Enteric viruses of humans and animals in aquatic environments: health risks, detection, and potential water quality assessment tools. *Microbiol Mol Biol Rev* **69**, 357–371.

Fuhrman, J.A., Liang, X. Noble, R.T. (2005) Rapid detection of enteroviruses in small volumes of natural waters by real-time quantitative reverse transcriptase PCR. *Appl Environ Microbiol* **71**(8): 4523–4530.

Fumian, T.M., Guimaraes, F.R., Pereira Vaz, B.J., da Silva, M.T., Muylaert, F.F., Bofill-Mas, S., Girones, R., Leite, J.P. and Miagostovich, M.P. (2010) Molecular detection, quantification and characterization of human polyomavirus JC from waste water in Rio De Janeiro, Brazil. *J Water Health* **8**, 438–445.

Gersberg, R.M., Rose, M.A., Robles-Sikisaka, R. and Dhar, A.K. (2006) Quantitative detection of hepatitis a virus and enteroviruses near the United States-Mexico border and correlation with levels of fecal indicator bacteria. *Appl Environ Microbiol* **72**, 7438–7444.

Gilgen, M., Wegmuller, B., Burkhalter, P., Buhler, H.P., Muller, U., Luthy, J. and Candrian, U. (1995) Reverse transcription PCR to detect enteroviruses in surface water. *Appl Environ Microbiol* **61**, 1226–1231.

Greenberg, H.B. and Estes, M.K. (2009) Rotaviruses: from pathogenesis to vaccination. *Gastroenterology* **136**, 1939–1951.

Gregory, J.B., Litaker, R.W. and Noble, R.T. (2006) Rapid one-step quantitative reverse transcriptase PCR assay with competitive internal positive control for detection of enteroviruses in environmental samples. *Appl Environ Microbiol* **72**, 3960–3967.

Gregory, J.B., Webster, L.F., Griffith, J.F. and Stewart, J.R. (2011) Improved detection and quantitation of norovirus from water. *Viro Methods* **172**, (1–2): 38–45.

Griffin, D.W., Donaldson, K.A., Paul, J.H. and Rose, J.B. (2003) Pathogenic human viruses in coastal waters. *Clin Microbiol Rev* **16**, 129–143.

Griffin, D.W., Gibson, C.J., 3rd, Lipp, E.K., Riley, K., Paul, J.H., 3rd and Rose, J.B. (1999) Detection of viral pathogens by reverse transcriptase PCR and of microbial indicators by standard methods in the canals of the Florida Keys. *Appl Environ Microbiol* **65**, 4118–4125.

Guttman-Bass, N. and Nasser, A. (1984) Simultaneous concentration of four enteroviruses from tap, waste, and natural waters. *Appl Environ Microbiol* **47**, 1311–1315.

Hamza, I.A., Jurzik, L., Stang, A., Sure, K., Uberla, K. and Wilhelm, M. (2009) Detection of human viruses in rivers of a densly-populated area in Germany using a virus adsorption elution method optimized for PCR analyses. *Water Res* **43**, 2657–2668.

Haramoto, E., Katayama, H., Oguma, K., Yamashita, H., Tajima, A., Nakajima, H. and Ohgaki, S. (2006) Seasonal profiles of human noroviruses and indicator bacteria in a wastewater treatment plant in Tokyo, Japan. *Water Sci Technol* **54**, 301–308.

Haramoto, E., Kitajima, M., Katayama, H. and Ohgaki, S. (2010) Real-time PCR detection of adenoviruses, polyomaviruses, and torque teno viruses in river water in Japan. *Water Res* **44**, 1747–1752.

Havelaar, A.H., Pot-Hogeboom, W.M., Furuse, K., Pot, R. and Hormann, M.P. (1990) F-specific RNA bacteriophages and sensitive host strains in faeces and wastewater of human and animal origin. *J Appl Bacteriol* **69**, 30–37.

He, J. and Jiang, S. (2005) Quantification of enterococci and human adenoviruses in environmental samples by real-time PCR. *Appl Env Micro* **71**, 2250–2255.

Hijnen, W.A., Beerendonk, E.F. and Medema, G.J. (2006) Inactivation credit of UV radiation for viruses, bacteria and protozoan (oo)cysts in water: a review. *Water Res* **40**, 3–22.

Hirsch, H.H. and Steiger, J. (2003) Polyomavirus BK. *Lancet Infect Dis* **3**, 611–623.

Hogan, T.F., Padgett, B.L., Walker, D.L., Borden, E.C. and McBain, J.A. (1980) Rapid detection and identification of JC virus and BK virus in human urine by using immunofluorescence microscopy. *J Clin Microbiol* **11**, 178–183.

Hsu, B.M., Chen, C.H. and Wan, M.T. (2008) Prevalence of enteroviruses in hot spring recreation areas of Taiwan. *FEMS Immunol Med Microbiol* **52**, 253–259.

Hundesa, A., Maluquer de Motes, C., Albinana-Gimenez, N., Rodriguez-Manzano, J., Bofill-Mas, S., Sunen, E. and Rosina Girones, R. (2009) Development of a qPCR assay for the quantification of porcine adenoviruses as an MST tool for swine fecal contamination in the environment. *J Virol Methods* **158**, 130–135.

Hundesa, A., Maluquer de Motes, C., Bofill-Mas, S., Albinana-Gimenez, N. and Girones, R. (2006) Identification of human and animal adenoviruses and polyomaviruses for determination of sources of fecal contamination in the environment. *Appl Environ Microbiol* **72**, 7886–7893.

Jiang, S., Noble, R. and Chu, W. (2001) Human adenoviruses and coliphages in urban runoff-impacted coastal waters of Southern California. *Appl Environ Microbiol* **67**, 179–184.

Jiang, S.C. (2006) Human adenoviruses in water: occurrence and health implications: a critical review. *Environ Sci Technol* **40**, 7132–7140.

Jiang, S.C. and Chu, W. (2004) PCR detection of pathogenic viruses in southern California urban rivers. *J Appl Microbiol* **97**, 17–28.

Jiang, S.C., Chu, W. and He, J.W. (2007a) Seasonal detection of human viruses and coliphage in Newport Bay, California. *Appl Environ Microbiol* **73**, 6468–6474.

Jiang, S.C., Chu, W., Olson, B.H., He, J.W., Choi, S., Zhang, J., Le, J.Y. and Gedalanga, P.B. (2007b) Microbial source tracking in a small southern California urban watershed indicates wild animals and growth as the source of fecal bacteria. *Appl Microbiol Biotechnol* **76**, 927–934.

Jimenez-Clavero, M.A., Escribano-Romero, E., Mansilla, C., Gomez, N., Cordoba, L., Roblas, N., Ponz, F., Ley, V. and Saiz, J.C. (2005) Survey of bovine enterovirus in biological and environmental samples by a highly sensitive real-time reverse transcription-PCR. *Appl Environ Microbiol* **71**, 3536–3543.

Katayama, H., Haramoto, E., Oguma, K., Yamashita, H., Tajima, A., Nakajima, H. and Ohgaki, S. (2008) One-year monthly quantitative survey of noroviruses, enteroviruses, and adenoviruses in wastewater collected from six plants in Japan. *Water Res* **42**, 1441–1448.

Katayama, H., Shimasaki, A. and Ohgaki, S. (2002) Development of a virus concentration method and its application to detection of enterovirus and norwalk virus from coastal seawater. *Appl Environ Microbiol* **68**(3): 1033–1039.

Khalil, S.M. and Stoner, G.L. (2001) *Human Polyomaviruses: Molecular and Clinical Perspectives.* New York: Wiley-Liss Inc.

Khetsuriani, N., Lamonte-Fowlkes, A., Oberst, S. and Pallansch, M.A. (2006) Enterovirus surveillance-United States, 1970-2005. *MMWR Surveill Summ* **55**, 1–20.

Kramer, M.H., Herwaldt, B.L., Craun, G.F., Calderon, R.L. and Juranek, D.D. (1996) Surveillance for waterborne-disease outbreaks--United States, 1993-1994. *MMWR CDC Surveill Summ* **45**, 1–33.

Kuniholm, M.H., Purcell, R.H., McQuillan, G.M., Engle, R.E., Wasley, A. and Nelson, K.E. (2009) Epidemiology of Hepatitis E Virus in the United States: Results from the Third National Health and Nutrition Examination Survey, 1988-1994. *J Infect Dis* **200**, 48–56.

Kunitake, T., Kitamura, T., Guo, J., Taguchi, F., Kawabe, K. and Yogo, Y. (1995) Parent-to-child transmission is relatively common in the spread of the human polyomavirus JC virus. *J Clin Microbiol* **33**, 1448–1451.

Lee, G. and Lee, C. (2008) Molecular detection and characterization of human enteroviruses in Korean surface water. *J Microbiol* **46**, 319–324.

Lee, S.H., Lee, C., Lee, K.W., Cho, H.B. and Kim, S.J. (2005) The simultaneous detection of both enteroviruses and adenoviruses in environmental water samples including tap water with an integrated cell culture-multiplex-nested PCR procedure. *J Appl Microbiol* **98**, 1020–1029.

Lee, S.H., Levy, D.A., Craun, G.F., Beach, M.J. and Calderon, R.L. (2002) Surveillance for waterborne-disease outbreaks--United States, 1999-2000. *MMWR Surveill Summ* **51**, 1–47.

Lee, T.W. and Kurtz, J.B. (1982) Human astrovirus serotypes. *J Hyg (Lond)* **89**, 539–540.

Lemarchand, K., Masson, L. and Brousseau, R. (2004) Molecular biology and DNA microarray technology for microbial quality monitoring of water. *Crit Rev Microbiol* **30**, 145–172.

Leskinen, S., Brownell, M., Lim, D. and Harwood, V.J. (2010) Hollow-Fiber ultrafiltration and PCR detection of human-associated genetic markers from various types of surface water in Florida. *Appl Env Micro* **76**, 4116–7.

Levy, D.A., Bens, M.S., Craun, G.F., Calderon, R.L. and Herwaldt, B.L. (1998) Surveillance for waterborne-disease outbreaks--United States, 1995-1996. *MMWR CDC Surveill Summ* **47**, 1–34.

Ley, V., Higgins, J. and Fayer, R. (2002) Bovine enteroviruses as indicators of fecal contamination. *Appl Environ Microbiol* **68**, 3455–3461.

Lipp, E.K., Futch, J.C. and Griffin, D.W. (2007) Analysis of multiple enteric viral targets as sewage markers in coral reefs. *Mar Pollut Bull* **54**, 1897–1902.

Maluquer de Motes, C., Clemente-Casares, P., Hundesa, A., Martin, M. and Girones, R. (2004) Detection of bovine and porcine adenoviruses for tracing the source of fecal contamination. *Appl Environ Microbiol* **70**, 1448–1454.

Markowitz, R.B., Thompson, H.C., Mueller, J.F., Cohen, J.A. and Dynan, W.S. (1993) Incidence of BK virus and JC virus viruria in human immunodeficiency virus-infected and -uninfected subjects. *J Infect Dis* **167**, 13–20.

McQuaig, S.M. (2009) The development of a human polyomavirus quantitative PCR assay to assess viral persistence, water quality, and human health risks *Dissertation, University of South Florida.*

McQuaig, S.M., Scott, T.M., Harwood, V.J., Farrah, S.R. and Lukasik, J.O. (2006) Detection of human derived fecal pollution in environmental waters using a PCR based human polyomavirus assay. *Appl Environ Microbiol* **72**, 7567–7574.

McQuaig, S.M., Scott, T.M., Lukasik, J.O., Paul, J.H. and Harwood, V.J. (2009) Quantification of human polyomaviruses JC Virus and BK Virus by TaqMan quantitative PCR and comparison to other water quality indicators in water and fecal samples. *Appl Environ Microbiol* **75**, 3379–3388.

Moore, A.C., Herwaldt, B.L., Craun, G.F., Calderon, R.L., Highsmith, A.K. and Juranek, D.D. (1993) Surveillance for waterborne disease outbreaks--United States, 1991-1992. *MMWR CDC Surveill Summ* **42**, 1–22.

Murray, P., Rosenthal, K., Kobayashi, G. and Pfaller, M. (2002) *Medical Microbiology*. St. Louis: Mosby-Year Book, Inc.

Neefe, J.R., Baty, J.B., Reinhold, J.G. and Stokes, J. (1947) Inactivation of the Virus of Infectious Hepatitis in Drinking Water. *Am J Public Health Nations Health* **37**, 365–372.

Oberste, M.S., Maher, K., Kilpatrick, D.R. and Pallansch, M.A. (1999) Molecular evolution of the human enteroviruses: correlation of serotype with VP1 sequence and application to picornavirus classification. *J Virol* **73**, 1941–1948.

Olszewski, J., Winona, L. and Oshima, K.H. (2005) Comparison of 2 ultrafiltration systems for the concentration of seeded viruses from environmental waters. *Can J Microbiol* **51**, 295–303.

Patel, M.M., Hall, A.J., Vinje, J. and Parashar, U.D. (2009) Noroviruses: a comprehensive review. *J Clin Virol* **44**, 1–8.

Paul, J.H., Jiang, S.C. and Rose, J.B. (1991) Concentration of viruses and dissolved DNA from aquatic environments by vortex flow filtration. *Appl Environ Microbiol* **57**, 2197–2204.

Payment, P., Berube, A., Perreault, D., Armon, R. and Trudel, M. (1989) Concentration of *Giardia lamblia* cysts, *Legionella pneumophila, Clostridium perfringens*, human enteric viruses, and

coliphages from large volumes of drinking water, using a single filtration. *Can J Microbiol* **35**, 932–935.
Payment, P., Larose, Y. and Trudel, M. (1979) Polioviruses and other enteroviruses in urban sewage from Laval (Canada): presence of non-vaccinal strains of poliovirus. *Can J Microbiol* **25**, 1305–1309.
Phan, T.G., Khamrin, P., Quang, T.D., Dey, S.K., Takanashi, S., Okitsu, S., Maneekarn, N. and Ushijima, H. (2007) Detection and genetic characterization of group A rotavirus strains circulating among children with acute gastroenteritis in Japan. *J Virol* **81**, 4645–4653.
Pianetti, A., Baffone, W., Citterio, B., Casaroli, A., Bruscolini, F. and Salvaggio, L. (2000) Presence of enteroviruses and reoviruses in the waters of the Italian coast of the Adriatic Sea. *Epidemiol Infect* **125**, 455–462.
Pina, S., Puig, M., Lucena, F., Jofre, J. and Girones, R. (1998) Viral pollution in the environment and in shellfish: human adenovirus detection by PCR as an index of human viruses. *Appl Environ Microbiol* **64**, 3376–3382.
Polo, C., Perez, J.L., Mielnichuck, A., Fedele, C.G., Niubo, J. and Tenorio, A. (2004) Prevalence and patterns of polyomavirus urinary excretion in immunocompetent adults and children. *Clin Microbiol Infect* **10**, 640–644.
Pusch, D., Oh, D.Y., Wolf, S., Dumke, R., Schroter-Bobsin, U., Hohne, M., Roske, I. and Schreier, E. (2005) Detection of enteric viruses and bacterial indicators in German environmental waters. *Arch Virol* **150**, 929–947.
Rafique, A. and Jiang, S.C. (2008) Genetic diversity of human polyomavirus JCPyV in Southern California wastewater. *J Water Health* **6**, 533–538.
Rajal, V.B., McSwain, B.S., Thompson, D.E., Leutenegger, C.M. and Wuertz, S. (2007) Molecular quantitative analysis of human viruses in California stormwater. *Water Res* **41**, 4287–4298.
Rigotto, C., Kolesnikovas, C.K., Moresco, V., Simoes, C.M. and Barardi, C.R. (2009) Evaluation of HA negatively charged membranes in the recovery of human adenoviruses and hepatitis A virus in different water matrices. *Mem Inst Oswaldo Cruz* **104**, 970–974.
Rose, M.A., Dhar, A.K., Brooks, H.A., Zecchini, F. and Gersberg, R.M. (2006) Quantitation of hepatitis A virus and enterovirus levels in the lagoon canals and Lido beach of Venice, Italy, using real-time RT-PCR. *Water Res* **40**, 2387–2396.
Schwartzbrod, L., Pommepuy, M., Albert, M., Dupray, E., Boher, S., Beril, C. and Guichaoua, C. (1991) Influence of sewage discharge on the pollution of an estuary (Morlaix, France). *Zentralbl Hyg Umweltmed* **192**, 238–247.
Shah, K.V. (1996) Polyomaviruses. Philadelphia: Lippincott–Raven.
Shah, K.V., Daniel, R.W. and Warszawski, R.M. (1973) High prevalence of antibodies to BK virus, an SV40-related papovavirus, in residents of Maryland. *J Infect Dis* **128**, 784–787.
Shields, P.A. and Farrah, S.R. (1986) Concentration of viruses in beef extract by flocculation with ammonium sulfate. *Appl Environ Microbiol* **51**, 211–213.
Sobsey, M.D. and Glass, J.S. (1984) Influence of water quality on enteric virus concentration by microporous filter methods. *Appl Environ Microbiol* **47**, 956–960.
Thurston-Enriquez, J.A., Haas, C.N., Jacangelo, J. and Gerba, C.P. (2003a) Chlorine inactivation of adenovirus type 40 and feline calicivirus. *Appl Environ Microbiol* **69**, 3979–3985.
Thurston-Enriquez, J.A., Haas, C.N., Jacangelo, J., Riley, K. and Gerba, C.P. (2003b) Inactivation of feline calicivirus and adenovirus type 40 by UV radiation. *Appl Environ Microbiol* **69**, 577–582.
Tsai, Y.L., Sobsey, M.D., Sangermano, L.R. and Palmer, C.J. (1993) Simple method of concentrating enteroviruses and hepatitis A virus from sewage and ocean water for rapid detection by reverse transcriptase-polymerase chain reaction. *Appl Environ Microbiol* **59**, 3488–3491.
Tsai, Y.L., Tran, B., Sangermano, L.R. and Palmer, C.J. (1994) Detection of poliovirus, hepatitis A virus, and rotavirus from sewage and ocean water by triplex reverse transcriptase PCR. *Appl Environ Microbiol* **60**, 2400–2407.
U.S. Environmental Protection Agency (1995) *USEPA Manual of Methods for Virology* No. EPA 600/4-84/013 (N15). Washington, D.C.: U.S. Environmental Protection Agency.

van Heerden, J., Ehlers, M.M., Heim, A. and Grabow, W.O. (2005) Prevalence, quantification and typing of adenoviruses detected in river and treated drinking water in South Africa. *J Appl Microbiol* **99**, 234–242.

Vanchiere, J.A., White, Z.S. and Butel, J.S. (2005) Detection of BK virus and simian virus 40 in the urine of healthy children. *J Med Virol* **75**, 447–454.

Vantarakis, A. and Papapetropoulou, M. (1999) Detection of enteroviruses, adenoviruses and hepatitis A viruses in raw sewage and treated effluents by nested-PCR. *Water, Air, and Soil Poll* **114**, 85–93.

Villar, L.M., de Paula, V.S., Diniz-Mendes, L., Lampe, E. and Gaspar, A.M. (2006) Evaluation of methods used to concentrate and detect hepatitis A virus in water samples. *J Virol Methods* **137**, 169–176.

Walter, J.E. and Mitchell, D.K. (2003) Astrovirus infection in children. *Curr Opin Infect Dis* **16**, 247–253.

Wetz, J.J., Lipp, E.K., Griffin, D.W., Lukasik, J., Wait, D., Sobsey, M.D., Scott, T.M. and Rose, J.B. (2004) Presence, infectivity, and stability of enteric viruses in seawater: relationship to marine water quality in the Florida Keys. *Mar Pollut Bull* **48**, 698–704.

White, M.K., Gordon, J., Reiss, K., Del Valle, L., Croul, S., Giordano, A., Darbinyan, A. and Khalili, K. (2005) Human polyomaviruses and brain tumors. *Brain Res Brain Res Rev* **50**, 69–85.

Wolf, S., Hewitt, J. and Greening, G.E. A viral tool box of multiplex qPCR assays for tracking the sources of fecal contamination. *Appl Environ Microbiol* 2010; **76**, 1388–1394.

Wong, M., Kumar, L., Jenkins, T.M., Xagoraraki, I., Phanikumar, M.S. and Rose, J.B. (2009) Evaluation of public health risks at recreational beaches in Lake Michigan via detection of enteric viruses and a human-specific bacteriological marker. *Water Res* **43**, 1137–1149.

Yoder, J.S., Blackburn, B.G., Craun, G.F., Hill, V., Levy, D.A., Chen, N., Lee, S.H., Calderon, R.L. and Beach, M.J. (2004) Surveillance for waterborne-disease outbreaks associated with recreational water--United States, 2001-2002. *MMWR Surveill Summ* **53**, 1–22.

Yoder, J.S., Hlavsa, M.C., Craun, G.F., Hill, V., Roberts, V., Yu, P.A., Hicks, L.A., Alexander, N.T., Calderon, R.L., Roy, S.L. and Beach, M.J. (2008) Surveillance for waterborne disease and outbreaks associated with recreational water use and other aquatic facility-associated health events--United States, 2005-2006. *MMWR Surveill Summ* **57**, 1–29.

Zaitlin, M. (1999) Tobacco mosaic virus and its contributions to virology. *ASM News* **65**, 675–680.

Zhong, S., Zheng, H.Y., Suzuki, M., Chen, Q., Ikegaya, H., Aoki, N., Usuku, S., Kobayashi,N., Nukuzuma, S., Yasuda, Y., Kuniyoshi, N., Yogo, Y. and Kitamura, T. (2007) Age-related urinary excretion of BK polyomavirus by nonimmunocompromised individuals. *J Clin Microbiol* **45**, 193–198.

Chapter 6
Phage Methods

Juan Jofre, Jill R. Stewart, and Willie Grabow

Abstract Bacteriophages infecting enteric bacteria have attractive features for tracking sources of fecal pollution in water. Some of these phages remain viable in water environments in numbers suitable for tracking purposes. Currently, F-specific RNA coliphages and *Bacteroides* phages seem to have the greatest potential. Standardized methods are available for the easy, sensitive, and inexpensive detection of these phages by culture assays or molecular techniques. F-RNA coliphages typically infect *Escherichia coli*, as well as some closely related species, through their sex pili. These phages are subdivided into four genogroups, of which groups I and IV predominate in animal feces and groups II and III in human feces. Unfortunately, the survival of these groups in water environments seems to differ. Certain strains of various species of *Bacteroides* detect phages typically present in human feces. Emerging data indicate that certain strains of *Bacteroides* may detect phages specific for animals, and possibly even different species of animals. A major weakness of these anaerobic host bacteria is that different strains seem to be required to detect phages specific for humans or animals in different geographical areas. Sound progress is being made with computer programs in which data on phages, bacteria, and chemical indicators are processed together with details on geographical and other variables for fecal source tracking. These tools are due to play an important role in new strategies for water-quality management.

Keywords Bacteriophages • Coliphages • F-specific • *Bacteroides*

6.1 Introduction

Infectious diseases are the most important concern about the quality of water (WHO 2004). Unsafe water together with childhood underweight, unsafe sex, alcohol use, and high blood pressure are considered the group of five leading risk

J. Jofre (✉)
Department of Microbiology, School of Biology, University of Barcelona,
Avinguda Diagonal 645, 00028 Barcelona, Spain
e-mail: jjofre@ub.edu

C. Hagedorn et al. (eds.), *Microbial Source Tracking: Methods, Applications, and Case Studies*, DOI 10.1007/978-1-4419-9386-1_6,

factors responsible for one quarter of all deaths and one fifth of all disability-adjusted life years (DALYs) in the world (Stevens et al. 2009). The spectrum of microorganisms potentially present in water environments that may be associated with adverse health effects is so wide, diverse, and complex that it is practically impossible to asses or monitor health risks by analyzing water for the presence of all pathogens that may be present.

The pioneering observations of John Snow on fecal pollution being the cause of waterborne diseases initiated the development of strategies for water-quality management toward the end of the nineteenth century. These strategies were based on the detection of fecal pollution by testing water for the presence of microbes of the gastrointestinal flora of humans and animals that typically occur in feces, such as coliform bacteria and enterococci (Ashbolt et al. 2001). These tools proved most valuable for water-quality assessment and monitoring and made a major contribution to the control of waterborne diseases.

However, increasing awareness of the shortcomings of fecal bacteria as indicators of the presence and behavior of many intestinal pathogens, notably viruses and protozoan parasites, in various raw and treated water environments attracted attention to bacteriophages (phages) as alternative indicators for fecal pollution. Bacteriophages are viruses that infect bacteria. They were independently discovered by Twort (1915) and D'Herelle (1917). Guelin (1948) was the first to advocate bacteriophages as an indicator of fecal contamination. Shortly thereafter, Romanian researchers showed that bacteriophages infecting *Salmonella typhi* and *E. coli* correlated with environmental pollution in groundwater (Cornelson et al. 1956; Sechter et al. 1957). Bacteriophages that infect an *E. coli* host are often termed "coliphages," although some strains are also capable of infecting other host species. Early work included the development of techniques for the detection of phages that infect *Serratia marcescens* because these hosts are typical inhabitants of the gastrointestinal tract (Coetzee 1962). They do not readily multiply or produce phages in other environments, which implies that their phages are relatively specific indicators of fecal pollution compared to many bacterial indicators that may also multiply in water environments and soil. Although *S. marcescens* phages had valuable features, attention eventually focused on phages of other enteric bacteria such as *E. coli* (coliphages) and *Bacteroides fragilis* (IAWPRC 1991; Grabow 2001). Phages of typical intestinal host bacteria may indicate fecal pollution as reliably as their hosts for a number of reasons.

Phages share many structural features with enteric viruses, which implies that they may reflect the behavior and resistance to treatment process of viruses much closer than bacterial indicators. Another valuable feature of phages is that they are detectable by simple and inexpensive techniques that yield results in a relatively short period of time. Also, phages do not constitute a health risk to laboratory workers. Typically, bacteriophages are detected by their effects on the host bacteria that they infect.

Numbers of phages are generally determined by direct quantitative plaque assays, the principles of which were designed as early as 1936 by Gratia (Adams 1959).

The presence of phages in a given volume of sample can also be determined by the qualitative presence–absence enrichment test. Enrichment is accomplished by adding host bacterium and nutrients to a sample and then incubating under conditions that permit infection of the bacteria and replication of the phages present in the sample. The number of phages increases to the point where they are readily detectable by plaque assay or spot test on a lawn of the host strain (Adams 1959). Enrichment of multiple tube serial dilutions allows estimating numbers of phages by "quantal" methods, as for example the most probable number (MPN) procedure. Phages have been successfully recovered from various sources by means of methods based on the principles of those used for human enteric viruses. However, some methods used for enteric viruses, notably those based on adsorption–elution principles, are not suitable for the recovery of many bacteriophages (Grabow 2001). For sample volumes ranging from 100 to 1,000 mL, two methods arise as the most recommendable. For water with low turbidity, Sobsey et al. (1990) developed a simple, inexpensive, and practical procedure for the recovery and detection of F-specific RNA phages using mixed cellulose and acetate membrane filters with a diameter of 47 mm and a pore size of 0.45 μm; this method has been slightly modified by Mendez et al. (2004) and has an excellent performance for up to 1 L for somatic coliphages, F-specific RNA phages, and bacteriophages of *B. fragilis*.

Although water analysis reflected by a variety of indicator bacteria and phages is most valuable, more details on pathogens potentially present would be important for a number of reasons. For instance, the distinction between fecal pollution of human and animal origin would greatly facilitate assessment of health risks constituted by fecal pollution. This relates to the host specificity of many pathogens. Some pathogens known as zoonotic pathogens infect and cause clinical disease in both humans and animals. These include bacteria such as species of *Salmonella* and *Campylobacter* and pathogenic strains of *E. coli*, and protozoan parasites such as species of *Cryptosporidium* and *Giardia*. Enteric viruses are typically host specific and species or strains that cause infection and clinical disease in humans rarely if ever cause disease in animals; one exception to the rule may be the hepatitis E virus, which seems to have zoonotic features (Grabow 2007). Other pathogens specific to humans include bacteria such as *Vibrio cholerae*, *Salmonella enterica* serovar Typhi, and *Shigella* species, and protozoan parasites such as species of *Acanthamoeba*, *Cyclospora*, and *Entamoeba*. In view of the host specificity of many pathogens, sewage of human origin is generally considered to constitute a higher health risk to humans than wastewater of animal origin.

Apart from information on health risks, the distinction between fecal pollution of human and animal origin greatly assists remedial action when pollution is detected. Details on the origin of fecal pollution are also of fundamental importance in pollution abatement and management of the quality of water resources, as well as in epidemiological studies on waterborne diseases.

Interest in technology and expertise for distinction between fecal pollution of human and animal origin based on microbiological methods is growing, and the term "microbial source tracking" (MST) became established (Scott et al. 2002). There are also chemical methods for fecal source tracking, but these are not addressed in this chapter. Microbiological methods include genotypic and phenotypic techniques for detailed identification of microorganisms to establish host specificity.

An important tool in MST is direct identification of selected host-specific pathogens such as species or strains of enteric viruses that are specific for humans or animals. This can be accomplished by means of molecular techniques. One restriction to this approach is that numbers of viruses in some water environments may be low and not readily detectable.

This chapter addresses the application of phages for MST. In addition to valuable features as indicators for fecal pollution and the survival of intestinal organisms in water treatment and disinfection processes referred to above, some phages are also remarkably specific for the feces of selected humans and animals. This is related to the gastrointestinal tracts populated by the host bacteria of the phages. Attractive features of phages include ease of detection and typical longer survival of phages in water environments than their host bacteria. Early approaches such as using phages of host bacteria such as *S. marcescens* to specifically indicate fecal pollution were followed up by evidence that F-RNA coliphages and phages that use specific strains of *B. fragilis* as hosts are valuable tools for MST. Attention focuses on methods for the detection of host-specific phages. Attractive possibilities for further improvement of detection methods and application in MST are outlined.

6.2 F-Specific RNA Bacteriophages

6.2.1 General Information

Bacteriophages that infect an *E. coli* host are often termed "coliphages," although some strains are also capable of infecting other host species. "Male-specific" also named F-specific bacteriophages infect their hosts through receptors on F pili, while "somatic coliphages" infect their hosts through receptors on the cell wall (Fig. 6.1). Six taxonomic groups of bacteriophages are conventionally recognized. Four of these, Myoviridae, Microviridae, Siphoviridae (or Styloviridae), and Podoviridae, are somatic bacteriophage families. Leviviridae and Inoviridae comprise male-specific bacteriophage families that contain RNA and DNA genomes, respectively. Additionally, F-RNA bacteriophages (Leviviridae) have been further subdivided into four genogroups, which are primarily associated with either human (groups II and III) or animal (groups I and IV) origins (Furuse 1987). Modern genomic analysis is further identifying intermediate and possibly additional genogroups (Vinjé et al. 2004).

The F-specific RNA bacteriophages (also termed F-RNA coliphages) are morphologically similar to Enteroviruses, Caliciviruses, Astroviruses, and Hepatitis A and E viruses. They have an icosahedral shape, small size (~24 nm diameter) and

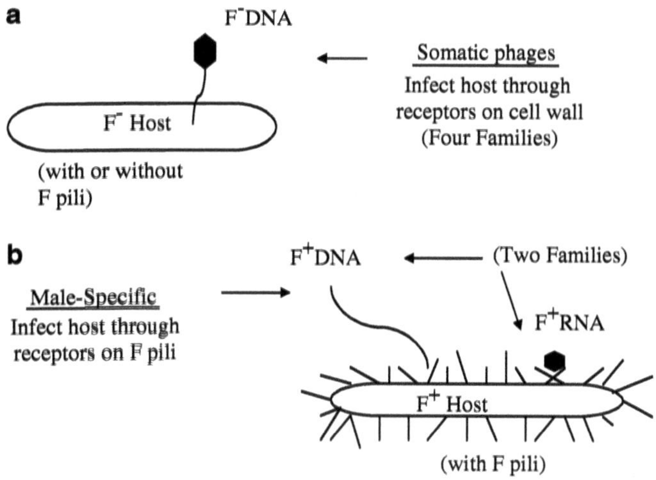

Fig. 6.1 Sites of infection for (**a**) somatic and (**b**) male-specific phages

possess single-stranded RNA surrounded by a protein coat. Owing to their morphological similarity, F-RNA coliphages and the aforementioned enteric viruses are expected to exhibit similar persistence and survivability in the environment and through treatment processes (Havelaar et al. 1993).

Initial classifications of RNA phages were based on serological typing. This technique measures phage neutralization (inhibition of infectivity) by serum raised against the same or another phage. One of the first attempts at serological typing was a study on 30 isolates from which three serogroups were deduced (Watanabe et al. 1967). Serogroup III was then divided into three subgroups (Miyake et al. 1968), and two additional serogroups were reported (Sakurai et al. 1968; Miyake et al. 1969). These last two serogroups, based on the coliphages SP and FI, were eventually made into serogroup IV with two subgroups. Miyake et al. (1971) were the first to separate RNA phages into four major groups (I–IV) based on template specificity of RNA replicases, physicochemical parameters, and serological typing data. These four groups are still in use today.

6.2.2 *Detection Methods*

Standard methods for detection of bacteriophages have been established by the International Organization for Standardization (ISO) and by the US Environmental Protection Agency (US EPA). All of these methods involve incubating coliphages in a nutrient medium with host bacteria to test for plaque formation. Plaques are zones of lysis, or clearings, observed on a bacterial host lawn. The quantity of coliphages in a sample is typically expressed as plaque-forming units (PFU) per a given volume of sample.

ISO Methods for somatic coliphages (ISO 2000) and F-specific RNA bacteriophages (ISO 1995) are available. Both include a double layer agar procedure for quantification and a presence–absence test that can be adapted to the MPN. Methods approved by the US EPA include the two two-step enrichment procedure (US EPA 2001a) and the single agar layer assay (US EPA 2001b). The single agar layer and the double agar layer methods are plaque assay methods used to enumerate coliphages in volumes up to 100 mL (Grabow and Coubrough 1986). The two-step enrichment procedure involves a liquid culture enrichment of 1 L samples. The procedure was originally validated for presence/absence analysis but can be adapted to "quantal" methods such as the MPN test using multiple sample volumes (Kott et al. 1974; Sobsey et al. 2004). Rapid methods are also being developed using 2- to 5-h enrichment steps followed by latex agglutination (Love and Sobsey 2007) or bioluminescence relating to host lysis (Guzmán et al. 2009).

Molecular approaches to bacteriophage detection have also been developed with promise for detecting phages in hours, without the need for culture. Reverse transcription-polymerase chain reaction (RT-PCR) assays have been developed for MS2, a prototypical F-RNA bacteriophage (O'Connell et al. 2006). Multiplexed RT-PCR assays have also been published to simultaneously detect F-RNA and F-DNA coliphages, coupled with a reverse line blot hybridization technique to genotype male-specific coliphages (Vinjé et al. 2004). More recently, multiplex RT-PCR assays have been introduced to distinguish the four genogroups of F-RNA coliphages (Ogorzaly and Gantzer 2006; Kirs and Smith 2007; Friedman et al. 2009).

6.2.3 *Occurrence in the Water Environment*

Male-specific coliphages have been consistently isolated from treated and untreated wastewaters including domestic, hospital, and slaughterhouse wastewaters (Funderburg and Sorber 1985; Nieuwstad et al. 1988; Hill and Sobsey 1998; Harwood et al. 2005) and surface waters polluted with sewage (Havelaar et al. 1993; Hill and Sobsey 1998; Contreras-Coll et al. 2002; Lucena et al. 2003; Lodder and de Roda 2005).

However, F-RNA coliphages appear to be of low prevalence in feces (Havelaar et al. 1986; Cornax et al. 1994). F-RNA coliphages are reportedly isolated in less than 10% of human feces samples and at variable rates in nonhuman animal feces samples (Havelaar et al. 1986; Calci et al. 1998; Schaper et al. 2002a; Long et al. 2005; Blanch et al. 2006).

Further research is necessary to determine whether the consistently higher concentrations of coliphages in sewage relative to feces are the result of direct environmental input or multiplication. However, environmental multiplication appears unlikely for F-RNA coliphages. Approximately 25% of known wild-type strains of *E. coli* carry F plasmids and are susceptible to F-RNA phage infection. Production of F pili on these bacteria is also temperature-dependent. They are not produced below 25°C (Novotny and Lavin 1971), a limitation that disfavors phage multiplication in the environment to any appreciable extent. Woody and Cliver (1995) concluded that pili formation ceases below 25°C and that Qβ, a prototypical F-RNA bacteriophage, did not replicate in 20°C or 22°C batch cultures.

Vaughn and Metcalf (1975) reported coliphage replication in estuarine waters seeded with host bacteria. However, laboratory observations such as this do not necessarily mean that coliphages commonly multiply in the environment. Several studies have shown that coliphages do not replicate below a bacterial host density of 10^4 CFU/mL (Wiggins and Alexander 1985; Woody and Cliver 1997). Woody and Cliver (1997) further demonstrated that F-RNA coliphage cannot replicate in nutrient-poor environments. Cornax et al. (1991) asserted that the low survivability of the *E. coli* bacterial host in marine environments does not support the replication of coliphages. A study of coliphage survival and reproducibility in tropical waters found that coliphages survived for extended periods of time but that neither sewage nor laboratory phage strains were able to replicate (Hernandez-Delgado and Toranzos 1995). The researchers showed that tested bacterial hosts were not permissive to phage replication at host densities of 10^4, 10^3, or 10^2 CFU/mL. These and other data reviewed by Jofre (2009) show that the limited conditions under which replication occurs most likely limits multiplication to areas with direct, fresh fecal input and a sufficiently high host density.

Given the presence of coliphages in sewage, the IAWPRC Study Group on Health Related Microbiology (1991) suggested that bacteriophages may be a more appropriate index of sewage contamination than fecal contamination. Havelaar (1993) further pointed out that indirect fecal input via sewage is more common than direct fecal inputs into water bodies, supporting the idea that coliphages could be a valuable microbial indicator.

6.2.4 F-RNA Coliphages Typing for Microbial Source Tracking

Most of the initial phage ecology studies were conducted in Asia, where researchers serologically typed thousands of RNA phage isolates from various sources. During the course of these studies, the researchers noted two trends in the distribution of RNA phages. One trend related to the geographical distribution of the phages and the other related to preferential distribution of RNA phage groups in animal hosts. Group II and III F-RNA bacteriophages were predominantly observed in humans, while groups I and IV F-RNA bacteriophages predominated in animals (Dhillon and Dhillon 1974; Furuse et al. 1975, 1978, 1981; Osawa et al. 1981). Subsequent research in the Netherlands and elsewhere confirmed the general association of F-RNA serogroups with source (Havelaar et al. 1986). These apparent ecological niches form the basis of typing F-RNA coliphages for MST.

In the 1990s, a genotyping system was developed to type the F-RNA coliphages based on hybridization of oligonucleotides with sections of the phage RNA (Hsu et al. 1995; Beekwilder et al. 1996). Both studies found that genotyping and serotyping results are analogous and that genotyping can be used to type isolates with ambiguous serotypes. This system has since been used to genotype F-RNA bacteriophages for MST in a number of applications (Griffin et al. 2000; Brion et al. 2002; Cole et al. 2003; Stewart-Pullaro et al. 2006).

As with all other microbial source methods validated to date, association of F-RNA genotypes with source is not absolute. Group II and III genotypes have been identified from a small number of animal-source samples (Schaper et al. 2002b; Stewart et al. 2006). More frequently, group I genotypes have been isolated from municipal sewage (Dhillon and Dhillon 1974; Griffin et al. 2000; Stewart-Pullaro et al. 2006). It is not clear whether the association of group I F-RNA coliphages with sewage is the result of animal wastes in the sewage, or if there is also a human origin of genogroup I.

Schaper et al. (2002b) found a statistically significant relationship for the traditional source-specific associations of F-RNA bacteriophages, that is for the association between genotypes II and III with human sources and between genotypes I and IV with animal sources. Although sources may not be absolute, a probabilistic approach may be useful for managers who need tools to identify sources of fecal contamination. If the probability were known that a certain viral strain is associated with a specific source (i.e., humans), identification of the strain could be extremely useful in a management context. Managers could then make informed decisions regarding risks, use, and remediation.

Several studies have reported differential survival within and among F-RNA bacteriophage genogroups (Brion et al. 2002; Schaper et al. 2002a). In general, group I F-RNA bacteriophages appear to persist longer under environmental stress (e.g., extreme pHs and temperatures, chlorine, high salinity), while group IV is the least resistant to environmental stress. Furthermore, enrichment of coliphages from water samples appears to decrease the representativeness of coliphage diversity within a sample, with group I strains growing to mask the presence of other genogroups (Stewart-Pullaro et al. 2006). While the enrichment method is the most sensitive for culture of F-RNA bacteriophage (Sobsey et al. 2004), plating techniques may be preferable for isolating coliphage strains that will subsequently be subject to typing. Molecular methods (e.g., Friedman et al. 2009) are also recommended for evaluation of representative genotypes without an enrichment bias.

In large methods comparison studies, F-RNA coliphage typing has shown promise for identifying waters contaminated with human-source sewage. One early study demonstrated that F-RNA coliphage methods reliably identified human-source contamination in blind samples seeded with sewage (Griffith et al. 2003). Coliphage methods and other viral detection approaches also demonstrated the lowest false-positive rates among tested methods, defined as the percentage of samples not containing human fecal material that were incorrectly identified as containing a human source (Noble et al. 2003). Another large, multilaboratory MST study conducted in the European Union identified somatic coliphages and phages infecting *Bacteroides thetaiotaomicron* as central parameters to predictive models capable of identifying sources of recent fecal pollution. The distribution of F-RNA genotypes was also among the best predictors (Blanch et al. 2006).

Nucleotide sequencing of F-RNA coliphages, although not widely adopted yet, will likely prove to be a very useful approach for MST (Stewart et al. 2006;

Lee et al. 2009). Lytic RNA viruses have the highest mutation rates observed in nature, owing largely to a lack of proofreading abilities for the virus-coded RNA polymerase (Drake 1993; Drake et al. 1998). By identifying unique sequences from fecal pollution sources, sequencing can be used to link coliphages in contaminated waters to their origins. As sequencing becomes more practical, economical, and high throughput, this approach shows great promise in convincingly identifying specific sources of contamination.

6.3 Bacteriophages Infecting *Bacteroides*

6.3.1 General Information

Bacteriophages infecting strains of *B. fragilis, B. thetaiotaomicron, B. ruminicola,* and *B. ovatus* have been detected in feces and wastewater (Booth et al. 1979; Masanori et al. 1985; Kai et al. 1985; Tartera and Jofre 1987; Klieve et al. 1991; Payan et al. 2005). Bacteriophages infecting *B. fragilis* have not been reported to replicate outside the gut probably because of the special necessities, including anaerobiosis and nutrient requirements, of the host strain to support phage replication (Tartera and Jofre 1987).

All bacteriophages infecting different *Bacteroides* species described so far are tailed. The great majority has the morphology of Siphoviridae (Queralt et al. 2003; Booth et al. 1979; Masanori et al. 1985; Klieve et al. 1991; Payan 2006), with flexible tails. Those with slightly curved tails are the most abundant, but phages with curved and curly tails are not rare.

The genome of the few *Bacteroides*-infecting phages that have been studied consists of double-stranded DNA, corresponding to that of Siphoviridae (Kory and Booth 1986; Klieve et al. 1991; Puig and Gironés 1999; Hawkins et al. 2008). Recently, the first genome sequence of a phage (ATCC 51477-B1–B40-8) infecting *B. fragilis* HSP40 has been completed. It has 44,929 base pairs with a G+C content of 38.7% and 46 putative reading frames (Hawkins et al. 2008).

Phages infecting *Bacteroides* are supposed to infect the host through the cell wall as tailed bacteriophages do. They are somatic bacteriophages (Fig. 6.1) with receptors identified as cell-wall proteins (Puig et al. 2001). The amount of capsule of the host seems to play a role in phage infectivity, presumably by regulating accessibility of the receptor sites on the bacterial surface (Booth et al. 1979; Klieve et al. 1991).

Most *Bacteroides* phages have a narrow host range (Keller and Traub 1974; Cooper et al. 1984; Kai et al. 1985; Kory and Booth 1986; Tartera and Jofre 1987; Payan 2006). The reasons for this narrow host range of *Bacteroides* phages are not fully understood. A potential explanation is that phages infecting anaerobic bacteria might have coevolved with the animal hosts more separately than in the case of facultative bacteria of the intestinal microbiota. In fact, a great degree of very specific interactions between *Bacteroides* and the animal host have been described for humans and *Bacteroides* (Xu et al. 2003).

Strains of *Bacteroides* spp. differ in the numbers of phages that they recover from sewage or sewage-polluted waters (Puig et al. 1999). They differ also in their capability to detect bacteriophages in the fecal material of different animal species, including humans, and hence in the capability of discerning the fecal source that contaminates a given sample. Thus, strain RYC2056 and VPI3625 of *B. fragilis* detect phages both in human and nonhuman fecal wastes (Kator and Rhodes 1992; Puig et al. 1999; Blanch et al. 2006), whereas *B. fragilis* HSP40 (Tartera and Jofre 1987; Tartera et al. 1989; Grabow et al. 1995), *B. thetaiotaomicron* GA17 (Payan et al. 2005; Blanch et al. 2006), and *B. fragilis* GB124 (Payan et al. 2005; Ebdon et al. 2007) detect phages mostly in human fecal wastes (much lower numbers if any in nonhuman fecal wastes). Preliminary results indicate that obtaining host strains detecting preferably phages in feces of a given animal species, i.e., pigs, is also possible (Payan 2006). Nondiscriminant strains such as RYC2056 seem to detect similar numbers of phages in sewage around the world (Puig et al. 1999; Lucena et al. 2003, 2004; Blanch et al. 2006; McLaughlin and Rose 2006), whereas those that are able to identify the fecal source such as *B. fragilis* HSP40 and HB13, *B. ovatus* GB124 and *B. thetaiotaomicron* GA17 have restricted geographical areas of application because of their low numbers in human fecal wastes in some geographical areas (Kator and Rhodes 1992; Chung et al. 1998; Puig et al. 1999; Payan et al. 2005; McLaughlin and Rose 2006). However, an easy method to isolate *Bacteroides* hosts convenient for a given geographical area has been described (Payan et al. 2005). At present, discerning hosts for Southern Europe (GA17), Great Britain (GB124), and Hawaii (HB73) are available. Table 6.1 summarizes information available about this topic.

Bacteriophages infecting *B. fragilis* have been reported to be quite resistant to both natural and anthropogenic stressors. Their resistance to stressors such as heat, UV radiation, and chemical disinfectants surpasses that of the traditional bacterial indicators and is similar to that of the most resistant viruses and other groups of bacteriophages. Bacteriophages infecting *Bacteroides* show first-rate inactivation kinetics to chlorine (Duran et al. 2003; Baert et al. 2009), UV irradiation (Sommer et al. 1998), and pasteurization (Mocé-Llivina et al. 2003). Persistence measured through microcosm experiments in the laboratory or through "in situ" inactivation experiments confirms that phages infecting *Bacteroides* rank among the more persistent under environmental conditions. Indeed, slow dieoff has been observed in freshwater (Duran et al. 2002), sea water (Kator and Rhodes 1992; Chung and Sobsey 1993; Mocé-Llivina et al. 2005; McLaughlin and Rose 2007), sea sediments (Chung and Sobsey 1993), water distribution pipe biofilms (Storey and Ashbolt 2001), and fomites (Abad et al. 1994).

Concurrently, data indicate that naturally occurring phages infecting *Bacteroides* survive primary and secondary wastewater treatments similar to other indicators (Sun et al. 1997; Lucena et al. 2004) and that they persist through tertiary treatments including UV irradiation and/or chemical disinfection (Chung et al. 1998; Gantzer et al. 1998; McLaughlin and Rose 2007; Costán-Longares et al. 2008). These phages also show very good survival to drinking water chlorination (Jofre et al. 1995; Sun et al. 1997), and they accumulate in raw sewage sludges and survive sludge treatments (Lasobras et al. 1999; Guzmán et al. 2007).

Table 6.1 Summarized information about *Bacteroides* host strains. Values are per 100 mL

Host strain	Fecal source discerning capability	Applicable in	Nonapplicable in	Levels in municipal wastewaters	Levels in slurries and wastewaters of abattoirs	References
B. fragilis, RYC2056	NO	Europe, USA, South America, and South Africa	–	10^4–5×10^4	10^3–10^4	Puig et al. (1999), Blanch et al. (2006), Lucena et al. (2004), McLaughlin and Rose (2007)
B. fragilis, VPI3625	NO	USA	–	10^4	–	Chung et al. (1998), Kator and Rhodes (1992)
B. fragilis, HSP40	YES	Southern Europe, Israel, and South Africa	USA, Great Britain, Scandinavia	5×10^3–10^4	<10	Tartera and Jofre (1987), Tartera et al. (1989), Grabow et al. (1993, 1995), Armon (1993), Sun et al. (1997), Kator and Rhodes (1992), McLaughlin and Rose (2007), Cornax et al. (1990)
B. tethaioataomicron, GA17	YES	Southern Europe	Great Britain	5×10^4–5×10^5	$<10^2$	Blanch et al. (2006), Vijayavel et al. (2010)
B. ovatus, GB124	YES	Great Britain	–	10^4	<10	Payan et al. (2005), Ebdon et al. (2007)
B. fragilis, HB13	YES	Spain	–	5×10^4–5×10^5	<10	Payan et al. (2005), Payan (2006)
Bacteroides spp., HB73	YES	Hawaii	–	10^5	<10	Vijayavel et al. (2010)

Persistence of *B. fragilis* bacteriophages relative to traditional fecal indicators results in proportions between *Bacteroides* bacteriophages and fecal indicators in receiving waters, which are different to the proportions in recently contaminated matrixes such as raw sewage. In temperate and temperate to warm climates, the variation of proportions, with the exception of the one with spores of sulfite-reducing clostridia, favors bacteriophages of *Bacteroides* and the change is more evident with increasing space and time from the source of pollution (Lucena et al. 2003, 1996; Chung et al. 1998; Duran et al. 2002; Mocé-Llivina et al. 2005).

6.3.2 Detection Methods

Standardized plaque assay and enrichment methods for bacteriophages infecting *B. fragilis* are available (ISO 2001). The cost, considering material and labor, for *B. fragilis* phages testing is about twice as much as testing for conventional fecal indicator bacteria.

Methods described for concentrating F-RNA bacteriophages are also suitable to concentrate phages infecting *B. fragilis* (Mendez et al. 2004). Thus, with the numbers of bacteriophages reported in different water environments (see Sect. 6.2.3), the source tracking method with phages infecting *Bacteroides* is applicable to waters with 10^2 CFU of *E. coli*/100 mL. Molecular methods for the detection of bacteriophages infecting *B. fragilis* are feasible (Puig et al. 2000), though further investigation about these procedures is needed. The availability of the sequence of the genome of a *B. fragilis* infecting phage (Hawkins et al. 2008) will certainly facilitate this need.

6.3.3 Occurrence in Water Environments

Reports on the occurrence in fecal sources of bacteriophages infecting strains of *B. fragilis* with fecal source distinction capabilities are only available for strain HSP40 (Tartera and Jofre 1987; Grabow et al. 1995). The percentage of human fecal samples containing *B. fragilis* were low, 10 and 14%, respectively, in the two studies, but absent in all animal samples tested. In addition, numbers of phages determined in fecal samples were highly variable (Tartera and Jofre 1987).

Phages infecting GB124 have been detected in the effluents of sewage treatment works receiving wastewater from a population as small as 28 people and also in those of small communities (Ebdon et al. 2007). Phages infecting GA17 have also been identified in the sewage from a small-to-medium-sized hospital (Blanch et al. 2006). Concentrations of bacteriophages infecting *Bacteroides* range from 5×10^3 to 5×10^5 PFU/100 mL in municipal sewage when appropriate host strains are used. Strains GA17, HB13, and HB73 provide the highest values. Interestingly, no variations in the numbers of phages detected using strains HSP40 and GA17 have been reported since the first determinations were performed, 1987 for HSP40 and 2005 for GA17 (data unpublished).

Phages infecting HSP40 (Tartera and Jofre 1987; Sun et al. 1997; Gantzer et al. 1998), GA17 (Payan 2006) and GB124 (Ebdon et al. 2007) were detected in 100% of standard secondary effluents tested with values ranging from 200 to 50,000 PFU/100 mL. These values depended on the quality of the effluent. The proportions between the numbers of these phages and the numbers of other bacterial and bacteriophage indicators were similar to those of raw sewage. In addition, values of bacteriophages infecting GA17 in raw sewage sludge exceed 10^4 PFU/g of dry matter and as in secondary effluent they keep the proportions with other indicator bacteriophages (Payan 2006).

In 63 reclaimed water samples tested with an average value of *E. coli* of 10 CFU/100 mL (57% positive samples), indicating tertiary effluents of excellent quality, phages infecting strain GA17 were detected in 59% of the samples with an average value of 0.7 PFU/100 mL (Costán-Longares et al. 2008).

Regarding receiving waters, phages infecting GA17 (Payan 2006) and GB127 (Ebdon et al. 2007) have been detected in river water moderately (about 10^3 *E. coli*/100 mL) impacted by fecal contaminants of human origin. Phage concentrations in these waters slightly exceeded 10^2 PFU/100 mL. Phages infecting strain GA17 were detected in 11 of 17 samples of a bathing area impacted exclusively with secondary effluents of municipal wastewater and *E. coli* levels well below 100 CFU/100 mL. Only 9 of 17 samples corresponded to *E. coli* concentrations greater than 0 CFU/100 mL (Mocé-Llivina 2004), whereas 100% of samples exceeding 100 *E. coli*/100 mL were positive for phages infecting GA17 (from 2 to 128). In all these cases, the proportions of *Bacteroides* phages to other indicators remained similar or greater than in raw sewage.

Phages infecting HSP40 have been detected in sea (Lucena et al. 1996) and river sediments (Tartera and Jofre 1987) and groundwater (Lucena et al. 1996) with variable values, but always with ratios to other fecal indicators greater than those of sewage.

6.3.4 Source Tracking Studies with Phages Infecting Bacteroides

Two reports describe research in which the aptness of phages infecting *Bacteroides* for MST was tested. The first report compared the occurrence and levels of different potential source trackers, including bacteriophages infecting strain GA17 of *B. tethaiaotaomicron* and fecal indicators, in waters (municipal sewage, slaughterhouse wastewater and slurries) highly contaminated with human or animal feces. The pair somatic coliphages and bacteriophages infecting strain GA17 or the pair fecal coliforms and bacteriophages infecting strain GA17 was highly predictive of the fecal source (Blanch et al. 2006).

In the second report, Ebdon et al. (2007) analyzed bacteriophages infecting strain GB124, somatic coliphages, and bacterial indicators in 306 river water, effluents of municipal wastewater treatment plants, and animal fecal samples. Bacteriophages capable of infecting strain GB124 were present in all municipal wastewaters but were not detected in fecal samples from animals. These phages were detected at significantly higher levels in river water downstream from a wastewater treatment plant than in water at the origin of the river allegedly impacted by diffuse pollution of nonhuman origin. The ratio somatic coliphages: bacteriophages infecting strain GB124 were also higher in water downstream of a dairy farm despite the presence of high levels of fecal indicator bacteria and somatic coliphages at these sites.

6.4 General Conclusions

Details summarized in this chapter show that phages are valuable tools for MST. Well-defined procedures are in place for the meaningful application of phages, notably *B. fragilis* phages and F-RNA specific bacteriophages, to distinguish between fecal pollution of human and animal origin. In spite of the promising prospects, there are some dilemmas that need to be elucidated, as for example which are the minimum volumes of sample required to detect bacteriophages and why there are reports of situations where the phages should have been present in the samples but the investigators could not find them. It is clear that the full potential of phages for MST remains to be explored. It seems possible that host bacteria that specifically colonize the gastrointestinal tract of humans or animals, and possibly even different species of animals, may likewise have phages that can be applied in MST. Recent developments in PCR based methods for family-specific detection of somatic coliphages (Lee 2009) or the recently described method for detecting phages infecting *Enterococcus faecalis.* (Bonilla et al. (2009) will very likely provide additional tools for MST. It is evident that no single indicator will be able to ideally meet all the expectations of MST. The answer lies in combinations of indicators, each of which contributes information to a comprehensive picture. Such batteries may consist of a variety of indicators, including a selected spectrum of bacteria, phages, and enteric viruses. In view of features such as host specificity, small size, genome structure, and resistance to unfavorable conditions including treatment and disinfection processes, phages may play a valuable role in such MST programs. The data generated by test batteries can be processed in computer programs for assessment in terms of data profiles on the host specificity and other relevant details on microbial indicators, supplemented by data from chemical analyses. The libraries could even compensate for variables such as the geographical incidence of certain indicators. The detection of microbial indicators can be greatly facilitated by means of molecular techniques that are highly specific, rapid, and cost-effective. Such computer models for MST are feasible, and progress along these lines has been reported (Scott et al. 2002; Blanch et al. 2006; Stewart et al. 2006; Lee et al. 2009). These tools may be able to distinguish not only between human and animal fecal pollution but also between fecal pollution from different animals, at least categories such as cattle, pigs, and poultry. In addition, the models may give meaningful indications of quantities of fecal pollution from various sources, as well as the distance of the discharge site from the sampling point. Basic information and technologies are in place to take microbial MST even further by means of gene chip, microarray, and biosensor techniques that may enable automatic in-line continuous monitoring systems. The MST toolbox with exciting possibilities is due to play a major role in future management strategies to comply with increasing water-quality demands.

References

Abad FX, Pinto RM, Bosch A (1994) Survival of enteric viruses on environmental fomites. Appl Environ Microbiol 60:3704–3710

Adams MH (1959) Bacteriophages. Interscience Publishers Inc, New York

Ashbolt NJ, Grabow WOK, Snozzi M (2001) Chapter 13: Indicators of microbial water quality. In Fewtrell L and Bartram J (ed) Water Quality Guidelines: Guidelines, Standards and Health. World Health Organization Water Series. IWA Publishing, London

Armon R (1993) Bacteriophages monitoring in drinking water: do they fulfill the index or indicator function. Wat Sci Technol 27:463–467

Baert L, Vandekinderen I, Devlieghere F et al (2009) Efficacy of sodium hypochlorite and peroxacetic acid to reduce murine norovirus 1, B40-8, *Listeria monocytogenes*, and *Escherichia coli* O157:H7 on shredded iceberg lettuce and in residual water. J Food Protect 72:1047–1054

Beekwilder J, Nieuwenhuizen R, Havelaar AH, van Duin J (1996) An oligonucleotide hybridization assay for the identification and enumeration of F-specific RNA phages in surface water. J Appl Bacteriol 80:179–186

Blanch AR, Belanche-Munoz L, Bonjoch X et al (2006) Integrated analysis of established and novel microbial and chemical methods for microbial source tracking. Appl Environ Microbiol 72:5915–5926

Bonilla N, Santiago T, Davila C et al (2009) Enterophages, a group of phage infecting *Enterococcus faecalis* and their potential as alternative indicators of human fecal contamination. 15th Health Related Water Microbiology Symposium Abstracts IWA), page 68

Booth SJ, Van Tassell RL, Johnson JL, Wilkins TD (1979) Bacteriophages of *Bacteroides*. Rev Infect Dis 2:325–334

Brion GM, Meschke JS, Sobsey MD (2002) F-specific RNA coliphages: occurrence, types, and survival in natural waters. Water Res 36:2419–2425

Calci KR, Burkhardt III W, Watkins WD, Rippey SR (1998) Occurrence of male-specific bacteriophage in feral and domestic animal wastes, human feces, and human-associated wastewaters. Appl Environ Microbiol 64:5027–5029

Chung H, Jaykus L-A, Lovelance G, Sobsey, MD (1998) Bacteriophages and bacteria as indi-cators of enteric viruses in oysters and their harvest waters. Wat Sci Technol 38:37–44

Chung H, Sobsey, MD (1993) Comparative Survival of Indicator Viruses and Enteric Viruses in Seawater and Sediment. Wat Sci Technol 27:425–428

Coetzee OJ (1962) Bakteriophagen als Indikator fekaler Wasser-verunreinigung. Gesundheits-Ingenieur 12:371–372

Cole D, Long SC, Sobsey MD (2003) Evaluation of F+ RNA and DNA coliphages as source-specific indicators of fecal contamination in surface waters. Appl Environ Microbiol 69:6507–6514

Contreras-Coll N, Lucena F, Mooijman K et al (2002) Occurrence and levels of indicator bacteriophages in bathing waters through Europe, Water Res 36:4963–4974

Cooper SW, Szymczak EG, Jacobus NV, Tally FP (1984) Differentiation of *Bacteroides ovatus* and *Bacteroides tethaiotaomicron* by means of bacteriophage. J Clin Microbiol 20:1122–1125

Cornax R, Moriñigo MA, Paez IG et al (1990) Application of direct plaque assay for detection and enumeration of bacteriophages infecting *Bacteroides fragilis* from contaminated water samples. Appl Environ Microbiol 56:3170–3173

Cornax R, Morinigo MA, Balebona MC et al (1991) Significance of several bacteriophage groups as indicators of sewage pollution in marine waters. Water Res 25:675–678

Cornax R, Morinigo MA, Gonzalez-Jaen F et al (1994) Bacteriophages presence in human faeces of healthy subjects and patients with gastrointestinal disturbances. Zentralbl Bakteriol 281:214–224

Cornelson DA, Secter I, Zamfir GH et al (1956) Cercetari asupra prezentei si semnificantiei bvacteriofagilor enterici in apa de fintina. Igiena 9:317

Costán-Longares A, Montemayor M, Payán A et al (2008) Microbial indicators and pathogens: removal, relatioships and predictive capabilities in water reclamation facilities. Water Res 42:4439–4448

D'Herelle F (1917) Sur un microbe invisible antagonistic des bacilles dysenteriques. C R Acad Sci Paris 165:373–375

Dhillon EK, Dhillon TS (1974) Synthesis of indicator strains and density of ribonucleic acid-containing coliphages in sewage. Appl Microbiol 27:640–647

Drake JW (1993) Rates of spontaneous mutation among RNA viruses. Proc Natl Acad Sci USA 90:4171–4175

Drake JW, Charlesworth B, Charlesworth B, Crow JF (1998) Rates of spontaneous mutation. Genetics 148:1667–1686

Duran AE, Muniesa M, Méndez X et al (2002) Removal and inactivation of indicator bacteriophages in fresh waters. J Appl Microbiol 92:338–347

Duran AE, Muniesa M, Mocé-Llivina L et al (2003) Usefulness of different groups of bacteriophages as model micro-organisms for evaluating chlorination. J Appl Microbiol 95:29–37

Ebdon J, Muniesa M, Taylor H (2007) The application of a recently isolated strain of *Bacteroides* (GB-124) to identify human sources of faecal pollution in a temperate river catchment. Water Res 41:3683–3690

Friedman SD, Cooper EM, Casanova L et al (2009) A reverse transcription-PCR assay to distinguish the four genogroups of male-specific (F+) RNA coliphages. J Virol Methods 159:47–52

Funderburg SW, Sorber CA (1985) Coliphages as indicators of enteric viruses in activated sludge. Water Res 19:547–555

Furuse, K (1987) Distribution of coliphages in the environment: general considerations In: Goyal SM, Gerba CP, Bitton, G (ed.) Phage Ecology. Wiley, New York

Furuse K, Ando A, Watanabe I (1975) Isolation and grouping of RNA phages. A survey in the islands in the adjacent seas of Japan. J Keio Med. Soc 52:259–263

Furuse K, Ando A, Osawa S, Watanabe I (1981) Distribution of ribonucleic acid coliphages in raw sewage from treatment plants in Japan. Appl Environ Microbiol 41:1139–1143

Furuse K, Sakurai T, Hirashima A et al (1978) Distribution of ribonucleic acid coliphages in south and east Asia. Appl Environ Microbiol 35:995–1002

Gantzer C, Maul A, Audic JM, Schwartzbrod L (1998) Detection of infectious enteroviruses, enterovirus genomes, somatic coliphages, and *Bacteroides fragilis* phages in treated wastewater. Appl Environ Microbiol 64:4307–4312

Grabow WOK (2001) Bacteriophages: Update on application as models for viruses in water. Water SA 27:251–268

Grabow WOK (2007) Overview of Health-Related Water Virology. In: Human Viruses in Water. Bosch A (ed) Elsevier Science Series: Perspectives in Medical Virology, 17. Elsevier, London

Grabow WOK, Coubrough P (1986) Practical direct plaque assay for coliphages in 100-ml samples of drinking water. Appl Environ Microbiol 52:430–433

Grabow WOK, Holtzhausen CS, De Villiers CJ (1993) Research on Bacteriophages as Indicators of Water Quality 1990-1992. Water Research Commission Report n°321/1/93. Water Research Commission, Pretoria

Grabow WOK, Neubrech TE, Holtzhausen,CS, Jofre, J (1995) *Bacteroides fragilis* and *Escherichia coli* bacteriophages. Excretion by humans and animals. Wat Sci Technol 31:223–230

Griffin DW, Stokes R, Rose JB, Paul III JH (2000) Bacterial indicator occurrence and the use of an F(+) specific RNA coliphage assay to identify fecal sources in Homosassa Springs, Florida. Microb Ecol 39:56–64

Griffith JF, Weisberg SB, McGee CD (2003) Evaluation of microbial source tracking methods using mixed fecal sources in aqueous test samples. J Water Health 1:141–151
Guelin A (1948) Etude quantitative de bacteriophages typhiques. VI. Dans les eaux. Ann Inst Pasteur 79:447
Guzmán C, Costan-Longares A, Lucena F, Jofre J (2009) Detection of somatic coliphages through a bioluminescence assay measuring phage mediated release of adenylate kinase and adenosine 5'-triphosphate. J Virol Methods 161:107–113
Guzmán C, Jofre J, Montemayor M, Lucena, F (2007) Occurrence and levels of indicators and selected pathogens in different sludges and biosolids. J Appl Microbiol 103:2420–2429
Harwood VJ, Levine AD, Scott TM et al (2005) Validity of the indicator organism paradigm for pathogen reduction in reclaimed water and public health protection. Appl Environ Microbiol 71:3163–3170
Havelaar AH (1993) Bacteriophages as models of human enteric viruses in the environment. ASM News 59:614–619
Havelaar AH, Furuse K, Hogeboom WM (1986) Bacteriophages and indicator bacteria in human and animal faeces. J Appl Bacteriol 60:255–262
Havelaar AH, van Olphen M, Drost YC (1993) F-specific RNA bacteriophages are adequate model organisms for enteric viruses in fresh water. Appl. Environ. Microbiol. 59:2956–2962
Hawkins SA, Layton AC, Ripp S et al (2008) Genome sequence of the *Bacteroides fragilis* ATCC 51477-B1. Virol J 5:97–101
Hernandez-Delgado EA, Toranzos GA (1995) In situ replication studies of somatic and male-specific coliphages in a tropical pristine river. Wat Sci Technol 31:247–250
Hill VR, Sobsey MD (1998) Microbial indicator reductions in alternative treatments systems for swine wastewater. Wat Sci Technol 38:119–122
Hsu FC, Shieh YS, van Duin J et al (1995) Genotyping male-specific RNA coliphages by hybridization with oligonucleotide probes. Appl Environ Microbiol 61:3960–3966
IAWPRC (1991) Bacteriophages as model viruses in water quality control. Water Res 25:529–545
ISO (1995) ISO 10705-1: Water quality. Detection and enumeration of bacteriophages -part 1: Enumeration of F-specific RNA bacteriophages. International Organization for Standardization, Geneva
ISO (2000) ISO 10705-2: Water quality. Detection and enumeration of bacteriophages-part 2: Enumeration of somatic coliphages. International Organization for Standardization, Geneva
ISO (2001) ISO 10705-4: Water quality. Detection and enumeration of bacteriophages-part 4: Enumeration of bacteriophages infecting *Bacteroides fragilis*. International Organization for Standardization, Geneva
Jofre, J (2009) Is the replication of somatic coliphages in water environments significant? J Appl Microbiol 106:1059–1069
Jofre J, Ollé E, Ribas F et al (1995) Potential usefulness of bacteriophages that infect *Bacteroides fragilis* as model organisms for monitoring virus removal in drinking water treatment plants. Appl Environ Microbiol 61:3227–3231
Kai M, Watanabe S, Furuse K, Ozawa A (1985) *Bacteroides* bacteriophages isolated from human faeces. Microbiol Immunol 29:895–899
Kator H, Rhodes M (1992) Evaluation of *Bacteroides fragilis* bacteriophages, a candidate human-specific indicator of fecal contamination from shellfish growing waters. Final Report. Virginia Institute of Marine Sciences, College of Williams and Mary, Gloucester Point, Virginia, USA
Keller R, Traub N (1974) The characterization of *Bacteroides fragilis* bacteriophage recovered from animal sera: observations on the nature of *Bacteroides* bacteriophage carrier cultures. J Gen Virol 29:179–189
Kirs M, Smith DC (2007) Multiplex quantitative real-time reverse transcriptase PCR for F+-specific RNA coliphages: a method for use in microbial source tracking. Appl Environ Microbiol 73:808–814
Klieve AV, Gregg K, Bauchop T (1991) Isolation and characterization of lytic phages from *Bacteroides ruminicola* ss *brevis*. Curr Microbiol 23:183–187

Kory MM, Booth SJ (1986) Characteristics of *Bacteroides fragilis* bacteriophages and comparison of their DNAs. Curr Microbiol 14:199–203

Kott Y, Roze N, Sperber S, Betzer N (1974) Bacteriophages as viral pollution indicators. Water Res 8:165–177

Lasobras J, Dellundé J, Jofre J, Lucena F (1999) Occurrence and levels of phages proposed as surrogate indicators of enteric viruses in different types of sludges. J Appl Microbiol 86:723–729

Lee HS (2009) Somatic coliphage families as potential indicators of enteric viruses in water and methods for their detection. Ph. D. Thesis. University of North Carolina, Chapel Hill, NC

Lee JE, Lim MY, Kim SY et al (2009) Molecular characterization of bacteriophages for microbial source tracking in Korea. Appl Environ Microbiol 75:7107–7114

Lodder WJ, de Roda Husman AM (2005) Presence of noroviruses and other enteric viruses in sewage and surface water in the Netherlands. Appl Environ Microbiol 71:1453–1461

Long SC, El-Khoury SS, Oudejans SJG et al (2005) Assessment of sources and diversity of male-specific coliphages for source tracking. Environ Eng Sci 22:367–377

Love DC, Sobsey MD (2007) Simple and rapid F+ coliphage culture, latex agglutination, and typing assay to detect and source track fecal contamination. Appl Environ Microbiol 73:4110–4118

Lucena F, Araujo R, Jofre J (1996) Usefulness of bacteriophages infecting *Bacteroides fragilis* as index microorganisms of remote faecal pollution. Water Res 30:2812–2816

Lucena F, Duran AE, Morón A et al (2004) Reduction of bacterial indicators and bacteriophages infecting faecal bacteria in primary and secondary wastewater treatments. J Appl Microbiol 97:1069–1076

Lucena F, Mendez X, Morón A et al (2003) Occurrence and densities of bacteriophages proposed as indicators and bacterial indicators in river waters from Europe and South America. J Appl Microbiol 94:808–815

Masanori K, Watanabe S, Furuse K, Ozawa A (1985) *Bacteroides* bacteriophages isolated from human faeces. Microbiol Immunol 29:895–899

McLaughlin MR, Rose JB (2006) Application of *Bacteroides fragilis* phages as an alternative indicator of sewage pollution in Tampa Bay, Florida. Estuaries and Coasts 29:246–256

Mendez J, Audicana A, Isern A et al (2004) Standardised evaluation of the performance of a simple membrane filtration-elution method to concentrate bacteriophages from drinking water. J Virol Methods 117:19–25

Miyake T, Haruna I, Shiba T et al (1971) Grouping of RNA phages based on the template specificity of their RNA replicases. Proc Natl Acad Sci USA 68:2022–2024

Miyake T, Shiba T, Sakurai T, Watanabe I (1969) Isolation and properties of two new RNA phages SP and FI. Jpn J Microbiol 13:375–382

Miyake T, Shiba T, Watanabe I (1968) Subgroups in group 3 of RNA phages. Jpn J Microbiol 12:167–170

Mocé-Llivina L (2004) Avenços metodològics en la detecció de virus entèrics en aigües. Ph. D. Thesis. University of Barcelona, Barcelona, Spain

Mocé-Llivina L, Lucena F, Jofre J (2005) Enteroviruses and bacteriophages in bathing waters. Appl Environ Microbiol 71:6838–6844

Mocé-Llivina L, Muniesa M, Pimenta-Vale H et al (2003) Survival of bacterial indicator species and bacteriophages after thermal treatment of sludge and sewage. Appl Environ Microbiol 69:1452–1456

Nieuwstad TJ, Mulder EP, Havelaar AH, van Olphen M (1988) Elimination of microorganisms from wastewater by tertiary precipitation followed by filtration. Water Res 22:1389–1397

Noble RT, Allen SM, Blackwood AD et al (2003) Use of viral pathogens and indicators to differentiate between human and non-human fecal contamination in a microbial source tracking comparison study. J Water Health 1:195–207

Novotny CP, Lavin K (1971) Some effects of temperature on the growth of F-pili. J Bacteriol 107:671–682

O'Connell KP, Bucher JR, Anderson PE et al (2006) Real-time fluorogenic reverse transcription-PCR assays for detection of bacteriophage MS2. Appl Environ Microbiol 72:478–483

Ogorzaly L, Gantzer C (2006) Development of real-time RT-PCR methods for specific detection of F-specific RNA bacteriophage genogroups: application to urban raw wastewater. J Virol Methods 138:131–139

Osawa S, Furuse K, Choi MS et al (1981) Distribution of ribonucleic acid coliphages in Korea. Appl Environ Microbiol 41:909–911

Payan A (2006) Bacteriófagos como modelo en el origen de la contaminación fecal y en aguas regeneradas. Ph D Thesis. University of Barcelona, Barcelona, Spain

Payan A, Ebdon J, Taylor H et al (2005). Method for Isolation of *Bacteroides* bacteriophages host strains suitable for tracking sources of fecal pollution in water. Appl Environ Microbiol 71:5659–5662

Puig A, Araujo R, Jofre J, Frias-Lopez J (2001) Identification of cell wall proteins of *Bacteroides fragilis* to which bacteriophages B40-8 binds specifically. Microbiol (UK) 147:281–288

Puig M, Gironés R (1999) Genomic structure of phage B40-8 of *Bacteroides fragilis.* Microbiol (UK) 145:1661–1670

Puig M, Jofre J, Gironés R (2000) Detection of phage infecting *Bacteroides fragilis* HSP40 using a specific DNA probe. J Virol Methods 88:163–173

Puig A, Queralt N, Jofre J, Araujo R (1999) Diversity of *Bacteroides fragilis* strains in their capacity to recover phages from human and animal wastes and from fecally polluted wastewater. Appl Environ Microbiol 65:1772–1776

Queralt N, Jofre J, Araujo R, Muniesa, M (2003) Homogeneity of the morphological groups of bacteriophages infecting *Bacteroides fragilis* strain HSP40 and strain RYC2056. Curr Microbiol 46:163–168

Sakurai T, Miyake T, Shiba T, Watanabe I (1968) Isolation of a possible fourth group of RNA phage. Jpn J Microbiol 12:544–546

Schaper M, Duran AE, Jofre J (2002a) Comparative resistance of phage isolates of four genotypes of F-specific RNA bacteriophages to various inactivation processes. Appl Environ Microbiol 68:3702–3707

Schaper M, Jofre J, Uys M, Grabow WOK (2002b) Distribution of genotypes of F-specific RNA bacteriophages in human and non-human sources of faecal pollution in South Africa and Spain. J Appl Microbiol 92:657–667

Scott TM, Rose JB, Jenkins TM et al (2002) Microbial source tracking: Current methodology and future directions. Appl Environ Microbiol 68:5796–5803

Sechter I, Birzu AL, Scherzer P et al (1957) Contributii la studiul tulpinilor test pentru cercetarea bacteriofagilor tifici Vi in ape. Medicina 9:315

Sobsey MD, Schwab K, Kellogg et al (1990) Simple membrane filter method to concentrate and enumerate male-specific RNA coliphages. J Am Water Works Ass 82:52–59

Sobsey MD, Yates MV, Hsu FC et al (2004) Development and evaluation of methods to detect coliphages in large volumes of water. Wat Sci Technol 50:211–217

Sommer R, Haider T, Cabaj A et al (1998) Time dose reciprocity in UV disinfection of water. Wat Sci Technol 12:145–150

Stevens G, Mascarenhas M, Mathers C (2009) Global health risks: progress and challenges. Bull World Health Org 87:646

Stewart-Pullaro J, Daugomah JW, Chestnut DE et al (2006) F+ RNA coliphage typing for microbial source tracking in surface waters. J Appl Microbiol 101:1015–1026

Stewart JR, Vinjé J, Oudejans SJ et al (2006) Sequence variation among group III F-specific RNA coliphages from water samples and swine lagoons. Appl Environ Microbiol 72:1226–1230

Storey MV, Ashbolt NJ (2001) Persistence of two model enteric viruses (B40-8 and MS2 bacteriophages) in water distribution pipe biofilms. Wat Sci Technol 43:133–138

Sun ZP, Levi Y, Kiene L et al (1997) Quantification of bacteriophages of *Bacteroides fragilis* in environmental water samples of Seine river. Water, Air and Soil Pollut 96:175–183

Tartera C, Jofre J (1987) Bacteriophage active against *Bacteroides fragilis* in sewage polluted waters. Appl Environ Microbiol 53:1632–1637

Tartera C, Lucena F, Jofre J (1989) Human origin of *Bacteroides fragilis* bacteriophages present in the environment. Appl Environ Microbiol 55:2696–2701

Twort FW (1915) An investigation on the nature of ultra-microscopic viruses. Lancet 2:1241
US EPA (2001a) Method 1601. Male-specific (F+) and somatic coliphage in water by two-step enrichment procedure, vol. 821-R-01-030. EPA Office of Water, Washington, DC
US EPA (2001b) Method 1602. Male-specific (F+) and somatic coliphage in water by single agar layer (SAL) procedure, vol. 821-R-01-029. EPA Office of Water, Washington, DC
Vaughn JM, Metcalf TG (1975) Coliphages as indicators of enteric viruses in shellfish and shellfish raising estuarine waters. Water Res 9:613–616
Vijayavel K, Fujioka RS, Ebdon J, Taylor H (2010) Isolation and characterization of *Bacteroides* host strain HB-73 used to detect sewage specific phages in Hawaii. Water Res 44:3714–3724
Vinjé J, Oudejans SJ, Stewart JR et al (2004). Molecular detection and genotyping of male-specific coliphages by reverse transcription-PCR and reverse line blot hybridization. Appl Environ Microbiol 70:5996–6004
Watanabe I, Miyake T, Sakurai T et al (1967) Isolation and grouping of RNA phages. Proc Jpn Acad 43:204
Wiggins BA, Alexander M (1985) Minimum bacterial density for bacteriophage replication: implications for significance of bacteriophages in natural ecosystems. Appl Environ Microbiol 49:19–23
Woody MA, Cliver DO (1995) Effects of temperature and host cell growth phase on replication of F-specific RNA coliphage Q beta. Appl Environ Microbiol 61:1520–1526
Woody MA, Cliver DO (1997) Replication of coliphage Q beta as affected by host cell number, nutrition, competition from insusceptible cells and non-FRNA coliphages. J Appl Microbiol 82:431–440
WHO (2004) Guidelines for Drinking-water Quality, Third Edition, Volume 1: Recommendations. World Health Organization, Geneva
Xu J, Bjursell MK, Himrod J et al (2003) A genomic view of the human-*Bacteroides tethaiotaomicron* simbiosis. Science 299:2074–2076

Chapter 7
Pathogenic Protozoa

Joseph A. Moss and Richard A. Snyder

Abstract Waterborne diseases associated with polluted recreational and potable waters have been scientifically documented for over a century. *Cryptosporidium* and *Giardia*, as well as the Microsporidia, are known to be key causative agents of a majority of the reported cases of gastrointestinal disease worldwide. Concern regarding these protozoan pathogens has led to the continual design and application of various methodologies directed at their detection. We review and provide a guide to the literature on current background information on pathogenic protozoa and on new, innovative, and advanced methods. Molecular diagnostics are shifting the approach to evaluating and monitoring potable and environmental samples. A variety of techniques have been tested for use in protozoan-specific pathogen assays, some more sensitive, facile, or cost-effective than others. We also propose a set of standards to serve as guidance for method development for monitoring efforts that would make comparison of emerging techniques more meaningful, including addressing standard quality control and quality assurance steps in the testing and reporting of new methods.

Keywords *Cryptosporidium Giardia* Microsporidia

7.1 Introduction

Waterborne diseases associated with polluted recreational and potable waters have been scientifically documented for over a century. Over the past 30 years, pathogenic enteric protozoa have been increasingly implicated in compromising the health of millions of people, mostly in developing nations. These protozoan

R.A. Snyder (✉)
Center for Environmental Diagnostics and Bioremediation,
University of West Florida, Pensacola, FL, USA
e-mail: rsnyder@uwf.edu

C. Hagedorn et al. (eds.), *Microbial Source Tracking: Methods, Applications, and Case Studies*, DOI 10.1007/978-1-4419-9386-1_7,

agents contribute significantly to the staggering caseload of diarrheal disease morbidity encountered in these regions yet also are significant concerns in industrialized countries despite improved sanitation. Major diseases associated with waterborne protozoan include amoebiosis, cryptosporidiosis, giardiasis, and microsporidiosis.

Significant progress has been achieved in improving water quality, though many countries suffer from persistent waterborne pathogenic protozoans and will continue to encounter problems in the future. Human enteric protozoans may be found widespread in environmental waters impacted by agriculture, human wastewater, and stormwater runoff. Pathogenic enteric protozoa of primary concern for human health include those from the genera *Cryptosporidium*, *Encephalitozoon*, *Enterocytozoan*, and *Giardia* that are human-specific or zoonotic. Additional protozoan pathogens, such as *Isospora belli*, and *Cyclospora cayetanensis*, have been associated with waterborne outbreaks (Khan 2008). This chapter reviews progress in technology and methodologies designed for the detection and monitoring of waters for human use for contamination by *Cryptosporidium* spp., *Giardia* spp., and Microsporidia.

7.1.1 *Cryptosporidium*

Though only first recognized as a waterborne agent in 1985 (D'Antonio et al. 1985), cryptosporidiosis from *Cryptosporidium* spp. infection is regarded a serious condition and is potentially fatal in immunocompromised individuals. There are currently 16–18 named species with over 40 genotypes of *Cryptosporidium* recognized in the literature (Alvarez-Pellitero et al. 2004; Xiao and Ryan 2004, 2008; Fayer et al. 2005; Fayer 2008; Xiao and Fayer 2008; Xiao 2010). Of these, there are essentially 5–7 *Cryptosporidium* spp. (*canis*, *felis*, *hominis*, *meleagridis*, *muris*, *parvum*, and *suis*) that have been attributed to most human infections (Cacciò et al. 2005). Two genotypes of *C. parvum* (Type I and II; *parvum* and *hominis*) are considered to be accountable for most cases of cryptosporidiosis. Other species and genotypes of *Cryptosporidium* have been sporadically detected in infected persons and warrant further investigation (Robinson et al. 2008; Chalmers et al. 2009).

A number of excellent reviews are available covering these organisms. The historical taxonomy has been integrated with more modern molecular information for *Cryptosporidium* and *Giardia* by Xiao and Fayer (2008). A thorough review of *Cryptosporidium* taxonomy, its historical perspective, and public health implications is discussed in Xiao et al. (2000, 2004b) and is not a focus of this review. Additionally, a recent and thorough compilation of *Cryptosporidium* historical taxonomy, genomics, life stages, and morphology is found in Xiao and Cama (2006) "*Cryptosporidium* and cryptosporidiosis." Transmission dynamics and epidemiology, sources of contamination in foodborne transmission, outbreaks and governmental regulations concerning drinking water, recreational waters, and waste management were also compiled by Fayer (2008). Information pertaining to morphology, survival, as well as regulatory measures and directives, can be found in Smith and Thompson (2001) and Harrington (2008).

7.1.2 *Giardia*

Giardia duodenalis (syn. *G. lamblia*, *G. intestinalis*) is arguably the most widespread protozoan causing diarrhea with 200 million symptomatic individuals worldwide (WHO 2006). *G. duodenalis* infections can be asymptomatic, and prevalence is 2–5% in industrialized countries and 20–30% in developing countries. *G. duodenalis* is currently categorized into seven genotypes or assemblages (Thompson and Monis 2004; Sprong et al. 2009). Only two of these genetic clusters have been detected in humans, but both also infect other mammals. *Giardia* isolates recovered from humans and many other mammalian species fall into one of the two major genotypic assemblages, A or B (Sprong et al. 2009). Molecular analyses have shown that the genetic distance separating these assemblages exceeds that used to delineate other species of protozoa and have demonstrated the existence of genetic subgroups within each assemblage.

Livestock, dogs, and birds may be significant reservoirs for human infection (Castro-Hermida et al. 2009; Little et al. 2009; Plutzer and Tomot 2009), although in a study in New Zealand, cryptosporidiosis was found to be correlated with rural areas and livestock, whereas giardiasis was associated more with human to human contact (Snel et al. 2009). Both *Cryptosporidium* spp. and *Giardia* cysts are known to survive wastewater treatment processes and are present in effluent and sludge (Cheng et al. 2009). A recent comprehensive book covering both *Giardia* and *Cryptosporidium* can be found (Ortega-Pierres et al. 2009).

7.1.3 *Microsporidia*

Owing to increasingly sophisticated tools for detecting agents of disease, Microsporidia are considered emerging human enteric pathogens of concern (Didier et al. 2004; Mathis et al. 2005). The phylum Microsporidia consists of more than 140 genera, with more than 1,200 described species (Wittner and Weiss 1999; Weiss 2001). Microsporidia can cause disease in immunocompetent but more often in immunocompromised humans (Cotte et al. 1999). Microsporidia infections have been reported worldwide (Leder et al. 1998; Franzen and Müller 1999; Didier and Weiss 2006) with 14 species in eight genera reported as human pathogens (Didier 2005). Microsporidian genera more commonly recovered from humans include the following: *Brachiola*, *Encephalitozoon*, *Enterocytozoon*, *Nosema*, *Pleistophora*, *Trachipleistophora*, *and Vittaforma*. Of these, the *Encephalitozoon* species (*E. cuniculi*, *hellem*, and *intestinalis*) are most often reported as opportunistic pathogens in human immunodeficiency virus (HIV)-infected individuals as well as other immunocompromised individuals (Wittner and Weiss 1999; Weiss 2001).

The species *Enterocytozoon bieneusi* and *E. intestinalis* have been found to be responsible for gastrointestinal disease worldwide (Hartskeerl et al. 1995; Didier 2005; WHO 2006). However, *E. bieneusi* is more commonly detected in patients with HIV infection (Leelayoova et al. 2006; Espern et al. 2007; Stark et al. 2009). Over 30 various genotypes of *E. bieneusi* have been reported to infect humans. Subject matter pertaining to morphology, disease, and treatment, beyond the scope of this chapter, has been well covered elsewhere (Wittner 1999; Omalu et al. 2006). Information concerning phylogeny, taxonomy, and life cycles of the Microsporidia can be found in reviews by Weiss (2001) and Franzen and Müller (1999). An excellent review on the morphology, life cycle, taxonomy, and historical research on the Microsporidia is available (Franzen 2008).

E. bieneusi, a genetically diverse protozoan, is composed of over 80 host-specific and zoonotic genotypes. *E. bieneusi* has been detected in a variety of wild and domestic mammals (Sulaiman et al. 2003) and chickens (Reetz et al. 2002). These pathogenic intestinal protozoa have been reported for clinically important infections in the developed and the developing world and are becoming more and more prevalent in USA. *E. bieneusi*, now recognized as a "true pathogen," is acknowledged to be the most frequent microsporidial infection of humans.

Cryptosporidium and *Giardia*, as well as Microsporidia, are known to be ubiquitous, with numerous cases of outbreaks reported globally (Smith 1999; Davis 2009). Information regarding global cases for *Cryptosporidium* and *Giardia,* as well as that of other protozoan pathogens, can be found in reviews by Savioli et al. (2006) and Karanis et al. (2007). While we know that *C. parvum*, *C. hominis*, and *G. duodenalis* are the species of parasites responsible for the vast majority of cases, information is sparce regarding their pathobiology. All three have been documented in zoonotic transmissions (Thompson 2000; Widmer et al. 2002). *C. parvum* has been found in cattle, goats, horses, and sheep (Mueller-Doblies et al. 2008; Smith et al. 2004; Quilez et al. 2008). *G. duodenalis* has been found in cattle, dogs, and waterfowl (Thompson 2004; Traub et al. 2004; Feng et al. 2008a; Plutzer and Tomot 2009; Winkworth et al. 2009). *Encephalitozoon* species (*intestinalis, cuniculi, hellem*) have various reservoirs including rabbits, birds, and dogs, to name a few (Graczyk et al. 2008; Santaniello et al. 2009; Snowden et al. 2009). Various genotypes of *E. cuniculi* and *E. bieneusi* have been detected in swine (Reetz et al. 2009). *E. bieneusi* has recently been detected in the feces of rhesus macaques, cats, and cattle. Current information pertaining to known and suspected reservoirs for *E. bieneusi* can be found in Dengjel et al. (2001), Rinder et al. (2000), Leelayoova et al. (2009), Curry (1999), and Didier et al. 2004. Zoonotic potential and the occurrence of zoonotic transmission of *Cryptosporidium* and *Giardia* (Stantic-Pavlinic et al. 2003; Goh et al. 2004; Thompson 2004; Traub et al. 2004; Hunter and Thompson 2005; Feng et al. 2008a) and Microsporidia (Rinder et al. 2000; Dengjel et al. 2001; Sulaiman et al. 2003; Cama et al. 2007) have been and continue to be well documented.

Transmissions of protozoan parasites through potable and recreational water are the primary avenues for human infection (Fayer et al. 2000; Smith and Grimason 2003); however, their detection in environmental and recreational waters is highly problematic. The ability to monitor effectively is difficult due to the dilution that occurs as pathogens are disseminated from terrestrial sources to aquatic ecosystems. The infective dose (ID50) to human beings is considered to be small relative to numbers in source material, for *G. intestinalis* (ID50 = 25–100 cysts; Rendtorff 1954, 1979) and *C. parvum* (ID50 = 10–132 oocysts, DuPont et al. 1995; Okhuysen et al. 1999; Heitman et al. 2002). For Microsporidia, the minimal infectious dose is still undetermined, though nonhuman trials indicate the infectious dose to be low (Didier et al. 2004). Infectious dose from potable and recreational waters is a function of both consumption volume and pathogen concentration. Clearly, a lower concentration in drinking waters may provide an equivalent dose as a much higher concentration in recreational waters.

Human-enteric protozoan species have been continually reported to be present in potable (Dowd et al. 2003), environmental (Dowd et al. 1998), and wastewater (Robertson et al. 2006; Lee et al. 2007) samples. Widespread awareness that immunocompromised individuals are susceptible to more zoonotic species and genotypes of *Cryptosporidium*, *Giardia*, and Microsporidia (Morgan et al. 2000; Pieniazek et al. 1999; Guyot et al. 2001; Weitzel et al. 2001; Xiao et al. 2001a; McOliver et al. 2009) exacerbates species-specific monitoring and assessment of risk, emphasizing the necessity for methods that are in their totality more inclusive, reliable, and reproducible.

7.2 Detection Methods

Development of protocols for pathogen detection in various matrices are continuously evolving. Detection of pathogens in potable and recreational water, as well as monitoring of sewage effluent, has been of increasing importance. This concern has led to the continual design and application of various methodologies, some more sensitive, facile, or cost-effective than others, directed at their detection. Though soon to be discarded, the US Environmental Protection Agency Enhanced Surface Water Treatment Rule (LT2 rule) presently mandates the use of Method 1623 for monitoring source waters and drinking water. The method consists of filtration, concentration, immunomagnetic separation (IMS), fluorescent antibody and 4′, 6′-diamidino-2-phylindole (DAPI) counter staining, and microscopic detection and enumeration. However, the use of this technique is hampered by its labor-intensive, time-consuming nature (LeChevallier et al. 1995), its inability to provide an insight into infectivity (LeChevallier et al. 2003), and its difficulty in distinguishing between human- and animal-specific *Cryptosporidium* and *Giardia* species (Pedraza-Díaz et al. 2000; Jiang et al. 2005b). This technique may also incur problems with false positives (Jiang et al. 2005a) and either low and/or variable rates of recovery (Francy et al. 2004). At present, no standardized method for detection of Microsporidia in water samples currently exists.

New, innovative, and advanced methods involving molecular diagnostics are shifting the approach to evaluating and monitoring potable and environmental samples. Advanced sequencing machines, combined with novel sequencing and computation strategies, have greatly accelerated the accumulation of genetic sequences for development of novel and enhanced PCR assays. The genome sequence of *G. duodenalis* (McArthur et al. 2000), *E. cuniculi* (Katinka et al. 2001) and more recently that of *C. parvum* (Abrahamsen et al. 2004) and *C. hominis* (Xu et al. 2004) have become available, allowing researchers to identify loci suitable for detection, taxonomic delineation, and epidemiologic study purposes. While PCR was almost immediately employed in research and clinical laboratories to study a variety of infectious diseases, it has been only recently that PCR methodologies have become more routine and user-friendly so that they may be implemented regularly for monitoring purposes. The US EPA has now recognized the use of alternative techniques as valid complements and/or replacements to microscopy (EPA 2004a).

Fast, sensitive, and cost-effective protozoan pathogen detection methods are essential for human health risk monitoring efforts. Molecular methodologies have addressed many of the limitations of conventional microscopy-based methods. Advantages of PCR-related methodologies include greater sensitivity (Rochelle et al. 1997) relatively low cost, rapid analysis of multiple samples, and the ability to differentiate pathogen species and strains. PCR methods for the detection of protozoan pathogens in environmental water samples have been recently reviewed (Wiedenmann et al. 1998; Bouzid et al. 2008).

In order to provide sensitivity as well as more reliability for taxonomic delineation, the nested-PCR approach has been found to be quite sensitive. Nested PCR has been reported to have exceptional sensitivity in ideal conditions (Sturbaum et al. 2001; Yu et al. 2009b). The nested approach has been reported to have attained detection limits as low as one cyst (Miller and Sterling 2007) and oocyst (Sturbaum et al. 2001; Hashimoto et al. 2006). Comparable detection capabilities for *Cryptosporidium* have been shown using the nested approach on slide preparations (Nichols et al. 2006a, b; Sunnotel et al. 2006). Nested PCR sensitivity comes with its own set of problems associated with its methodology, primarily a high susceptibility to contamination from sample manipulation and further amplification of contamination from a previous PCR (Álvarez-Nava et al. 2004; Calderaro et al. 2006). The nested PCR is also more expensive, and running two separate reactions per sample trial increases the time involved. Additionally, confirmation reactions and resampling of positives are also more expensive.

Diagnosis of certain protozoan pathogens to the species/genotype level is a challenge but is fundamental for the evaluation of public health. Molecular epidemiology studies have shown that multiple genotypes of protozoan pathogens can circulate within a geographic region (Xiao et al. 2001a–c; Feng et al. 2009a). The nested PCR reaction method has been combined with RFLP for differentiation of

Cryptosporidium species and genotypes present in samples and fixed slides (Nichols et al. 2006a, b). Numerous studies have been published regarding the utility of this combination for a variety of organisms (Gibbons and Awad-El-Kariem 1999; Xiao et al. 2004a, b; Coupe et al. 2005).

A main goal of subtyping is for contamination source tracking during outbreaks (Xiao et al. 2009). Over the last 15 years, genetic characterization of *Cryptosporidium*, *Giardia*, and Microsporidia, as well as that of other protozoans from various sources has led to the development of many genotyping and subtyping tools. Subtyping of intragenotypic genetic diversity of *C. parvum* and *C.hominis* has provided a possible means of tracking their source of contamination during outbreaks (Cacciò et al. 2000, 2001; Feng et al. 2000; Mallon et al. 2003a, b) and for characterizing their relative population structures and transmission dynamics (Widmer et al. 2008). Subtyping allows contamination source tracking during outbreaks (WHO 2006; Thompson and Monis 2004) The utility of subtyping tools for the analysis of samples from clinical, foodborne, and waterborne outbreaks of cryptosporidiosis has been demonstrated (Sulaiman et al. 2001; Glaberman et al. 2002; Leoni et al. 2003, 2007; Cama et al. 2008).

The SSU rRNA gene is a commonly used target for genotyping *Cryptosporidium* (Jiang et al. 2005a, b; Robertson et al. 2006; Ruecker et al. 2007; Plutzer et al. 2008; Lv et al. 2009), *Giardia* (Berrilli et al. 2004; Robertson et al. 2006; Xiao et al. 2006; Santín et al. 2007; Giangaspero et al. 2009; Lim et al. 2009) and Microsporidia (Asakura et al. 2006). SSU rRNA contains genus-specific information and is present in high copy number per cell. The SSU rRNA gene has also been shown useful for detection and subtyping of various other (*Blastocystis*, etc.) protozoan pathogens (Leelayoova et al. 2008).

The outer wall protein (COWP) gene has also been used to discriminate *Cryptosporidium* (Leone et al. 2009; Robinson et al. 2008; Giangaspero et al. 2009) species. Subtyping through use of the glycoprotein precursor gene GP60 has also shown promise (Strong et al. 2000; Sulaiman et al. 2001; Peng et al. 2001, 2003; Leav et al. 2002; Alves et al. 2003; Cama et al. 2008; Lobo et al. 2008). A report linking the GP60 subtype with strain virulence may be beneficial for future risk assessments (Cama et al. 2007). Additional studies have included use of the conserved heat-shock protein genes HSP70 (Pennil et al. 2008) and HSP90 (Feng et al. 2008b, 2009b). A polymorphic genetic marker in a structural gene (the thrombospondin-related adhesive protein C1 [TRAP-C1]) has also been utilized (Spano et al. 1998; Muthusamy et al. 2006; Ajjampur et al. 2007). Different gene targets have been used for investigating *Giardia* genotypes, including the β-giardin gene (Caccio et al. 2002) the glutamate dehydrogenase (*gdh*) gene (Read et al. 2004), and the small subunit (SSU)-rRNA gene (Hopkins et al. 1997).

Methods utilizing nested-PCR and multilocus sequence typing (Cama et al. 2006; Xiao et al. 2006) can be valuable for a more expansive knowledge of species and genotypes present in the environment and their respective transmission dynamics (Zhou et al. 2003). Multilocus sequence typing has been shown to be helpful in the characterization of isolates (Cama et al. 2006; Gatei et al. 2006) and the constituents of mixed populations (Mallon et al. 2003b; Lasek-Nesselquist et al. 2008). Comparable methodologies based on subtyping research have also made use of microsatellite and minisatellite length polymorphisms for distinguishing between *C. hominis* and *C. parvum* (Mallon et al. 2003a; Tanriverdi and Widmer 2006) and have also shown insight into population structures of *C. parvum* and *C. hominis* (Tanrıverdi et al. 2008). An earlier but thorough listing of various primer pairs and hybridization probes used for detection and species differentiation of human Microsporidia and their respective target organisms and genes can be found in Franzen and Müller (1999). A more recent listing of diagnostic primers (with corresponding anneal temps, amplicon size, etc.) for various species of the Microsporidia has been provided by Ghosh and Weiss (2009).

Real-time PCR (qPCR) applications for detection of protozoan pathogens have been emerging, with the promise of semiquantitative molecular-specific detection. Target detection via use of a fluorescent probe during reaction cycling eliminates post-PCR processing, thereby preventing potential PCR product carryover contamination. Results may be obtained within 30–90 min, pending the instrumentation and application utilized. Low detection (<10 oocysts) sensitivity for purified and spiked samples has been reported (Fontaine and Guillot 2003). Clinical studies have shown the methodology to provide accurate and reproducible relative quantification of pathogens (Hester et al. 2002) and of gene copies (Fang et al. 2002). Quantitative-PCR reactions using intercalating and surface-binding dyes have been tried (Tanriverdi et al. 2002; Ramirez and Sreevatsan 2006); however, these methods may have decreased specificity due to detection of primer dimers and nonspecific products (Heid et al. 1996).

Real-time PCR (qPCR) is becoming a method of choice for detection and quantification of gene target sequences (Walker 2002) and has shown potential when applied to detection of *Cryptosporidium* and *Giardia* in fecal and water samples (Higgins et al. 2001; Fontaine and Guillot 2002; Limor et al. 2002; Amar et al. 2003, 2004; Guy et al. 2003; Verweij et al. 2003, 2004; Stroup et al. 2006; Sunnotel et al. 2006; Haque et al. 2007; Helmy et al. 2009; Alonso et al. 2011; Basque et al. 2011; Hadfield et al. 2011). Several investigations in clinical settings have reported qPCR methods for Microsporidia (Blessmann et al. 2002; Menotti et al. 2003a, b; Hester et al. 2002; Wolk et al. 2002; Lejeune et al. 2005; Espern et al. 2007; Verweij et al. 2007; Polley et al. 2011). However, assays optimized for clinical use may not be well suited for environmental samples due to specificity problems, hybridization with competitive DNA, and inhibitory compounds in the matrix. Real-time PCR has been utilized for various other protozoans including *Balamuthia* (Kiderlen et al. 2008); *Toxoplasma gondii* (Jauregui et al. 2001; Lin et al. 2000), *C. cayetanensis* (Varma et al. 2002), *Entamoeba histolytica* (Roy et al. 2005), and *Isospora belli* (ten Hove et al. 2008).

Multiplex real-time PCR is defined by the use of multiple fluorogenic oligoprobes for the detection of amplicons that have been produced by one or more primer pairs. Though problematic to develop, a few multiplex assays have been developed for use with protozoans (Guy et al. 2003; Qvarstrom et al. 2006; Verweij et al. 2007; Stark et al. 2011). A thorough review of the qPCR, how it works, variations of the approach, and where it was applied in the past is available (Mackay 2004).

7.3 Other Detection Methods

A variety of techniques have been tested for use in protozoan-specific pathogen assays. Fluorescent in situ hybridization (FISH)-based methods have been used for specific microscopical detection of *C. parvum* and *G. lamblia* (Deere et al. 1998; Vesey et al. 1998; Dorsch and Veal 2001; Graczyk et al. 2003a). FISH has also been directed against the small subunit or intergenic regions of Microsporidia rRNA to detect *E. bieneusi* and *E. hellem* (Hester et al. 2000; Velásquez et al. 1999) in clinical diagnostic studies. Graczyk et al. (2007a–c) used FISH for detection of Microsporidia. FISH has also been implemented for detection of various protozoan pathogens in recreational waters (Graczyk et al. 2007a, b) as well as detection of protozoa in crustaceans and bivalves (Graczyk et al. 2007c, d). While FISH is desirable for specific and quantitative analysis, it has several drawbacks. It is time-consuming, technically challenging, and less suitable for environmental samples that contain background material and low pathogen numbers. Nevertheless, when applicable, FISH may prove particularly valuable for environmental samples wherein discrimination of live vs. dead organisms is important, which can be provided using probes directed at RNA.

Recently, oligonucleotide microarrays have been used effectively, mostly in clinical settings, for the detection of pathogens. An advantage of this detection approach is that it combines PCR amplification strategies with subsequent hybridization to oligonucleotide probes specific for multiple target sequences. This provides the potential for high-throughput pathogen detection and genotyping. Using immobilized, specific oligonucleotides to monitor and screen purified nucleotide mixtures allows for rapid and simultaneous identification of many different target microorganisms in a single assay. With increasing confidence in genetic tests for identification and differentiation of pathogens, and multiple protozoan species present in most environmental samples, the use of microarrays would appear to be preferable for the detection of waterborne protozoan parasites. One clinical study reports the development of an oligonucleotide microarray to simultaneously detect *E. cuniculi, hellem,* and *intestinalis*, as well as *E. bieneusi* from fecal samples (Wang et al. 2005). Additionally, a similar method has also been designed for simultaneous detection and genotyping of *E. histolytica*, *E. dispar*, *G. lamblia*, and *C. parvum* (Wang et al. 2004).

There has been an increase in the number of studies reporting development and/or improvement of oligonucleotide microarrays for microbial diagnostics for bacterial and viruses, but use of the technology in routine diagnostics is still constrained by a variety of factors (Call 2005; Chandler and Jarrell 2005). One drawback when monitoring environmental isolates using microarrays arises due to the variability in relative abundance of different pathogens, resulting in variability in the signal-to-noise ratio. In addition, microarray hybridizations are not optimized for a single probe reaction, but represent a compromise to accommodate the multiple targets and their potential differences in optimized conditions. Currently, the purchase or construction of DNA microarrays is prohibitively costly for routine application. Various reviews concerning recent applications of microarrays for detection and characterization of pathogens can be found in Kostrzynska and Bachand (2006) and Heller (2002).

An additional method, nucleic-acid sequence-based amplification (NASBA) has shown some promise, demonstrating rapid and sensitive detection when applied to *C. parvum* (Baeumner et al. 2001), but its application on potable and complex environmental and wastewater samples has yet to be verified fully. Additional information regarding common methodologies and less frequently used technologies for detection of waterborne parasites can be found in Lazcka et al. (2007) and Bouzid et al. (2008).

7.4 Collection Methods

Waterborne protozoa are usually detected from large-volume (10–1,000 L) water samples by filter-based concentration methods as per US EPA standard methods. However, the methods reported in the research literature for the collection of water samples have not been standardized. The collection methodology used should complement the intended detection method. Several collection methods have been more widely used than others.

A number of products, some more effective for different applications, are available for the concentration of protozoan pathogens from environmental and potable waters. Several filters have been accepted for use in methods 1622 (EPA 2001a; EPA 821-R-01-026) and 1623 (EPA 2001b; EPA 821-R-01-025; EPA 2005, 815-R-05-002) and the UK Drinking Water Inspectorate (DWI) including the Whatman CrypTest cartridge filter, the Pall Gelman Envirochek HV capsule for *Cryptosporidium*, and the IDEXX (2000) Filta-Max system. Standard test validation studies have not been performed for the Microsporidia or other protozoan pathogens, and no standardized filter or filtration method has been developed for the Microsporidia or other protozoans other than *Cryptosporidium* and *Giardia*.

Envirocheck capsule filters (Pall Life Sciences 2000) have been commonly used for conventional studies since the early 1990s in combination with IMS (Hallier-Soulier and Guillot 1999; Xiao et al. 2001a–c; Lemarchand and Lebaron 2003; Peng et al. 2003; Costán-Longaresa et al. 2008; Keeley and Faulkner 2008). FiltaMax™ filter capsules, though not as regularly used, have also been evaluated (McCuin and Clancy 2003) and applied in various studies (Li et al. 2005, 2009; Ortega et al. 2009). Both products are designed for field filtration with subsequent lab backflushing of the filters to elute target cysts (Idexx requires additional equipment for elution). Filter capsules used for the Crypto–Giardia standard method (1623) are relatively expensive, generally \$100/sample for the filter alone.

Other current, yet less often used, filtration methods involve flatbed membranes (Thurston et al. 2001; Cox et al. 2003) and tangential flow (HemoFlow) ultrafiltration units (Fresenius Medical Care, Homberg, Germany) (Simmons et al. 2001). Flatbed membranes have also been used for detection of Microsporidia as well as additional protozoan pathogens. Though dependent on the method of detection and target organism used, membrane filtration has been shown to result in higher recovery and greater detection than filter capsule methods (Hsu et al. 2001). Use of a disposable hollow-fiber, ultrafiltration system resulted in sample acquisition with equal or better capture efficiency than capsule filters (Simmons et al. 2001; Kuhn and Oshima 2002). A recent study compared the recovery of *Cryptosporidium* oocysts and *Giardia* cysts from raw waters using these products and their respective concentration/elution method (Ferguson et al. 2004). These authors found recoveries using FiltaMax™ and Envirochek™ HV capsule filters were lower when compared to use of flatbed and HemoFlow methods. Additional data from du Preez et al. (2003) indicated that the flatbed technique recovery for *Cryptosporidium* and *Giardia* was comparable or greater than that with Envirochek™ and Filta-Max™ cartridges.

To circumvent losses accrued from backwashing, some studies have capitalized on the rigidity and solvent resistance of protozoans in an acetone-dissolution method (ADM) as a means for concentrating parasites prior to immunofluorescence (IF) microscopy (Aldom and Chagla 1995; Graczyk et al. 1997a, b; Franco et al. 2001). The methodology has been routinely used as a standard procedure in Japan (Japanese Ministry of Health and Welfare 1998) and has also shown value as a preliminary step to gDNA extraction (Udeh et al. 2000, 2007e). The method has been shown to be successful for the collection and detection of *C. parvum* oocysts, *G. lamblia* cysts, as well as Microsporidian spores when combined with FISH (Graczyk et al. 2007a–e).

Other methods for the concentration of target protozoa include continuous flow centrifugation (CFC) (Higgins et al. 2003; Hoffman et al. 2007) and continuous separation channel centrifugation (Borchardt and Spencer 2002; Borchardt et al. 2009). Reported recovery efficiencies ranged from 38 to 75% (tap water) and 78–90% (ultrapure water) (Hoffman et al. 2007). CFC, however, has been reported to have low to moderate recoveries (44–63%) using *E. intestinalis* spores spiked into tap water, natural, and artificially created water samples (NCER/R828041). Other limitations include lengthy process times (~150 mL/min), a low number of samples that can be run simultaneously, and clogging of separation channels when using samples with higher turbidities and larger volumes. Detection sensitivities have not been thoroughly examined for complex environmental matrices. A review covering traditional methods can be found (Smith 1999), and information covering more recent retention methods has been published (Zarlenga and Trout 2004).

There is also continual interest in alternate methods for the collection of protozoan pathogens that circumvent filtration methods. Attention has been directed toward utilizing filter-feeding bivalves (Gómez-Bautista et al. 2000; Gómez-Cousoa et al. 2004; Graczyk et al. 2003b, 2006; Miller et al. 2005, 2006; Lucy et al. 2008) as well as biofilms (Searcy et al. 2006; Packman et al. 2007; Helmi et al. 2008; Wolyniak et al. 2009) as sentinel indicators used for entrapping protozoan pathogens. These types of samples are by necessity qualitative (presence/absence) indicators as volumes would be unknown.

The sensitivity and accuracy of detection in environmental matrices are influenced by a variety of factors, including but not limited to filtration effectiveness, sampling replication, presence of inhibitors or masking substances, the fidelity of polymerases, and quality of reagents. Quite often, especially dealing with environmental samples, there is a great deal of organic and inorganic nontarget material collected, which impede the detection capabilities of both microscopic and molecular techniques, especially in fresh waters where contamination with protozoans is more prevalent.

One of the main problems encountered in microscopy-based or molecular-based detection analyses is the separation of these matrix components that interfere with pathogen recovery and the sensitivity of detection. Samples often contain substances (particulates, proteins, polysaccharides, humic acids, etc.) that not only complicate pathogen detection via microscopy (Nieminski et al. 1995; Feng et al. 2003) but also inhibit PCR (Jiang et al. 2005a, b). Various techniques has focused on alleviating or overcoming these shortcomings.

One such technique, IMS, began development in the mid-late 1990s (Wright et al. 1994; Enroth and Engstrand 1995; Campbell and Smith 1997) to aid specifically in the separation of pathogens from various food, fecal, and water sources. In short, the methodology entails the use of paramagnetic beads coated with antibodies specific to surface antigens of target organisms to separate and concentrate organisms from a heterogeneously mixed sample. Studies involving detection of *Cryptosporidium* and *Giardia* have used the IMS procedure prior to cell culture, IF, and DNA extraction to remove nontarget organisms and PCR inhibitors present in samples (Mahbubani et al. 1998; Hallier-Soulier and Guillot 1999, 2000; Xiao et al. 2000; Jellison et al. 2002; Rimhanen-Finne et al. 2002; Sturbaum et al. 2002; Ward et al. 2002; LeChevallier et al. 2003; Nichols et al. 2006; Coupe et al. 2006). With respect to studies involving detection of *C. parvum*, the use of IMS as a prep to PCR has been reported to be sensitive and discriminating (Johnson et al. 1995; Wiedenmann et al. 1998; Kostrzynska et al. 1999; Hallier-Soulier and Guillot 1999, 2000). Various authors report the use of IMS as an effective pre-PCR step for not only the isolation of target pathogens but also the reduction of Taq polymerase inhibitors (Hallier-Soulier and Guillot 2000; Fontaine and Guillot 2003).

IMS has been developed for the Microsporidia (Enriquez et al. 1997; Accoceberry et al. 1999), and its effectiveness has been evaluated in comparative studies (Hoffman et al. 2001, 2007) and applied for use in water monitoring (Sorel et al. 2003). Though separation by IMS showed good recovery (78–90%) using ultrapure water, concerns regarding specificity and availability of antibodies have been made (Hoffman et al. 2001, 2007). An additional concern commonly reported with the use of IMS on *Cryptosporidium* and *Giardia* is that of decreased efficiency with increased turbidity (DiGiorgio et al. 2002). Particulate matter associated with turbid samples can inhibit or interfere with antibody specificity and adherence (Hoffman et al. 2007; Zhang et al. 2007). Various matrix factors thought to influence recovery have been reported, including pH and the presence of clay and metals (Connell et al. 2000; McCuin et al. 2001; McElroy et al. 2001; Kuhn et al. 2002). Unequal performance of IMS antibodies available has also been reported (Bukhari et al. 1998; Rochelle et al. 1999; Hoffman et al. 2001). Additionally, and perhaps most importantly, the cost, time, training, and necessary equipment required to successfully perform IMS methodologies make them inappropriate for routine water monitoring in many cases (Bukhari et al. 1998; Royer et al. 2002). An IMS assay for other known pathogens (e.g., *Cyclospora* and *Isospora*) is not available.

The presence of PCR inhibitors (Wilson 1997), common in complex samples, has been a consistent dilemma associated with many molecular-based assays of environmental samples including those in protozoan pathogen studies (Sluter et al. 1997; Loge et al. 2002). To alleviate problems associated with PCR inhibition, a few studies have compared various extraction methods, both traditional methods and extraction kits, and treatments (PVPP, BSA, etc.) to maximize DNA yields while reducing inhibitors (Guy et al. 2003; Subrungruan et al. 2004; Jiang et al. 2005b; Nikaeen and Makimura 2007; Babaei et al. 2011). Additional information and similar comparisons can be found in Xiao et al. (2006), Yu et al. (2009a), and Smith and Nichols (2010).

7.5 Concerns

Evaluation of new monitoring approaches for protozoan pathogens of fecal origin, whether dealing with collection, purification, or detection, is difficult without consistent and realistic tests of their implementation. Initial methods development requires the use of ideal controlled conditions, but the methods ultimately need to be optimized and tested under the conditions and with sample matrices that reflect their ultimate use. Optimization of procedures for ideal laboratory conditions is not the same as optimizations for field or point-of-use conditions, and in many instances, what is possible in the laboratory is not transferable to field samples. Many of the reported detection methodologies in the research literature are not amenable to real-life situations, either due to problems incurred by cost, facility, repeatability, or versatility.

Comparative studies of target genes and the respective primers developed for their detection have been studied for *Cryptosporidium* spp. (Rochelle et al. 1997; Chalmers et al. 2005; Leetz et al. 2007; Yu et al. 2009a, b), *Giardia* spp. and the Microsporidia. Many of these studies, though important for advancement of PCR and phylogenetic studies, do not address reaction efficacy in samples derived from environmental media containing competitive DNA and PCR inhibitors. As an example, in some cases dilution of extracted DNA from environmental sources can improve PCR reaction efficiency. Reported detection sensitivities with single organism target DNA under ideal conditions and without regard to their use in complex and inhibitory environmental matrices are difficult to evaluate for point-of-use purposes.

The use of the term "detection sensitivity" is in itself a source of confusion, since it is sample matrix and context dependent. While some reported "detection limits" may indeed be impressive, they may not translate well to environmental samples containing both a diversity of background DNA and inhibitory organic substrates that may require different optimization strategies. Also, appropriate lower detection limits determined using established quality assurance and quality control procedures are less concerned with the minimal amount that was detected, as often reported, than a repeatable lower detection limit with defined confidence limits.

Clearly, there is a difference between basic research and vetting methods for regulatory use, but to facilitate research to a common goal we would suggest establishing standard criteria for the development and evaluation of methodologies in the basic research phase of development. There is an established procedure for EPA approval of detection methods (EPA 2004b; EPA-821-B-03 004). However, what is envisioned are criteria oriented toward basic research that would allow objective comparative assessment between methods prior to advancing to standard method validation. Without this frame work, screening methods for promising regulatory trial evaluations will continue to be problematic. Standard test development criteria not only would act as a guide for establishing detection limits and for accurately comparing methodologies but will also act as an impetus for resolving site-specific differences that may be required. It may also act as a catalyst for additional discussion and collaboration with studies regarding clinical diagnoses and information gaps in pathogen genealogy.

In order to advance pathogen detection capabilities and their use in potable and environmental matrices, we suggest the following:

(a) A standard turbidity testing criteria for evaluating and comparing protocol sensitivities for use in environmental matrices.
(b) Use of quality assurance and quality control measures (i.e., NELAC) for defining accuracy and precision of methods and lower detection limits.
   - Use of appropriate controls.
   - Use of statistics.
(c) Cost-effectiveness as an evaluation criterion.
(d) Ease of use and training as an evaluation criterion.

Foremost, the definition of the term "detection limit" must be clear and unambiguous. Most importantly, an assay's "detection limit" must encompass sample collection, processing, and detection. With regard to finished, environmental and recreational waters, reported detection sensitivities need to be accompanied by some measure of water quality such as turbidity (i.e., <2 for potable, 2–9 for environmental, >9 for WWTP). Accordingly, for the estimation of an assay's sensitivity, volume of water processed needs to be reported with some minimum for practical monitoring purposes, relating to infectious dose estimates and average ingested/exposure amount. Obviously, ingested volume would be different for drinking water vs. recreational waters.

Guidelines for evaluating and comparing methodologies using potable water or environmental matrices are listed below (Table 7.1). Minimum volume requirements

Table 7.1 Evaluation Guidelines

Matrix investigated	Potable	Environmental	WWTP/sludge
Turbidity range (NTUs)	<2	2–9	>9
Volume (minimum liters)	20	10	<5
Detection limit ((oo)cyst/spore per liter)	4	10	25

are based on the capabilities of current methods used for the capture and concentration of protozoan pathogens in matrices of varying turbidity. Clearly, when possible, investigators should strive to exceed these minimum requirements. Pathogen quantities used for spiking should be adjusted accordingly to meet the detection limit (i.e., cysts/L) designated for each range of turbidity.

Detection limit criterion (Table 7.1) for potable water is based on recent reported detection capabilities using molecular-grade or tap water. The detection limit criterion listed (10; Table 7.1) for environmental (>2 NTUs) samples is established from reported ID50s, recreational ingesting estimates, and limitations associated with increased inhibition in higher turbidities. Estimates of average volumes of water ingested during recreational swimming reported by Dang (1996) and Coupe et al. (2006) have been approximated at 50 and 76 μL, respectively. Taken into account the infective dose (ID50) for *G. intestinalis* (25–100 cysts; Rendtorff 1954, 1979) and *C. parvum* (10–132 oocysts, DuPont et al. 1995; Okhuysen et al. 1999; Atwill et al. 2001; Heitman et al. 2002), the acceptable limits of detection for *Giardia* and *Cryptosporidium* fall between ~329–2,000 cysts/L and ~132–2,640 oocysts/L, respectively. However, a conservative approach should assume an infective dose to be 1 (oo)cyst. Detection limit criteria for environmental samples are set at 10, which is readily attainable and considerably lower than 13/L (which is based on the lowest reported infectivity (one cyst) and avg. vol. of recr. water ingested (65 μL)). For Microsporidia, the minimal infectious dose is still uncertain, yet assumed to be low (Didier et al. 2004). Until a more certain ID50 is ascertained for various members of the Microsporidia, investigations involving Microsporidia should adhere to criteria set for *Cryptosporidium* and *Giardia.*

Cost and time investments in assays are of essential importance for both microbial source tracking and environmental risk assessment if an assay is to be employed on a routine basis. A reliable method for monitoring and risk assessment should be sensitive, economical, straightforward, and repeatable. No one method may be applicable for all applications. Epidemiologic studies implementing various assays for pathogen subtyping (Nested-PCR, sequencing) require longer assay times (8–16 h). However, the impetus for method development to a large degree has been for the development and use of methods for outbreak purposes. For this type of application, methodologies having complete assay times of 3–4 h (1–20 samples) may be needed. Recent advances in molecular equipment and methodologies make these time constraints increasingly attainable if the collection and sample processing steps can be expedited.

Since large outbreaks can occur and have occurred, it can be the case that extraction of DNA from hundreds of field samples at a time may be necessary, including the need for multiple samples to increase the accuracy of detection (Xiao et al. 2006). The current expense in USA (EPA-approved labs) for the evaluation of *Cryptosporidium* and *Giardia* in water samples exceeds $400/sample, most of which is incurred from materials and equipment (filtration capsules; IMS kits). This reality has been the bane of monitoring initiatives (Bouzid et al. 2008; Fayer and Xiao 2008) and is due in part to both the government and scientific community that have been too receptive of the technologies that entail the use of specialized expensive equipment and highly trained personnel.

In addition to a set of standards for guidance during method development and comparisons use, investigators should adhere to use of appropriate controls needed for both comparative and monitoring efforts. Some common items as well as some novel ideas are listed below with regard to routine use and comparisons of PCR-related methodologies for pathogen detection: some are specified in method 1623 and have been adapted for use in PCR-related detection methodologies.

7.5.1 Sampling

Matrix spike – A known quantity of organisms to a specified amount of sample matrix for which an independent estimate of target analyte concentration and assay capability. In order to estimate the overall detection sensitivity of a developed methodology, matrix spikes will be required. The overall detection capabilities of newly developed methodologies should be evaluated against concentration of target organisms in matrix spikes, conforming to the predetermined "detection limit" criteria. This includes studies evaluating newly developed primers. For monitoring purposes, a matrix spike should be used to determine the effect of the matrix on a method's detection capability. All investigations will need to follow a common protocol for spiking to avoid discrepancies in analyte loss prior to collection.

Method blank (negative control) – A method blank (i.e., blank filter) must be run for every trial series to demonstrate independence from contamination during purification, gDNA extraction, and PCR setup and application.

Internal positive control (IPC) – To ensure acceptable execution of the gDNA extraction method used, an IPC should be used with each trial series. A known and consistent amount (e.g., 100 oocysts) of target organisms should be added during the processing stage, subjected to gDNA extraction and subsequent molecular methodology, in parallel with unknowns and method blanks. This step is critical not only for comparative studies but also for aiding in resolving any problems (i.e., method spike does not work) incurred during sample processing. The analyst will be able to elucidate if the reason for the Matrix spike failure was due to sample processing prior to gDNA extraction or if it was the extraction method itself. Determination of the appropriate number of replicate IPC-reactions executed as well as the criteria (i.e., cycle threshold) used to demarcate a successful extraction will be necessary.

7.5.2 PCR

No template control (NTC) – To ensure purity of PCR reagents and the occurrence of false positives during the molecular methodology used, NTCs must be run with every PCR reaction series. The minimal amount of negative controls used and what constitutes (e.g., mol. grade water) an acceptable negative control should be established. NTC volumes should be consistent with the template volume specified in the PCR protocol.

Positive control – To ensure the vitality of the PCR reagents used and to confirm the credibility of the results, positive controls must be included in each reaction series. Acceptable positive control type (purified recombinant DNA, DNA ligated into plasmids, etc.), the number of positive controls per trial and the respective % necessary to be positive, and criteria as to what constitutes an acceptable positive (Ct value, reaction efficiency) needs to be established.

Unknowns – A minimum of ten replicate PCR reactions should be run for each unknown to account for the probability of missing target sequences during preparation of aliquots of potential template DNA from gDNA extraction samples.

Additional regulatory controls and conditions pertain to the application of standards for relative quantification of target pathogens, the minimal amount of replicate reactions necessary for each unknown and standard, the use of acceptable statistical methods, and matters concerning use of IPCs (EPA 2004a). The establishment of the criteria listed will take time considering the number of methods currently available; some initial incongruity will most likely ensue. However, once established these guidelines would provide a template for researchers to follow and allow for objective comparisons.

7.6 Conclusion

The ultimate objective of determining the microbiological quality of water is to identify and then minimize the public health risk from consuming contaminated drinking water and from exposure to contaminated recreational waters. The microbial safety of water for human contact relies on indicator species providing a probability of health risk as a cost-effective compromise. This is because no single technique can measure all types of pathogens, or even all types of bacteria, viruses, or protozoa. However, the indicator species have been demonstrated to be unreliable in determining the presence of protozoan pathogens, which normally do not correlate with bacterial or physicochemical parameters (Harwood et al. 2005) Standardized molecular detection protocols for protozoan pathogens, especially for outbreak investigations, are needed to adequately define human health risks.

This chapter has provided a review of recent research in the development of detection methodologies for protozoan pathogens of fecal contamination origin, and has suggested how research might be better directed to that end. Advances in molecular technologies will inevitably continue to provide increased sensitivity and specificity in detection, yet development of these methods for regulatory assays will require a different perspective than the typical bench molecular biologist may take. Continual evaluative efforts, more common than previously applied, will be necessary if we want not only to keep standards consistent with capabilities of available technologies but also to continually foster scientific advancement by giving direction as to what is now currently acceptable and attainable.

References

Abrahamsen MS, Templeton TJ, Enomoto S, Abrahante JE, Zhu G, Lancto CA, Deng M, Chang C, Widmer G, Tzipori S, Buck GA, Xu P, Bankier AT, Dear PH, Konfortov BA, Spriggs HF, Iyer L, Anantharaman V, Aravind L, Kapur V(2004) Complete Genome Sequence of the Apicomplexan, *Cryptosporidium parvum.* Science 304:441–445

Accoceberry I, Thellier M, Desportes-Livages I, Achbarou A, Biligui S, Danis M, Datry A (1999) Production of monoclonal antibodies directed against the microsporidium *Enterocytozoon bieneusi.* J Clin Microbiol 37:4107–4112

Ajjampur SSR, Gladstone BP, David Selvapandian, Jaya Prakash Muliyil, Honorine Ward, and Gagandeep Kang (2007) Molecular and Spatial Epidemiology of Cryptosporidiosis in Children in a Semiurban Community in South India. J Clin Microbiol 45:915–920

Aldom JE, Chagla AH (1995) Recovery of *Cryptosporidium* oocysts from water by a membrane filter dissolution method. Lett Appl Microbiol 20:186–187

Alonso JL, Amorós I, Cañigral I (2011) Development and evaluation of a real-time PCR assay for quantification of *Giardia and Cryptosporidium* in sewage samples. Appl Microbiol Biotechnol 89:1203–1211

Álvarez-Nava F, Soto M, Fernández E, Zenteno JC (2004) Nested polymerase chain reaction—not ready for gold standard status yet! Fertil Steril 82:1471

Alvarez-Pellitero P, Quiroga MI, Sitjà-Bobadilla A, Redondo MJ, Palenzuela O, Padrós F, Vázquez S, Nieto JM (2004) *Cryptosporidium scophthalmi* n. sp. (Apicomplexa: Cryptosporidiidae) from cultured turbot *Scophthalmus maximus.* Light and electron microscope description and histopathological study. Dis Aquat Org 62:133–145

Alves M, Xiao L, Sulaiman I, Lal AA, Matos O, Antunes F (2003) Subgenotype analysis of *Cryptosporidium* isolates from humans, cattle, and zoo ruminants in Portugal. J Clin Microbiol 41:2744–2747

Amar CF, Dear PH, McLauchlin J (2003) Detection and genotyping by real-time PCR/RFLP analyses of *Giardia duodenalis* from human faeces. J Med Microbiol 52:681–683

Amar CFL, Dear PH, McLauchlin J (2004) Detection and identification by real time PCR/RFLP analyses of *Cryptosporidium* species from human faeces. Letters Appl Microbiol 38:217–222

Asakura T, Nakamura S, Ohta M, Une Y, Furuya K (2006) Genetically unique microsporidian *Encephalitozoon cuniculi* strain type III isolated from squirrel monkeys. Parasitol Int 55:159–162

Babaei Z, Oormazdi H, Rezaie S, Rezaeian M, Razmjou E (2011) Giardia intestinalis: DNA extraction approaches to improve PCR results. Exp Parasitol 128:159–162

Baeumner AJ, Humiston MC, Montagna RA, Durst RA (2001) Detection of Viable Oocysts of *Cryptosporidium parvum* Following Nucleic Acid Sequence Based Amplification. Anal Chem 73:1176–1180

Baque RH, Gilliam AO, Robles LD, Jakubowski W, Slifko TR (2011) A Real Time RT-PCR method to detect viable giardia lamblia cysts in environmental waters. Water Res doi: 10.1016/j.watres.2011.03.032

Berrilli F, Di Cave D, De Liberato C, Franco A, Scaramozzino P, Orecchia P (2004) Genotype characterisation of *Giardia duodenalis* isolates from domestic and farm animals by *SSU-rRNA* gene sequencing. Vet Parasitol 122:199

Blessmann J, Buss H, Nu PA, Dinh BT, Ngo QTV, Van AL, Alla, Jackson T, Ravdin JI, Tannich E (2002) Real-time PCR for detection and differentiation of *Entamoeba histolytica* and *Entamoeba dispar* in fecal samples. J Clin Microbiol 40:4413–4417

Borchardt MA, Spencer SK (2002) Concentration of *Cryptosporidium*, microsporidia and other waterborne pathogens by continuous separation channel centrifugation. J Appl Microbiol 92:649–656

Borchardt MA, Spencer SK, Bertz PD, Ware MW, Dubley JP, Lindquist HD (2009) Concentrating *Toxoplasma gondii* and *Cyclospora cayetanensis* from surface water and drinking water by continuous separation channel centrifugation. J Appl Microbiol 107:1089–1097

Bouzid M, Steverding D, Tyler KM (2008) Detection and surveillance of waterborne protozoan parasites. Curr Opin Biotechnol 19:3302–3306

Bukhari Z, McCuin RM, Fricker CR, Clancy JL (1998) Immunomagnetic separation of *Cryptosporidium parvum* from Source Water Samples of Various Turbidities. Appl Environ Microbiol 64:4495–4499

Caccid S, Homan W, Camilli R, Traldi G, Kortbeek T, Pozio E (2000) A microsatellite marker reveals population heterogeneity within human and animal genotypes of *Cryptosporidium parvum*. Parasitol 120:237–244

Caccid S, Spano F, Pozio E (2001) Large sequence variation at two microsatellite loci among zoonotic (genotype C) isolates of *Cryptosporidium parvum*. Int J Parasitol 31:1082–1086

Caccio SM, De Giacomo M, Pozio E (2002) Sequence analysis of the beta-giardin gene and development of a polymerase chain reaction-restriction fragment length polymorphism assay to genotype *Giardia duodenalis* cysts from human faecal samples. Int J Parasitol 32:1023–1030

Caccid S, Thompson RC, McLauchlin J, Smith HV (2005) Unravelling *Cryptosporidium* and *Giardia* epidemiology. Trends Parasitol 21:430–437

Calderaro A, Piccolo G, Gorrini C, Peruzzi S, Zerbini L, Bommezzadri S, Dettori G, Chezzi C (2006) Comparison between two real-time PCR assays and a nested-PCR for the detection of *Toxoplasma gondii*. Acta Biomed 77:75–80

Call DR (2005) Challenges and Opportunities for Pathogen Detection Using DNA Microarrays Crit. Rev. Microbiol 31:91–99

Cama VA, Arrowood MJ, Ortega YR, Xiao L (2006) Molecular Characterization of the *Cryptosporidium parvum* IOWA Isolate Kept in Different Laboratories. J Euk Microbiol 53:40–42

Cama VA, Pearson J, Cabrera L, Pacheco L, Gilman R, Meyer S, Ortega Y, Xiao L (2007) Transmission of *Enterocytozoon bieneusi* between a Child and Guinea Pigs. J Clin Microbiol 45:2708–2710

Cama VA, Bern C, Roberts J, Cabrera L, Sterling CR, Ortega Y, Gilman RH, Xiao L (2008) *Cryptosporidium* Species and Subtypes and Clinical Manifestations in Children, Peru. Emerg Infect Dis. 14:1567–1574

Campbell AT, Smith HV (1997) Immunomagnetisable separation of *Cryptosporidium parvum* oocysts from water samples. Water Science and Technology 35:397–401

Castro-Hermida, JA, Garcia-Presedo I, Aimeida A, Gonzàlez-Warieta M, Correia Da Costa M and Mezo M. 2009. Detection of *Cryptosporidium* spp. And *Giardia duodenalis* in surface water: a health risk for humans and animals. Water Research 43:4133–4142

Chalmers RM, Ferguson C, Caccid S, Robin B, Gasser RB, Youssef G, EL-Osta A, Heijnen L, Xiao L, Elwin K, Hadfield S, Sinclair M, Stevens M (2005) Direct comparison of selected methods for genetic categorisation of Cryptosporidium parvum and Cryptosporidium hominis species. Int J Parasitol 35:397–410

Chalmers RM, Robinson G, Elwin K, Hadfield SJ, Xiao L, Ryan U, Modha D, Mallaghan C (2009) *Cryptosporidium* Rabbit Genotype, a Newly Identified Human Pathogen. Emerg Infect Dis 15:829–830

Chandler D P, Jarrell AE (2005) Taking arrays from the lab to the field: trying to make sense of the unknown. BioTechniques 38:591–600

Cheng H-W, Lucy FE, Graczyk TK, Broaders MA, Tamang L, and Connolly M. 2009. Fate of *Cryptosporidium parvum* and *Cryptosporidum hominus* oocysts and *Giardia duodenalis* cysts during secondary wastewater treatments. Parasitol. Res. 105:689–696

Connell K, Rodgers CC, Shank-Givens HL, Scheller J, Pope ML, Miller K (2000) Building a better protozoa data set. J Am Water Works Assoc 92:30–43

Cotte L, Rabodonirina M, Chapuis F, Bailly F, Bissuel F, Raynal C, Gelas P, Persat F, Piens M-A, Trepo C (1999) Waterborne Outbreak of Intestinal Microsporodiosis in Persons with and without Human Immunodeficiency Virus Infection. J Infect Dis 180:2003–2008

Coupe S, Sarfati C, Hamane S, Derouin F (2005) Detection of *Cryptosporidium* and Identification to the Species Level by Nested PCR and Restriction Fragment Length Polymorphism J Clin Microbiol 43:1017–1023

Coupe S, Delabre K, Pouillot R, Houdart S, Santillana-Hayat M, Derouin F (2006) Detection of *Cryptosporidium, Giardia* and *Enterocytozoon bieneusi* in surface water, including recreational areas: a one-year prospective study. FEMS Immun Med Microbiol 47:351–359

Costán-Longaresa A, Montemayora M, Payána A, Méndeza J, Jofrea J, Mujeriegob R, Lucenaa F (2008) Microbial indicators and pathogens: Removal, relationships and predictive capabilities in water reclamation facilities. Wat Res 42:4439–4448

Cox P, Fisher I, Kastl G, Jegatheesan V, Warnecke M, Angles M, Bustamante H, Chiffings, T, Hawkins PR (2003) Sydney 1998 — lessons from a drinking water crisis. J. Am Water Works Assoc 95:147–161

Curry A (1999) "Human microsporidial infection and possible animal sources," Curr Opin Infect Dis 12:473–480

Dang W (1996) The swimmer exposure assessment model (SWIMODEL) and its use in estimating risks of chemical use in swimming pool. EPA internal guidance document. U.S. Environmental Protection Agency, Washington DC, USA

D'Antonio RG, Winn RE, Taylor JP, Gustafson TL, Current WL, Rhodes MM, Gary Jr GW, Zajac RA (1985) A waterborne outbreak of cryptosporidiosis in normal hosts. Annals Internal Med 103:886–888

Davis J P (2009) The massive Waterborne Outbreak of Cryptosporidium infection, Milwaukee, Wisconsin, 1993. Book entitled Outbreak Investigation Around the World; M.S. Dworkin (ed). Jones and Bartlett Pub, Sudbury

Deere D, Vesey G, Milner M, Williams K, Ashbolt N, Veal DA (1998) Rapid method for fluorescent in situ ribosomal RNA labeling of *Cryptosporidium parvum*. J Appl Microbiol 85:807–818

Dengjel B, Zahler M, Hermanns W, Heinritzi K, Spillmann T, Thomschke A, Löscher T, Gothe R, Rinder H (2001) Zoonotic Potential of *Enterocytozoon bieneusi*. J Clin Microbiol 39:4495–4499

Didier ES (2005) Microsporidiosis: An emerging and opportunistic infection in humans and animals Acta Tropica 94:61–76

Didier ES, Stovall ME, Green LC, Brindley PJ, Sestak K, and Didier PJ (2004) Epidemiology of microsporidiosis: source and modes of transmission. Vet Parasitol 126:145–166

Didier ES, Weiss, LM (2006) Microsporidiosis: current status Current Opinion in Infectious Diseases 19:485–492 doi: 10.1097/01.qco.0000244055.46382.23

DiGiorgio CL, Gonzalez DA, Huitt CC (2002) *Cryptosporidium* and *Giardia* Recoveries in Natural Waters by Using Environmental Protection Agency Method 1623. Appl Environ Microbiol 68:5952–5955

Dorsch MR, Veal DA (2001) Oligonucleotide probes for specific detection of *Giardia lamblia* cysts by fluorescent in situ hybridization. J Appl Microbiol 90:836–842

Dowd SE, Gerba CP, Pepper IL (1998) Confirmation of the Human-Pathogenic Microsporidia *Enterocytozoon bieneusi*, *Encephalitozoon intestinalis*, and *Vittaforma corneae* in Water. Appl Environ Microbiol 64:3332–3335

Dowd SE, Eliopolus JD, Gerba CP, Naranjo J, Klein R, Lopez B, deMejia M, Mendoza CE, Pepper IL (2003) Confirmed detection of *Cyclospora cayentanensis, Encephalitozoon intestinalis* and *Cryptosporidium parvum* in water used for drinking. J Water Health 1:117–123

DuPont HL, Chappell CL, Sterling CR, Okhuysen PC, Rose JB, Jakubowski W (1995) The infectivity of *Cryptosporidium parvum* in healthy volunteers. New Engl J Med 332:855–859

du Preez, M., Idema G, Ramgopaul N (2003) Validation of a flat-bed membrane concentration method for *Cryptosporidium* oocysts and *Giardia* cysts using USEPA method 1623 as a reference. IWA Health Related Water Microbiology, Cape Town. pp 152–153

DWI. (1999) Standard operating protocol for the monitoring of *Cryptosporidium* oocysts in treated water supplies to satisfy water supply (water quality) amendment regulations 1999, SI No. 1524, Drinking Water Inspectorate. (Available from http://www.dwi.gov.uk/regs/crypto/index.htm)

Enriquez FJ, Ditrich O, Palting JD, Smith K (1997) Simple diagnosis of *Encephalitozoon* sp. microsporidial infections by using a panspecific antiexospore monoclonal antibody. Journal of Clinical Microbiology 35:724–729

Enroth H, Engstrand L (1995) Immunomagnetic separation and PCR for detection of *Helicobacter pylori* in water and stool specimens. J Clin Microbiol 33:2162–2165

Espern A, Morio F, Miegeville M, Illa H, Abdoulaye M, Meyssonnier V, Adehossi E, Lejeune A, Cam PD, Besse B, Gay-Andrieu F (2007) Molecular Study of Microsporidiosis Due to *Enterocytozoon bieneusi* and *Encephalitozoon intestinalis* among Human Immunodeficiency Virus-Infected Patients from Two Geographical Areas: Niamey, Niger, and Hanoi, Viet J Clin Microbiol 45:2999–3002

Fayer R, Xiao L (2008) *Cryptosporidium* and *cryptosporidiosis* (second ed). Fayer, R. and Xiao, L. (eds.), CRC Press and IWA Publishing, Boca Raton 251 pp 251

Fang Y, Wu W-H, Pepper JL, Larsen JL, Salvatore A, Marras E, Nelson, EA, Epperson WB, Christopher-Hennings J (2002) Comparison of Real-Time, Quantitative PCR with Molecular Beacons to Nested PCR and Culture Methods for Detection of *Mycobacterium avium* subsp. *paratuberculosis* in Bovine Fecal Samples. J Clin Microbiol 40:287–291

Fayer R (2008) Fayer, R., Xiao, L. (Eds.), *Cryptosporidium* and cryptosporidiosis (second ed). CRC Press and IWA Publishing, Boca Raton 251 pp 251

Fayer R, Morgan U, Upton SJ (2000) Epidemiology of *Cryptosporidium*: transmission, detection and identification. Int. J. Parasitol. 30:1305–1322

Fayer R, Santin M, Xiao L (2005) *Cryptosporidium bovis* N. Sp. (Apicomplexa: Cryptosporidiidae) in Cattle (*Bos taurus*) J Parasitol 91:624–629

Feng X, Rich SM, Akiyoshi D, Tumwine JK, Kekitiinwa A, Nabukeera N, Tzipori S, Widmer G (2000) Extensive Polymorphism in *Cryptosporidium parvum* Identified by Multilocus Microsatellite Analysis. Appl Environ Microbiol 66:3344–3349

Feng YY, Ong SL, Hu JY, Song LF, Tan XL, Ng WJ (2003) Effect of Particles on the Recovery of *Cryptosporidium* Oocysts from Source Water Samples of Various Turbidities. Appl Environ Microbiol 69:1898–1903

Feng Y, Ortega Y, Cama V, Terrell J, Xiao L (2008a) High intragenotypic diversity of *Giardia duodenalis* in dairy cattle on three farms. Parasitol Res 103:87–92

Feng YY, Dearen T, Cama V, Xiao L (2008b) Development of a genotyping tool for species differentiation of human-pathogenic *Cryptosporidium* based on the 90-kilodalton heat shock protein gene, abstr. PO67. Abstr. 10th Int. Wkshps. Opportun. Protists, Boston, MA, 28 to 31 May 2008

Feng Y, Li N, Duan L, Xiao L (2009a) *Cryptosporidium* Genotype and Subtype Distribution in Raw Wastewater in Shanghai, China: Evidence for Possible Unique *Cryptosporidium hominis* Transmission. J Clin Microbiol 47:153–157

Feng Y, Theresa Dearen, Vitaliano Cama, and Lihua Xiao (2009b) 90-Kilodalton Heat Shock Protein, Hsp90, as a Target for Genotyping *Cryptosporidium* spp. Known To Infect Humans. Euk Cell 4:478–482

Ferguson C, Kaucner C, Krogh M, Deere D, Warnecke M (2004) Comparison of methods for the concentration of *Cryptosporidium* oocysts and *Giardia* cysts from raw waters. Can J Microbiol 50:675–682

Fontaine M, Guillot E (2002) Development of a TaqMan quantitative PCR assay specific for *Cryptosporidium parvum*. FEMS Microbiol Lett 214:13–17

Fontaine M, Guillot E (2003) An immunomagnetic separation–real-time PCR method for quantification of *Cryptosporidium parvum* in water samples. J Microbiol Meth 54:29–36

Franco RMB, Rocha-Eberhardt R, Cantusio-Neto R (2001) Occurrence of *Cryptosporidium* oocysts and *Giardia* cysts in raw water from the Atibaia river, Campinas, Brazil. Rev Inst Med Trop S Paulo 43:109–111

Francy DS, Simmons III OD, Ware MW, Granger EJ, Sobsey MD, Schaefer III FW (2004) Effects of Seeding Procedures and Water Quality on Recovery of *Cryptosporidium* Oocysts from Stream Water by Using U.S. Environmental Protection Agency Method 1623. Appl Environ Microbiol 70:4118–4128

Franzen C (2008) Microsporidia: A Review of 150 Years of Research. Open Parasitol J 2:1–34

Franzen C, Müller A (1999) Molecular Techniques for Detection, Species Differentiation, and Phylogenetic Analysis of Microsporidia. Clin Microbiol Rev 12:243–285

Gatei W, Hart CA, Gilman RH, Das P, Cama V, Xiao L (2006) Development of a Multilocus Sequence Typing Tool for *Cryptosporidium hominis*. J Eukaryot Microbiol 53:43–48

Ghosh K, Weiss LM (2009) Review Article: Molecular Diagnostic Tests for Microsporidia Interdisciplinary Perspectives on Infectious Diseases. Volume 2009, Article ID 926521, pp 13

Giangaspero A, Cirillo R, Lacasella V, Lonigro A, Marangi M, Cavallo P, Berrilli F, Di Cave D, Brandonisio O (2009) *Giardia* and *Cryptosporidium* in inflowing water and harvested shellfish in a Lagoon in Southern Italy Parasitol Int 58:12–17

Gibbons CL, Awad-El-Kariem FM (1999) Nested PCR for the Detection of *Cryptosporidium parvum* Parasitology Today Volume 15, Issue 8, 1 August 1999, pp 345

Glaberman S, Moore JE, Lowery CL, Chalmers RM, Sulaiman I, Elwin K, Rooney PJ, Millar BC, Dooley JSG, Lal AA, Xiao L (2002) Three Drinking-Water–Associated Cryptosporidiosis Outbreaks, Northern Ireland Emerg Infect Dis 8:631–633

Goh S, Reacher M, Casemore DP, Verlander NQ, Chalmers R, Knowles M, Williams J, Osborn K, Richards S (2004) Sporadic Cryptosporidiosis, North Cumbria, England, 1996–2000. Emerg Infect Dis 10:1007–1015

Gómez-Bautista M, Ortega-Mora LM, Tabares E, López-Rodas V, Costas E (2000) Detection of infectious *Cryptosporidium parvum* oocysts in mussels (*Mytilus galloprovincialis*) and cockles (Cerastoderma edule). Appl Environ Microbiol 66:1866–1870

Gómez-Cousoa H, Freire-Santosa F, Amarb CFL, Grantb KA, Williamsonc K, Ares-Mazása ME, McLauchlinb J (2004) Detection of *Cryptosporidium* and *Giardia* in molluscan shellfish by multiplexed nested-PCR. Int J Food Microbiol 91:279–288

Graczyk TK, Cranfield MR, Fayer R (1997a) Recovery of waterborne oocysts of *Cryptosporidium* from water samples by the membrane-filter dissolution method. Parasitol Res 83:121–125

Graczyk TK, Fayer R, Cranfield MR, Owens R (1997b) *Cryptosporidium parvum* oocysts recovered from water by the membrane filter dissolution method retain their infectivity. J. Parasit. 83:111–114

Graczyk TK, Grimes BH, Knight R, Da Silva AJ, Pieniazek NJ, Veal DA (2003a) Detection of *Cryptosporidum* and *Giardia lamblia* carried by synanthropic flies by combined fluorescent in situ hybridization and a monoclonal antibody. Am J Trop Med Hyg 68:228–232

Graczyk TK, Conn DB, Marcogliese DJ, Graczyk H, de Lafontaine Y (2003b) Accumulation of human waterborne parasites by zebra mussels (*Dreissena polymorpha*) and Asian freshwater clams (*Corbicula fluminea*). Parasitol Res 89:107–112

Graczyk TK, Girouard AS, Tamang L, Nappier SP, Schwab KJ (2006) Recovery, bioaccumulation, and inactivation of human waterborne pathogens by the Chesapeake Bay non-native oyster, *Crassostrea ariakensis*. Appl Environ Microbiol 72:3390–3395

Graczyk TK, Lucy FE, Tamang L, Miraflor A (2007a) Human enteropathogen load in activated sewage sludge and corresponding sewage sludge-end products. Appl Environ Microbiol 73:201–215

Graczyk TK, Sunderland D, Tamang L, Lucy FE, Breysse PN (2007b) Bather density and levels of *Cryptosporidium*, *Giardia*, and pathogenic microsporidian spores in recreational bathing water. Parasitol Res 101:1729–1731

Graczyk TK, Sunderland D, Tamang L, Shields TM, Lucy FE, Breysee PN (2007c) Quantitative evaluation of the impact of bather density on levels of human-virulent microsporidian spores in recreational water. Appl Environ Microbiol 73:4095–4099

Graczyk TK, McOliver C, Silbergeld EK, Tamang L, Roberts JD (2007d) Risk of handling as a route of exposure to infectious waterborne *Cryptosporidium parvum* oocysts by Atlantic blue crabs (*Callinectes sapidus*). Appl Environ Microbiol 73:4069–4070

Graczyk TK, Lewis EJ, Glass G, Dasilva AJ, Tamang L, Girouard AS, Curriero FC (2007e) Quantitative assessment of viable *Cryptosporidium parvum* load in commercial oysters (*Crassostrea virginica*) in the Chesapeake Bay. Parasitol Res 100:247–253

Graczyk TK, Majewska AC, Schwab KJ (2008) The role of birds in dissemination of human waterborne enteropathogens. Trends Parasitol 24:55–59

Guy RA, Payment P, Krull UJ, Horgen PA (2003) Real-Time PCR for quantification of *Giardia* and *Cryptosporidium* in environmental water samples and sewage. Appl Environ Microbiol 69:5178–5185

Guyot K, Follet-Dumoulin A, Lelièvre E, Sarfati C, Rabodonirina M, Nevez G, Cailliez JC, Camus D, Dei-Cas E (2001) Molecular Characterization of *Cryptosporidium* Isolates Obtained from Humans in France. J Clin Microbiol 39:3472–3480

Hadfield SJ, Robinson G, Elwin K, Chalmers RM (2011) Detection and differentiation of *Cryptosporidium* spp. in human clinical samples by use of real-time PCR. J Clin Microbiol 49:918–924

Hallier-Soulier S, Guillot E (1999) An immunomagnetic separation polymerase chain reaction assay for rapid and ultra-sensitive detection of *Cryptosporidium parvum* in drinking water. FEMS Microbiology Letters 176:285–289

Hallier-Soulier S, Guillot E (2000) Detection of cryptosporidia and *Cryptosporidium parvum* oocysts in environmental water samples by immunomagnetic separation-polymerase chain reaction. J Appl Microbiol 89:5–10

Haque R, Roy S, Siddique A, Mondal U, Rahman SMM, Mondal D, Houpt E, Petri Jr WA (2007) Real-Time PCR Assay for Detection of *Entamoeba histolytica, Giardia intestinalis*, and *Cryptosporidium* spp. Am J Trop Med Hyg 76:713–717

Hartskeerl R A, Van Gool T, Schuitema AR, Didier ES, Terpstra WJ (1995) Genetic and immunological characterization of the microsporidian *Septata intestinalis* Cali, Kotler and Orenstein, 1993: reclassification to *Encephalitozoon intestinalis*. Parasitol 110:277–285

Harrington BJ (2008) Microscopy of 4 Pathogenic Enteric Protozoan Parasites: A Review. LabMedicine 39:231–238

Harwood VJ, Levine AD, Scott TM, Chivukula V, Lukasik J, Farrah SR, Rose JB (2005) Validity of the Indicator Organism Paradigm for Pathogen Reduction in Reclaimed Water and Public Health Protection. Appl Environ Microbiol 71:3163–3170

Hashimoto A, Sugimoto H, Morita S, Hirata T (2006) Genotyping of single *Cryptosporidium* oocysts in sewage by semi-nested PCR and direct sequencing. Water Research 40:2527–2532

Heid, CA, Stevens J, Livak HJ (1996) Real time quantitative PCR. Genome Res. 6:986–994

Heitman TL, Frederick LM, Viste JR, Guselle NJ, Morgan UM, Thompson RCA, Olson ME (2002) Prevalence of *Giardia* and *Cryptosporidium* and characterization of *Cryptosporidium* spp. isolated from wildlife, human, and agricultural sources in the North Saskatchewan River Basin in Alberta, Canada. Can J Microbiol 48:530–541

Heller MJ (2002) DNA microarray technology: devices, systems, and applications. Annual Rev Biomed Engineer 4:129–153

Helmi K, Skraber S, Gantzer C, Willame R, Hoffmann L, Cauchie H-M (2008) Interactions of *Cryptosporidium parvum*, *Giardia lamblia*, Vaccinal Poliovirus Type 1, and Bacteriophages _X174 and MS2 with a Drinking Water Biofilm and a Wastewater Biofilm. Appl Environ Microbiol 74:2079–2088

Helmy MMF, Abdel-Fattah HS, Rashed L (2009) Real-Time Pcr/rflp Assay to Detect *Giardia intestinalis* Genotypes in Human Isolates with Diarrhea in Egypt. J Parasitol 95:1000–1004

Hester JD, Lindquist HDA, Bobst AM, and Schaefer III FW (2000) Fluorescent *in situ* detection of *Encephalitozoon hellem* spores with a 6-carboxyfluorescein-labeled ribosomal RNA-targeted oligonucleotide probe. J Euk Microbiol 47:299–308

Hester JD, M. Varma, A. M. Bobst, M. W. Ware, H. D. A. Lindquist and F. W. Schaefer, III (2002) Species-specific detection of three human-pathogenic microsporidial species from the genus *Encephalitozoon* via fluorogenic 5′ nuclease PCR assays Mol Cell Probes 16:435–444

Higgins JA, Fayer R, Trout JM, Xiao L, Lal AA, Kerby S, Jenkins MC (2001) Real-time PCR for the detection of *Cryptosporidium parvum.* J Microbiol Meth 47:323–327

Higgins JA, Trout JM, Fayer R, Shelton D, Jenkins MC (2003) Recovery and detection of *Cryptosporidium parvum* oocysts from water samples using continuous flow centrifugation. Water Res. 37:3551–3560

Hoffman RM, Borchardt M, Marshall MM, Sonzogni WC 2001. Final Report: FDP, Development of Detection and Viability Methods for Waterborne Microsporidia. NCER/Grant number: R828041

Hoffman RM, Wolk DM, Spencer SK, Borchardt MA (2007) "Development of a method for the detection of waterborne microsporidia," *Journal of Microbiological Methods*. 70:312–318

Hopkins, RM, Meloni BP, Groth DM, Wetherall JD, Reynoldson JA, Thompson RC (1997) Ribosomal RNA sequencing reveals differences between the genotypes of *Giardia* isolates recovered from humans and dogs living in the same locality. J. Parasitol. 83:44–51

Hsu BM, Huang C, Hsu YF, Jiang GY, Hsu CL 2001. Evaluation of two concentration methods for detecting *Giardia* and *Cryptosporidium* in water, Water Res 35:419–424

Hunter PR, Thompson RCA 2005. The zoonotic transmission of *Giardia* and *Cryptosporidium*. Int J Parasitol 35:1181–1190

IDEXX. 2000. Filta-Max — *Cryptosporidium* and *Giardia* collection and recovery. IDEXX, Westbrook, Maine

Jiang J, Alderisio KA, Xiao L (2005) Distribution of *Cryptosporidium* Genotypes in Storm Event Water Samples from Three Watersheds in New York. Appl Environ Microbiol 71:4446–4454

Jiang J, Alderisio KA, Singh A, Xiao L (2005) Development of Procedures for Direct Extraction of *Cryptosporidium* DNA from Water Concentrates and for Relief of PCR Inhibitors. Appl Environ Microbiol 71:1135–1141

Japanese Ministry of Health and Welfare (1998) Tentative guidelines for protection of water supplies from *Cryptosporidium* pollution

Jauregui LH, Higgins J, Zarlenga D, Dubey JP, Lunney JK (2001) Development of a real-time PCR assay for detection of *Toxoplasma gondii* in pig and mouse tissues. J Clin Microbiol 39:2065–2071

Jellison KL, Hemond HF, Schauer DB (2002) Sources and Species of *Cryptosporidium* Oocysts in the Wachusett Reservoir Watershed. Appl Environ Microbiol 68:569–575

Johnson DW, Pieniazek NJ, Griffin DW, Misener L, Rose JB (1995), Development of a PCR protocol for sensitive detection of Cryptosporidium oocysts in water samples. Appl Environ Microbiol 61:3849–3855

Karanis P, Kourenti C, Smith H (2007) Waterborne transmission of protozoan parasites: A worldwide review of outbreaks and lessons learnt. J Water Health 5:1–38

Katinka, MD, Duprat S, Cornillot E, Metenier G, Thomarat F, Prensier G, Barbe V, Peyretaillade E, Brottier P, Wincker P, Delbac F, El Alaoui H, Peyret P, Saurin W, Gouy M, Weissenbach J, Vivares CP (2001) Genome sequence and gene compaction of the eukaryote parasite *Encephalitozoon cuniculi.* Nature 414:450–453

Keeley A, Faulkner BR (2008) Influence of land use and watershed characteristics on protozoa contamination in a potential drinking water resources reservoir. Wat Res 42:2803–2813

Khan NA (2008) Emerging Protozoan Pathogens. N. A. Khan (ed.) 2008.Taylor and Francis Publishing Co., London, U.K.

Kiderlen AF, Elke Radam E, Lewin A (2008) Detection of *Balamuthia mandrillaris* DNA by real-time PCR targeting the RNase P gene. BMC Microbiol 8:210

Kostrzynska M, Sankey M, Haack E, Power C, Aldom JE, Chagla AH, Unger S, Palmateer G, Lee H, Trevors JT, De Grandis SA (1999) Three sample preparation protocols for polymerase chain reaction based detection of *Cryptosporidium parvum* in environmental samples. J Microbiol Meth 35:65–71

Kostrzynska M, Bachand A (2006) Application of DNA microarray technology for detection, identification, and characterization of food-borne pathogens. Can J Microbiol 52:1–8

Kuhn, RC, Oshima, KH (2002) Hollow-fiber ultrafiltration of *Cryptosporidium parvum* oocysts from a wide variety of 10-L surface water samples. Can J Microbiol 48:542–549

Kuhn RC, Rock CM, Oshima KH (2002) Effects of pH and magnetic material on immunomagnetic separation of Cryptosporidium oocysts from concentrated water samples. Appl Environ Microbiol 68:2066–2070

Lasek-Nesselquist E, Bogomolni AL, Gast RJ, Welch DM, Ellis JC, Sogin ML, Moore MJ (2008) Molecular characterization of *Giardia intestinalis* haplotypes in marine animals: variation and zoonotic potential. Dis Aquat Org 81:39–51

Lazcka O, Del Campo FJ, Muñoz FX (2007) Pathogen detection: A perspective of traditional methods and biosensors. Biosens Bioelectron 22:1205–1217

Leav, B.A., Mackay, M.R., Anyanwu, A., O'Connor, R.M., Cevallos, A.M., Kindra, G., Rollins, N.C., Bennish, M.L., Nelson, R.G., Ward, H.D., 2002. Analysis of sequence diversity at the highly polymorphic Cpgp40/15 locus among *Cryptosporidium* isolates from human immunodeficiency virus-infected children in South Africa. Infection and Immunity 70:3881–3890

LeChevallier MW, Norton WD, Siegel JE, Abbaszadegan M (1995) Evaluation of the immunofluorescence procedure for detection of *Giardia* cysts and *Cryptosporidium* oocysts in water. Appl Environ Microbiol 61:690–697

LeChevallier MW, Di Giovanni G D, Clancy JL, Bukhari Z, Bukhari S, Rosen JS, Sobrinho J, Frey MM (2003) Comparison of Method 1623 and Cell Culture-PCR for Detection of *Cryptosporidium* spp. in Source Waters. Appl Environ Microbiol 69:971–979

Leder K, Ryan N, Spelman D, Crowe SM (1998) Microsporidial Disease in HIV-infected Patients: A Report of 42 Patients and Review of the Literature. Scand J Inf Dis 30:331–338

Lee SH, Lee CH, Kim YH, Do JH, Kim SH (2007) Occurrence of *Cryptosporidium* and *Giardia* cysts in the Nakdong River and their removal during water treatment. J Water Health 5:163–169

Leelayoova S, Subrungruang I, Suputtamongkol Y, Worapong J, Petmitr PC, Mungthin M (2006) Identification of genotypes of *Enterocytozoon bieneusi* from stool samples from human immunodeficiency virus-infected patients in Thailand. J Clin Microbiol 44:3001–3004

Leelayoova S, Siripattanapipong S, Thathaisong U, Naaglor T, Taamasri P, Piyaraj P, Mungthin M (2008) Drinking water: a possible source of *Blastocystis* spp. subtype 1 infection in school children of a rural community in central Thailand. Am J Trop Med Hyg 79:401–406

Leelayoova S, Piyaraj P, Subrungruang I, Pagornrat W, Naaglor T, Phumklan S, Taamasri P, Suwanasri J, Mungthin M (2009) Genotypic characterization of *Enterocytozoon bieneusi* in specimens from pigs and humans in a pig farm community in central Thailand. J Clin Microbiol 47:1572–1574

Leetz AS, Sotiriadou I, Ongerth J, Karanis P (2007) An evaluation of primers amplifying DNA targets for the detection of *Cryptosporidium* spp. using *C. parvum* HNJ-1 Japanese isolate in water samples. Parasitol Res 101:1432–1955

Lejeune A, Espern A, Phung DC, Nguyen TC, Miegeville M (2005) Presentation of the first *Enterocytozoon bieneusi* intestinal microsporidia case in an HIV patient, Hanoi, Vietnam. Med. Mal. Infect. 35:425–426

Lemarchand K, Lebaron P (2003) Occurrence of *Salmonella* spp. and *Cryptosporidium* spp. in a French coastal watershed: relationship with fecal indicators. FEMS Microbiology Letters 218:203–209

Leone A, Ripabelli G, Sammarco ML, Grasso GM (2009) Detection of *Cryptosporidium* spp. from human faeces by PCR-RFLP, cloning and sequencing. Parasitol Res 104:583–587

Leoni F, Gallimore CI, Green J, McLauchlin J (2003) Molecular epidemiological analysis of *Cryptosporidium* isolates from humans and animals by using a heteroduplex mobility assay and nucleic acid sequencing based on a small double-stranded RNA element. J Clin Microbiol 41:981–992

Leoni F, Mallon ME, Smith HV, Tait A, McLauchlin J (2007). Multilocus analysis of *Cryptosporidium hominis* and *Cryptosporidium parvum* isolates from sporadic and outbreak-related human cases and *C. parvum* isolates from sporadic livestock cases in the United Kingdom. J Clin Microbiol 45:3286–3294

Li D, Craik SA, Smith DW, Belosevic M (2005) The Infectivity of *Giardia lamblia* Cysts in Wastewater Effluents Before and After UV treatment Proc Wat Env Fed Dis 11:983–993

Li D, Craik SA, Smith DW, Belosevic M (2009) Infectivity of *Giardia lamblia* cysts obtained from wastewater treated with ultraviolet light. Wat. Res. 43:3037–3046

Lim Y, Ramasame SD, Mahdy M, Yusoff W, Sulaiman W, Smith HV (2009) Detection and molecular characterization of *Giardia* isolated from recreational lake water in Malaysia. Parasitol Res 106:289–291

Limor JR, Lal AA, Xiao L (2002) Detection and Differentiation of *Cryptosporidium* Parasites That Are Pathogenic for Humans by Real-Time PCR. J Clin Microbiol 40:2335–2338

Lin MH, Chen TC, Kuo TT, Tseng CC, Tseng CP (2000) Real-time PCR for quantitative detection of *Toxoplasma gondii*. J Clin Microbiol 38:4121–4125

Little SE, Johnson EM, Lewis D, Jaklitsch RP, Payton ME, Blagburn BL, Bowman DD, Moroff S, Tams T, Rich L, and Aucoin D. 2009. Prevalence of intestinal parasites in pet dogs in the United States. Veterinary Parasitology 166:144–152

Lobo ML, Xiao L, Antunes F, Matos O (2008) Occurrence of *Cryptosporidium*, *Giardia*, and microsporidia in water supplies in Portugal, abstr. PO28. Abstr. 10th Int. Wkshps. Opportun. Protists, Boston, MA, 28 to 31 May 2008

Loge FJ, Thompson DE, Call DR (2002) PCR detection of specific pathogens in water: a risk-based analysis. Environ Sci Technol 36:2754–2759

Lucy FE, Graczyk TK, Tamang L, Miraflor A, Minchin D (2008) Biomonitoring of surface and coastal water for *Cryptosporidium, Giardia*, and human-virulent microsporidia using molluscan shellfish. Parasitol Res 103:1369–1375

Lv C, Zhang L, Wang R, Jian F, Zhang S, Ning C, Wang H, Feng C, Wang X, Ren X, Qi M, Xiao L (2009) Prevalence and Molecular Characterization of *Cryptosporidium* spp. in Wild, Laboratory, and Pet Rodents in China. Appl Environ Microbiol 75:7692–7699

Mackay IM (2004) Real-time PCR in the microbiology laboratory. Clin Microbiol Infect 10:190–212

Mahbubani MH, Schaefer FW III, Jones DD, Bej AK (1998) Detection of *Giardia* in environmental waters by immuno-PCR amplification methods. In Microbiol 36:107–113

Mallon M, MacLeod A, Wastling JM, Smith H, Tait A (2003a) Multilocus genotyping of *Cryptosporidium parvum* Type 2: population genetics and sub-structuring, Infect Genet Evol 3:207–218

Mallon M, MacLeod A, Wastling J, Smith H, Reilly B, Andy Tait A (2003b) Population Structures and the Role of Genetic Exchange in the ZoonoticPathogen *Cryptosporidium parvum*. J Mol Evol 56:407–417

Mathis A, Weber R, and Deplazes P (2005) Zoonotic potential of microsporidia. Clin Microbiol Rev 18:423–445

McArthur AG et al. (2000) The *Giardia* genome project database. FEMS Microbiol Lett 189:271–273

McCuin RM, Bukhari Z, Sobrinho J, Clancy JL (2001) Recovery of *Cryptosporidium* oocyst and *Giardia* cysts from source water concentrates using immunomagnetic separation. J Microbiol Meth 45:69–76

McCuin RM, Clancy JL (2003) Modifications to United States Environmental Protection Agency methods 1622 and 1623 for detection of Cryptosporidium oocysts and Giardia cysts in water. Appl Environ Microbiol 69:267–274

McElroy W, Cabello E, Pillai SD (2001) Efficacy of an immuno-magnetic separation system for recovering Cryptosporidium oocysts from soils. J Rap Meth Autom Microbiol 9:63–70

McOliver CC, Lemerman HB, Silbergeld EK, Moore RD, Graczyk TK (2009) Risks of Recreational Exposure to Waterborne Pathogens Among Persons With HIV/AIDS in Baltimore, Maryland. AJPH 99:1116–1122

Menotti J, Cassinat B, Porcher R, Sarfati C, Derouin F, Molina J-M (2003a) Development of a real-time polymerase-chain reaction assay for quantitative detection of *Enterocytozoon bieneusi* DNA in stool specimens from immunocompromised patients with intestinal microsporidiosis. J Infect Dis 187:1469–1474

Menotti J, Cassinat B, Sarfati C, Liguory O, Derouin F, Molina J-M (2003b) "Development of a real-time PCR assay for quantitative detection of *Encephalitozoon intestinalis* DNA," J Clin Microbiol 41:1410–1413

Miller WA, Atwill ER, Gardner IA, Miller MA, Fritz HM, Hedrick RP, et al. (2005) Clams (*Corbicula fluminea*) as bioindicators of fecal contamination with *Cryptosporidium* and *Giardia* spp. in freshwater ecosystems in California. Int J Parasitol 35:673–684

Miller WA, Gardner IA, Atwill ER, Leutenegger CM, Miller MA, Hedrick RP, Melli AC, Barnes NM, Conrad PA (2006) Evaluation of methods for improved detection of *Cryptosporidium* spp. in mussels (*Mytilus californianus*). J Microbiol Meth 65:367–379

Miller KM, Sterling CR (2007) Sensitivity of Nested PCR Detection for Low Numbers 1 of *Giardia lamblia* Cysts Appl Environ Microbiol 73:5949–5950

Morgan U, Weber R, Xiao L, Sulaiman I, Thompson RC, Ndiritu W, Lal A, Moore A, Deplazes P (2000) Molecular Characterization of *Cryptosporidium* isolates obtained from human immunodeficiency virus-infected individuals living in Switzerland, Kenya, and the United States. J Clin Microbiol 38:1180–1183

Mueller-Doblies DM, Elwin GK, Smith RP, Clifton-Hadley FA, Chalmers RM (2008) Distribution of *Cryptosporidium* species in sheep in the UK. Vet Parasitol 154:214–219

Muthusamy D, Rao SS, Ramani S, Monica B, Banerjee I, Abraham OP, Mathai DC, Primrose B, Muliyil J, Wanke CA, Ward HD, Kang G (2006) Multilocus Genotyping of *Cryptosporidium* sp. Isolates from Human Immunodeficiency Virus-Infected Individuals in South India. J Clin Microbiol 44:632–634

Nichols RAB, Campbell BM, Smith HV (2006) Molecular Fingerprinting of *Cryptosporidium* Oocysts Isolated during Water Monitoring. Appl Environ Microbiol 72:5428–5435

Nichols GL, Chalmers RM, Sopwith W, Regan M, Hunter CA, Grenfell P, Harrison F, Lane C (2006) Cryptosporidiosis: A report on the surveillance and epidemiology of *Cryptosporidium* infection in England and Wales.. Drinking Water Directorate Contract Number DWI 70/2/201. Drinking Water Inspectorate, U.K., 142

Nieminski EC, Schaefer III FW, Ongerth JE (1995) Comparison of Two Methods for Detection of *Giardia* Cysts and *Cryptosporidium* Oocysts in Water. Appl Environ Microbiol 61:1714–1719

Nikaeen M, Makimura K (2007) Evaluation of QIAamp DNA mini kit for removing of inhibitors in detection of *Cryptosporidium parvum* oocysts in water samples by a nested– PCR assay *International Journal of Environmental Science and Technology.* 4:.241–246

Okhuysen PC, Chappell CL, Crabb JH, Sterling CR, DuPont HL,(1999) Virulence of three distinct *Cryptosporidium parvum* isolates for healthy adults. J Infect Dis 180:1275–1281

Omalu ICJ, Duhlinska DD, Anyanwu GI, Parn VA, Inyama PU (2006) Human Microsporidial Infections. Online J Health Allied Scs 3:2

Ortega C, Solo-Gabriele HM, Abdelzaher A, Wright M, Deng Y, Stark LM (2009) Correlations between microbial indicators, pathogens, and environmental factors in a subtropical Estuary. Mar. Poll. Bull. 58:1374–1381

Ortega-Pierres G, Cacciò SM, Fayer R, Mank TG, Smith HV, Thompson RCA (Eds.) (2009) Giardia and *Cryptosporidium*: from molecules to disease. CABI, Cambridge, Massachusetts. xv þ 502 p. ISBN: 978-1-84593-391-3

Pall Life Sciences. 2000. Envirochek HV sampling capsule 02/00, xk, GN00.0120. PALL Life Sciences, East Hills, New York

Peng MM, Meshnick SR, Cunliffe NA, Thindwa BD, Hart, CA, Broadhead RL, Xiao L., 2003. Molecular epidemiology of cryptosporidiosis in children in Malawi. Journal of Eukaryotic Microbiology 50 (Suppl), 557–559

Packman AI, Lau BLT, Harter T, Atwill ER (2007) Processes affecting the transport of *Cryptosporidium parvum* and other persistent pathogens in surface- and ground-waters., Eos Trans. AGU, 88(52), Fall Meet. Suppl., Abstract H33F-1701

Pedraza-Díaz S, Amar C, McLauchlin J (2000) The identification and characterisation of an unusual genotype of *Cryptosporidium* from human faeces as *Cryptosporidium meleagridis.* FEMS Microbiol Lett 189:189–194

Peng MM, Matos O, Gatei W, Das P, Stantic-Pavlanic M, Bern C, Sulaiman IM, Glaberman S, Lal AA, Xiao L (2001) A Comparison of *Cryptosporidium* Subgenotypes from Several Geographic Regions J Euk Microbiol 48:28–31

Pennil CC, Clarke ME, Giddings CW, McEvoy JM (2008) Source of *Cryptosporidium* in a rural watershed, abstr. PO68. Abstr. 10th Int. Wkshps. Opportun. Protists, Boston, MA, 28 to 31 May 2008

Pieniazek NJ, Bornay-Llinares FJ, Slemenda SB, da Silva AJ, Moura IN, Arrowood MJ, Ditrich O, and Addiss DG 1999 New *Cryptosporidium* genotypes in HIV-infected persons. Emerging Infectious Diseases 5:444–449

Plutzer J, Karanis P, Domokos K, Torokne A, Marialigeti K (2008) Detection and characterisation of *Giardia* and *Cryptosporidium* in Hungarian raw, surface and sewage water samples by IFT, PCR and sequence analysis of the SSUrRNA and GDH genes. Int J Hyg Environ Health 211:524–533

Plutzer, J. and Tomot, B. 2009. The role of aquatic birds in the environmental dissemination of human pathogenic Giardia duodenalis cysts and *Cryptosporidium* oocysts in Hungary. Parasitology International 58:227–231

Polley SD, Boadi S, Watson J, Curry A, Chiodini PL (2011) Detection and species identification of microsporidial infections using SYBR Green real-time PCR J Med Microbiol 60:459–466
Quilez JE, Torres RM, Chalmers SJ, Hadfield, E. Del Cacho, Sanchez-Acedo C (2008) Genotype and subtype characterization of *Cryptosporidium* in lambs and goad kids in Spain. Appl Environ Microbiol 74:6026–6031
Qvarstrom Y, Visvesvara GS, Sriram R, Silva AJ (2006) Multiplex real-time assay for simultaneous detection of *Acancthamoeba spp. Balamuthia mandrillaris, and Naegleria fowleri.* J Clin Microbiol 44:3589–3595
Ramirez N, Sreevatsan S (2006) Development of a sensitive detection system for *Cryptosporidium* in environmental samples. Vet Parasitol 136:201–213
Read CM, Monis PT, Thompson RC (2004) Discrimination of all genotypes of *Giardia duodenalis* at the glutamate dehydrogenase locus using PCR-RFLP. Infect Genet. Evol 4:125–130
Reetz J, Rinderb H, Thomschkeb A, Mankea H, Schwebsa M, Bruderek A (2002) First detection of the microsporidium *Enterocytozoon bieneusi* in non-mammalian hosts (chickens). Int J Parasitol 32:785–787
Reetz J, Karsten Nöckler, Sabine Reckinger, Maria Margarida Vargas, Wolf Weiske and Alessandro Broglia (2009) Identification of *Encephalitozoon cuniculi* genotype III and two novel genotypes of *Enterocytozoon bieneusi* in swine. Parasitol Int 58:285–292
Rendtorff RC. (1954) The experimental transmission of human intestinal protozoan parasites. II. *Giardia lamblia* cysts given in capsules. Am J Hyg 59:209–220
Rendtorff RC (1979) The experimental transmission of *Giardia lamblia* among volunteer subjects, pp. 64–81. *In* W. Jakubowski and J. C., Hoff (ed.), Waterborne transmission of giardiasis. EPA-600/9-79-001. U.S. Environmental Protection Agency, Cincinnati, OH
Rimhanen-Finne R, Hörman A, Ronkainen P, Hänninen M-L (2002) An IC-PCR method for detection of *Cryptosporidium* and *Giardia* in natural surface waters in Finland. J Microbiol Meth 50:299–303
Rinder H, Thomschke A, Dengjel B, Gothe R, Löscher T, Zahler M (2000) Close genotypic relationship between *Enterocytozoon bieneusi* from humans and pigs and first detection in cattle. J Parasitol 86:185–188
Ruecker NJ, Braithwaite SL, Topp E, Edge T, Lapen DR, Wilkes G, Robertson W, Medeiros D, Sensen CW, Neumann NF (2007) Tracking host sources of *Cryptosporidium* spp. in raw water for improved health risk assessment. Appl Environ Microbiol 73:3945–3957
Robertson LJ, Forberg T, Hermansen L, Gjerde BK, Alvsvåg JO, Langeland N. (2006) *Cryptosporidium parvum* Infections in Bergen, Norway, during an Extensive Outbreak of Waterborne Giardiasis in Autumn and Winter 2004. Appl Environ Microbiol 72:2218–2220
Robinson G, Elwin K, Chalmers RM (2008) Unusual *Cryptosporidium* Genotypes in Human Cases of Diarrhea Emerg Infect Dis. 2008 November; 14(11):1800–1802
Rochelle PA, De Leon R, Stewart MH, Wolfe RL (1997) Comparison of primers and optimization of PCR conditions for detection of *Cryptosporidium parvum* and *Giardia lamblia* in water. Appl. Environ. Microbiol 63:106–114
Rochelle PA, De Leon R, Johnson A, Stewart MH, Wolfe RL (1999) Evaluation of Immunomagnetic Separation for Recovery of Infectious *Cryptosporidium parvum* Oocysts from Environmental Samples. Appl Environ Microbiol 65:841–845
Roy S, Kabir M, Mondal D, Ali IKM, Petri WA Jr, Haque R (2005) Real-time PCR assay for the diagnosis of *Entamoeba histolytica* Infection. J Clin Microbial 43:2168–2172
Royer M, Xiao L, Lal AA (2002) Animal Source Identification Using A *Cryptosporidium* DNA Characterization Technique. EPA/600/R-03/047
Santaniello A, Ludovico Dipineto, Laura Rinaldi, Lucia Francesca Menna, Giuseppe Cringoli and Alessandro Fioretti (2009) Serological survey of *Encephalitozoon cuniculi* in farm rabbits in Italy. Res Vet Science 87:67–69
Santín M, Trout JM, Fayer R (2007) Prevalence and molecular characterization of *Cryptosporidium* and *Giardia* species and genotypes in sheep in Maryland. Vet Parasitol 146:17–24
Savioli L, Smith H, Thompson A (2006) *Giardia* and *Cryptosporidium* join the 'Neglected Diseases Initiative' TRENDS in Parasitol 22 No.5

Searcy KE, Packman AI, Atwill ER, Harter T (2006) Capture and Retention of *Cryptosporidium parvum* Oocysts by *Pseudomonas aeruginosa* Biofilms. Appl Environ Microbiol 72:6242–6247

Simmons III OD, Sobsey MD, Heaney CD, Schaefer III FW, Francy DS (2001) Concentration and Detection of *Cryptosporidium* Oocysts in Surface Water Samples by Method 1622 Using Ultrafiltration and Capsule Filtration Appl Environ Microbiol 67:1123–1127

Sluter SD, Tzipori S, Widmer G (1997) Parameters affecting polymerase chain reaction detection of waterborne *Cryptosporidium parvum* oocysts. Appl Microbiol Biotechnol 48:325–330

Smith HV (1999) Detection of parasites in the environment. Parasitol 117:113–141

Smith M, Thompson KC (2001) *Cryptosporidium*: the analytical challenge. Smith/Thomspon (eds.) Royal Society of Chemistry (Great Britain)

Smith, JJ, Gunasekera TS, Barardi CR, Veal D, Vesey G (2004) Determination of *Cryptosporidium parvum* oocyst viability by fluorescence in situ hybridization using a ribosomal RNA-directed probe. J Appl Microbiol 96:409–417

Smith HV, Grimason AM (2003) *Giardia and Cryptosporidium* in water and wastewater. In The Handbook of Water and Wastewater Microbiology (ed. D. Mara & N. Horan), Elsevier Science, Oxford, pp 619–781

Smith HV, Nichols RAB (2010) *Cryptosporidium*: Detection in water and food. Exp Parasitol 124:61–79

Snel SJ, Baker MG, Kamalesh V, French N, Learmonth J (2009) A tale of two parasites: the comparative epidemiology of *cryptosporidiosis* and *giardiasis*. Epidemiol Infect 137:1641–1650

Snowden KF, Lewis BC, Hoffman J, Mansell J. (2009) *Encephalitozoon cuniculi* Infections in Dogs: A Case Series. J Amer Anim Hosp Assoc 45:225–231

Sorel N, Guillot E, Thellier M, Accoceberry I, Datry A, Mesnard-Rouiller L, Miégeville M (2003) Development of an immunomagnetic separation-polymerase chain reaction (IMS-PCR) assay specific for *Enterocytozoon bieneusi* in water samples. J Appl Microbiol 94:273–279

Spano F, Putignani L, Naitza S, Puri C, Wright S, Crisanti A (1998) Molecular cloning and expression analysis of a *Cryptosporidium parvum* gene encoding a new member of the thrombospondin family. Mol. Biochem Parasitol. 92:147–162

Sprong H, Cacciò, SM, van der Giessen JWB,(2009) ZOOPNET network and partners. Identification of zoonotic genotypes of Giardia duodenalis. PLoS neglected tropical diseases 3:e558. doi:10.1371

Stantic-Pavlinic M, Xiao L, Glaberman S, Lal AA, Oražen T, Rataj-Verglez A, Logar J, Berce I (2003) Cryptosporidiosis associated with animal contacts. Wiener Klinische Wochenschrift 115:125–127

Stark D, van Hal S, Barratt J, Ellis J, Marriott D, Harkness J (2009) Limited genetic diversity among genotypes of *Enterocytozoon bieneusi* strains isolated from HIV-infected patients from Sydney, Australia. J Med Microbiol 58:355–357

Stark D, Al-Qassab SE, Barratt JLN, Stanley K, Roberts T, Marriott D, Harkness J, Ellis JT (2011) Evaluation of Multiplex Tandem Real-Time PCR for Detection of *Cryptosporidium* spp., *Dientamoeba fragilis, Entamoeba histolytica*, and *Giardia intestinalis* in Clinical Stool Samples. J Clin Microbiol 49:257–262

Strong WB, Gut J, Nelson RG (2000) Cloning and sequence analysis of a highly polymorphic *Cryptosporidium parvum* gene encoding a 60-kilodalton glycoprotein and characterization of its 15-and 45-kilodalton zoite surface antigen products. Infect Immun 68:4117–4134

Stroup SE, Roy S, Mchele J, Maro V, Ntabaguzi S, Siddique A, Kang G, Guerrant RL, Kirkpatrick BD, Fayer R, Herbein J, Ward H, Haque R, Houpt ER (2006) Real-time PCR detection and speciation of *Cryptosporidium* infection using Scorpion probes. J Med Microbiol 55:1217–1222

Sturbaum G, Reed C, Hoover PJ, Jost BH, Marshall MM, Sterling CR (2001) Species-specific, nested PCR-restriction fragment length polymorphism detection of single *Cryptosporidium parvum* oocysts, Appl Environ Microbiol 67:2665–2668

Sturbaum GD, Klonicki PT, Marshall MM, Jost BH, Clay BL, Sterling CR. (2002). Immunomagnetic Separation (IMS)-Fluorescent Antibody Detection and IMS-PCR Detection of Seeded *Cryptosporidium parvum* Oocysts in Natural Waters and Their Limitations. Appl Environ Microbiol 68:2991–2996

Subrungruan IM. Mungthi P. Chavalitshewinkoon-Petmit R. Rangsi T. Naaglor and S. Leelayoova (2004) "Evaluation of DNA extraction and PCR methods for detection of *Enterocytozoon bienuesi* in stool specimens," J Clin Microbiol 42:3490–3494

Sunnotel O, Snelling WJ, Xiao L, Moule K, Moore JE, Millar B, Dooley JSG, Lowery CJ (2006) Rapid and Sensitive Detection of Single *Cryptosporidium* Oocysts from Archived Glass Slides. J Clin Microbiol 44:3285–3291

Sulaiman IM. Lal AA. and L. Xiao. 2001. A population genetic study of the *Cryptosporidium parvum* human genotype parasites. J. Eukaryot. Microbiol. 48:24–27

Sulaiman IM, Fayer R, Lal AA, Trout JM, Schaefer FW III, Xiao L (2003). Molecular Characterization of Microsporidia Indicates that Wild Mammals Harbor Host-Adapted *Enterocytozoon* spp. as well as Human-Pathogenic *Enterocytozoon bieneusi*. Appl Environ Microbiol 69:4495–4501

Tanriverdi S, Tanyeli A, Baslamisli F, Köksal F, Kilinc Y, Feng X, Batzer G, Tzipori S, and Widmer G (2002) Detection and genotyping of oocysts of *Cryptosporidium parvum* by real-time PCR and melting curve analysis. J Clin Microbiol 40:3237–3244

Tanriverdi S, Widmer G (2006) Differential evolution of repetitive sequences in *Cryptosporidium parvum* and *Cryptosporidium hominis*. Infect Genet Evol 6:113–122

Tanrıverdi S, Grinberg A, Chalmers RM, Hunter PR, Petrovic Z, Akiyoshi DE, London E, Zhang L, Tzipori S, Tumwine JK, Widmer G (2008) Inferences about the Global Population Structures of *Cryptosporidium parvum* and *Cryptosporidium hominis*. Appl Environ Microbiol 74:7227–7234

ten Hove R, van Lieshout L, Brienen E, Perez M, Verweij J (2008) Real-time polymerase chain reaction for detection of *Isospora belli* in stool samples Diag Microbiol Infect Dis 61:280–283

Thompson RCA (2000) Giardiasis as a re-emerging infectious disease and its zoonotic potential. Int J Parasitol 30:1259–1267

Thompson RCA (2004) The zoonotic significance and molecular epidemiology of *Giardia* and giardiasis. Vet Parasitol 126:15–34

Thompson RCA, Monis PT (2004) Variation in Giardia: implications for taxonomy and epidemiology. Adv Parasitol 58:69–137

Thurston JA, Gerba CP, Foster KE, Karpiscak MM (2001) Fate of indicator microorganisms, *Giardia* and *Cryptosporidium* in subsurface flow constructed wetlands. Wat Res 35:1547–1551

Traub RJ, Monis PT, Robertson I, Irwin P, Mencke N (2004) Epidemiological and molecular evidence supports the zoonotic transmission of *Giardia* among humans and dogs living in the same community. Parasitol 128:253–262

Udeh P,Veenstra J, Abraham AJ, John GH (2000) Quantitative polymerase chain (QPCR) reaction using the MIMIC approach to estimate *Cryptosporidium parvum* oocysts, an intestinal pathogen, in municipal water treatment sludge samples. Mol Cell Probes 14:121–126

U.S. Environmental Protection Agency (2001a) USEPA method 1622: *Cryptosporidium* in water by filtration/IMS/FA. EPA 821-R-01–026. Office of Water, U.S. Environmental Protection Agency, Washington, D.C.

U.S. Environmental Protection Agency (2001b) USEPA method 1623: *Cryptosporidium* and *Giardia* in water by filtration/IMS/FA. EPA 821-R-01–025. Office of Water, U.S. Environmental Protection Agency, Washington, D.C.

U.S. Environmental Protection Agency (2004a) Quality Assurance/Quality Control Guidance for Laboratories Performing PCR Analyses on Environmental Samples October 2004 http://www.epa.gov/nerlcwww/qa_qc_pcr10_04.pdf

U.S. Environmental Protection Agency (2004b). EPA Microbiological Alternate Test Procedure (ATP) Protocol for Drinking Water, Ambient Water, and Wastewater Monitoring Methods (2004) EPA- 821-B-03 004 http://www.epa.gov/waterscience/methods/atp/EPA821B03004.pdf

U.S. Environmental Protection Agency, (2005). Method 1623: Cryptosporidium and Giardia in water by filtration/IMS/FA. Office of Water. U.S. Government Printing Office, Washington, DC. EPA 815-R-05-002

Varma M, Hester JD, Schaefer, III FW, Ware MW, Lindquist HDA (2002) Detection of *Cyclospora cayetanensis* using a quantitative real-time PCR assay. J Microbiol Meth 53:27–36

Velásquez JN, Carnevale S, Labbé JH, Chertcoff A, Cabrera MG, and Oelemann W (1999) In situ hybridization: a molecular approach for the diagnosis of the microsporidian parasite *Enterocytozoon bieneusi*. Human Path 30:54–58

Verweij JJ, Schinkel J, Laeijendecker D, van Rooyen MAA, van Lieshout L, Polderman AM (2003) Real-time PCR for the detection of *Giardia lamblia*. Mol Cell Probes 17:223–225

Verweij JJ, Blange RA, Templeton K, Schinkel J, Brienen EAT, van Rooyen MAA, van Leishout L, Polerman AM (2004) Simultaneous detection of *Entamoeba histolytica, Giardia lamblia,* and *Cryptosporidium parvum* in fecal samples using multiplex real-time PCR. *J Clin Microbiol* 42:1220–1223

Verweij J, ten Hove R, Brienen E, van Lieshout L (2007) Multiplex detection of *Enterocytozoon bieneusi* and *Encephalitozoon* spp. in fecal samples using real-time PCR Diag Microbiol Infect Dis 57:163–167

Vesey G, Ashbolt N, Fricker EJ, Deere D, William KL, Veal DA, Dorsch M (1998) The use of a ribosomal RNA targeted oligonucleotide probe for fluorescent labelling of viable *Cryptosporidium parvum* oocysts. J Appl Microbiol 85:429–440

Walker NJ (2002) Tech.Sight. A technique whose time has come. Science 296:557–559

Wang Z, Vora GJ, Stenger DA (2004) Detection and Genotyping of *Entamoeba histolytica, Entamoeba dispar, Giardia lamblia,* and *Cryptosporidium parvum* by Oligonucleotide Microarray J Clin Microbiol 42:3262–3271

Wang Z, Orlandi PA, Stenger DA (2005) Simultaneous Detection of Four Human Pathogenic Microsporidian Species from Clinical Samples by Oligonucleotide Microarray. J Clin Microbiol 43:4121–4128

Ward PI, Deplazes P, Regli W, Rinder H, Mathis A (2002) Detection of eight *Cryptosporidium* genotypes in surface and waste waters in Europe. Parasitol 124:359–368

Weiss LM (2001) Microsporidia: emerging pathogenic protists. Acta Tropica 78:89–102

Weitzel T, Wolff M, Dabanch J, Levy I, Schmetz C, Visvesvara GS, Sobottka I (2001) Dual Microsporidial Infection with *Enchephalitozoon cuniculi* and *Enterocytozoon bieneusi* in an HIV-Positive Patient. Infection 29:237–239

WHO: *Guidelines for drinking-water quality. Vol 1 Recommendations*. Edited by. Geneva: World Health Organization; 2006 [Guidelines for drinking-water quality, vol 1.]

Widmer G, Lin J, Kapur V, Feng X, Abrahamsen MS (2002) Genomics and genetics of *Cryptosporidium parvum*: the key to understanding cryptosporidiosis. Microb Infect 4:1081–1090

Widmer G, Grinberg A, Tanriverdi S, Chalmers RM, Hunter PR, Petrovic Z, Akiyoshi DE, London E, Zhang L, Tumwine JK, Tzipori S (2008) Multilocus genotype analysis of the global phylogeography and population structure of *Cryptosporidium parvum* and *Cryptosporidium hominis*, abstr. PL56. Abstr. 10th Int. Wkshps. Opportun. Protists, Boston, MA, 28 to 31 May 2008

Wiedenmann A, Krüger P, Botzenhart K (1998) PCR detection of *Cryptosporidium parvum* in environmental samples—a review of published protocols and current developments. J. Ind Microbiol Biotech 21:150–166

Wilson IG (1997) Inhibition and facilitation of nucleic acid amplification. Appl Environ Microbiol 63:3741–3751

Winkworth CL, James J. Learmonth, Christoph D. Matthaei, Colin R. Townsend (2009) Molecular characterization of *Giardia* from calves and humans in a recently intensified farming region Appl. Environ. Microbiol. doi:10.1128/AEM.00232-08

Wittner M (1999) *The Microsporidia And Microsporidiosis*. Wittner M., L. M. Weiss (eds.) ASM Press, pp 572

Wittner M, Weiss LM (1999) The microsporidia and microsporidiosis. *Am. Soc. Microbiol.* Washington, D.C.

Wolk DM, Schneider SK, Wengenack NL, Sloan LM, Rosenblatt JE (2002) Real-time PCR method for detection of *Encephalitozoon intestinalis*. J Clin Microbiol 40:3922–3928

Wolyniak EA, Hargreaves BR, Jellison KL (2009) Retention and Release of *Cryptosporidium parvum* Oocysts by Experimental Biofilms Composed of a Natural Stream Microbial Community. Appl Environ Microbiol 75:4624–4626

Wright D J, Chapman PA, Siddons CA (1994) Immunomagnetic separation as a sensitive method for isolating *Escherichia coli* O 157 from food samples. Epidemiol infect 113:31–39

Xiao L (2010) Molecular epidemiology of cryptosporidiosis: An update, *Experimental Parasitology* 124:80–89

Xiao L, Morgan UM, Fayer R, Thompson RC, Lal AA (2000) *Cryptosporidium* systematices and implications for public health. Parasitol Today 16:287–292

Xiao L, Singh A, Limor J, Graczyk TK, Gradus S, Lal AA (2001) Molecular Characterization of *Cryptosporidium* Oocysts in Samples of Raw Surface Water and Wastewater. Appl Environ Microbiol 67:1097–1101

Xiao L, Bern C, Limor J, Sulaiman I, Roberts J, Checkley W, Cabrera L, Gilman RH, Lal AA (2001a) Identification of 5 types of *Cryptosporidium* parasites in children in Lima, Peru. J Infect Dis 183:492–497

Xiao L, Limor J, Bern C, Lal AA (2001b) Tracking *Cryptosporidium parvum* by sequence analysis of small double-stranded RNA. Emerg Infect Dis 7:141–145

Xiao L, Cama V (2006) *Cryptosporidium* and Cryptosporidiosis, Foodborne Parasites; Springer; US. P57-108

Xiao L, Ryan, UM (2004) Cryptosporidiosis: an update in molecular epidemiology. Current Opin Infect Dis 17:483–490

Xiao L, Lal AA, Jiang J (2004a) Detection and Differentiation of Cryptosporidium Oocysts in Water by PCR-RFLP Methods in Molecular Biology 268:163–176

Xiao L, Fayer R, Ryan U, Upton SJ (2004b). *Cryptosporidium* Taxonomy: Recent Advances and Implications for Public Health. Clin Microbiol Rev 17:72–97

Xiao L, Singh A, Alderisio K (2006) Development and Standardization of a *Cryptosporidium* Genotyping Tool for Water Samples. AWWA Res Foun pp 116

Xiao L, Fayer R (2008) Molecular characterisation of species and genotypes of *Cryptosporidium* and *Giardia* and assessment of zoonotic transmission Int J Parasitol 38:1239–1255

Xiao L, Ryan UM (2008) Molecular epidemiology. In: Fayer, R., Xiao, L. (Eds.), *Cryptosporidium* and Cryptosporidiosis. CRC Press and IWA Publishing, Boca Raton, FL, pp 119–171

Xiao L, Michele C. Hlavsa, Jonathan Yoder, Christina Ewers, Theresa Dearen, Wenli Yang, Randall Nett, Stephanie Harris, Sarah M. Brend, Meghan Harris, Lisa Onischuk, Amy L. Valderrama, Shaun Cosgrove, Karen Xavier, Nancy Hall, Sylvia Romero, Stephen Young, Stephanie P. Johnston, Michael Arrowood, Sharon Roy, and Michael J. Beach (2009) Subtype analysis of *Cryptosporidium* specimens from sporadic cases in four U.S. states in 2007: the wide occurrence of one *Cryptosporidium hominis* subtype and a case report of an infection with the *Cryptosporidium* horse genotype J. Clin. Microbiol 47:3017–3020

Xu P, Widmer G, Wang Y, Ozaki LS, Alves JM, Serrano MG, Puiu D, Manque P, Akiyoshi D, Mackey AJ, Pearson WR, Dear PH, Bankier AT, Peterson DL, Abrahamsen MS, Kapur V, Tzipori S, Buck GA (2004) The genome of *Cryptosporidium hominis*. Nature 431:1107–1112

Yu X, Van Dyke MI, Portt A, Huck PM (2009) Development of a direct DNA extraction protocol for real-time PCR detection of *Giardia lamblia* from surface water. Ecotoxicol 18:661–668

Yu J-R, Lee S-U, Park W-Y (2009) Comparative Sensitivity of PCR Primer Sets for Detection of *Cryptosporidium parvum* Korean J Parasitol. 47:293–297

Zarlenga DS, Trout JM (2004) Concentrating, purifying and detecting waterborne parasites. Vet Parasitol 126:195–217

Zhang T, Hu H-Y, Zong, Z-S (2007) Factors Affecting the Determination of *Cryptosporidium* and *Giardia* in Water Samples by Immunomagnetic Separation. J Environ Health 2:56–58

Zhou L, Singh A, Jiang J, and Xiao L 2003 Molecular Surveillance of *Cryptosporidium* spp. in Raw Wastewater in Milwaukee: Implications for Understanding Outbreak Occurrence and Transmission Dynamics. J Clin Microbiol 41:5254–5257

Chapter 8
Chemical-Based Fecal Source Tracking Methods

Charles Hagedorn and Stephen B. Weisberg

Abstract A useful source tracking approach that has emerged over the past few years involves detection of human-specific chemicals found in wastewaters and septage. Chemicals specific to human wastewaters offer some potential advantages over biologically based methods: they are generally faster to prepare and analyze; they are more source-specific because they are not confounded by regrowth in the environment; and chemicals may be more geographically and temporally stable. However, chemicals have received less scrutiny to date, often require specialized equipment, and sample processing is usually more expensive. Additionally, many chemicals specific to human waste streams may occur at concentrations low enough to be diluted below detection limits once the waste stream enters the ambient environment. This chapter describes the different classes of wastewater chemicals and explores the advantages, disadvantages of each as potential source tracking candidates. While no single chemical has emerged as the best, there are several viable candidates for source tracking applications. For initial water sample screenings, optical brighteners (OBs) in detergents have shown considerable promise. Although not as sensitive as most microbial assessments, OBs can be measured with a handheld fluorometer, providing near real-time and relatively inexpensive tracking of signals in the field if the sample contains an OB concentration large enough to produce a measurable signal. Fecal sterols and stanols have been found to work best for source attribution, correlating with fecal indicator bacteria and providing discrimination between humans and several other fecal sources.

Keywords Chemical tracking • Fecal pollution • Fluorometry • Optical brighteners • Pharmaceuticals • Fecal sterols and stanols • Caffeine • Recreational water quality • Microbial source tracking

---

C. Hagedorn (✉)
Department of Crop and Soil Environmental Sciences,
401 Price Hall, Virginia Tech, Blacksburg, VA 24061, USA
e-mail: chagedor@vt.edu

C. Hagedorn et al. (eds.), *Microbial Source Tracking: Methods, Applications, and Case Studies*, DOI 10.1007/978-1-4419-9386-1_8,

8.1 Introduction

There is a substantial body of information and guidance on MST that addresses microbiologically based methods and approaches (this book, for example). However, chemicals that are specific to human wastewater have received much less scrutiny to date but offer several potential advantages (and disadvantages), as described in a recent review by Hagedorn and Weisberg (2009). Some chemicals require less sample preparation and analysis time that microbial methods, while others have very detailed and time-consuming procedures, but most are very specific to human-origin pollution. There may be fewer issues with geographic and temporal variations that have been found to be impediments to many microbial methods, assuming that chemicals can be identified that are both widely used and readily detectable (Rangdale et al. 2003). Additionally, chemical methods are not subject to some of the problems that affect microbe-based approaches such as persistence in the environment and possible regrowth (Seurinck et al. 2005; Wu et al. 2008).

In the largest study performed to date, 110 chemicals were quantified from wastewater effluent samples collected across USA, and 78 were detected at least once in a survey of treatment plants (Glassmeyer et al. 2005). These compounds included food additives, pharmaceuticals, hygienic products, cosmetics, fecal stanols/sterols (the metabolic by-products of cholesterols), whitening agents and other compounds prevalent in household detergents, caffeine, and a variety of household chemicals such as plasticizers, biocides, and wood preservatives. While results suggested that as many as 35 of these compounds might prove useful as indicators of anthropogenic pollution, five of them, including two pharmaceuticals (carbamazepine and diphenhydramine), one stanol (coprostanol), fluorescent whitening agents from detergents (OBs), and caffeine demonstrated the greatest potential (most widespread and present in sufficient concentrations). Nakada et al. (2008) examined 13 pharmaceuticals and personal-care products as indicators of sewage contamination in riverine, coastal, and groundwaters in Japan. They concluded that crotamiton and carbamazepine could serve as conservative markers for urban riverine and coastal environments. Although both studies demonstrated the possibility of using chemical analytes to assess human-origin pollution, neither examined the relationship between fecal indicator bacteria (FIB) density and chemical presence and/or concentration.

This chapter describes the relative advantages and disadvantages of specific chemicals within the following groups of compounds that appear to have the greatest potential (Table 8.1): (1) pharmaceuticals and personal-care products, (2) fecal sterols/stanols, (3) optical brighteners (OBs)/fluorescent whitening agents, (4) caffeine, and (5) miscellaneous analytes. Discussion of the relationship of these chemicals to FIB and establishing performance criteria for chemical methodology are also included (also see Chap. 2).

Table 8.1 Advantages and disadvantages of different classes of chemical source tracking compounds

Advantages	Disadvantages
Pharmaceuticals and personal-care products	
Synthetic chemicals specific to human wastewater, no natural sources or known analogs	Extraction, elution, and concentration procedures not yet standardized for many compounds
Sensitivity, accuracy, and repeatability will largely depend on the sample matrix. Very small amounts can be accurately measured	Sensitivity of detection in large bodies of water could be a problem due to dilution. The target compound has to be used heavily enough so that it can be reliably detected
Temporal and geographic applicability not a problem as long as the target compound is being used in the area studied	High costs of analytical equipment and trained personnel dedicated to the equipment
	Obtaining results quickly will depend on the schedule of the analytical lab
	Correlation to health risks and FIB, plus environmental degradation rates or persistence in different matrices (and possible interference with detection) are unresolved
Fecal sterols/stanols	
Coprostanol, a 5β-stanol, is reasonably specific to humans	Detection methods not yet standardized
Detection methods that provide adequate sensitivity, accuracy, and repeatability have been reported	Levels of sensitivity required for detection in larger bodies of water not yet established
Temporal and geographic applicability should not be problematic	There may be natural sources or analogs of the target stanols in trace amounts in soils and sediments; these may negatively affect specificity, sensitivity, accuracy, and repeatability
Several ratios for different fecal sources based on a comparison of sterol/stanol concentrations have been proposed	High costs of analytical equipment and trained personnel dedicated to the equipment. Considerable experience is required to accurately identify different peaks
Other sterols/stanols may be specific and useful for other animal sources	Obtaining results quickly will depend on the schedule of the analytical lab
	Correlation to health risks and FIB, plus environmental degradation rates or persistence in different matrices are unresolved
Optical brighteners/fluorescent whitening agents	
Dye-free cotton pads	
The simplest and most cost effective approach	Pads must be placed in locations where OB concentrations are high
	Specificity and sensitivity are low, no quantification possible
	Other fluorescent chemicals may be present in the environment, resulting in false positives

(continued)

Table 8.1 (continued)

Advantages	Disadvantages
Direct measurement	
Very sensitive levels of detection. Accuracy and repeatability are very good	High costs of analytical equipment and trained personnel dedicated to the equipment
	Obtaining results quickly will depend on the schedule of the analytical lab
	Correlation to health risks and FIB, plus environmental degradation rates or persistence in different matrices are unresolved
Fluorometry	
Relatively easy to use; has a reasonably high degree of sensitivity, accuracy, and repeatability	Requires subsequent exposure to UV light for validation, but same-day results are possible
Initial results can be obtained very quickly	Dilution of OBs below detection limits in large bodies of water can occur
Instrument costs not an issue, an inexpensive handheld fluorometer works well	Potential interference from unknown compounds
No need to send samples to an analytical lab, highly trained technical personnel not required	Correlation to health risks, plus environmental degradation rates or persistence in different matrices are unresolved
A few good correlations with FIB have been reported	
Caffeine	
Very sensitive levels of detection. Accuracy and repeatability are very good	Same disadvantages as listed above for Pharmaceuticals and Personal-Care Products, except that good extraction, elution, and concentration procedures are available for caffeine
Temporal and geographic applicability not a problem as long as the target compound is being used in the area studied	In certain areas, there may be different plant species that secrete caffeine

8.2 Pharmaceuticals and Personal-Care Products

The application of pharmaceutical chemicals to detect sewage-based human pollution has received considerable scrutiny since described by Buser et al. (1999). Chemicals used in the pharmaceutical industry have been examined as indicators of human wastewater pollution (Leeming and Nichols 1996), as impacting aquatic ecosystems, including acute and chronic toxicity (Brun et al. 2006; Fent et al. 2006), and affecting growth and reproduction (Binzcik et al. 2004). The successful detection of a variety of pharmaceutical chemicals has been demonstrated in freshwaters (Hilton and Thomas 2003), seawater (Weigel et al. 2004), estuaries (Thomas and Hilton 2004), sediments (Bolz et al. 2001), sewage sludge (Temes et al. 2002), and wastewater effluents (Roberts and Thomas 2006).

Buser et al. (1999) detected concentrations of ibuprofen in wastewater influents at concentrations as high as 3 μg/L, as well as the urinary metabolites hydroxy-ibuprofen and carboxy-ibuprofen at even greater concentrations. Even though the effective degradation of ibuprofen and its corresponding metabolites during waste treatment processes was shown to be greater than 95%, sufficient detectable amounts remained, suggesting the potential for use as a human-specific chemical marker. A study of 24 pharmaceuticals and personal-care products in treated effluents and nearby wells and creeks in northwestern Washington found sixteen compounds that were consistently detected in effluents, while only caffeine, nicotine, and metformin, a drug used to treat diabetes, were detected (all at concentrations of 25 μg/L or less) in samples from nearby creeks and wells (Johnson et al. 2004). The investigators concluded that additional testing of these three chemicals as potential anthropogenic markers appeared warranted.

Other chemicals that have been investigated include carbamazepine, a drug used in the treatment of epilepsy and bipolar disorders (Clara et al. 2004), propranolol, a commonly prescribed β-blocker (Fono and Sedlack 2005), and triclosan, a biocide found in an increasing number of consumer products, such as soaps, deodorants, toothpastes, mouthwashes, and cleaning supplies (Weigel et al. 2004). Polycyclic musks, chemicals used as fragrance ingredients in household products including soaps, detergents, cosmetics, perfumes, and air fresheners, have also been assessed as potential anthropogenic markers, along with *N,N*-diethyl-*meta*-toluamide (abbreviated DEET), the most common active ingredient in insect repellents (Buerge et al. 2003a). While detection proved successful with all of these compounds, low abundance in both treated and untreated wastewaters appeared problematic for real-world source tracking applications (Buerge et al. 2003b).

Advantages of pharmaceuticals include the specificity of many of the synthetic chemicals to human wastewater, as there are no natural sources or known analogs of these compounds (Table 8.1). Accuracy and repeatability should not be issues with the current methods available for measuring pharmaceuticals, although extraction methods (e.g., solid phase or continuous liquid-liquid phase) followed by appropriate elution and concentration procedures are not yet standardized for many candidate pharmaceuticals. Sensitivity of detection in large bodies of water could be an issue, and the target compound has to be used heavily enough within a given locality so that it can be reliably detected. This may limit application to where the target compounds are more prevalent, such as larger cities rather than small or rural communities, and country of origin.

Other disadvantages include the cost of analytical equipment necessary to analyze samples with liquid or gas chromatography or mass spectrometry, and costs associated with trained personnel dedicated to the equipment. Such considerations can result in high per sample costs and the quickness of obtaining results will depend on the schedule of the analytical lab. Lastly, correlation to both health risks and FIB, and an understanding of environmental degradation rates or persistence, especially in different matrices (and possible interference with detection), remain unresolved at this time for all candidate pharmaceuticals (Table 8.1).

8.3 Fecal Sterols/Stanols

Assessing fecal sterols and stanols in attempts to identify human-origin pollution has received notable attention around the world, including Antarctica (Edwards et al. 1998), Brazil (da Costa and Carreira 2005), China (Chan et al. 1998), France (Tolosa et al. 2003), Greece (Bull et al. 2003), Japan and Vietnam (Isobe et al. 2004), New Zealand (Gregor et al. 2002), Spain (Maldonaldo et al. 1999), UK (Elhmmali et al. 2000), and USA (LeBlanc et al. 1992; Noblet et al. 2004). Much of the fecal sterol/stanol research to date has been conducted in Australia and New Zealand (Leeming and Nichols 1996; Devane et al. 2006; Chap. 22). Fecal sterols are the only class of compounds in this chapter where there is potential for identification of other sources in addition to humans.

Fecal stanols are formed in the guts of animals from the metabolism of sterols, with the metabolic end products varying in concentration between animal groups based on diet and intestinal flora (Leeming et al. 1996). The sterol cholesterol is largely metabolized to the 5β-stanol (coprostanol) in humans, representing approximately 60% of the total stanols within the human gut (MacDonald et al. 1983). In comparison, the quantity of coprostanol present in the feces of nonhuman animals is typically much lower (Bull et al. 2002). The metabolism of cholesterol by microorganisms within the environment usually generates cholestanol.

Fecal material from animals such as cattle, horses, and sheep contains a greater relative proportion of 5β-campestanol and 5β-stigmastanol, metabolic products of the sterols campesterol and sitosterol abundant in the diet of herbivorous animals (Leeming et al. 1996). By contrast, cholesterol is dominant in dog feces, while bird feces contain mainly cholesterol and sitosterol (Leeming et al. 1997), likely due to the absence of specific intestinal bifidobacteria necessary to utilize sterols. Leeming et al. (1996) determined that while coprostanol and 5β-campestanol/5β-stigmastanol production is not exclusively limited to humans and herbivores, respectively, the predominance of formation within these groups is significant enough to serve as potential source-specific markers.

Several ratios of stanol concentrations have been proposed as representative of different fecal sources. Quantification of the 5α- and 5β-stanol isomers led to the creation of a $5\beta/(5\alpha+5\beta)$ ratio by Grimalt et al. (1990), with a ratio greater than 0.7 believed to be indicative of human fecal pollution. Evershed and Bethell (1996) proposed a ratio of coprostanol to 5β-stigmastanol to separate human and ruminant pollution, with ratios above 1.5 considered positive for human fecal contamination. Leeming et al. (1997) suggested that if coprostanol/(coprostanol + 5β-stigmastanol) was greater than 0.73, pollution may be as much as 100% human in origin; if the ratio was less than 0.28, then herbivores could be responsible for up to 100% of the fecal pollution. Shah et al. (2007) reported that fecal sterol ratios were effective at identifying which mixtures contained a human contribution, but could not accurately determine the percent contributions of the different sources (no consensus on ratios).

A few field studies have successfully applied stanol ratios as a source identification approach. Noblet et al. (2004) used the $5\alpha/(5\beta+5\alpha)$ ratio to determine that sewage was unlikely to be a major fecal contributor in a California creek and were able to identify birds as the likely main contributor based on a suite of fecal steroids. Gregor et al. (2002) found that effluent from septic tanks contained higher concentrations of fecal stanols than treatment plant effluents. LeBlanc et al. (1992) used coprostanol detected in surface waters and sediments to trace sewage pollution in discharges into the Providence River, Rhode Island. Concentrations of 0.02–0.22 μg/L were found in surface waters 0.22–0.33 μg/L in sediments, with values decreasing away from the mouth of the Providence River. Ahmed et al. (Table 21.2) listed ten sterol ratios and the corresponding source interpretations for each, with several of these ratios allowing discrimination between human, herbivore, and avian sources. These authors described two studies in different locations in New Zealand where the ten ratios were evaluated, along with other MST methods, and the results met expectations regarding possible sources at each location (Tables 21.13 and 21.14).

As with all of the chemicals described in this chapter, there are a number of impediments that limit widespread adoption of the fecal sterol approach. One concern is with sample processing, including both the cost of specialized equipment and trained personnel. Also, there are no standard methods for analysis of 5β-stanols, hindering comparison of results among studies and different labs (Bull et al. 2002). Many stanols/sterols produce very similar analytic peaks that require considerable experience to separate (Reeves and Patton 2001; Table 8.1). There are also sensitivity concerns for sterol/stanol detection, particularly in larger water bodies where dilution can be extensive (Pond et al. 2004; Table 1). Large sample quantities (as much as 4.0 L/sample) may be required to detect stanols, creating difficulties for projects with lots of samples and frequent collection schedules. Also, the use of coprostanol alone as a biomarker of human fecal pollution (Murtaugh and Bunch 1967; Leeming et al. 1996) can lead to false source identification, as coprostanol is also present in the feces of some other animals and small amounts can be generated from cholesterol in anaerobic sediments (Mudge and Gwyn Lintern 1999). Lastly, there are a few additional issues that should be considered with this approach because various other sterols, in addition to coprostanol, can be found in humans and animals, and these can vary based on diet, variation in digestive tracts, and the diversity of gut microflora.

Although a direct relationship between pathogenic organisms and fecal sterols has yet to be established (Scott et al. 2002), several studies have found a correlation between sterols and FIB. Coprostanol concentrations of 60 and 400 ng/L were found to correlate with elevated fecal coliforms (FCs) concentrations of 150 and 1,000 cfu/100 mL, respectively, and enterococci concentrations of 35 and 230 cfu/100 mL, respectively, in sewage-polluted Tasmanian waters (Leeming et al. 1997). Isobe et al. (2004) found a strong correlation between *Escherichia coli* densities and coprostanol concentrations, although seasonal variations were observed. However, measurements of coprostanol concentrations can lead to conflicting conclusions due to its apparent ubiquity in trace amounts in soils and sediments (Bull et al. 2002). It also has a tendency to degrade under aerobic conditions, with a half-life of about 10 days at 20°C (Isobe et al. 2004). In a European methods comparison study, β-stanols, coprostanol, and 24-ethylcoprostanol were significantly different in human and nonhuman samples; however, the number of incorrect classifications among individual water samples prevented the study authors from recommending their use as an exclusive detection method (Blanch et al. 2006).

8.4 Optical Brighteners/Fluorescent Whitening Agents

OBs, also known as fluorescent whitening agents, are organic compounds used in household detergents to whiten clothing that absorb long-wave ultraviolet light (365 nm) and transmit strongly within the blue portion (violet [415 nm] to indigo [435 nm]) of the visible spectrum. This transmittance of blue light from the OBs that bind to clothing in a wash helps to balance the natural yellow color of cotton fabrics, making them appear bright white. Anywhere from 25 to 95% of the OBs used in a wash cycle are bound to clothing during the wash (Poiger et al. 1998), while the remainder is discharged into wastewater. The mixing of wastewater and gray water in household plumbing systems allows for the detection of OBs in both septic systems (Close et al. 1989; Boving et al. 2004) and untreated sewage (Poiger et al. 1998). In addition, OBs are used in a variety of other home products, including toilet papers and dishwashing detergents (Hagedorn et al. 2005a), providing additional OB input into wastewaters. There are no known natural sources of OBs, so their detection is indicative of human wastewater (Table 8.1).

There are currently three methods for detecting OBs in water and each has advantages and disadvantages. The simplest and most cost-effective approach is to place dye-free cotton pads in environmental waters for a period of time (2–3 days), allowing for any OBs present to bind to the fibers in the pads. With subsequent exposure to UV light, the pads will fluoresce if OBs are present in the water (Dixon et al. 2005). The major disadvantages of this method are low specificity and inability to quantify results (no threshold concentration can be determined). Because numerous other chemicals may be present in the environment that will fluoresce under UV light (Dickerson et al. 2007), false positives can be a serious issue. The second method employs high-performance liquid chromatography or mass spectrometry (Shu and Ding 2005) and offers high sensitivity but is greatly hindered by high per sample costs based on the specialized equipment and technicians needed. The third method is the use of a fluorometer, an instrument that varies in initial cost ($2,000–20,000), but is relatively easy to use and has a reasonably high degree of sensitivity, accuracy, and repeatability (Hagedorn et al. 2005b).

OBs have been successfully detected in numerous watersheds in USA (Hagedorn et al. 2003; McDonald et al. 2006; Dickerson et al. 2007), Japan (Hayashi et al. 2002), and New Zealand (Close et al. 1989; Gilpin et al. 2002). Dickerson et al. (2007) found difficulties in detecting OBs in river and bay waters resulting from dilution but was successful in tracing OBs back to sewer infrastructure problems in near-shore waters. Some fluorometers possess the additional advantage of being able to operate in continuous-flow mode, allowing the recoding of continuous readings along shorelines or within sections of a waterbody (Hagedorn et al. 2005b). A fluorometer in continuous-flow mode successfully detected nonpoint sources in embayments in eastern Virginia; where the contamination was from faulty septic systems and could be traced back to individual homes (Hagedorn et al. 2003). Ahmed et al. (Table 21.12) described a study in New Zealand where OBs were evaluated, along with fecal stanols and microbe-based MST methods, and the results met expectations regarding presence/absence of human sources.

Often high fluorometric readings correspond to high bacterial counts (Hagedorn et al. 2005b; McDonald et al. 2006; Dickerson et al. 2007); however, this has not been the case for all studies (Close et al. 1989; Wolfe 1995). Many of the high OB/low bacteria situations are likely the result of unidentified chemicals that absorb and transmit light at wavelengths similar to that of OBs, such as diesel fuel, organic matter, or certain compounds produced by algae (Hagedorn et al. 2005a). It is possible to separate such non-OB sources by exposing samples to UV light (Hartel et al. 2007). OBs are known to degrade under UV light (Kramer et al. 1996) and prolonged exposure of water samples to UV light (varying from 30 min to 4 h based on UV intensity) will substantially reduce OB fluorescence. The change in fluorometric readings, both before and after UV exposure, represents the approximate amount of OBs present in the water sample and can serve as a confirmatory test that the fluorescent compound in the water was due to OBs and not materials that fluoresce under the same conditions (Hartel et al. 2007). Cao et al. (2009) found that using UV exposure to differentiate OB fluorescence from natural organic matter in southern California proved valid, although the method failed to detect half of the detergents tested in natural stream water due to conservative thresholds placed on the method. This problem was surmounted by a protocol modification that improved sensitivity based on differences in the shape of photodecay curves between OBs and natural organic matter, resulting in detection of all detergents in sewage at 1:10 dilution and septage at 1:100 dilution.

The biggest advantages of OBs is the relative simplicity of the method, that results can be obtained quickly (automation is very feasible), and the ubiquity with which OBs are used in wash waters on a global scale (Hagedorn et al. 2005b). The drawbacks include dilution of OBs below detection limits in large water bodies and the potential for interference from unknown compounds. Fluorometry can be recommended as a means to trace point sources of human fecal contamination or as a supplemental detection method with other source tracking methods (Dickerson et al. 2007), but a study comparing both fluorometry results and mass spectrometry concentrations of OBs in different matrices over time is needed (Table 8.1).

8.5 Caffeine

Caffeine (1,3,7-trimethylxanthine) is another potential chemical indicator of human-origin pollution. While present in over 60 plants, most caffeine-containing species are nonnative to USA, with the exception of the yaupon holly (*Ilex vomitoria*) found in southeastern states (Peeler et al. 2006). As an ingredient in coffee, tea, soda, energy drinks, and chocolates, the average person globally consumes about 70 mg of caffeine per day (Buerge et al. 2003b). With even higher consumption levels in USA (210 mg/day) and UK (440 mg/day), the obvious potential as a human source chemical marker has been described in detail (Seiler et al. 1999; Siegener and Chen 2002). While only about 3% of caffeine ingested is not metabolized and is excreted in the urine, disposal of unconsumed coffee, tea, soft drinks, or other caffeinated beverages may also represent significant quantities in wastewater (Seiler et al. 1999).

The chief metabolite of the caffeine alkaloid in the human body is paraxanthine, or 1,7-dimethylxanthine, a dimethyl derivative of xanthine. Like caffeine, paraxanthine is a psychoactive central nervous system (CNS) stimulant. It possesses potency roughly equal to that of caffeine and is likely involved in the mediation of the effects of caffeine itself. Paraxanthine is not produced by plants and is only observed in nature as a metabolite of caffeine in humans but has been poorly studied and its suitability as a human source indicator is unknown. The compound is produced from caffeine breakdown and roughly 84% is demethylated at the 3-position to yield paraxanthine (Buerge et al. 2003b; Siegener and Chen 2002).

In wastewater influent and effluent, caffeine has been detected at 37 and 4.0 μg/L, respectively (Paxéus and Schröder 1996), indicating the relative high removal efficiency (80–90%) of wastewater treatment and the challenge of using caffeine for detecting wastewater sources. In an environment with less treatment such as septic tanks, caffeine has been detected at concentrations between 100 and 120 μg/L (Seiler et al. 1999). Caffeine is generally extracted using liquid–liquid extraction or solid-phase extraction (Rangdale et al. 2003), with detection by gas chromatography-mass spectrometry (GC-MS) or by high-performance liquid chromatography-mass spectrometry (HPLC-MS).

Caffeine has been identified in freshwater rivers and lakes (Buerge et al. 2003a; Peeler et al. 2006), groundwater (Seiler et al. 1999; Chen et al. 2002), marine waters (Siegener and Chen 2002; Peeler et al. 2006), and stormwater outfalls (Sankararamakrishnan and Guo 2005). Siegener and Chen (2002) reported the successful use of caffeine in tracing human pollution to specific tributaries in Boston Harbor. However, caffeine has also been detected in presumably uncontaminated marine water samples collected in the Tromsø-Sound, Norway (Weigel et al. 2004). Peeler et al. (2006) successfully detected caffeine levels in small Georgia creeks with known anthropogenic sources. As these creeks drained into larger rivers concentrations were diluted below detectable levels. In addition, caffeine was detected in nearby nonimpacted wetlands containing populations of yaupon hollies (plants that contain caffeine), raising questions concerning the abundance of natural caffeine in surface waters, but failure to detect caffeine in rural creeks in other areas suggests that the wetlands detection in Georgia may be atypical. Finally, Buerge et al. (2003b) reported difficulties correlating levels of caffeine with FIBs, probably because it has a half-life of approximately 12 days when exposed to natural light.

8.6 Miscellaneous Compounds

There are other metabolites that might serve as anthropogenic markers, but very few have been adequately evaluated. Given the reports described in this section, it is very likely that additional new chemical analytes will be discovered in the future that could be very useful in source tracking. Katz et al. (2009) examined 78 wastewater compounds in sewage applied to sprayfields in Florida and reported that carbamazepine (an anticonvulsant drug) was detected in groundwater at two locations. While this study did not evaluate any of the compounds as possible anthropogenic agents, it did provide a long list of analytes that should be tested for this purpose, including fragrances and flavorants, flame retardants, fuel-related chemicals, plasticizers, and solvents and preservatives. Another example is cotinine, the most common metabolite of nicotine – readily detected as a by-product of tobacco use. There is very little information available to date, but cotinine was found in larger quantities than nicotine in one study (Glassmeyer et al. 2005). Like many other chemicals, detection of cotinine may be limited to larger communities or regions where tobacco use is more prevalent. Another novel group of compounds of unknown potential are artificial sweeteners. Buerge et al. (2009) is the only study so far that examined different artificial sweeteners and they reported that acesulfame might be an ideal chemical marker of domestic wastewater.

One final example is the bile acids (Tyagi et al. 2009). Normal human feces contain more than 20 different bile acids that are formed from the primary bile acids, cholic acid (CA), and chenodeoxycholic acid (CDOCA, Bull et al. 2002). The two primary bile acids are formed in the liver from cholesterol and secreted with the bile to the intestine. The microorganisms present in the intestine transform primary bile acids (CDOCA and CA) to secondary bile acids (some 20+ different acids). Most of the secondary bile acids are absorbed in the intestine and returned to the liver, while a small but significant fraction is excreted in the feces. Source differentiation is possible based on the profile of the secondary bile acids. Ruminant animals (e.g., bovines) produce predominantly deoxycholic acid, while nonruminants (e.g., canines and humans) produce significant quantities of lithocholic acid. The absence of deoxycholic acid and the presence of hyocholic acids in porcine fecal material enable it to be distinguished from human and canine contamination. Tyagi et al. (2009) reported that combinations of fecal stanols and bile acids provided reliable and specific biomarkers for chickens, cows, horses, and pigs, suggesting that such combinations could be used to identify sources and occurrence of fecal matter in water and sediments.

8.7 Relationship of Chemicals to Fecal Indicator Bacteria

One of the main issues with both microbes and chemicals that are used in source tracking is to find predictive relationships with the FIBs that are monitored for regulatory purposes. Noblet et al. (2004) tested a suite of fecal sterols as chemical markers to determine if sewage was a significant source of FIB in the Lower Santa Anna River watershed in California. They found that bird fecal sterols were better correlated with FIB than typical sewage sterols, and their results indicated that sewage was not an appreciable source of FIB in the study area. Sankararamakrishnan and Guo (2005) reported that caffeine, anionic surfactants, fluoride, and fluorescent whitening agents were readily detectable in stormwater outfalls in Deal Lake (New Jersey) and were strongly correlated with FC counts, further indicating the human source of the pollution in stormwater. Wu et al. (2008) found that caffeine was a better indicator than a variety of human pharmaceutical compounds in detecting human-origin pollution in Marina Bay, Singapore, and caffeine concentrations were reasonably correlated with FC densities. However, all three of the previous reports were modest regarding the size of the study area, the numbers of compounds tested and samples collected, and the time frame of the project (assessment under different environmental conditions over time).

Haack et al. (2009) conducted the largest and most complete study to date where both a wide array of wastewater chemicals and selected bacterial pathogen genes were measured and compared against FIB. Only the *E. coli eae*A toxin gene was positively correlated with FIB concentrations, and some of the chemical and gene-based indicators were present in waters where FIB standards were met. They concluded that the best approach at present was to use multiple water-quality indicators with different levels of environmental persistence and fate.

8.8 Performance Criteria for Chemical-Based Methods

Method performance criteria that are standardized as much as possible should be used to assess procedures for measuring chemicals within each class of compounds (Stoeckel and Harwood 2007). The seven performance criteria listed here for chemicals were selected by combining (and largely mirroring) those suggested for microbial-based methods (USEPA 2007), plus the method performance approach that is widely used in water monitoring (APHA 2005), and the criteria described in a large European study (Blanch et al. 2006). See Chap. 2 for a detailed discussion of performance criteria for microbiological-based methods that should also be applied to chemical methods. The following section briefly describes issues within each criterion that relate mainly (but not exclusively) to chemical measurements.

Specificity – All compounds of choice must be specific to human waste streams, having a minimum of alternative sources that would cause the indicator to occur where human fecal material is absent, and should be present whenever there is a human fecal source.

Temporal and geographic applicability – Chemical indicators should produce a comparable signal across as large a geographic range as possible (although such applicability can be affected by varying uses of chemicals among locales, regions, and countries), and the signal should not be seasonal, or seasonality of its presence should be understood.

Sensitivity and accuracy – These two parameters are typically quantified in terms of false positives and false negatives, best addressed through challenge samples. Sensitivity is the analytical detection limit of a method (the smallest amount detectable using the method), or the lowest quantity of a substance that can be distinguished from the absence of that substance (a blank). Accuracy is the ability to measure the target chemical in an unbiased manner and is usually defined with regards to a positive success rate.

Repeatability – All methods of choice should measure analyte concentrations precisely over repeated times. Repeatability both within a laboratory and across labs (precision) is critical, as is the expected or allowable degree of variability. The desired precision level must meet the needs of regulatory decisions.

Matrix-independence – Waters that are saline, turbid, or contain a high organic content have the potential to interfere with chemical measurement methods. Tannic and humic acids from decaying plant materials and low levels of residual chlorine from waste treatment facilities are also concerns. Methods that require filtration might be susceptible to interference from high suspended solid loads.

Practicality – Sampling logistics and costs (labor, materials, capital, and training) often governs the analytical method of choice. Cost concerns can be important when large numbers of samples or large sample volumes are needed. It is always possible to measure lower concentrations with high-volume collection strategies, but it is preferable for chemicals of choice to be present at high enough density to be easily detected in sample volumes that are convenient to collect and transport to a laboratory. Also, many chemical measurement technologies may require large initial investments if the needed equipment and personnel are not already in place.

Rapidity of results – Speed of the analytical method is an important characteristic, particularly when warning decisions are involved and potential human exposure might occur during the laboratory analysis period. Although speed can be constrained by backlogs in the analytical lab, and technicians may not be familiar with the required procedures, automated real-time measurement is very feasible for OBs.

8.9 Conclusions

A number of chemicals are potentially good markers of human fecal sources, having been found to be more source-specific and geographically stable than most microbial markers reported to date. However, most comparison studies have only evaluated source specificity and not addressed performance characteristics of the measurement methods. Where performance has been examined, measurement sensitivity has been found to be an impediment, and both temporal and geographic applicability for most is unknown. Many chemicals that are consistently present in waste streams occur at low densities that quickly become diluted below measurement detection limits in the ambient environment (however, the same is true for microbe-based methods as well). Many of the methods also have a higher cost, and most require specialized measurement equipment such as mass spectrometers and trained, dedicated personnel not often available at local laboratories.

Based on these shortcomings, it is unlikely that chemical measures will supplant microbial source identification methods, but there are two types of compounds that can be used now. The first are a small class of chemicals that are so highly specific to human sources that they can be important contributors when certainty about human sources are critical, such as in drinking water applications. These chemicals, based on results presented in this chapter, include two pharmaceuticals (carbamazepine and diphenhydramine), one stanol (coprostanol), and caffeine (although most analytes have yet to be thoroughly tested). The second is using OBs as a screening tool. OBs are not as sensitive as microbial measures but have been found to be sufficiently sensitive to identify spatial patterns in the ambient environment when human fecal contribution is substantial. One big advantage of OBs is that they can be measured with a handheld fluorometer, providing for near real-time and almost cost-free tracking of signals in the field. The US Environmental Protection agency is committed to developing new recreational water-quality criteria, based on a variety of novel indicators, for coastal waters by 2012 (Boehm et al. 2009). Using multiple water-quality indicators with different levels of environmental persistence and fate, including appropriate chemical markers, will yield greater confidence in fecal pollution assessment and regulatory decisions.

References

APHA (2005) Standard methods for the examination of water and wastewater, 21st ed. American Public Health Association/American Water Works Association/Water Environment Federation, Washington, D.C

Binzcik GA, Gates JL, Gray Jr. LE, et al (2004) Endocrine-disrupting effects of cattle feedlot effluent on an aquatic sentinel species, the fathead minnow. Environ. Health Perspectives 112: 353–358

Blanch AR, Belanche-Munoz L, Bonjoch X, et al (2006) Integrated analysis of established and novel microbial and chemical methods for microbial source tracking. Appl. Environ. Microbiol. 72: 5915–5926

Boehm AB, Ashbolt NJ, Colford JM, et al (2009) A sea change ahead for recreational water quality criteria. J. Water & Health. 7: 9–20

Bolz U, Hagenmaiser H & Komer W (2001) Phenolic xenoestrogens in surface water, sediments, and sewage sludge from Baden-Wurttemberg, south-west Germany. Environ. Pollut. 115: 291–301

Boving TB, Meritt DL & Boothroyd JC (2004) Fingerprinting sources of bacterial input into small residential watersheds: Fate of fluorescent whitening agents. Environ. Geol. 46: 228–232

Brun GL, Bernier M, Losier R, et al (2006) Pharmaceutically active compounds in Atlantic Canadian sewage treatment plant effluents and receiving waters, and potential for environmental effects as measured by acute and chronic aquatic toxicity. Environ. Toxicol. Chem. 25: 2163–2176

Buerge IJ, Buser HR, Muller MD, et al (2003a) Behavior of the polycyclic musks HHCB and AHTN in lakes, two potential anthropogenic markers for domestic wastewater in surface waters. Environ. Sci. Technol. 37: 5636–5644

Buerge IJ, Poiger T, Muller MD, et al (2003b) Caffeine, an anthropogenic marker for wastewater contamination of surface waters. Environ. Sci. Technol. 37: 691–700

Buerge IJ, Buser H-R, Kahle M, et al (2009) Ubiquitous occurrence of the artificial sweetener acesulfame in the aquatic environment: An ideal chemical marker of domestic wastewater in groundwater. Environ. Sci. Technol. 43: 4381–4385

Bull ID, Lockheart MJ, Elhmmali MM, et al (2002) The origin of faeces by means of biomarker detection. Environ. Int. 27: 647–654

Bull ID, Elhmmali MM, Roberts DJ, et al (2003) The application of steroidal biomarkers to track the abandonment of a roman wastewater course at the Agora (Athens, Greece). Archaeometry 45: 149–161

Buser HR, Poiger T & Muller MD (1999) Occurrence and environmental behavior of the chiral pharmaceutical drug ibuprofen in surface waters and in wastewater Environ. Sci. Technol. 33: 2529–2535

Cao Y, Griffith JF & Weisberg SB (2009) Evaluation of optical brightener photodecay characteristics for detection of human fecal contamination. Water Res. 43: 2273–2279

Chan KH, Lam MHW, Poon KF, et al (1998) Application of sedimentary fecal stanols and sterols in tracing sewage pollution in coastal water. Water Res. 32: 225–235

Chen Z, Pavelic P, Dillon P, et al (2002) Determination of caffeine as a tracer of sewage effluent in natural waters by on-line solid-phase extraction and liquid chromatography with diode-array detection. Water Res. 36: 4830–4838

Clara M, Strenn B & Kreuzinger N (2004) Carbamazepine as a possible anthropogenic marker in the aquatic environment: investigations on the behaviour of carbamazepine in wastewater treatment and during groundwater infiltration. Water Res. 38: 947–954

Close ME, Hodgson LR & Tod G 1989 Field evaluation of fluorescent whitening agents and sodium tripolyphosphate as indicators of septic tank contamination in domestic wells. N. Z. J. Mar. & Fresh. Res. 23: 563–568

da Costa RL & Carreira RS (2005) A comparison between faecal sterols and coliform counts in the investigation of sewage contamination in sediments. Braz. J. Oceanogr. 53: 157–167

Devane M, Saunders D, Gilpin B (2006) Faecal sterols and fluorescent whiteners as indicators of the source of faecal contamination. Chemistry in New Zealand 70: 74–77

Dickerson JW Jr., Hagedorn C & Hassall A (2007) Detection and remediation of human-origin pollution at two public beaches in Virginia using multiple source tracking methods. Water Res. 41: 3758–3770

Dixon LK, Taylor HM, Staugler E, et al (2005) Development of a fluorescence method to detect optical brighteners in the presence of varying concentrations of fluorescent humic substances: identifying regions in influenced by OSTDS in the estuarine waters of Charlotte Harbor. Mote Marine Laboratory Technical Report No. 1045

Edwards DD, McFeters GA & Venkatesan MI (1998) Distribution of *Clostridium perfringens* and fecal sterols in a benthic coastal marine environment influenced by the sewage outfall from McMurdo Station, Antarctica. Appl. Environ. Microbiol. 64: 2596–2600

Elhmmali MM, Roberts DJ & Evershed RP (2000) Combined analysis of bile acids and sterols/stanols from riverine particulates to assess sewage discharges and other fecal sources. Environ. Sci. Technol. 34: 39–46

Evershed RP & Bethell PH (1996) Application of multimolecular biomarker techniques to the identification of faecal material in archaeological soils and sediments. ACS Symp. Ser. 625: 157–172

Fent K, Weston AA & Caminada D (2006) Ecotoxicology of human pharmaceuticals. Aquat. Toxicol. 76: 122–159

Fono LJ & Sedlack DL (2005) Use of the chiral pharmaceutical propranolol to identify sewage discharges into surface waters. Environ. Sci. Technol. 39: 9244–9252

Gilpin BJ, Gregor JE & Savill MG (2002) Identification of the source of faecal pollution in contaminated rivers. Water Sci. Technol. 46: 9–15

Glassmeyer ST, Furlong ET, Kolpin DW, et al (2005) Transport of chemical and microbial compounds from known wastewater discharges: potential for use as indicators of human fecal contamination. Environ. Sci. Technol. 39: 5157–5169

Gregor J, Garrett N, Gilpin B, et al (2002) Use of classification and regression tree (CART) analysis with chemical faecal indicators to determine sources of contamination. N. Z. J. Marine & Freshwater Res. 36: 387–398

Grimalt JO, Fernandez P, Bayona JM, et al (1990) Assessment of fecal sterols and ketones as indicators of urban sewage inputs to coastal waters. Environ. Sci. Technol. 24: 357–63

Haack SK, Duris JW, Fogarty LR, et al (2009) Comparing wastewater chemicals, indicator bacteria concentrations, and bacterial pathogen genes as fecal pollution indicators. J. Environ. Qual. 38: 248–258

Hagedorn C, Reneau Jr. RB, Saluta M, et al (2003) Impact of onsite wastewater systems on water quality in coastal regions. Virginia Coastal Resources Management Program Memorandum of Agreement 50312-01-13-PT. Virginia Dept. of Conservation and Recreation, Richmond, VA

Hagedorn C, Saluta M, Hassall A, et al (2005a) Fluorometric detection of optical brighteners as an indicator of human sources of water pollution. Part I. Description and detection of optical brighteners. Crop & Soil Environmental News. online http://filebox.vt.edu/users/chagedor/biol_4684/BST/OB%20Article-I.pdf (accessed 6/24/09)

Hagedorn C, Saluta M, Hassall A, et al (2005b) Fluorometric detection of optical brighteners as an indicator of human sources of water pollution. Part II. Development as a source tracking methodology in open waters. Crop & Soil Environmental News. online http://filebox.vt.edu/users/chagedor/biol_4684/BST/OB%20Article-II.pdf (accessed 6/24/09)

Hagedorn, C., and S.B. Weisberg. 2009. Development of chemical-based fecal source tracking methods. Rev. Environ. Sci. Bio/Technol. 8: 275–287

Hartel PG, Hagedorn C, McDonald JL, et al (2007) Exposing water samples to ultraviolet light improves fluorometry for detecting human fecal contamination. Water Res. 41: 3629–3642

Hayashi Y, Managaki S & Takada H (2002) Fluorescent whitening agents in Tokyo Bay and adjacent rivers: their application as anthropogenic molecular markers in coastal environments. Environ. Sci. Technol. 36: 3556–3563

Hilton HJ & Thomas KV (2003) Determination of selected human pharmaceutical compounds in effluent and surface water samples by high-performance liquid chromatography-electrospray tandem mass spectrometry. J. Chromatogr A. 1015: 129–141

Isobe KO, Tarao M, Chiem NH, Minh LY, et al (2004) Effect of environmental factors on the relationship between concentrations of coprostanol and fecal indicator bacteria in tropical (Mekong Delta) and temperate (Tokyo) freshwaters. Appl. Environ. Microbiol. 70: 814–821

Johnson A, Carey B & Golding S (2004) Results of a screening analysis for pharmaceuticals in wastewater treatment plant effluents, wells, and creeks in the Sequim-Dungeness area. Environmental Assessment Program, Olympia, Washington 98504-7710. Publication No: 04-03-051

Katz BG, Griffin DW & Davis JH (2009) Groundwater quality impacts from the land application of treated municipal wastewater in a large karstic spring basin: Chemical and microbiological indicators. Sci. of the Total Environment 407: 2872–2886

Kramer JB, Canonica S & Hoigne J (1996) Degradation of fluorescent whitening agents in sunlit natural waters. Environ. Sci. Technol. 30: 2227–2234

LeBlanc LA, Latimer JS, Ellis JT, et al (1992) The geochemistry of coprostanol in waters and surface sediments from Narragansett Bay. Estuarine, Coastal and Shelf Sci. 34: 439–458

Leeming R & Nichols PD (1996) Concentration of coprostanol that corresponds to existing bacterial indicator guideline limits. Water Res. 30: 2997–3006

Leeming R, Ball A, Ashbolt N, et al (1996) Using faecal sterols from humans and animals to distinguish faecal pollution in receiving waters. Water Res. 30: 2893–2900

Leeming R, Latham V, Rayner M, et al (1997) Detecting and distinguishing sources of sewage pollution in Australian inland and coastal waters and sediments. ACS Symp. Ser. 671: 306–319

MacDonald IA, Bokkenheuser VD, Winter J, et al (1983) Degradation of fecal sterols in the human gut. J. Lipid Res. 24: 675–694

Maldonaldo C, Dachs J & Bayona JM (1999) Trialkylamines and coprostanol as tracers of urban pollution in waters from enclosed seas: The Mediterranean and Black Sea. Environ. Sci. Technol. 33: 3290–3296

McDonald JL, Hartel PG, Gentit LC, et al (2006) Identifying sources of fecal contamination inexpensively with targeted sampling and bacterial source tracking. J. Environ. Qual. 35: 889–897

Mudge, S.M., Lintern, D., (1999) Comparison of sterol biomarkers for sewage with other measures in Victoria Harbour, BC, Canada. Estuarine. Coastal and Shelf Science, 48: 27–38

Murtaugh, J.J., Bunch, R.L., (1967) Sterols as a measure of fecal pollution. J. Wat. Pollut. Control Fed., 39: 404–409

Nakada N, Kiri K, Shinohara H, et al (2008) Evaluation of pharmaceuticals and personal care products as water-soluble markers of sewage. Environ. Sci. Technol. 42: 6347–6353

Noblet JA, Young DL, Zeng EY, et al (2004) Use of fecal steroids to infer the sources of fecal indicator bacteria in the Lower Santa Ana River Watershed, California: sewage is unlikely a significant source. Environ. Sci. Technol. 38: 6002–6008

Paxéus N & Schröder HF (1996) Screening for non-regulated organic compounds in municipal wastewater in Göteborg, Sweden. Water Sci. Technol. 33: 9–15

Peeler KA, Opsahl SP & Chanton JP (2006) Tracking anthropogenic inputs using caffeine, indicator bacteria, and nutrients in rural freshwater and urban marine systems. Environ. Sci. Technol. 40: 7616–7622

Pond KR, Rangdale R, Meijer WG, et al (2004) Workshop Report: Developing pollution source tracking for recreational and shellfish waters. Environ. Forensics 5: 237–247

Poiger T, Field JA, Field TM, et al (1998) Behavior of fluorescent whitening agents during sewage treatment. Water Res. 32: 1939–1947

Rangdale RE, Meijer W, Rincé A, et al (2003) Development of methods for pollution source tracking-consolidated literature review. Completed for Environment Agency North West Region INTERREG IIIB Atlantic Area Programme Improved Coastal & Recreational Waters (ICReW)

Reeves AD, Patton D (2001) Measuring change in sterol input to estuarine sediments. Physics and Chemistry of the Earth Part B: Hydrology Oceans and Atmosphere 26: 753

Roberts PH & Thomas KV (2006) The occurrence of selected pharmaceuticals in wastewater effluent and surface waters of the lower Tyne catchment. Sci. Total Environ. 56: 143–53

Sankararamakrishnan N & Guo Q (2005) Chemical tracers as indicator of human fecal coliforms at storm water outfalls. Environ. Intern. 31: 1133–1140

Scott TM, Rose JB, Jenkins TM, et al (2002) Microbial source tracking: current methodology and future directions. Appl. Environ. Microbiol. 68: 5796–5803

Seiler RL, Zaugg SD, Thomas JM, et al (1999) Caffeine and pharmaceuticals as indicators of wastewater contamination in wells. Ground Water 37: 405–410

Seurinck S, Verstraete W & Siciliano SD (2005) Microbial source tracking for identification of fecal pollution. Rev. Environ. Sci. Bio/Technol. 4: 19–37

Siegener R & Chen RF (2002) Caffeine in Boston Harbor seawater. Mar. Pollut. Bull. 44: 3–387

Shah VG, Dunstan RH, Geary PM, et al (2007) Evaluating potential applications of fecal sterols in distinguishing sources of faecal contamination from mixed faecal samples. Water Res. 41: 3667–3674

Shu WC & Ding WH (2005) Determination of fluorescent whitening agents in laundry detergents and surface waters by solid-phase extraction and ion-pair high-performance liquid chromatography. J. Chromatography A 1–2: 218–223

Stoeckel DM & Harwood VJ (2007) Performance, design, and analysis in microbial source tracking studies. Appl. Environ. Microbiol. 73: 2405–2415

Temes TA, Andersen H, Gilberg D, et al (2002) Determination of estrogens in sludge and sediments by liquid extraction and GC/MS/MS. Anal. Chem. 74: 3498–3504

Thomas KV & Hilton MJ (2004) The occurrence of selected human pharmaceutical compounds in UK estuaries. Mar. Pollut. Bull. 49: 436–444

Tolosa I, LeBlond N, Copin-Montégut C, et al (2003) Distribution of sterol and fatty alcohol biomarkers in particulate matter from the frontal structure of the Alboran Sea (SW Mediterranean Sea). Marine Chem. 82: 161–183

Tyagi Punam, Dwayne R Edwards & Mark S Coyne (2009) Fecal sterol and bile acid biomarkers: runoff concentrations in animal waste-amended pastures. Water Air Soil Pollut. 198: 45–54

USEPA (2007) Experts Scientific Workshop on Critical Research Needs for the Development of New or Revised Recreational Water Quality Criteria. EPA 823-R-07-006; June 2007, online http://www.epa.gov/waterscience/criteria/recreation/experts/index.html

Weigel S, Berger U, Jensen E, et al (2004) Determination of selected pharmaceuticals and caffeine in sewage and seawater from Tromsø/Norway with emphasis on ibuprofen and its metabolites. Chemosphere 56: 583–592

Wolfe TM (1995) A comparison of fecal coliform densities and fluorescent intensities in Murrells Inlet, a highly urbanized estuary and in North Inlet, a pristine forested estuary. M.S. thesis. University of South Carolina, Columbia

Wu J, Yue J, Hu R, et al (2008.) Use of caffeine and human pharmaceutical compounds to identify sewage contamination. Proc. World Acad. Sci. Eng. & Tech. 34: 2070–3740

Chapter 9
Statistical Approaches for Modeling in Microbial Source Tracking

Lluís A. Belanche and Anicet R. Blanch

Abstract Microbial source tracking (MST) concerns the definition of new indicators and appropriate detection methods, the identification of host-specific indicators of fecal pollution, and ultimately the development of useful and reliable predictive models for practical deployment. Optimal predictive models should be designed using proper statistical and computational tools for the analysis of the available data samples. A further requirement is found in the determination of appropriate sets of predictors (indicators, tracers) for developing accurate and low-cost MST solutions. This chapter briefly reviews some of these modeling tools, and their use and feasibility in providing more accurate MST-based results. It also evaluates the potential of established and new algorithmic methods to the identification of fecal pollution sources.

Keywords Predictive models • Fecal pollution • Microbial indicators

9.1 Introduction

Microbial source tracking (MST) became a defined knowledge topic and research field in the early 1980s as a result of social, legal, and regulatory pressures (Malakoff 2002). Starting in the 1970s, most research in MST was aimed at defining new indicators and appropriate detection methods. MST studies were focused mainly on the use of selective media in the attempt to identify host-specific indicators of fecal pollution in water (Geldreich 1976; Mara and Oragui 1981; Osawa et al. 1981; Allsop and Stickler 1985). The feasibility of the approaches developed was evaluated by calculating the percentage of correct classification using water

A.R. Blanch (✉)
Department of Microbiology, University of Barcelona, Avda. Diagonal 645, Barcelona, Spain
e-mail: ablanch@ub.edu

C. Hagedorn et al. (eds.), *Microbial Source Tracking: Methods, Applications, and Case Studies*, DOI 10.1007/978-1-4419-9386-1_9,

samples with known sources of fecal pollution. Similar conceptual approaches are still published today to develop new methods (Layton et al. 2006; Reischer et al. 2007) or to suggest new indicators for fecal source detection (Martellini et al. 2005; Payán et al. 2005).

Different methodologies proposed during the last few years have mainly been based on the requirements of the microorganisms to be cultivated and dependence on a prior developed library of reference. According to these characteristics, methods have been classified as library-dependent or library-independent and culture-dependent or culture-independent (Simpson et al. 2002; Scott et al. 2005; Stoeckel and Harwood 2007; Blanch et al. 2008). Moreover, the arrival of molecular techniques based on nucleic-acid amplification intensified the search for new methods and new targets specific to fecal sources that could avoid the dependency on libraries and local approaches.

Problems with comparability of methods and results, and the significance of databases, soon appeared. It was observed that the standardization of methods, data sharing, and the comparison of approaches covering wider geographical and climatic areas would be essential to make further progress in the definition of reliable MST indicators and methods. Several collaborative studies were performed by independent laboratories based on analyzing water samples spiked with fecal material from a known origin. Some laboratories used genotypic or phenotypic library-dependent techniques (Harwood et al. 2003; Griffith et al. 2003). Although genotypic methods performed better than phenotypic methods, it was observed that the internal accuracy of the libraries did not correspond to the accuracy of source prediction. Library-based methods showed a high degree of variability across assayed statistical methods. A careful examination of libraries, the removal of nuisance variables, and a quest for methods capable of better discrimination were recommended (Ritter et al. 2003). Other collaborative studies were performed using genotype-based and library-independent methods or culture-independent and library-independent genotypic methods (Field et al. 2003; Myoda et al. 2003; Noble et al. 2003), all of them showing a high false-result rate. Most of these collaborative studies were performed using experimental prepared samples or spiked samples with analyzed indicators. Consequently, their performance in real scenarios using environmental water samples was unknown. It was also identified that comparative and integrated studies between independent research groups needed to follow standardized procedures to avoid differences in implementation (Stewart et al. 2003).

Interlaboratory studies have been performed using standardized procedures, point-source polluted water samples, and the combined use of proposed MST host-specific indicators with other nondiscriminant traditional microbial indicators (Blanch et al. 2004). These studies concluded that no single MST indicator was best able to determine the source of fecal pollution. Moreover, the effects of the load of fecal material, persistence of indicators in the environment, dilution effects of water bodies, and the mixture of fecal origins should all be considered, since they could impair the relative feasibility of each method examined. Therefore, MST methods need to be adapted and optimized to minimize modeling errors and improve their general accuracy and reliability.

At present, there is widespread agreement that no single microbial or chemical indicator (or parameter) is able to best determine sources of fecal pollution. In consequence, a subset of indicators should be carefully selected based on appropriate statistical analyses. In particular, combinations of at least two parameters (one discriminating and another nondiscriminating indicator) are needed to accurately differentiate between two distinct fecal pollution sources. The discriminating indicator identifies the source, and the universal (nondiscriminating) indicator provides information on the fecal load (Blanch et al. 2006). This distinction is an innovative contribution with respect to the pairs given in other studies, such as combinations of sterols (Leeming et al. 1996) or ratios between traditional bacterial indicators (Geldreich 1976). It is envisaged that such combinations may offer advantages when analyzing challenging characteristics within sets of data samples (such as diluted, aged, and mixed observations).

There has been a lack of consensus regarding the most appropriate statistical analysis to determine optimal predictive models and appropriate sets of variables (indicators) for developing useful MST solutions. Algorithmic methods have been borrowed from other modeling disciplines, such as Machine Learning (ML), an active field of research within computer science with strong bridges to statistics. In ML, the aim is to obtain computer algorithms capable of learning from experience which will be the base for the modeling (Mitchell 1997). Specifically, neural networks – which have been used successfully in many disciplines – have provided superior predictive power in nonlinear relationships compared to some traditional statistical methods when applied to multiparameter MST studies (Brion and Lindgireddy 2000).

Recent ML methods proved to be useful for the development of MST predictive models, achieving 100% success in distinguishing between wastewaters of human and nonhuman origin (Blanch et al. 2006; Belanche and Blanch 2008). These approaches pursued the formulation of new predictive models for MST, including determination of the most discriminating indicators among those that show broad and consistent geographical stability. The overall goal is the development of models using the minimum number of variables possible, as well as solutions with different degrees of accuracy and technical cost, depending on the presence/absence of certain user-defined variables.

This chapter aims at a critical review of the strengths and weaknesses of widely used statistical methods that have been applied to evaluate the feasibility of MST methodologies. Recently introduced ML methods (some of which have only just been applied to this research field, if at all) are also reviewed, and their potential for the future definition of reliable MST models is assessed. A major aim is to stimulate biologists to work together with their local statisticians and computer scientists to come up with better solutions to the problem of determining the origins of fecal pollution in water. This collaboration will foster research and hopefully find new and useful approaches for MST.

9.2 Modeling Approaches for MST

Most of the modeling approaches used in MST make use of traditional techniques, such as discriminant analysis, nearest neighbors, or maximum and average similarity (Wiggins 1996; Harwood et al. 2000; Ritter et al. 2003; Blanch et al. 2004), though a few use artificial neural networks (ANNs), most notably the multilayer perceptron or MLP (Brion and Lingireddy 2003). Among the unsupervised methods, clustering variants, principal components analysis (PCA) and factor analysis (FA) still find good use among practitioners because of the possibilities they bring for dimension reduction and data visualization. For instance, PCA and FA were used to identify pollution sources (among them, fecal pollution) in Hong Kong coastal waters (Zhou et al. 2007), and PCA was used for the successful identification of fecal pollution (human from other animals) in receiving waters using fecal sterol profiles (Leeming et al. 1996).

Discriminant analysis with equal priors and a pooled covariance matrix was used by Wiggins (1996) to achieve up to 95% accuracy for two-class classification (human vs. nonhuman) of the fecal pollution source, and 84% accuracy for four-class classification (human, cattle, poultry, and wild animal). This work is also interesting in that it suggests the use of different models for different tasks, guided by available knowledge, and the recognition that larger data samples will be needed when geographical or temporal variation are an issue. Different results in the human vs. nonhuman task were obtained by discrimination based on *E. coli* ribotype profiles in estuarine waters using equal priors and separate covariance matrices for each class (Parveen et al. 1999).

An MLP model was used by Neelakantan et al. (2001) for predicting background or above-background concentrations of waterborne encysted protozoa in the Delaware River at Trenton, NJ (USA) and for identifying encysted protozoan-laden sources in surface waters from urban or agricultural runoff using water-quality indicators at Jacobson Reservoir #4, KY (USA). This work was expanded into a multiyear study performed at several nearby sites, with the multiple aim of distinguishing human sewage from animal-impacted runoff, fresh from aged runoff, and fresh agricultural from urban runoff, by means of separate modeling tasks that were hierarchically cascaded (Brion et al. 2002).

There are few thorough comparisons of methods and modeling approaches to MST (Ritter et al. 2003; Stoeckel and Harwood 2007). Such comparisons entail several inherent difficulties that may hinder their usefulness. The working hypotheses of every modeling approach, the different ways of overfitting control, the use of different resampling procedures, and the existence of varying numbers of free parameters to be determined, among others, make a fair and principled comparison a delicate task. Indeed, the presence of all these factors partly explains the high degree of variability found in experimental studies. Another partial explanation can be found in the differences among the samples used for building the models: number of classes (known groups) and their distribution, sample size and number, and importance of the measured variables make the analysis of different data samples a difficult task that may require at least distinct, if not separate, modeling approaches.

These problems are more acute when MST approaches are designed using a low number of observations or a low ratio number of observations vs. number of variables. In this sense, few authors perform and report explicit variable selection studies, and most of these are limited in their extent. Neelakantan et al. (2001) perform a selection with expert eyes, looking for good combinations of predictors; however, in the end, all the predictors are used. In other studies, only one step of backward elimination is carried out, removing every possible variable at a time and keeping the best model that is found (Brion et al. 2002).

When the age of fecal pollution is also an issue, identification becomes more complicated and thus the availability of reliable decay rates is of primary importance to MST (Bonjoch et al. 2009). In surface waters impacted by runoff and point sources, the use of ratios between parameters can help in determining the origin of fecal contamination (Nieman and Brion 2003). Persistence studies of each selected parameter are needed to provide complementary information (Pote et al. 2009; Bell et al. 2009). These studies need to address the effects of environmental aspects such as temperature, solar radiation, salinity, pH, chemical pollutants, water filtration, turbidity, starvation, predation, and presence of heavy metals, among others.

9.2.1 Model Fitting Issues

It is known that the process of creating (inducing) predictive models suffers from some or all of several sources of error:

(a) The modeling error inherent in fitting a model that deviates substantially from the true model (called the *bias* of the model).
(b) The modeling error due to using a specific (and finite) data sample (called the *variance* of the model).
(c) The modeling error due to ignorance of the factors affecting the true model, mainly the effect of unmeasured relevant variables.

In addition, there is the phenomenon of inherent *random variation*. For instance, some methods for pathogen or indicator detection or enumeration are costly and time-consuming, and on many occasions surrogate methods have to be used. Although these water-quality parameters may be more easily measured, they may not be perfectly correlated with the presence of fecal sources, thereby introducing process modeling errors. Another example of inherent variation is found in the analytical uncertainty of microbial methods. These sources of variation are gathered in the joint data distribution and usually treated in practice as "noise" in the data. Last but not least, there is *model selection* uncertainty due to the method used for picking a model among a set of candidates (e.g., via resampling). This implies that the error of a model cannot be measured exactly.

In order to minimize the expected prediction error (EPE) of a regression model, the square error (used to measure discrepancies in the predictions of desired true values) can be derived from the assumption of homoscedastic normality of an additive error term (Duda et al. 2001). As it turns out, the EPE can be expressed as the sum of three terms: the first two are the (squared) bias and the variance of the model (their sum is known as the *reducible* error); the third is the aforementioned random variation of the process, and represents the *irreducible* error. This error is independent of the model and is the minimum achievable error (called the Bayes error). A similar bias-variance decomposition holds for classification tasks (Domingos 2000): in this case, the true error of a model is the probability that the model misclassifies an observation drawn at random according to the underlying data distribution.

When a model is too flexible with respect to the available data, unless this flexibility is kept under control by the modeler, the fitting of the model shows a tendency to memorize the data or fit the noise in the data to a point that is deleterious for generalization (note the true error is also known as the generalization error). This phenomenon, sometimes caused by an excess of parameters, is known as *overfitting*. It is perhaps the most notable concern that is common to all modeling methods, since no method is entirely safe from overfitting the data. More formally, a model overfits a data sample if some other possible model fits the same data less well but will actually perform better over the entire distribution of observations.

The error that a model commits in the same data sample that was used to fit it is known as the *apparent* error. Ideally, one should build a model such that its apparent error across different data samples is roughly equal to its true error (and its true error is just above the Bayes error). If the apparent error is lower than the true error, one is most likely in dealing with an overfitted model. Figure 9.1 provides a simple graphical illustration of overfitting.

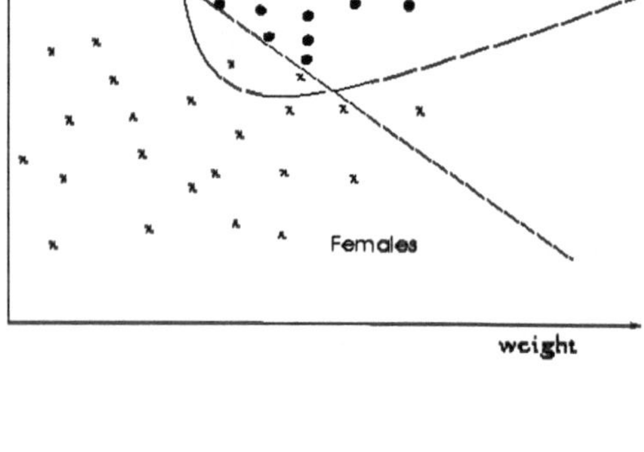

Fig. 9.1 Scatter plot of pairs (weight, height) corresponding to male and female insects of a certain species. The two groups are normally distributed with unequal covariance matrices and different priors. The Bayes optimal solution, as given by QDA - see 9.2.7– is indicated by a parabola. The linear model (a *straight line*) overfits the data, showing a lower modeling error (the apparent error) on the training data

9.2.2 Model Selection and Calibration

Model selection is concerned with the process of finding the optimal model for a set of observations among a set of candidate models. Model calibration (or validation) refers to the acquirement of a reliable estimation of the true error of a selected model. Estimating true model performance is a key issue that has recently been pointed out in the MST literature (Stoeckel and Harwood 2007; Santo Domingo et al. 2007; Gronewold et al. 2009). In order for such calibration to be possible, the estimation of the true error should be carried out in a sample that is independent of the model. If one had access to an unlimited number of observations, the matter would have a straightforward answer: choose the model that provides the lowest error rate on the entire population. However, in real applications only a finite set of observations is available. This has lead to a family of *resampling* methods whose goal is to make better use of the available data. These methods are very useful for assessing how a predictive model that can be the result of a complex modeling process will perform in practice.

Common procedure entails using a first set (called the training set or TR set) to fit the model and a second, independent validation set (or VA set), to assess the performance of a fully trained model. In the *resubstitution* method, VA equals TR (i.e., the validation error is the apparent error). Note that using the entire dataset to select the "optimal" model and then estimating the error rate on the same data will normally overfit the data. Therefore, this error is known to be an optimistically biased assessment of how well the model will fit an independent dataset. Cross-validation (CV) is a very popular technique that can be used to compensate for an optimistic apparent error rate, as explained below.

In the *holdout* method, a subset of observations is chosen randomly from the initial sample to form the VA set and the remaining observations are retained as the TR set; a popular choice is 2/3 for TR and 1/3 for VA. The rationale for this procedure is that a second, independent sample is unlikely to show the same random fluctuations as the TR sample and thus it guards against possible false data regularities that could be present in the TR sample. The error estimation is pessimistically biased, since not all the data used were available to fit the model. The method also exhibits Monte Carlo variation: the results will vary if the analysis is repeated with different random splits. Therefore, a great deal of uncertainty is associated with using a *single* TR/VA partition. To remedy this, *random subsampling* performs several data splits of the entire dataset. For each such split, the model is fitted to the training data, and predictive accuracy is assessed using the validation data. The estimate of the true error is obtained as the average of the separate estimates, which can be significantly better than the holdout estimate. The advantage of this method (over k-fold CV, see below) is that the proportion of the TR/VA splits (called *folds*) is not dependent on the number of splits. A disadvantage is that some observations may never be selected in the validation subsample, whereas others may be selected more than once. In other words, validation subsets may overlap. If the data sample is large, this method is fine. However, if the TR set is small, the estimated parameters will be unreliable and the chances of overfitting will be high. On the contrary, if the VA set is small, the estimated error will have low significance. Thus, there is a trade-off between both sample sizes.

The generic goal of CV is to estimate the expected error of a model in a dataset that is independent of the data that were used to train the model. One round of k-fold CV (or k-CV) involves partitioning the sample into k complementary subsets, systematically performing the modeling on the union of k–1 such subsets and checking the obtained model on the remaining subset (acting as a VA set). To reduce variability, again multiple rounds can be performed using different partitions, and the results averaged over the rounds. The advantage of this method over repeated random subsampling is that all observations are used for both TR and VA, and each observation is used for VA exactly once. In *stratified* k-CV, in the regression case, the splits are carried out so that the mean response value is approximately equal in all the subsets. In the classification case, each subset contains roughly the same proportions of class labels, which is particularly necessary in severely unbalanced class distributions.

As the name suggests, *leave-one-out* CV (LOOCV, also known as the *jackknife* method) involves using a single observation from the original sample as the VA data, and the remaining ones as the TR data. This is repeated such that each observation in the sample is used once as the VA data. Note this amounts to k-CV with k being equal to the number of observations in the data sample. LOOCV is often computationally expensive because of the large number of times the training process is repeated. In some cases, such as least squares regression, k-CV can be sped up significantly by precomputing values that are needed repeatedly in the training, or by using fast updating rules such as the Sherman–Morrison formula. In particular, in linear regression the LOOCV error has a closed-form expression known as the prediction residual error sum of squares or PRESS (McCullagh and Nelder 1989).

The result of k-CV is an estimation of the error if only a fraction $(k-1)/k$ of the available data is used. This error is, thus, expected to be *conservative* (larger than the error obtained if the entire sample was used). This sampling bias can be reduced by increasing k. How many folds are thus needed? With a large number of folds, the bias of the true error rate estimator will be small (the estimator will be very accurate on average), but the variance will be large; also, the computational time will be large as well (many experiments). On the contrary, with a small number of folds, the variance of the estimator will be small, but the bias will be large; moreover, computation time is lower. LOOCV is almost unbiased, but its variance can be large. For this reason, if several models are compared based on the results of k-CV, the model with the lower estimated error may not actually be the best.

Bootstrap resampling techniques permit mean performance estimates and their variability to be obtained, as well as confidence intervals (CI) for generalization error. Assuming one has a dataset $S=\{(x_i,y_i)\}_{i=1},\ldots,_p$, *bootstrap samples* $S^{*1},\ldots,S^{*B}$ of size p are drawn by sampling S with replacement and refit a model to each of the S^{*b}, $b=1$,..., B. The LOOCV estimate of prediction error is, for the classification case:

$$Err^* = \frac{1}{p}\sum_{i=1}^{p}\frac{1}{S^{-i}}\sum_{b\in S^{-i}}\Im[y_i \neq f_b^*(x_i)],$$

where $\Im[z]$ is 1 when z is true and 0 otherwise, S^{-i} is the set of indexes of the bootstrap samples that do not contain observation x_i and f_b^* is the model developed in S^{*b}. Thus, the bootstrap can help in reducing the variance of the LOOCV estimate. Similarly, the resubstitution error is estimated as:

$$\overline{e}^* = \frac{1}{p}\sum_{i=1}^{p}\frac{1}{S^{i}}\sum_{b\in S^{i}}\Im[y_i \neq f_b^*(x_i)],$$

where $S^i = \{1,\ldots,B\}\setminus S^{-i}$. In both cases, if an index set is empty, the term is skipped and the formula is normalized accordingly. The 0.632-*bootstrap* estimate is formed as $0.368.\overline{e}^* + 0.632.Err$ (Efron and Tibshirani 1997). Intuitively, this formula pulls the LOOCV bootstrap estimate down toward the training error, thereby reducing its likely upward bias (Hastie et al. 2001). Further, CIs can be obtained by the *percentile* method, as follows: let θ^{*b}, $b=1$,..., B, denoting the bootstrap distribution of a statistic θ on the samples S^{*b}. The α-level CI is constructed by ordering these values in ascending order and choosing critical value observations θ^{*L} and θ^{*U} as the end points of the CI, such that $\Pr\{\theta^{*L}\leq\theta\leq\theta^{*U}\}=1-\alpha$. For instance, for $B=1{,}000$, observations 26 and 975 are the end points of the 95% CI.

9.2.3 Selection of the Relevant Variables

ML methods can be complemented with feature subset selection (FSS) techniques (Guyon and Elisseeff 2003; Liu and Motoda 1998) for the selection of the best parameters in a MST model (often called *variables* or *predictors* in statistics and *features* in the ML literature). The correct selection of a subset of the available parameters is a ubiquitous problem in an inductive setting and its importance is beyond doubt. First, it can make the task of data visualization and understanding easier by eliminating irrelevant features which can mislead the interpretation of the data. Second, when many *nonuseful* features (irrelevant or redundant) are present, modeling methods are prone to finding false regularities in the data and therefore are more likely to overfit. There are also the *scientific* benefits associated with a smaller number of features: a reduced measurement cost and hopefully a better understanding of the domain. There is a long tradition of statistical modeling tools for the purpose of identifying a useful set of variables – especially in regression, see Hocking (1976). In MST, the objective is to find the most prominent or appropriate combination of water-quality indicators that yield the highest discriminatory power.

Feature selection can be seen as a *search* problem, where each state in the search space corresponds to a subset of features. In the ML literature, a wide family of suboptimal algorithms depart from an initial solution and iteratively add or delete features by locally optimizing the error function. Three commonly used methods are forward selection (adds variables iteratively starting from the empty set), backward elimination (removes variables iteratively starting from the full set), and stepwise regression (adds and removes variables). More sophisticated methods work by combining $l \geq 0$ forward and $r \geq 0$ backward steps, such as the plus l – take away r (Stearns 1976) or the floating search (Pudil et al. 1994), at an increase in computational cost.

Relief (Kira and Rendell 1992) is a general algorithm that works by repeatedly and randomly choosing an instance x (observations are usually called instances or examples in ML) and finding its *near hit* and its *near miss*. The former is the closest instance to x among all the instances in the same class of x. The latter is the closest instance to x among all the instances in a different class. The underlying idea is that a feature is more relevant to x the more it separates x and its near miss, and the least it separates x and its near hit. The result is a weighted version of the original feature set. The most important advantage of Relief is the rapid assessment of irrelevant features with a sound approach; however, it is not good at discriminating between redundant features.

These selection methods, if used in a principled way, can be a very valuable tool for the data analyst. However, a word of warning is due against *exhaustive* search: since this procedure systematically snoops through many models (possibly thousands of them), there is a great chance that the best-fit hypothesis will overfit the data. This is likely to happen even when using a safe resampling method such as k-CV. Thus, it is strongly advised that any special knowledge the analyst may have about the data be used to filter out vast numbers of unfeasible or undesirable models.

9.2.4 Recommended Practices

CV is a generally applicable way to predict the performance of a model and can also be used to decide among a number of competing models (*model selection*). These models can be different because of the modeling procedures, the choices of free parameters of the models, the selection of variables, or a combination thereof. The question remains: how to optimize for the best model in the available data sample and at the same time get an unbiased estimate of the generalization error? The answer is that if model selection and estimations of the true error are to be computed simultaneously, a *three-way* disjoint split of the available data is strongly recommended (Ripley 1996): a first set (the TR set) used to fit the parameters of the model, a second, independent validation set (the VA set), used to tune the free parameters of the model or to choose among a number of competing models (both tasks are equivalent in this setting), and finally a third set, independent of the previous two, and called a test set (or TE set), used only to assess the performance of a completely specified and selected model. Why separate TE and VA sets? This is because the error estimation of the final model on VA data will be biased downward, since this set *has* been used to select the final model. If, as is often the case, the available data size is small to medium, resampling methods for multiple splits are needed, for the reasons explained above.

Suppose one has a data sample S and wish to choose a model among a set M of candidates and obtain a reliable error estimation of the chosen model. These candidate models are determined by different values $\theta_1, \ldots, \theta_t$ of a free parameter θ. As a simple example, consider the problem of selecting the number h of hidden neurons in a one-hidden-layer MLP (a model selection problem). In this case, $\theta_i = h_i$, being h_i, $i = 1, \ldots, t$ the desired values to be tested.

The idea is to perform two nested CV loops (sometimes called *double* CV). The outline of the whole procedure is as follows:

Select reasonable (best judgement) values for positive integers k_1 and k_2

Perform an *outer* loop of k_1-CV:

1. Partition the available data S into k_1 subsets S_i of equal size
2. For every subset S_i, perform an *inner* loop of k_2-CV in $S \backslash S_i$:

   (a) Partition $S \backslash S_i$ into k_2 subsets S_j of equal size
   (b) Repeat for all the subsets S_j, $j = 1, \ldots, k_2$:

      - Fit all the candidate models in M to $(S \backslash S_i) \backslash S_j$
      - Evaluate the fitted models in S_j (acts as a VA set)

   (c) Call V the average VA error of every model

3. Choose the best model as the one having the lowest V error on *average*
4. Refit this model in every $S \backslash S_i$, $i = 1, \ldots, k_1$; call the obtained models m_i
5. Evaluate every m_i in S_i (acts as a TE set for m_i); call T the obtained average TE error; this error is an estimation of the true error

Refit the model in S and call it m; return as final result the pair (m, T)

After assessing the final model on the TE sets (which should be the main reported result), the decision on the chosen model cannot be rectified. Note that the final error estimation is conservative (which is good), for the reasons explained above.

In practice, the choice of the number k of folds in k-CV depends on the size of the dataset. For medium-sized datasets, $k = 3$ is a sensible choice. For small datasets, we may well need to use LOOCV to fit the models on as many observations as possible. Popular intermediate choices are $k = 5$ or $k = 10$. If the dataset is very small, it will be worth considering using the bootstrap instead.

The described procedure only yields meaningful results if the different samples used for TR, VA and TE are drawn from the same population, which should be guaranteed if they are obtained by repeated partitions of the original data sample. In many applications, however, the structure of the process being studied evolves over time. This means that both the induced model and its error estimation may not be useful for predictions in the distant future. A possible workaround solution is to gather new data and perform a new modeling run from time to time.

9.2.5 Assessing Final Model Performance

Once the final model has been selected, the question of how to assess and report its performance remains. Given that the used TE set is a random sample, the error of the selected model should be regarded as a random variable and the obtained value a random value, with its associated uncertainty. Therefore, it is necessary to reflect this uncertainty in the reported performance estimation of the chosen model. This can be quantified by the use of error bars.

The following is a procedure useful for classification tasks, as described by Mitchell (1997), where the reader will find additional information and statistical justification. Given a model h and a data sample S of size n (which, for this example, is indeed the TE set), the quantity "number of errors in S committed by h" follows a binomial distribution $B(n, Pe)$, where Pe is the true error probability of h. We are interested in using the precise value that this random variable takes in S to quantify the uncertainty in Pe. Let m be the number of errors in S committed by h and let $\hat{P}e = m / n$, which is the maximum-likelihood estimate for Pe (Duda et al. 2001). The larger n is, the more precise the estimate of Pe will be (and the narrower the CIs). If $n \cdot \hat{P}e \cdot (1 - \hat{P}e) >= 5$ (or, at least, $n \geq 30$ and $\hat{P}e$ is not very close to 0 or to 1), the binomial can be reliably approximated by a Gaussian. In these conditions, a $N\%$ CI for Pe is:

$$Pe \in \left[\hat{P}e \pm z_N \sqrt{\frac{\hat{P}e(1 - \hat{P}e)}{n}} \right].$$

For example, for $N=95\%$ (level of significance 0.05), $z_N = 1.96$.

9.2.6 Assessing the Differences in Error of Two Models

There are at least two reasons for estimating the error of a model on a problem: to ascertain if it performs well enough to be useful and to be able to compare it with another model. With the latter goal in mind, imagine comparing two models h_1 and h_2 that are being tested on a common (test) data sample S of size n. Let $Pe(h_1), Pe(h_2)$ be the true errors of h_1 and h_2, respectively and define $d = Pe(h_1) - Pe(h_2)$. Let again $\hat{P}e(h_1), \hat{P}e(h_2)$ be maximum-likelihood estimates for $Pe(h_1), Pe(h_2)$ and define $\hat{d} = \hat{P}e(h_1) - \hat{P}e(h_2)$. It turns out that $\hat{d}$ is an unbiased estimator of d and that, provided the corresponding conditions are met, it follows a Gaussian distribution, with mean d and variance equal to the sum of the variances of the error estimates $\hat{P}e(h_1)$ and $\hat{P}e(h_2)$ (which are $\hat{P}e(h_1)(1 - \hat{P}e(h_1)) / n$ and $\hat{P}e(h_2)(1 - \hat{P}e(h_2)) / n$, respectively). Therefore, a $N\%$ CI for d is:

$$d \in \left[\hat{d} \pm z_N \sqrt{\frac{\hat{P}e(h_1)\left[(1 - \hat{P}e(h_1)\right] + \hat{P}e(h_2)\left[(1 - \hat{P}e(h_2)\right]}{n}} \right].$$

It should be noted that these CIs are approximate in that they involve making a number of assumptions. Nevertheless, they are much better than "nothing at all" and surely provide useful guidance for interpreting the true utility of induced models.

9.2.7 Review of Modeling Techniques

While it seems obvious that, for a certain task, some modeling techniques are superior, in the absence of prior information about the problem there is no reason to prefer one method over another. Different models and fitting methods entail different working assumptions, which are sometimes neglected (often consciously) for the sake of practical performance. Most importantly, different methods have different (parameterized) hypothesis spaces and inductive biases (preference of some hypotheses over others) and distinct ways of overfitting control (both in nature and number). In these conditions, it is no surprise that different methods yield different results from the same or similar problems (Lasalde et al. 2005).

Among the modeling techniques, linear models are often preferred chiefly due to their analytical tractability, interpretability, and ease and reliability of the optimization process. Uniqueness of the solution (though not specific to linear models) is also an issue because it allows the tedious process of carrying out different execution runs to be avoided. Another useful distinction is that of parametric vs. nonparametric methods. In the former, the model has a certain fixed (parametric) functional form that is completely specified when the parameter set is estimated from the data; an example is found in Gaussian discriminant analysis methods. The main drawback is that the assumed form may not be adequate to model the available data sample. In the latter, no such assumptions are taken; however, typically the number of parameters grows with sample size. A compromise is found in semiparametric models (e.g., neural networks), where the assumed form is fixed but very flexible and the number of parameters can be varied independently of sample size.

Linear and quadratic discriminant analyses or LDA/QDA (Duda et al. 2001) are widely used parametric methods. These methods assume that the class distributions are multivariate Gaussians. The theory also assumes knowledge of population parameters (means, covariances and priors for every class). If this information is not available, maximum-likelihood estimates can be used, although in this case the Bayesian optimality properties are no longer valid. With LDA, all classes are assumed to have the same covariance matrix. QDA does not need such an assumption; however, the number of parameters to be estimated from the data available for each class is much higher, entailing lower statistical significance. A compromise is found in regularized discriminant analysis where a weighted combination of the equal and individual covariance matrices is used. In LDA/QDA, an observation is classified into a class if the squared distance (also called the Mahalanobis distance) of the observation to the class center (its mean) is the minimum. These are attractive methods because they need no parameter tuning and allow prior probabilities as well as costs for the different types of misclassifications to be specified. Moreover, their limited complexity (quadratic at most) is a solid guard against overfitting the data.

One highly practical simplification is the so-called Naïve Bayes classifier, which makes the assumption that the variables are class-conditionally independent (this assumption is not as rigid as assuming independent variables). Thus, only the univariate densities need to be estimated (usually assumed Gaussian). An attractive feature of this method is that variables of mixed data types (continuous and categorical) can be combined with ease. Despite its simplicity, Naïve Bayes has been shown to have comparable performance in some domains to more sophisticated methods, such as ANNs or decision trees (DTs).

The *k*-Nearest Neighbors (*k*NN) is a very intuitive nonparametric technique that classifies new observations based on their distance to observations in the training set. Given an unlabelled observation *x*, *k*NN finds the *k*-closest labeled observations to *x* and predicts the class of *x* as the majority class within these observations. The method only requires the specification of an integer $k>0$ and a metric to measure distances; it has the advantages of being analytically tractable and straightforward to implement. On the negative side, there are the large storage requirements and computationally intensive recall. Practical methods for optimal choice of *k* have apparently not been given. The choice of *k* determines the bias/variance trade-off in the estimator – smaller values having in general less bias but higher variance. Moreover, *k* must also be certainly lower than the size of the smaller class – which is very relevant when one class is much smaller than the rest. Large values of *k* offer smoother decision regions and may provide probabilistic information about the decisions. However, too large a value of *k* is detrimental, since it destroys the locality of the estimation, ultimately leading to underfitting the data. More often than not, a plausible value for *k* is found by CV.

Logistic regression (LR) has a long tradition in modeling relationships between predictor variables and a categorical response variable. LR obtains maximum-likelihood estimates of the parameters using an iterative reweighed least squares algorithm (McCullagh and Nelder 1989). It offers a number of tests for assessing model performance, as well as to determine the significance of a given variable in the model. Moreover, if used with the *logit* link, it provides a natural interpretation of the estimated coefficients. In addition, LR does not require normally distributed variables. It also has limitations, requiring that observations be independent and that the predictor variables be linearly related to the logit of the response variable.

ANNs (Bishop 1995; Haykin 1994; Hertz et al. 1991; Hecht-Nielsen 1990) are information processing structures evolved as an abstraction of known principles of how the brain works. The computing elements, called neurons, are linked to one another with a certain strength, called weight. In their simplest form, each unit computes a function of its inputs – which are either the outputs from other units or external signals – influenced by the weights of the links conveying these inputs. The network is said to learn when the weights of all the units are adapted to represent the information present in a data sample, in an optimal sense given by an error function. ANNs provide a general method for learning nonlinear mappings from a sample and have been used with considerable success in many application domains

(Paliwal and Kumar 2009). They have become increasingly popular in MST because of their great flexibility in capturing complex relations with a limited amount of data (Brion and Lingireddy 2003).

ANNs implement a semiparametric approximating function and are capable of representing generic classes of functions (as continuous or integrable functions) to an arbitrary degree of accuracy. Known drawbacks are associated with difficulties in the training process: the possibility of getting stuck in local optima of the error function, oscillations and network paralysis and poor generalization. The latter can be due to several causes: insufficient training data, an excessive number of hidden units, lengthy training times, or a badly chosen preprocessing technique. All of them lead to overfitting of the training data, in which the ANN fits the training set merely as an interpolation task. Finally, ANNs are usually not amenable to inspection: it is generally arduous to interpret the knowledge learned, especially in large networks or with a high number of parameters.

In the ANN context, the VA set is sometimes used for *early stopping*: in this technique, the network is trained (in the TR set) and the error in the VA set is monitored; when this latter error exhibits a clear tendency to grow, the training is stopped and the network having the lowest error in the VA set is returned. In the case of having a small data sample that we do not wish to split, or of having no extra VA set, k-CV is recommended, as follows (Mitchell 1997): a different network is trained in every set of $k-1$ folds in such a way that every training process uses the remaining fold as a VA set (in the early stopping sense). The result is a collection of k estimations of the optimal number of training iterations, which is then averaged. Then a new network is trained using the full sample for this average number of iterations. It is important to note that this procedure is tantamount to performing model selection: the "competing" models are networks at different training stages.

Another way of preventing overfitting in ANNs is *regularization*, in which a penalty term is added to the error function, to keep network weights near zero unless they yield significant reductions in error. The impact of this term is regulated by a user-defined parameter: the higher this value is, the more those weights with large values (positive or negative) are penalized. This method has the advantage that no separate VA set is needed (at least for training purposes); however, a loop of k-CV may be necessary for estimating a sensible value for this parameter. It turns out that this method is essentially equivalent to performing a weight decay, in which each weight is decreased by a small factor during training (Bishop 1995).

In DTs, a nonparametric method greatly popularized by Quinlan (1993), the internal nodes of the tree are questions about the possible values of a variable and the leaves are decision nodes labeled with the class. The trees are grown in an iterative top–down process. At every step, the remaining set of observations is split according to the variable that most reduces the uncertainty (e.g., measured by entropy) of this set with respect to the classes. Once it has finished, test observations are classified through a sequence of questions, one per internal node. The main benefit of DTs over, say, neural networks is interpretability. On the one hand, the decision process for a test observation can be expressed as the conjunction of decisions taken along the path from the root to the corresponding leaf node. On the other hand, the classes themselves can be characterized as logical descriptions using conjunctions and disjunctions (Duda et al. 2001); in geometrical terms, this corresponds to a union of hyperboxes (this is the generalization of a parallelepiped in higher dimensions).

The most important question with DTs concerns deciding when to stop splitting, a delicate issue directly related to overfitting and to correct generalization. One technique is to rely on a VA set, in a way conceptually similar to early stopping in neural networks. Another method is based on finding a compromise between tree size and overall leaf uncertainty; this method is reminiscent of regularization, in which the penalty term is here the size of the tree. A very popular alternative to stopping splitting is pruning: the tree is grown fully and then pairs of neighbouring nodes (having a direct common ancestor in the tree) are considered for joining (the inverse operation of splitting), if the increase in uncertainty is small. Whatever the case, trees are considered quite unstable learners: the addition or removal of a small set of observations can result in an entirely different fully grown tree.

Support Vector Machines (SVMs) (Vapnik 1998) have gained much popularity in the last decade because of their firm theoretical results and the excellent performance in some difficult two-class classification problems. Intuitively, given a set of observations, a so-called linear SVM finds the hyperplane leaving the largest possible fraction of observations of the same class on the same side (thus classifying these observations well), while maximizing the distance (called the *margin*) of either class from the hyperplane. This hyperplane minimizes the risk of misclassifying unseen observations, since the margin is theoretically related to the generalization ability: a larger margin implying a better generalization ability (Vapnik 1998). The interested reader may consult Burges (1998) or Christianini and Shawe-Taylor (2000) for a thorough introduction. SVMs also rely on nonlinear data preprocessing to represent the observations in a dimension that can be much higher than the original (it could be infinite as well). With an appropriate mapping (which is implicit in the choice of a so-called *kernel* function), data from two classes can be made almost linearly separable. If they are not, the SVM has a means of controlling the trade-off between forgiving misclassified observations and maximizing the margin. This trade-off can be regarded as a form of regularization via a user-defined parameter, a larger value corresponding to the assignment of a higher penalty to classification errors.

9.2.8 New Algorithmic Methods

The need for reliable predictive models is apparent in that they would allow for *proactive* or preventive policies, rather than reactive water-quality approaches (Brion and Lindgireddy 2000). In this sense, ML algorithms have made significant inroads and proven to be of great practical importance in a variety of application domains. There are many exciting developments in algorithmic modeling that can make important contributions to the statistician's tool kit, including SVMs (reviewed in 9.2.7), Bayesian methods (Carlin and Thomas 2008), and especially ensemble methods (Dietterich 2000). These are algorithms that construct a set of models and then classify new observations by taking a vote on their predictions. In *bagging*, a bag of models is built as a result of subsampling the TR set via the bootstrap. When a prediction is needed, the class that has the majority vote is the bagged response; this has been shown to reduce variance (Breiman 1996). *Boosting* takes a different resampling approach to bagging – which maintains a constant uniform probability for the selection of each observation. In boosting, this probability is adapted over time by biasing the resampling toward the mistakes. The individual classifiers are built sequentially and observations that are mislabelled by previous classifiers are chosen more often. Boosting is based on the concept of a "weak learner," an algorithm that performs slightly better than chance: these can be converted into a strong learner by changing the distribution of training observations (Freund and Schapire 1996). *Random Forests* builds multiple DTs again via the bootstrap, each tree using different (small) subsets of variables, and returns the majority class among individual trees (Breiman 2001).

9.3 Conclusions and Future Directions

Natural systems are complex and multifactorial, and the same is also applicable to MST. The ability to make precise and significant statements about the behavior of fecal polluted environmental waters decreases as the complexity of the scenario increases. This statement is correct until a threshold is achieved beyond which precision and significance of the developed models become mutually exclusive characteristics (Kosko 1993).

The particular problems faced include independence of geographical location (or at least portability under similar watershed conditions), nature of the dominant fecal pollution contributions (anthropogenic or nonanthropogenic), persistence of indicators and their measured parameters, effects of dilution in watersheds, and presence of complex mixtures from several distinct animal species.

In MST, a need has been pointed out by Brion and Lindgireddy (2000) for new modeling approaches and tools that take the best from several disciplines (microbiology, epidemiology, ecology, statistics, and computer science, among others). Moreover, the used statistical and algorithmic methods should account for the limitations already found in the development of predictive models. In this vein, the recent predictive solutions mentioned in the introduction to this chapter, using only two variables and a linear separation, are first steps in this direction (Belanche and Blanch 2008). These results – together with those emerging from EPA-backed studies – make clear that new approaches to MST are required to solve, in a step-by-step fashion, the different factors that certainly influence and limit the successful identification of fecal pollution sources. These factors are addressed one by one, following a deliberative order:

1. Abundance in fecal sources. The parameters must be numerous in samples with high fecal load, if dilution is not to affect the validity of the approach.
2. More than one indicator. No single indicator can determine the origin of fecal sources in all situations. At least two parameters – one which discriminates sources and one which does not – are required to create predictive models capable of distinguishing fecal pollution sources.
3. Prevalence in the environment. The concentrations (densities) of the selected indicators should be detectable by the respective method of measurement for any matrix of water analyzed.
4. Persistence in the environment. The persistence in the environment and the resistance to water treatments of the different indicators used in predictive models should be similar. No seasonal variations should occur for the selected parameters. Short persistence in the environment, as is the case for anaerobic bacteria (Fisdal et al. 1985; Bonjoch et al. 2009), might hinder their use as tracers of fecal sources. Persistence in the environment and resistance to inactivating treatments may also affect the performance of approaches based on proportions between indicators.
5. Resistance to inactivating water treatments. The fate of microbial parameters in wastewater treatments affects their suitability for tracking fecal sources, since secondary effluents from wastewater treatment plants is – at least in developed countries – the most important contributor of fecal contamination of human origin in receiving water bodies. Low levels of persistence and resistance in some cases (e.g., for anaerobic bacteria) and mismatched persistence and resistance in others imply that these tracers might be unfeasible for MST.
6. Universal feasibility. MST indicators should be independent of geography, climate, or dietary habits. The extent of the geographical areas covered by different procedures as well as the cost of adapting them to different areas needs to be investigated and evaluated.
7. Mixture of fecal sources. The need will arise to find host-specific indicators that can differentiate complex mixtures of several fecal pollution sources (human vs. nonhuman animal species). The combination of distinct discriminating indicators for different fecal sources could provide the relative contribution to the total fecal load from each source in predictive models.
8. Low cost and ease of deployment. The indicators and their parameters should be accessible without incurring large economic or logistic costs.
9. Possibility of developing ad hoc solutions. Solutions should be developed in accordance to specific users' needs (e.g., the use of own tracers) and changing conditions (e.g., local seasonality, solar insulation, water temperature or rainfall), which affect tracer die-off and growth.

In order for these ambitious goals to be met, biologists should join efforts with statisticians and computational experts to share their accumulated knowledge. This synergy is a must for the development of new systems capable of predicting water-quality conditions and for evaluating problems of fecal pollutant loading situations. The ultimate goal is the achievement of "universal" systems for MST, in the sense of having all or most of the above included in any one system.

References

Allsop K, Stickler DJ (1985) An assessment of *Bacteroides fragilis* group organisms as indicators of human faecal pollution. Journal of Applied Bacteriology 58:95–99.

Belanche L, Blanch AR (2008) Machine learning methods for microbial source tracking. Environmental Modeling & Software 23:741–750.

Bell A, Layton AC, McKay L et al (2009) Factors influencing the persistence of fecal *Bacteroides* in stream water. J Environ Qual. 38:1224.

Bishop C (1995) Neural Networks for Pattern Recognition. Clarendon Press, New York.

Blanch AR, Belanche-Munoz L, Bonjoch X et al (2004) Tracking the origin of faecal pollution in surface water: an ongoing project within the European Union research programme. J. Water Health 2:249–260.

Blanch AR, Belanche-Munoz L, Bonjoch X et al (2006) Integrated analysis of established and novel microbial and chemical methods for microbial source tracking. Appl Environ Microbiol 72:5915–5926.

Blanch AR, Lucena F, Payan A et al (2008) Minimal requirements for parameters to be used in the development of predictive models for microbial source tracking: somatic coliphages and phages infecting *Bacteroides* as examples. Journal of Environmental Detection 1:2–21.

Bonjoch X, Lucena F, Blanch AR (2009) The persistence of bifidobacteria populations in a river measured by molecular and culture techniques. J Appl Microbiol 107:1178–1185.

Breiman L (1996) Bagging predictors. Machine Learning 24:123–140.

Breiman L (2001) Random Forests. Machine Learning 45:5–32.

Brion GM, Lindgireddy S (2000) Identification of pollution sources via neural networks, p. 179-197. In: R. S. Govindaraju and Ramachandra Rao (eds.) Artificial Neural Networks in Hydrology. Kluwer Academic Publishers. The Netherlands.

Brion GM, Lingireddy S (2003) Artificial neural network modeling: a summary of successful applications relative to microbial water quality. Water Sci Technol. 47:235–40.

Brion GM, Neelakantan TR, Lingireddy S (2002) A neural-network-based classification scheme for sorting sources and ages of fecal contamination in water. Water Res. 36:3765–3774.

Burges CJC (1998) A Tutorial on Support Vector Machines for Pattern Recognition. Knowledge Discovery and Data Mining 2:121–167

Carlin BP, Thomas AL (2008) Bayesian methods for data analysis. Third Edition. Chapman & Hall/CRC, Boca Raton, p. 552.

Christianini N, Shawe-Taylor J (2000) An introduction to support vector machines. Cambridge University Press. Cambridge.

Dietterich TG (2000) Ensemble methods in machine learning. Multiple Classifier SystemsSpringer, Heidelberg. LNCS 1857:1–15.

Domingos PA (2000) Unified bias-variance decomposition and its applications. Proceedings of the International Conference on Machine Learning Stanford, CA:231–238.

Duda RO, Hart PE, Stork L (2001) Pattern classification. Wiley, New York. pp. 654.

Efron B, Tibshirani R (1997) Improvements on cross-validation: The .632+ Bootstrap Method. Journal of the American Statistical Association 92:548–560.

Field KG, Chern EC, Dick LK et al (2003) A comparative study of culture-independent, library-independent genotypic methods of faecal source tracking. Journal of Water and Health 1:181–194.

Fisdal L, Makai JS, La Croix SJ et al (1985) Survival and detection of *Bacteroides spp.* prospective indicator bacteria. Applied and Environmental Microbiology 49:148–150.

Freund Y, Schapire R (1996) Experiments with a new boosting algorithm. In: Proceedings International Conference on Machine Learning, Morgan Kaufmann: 148–146.
Geldreich EE (1976) Faecal coliforms and faecal streptococcus relationship in waste discharge and receiving waters. Critical Reviews in Environmental Control 6:349–368.
Griffith JF, Weisbert SB, McGee CD (2003) Evaluation of microbial source tracking methods using mixed faecal sources in aqueous test samples. Journal of Water and Health 1:141–151.
Gronewold AD, Qian S, Wolpert RL et al (2009) Calibrating and validating bacterial water quality models: A Bayesian approach. Water Res. 43:2688–2698.
Guyon I, Elisseeff A (2003) An introduction to variable and feature selection. J. Mach. Learn. Res. 3:1157–1182.
Harwood VJ, Whitlock J, Withington V (2000) Classification of antibiotic resistance patterns of indicator bacteria by discriminant analysis: use in predicting the source of faecal contamination in subtropical waters. Applied and Environmental Microbiology, 66:3698–3704.
Harwood VJ, Wiggins B, Hagedorn C et al (2003) Phenotypic library-based microbial source tracking methods: efficacy in the California collaborative study. Journal of Water and Health 1:153–166.
Hastie T, Tibshirani R, Friedman J (2001) The Elements of Statistical Learning. Springer-Verlag, New York. p. 533.
Haykin S (1994) Neural Networks: A Comprehensive Foundation. MacMillan, New York, pp.696.
Hecht-Nielsen R (1990) Neurocomputing. Addison-Wesley. Redwood City.
Hertz J, Krogh A, Palmer RG (1991) Introduction to the Theory of Neural Computation, Addison-Wesley. Redwood City
Hocking RR (1976) The Analysis and Selection of Variables in Linear Regression. Biometrics 32:1–49.
Kira K, Rendell LA (1992) The feature selection problem: Traditional methods and a new algorithm, p. 129-134. In: AAAI-92, Proceedings of the Ninth National Conference on Artificial Intelligence, AAAI Press.
Kosko B (1993) Fuzzy thinking: the new science of fuzzy logic. Hyperion, NY.
Lasalde C, Rodríguez R, Toranzos GA (2005) Statistical analyses: possible reasons for unreliability of source tracking efforts. Appl Environ Microbiol 71:4690–4695.
Layton A, McKay L, Williams D et al (2006) Development of *Bacteroides* 16S rRNA gene TaqMan-based real-time PCR assays for estimation of total, human, and bovine faecal pollution in water. Appl Environ Microbiol 72:4214–4224.
Leeming R, Ball A, Ashbolt N et al (1996) Using faecal sterols from humans and animals to distinguish faecal pollution in receiving waters. Water Research, 30:2893–2900.
Liu H, Motoda H (1998) Feature selection for knowledge discovery and data mining. The Springer International Series in Engineering and Computer Science, Vol.454, Springer, New York, pp. 244.
Malakoff D (2002) Microbiologists in the trail of polluting bacteria. Science 295:2352–2353.
Mara DD, Oragui JI (1981) Occurrence of *Rhodococcus coprophilus* and associated actinomycetes in feces, sewage and freshwater. Appl Environ Microbiol 51:85–93.
Martellini A, Payment P, Villemur R (2005) Use of eukaryotic mitochondrial DNA to differentiate human, bovine, porcine and ovine sources in fecally contaminated surface water. Water Research 39:541–548.
McCullagh P, Nelder JA (1989) Generalized linear models. Chapman & Hall, London, Second Edition.
Mitchell M (1997) Machine Learning. McGraw-Hill Higher Education, NY.
Myoda SP, Carson CA, Fuhrmann JJ et al (2003) Comparison of genotypic-based microbial source tracking methods requiring a host origin database. J. Wat. Health 1:167–180.
Neelakantan TR, Brion GM, Lingireddy S (2001) Neural network modeling of *Cryptosporidium* and *Giardia* concentrations in the Delaware River, USA. Water Sci Technol. 43:125–132.
Nieman J, Brion GM (2003) Novel bacterial ratio for predicting fecal age. Water Sci Technol. 47:45–49.
Noble RT, Allen SM, Blackwood AD et al (2003) Use of viral pathogens and indicators to differentiate between human and non-human faecal contamination in a microbial source tracking comparison study. Journal of Water and Health 1:195–209.

Osawa S, Furuse K, Watanabe I (1981) Distribution of ribonucleic acid coliphages in animals. Appl Environ Microbiol 41:164–168.
Paliwal M, Kumar UA (2009) Neural networks and statistical techniques: a review of applications. Expert Systems with Applications 36:2–17.
Parveen S, Portier KM, Robinson K et al (1999) Discriminant analysis of ribotype profiles of *Escherichia coli* for differentiating human and nonhuman sources of fecal pollution. Appl Environ Microbiol 65:3142–3147.
Payán A, Ebdon J, Taylor H et al (2005) Method for Isolation of *Bacteroides* bacteriophages host strains suitable for tracking sources of faecal pollution in water. Appl Environ Microbiol 71: 5659–5662.
Pote J, Haller L, Kottelat R et al (2009) Persistence and growth of fecal culturable bacterial indicators in water column and sediments of Vidy Bay, Lake Geneva, Switzerland. J Environ Sci 21: 62–69.
Pudil P, Novovicová J, Kittler J (1994) Floating search methods in feature selection Pattern Recognition Letters 15:1119–1125.
Quinlan JR (1993) C4.5: programs for machine learning. Morgan Kaufmann Publishers Inc., Los Altos, pp. 302.
Reischer GH, Kasper DC, Steinborn R et al (2007) A quantitative real-time PCR assay for the highly sensitive and specific detection of human faecal influence in spring water from a large alpine catchment area. Letters in Applied Microbiology 44:351–356.
Ripley BD (1996) Pattern Recognition and Neural Networks. Cambridge University Press, Cambridge, pp. 403.
Ritter KJ, Carruthers E, Carson CA et al (2003) Assessment of statistical methods used in library-based approaches to microbial source tracking. Journal of Water and Health 1:209–223.
Santo Domingo JW, Bambic DG, Edge TA et al (2007) Quo vadis source tracking? Towards a strategic framework for environmental monitoring of fecal pollution. Water Res. 41:3539–3552.
Scott TM, Jenkins TM, Lukasik J et al (2005) Potential use of a host associated molecular marker in *Enterococcus faecium* as an index of human faecal pollution. Environmental Science and Technology, 39:283–287.
Simpson JM, Santo Domingo JW, Reasoner DJ (2002) Microbial source tracking: state of the science. Environmental Science and Technology 36:5279–5288.
Stearns SD (1976) On selecting features for pattern classifiers. Proceedings of the International Conference on Pattern Recognition (ICPR 1976), Coronado.
Stewart JR, Ellender RD, Gooch JA et al (2003) Recommendations for microbial source tracking: lessons from a methods comparison study. Journal of Water and Health 1:225–231.
Stoeckel DM, Harwood VJ (2007) Performance, design, and analysis in microbial source tracking studies. Appl Environ Microbiol 73:2405–2415.
Vapnik V (1998) Statistical Learning Theory. Wiley, New York.
Wiggins BA (1996) Discriminant analysis of antibiotic resistance patterns in faecal streptococci, a method to differentiate human and animal sources of faecal pollution in natural waters. Appl Environ Microbiol 62:3997–4002.
Zhou F, Guo HC, Liu Y, et al (2007) Identification and spatial patterns of coastal water pollution sources based on GIS and chemometric approach. J Environ Sci 19:805–810.

Chapter 10
Mitochondrial DNA as Source Tracking Markers of Fecal Contamination

Jane Caldwell, Pierre Payment, and Richard Villemur

Abstract Guidelines for water-quality monitoring have traditionally focused on the use of bacterial indicators. However, efforts to effectively mitigate fecal contamination necessitate greater clarity in source recognition. Host (mammalian and avian) epithelial cells are shed in the gut lumen and expelled in feces. These cells have multiple numbers of mitochondria, an organelle with its own genome, containing species-specific DNA sequences. These properties make mitochondrial DNA sequences (mtDNA) excellent molecular targets as they are host-specific and robust. This chapter describes the development of molecular methods such as PCR, qPCR, PCR with consensus primers, and DNA microarrays to detect and quantify mtDNA in effluents, influents, and environmental surface waters. These assays represent a paradigm shift in source tracking by detecting DNA from the host rather than its fecal bacterial population. Future development to increase the sensitivity of the assays and ease sample processing of large volumes is warranted. Contamination by nonfecal sources such as skin, hair, and sputum of swimmers needs to be evaluated in the context of providing data for source tracking, i.e., presence of human activity impacting the site. The significance of meat carryover in human feces, waste from kitchen garbage disposals and abattoir or industrial manufacturing requires further study to assess their impact on species-specific source tracking.

Keywords Mitochondrial DNA • Real-time PCR • Rapid DNA extraction • Consensus PCR primers • Alternative molecular methods

---

J. Caldwell (✉)
USDA/ARS Food Science Research Unit, Department of Food, Bioprocessing, & Nutrition Sciences, NC State University, 323 Schaub Hall, Box 7624, Raleigh, NC 27695-7624, USA
e-mail: Jane.caldwell@ars.usda.gov

C. Hagedorn et al. (eds.), *Microbial Source Tracking: Methods, Applications, and Case Studies*, DOI 10.1007/978-1-4419-9386-1_10,

10.1 Introduction

Methods to determine the source of fecal contamination in environmental waters have evolved and multiplied over many years. The list of potential source identifiers and strategies to detect them is still growing. Most methods have been based on microbial characteristics. Other methods have detected chemicals associated with human wastes such as fecal sterols, caffeine, laundry brighteners, pharmaceuticals, and personal care products (Hagedorn and Weisberg 2009). Although useful, most have limitations when used alone. For example, culture-based, library-dependent microbial methods require huge numbers of isolates to generate a reference bank, making the cost of routine monitoring prohibitive. Furthermore, they have not been entirely reliable (Stoeckel and Harwood 2007). The culture-independent library-independent microbial methods are based on the detection of animal-specific nucleic acids harbored by intestinal bacteria (Sinton et al. 1998; Hagedorn et al. 1999, 2003). The normal animal gut biota is a collection of complex microorganisms that is estimated at 10^{11}–10^{12} microbial cells/g feces and 300–500 bacterial species (Guarner and Malagelada 2003). Defining host-specific microorganisms is difficult in view of this diversity. Although *Bacteroidetes* appears to be promising as an animal-specific indicator, very few host-specific bacterial genetic markers have been identified (Shanks et al. 2006, 2009; Stoeckel and Harwood 2007). Adding new and different types of specific markers to traditional microbial and chemical indicators would increase the proficiency of host identification in source tracking analyses.

Over 20 years ago, the research group of Pierre Payment attempted to use enzyme-linked immunosorbent assays (ELISA) to detect coproantibodies in wastewater as a source identifier. IgA antibodies are species-specific and are always present in feces. However, the sensitivity was too low and all attempts were unsuccessful (Payment, personal communication). In 2000, the advent of molecular DNA methods prompted him to collaborate with Richard Villemur, a molecular biologist specializing in microbial ecology. The working hypothesis was that host epithelial cells are shed in the gut lumen and expelled in feces; therefore, host cell nucleic acids should be targeted instead of the resident fecal microorganism DNA. Villemur had expertise in monitoring specific microbial species in heterogeneous environments such as soil and water. He first suggested targeting 18S ribosomal RNA gene sequences, but with little success. Promising results were obtained by changing the target to another genetic element with properties similar to ribosomal RNA genes: the mitochondrial genome.

Mitochondria are found in multiple numbers in all cells of eukaryotes, and each mitochondrion possesses its own genome in multiple copies. In addition, the mitochondrial genome contains species-specific sequences. These properties make mitochondrial DNA sequences (mtDNA) excellent targets in terms of specificity (host-specific) and sensitivity (multiple copies per cell). Therefore, the advantages of targeting mtDNA as a source tracking tool are substantial. Polymerase chain reaction (PCR) and DNA microarrays of mitochondrial DNA can be used to identify the animal contaminator directly, rather than the microbial species it may host. Since the host is directly detected, debates about indicator organism host relevance, cultivability, and viability are unnecessary. This chapter reviews the potential of mtDNA as a useful marker in source tracking of fecal contamination.

10.2 The Mitochondrial Genome

The mitochondrion is an organelle found in the cells of most eukaryotes and has a crucial role in aerobic production of ATP. It contains a genome, in general circular, that encodes proteins, ribosomal RNAs, and tRNAs. The number of mitochondria in a cell and the copy number of its genome in a mitochondrion can vary greatly and depends on the cell type and its physiological conditions (Alberts et al. 2002). In one example, the rat liver cell, the number of mitochondria ranges from 1,400 to 2,200, with 5–10 genomes per mitochondrion, which makes a total of 7,000–22,000 mitochondrial genomes per cell (Bogenhagen and Clayton 1974; Garcia-Rodriguez 2007).

The mitochondrial genomes of the vertebrates (such as birds, mammals, and fish) have a size varying from 16,000 to 17,000 bp with a very similar gene arrangement. In humans, the mitochondrial genome contains 13 protein-coding regions, 22 tRNA genes, and 2 ribosomal RNA genes (Fig. 10.1). Substantial differences in mtDNA can be found between closely related species. For instance, there are 9% nonidentities between human and chimpanzee mtDNA. These variations in sequence facilitate creation of primers that are species-specific for use in molecular techniques such as PCR and DNA microarrays.

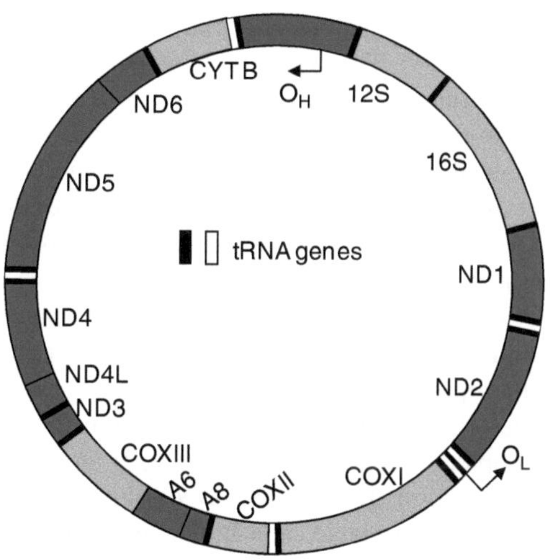

Fig. 10.1 The gene arrangement of the human mitochondrial genome. O_H and O_L: Origins of replication. Genes encode the 12S and 16S ribosomal RNAs, the subunits of NADH-coenzyme Q oxidoreductase (ND), cytochrome-c oxidase (COX), cytochrome-b (CYT B), and ATP synthase (A), and 22 tRNAs

10.3 Mitochondrial Sequences as Genetic Markers

Mitochondrial DNAs are used as typing markers in several fields. They have been adopted for barcoding almost all groups of higher animals (http://www.barcoding.si.edu/). A 648 nt region in the 5′ end of the cytochrome c oxidase subunit 1 mitochondrial region (COI) has emerged as the standard barcode region. Mitochondrial DNA is also used in human typing for forensic analysis (Hopwood et al. 1996; Andreasson et al. 2002; Budowle et al. 2003) using tissues such as bones, teeth, and hair shafts for DNA extraction. In food industry, PCR-based mtDNA analyses were used in the authentication of food and to trace contamination of other animals in the food products (Meyer and Candrian 1996; Lahiff et al. 2001; Zhang et al. 2007; Fujimura et al. 2008). The development of these molecular tools improves the monitoring of food quality and prevents fraudulent description of food content. The success of using mtDNA in a heterogeneous food matrix pointed to the potential of using the same in source tracking of fecal contamination.

10.4 Colonic Cells Shed in Feces

The epithelium of the gastrointestinal (GI) tract is composed of specialized cells involved in mucus and enzyme secretion, lubrication, food and water absorption, and host protection. Undifferentiated epithelium cells divide in the proliferation zones, then migrate and differentiate before being eliminated. Depending on the cell type and their location in the GI tract, the life span of these cells in humans is 2–8 days with a mean generation time of about 1–2 days (Lipkin 1973; Guyton 1976; Eastwood 1977; Guarner and Malagelada 2003).

It is estimated that there are 5×10^{10} human colonic epithelium cells and ca. 20% are replaced each day (Lipkin et al. 1963; Shorter et al. 1964; Potten et al. 1979; Mehl 1991). The precise mechanisms of elimination of these epithelial cells are not fully known. A substantial number of colonic cells are present in feces, suggesting their elimination by shedding. Methods to isolate exfoliated cells from feces have yielded between 1 and 20×10^6 viable cells/g fresh (wet weight) stool (Albaugh et al. 1992; Gireesh et al. 2004; Kamra et al. 2005).

Other studies have shown that apoptosis might play a central role in the regulation of the cell number of the epithelium (David 1967; Hall et al. 1994; Edelblum et al. 2006). Apoptosis bodies have been seen in the GI epithelium. The clearance of these bodies appears to occur by engulfment by adjacent epithelial cells or microphages, although shedding of the apoptosis bodies in the gut lumen is possible (Hall et al. 1994). It was shown that intact mitochondria are released from apoptotic bodies (Kerr et al. 1972) and that the mtDNA is less affected by DNA degradation than nuclear DNA during apoptosis (Tepper and Studzinski 1993). Shedding of viable cells and apoptosis bodies in the luminal gut, therefore, contributes to the release of mitochondria in feces.

Most studies on GI epithelium were performed on human and laboratory rodents. There are few studies of colonic cells in feces of other animal species, but one can assume that similar events occur in most vertebrates.

10.5 Host Mitochondrial DNA in Feces

Extraction of total DNA from feces allows the detection of specific genes from exfoliated colonic epithelial cells (Notarnicola et al. 2000). It has been reported that ca. 10^7 viable exfoliated colonic cells/g feces can be isolated (Albaugh et al. 1992; Gireesh et al. 2004; Kamra et al. 2005;). If one assumes 10^4 mtDNA / cell, this could represent up to 10^{11} mtDNA/g feces, which is comparable to the level of bacteria in feces (10^{11}–10^{12} bacteria/g; Guarner and Malagelada 2003). Therefore, mtDNA genes could be expected to give robust PCR signals similar to 16S rRNA bacteria genes due to high copy numbers.

Caldwell et al. (2007) determined the number of mtDNA copies in feces from 16 human volunteers, which ranged from 0.028 to 2.7×10^7 copies/g feces, and Schill and Mathes (2008) reported similar values at 1.4×10^7/g. The latter authors also measured this level with 11 bovine feces, which ranged from 1.69 to 10.6×10^7 mtDNA copies/g feces. This is 3–4 orders of magnitude less than the theoretical number. Hopwood et al. (1996) also reported a DNA quantity four orders of magnitude less than what was expected of total human DNA extracted from stool. The reason for these discrepancies might be the decrease in the number of mitochondria or the number of mitochondrial genomes per mitochondrion in the exfoliated cells during their passage in the feces, the overestimation of mitochondrial genomes per epithelial cell, and the extraction efficiency of colonic cell DNA from feces.

PCR-detectable mtDNA persists even after cellular death. Martellini et al. (2005) were able to detect human mtDNA by multiplex PCR up to 15 days in a wastewater sample after storage at room temperature. In another study, when samples were conserved at 4°C for months, Kortbaoui et al. (2009) were still able to detect mtDNA but only by nested PCR, suggesting a degradation of mtDNA. However, the persistence of exfoliated colonic cells and their mitochondria in diverse natural environments is unknown.

10.6 Mitochondrial DNA Protocols for Source Tracking

10.6.1 DNA Extraction

Centrifugation and filtration are two methods to concentrate suspended solids, including DNA, from water samples. Centrifugation is an efficient and rapid method to collect most of these solids (Martellini et al. 2005; Caldwell et al. 2007).

Filtration is also efficient, but filters can clog if the solid content of the sample is high. Villemur (personal communication) observed that some solids have a lower density, float to the surface, or remain in suspension after centrifugation. However, while centrifugation requires more expensive equipment and is usually performed in the laboratory, filtration requires minimal equipment and can be performed on site. The volume that can be filtered before clogging depends on the amount of suspended solids in the water and the type of filter used. Kortbaoui et al. (2009) found that even when 0.22 or 0.45-μm filters clogged after filtration of river water samples, there was enough biomass retained to obtain sufficient DNA for multiple PCR assays. Schill and Mathes (2008) also used filtration for fecal-contaminated samples. They added a supplemental step to recover the dissolved DNA that can be present in samples but not retained by the filters. The dissolved DNA in the filtrates was then concentrated by ultrafiltration.

All forms of PCR are enzymatic processes. Therefore, the sample matrix must not inhibit the amplification cascade, nor degrade the DNA during the reaction. Water samples can originate from vastly different environments; from pristine groundwater to highly charged, sediment-rich river flows or manure-contaminated agricultural runoffs. These samples can yield DNA of varying qualities due to other organic substances such as polysaccharides and humic acids, which can coextract with DNA and can inhibit PCR. To circumvent inhibition, dilution of the DNA at 1:10 or 1:100 can increase amplification but, paradoxically, decrease the sensitivity of the method. Caldwell and Levine (2009) developed a simple DNA extraction protocol from wastewater influent by centrifugation (9,000 × *g*, 15 min) followed by freeze/thawing, and heating the sample at 99°C, 300 rpm for 5 min. The crude DNA was suitable for real-time PCR (qPCR) with DNA concentrations ranging from 5 to 100 ng/μL. Internal amplification controls (IAC) were conducted with each sample, and no evidence of inhibitors was apparent (Caldwell, personal communication). Kortbaoui et al. (2009) and Schill and Mathes (2008) used commercial kits with mechanical homogenization using glass beads to extract DNA from material trapped on filters. The DNA was then purified with a silica-based matrix and eluted in a small volume. Supplemental steps to produce cleaner DNA include the polyvinylpolypyrrolidone spin column (Martellini et al. 2005) or gel filtration column that will trap organic contaminants. A commercial kit (Dynabeads SILANE genomic DNA from Invitrogen) using magnetic beads with silica-like surface characteristics was developed to purify genomic DNA from blood. This kit can be adapted for environmental DNA, allowing PCR amplifications of some of the most recalcitrant contaminated DNA samples (Villemur, personal communication).

10.6.2 PCR

More than two thousand complete mtDNA genomes are available at the National Center for Biotechnology Information genome Web page (2,061 sequences, January 2010) (www.ncbi.nlm.nih.gov), where more than 60% are from vertebrates.

Not only are mtDNA sequences from human, farm animals, and pets present but they are also from many wild animals such as Canada geese and white-tail deer that have been implicated in fecal contamination (Siewicki et al. 2007; Somarelli et al. 2007). Deriving primers specific for each animal of interest is performed by aligning sequences and locating unique stretches containing many single-nucleotide polymorphisms (SNP) for the chosen animals. There are sufficient divergences among species to design species-specific PCR primers for conventional, nested, and qPCR (both singleplex and multiplex) (Fig. 10.2). In nested PCR, the use of two

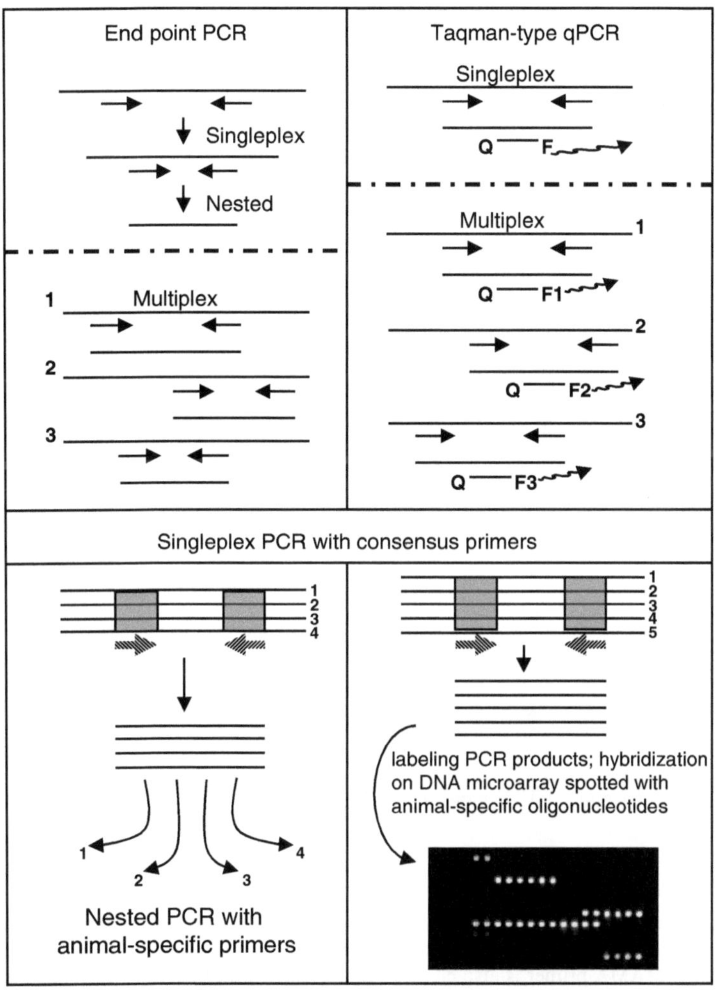

Fig. 10.2 PCR strategies. Each line represents mtDNA. Numbers represent different species. *Solid arrows* represent animal-specific primers. *Shaded arrows* represent consensus primers. *Shaded areas* represent consensus sequences where consensus primers were derived. *Q* Quencher. F1 to F3 represent different fluorochromes associated to the oligonucleotide probes

sets of primers increases the sensitivity of the method and enhances the specificity. Nonspecific amplifications are more likely with one set of primers. The diversity between mtDNA of different animal species allows the development of multiplex PCR, which can process up to four species simultaneously (Martellini et al. 2005; Caldwell et al. 2007). However, multiplex PCR is less sensitive than the singleplex and nested PCR and can be subjected to bias toward the most abundant mtDNA present in the samples (Caldwell et al. 2007).

Species specificity can be further enhanced by designing primers using the mismatch amplification mutation assay (MAMA) (Cebula et al. 1995), which utilizes primer ultimate and penultimate base pair mismatches to decrease the likelihood of nontarget amplification. The MAMA uses SNP in the ultimate 3′ position of the primer, plus a penultimate mismatch to distinguish between very similar sequences with PCR (Fig. 10.3). Caldwell et al. (2007) and Caldwell and Levine (2009) used this method to create mtDNA primers that did not cross-react between species.

Different regions of the mtDNA genome (ND2, ND5, CytB, 16S, COXII, COXIII, and A6–A8) were used to design specific primers targeting ten different animal species in the development of fecal source tracking markers (human, bovine, porcine, ovine, chicken, dog, cat, deer, goose, and horse) (Martellini et al. 2005; Caldwell et al. 2007; Schill and Mathes 2008; Baker-Austin et al. 2009; Caldwell and Levine 2009; Kortbaoui et al. 2009; Baker-Austin et al. 2010). In all

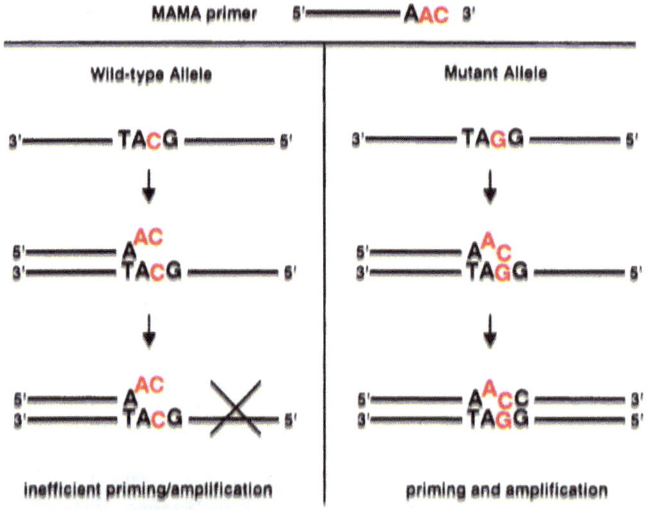

Fig. 10.3 Mismatch amplification mutation assay: The MAMA primer illustrated has an ultimate 3′ base (in *red*) that is complementary to the mutant allele being assayed. The penultimate 3′ base (in *red*) in the MAMA primer is also changed to mismatch with both the wild-type allele and the mutant allele. Upon annealing the MAMA primer to the template, a wild-type sequence will result in two mismatched bases, at the ultimate and penultimate sites, which will be inefficiently primed and/or amplified by DNA polymerase. A mutant sequence, alternatively, will have only a mismatch at the penultimate base, which allows for priming and amplification by DNA polymerase (Glaab and Skopek 1999)

cases, these primers generated very specific PCR products. Cross signals between species can be reduced by selection of primers, optimization of the PCR protocol, and determination of lower detection limits.

PCR methods have been used to detect mtDNA of specific animal species in different wastewater (WW) influents, feces, and surface waters. Martellini et al. (2005) and Kortbaoui et al. (2009) developed specific PCR primers for human, ovine, porcine, bovine, and chicken and were able to detect these species in wastewater influents, farm runoffs, swine waste, and rivers. One of the rivers sampled was at the mouth of an important watershed where several municipal and agricultural activities occurred. The river was sampled several times and all five species were detected at least once by nested PCR.

10.6.3 Real-Time PCR

Real-time PCR (qPCR) has several advantages over conventional PCR: it is rapid, does not require gels, and is quantitative. Multiplex qPCR can further increase speed of detection by amplifying and differentiating between as many as four amplicons simultaneously. Caldwell et al. (2007) used multiplex qPCR to detect human, porcine, and bovine mtDNA in WW influents, feces, and farm effluents. In a blind challenge test, the multiplex assay identified the sources of 10 out of 12 fecal-contaminated effluents with no false positives. By improving their DNA-extraction protocol with the use of heat treatment (99°C, 300 rpm, 5 min), the authors were able to identify the sources of samples 6 out of 6. In the Taqman-type qPCR approach, the oligonucleotide 24-mer probe recognizes sequence between the forward and reverse primers in the amplicon, which, as nested PCR, increases the specificity. Singleplex qPCR was used by Schill and Mathes (2008) targeting nine animals. Blind challenge test was also performed where 16 out of 20 samples were correctly identified. Three samples were below the level of detection. One sample, dog feces, generated dominant dog mtDNA with chicken.

10.6.4 Consensus PCR Primers and DNA Microarray Development

Instead of targeting a unique animal species, consensus mtDNA primers can be chosen to target a specific group, for instance poultry: chicken, turkey, quail, emu, ostrich, or wild birds such as the gull (*Larus* genus). When PCR products are obtained from these consensus primers, the detection of specific species can then proceed with nested species-specific PCR primers (Fig. 10.2). This has the advantage of limiting the number of assays. In screening several hundred water samples, it would be appropriate to determine first if a certain group of animals is present.

This would require less DNA, which is vital if the quantity of DNA is limited. Also, if the sequence of the mitochondrial genome of a species of interest is unavailable in the gene databases, the consensus PCR primers will amplify the unknown mtDNA, which can then be sequenced and added to the database.

Baker-Austin et al. (2009) developed consensus primers targeting *cytB* to amplify 124-bp products for bovine, ovine, caprine, and porcine mtDNA. No attempts were made to discriminate between species. Interestingly, they developed a multiplex PCR assay where these primers were used with two other sets of primers that targeted 16S rDNA sequence (positive control; indication of PCR inhibition) and *E. coli uidA* (indication of fecal contamination). They succeeded to correctly identify 80% of water samples amended with fecal slurry of different species. The 20% false-negative results were probably due to PCR inhibitors as no amplification was observed with the two other targeted genes.

The proof of concept for DNA microarrays was reported by Kortbaoui et al. (2009) (Fig. 10.2). They first found consensus sequences by aligning human, bovine, ovine, porcine, and chicken mtDNA to design three pairs of "universal" primers. These consensus sequences were derived from the 12S and 16S rRNA genes, and the tRNAmet-ND2 genes that amplified DNA fragments ranging from 235 to 506 bp. Flanked inside these universal sequences, specific sequences for each of the five animals was then found to derive 50-mer oligonucleotides that were spotted on nylon membranes (dot blot). With the consensus primers, up to five different animal mtDNA can be amplified in one reaction. Kortbaoui et al. (2009) performed a PCR assay with a DIG-labeled nucleotide, resulting with a DIG-labeled PCR product that was then used to probe the dot blot for discriminating between the different animal mtDNAs. When this approach was used to detect mtDNA in three river samples and three WW samples, exactly the same results were obtained with the nested PCR approach using the specific animal primers. With the availability of numerous mtDNA genomes, the consensus primers can be extended to a much broader number of species allowing the development of DNA microarrays.

10.6.5 PCR Assay Sensitivity

Martellini et al. (2005) inoculated different dilutions of freshly collected human cells from a cell line in autoclaved wastewater and showed by multiplex PCR that the limit of detection was between 10^1 and 10^2 cells/100 mL. It would have been 1–2 orders of magnitude lower with nested PCR (around one cell/100 mL). The same authors also showed by nested PCR that the limit of detection of human mtDNA was in the range of 0.1–1 pg total DNA/reaction even in the presence of 100 ng bacterial DNA. This corresponds to 0.016–0.16 nuclear genome (6 pg/human diploid genome). Considering that the number of mtDNA genomes/cell is in the range of 10^3–10^4 copies (Bogenhagen and Clayton 1974; Garcia-Rodriguez 2007), the level of detection by the nested PCR can be estimated between 10 and

1,000 mtDNA copies/reaction. Other studies using PCR and qPCR to amplify mtDNA reported the same level of sensitivity (Allen et al. 1998; Andreasson et al. 2002; Poe et al. 2007; Baker-Austin 2009). In singleplex qPCR assays, Schill and Mathes (2008) and Caldwell et al. (2007) detected mtDNA copies as low as 10^1, whereas 10^2 copies were detected in multiplex qPCR assays.

Caldwell et al. (2007) estimated that the sensitivity of their multiplex qPCR was 2×10^6 mtDNA copies/100 mL effluent or surface water, and assuming 1.1×10^7 copies/g feces, this represents 200 mg feces/100 mL. With their extraction protocol, Schill and Mathes (2008) achieved a higher sensitivity at 1.8 mg/100 mL (or 1.8×10^4 mtDNA copies/100 mL). These last authors correlated this amount of feces to 3,780 coliforms based on figures reported by Slanetz and Bartley (1957) (geometric mean of 2.1×10^6 coliforms/g feces; 20 human feces samples).

10.7 Carryover and Nonhost Mitochondrial DNA

Most humans are omnivorous and consume animal products: feces still contain nondegraded DNA from meat that can be detected by PCR or qPCR. Martellini et al. (2005) could not detect ovine, porcine, or bovine PCR signal from the feces of one human volunteer who had eaten these meat products the previous day. However, Caldwell et al. (2007) reported carryover signal for consumed beef in two out of four human volunteers. In the latter study, sixteen human volunteers provided fecal samples and completed a questionnaire concerning meals eaten for the previous 48 h. These feces were processed to determine if consumed meat products would provide mtDNA signal in human feces. After DNA was extracted using a commercial kit for stool and assayed by multiplex PCR, there were two instances of carryover mtDNA signal from meat consumed. The mtDNA signals collected corresponded to only the beef products consumed within the last 24 h and were 2×10^4 and 30×10^4 copies mtDNA/g feces. The human mtDNA signals for the same samples were 1×10^6 and 7×10^6 copies/g feces, respectively. Thus, the human mtDNA signal from human feces was 1–2 orders of magnitude greater than the consumed beef. There was no mtDNA carryover signal for consumers of pork. The difference between pork and beef could be the result of the cooking preference of each individual; pork meat is usually well cooked as opposed to beef, which can be served rare. Also, porcine primers were not as robust as bovine primers in multiplex PCR (Caldwell, personal communication).

It appears that carryover signal from consumed beef by humans must be considered when analyzing mitochondrial qPCR results. When human signal is detected in combination with bovine signal, it can mean human contamination alone. The order of magnitude of each mtDNA constituent should be considered to determine the source. Furthermore, PCR could conceivably detect mtDNA from nonfecal sources such as skin, hair, and sputum of swimmers, waste from kitchen garbage disposals and abattoir or industrial manufacturing wastewater such as plastics that use stearic acids derived from cattle. While these sources will confuse the interpretation of the data obtained, they still provide useful information on sources of pollution when correctly assessed in the general context. Host mDNA from skin is expected in recreational waters that are in use for bathing: their presence in itself provides indication on a level of risk due to the human presence (Gerba 2000).

Another nonfecal source is from the operators that can contaminate the samples during sampling, DNA extraction, and PCR. Inclusion of proper negative controls such as no-template control (NTC) should always be performed to rule out operator contamination. Good laboratory practices and quality control should rule out these errors.

10.8 Sources of Mitochondrial DNA in Sanitary Wastewaters

Sanitary or combined sewer WWs can contain mtDNA from nonhuman animals such as bovine, porcine, ovine, chicken, dog, and cat (Martellini et al. 2005; Caldwell and Levine 2009; Kortbaoui et al. 2009). Besides carryover in feces from consumed meats, the occurrence of mtDNA of these species is probably from food waste in sinks and animal feces flushed down commodes (Caldwell and Levine 2009). Wildlife mtDNA such as that of white-tail deer and Canada goose was found near detection limits and cited as spurious results from chimeric amplification of complex DNA mixtures (Caldwell and Levine 2009). Caldwell and Levine (2009) profiled domestic WW influent from two suburban treatment plants and concluded that for source tracking purposes, a combination of human (10^5 copies/mL) and dog mtDNA signal (10^2 copies/mL) were indicative of municipal domestic WW. In some samples, bovine signal was also detected at the same order of magnitude as human and was thought to originate from disposal of meats down drains as well as carryover from consumed meats in human feces. Schill and Mathes (2008) measured the concentration of human mtDNA of the inflows from one WW plant from a small community (10 total samples taken on two occasions), which averaged 7.2×10^4 copies/100 mL. Their figure is approximately 100 times lower than the measurement of two domestic WW influents (24 total samples collected weekly for 12 weeks) by Caldwell and Levine (2009). The difference cannot be attributed to the density of population that the WW plants were treating, which was similar.

Mitochondrial DNA concentrations were compared to other bacterial, chemical, and spectrophotometric parameters measured in domestic WW influents by Caldwell and Levine (2009). Human mtDNA was positively correlated with ammonia concentration and initial $OD_{600\ nm}$ reading, an indicator of the fecal loading of wastewaters. Bovine mtDNA was positively correlated with biological oxygen demand, final DNA concentration, initial and final humic acid concentrations, and final $OD_{600\ nm}$ at one WW treatment facility and with total suspended solids at two facilities. Fecal coliforms were not correlated with mtDNA concentrations of any species assayed, including human, bovine, swine, dog, and cat. Further study is needed to determine the use of mtDNA data with other source tracking indicators data to create profiles of residential septic systems, agricultural and wildlife sources of contamination.

10.9 Future Development

10.9.1 Enhancing Sensitivity

At present, all the various PCR protocols cited are restricted to water samples close to the contamination source, where the concentration of mtDNA will be high enough for detection. However, the level of sensitivity may not be sufficient for source tracking at low levels of fecal pollution. International guidelines and local regulations for recreational beaches suggest action levels for thermotolerant coliforms or other fecal bacteria indicators (WHO 2003). These levels are generally between 10 and 500 bacteria per 100 mL. Therefore, it will be imperative to increase the sensitivity of the mtDNA PCR methods by 1–2 orders of magnitude to be at the same sensitivity level as the microbial fecal indicators.

As described before, different protocols or commercial kits can be used to get clean environmental DNA. But even with the cleanest, highest-quality DNA, there remains one big limitation: high quantities of DNA (above 500–1,000 ng/reaction) inhibit amplification or generate spurious amplifications in qPCR (Schill and Mathes 2008). Using a rigorous DNA extraction protocol combining filtration and concentrated free-mtDNA from the filtrate, Schill and Mathes (2008) probably reached the absolute limit of PCR sensitivity with 1.8×10^4 mtDNA copies/100 mL water, the equivalent of 1.8 mg feces/100 mL or 3,780 coliforms in the water samples. Below this fecal concentration, PCR could not detect mtDNA because its proportion is too low relative to the total DNA in the environmental sample (which includes other DNA from bacteria, algae, plants and other sources). To overcome this limitation, mtDNA must be enriched from the total extracted DNA.

One enrichment approach is nucleic-acid sequence capture with magnetic beads (Fig. 10.4). Oligonucleotides representing species-specific or consensus mtDNA sequences are labeled with biotin and capture, by hybridization, the mtDNA from the DNA sample. Separation is then achieved with magnetic beads coated with streptavidin, which has very high affinity to biotin. Besides enrichment of mtDNA, this can eliminate PCR inhibitory substances, which can be washed from the beads. This method was successfully used to enrich mycobacterial genomic DNA from clinical specimens such as pleural fluid and feces samples (Millar et al. 1995; Mangiapan et al. 1996; Marsh et al. 2000). Mangiapan et al. (1996) were able to enrich mycobacterial DNA with a background of the sample DNA up to 750 μg and increase the sensitivity of their PCR protocol by 1–2 orders of magnitude. One disadvantage of enrichment of mtDNA may include impaired quantification, which can be important information for proper source tracking. Also, the enrichment may not capture all the mtDNA in the sample or favor one species over another due to primer bias. Another disadvantage is the larger volume of water that would have to be filtered (~2–3 L). However, this can be feasible with high-capacity filters.

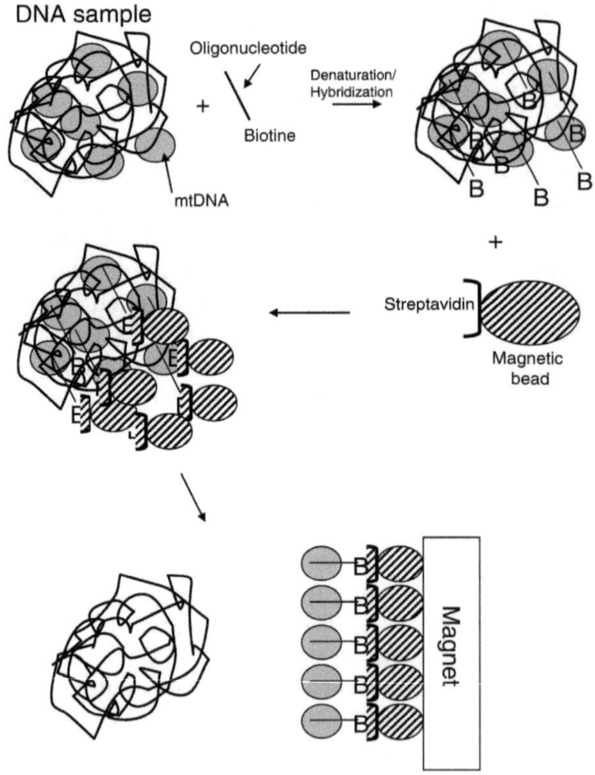

Fig. 10.4 mtDNA enrichment by magnetic beads

10.9.2 *Alternative Molecular Methods*

Processing water samples to source track fecal contamination for a limited number of species (e.g., human, some farm animals, or pets) remains a feasible approach, but one can envisage the need to screen an increasing number of species to accurately determine the source(s). Ideally, more than one specific sequence per animal should be used for better identification, preferentially with reactions performed in duplicate or triplicate. Including negative, positive, and internal amplification controls, this would require substantial numbers of PCR per water sample, which could be difficult to perform by hand. An increase in the number of species markers might require automation of molecular protocols.

Furthermore, the amount of DNA collected may not be sufficient for a very large assay as ~100–200 ng is needed per reaction to get the highest sensitivity. Although multiplex PCR can be performed, such protocols are usually less sensitive, limited to 4 specific markers and subject to primer bias. Different consensus primers representing classes of animals (such as birds, fish, farm-raised animals, wild animals, and domestic animals) can be used first for a broader screen. Upon positive results,

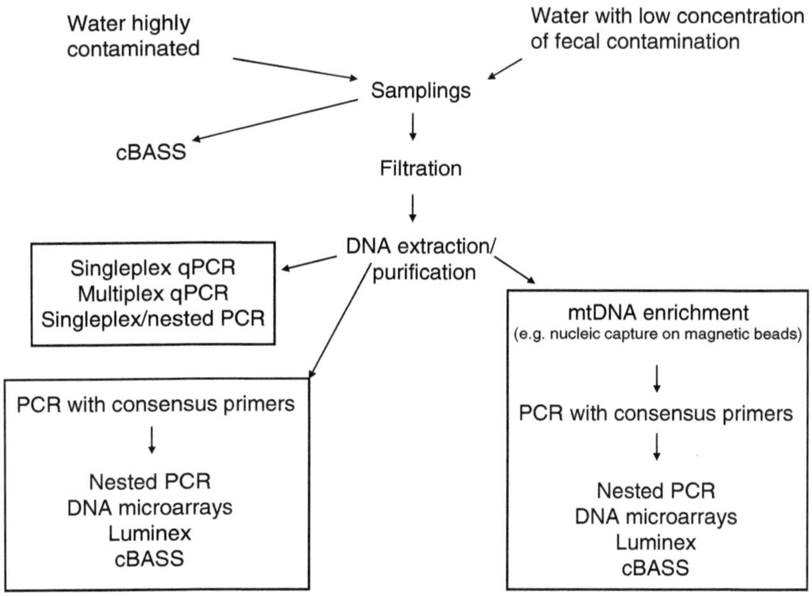

Fig. 10.5 Source tracking strategies with mtDNA

further PCR with specific-species primers can then be performed. However, even with this last strategy, excessive processing and handling would still be a burden for the operators.

One strategy is to develop consensus primers to amplify in one reaction all animal mtDNA. Consensus sequences representing 40–50 different animals can be designed (Villemur, personal communication). Segregation of animal species can then be carried out by different high-throughput technologies such as robotic PCR, DNA microarray, Luminex, and cBASS (Fig. 10.5).

Manipulating DNA samples, mixing PCR reactions, running PCR, and measuring PCR amplification in qPCR apparatus can be performed robotically. This also reduces operator bias. However, robotic platforms are expensive to operate and maintain. With the DNA microarray strategy, the PCR product resulting from the consensus primers is labeled with a fluorescent molecule and hybridized to a DNA microarray chip. Several hundreds, or thousands, of different mtDNA animal-specific oligonucleotides can be spotted on one array. Therefore, DNA microarrays can detect, in one set of manipulations, all the targeted animal species.

An emerging technology to process several markers simultaneously has been developed by Luminex Corporation. An array of up to 100 different animal-specific oligonucleotides can be designed. This technology uses polystyrene microspheres containing a mix of red and infrared fluorochromes. The different ratios of each fluorochrome yield specific spectral addresses. One hundred microsphere sets were designed, each with its own spectral address that can be discriminated by flow cytometry (Fig. 10.6). The microspheres contain at their surface carboxyl groups

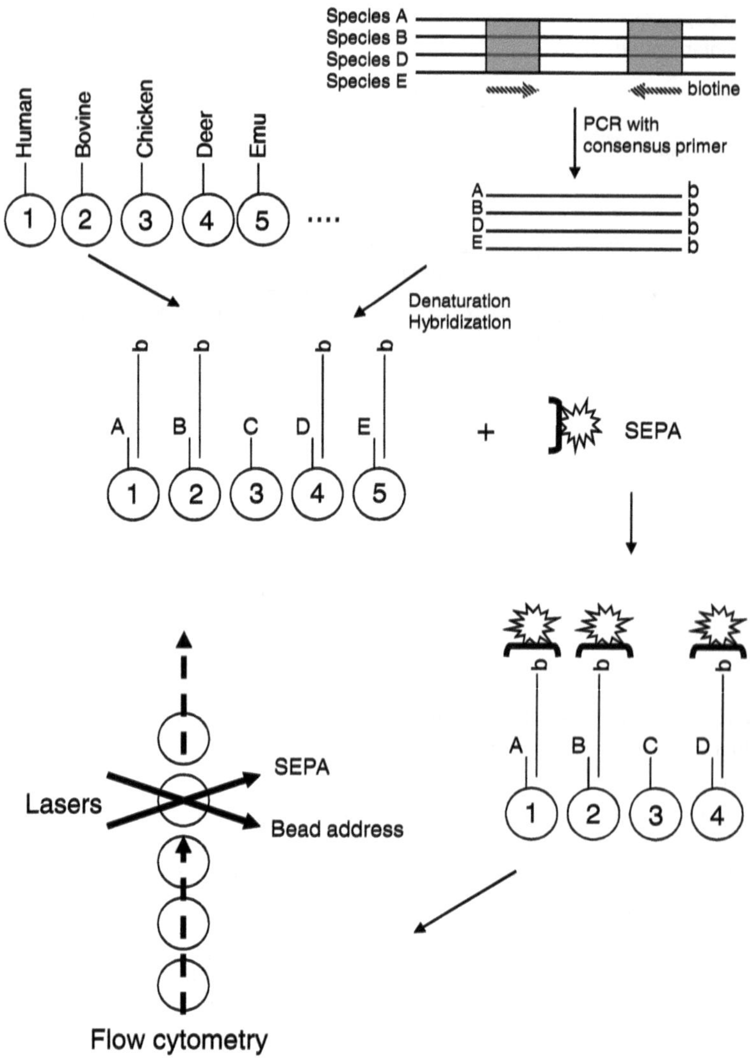

Fig. 10.6 Schematic summary of the Luminex technology. SEPA: streptavidin coupled to phycoerythrin

that can covalently bind oligonucleotides (Dunbar 2006). PCR is conducted with consensus primers, one of which is biotinylated. Next, the amplicons containing one, or more than one, mtDNA are mixed with the microsphere sets, each set coated with an animal-specific oligonucleotide. After liquid hybridization, hybridized PCR amplicons are labeled with steptavidin-conjugated with phycoerythrin (SEPA). The microsphere-captured PCR amplicons are then examined by flow cytometry with two lasers, one to detect SEPA (presence of mtDNA) and the other to detect the microsphere spectral addresses (the animal source). This can be

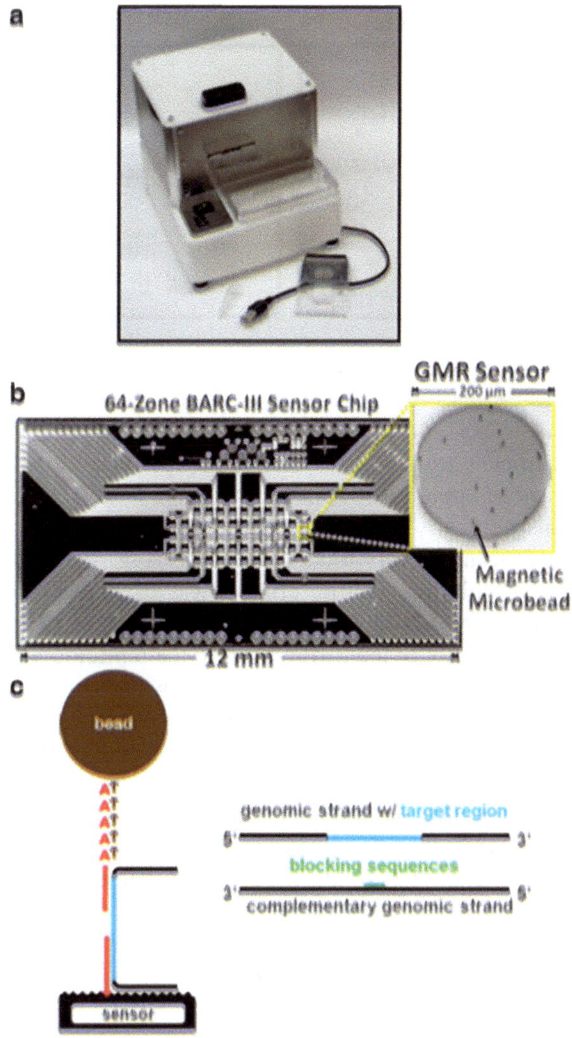

Fig. 10.7 (**a**) Picture of the compact Bead Array Sensor System (cBASS®). (**b**) Picture of the Bead ARray Counter (BARC) chip and up-close view of an individual GMR sensor with immobilized microbeads. (**c**) Cartoon depicting the 2-probe assay strategy used with fluidic force discrimination (FFD) hybridization assays. Also shown is the blocking strategy used for isolating the target sequence for detection

adapted for 96-well plates. For each well, up to 100 animal-specific sequences can be examined simultaneously. Assays for the detection of microorganisms showed the level of sensitivity on a single molecule per bead and to 100 microbial genomes per PCR reaction (see review by Dunbar 2006).

An alternative to PCR, the compact Bead Array Sensor System (cBASS®) utilizes Fluidic Force Discrimination (FFD) (Mulvaney et al. 2009) and DNA hybridization at room temperature. The cBASS (Fig. 10.7) is a field portable technology capable of multiplexed, femtomolar nucleic acid and attomolar toxin detection within minutes and without amplification (Mulvaney et al. 2007; Mulvaney et al. 2008; Tamanaha et al. 2008). This technology brings together electronics, fluidics,

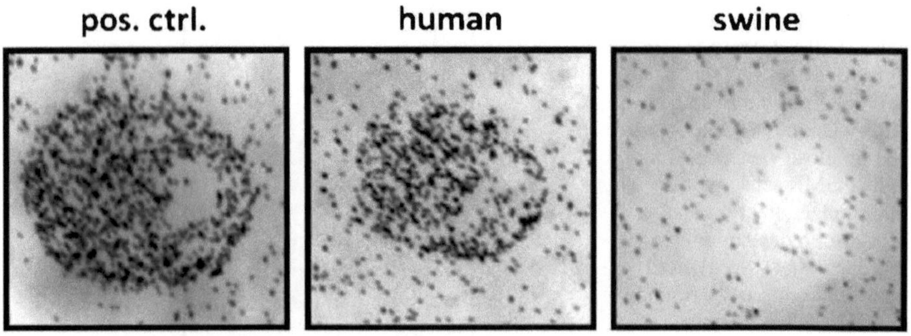

Fig. 10.8 Detection of the *NADH* gene in human mtDNA spiked into waste water

and magnetics for successful biological detection. Targets are captured over an array of magnetic sensors and then labeled with micromagnetic beads. Nonspecifically attached beads are removed with controlled fluidic forces. FFD assays result in very low backgrounds (Fig. 10.8) and few false positives, making cBASS® highly specific. cBASS was used to discriminate human, bovine, and swine mtDNA that were spiked in sterilized wastewater (Mulvaney et al. 2009). The sensitivity of the method was at the femtomolar level corresponding to around 1,000 mtDNA/μL with only 150 μL of water sample. However, this translates to 10^8 mtDNA copies/100 mL water, which is not as sensitive as the methods of Caldwell et al. (2007) or Schill and Mathes (2008). Most mtDNA is not free-floating in water but still inside cells or mitochondria, or in complexes with suspended particles. Therefore, this method would still require filtration or centrifugation and DNA extraction of the water samples. Alternatively, cBASS can be used to discriminate mtDNA from PCR products generated with consensus primers.

10.10 Conclusions

Detection of eukaryotic nucleic acids, specifically mitochondrial DNA, provides the most direct identification of host species that contribute fecal material to environmental waters. Molecular methods are already widely used in forensics. Due to popular media exposure via newspaper reports and television crime dramas, the general public, and therefore juries, are familiar with DNA evidence and have confidence in molecular assays. The specificity of PCR and DNA microarray methods allows investigators to point to the contaminator directly with minor ambiguities. DNA is relatively persistent in the environment, and when properly stored, can be frozen at –20°C or blotted on paper at room temperature for extended periods, creating an inexpensive library of past queries. Quantitative PCR assays are rapid and, at present, can identify up to four amplicons simultaneously. DNA microarray methods provide screening of large numbers of samples which reduce costs.

Looking at the challenges and issues, DNA becomes highly diluted in large bodies of water, so samples must be taken near the source. This limits these assays to point-source monitoring. PCR uses a very small sample size, 1–5 μL per reaction. Samples are usually larger than those used in molecular methods, i.e., 500–1,000 mL. This difference in scale requires reliable concentration methods. Inhibitors such as humic acid and fecal bile salts which co-elute with DNA can be easily eliminated by dilution, but this lowers the sensitivity of the assay. Finally, results must be put in perspective as they provide only evidence for the presence of the animal species whether fecal or nonfecal source of mtDNA such as hair and skin in water samples. Whether fecal or nonfecal their detection provides an indication of the source of the pollution and some level of risk associated with this presence.

The weight-of-evidence approach combines land-use analysis (Geographic Information Systems or mapping databases), particulate matter analysis (turbidity, particle counts), organic matter (TOC, DOC, UV254 absorbance), and indicator organisms (fecal coliforms, enterococci) to characterize water quality (Plummer and Long 2007). Land-use analysis can identify possible contamination sources and rely on mtDNA markers to verify point-source contributions. MtDNA analysis is the perfect complement to this tool box approach by identifying the source(s) directly with minor ambiguities related to carryover signal from host-consumed meats. New methods incorporating magnetic beads, DNA microarrays, microfluidics, and electronics will increase sensitivity; numbers of species assayed simultaneously and allow greater volumes to be monitored. Because they require relatively clean products, multiple temperatures and technically trained personnel, PCR and DNA microarrays have been limited to the laboratory. Our future challenge will be to create user-friendly assays that can be successfully performed and analyzed in the field by local water quality monitors.

References

Alberts B, Johnson A, Lewis J et al. (2002) Molecular Biology of the Cell. 4th edn. Garland Science, New York

Albaugh GP, Iyengar V, Lohani A et al. (1992) Isolation of exfoliated colonic epithelial-cells, a novel, noninvasive approach to the study of cellular markers. Int J Cancer 52:347–350

Allen M, Engstrom AS, Meyers S et al. (1998) Mitochondrial DNA sequencing of shed hairs and saliva on robbery caps: sensitivity and matching probabilities. J Forensic Sci 43(3):453–464

Andreasson H, Gyllensten U, Allen M (2002) Real-time DNA quantification of nuclear and mitochondrial DNA in forensic analysis. Biotechniques 33(2):402–411

Baker-Austin C, Rangdale R, Lowther J et al. (2010) Application of mitochondrial DNA analysis for microbial source tracking purposes in shellfish harvesting waters. Water Science & Technology 61(1):1–7

Baker-Austin C, Morris J, Lowther JA et al. (2009) Rapid identification and differentiation of agricultural faecal contamination sources using multiplex PCR. Lett Appl Microbiol 49(4):529–532

Bogenhagen D, Clayton DA (1974) The number of mitochondrial deoxyribonucleic acid genomes in mouse L and human HeLa cells. Quantitative isolation of mitochondrial deoxyribonucleic acid. J Biol Chem 249(24):7991–7995

Budowle B, Allard MW, Wilson MR et al. (2003) Forensics and mitochondrial DNA: applications, debates, and foundations. Annu Rev Genomics Hum Genet 4:119–141

Caldwell JM, Raley ME, Levine JF (2007) Mitochondrial multiplex real-time PCR as a source tracking method in fecal-contaminated effluents. Environ Sci Technol 41(9):3277–3283

Caldwell JM, Levine JF (2009) Domestic wastewater influent profiling using mitochondrial real-time PCR for source tracking animal contamination. J Microbiological Methods 77:17–22

Cebula TA, Payne WL, Feng P (1995) Simultaneous identification of strains of *Escherichia coli* Serotype O157:H7 and their shiga-like toxin type by mismatch amplification mutation assay-multiplex PCR. J Clin Microbiol 33:248–250

David H (1967) [On the mechanism of cell desquamation of the small intestine villi. (Electron microscopic studies)]. Virchows Arch Pathol Anat Physiol Klin Med 342:19–25

Dunbar SA (2006) Applications of Luminex xMAP technology for rapid, high-throughput multiplexed nucleic acid detection. Clin Chim Acta 363(1-2):71–82

Eastwood GL (1977) Gastrointestinal epithelial renewal. Gastroenterology 72(5 Pt 1):962–975

Edelblum KL, Yan F, Yamaoka T et al. (2006) Regulation of apoptosis during homeostasis and disease in the intestinal epithelium. Inflamm Bowel Dis 12(5):413–424

Fujimura T, Matsumoto T, Tanabe S et al. (2008) Specific discrimination of chicken DNA from other poultry DNA in processed foods using the polymerase chain reaction. Biosci Biotechnol Biochem 72:909–913

Garcia-Rodriguez LJ (2007) Basic properties of mitochondria. In: Pon LA, Schon EA (eds) Methods in cell biology: Mitochondria 2nd edn. Academic Press, CA

Gerba CP (2000) Assessment of enteric pathogen shedding by bathers during recreational activity and its impact on water quality. Quant. Microbiol. 2:55–68

Gireesh T, Nair PP, Sudhakaran PR (2004) Studies on the bioavailability of the provitamin A carotenoid, beta-carotene, using human exfoliated colonic epithelial cells. Br J Nutr 92(2):241–245

Glaab WE, Skopek TR (1999) A novel assay for allelic discrimination that combines the fluorogenic 5' polymerase chain reaction (TaqMan®) and mismatch amplification mutation assay. Mutation Research 430:1–12

Guarner F, Malagelada JR (2003) Gut flora in health and disease. The Lancet 361(9356):512–519

Guyton AC (1976) Textbook of medical physiology. 5th edn. W.B. Saunders Company, Philadelphia, PA

Hagedorn C, Crozier JB, Mentz KA et al. (2003) Carbon source utilization profiles as a method to identify sources of faecal pollution in water. J. Appl. Microbiol. 94(5):792–799

Hagedorn C, Robinson SL, Filtz JR et al. (1999) Determining sources of fecal pollution in a rural Virginia watershed with antibiotic resistance patterns in fecal streptococci. Appl Environ Microbiol 65(12):5522–5531

Hagedorn C, Weisberg SB (2009) Chemical-based fecal source tracking methods: current status and guidelines for evaluation. Rev Environ Sci Biotechnol 8(3):275–287

Hall PA, Coates PJ, Ansari B et al. (1994) Regulation of cell number in the mammalian gastrointestinal-tract - the importance of apoptosis. J Cell Sci 107:3569–3577

Hopwood AJ, Mannucci A, Sullivan KM (1996) DNA typing from human faeces. Int J Legal Med 108(5):237–243

Kamra A, Kessie G, Chen JH et al. (2005) Exfoliated colonic epithelial cells: Surrogate targets for evaluation of bioactive food components in cancer prevention. J Nutrition 135:2719–2722

Kerr JF, Wyllie AH, Currie AR (1972) Apoptosis: a basic biological phenomenon with wide-ranging implications in tissue kinetics. Br J Cancer 26(4):239–257

Kortbaoui R, Locas A, Imbeau M et al. (2009) Universal mitochondrial PCR combined with species-specific dot-blot assay as a source-tracking method of human, bovine, chicken, ovine, and porcine in fecal-contaminated surface water. Water Res 43:2002–2010

Lahiff S, Glennon M, O'Brien L et al. (2001) Species-specific PCR for the identification of ovine, porcine and chicken species in meta and bone meal (MBM). Mol Cell Probes 15(1):27–35

Lipkin M (1973) Proliferation and differentiation of gastrointestinal cells. Physiol Rev 53:891–915

Lipkin M, Bell B, Sherlock P (1963) Cell proliferation kinetics. In: The gastrointestinal tract of man. I. Cell renewal in colon and rectum. J. Clin. Invest. 42(6):767–776

Mangiapan G, Vokurka M, Schouls L et al. (1996) Sequence capture-PCR improves detection of mycobacterial DNA in clinical specimens. J Clin Microbiol 34(5):1209–1215

Marsh I, Whittington R, Millar D (2000) Quality control and optimized procedure of hybridization capture-PCR for the identification of *Mycobacterium avium* subsp. paratuberculosis in faeces. Mol Cell Probes 14(4):219–232

Martellini A, Payment P, Villemur R (2005) Use of eukaryotic mitochondrial DNA to differentiate human, bovine, porcine and ovine sources in fecally contaminated surface water. Water Res 39(4):541–548

Mehl LE (1991) A mathematical computer stimulation model for the development of colonic polyps and colon cancer. J. Surg. Oncol. 47(4):243–252

Meyer R, Candrian U (1996) PCR-based DNA analysis for the identification and characterization of food components. Food Science Technology-Lebensmittel-Wissenschaft & Technologie 29:1–9

Millar DS, Withey SJ, Tizard ML et al. (1995) Solid-phase hybridization capture of low-abundance target DNA sequences: application to the polymerase chain reaction detection of *Mycobacterium paratuberculosis* and *Mycobacterium avium* subsp. silvaticum. Anal Biochem 226(2):325–330

Mulvaney SP, Cole CL, Kniller MD (2007) Rapid, femtomolar bioassays in complex matrices combining microfluidics and magnetoelectronics. Biosens Bioelectron 23:191–200

Mulvaney SP, Myers KM, Sheehan PE (2008) Attomolar protein detection in complex sample matrices with semi-homogeneous fluidic force discrimination assays. Biosens Bioelectron 24(5):1109–1115

Mulvaney SP, Ibe CN, Caldwell JM et al. (2009) Detection of mitochondrial DNA with the compact bead array sensor system (cBASS®). Proc SPIE 7167:7167OV1–10

Notarnicola M, Cavallini A, Cardone R et al. (2000) K-ras and p53 mutations in DNA extracted from colonic epithelial cells exfoliated in faeces of patients with colorectal cancer. Digest Liver Dis 32:131–136

Plummer JD, Long SC (2007) Monitoring source water for microbial contamination: evaluation of water quality measures. Water Res 41:3716–3728

Poe BG, Navratil M, Arriaga EA (2007) Absolute quantitation of a heteroplasmic mitochondrial DNA deletion using a multiplex three-primer real-time PCR assay. Anal Biochem 362(2):193–200

Potten CS, Schofield R, Lajtha LG (1979) A comparison of cell replacement in bone marrow, testis and three regions of surface epithelium. Bioch Biophys Acta 560:281–299

Schill WB, Mathes MV (2008) Real-time PCR detection and quantification of mine potential sources of fecal contamination by analysis of mitochondrial cytochrome b targets. Environ Sci Technol 42(14):5229–5234

Shanks OC, Santo Domingo JW, Lamendella R et al. (2006) Competitive metagenomic DNA hybridization identifies host-specific microbial genetic markers in cow fecal samples. Appl Environ Microbiol 72(6):4054–4060

Shanks OC, Kelty CA, Sivaganesan M et al. (2009) Quantitative PCR for genetic markers of human fecal pollution. Appl Environ Microbiol 75(17):5507–5513

Shorter RG, Moertel CG, Titus JL et al. (1964) Cell kinetics. In: The jejunum and rectum of man. Amer J Digest Dis 9:760–763

Siewicki TC, Pullaro T, Pan W et al. (2007) Models of total and presumed wildlife sources of fecal coliform bacteria in coastal ponds. J Environ Manag 82:120–132

Sinton LW, Finlay RK, Hannah DJ (1998) Distinguishing human from animal faecal contamination in water: a review. New Zealand J Marine Freshwater Res 32(2):323–348

Slanetz LW, Bartley CH (1957) Numbers of enterococci in water, sewage, and feces determined by the membrane filter technique with an improved medium. J Bacteriol 74:591–595

Somarelli JA, Makarewicz JC, Sia R et al. (2007) Wildlife identified as major source of *Escherichia coli* in agriculturally dominated watersheds by BOX A1R-derived genetic fingerprints. J Environ Manag 82:60–65

Stoeckel DM, Harwood VJ (2007) Performance, design, and analysis in microbial source tracking studies. Appl Environ Microbiol 73(8):2405–2415

Tamanaha CR, Mulvaney SP, Rife JC (2008) Magnetic labeling, detection, and system integration. Biosens Bioelectron 24:1–13

Tepper CG, Studzinski GP (1993) Resistance of mitochondrial DNA to degradation characterizes the apoptotic but not the necrotic mode of human leukemia cell death. J Cell Biochem 52(3):352–361

WHO (2003) Guidelines for safe recreational water environments Vol 1 Coastal and fresh waters. World Health Organization, Geneva p. 862003

Zhang CL, Fowler MR, Scott NW et al. (2007) A TaqMan real-time PCR system for the identification and quantification of bovine DNA in meats, milks and cheeses. Food Control 18:1149–1158

Chapter 11
Community Analysis-Based Methods

Yiping Cao, Cindy H. Wu, Gary L. Andersen, and Patricia A. Holden

Abstract Microbial communities are each a composite of populations whose presence and relative abundance in water or other environmental samples are a direct manifestation of environmental conditions, including the introduction of microbe-rich fecal material and factors promoting persistence of the microbes therein. As shown by culture-independent methods, different animal-host fecal microbial communities are distinctive, suggesting that their community profiles can be used to differentiate fecal samples and to potentially reveal the presence of host fecal material in environmental waters. Cross-comparisons of microbial communities from different hosts also reveal relative abundances of genetic groups that can be used to distinguish sources. In increasing order of their information richness, several community analysis methods hold promise for MST applications: phospholipid fatty acid (PLFA) analysis, denaturing gradient gel electrophoresis (DGGE), terminal restriction fragment length polymorphism (TRFLP), cloning/sequencing, and PhyloChip. Specific case studies involving TRFLP and PhyloChip approaches demonstrate the ability of community-based analyses of contaminated waters to confirm a diagnosis of water quality based on host-specific marker(s). The success of community-based MST for comprehensively confirming fecal sources relies extensively upon using appropriate multivariate statistical approaches. While community-based MST is still under evaluation and development as a primary diagnostic tool, results presented herein demonstrate its promise. Coupled with its inherently comprehensive ability to capture an unprecedented amount of microbiological data that is relevant to water quality, the tools for microbial community analysis are increasingly accessible, and community-based approaches have unparalleled potential for translation into rapid, perhaps real-time, monitoring platforms.

Keywords Community analysis • Multivariate statistical method • Spatial source tracking • TRFLP • PhyloChip • MST on a chip

---

Y. Cao (✉)
Southern California Coastal Water Research Project,
3535 Harbor Blvd, Suite 110, Costa Mesa, CA 92626, USA
e-mail: yipingc@sccwrp.org

C. Hagedorn et al. (eds.), *Microbial Source Tracking: Methods, Applications, and Case Studies*, DOI 10.1007/978-1-4419-9386-1_11,

11.1 Introduction

11.1.1 Challenges in Water-Quality Diagnosis

Microbiological water-quality is a serious public health concern for drinking water, recreational swimming, shellfish consumption, and agricultural food production. The microbial pollutants of concern are pathogens, discharged with various fecal sources including human sewage and septage, domestic pet waste, and livestock manure. Pathogen-related coastal water-quality problems are recognized in a National Research Council (NRC) report (1993) where "pathogens and toxins that affect human health" and the "introduction of nonindigenous species" are listed among major US coastal environmental issues.

Historically, routine monitoring of environmental waters for all known bacterial and viral pathogens has been impractical; monitoring for all possible pathogens also has been impossible given that most of the human gut microbiome, from which fecal material originates, has not been identified (Eckburg et al. 2010). Instead, fecal indicator bacteria (FIB, Fig. 11.1) are used to infer the presence of fecal contamination (Eaton et al. 1998), the presence of pathogens, and the risk of disease (Fig. 11.2). While FIB, and thus the "chain of inference" for human health risks (Fig. 11.2), are accepted for safely monitoring drinking water, the scientific and regulatory communities increasingly regard FIB as flawed for monitoring fecal material in the environment. The historical basis and advantages for using FIB in environmental water-quality monitoring have been reviewed elsewhere (Wade et al. 2006). Early epidemiological studies showed a direct connection between swimmer illness and FIB (Fig. 11.2), but such studies were performed in coastal waters continuously receiving sewage (Cabelli et al. 1982). Now, most wastewaters in the United States are treated to either primary or secondary standards (National Research Council 1993), and except for acute short-term sewage spills, sources of human waste are diffuse, e.g., from leaking sewer pipes, cross-connections to storm

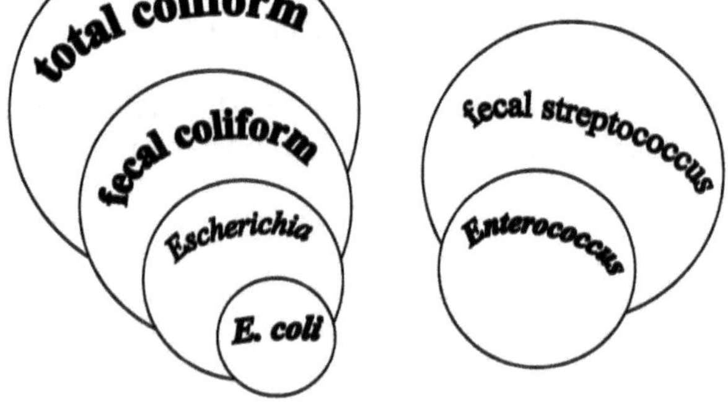

Fig. 11.1 Relatedness of the fecal indicator bacteria (FIB, shown in *bold*), cultured as indicators of microbiological water quality

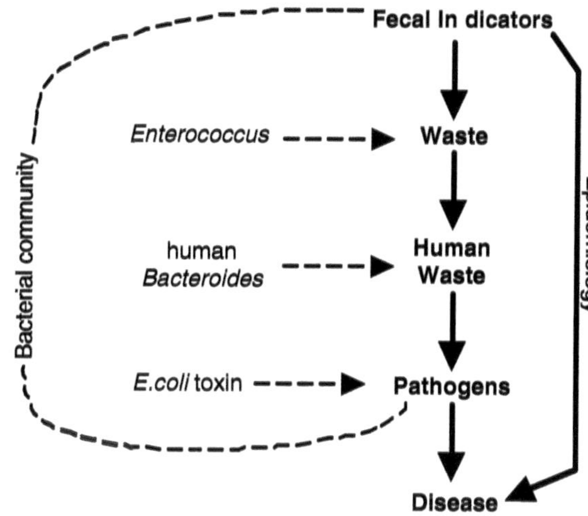

Fig. 11.2 The "chain of inference" that is the basis for traditional, cultivation-based microbiological water-quality monitoring. In *bold*, the inference is: If FIB are present, then waste is also present, and so on. Epidemiology has related FIB directly to disease (*right arrow*) for some cases. To the *left*, culture-independent, PCR-based assays (*dotted lines*) include *Enterococcus*, host-specific markers (e.g., human), and host-specific pathogen genes (e.g., *Escherichia coli*). Culture-independent, DNA-based whole microbial community analysis (*far left*) is comprehensive, encompassing most indicators along the chain and also bacterial pathogens, depending on the PCR primers employed

drains, or failed septic systems. The disadvantages in relying on FIB for diffuse, or nonpoint, fecal source monitoring arise primarily from the fact that FIB are nonspecific to sources, e.g., to human waste which is of concern as a carrier for human pathogens. While total maximum daily load (TMDL) compliance is a major driver for microbial source tracking (MST; U.S. Environmental Protection Agency 2005), another is the nonspecificity of FIB to human or other wastes. Other disadvantages of FIB, including the method's reliance on cultivation when FIB can become nonculturable in the environment (Leadbetter 1997), partly motivate the subject of this chapter: i.e., culture-independent microbial community analysis as a MST approach. Finally, and related to the nonspecificity of FIB to sources, some FIB can become indigenous, e.g., to plant surfaces (Hazen 1988) and to beach sands (Ishii et al. 2007), which results in false-positive indications of an increased risk of pathogen presence. These disadvantages contribute to the rate of error associated with FIB monitoring for predicting pathogen presence and protecting public health. Thus, while one MST objective may be discovering the sources of FIB, which is particularly pertinent to TMDL applications (U.S. Environmental Protection Agency 2005), another view, particularly in light of the fundamental problems with interpreting FIB environmental data, is that MST should be mainly motivated by discovering: (1) if fecal material is present, (2) the host sources of fecal material, and (3) the geographical origin of fecal material.

11.1.2 The MST Toolkit, Including Microbial Community Analysis

Because of the ubiquity of routine FIB monitoring, MST is often employed as a follow-up to determine if FIB signal the presence of fecal material. Frequently used in such follow-up studies, a stepwise MST study approach has been successful in some circumstances (Vogel et al. 2007), including a study revealing a leaking sanitary sewer line as an acute source of FIB near a beach (Boehm et al. 2003), and another discovering that aged sanitary sewers contribute diffuse contamination to storm drains discharging directly to coastal waters (Sercu et al. 2009). In such cases, study steps included: analyzing historical FIB time-course data in a spatial context, performing existing data and field system reconnaissance, nominating hypothetical sources based on sewer infrastructure and nearby features, designing a field sampling and sample analysis program to test the hypotheses, and employing one or more source-specific markers in sample analysis. This has been previously referred to as a "tiered approach" (Boehm et al. 2003), but the differences in decay rates between abundances of FIB and source-specific markers (Dick et al. 2010) means that FIB data may not provide a solid foundation upon which a tiered source diagnosis can be built. FIB data are useful for generally circumscribing the spatial emphases of MST, but not for reliably tracking sources of waste.

MST has been advanced significantly by the discovery of microbial host-specific waste markers, most notably DNA sequence encoding the 16S ribosomal RNA (rRNA) of a *Bacteroidetes* specific to human feces, as discovered by Bernhard and Field (2000b). The advantages of testing for this DNA sequence include its strong association with human waste and its culture-independence, coupled with quantitative polymerase chain reaction (qPCR) protocols (Seurinck et al. 2005) that enable quantifying waste markers relative to possible sources such as sewage (Seurinck et al. 2005; Sercu et al. 2009). Host-specific markers are reviewed more comprehensively elsewhere in this book (Chaps. 3–7) and are continuing to be an important, emerging component of the MST toolkit. There is still much work to be done in order to understand the true specificity of host-specific waste markers, the behavior of individual DNA-based waste markers under environmental conditions including their rates of decay relative to pathogens (Field and Samadpour 2007), and the potential for waste markers to become indigenous to the environment (a problem also noted for FIB). For these reasons, individual markers will generally not be used alone but will instead be employed within a suite of MST tools. Further, while a combination of bacterial host-specific markers (Chaps. 3 and 4), coupled with chemical markers (Chap. 8), host-related protozoa markers (Chap. 7), and assays for viral pathogens or host-specific bacteriophages indicators (Chaps. 5 and 6), can comprise of an effective suite of MST tools, a universally applicable and optimal tool combination has yet to be defined (Chap. 9). Meanwhile, the orientation of MST toward culture-independent waste markers suggests a potentially powerful use of DNA extracted from environmental samples: analysis of entire microbial communities which inevitably include pathogens, FIB, and other marker organisms

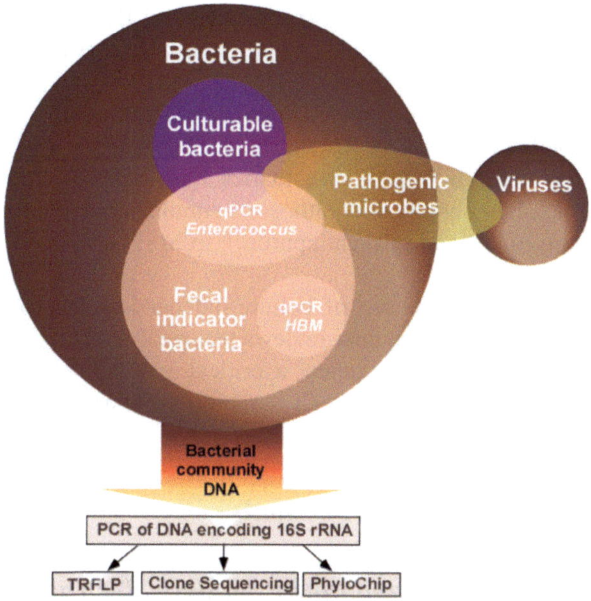

Fig. 11.3 Summary of microbial community analysis applied to MST. All bacteria (*Bacteria* and *Archaea*) along with *Eukarya* are recovered in environmental sampling; viruses, including coliphage, are recovered and analyzed separately (*right*). Bacteria, inclusive of culturable and nonculturable fecal indicators, can be analyzed holistically using total extractable DNA encoding 16S rRNA. The as-conceived associations of qPCR-based approaches assume that PCR primers are specific. By PhyloChip, bacterial pathogens, culturable or not, are automatically captured and can be specifically accounted for. Since pathogens are included, community analysis is conceptually highly responsive to public health concerns. Not shown, but conceptually similar, are other community analysis approaches such as DGGE, TGGE, and the analysis of membrane fatty acids by PLFA

(Fig. 11.3). This type of analysis could trace various environmental conditions, including biotic and abiotic perturbations.

Here, we use community analysis to refer to culture-independent characterization of the microbial community in a sample, i.e., what microbes, in what abundance, comprise of the community in a sample. Fecal input can alter microbial communities in the receiving environmental waters either directly or indirectly, i.e., by fecal source communities acting as inoculants, or by altering the environmental conditions of receiving water such as through changing water chemistry. Community analysis in MST has a strong potential to be useful for waste source assessment for the following reasons. First, gut microbial communities of various hosts vary significantly by host species and diet (Ley et al. 2008); therefore, microbial communities in feces would differ by host animal. Second, microbial communities are sensitive to perturbations and respond rapidly to environmental changes (Hwang et al. 2009), suggesting that their composition may reflect recent or ongoing contamination events. Third, the culture-independent analysis of microbial communities, by now a mature research endeavor (Liu and Jansson 2010), has successfully captured changes of microbial communities in the environment along gradients such as carbon availability (LaMontagne et al. 2003a) and proximity to hydrocarbon seeps (LaMontagne et al. 2004), and changes in microbial community due to perturbations such as inundation (Córdova-Kreylos et al. 2006), and vegetation and pollutant variations (Cao et al. 2006, 2008) in estuaries. Furthermore, community analysis detects many microbial signals and pathogens simultaneously to provide what is potentially a more robust and relevant approach to MST when compared to single indicators (Fig. 11.3).

This chapter examines the use of culture-independent microbial community analysis in MST. Briefly, this chapter reviews technical methods and data analysis approaches, provides two case studies, a critical evaluation of advantages and needs, and a view of community analysis in the future of MST.

11.2 Community Analysis Methods

11.2.1 Description of Community Analysis Techniques

A variety of holistic microbial community analysis techniques have been used in microbial ecology (for reviews, see Osborn and Smith 2005; Nocker et al. 2007). Culture-independent community analysis techniques differ on analysis targets and profiling techniques (Fig. 11.4). The analysis targets divide techniques into phenotypic and genotypic methods. The most common phenotypic method, phospholipid fatty acid (PLFA) analysis, targets phospholipids contained in the cell membranes of all living cells. Most genotypic methods are DNA-based, where total community DNA extracted from an environmental sample is analyzed directly through techniques such as "shotgun cloning," "shotgun sequencing," or via "DNA probe hybridization." Alternatively, polymerase chain reaction (PCR) is used to amplify specific genetic markers from community DNA, and the PCR products are subsequently analyzed by sequencing, hybridization, DNA melting behavior, or length polymorphism (Nocker et al. 2007).

Here, we describe methods that are most widely used. In particular, we focus on techniques that have the greatest potential to be used in MST studies. For each method, brief descriptions are provided regarding the method background, sample processing, and basic method-specific data processing. Multivariate data analysis is discussed in Sect. 11.3.

11.2.1.1 Phospholipid Fatty Acid (PLFA) Analysis

Phospholipids are essential membrane components that control cell permeability. Microbial fatty acids are largely linked to phospholipids, and in most cases, specific types of fatty acids predominate in a given taxon and are commonly associated with

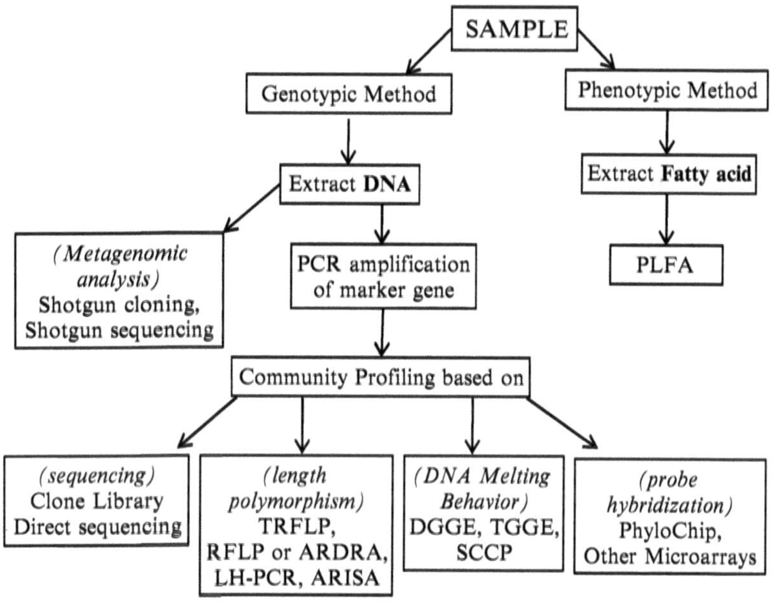

Fig. 11.4 Diagram to summarize microbial community analysis techniques which differ on the analysis target (**bold**) and target profiling technique (*italic in parentheses*). *PLFA* phospholipid fatty acid (analysis); *TRFLP* terminal restriction fragment length polymorphism; *RFLP* restriction fragment length polymorphism; *ARDRA* amplified ribosomal DNA restriction analysis; *LH-PCR* length heterogeneity polymerase chain reaction; *ARISA* automated ribosomal intergenic spacer analysis; *DGGE* denaturing gradient gel electrophoresis; *TGGE* temperature gradient gel electrophoresis; *SCCP* single strand conformation polymorphism

groups metabolizing similar substrates (Zelles 1999). Microorganisms also change membrane fatty acids composition in adaptation to environmental conditions including stressors (Loffhagen et al. 2004). Since PLFAs decompose quickly upon cell death, PLFA community analysis is typically regarded as reflecting viable or recently living cells (White 1994).

In performing PLFA analysis, PLFAs are first recovered from a sample by organic solvent extraction or solid-phase extraction (Zelles 1999). The extracted fatty acids and their derivatives are analyzed by gas chromatography to generate a PLFA profile that contains a list of fatty acids and their molar abundances. For data processing, the total mass of all PLFAs is often calculated as an indicator of total biomass, and the total mass of specific groups of PLFAs or mass ratios of certain PLFAs are also calculated as indicators of target groups of microorganisms or of environmental stresses (Zelles 1999). For example, several PLFAs are considered as biomarkers for members of the sulfate-reducer functional group; branched PLFAs are used as biomarkers for Gram-positive bacteria, and certain FAs can indicate heavy metal pollution (Pennanen et al. 1996; Córdova-Kreylos et al. 2006). Depending on downstream data analysis, the PLFA mass (i.e., absolute abundance) is also converted to percentage composition (i.e., relative abundance) so that biomass influences can be down-weighted and sample comparisons will be mainly based on community composition (Cao et al. 2006).

11.2.1.2 Terminal Restriction Fragment Length Polymorphism (TRFLP)

Terminal restriction fragment length polymorphism (TRFLP) produces genotypic fingerprints, or profiles, of the microbial community based on length polymorphism of PCR products of a specific marker gene, most frequently the gene encoding 16S rRNA, which contains regions that are conserved and regions that are variable among microorganisms. Additionally, other functional genes such as nitrite reductase genes (*nir*S, *nir*K) have also been used as marker genes for community analysis by TRFLP (Braker et al. 2001).

In TRFLP analysis (Liu et al. 1997), DNA is extracted from an environmental sample and genes from total community DNA (usually 16S rRNA genes) are amplified via PCR. One or both of the (forward and reverse) primers are labeled with a fluorescent dye. The resulting terminally labeled PCR products (amplicons) are digested with restriction enzymes that recognize specific cutting sites. These sites are located at different positions on the amplicons due to differences in gene sequences among microorganisms. Thus, the enzyme digestion generates fluorescently labeled terminal restriction fragments (TRFs) of various sizes (or lengths in base pairs). The sizes and abundances of the TRFs are determined using an automated DNA sequencer, where each TRF is represented as a peak on the electropherogram. The electropherogram is the graphical representation of the TRFLP profile, and the profile itself is a list of TRFs (a.k.a. peaks) in order of their base pair length and their abundance in relative fluorescent units. Both TRF height and area, for each peak, are provided in the profile dataset. Since the estimated sizes of TRFs from the same phylotype differ slightly due to run-to-run variability, TRFLP profiles are aligned across runs before further data analysis (Dunbar et al. 2001). Absolute abundances of TRFs are often converted to relative abundances, expressed as percentages of the total peak height or peak area for all TRFs in a sample. Normalization in this way is necessary to adjust for slight variations in the amount of DNA loaded onto the sequencer. If PCR bias is a concern, TRFLP data can be converted, prior to further analysis, to a simple binary format based on TRF presence or absence (Cao et al. 2006).

11.2.1.3 Denaturing Gradient Gel Electrophoresis (DGGE)

Denaturing gradient gel electrophoresis (DGGE) also generates community fingerprints based on profiling PCR products of a specific marker gene. However, the profiling is achieved based on separating the PCR products in a gel formulated with a chemical denaturant such as urea (Muyzer and Smalla 1998). Specially designed primers with a GC clamp are used in PCR so that the GC clamp can hold together the two separated strands of amplicon during denaturation. Briefly, as the PCR products are separated by electrophoresis, they denature in the gel due to exposure to the chemical denaturant. Denaturation converts easily electrophoresed double-stranded DNA into DNA whose migration is sterically hindered by its denatured content. Since sequence differences cause different melting behaviors, the timing and extent of denaturation of the PCR products differ along the gradient, which in turn determines how fast the various PCR products migrate in the gel. A community profile results from the pattern of separately migrated bands where each band represents a different organism or group. A similar technique is known as temperature gradient gel electrophoresis (TGGE) where a temperature gradient is used for denaturation.

After electrophoresis, the gel is stained with a nucleic acid-binding dye and photographed, resulting in an image of the banding pattern that is often imported into computer software for band density analysis (Esseili et al. 2008). Depending on the crispness of the separation, bands can also be excised, PCR-amplified, and sequenced for phylogenetic identification.

11.2.1.4 Cloning and Sequencing

A higher degree of phylogenetic resolution for community analysis is achieved by obtaining the actual sequence of a marker gene (or a metagenome, Fig. 11.4) through cloning followed by sequencing (Nocker et al. 2007) or direct sequencing (Shendure and Ji 2008). Here, a metagenome refers to the entire collection of genetic material recovered directly from environmental samples. During the cloning process, PCR products (or fragmented community DNA, i.e., the pieces of the metagenome) are inserted into plasmid vectors that are then transformed into *Escherichia coli*. The plasmids harboring specific PCR products (from specific organisms, presumably) are then multiplied by growing the transformed *E. coli* cells. The insertions are harvested by plasmid preparation and are subsequently sequenced. After quality check and trimming of the vector sequence (i.e., alignment), the marker gene sequences can be compared to sequence databases (Maidak et al. 2001) for phylogenetic identification. Computer programs used for sequence quality check, alignment, and comparison are widely available. Phylogenetic trees can be constructed to reveal similarities between sequences (http://rdp.cme.msu.edu/treebuilder/treeing.spr).

11.2.1.5 PhyloChip Microarray

Microarrays are high-throughput devices that allow for simultaneous detection of multiple DNA fragments (Bodrossy and Sessitsch 2004; Andersen et al. 2010). Detection is based on strand complementation and hybridization of the fluorophore-labeled target DNA with the probes representing known DNA sequences fixed on the array. After washing away the unbound target DNA fragments, the array is scanned at defined excitation wavelengths to image the bound, fluorophore-labeled DNA fragments. The locations of the probes with hybridized target DNA on the array indicate the presence of specific nucleic acid sequences in the query sample. The fluorescence intensity can also be used to quantify the relative abundance of the target when compared to the same probe on a separate array. However, different probes cannot be directly compared to each other due to differences in GC content and hybridization efficiency

The PhyloChip (G2, i.e., the second generation) is an Affymetrix (Santa Clara, CA)-platform microarray designed by researchers at the Lawrence Berkeley National Laboratory (Berkeley, CA). This high-density custom microarray encodes all known Archaea and Bacteria 16S rRNA sequences found in the 2004 public databases, and can identify up to 8,741 operational taxonomic units (OTUs) (Brodie et al. 2006; DeSantis et al. 2006, 2007). Currently, a G3 PhyloChip is being developed, and the probes are designed based on all 16S rRNA sequences available in the 2007 public databases. The G3 PhyloChip can detect up to ~60,000 OTUs and select pathogen specific genes. Additionally, the probes on the G2 and G3 PhyloChips are constantly being dynamically re-annotated based on the most current database information.

Sample processing includes extraction of the nucleic acids from an environmental microbial community, and amplification of the genomic DNA using universal bacterial and archaeal primers targeting the 16S rRNA gene (Fig. 11.5). The amplicons are then fragmented, biotin-labeled, denatured, and hybridized onto the microarray. Following overnight hybridization, the microarray is washed, stained with streptavidin-labeled fluorophore conjugate and an image is acquired using a confocal laser scanner. The response of individual oligonucleotide probes that make up a probe set is calculated to determine both the presence and relative concentration of defined bacterial and archaeal taxa. The data are subsequently normalized and are ready for downstream processing with uni-, multivariate, and phylogenetic analyses. Key features that set the PhyloChip apart from other similar technologies are the use of multiple oligonucleotide probes for every known

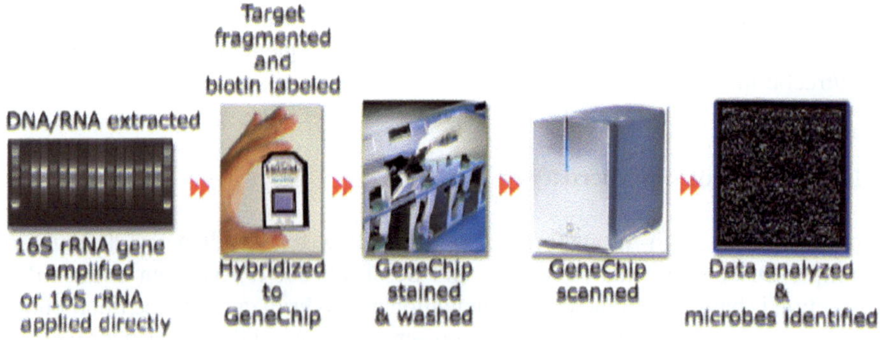

Fig. 11.5 PhyloChip sample processing schematic

category of Bacterial and Archaeal organisms for high confidence of detection and the pairing of a mismatch probe for every perfectly matched probe to minimize the effect of nonspecific hybridization. A strong linear correlation has been confirmed between microarray probe set intensity and concentration of OTU specific 16S rRNA gene copy number, allowing quantification over a wide dynamic range (Brodie et al. 2007; La Duc et al. 2009).

11.2.2 Applicability and Demonstration of Community Analysis Approaches for MST

Applicability of the various community analysis methods to specific types of questions differ because of their technological differences, state of method development and evaluation, and logistics. The demonstrated and potential usages of the methods for MST also vary.

PLFA analysis has been widely used to study changes in microbial community structure along environmental gradients or in response to environmental perturbations (Frostegard et al. 1993; Macalady et al. 2000; Kaur et al. 2005). Since phenotypic adaptation reflects the microorganisms' particular habitat, including host intestine and other sources, PLFA is potentially a useful tool for MST. As it mostly reflects living or recently living cells, PLFA may have the added benefit of detecting recent pollution sources, but not older ones. However, extensive knowledge about fatty acid patterns is generally required for interpreting the significance of specific fatty acids or fatty acid groups and for most efficient usage of PLFA data. Accessible databases for relating taxa or environmental stresses with fatty acid patterns are not available, and such interpretation often relies on a researcher's experience or familiarity with the PLFA literature. Furthermore, PLFA extraction methods influence the types of fatty acids recovered from a sample, and some extraction protocols may liberate fatty acids from nonliving organic matter, in which case PLFA composition could reflect more than the living microbial community (Zelles 1999). Nonetheless, these potentially confounding issues may be alleviated by commercial service laboratories that offer standardized PLFA analysis. Total PLFA abundance and PLFA profiles can be useful for tracking overall biomass and microbial community changes in MST studies without specifically focusing on individual fatty acids or fatty acid groups. For example, similarity and dissimilarity of the PLFA profiles from different samples were utilized to rule out kelp, but imply beach sand, as sources of fecal contamination to beach water (Izbicki et al. 2009).

For genotypic methods based on genes encoding 16S rRNA, abundant sequence data have been generated and are accessible in large databases such as the Ribosomal Database Project (RDP) (Maidak et al. 2001) and the Greengenes Database (greengenes.lbl.gov). Ever-growing databases are increasingly accessible because of developments in computational biology and bioinformatics that provide new and better tools for data handling. Performing most community analysis methods does require specialized equipment and expertise, but many genomic facilities can provide such services. However, it is important to be aware that, like culture-based or phenotypic methods, genotypic methods have their share of technical shortcomings arising from variations in DNA extraction efficiency, PCR bias, and/or sequencing accuracy and comprehensiveness.

Application of genotypic community analysis methods in MST has included: (1) identification of source-specific species as candidates for developing source-specific single indicators and (2) differentiation and/or tracking source of pollution based on the similarity of microbial community profiles from potential sources and sinks. Numerous studies have characterized microbial communities associated with human or animal feces using community analysis methods (Zoetendal et al. 2004). Although MST was not the objective, these studies provided abundant information regarding the host specificity of microbial communities and factors that affect such specificity (Ley et al. 2008); such studies also demonstrated the potential of community analysis for MST (Li et al. 2007). The following paragraphs discuss the application of TRFLP, cloning and sequencing, and PhyloChip in MST.

Since it is a high-throughput, sensitive, and reproducible approach whose data are readily amenable to quantitative statistical comparisons, TRFLP has been frequently used to analyze communities from a wide range of environments, including feces and digestive tracts of insects and mammals, and to characterize microbial community responses to environmental changes (for review, see Thies 2007; Schütte et al. 2008). Furthermore, because of its popularity in microbial ecology, an abundance of literature and many automated data processing software applications are available for adapting TRFLP for MST studies (Kent et al. 2003; Shyu et al. 2007).

TRFLP has been used successfully to develop source-specific single indicators. For example, highly reproducible, host-specific TRFLP patterns were identified in microbial communities from human and cow feces using primers specific to the *Bacteroides–Prevotella* group (Bernhard and Field 2000a), and subsequent cloning and sequencing of such source-specific TRFs led to designing of human- and cow-specific single indicator (q)PCR assays for MST (Bernhard and Field 2000b; Field et al. 2003a). More recently, TRFLP was used to find a poultry-specific *Brevibacterium* marker (Weidhaas et al. 2010). Studies also employed TRFLP to differentiate fecal sources based on overall community similarity. TRFLP was first shown to be successful in distinguishing deer fecal samples from sands while demonstrating high similarity between microbial communities in two discrete piles of deer fecal pellets (Clement et al. 1998). Using universal eubacterial primers, TRFLP analysis also clearly differentiated microbial communities from cattle feces, dog feces, and sewage (LaMontagne et al. 2003b). In a more comprehensive study, TRFLP was employed to analyze the *Bacteroides–Prevotella* community in multiple (10–50) fecal samples from each of nine host species (cattle, chicken, deer, dog, geese, horse, humans, pig, and seagulls) from different geographical locations and times of year (Fogarty and Voytek 2005). While no single TRF was identified as exclusive to a host species, and the previously identified cow- and human-specific TRFs (Bernhard and Field 2000a) were not resolved from their respective sources (Fogarty and Voytek 2005), the *Bacteroides–Prevotella* TRFLP community profiles were highly reproducible and much more similar within host species as compared to between host species. Attempts to identify sources using single TRFs or total community similarity for mixed-source samples, however, were less successful, perhaps due to biomass dominance from one source (Field et al. 2003b) or higher redundancy in TRFs from eubacterial primers (Liu et al. 1997). It is possible that a combination of TRFs, analyzed as a subset of the consortium, would be more useful for source identification in mixed-source samples. More recent studies utilized TRFLP to conduct source tracking in defined watersheds. Potential pollutant transport pathways were identified via similarity analysis of TRFLP community profiles from different sampling locations (Ibekwe et al. 2008). A case study using this approach will be discussed in a later section of this chapter.

DGGE has been widely used to assess diversity and to monitor dynamics of microbial communities (Ercolini 2004; Dorigo et al. 2005). Although DGGE analysis is often confined to PCR products with limited lengths (<400 bp), it can differentiate a single base pair difference in sequence fragments. Compared to expensive sequencers needed for TRFLP and sequencing, equipment for DGGE analysis is affordable for ordinary laboratories; however, DGGE is technically demanding. Methodological concerns such as inaccuracy and low reproducibility of band patterns, low sensitivity and lack of reliable quantification of band intensity, and in particular the need to optimize experimental conditions may also hinder its application (Dorigo et al. 2005). Nevertheless, DGGE applied to a 126 bp fragment of the β-D-glucuronidase gene (*uid*A) was successful in discriminating among *E. coli* phylotypes using DNA from cultured isolates, DNA from mixed culture-enriched *E. coli* populations, and community DNA extracted directly from environmental samples. Little difference in the DGGE patterns was observed for the latter two DNA sources, indicating that the culture enrichment step may be bypassed (Farnleitner et al. 2000). More recently, DGGE based on enriched *E. coli* cultures indicated similar *E. coli* populations for samples originating from the same sampling site (Sigler and Pasutti 2006). A more comprehensive study evaluated the applicability of 15 marker genes for use with DGGE for MST, and three genes (*mdh*, *phoE* and *uidA*) were identified to provide good discrimination among horses, pigs, and goats (Esseili et al. 2008). DGGE profiles from these three genes indicated greater *E. coli* population similarity (98–100%) between wastewater treatment plant (WWTP) effluents and downstream water samples, and lower similarity between upstream and downstream/effluent samples, providing strong evidence for a dominant contamination source from the WWTP. However, source attribution was less successful for contamination in a pond, presumably due to mixed sources from urban runoff in addition to goose feces deposition (Esseili et al. 2008).

Cloning followed by DNA sequencing has been a widely used tool in molecular microbial ecology from its inception. Although cloning and sequencing offers high phylogenetic resolution, the method is laborious, time consuming, and costly for routine usage. Also, rarely do clone libraries provide complete coverage of entire microbial communities. However, technology advancement in automation and parallel sequencing may greatly improve its speed and lower its cost while also improving its comprehensiveness (Shendure and Ji 2008). Since it provides actual sequence data, cloning followed by sequencing is often employed in developing and evaluating single indicator-based (q)PCR assays. For example, a library of genes encoding 16S rRNA extracted from gull feces revealed the abundance of a sequence closely related to *Catellicoccus marimammalium*, which was used successfully to develop a gull-specific, SYBR green (q)PCR assay (Lu et al. 2008). Libraries of *Bacteroidales* genes encoding 16S rRNA extracted from the feces of eight hosts revealed ruminant-, pig-, and horse-specific clusters of sequences, while human, dog, cat, and gull *Bacteroidales* communities shared greater similarities (Dick et al. 2005). The host-specific sequences were used to design PCR assays specific to pig and horse fecal matter. The analysis of other *Bacteroidales* clone libraries comprised of genes encoding 16S rRNA extracted from gull, goose, canine, raccoon, and sewage sources revealed concerns regarding instability of source identification assays against geographic or host individual differences (Jeter et al. 2009).

In addition to developing and evaluating single indicator assays, multiple clone libraries from potential contributing sources and environmental samples have also been developed to link pollution sources with environmental sinks. *Bacteroidales* clone library analysis revealed high similarity between cattle feces and water sample clone libraries, confirming cattle fecal pollution in a small watershed (Lamendella et al. 2007). However, clone libraries constructed from a horse manure pile and water samples from upstream and downstream of the manure pile using both universal eubacterial primers and *Bacteroides* group-specific primers showed little similarity between microbial communities from the manure pile and the downstream water samples, even though the water at 5 m downstream was visibly contaminated (Simpson et al. 2004). The authors offered two explanations: (1) downstream water was contaminated with the recently deposited surface material of the manure pile which harbored a different microbial community than the older interior manure pile from which the clone library had been constructed, (2) universal eubacterial primers do not offer sufficient sensitivity to detect manure pollution at the dilution level in the study sites. Direct shotgun cloning and sequencing of community DNAs (i.e., metagenomic analysis) was used to characterize a viral community in human feces (Breitbart et al. 2003); however, its direct application in MST has been limited at this time.

Phylogenetic microarrays such as the PhyloChip, which targets the currently known diversity within bacteria and archaea, have been employed to determine the composition of microbial communities in a number of different environments and conditions. When the PhyloChip microarray was applied to urban aerosols, the spatio-temporal distributions of known bacterial groups, including specific pathogens, were determined to be related to meteorologically driven transport processes as well as sources (Brodie et al. 2007). This microarray has been extensively validated and successfully used on a number of complex environmental samples, and the resulting findings have been confirmed by additional methods, including qPCR and 16S rRNA gene clone libraries (Brodie et al. 2006; Flanagan et al. 2007; Chivian et al. 2008; Tsiamis et al. 2008; Wrighton et al. 2008; Cruz-Martinez et al. 2009; DeAngelis et al. 2009; Sagaram et al. 2009; Sunagawa et al. 2009; Yergeau et al. 2009; Rastogi et al. 2010; Wu et al. 2010). Studies using split samples have confirmed that >90% of all 16S rRNA sequence types identified by the more expensive clone library method are also identified by the PhyloChip (DeSantis et al. 2007). In addition, the PhyloChip has demonstrated several-fold increases in detected microbial diversity over the clone library method and metagenomic sequencing with second-generation sequencers. One of the reasons for this is the high sensitivity of the PhyloChip, with the ability to detect organisms present at a proportional fraction of less than 10^{-4} abundance compared to the total sample (La Duc et al. 2009). Each sample analysis by the PhyloChip provides detailed information on microbial composition, and the highly parallel and reproducible nature of this array also allows tracking community dynamics over time and treatment. With no prior knowledge, specific microbial taxa may be identified in urban watersheds that are keys to human-associated fecal influence.

The PhyloChip is ideal for characterizing complex microbial communities, and its application for MST is currently being investigated. The comprehensiveness and sensitivity of the PhyloChip allows for better characterization of low-abundance organisms, leading to improved description of microbial diversity (La Duc et al. 2009; Sagaram et al. 2009). The reproducibility of the PhyloChip data on microbial community composition provides the opportunity to obtain results with high levels of statistical confidence (Brodie et al. 2006; DeSantis et al. 2007). A case study with PhyloChip-analyzed bacterial communities from an urban creek with known fecal pollution is discussed in a later section.

11.3 Multivariate Data Analysis, Interpretation, and Presentation

11.3.1 Why Multivariate Techniques?

Multivariate analysis involves simultaneous analysis of multiple, often correlated, variables. Multivariate analysis of community profiles has been developed and is routinely used by ecologists who study animals or plants, yet the application of such tools has been limited in microbial ecology, both in terms of frequency and choice of multivariate methods (Ramette 2007). However, multivariate tools are necessary to analyze the multivariate datasets generated from community analysis-based methods for MST.

Datasets generated by microbial community analysis methods usually contain rows representing samples or sites and columns representing OTUs. OTUs can be fatty acids from PLFA, TRFs from TRFLP, gel band identifiers from either DGGE or TGGE, or sequences or species from either clone libraries, or direct sequencing or PhyloChip analyses. Although each OTU could be treated as a single variable and analyzed by univariate statistical methods separately, separate univariate analyses not only are logistically difficult because there are hundreds to tens of thousands columns in a community profile but also are scientifically undesirable because microbial communities evolve and adapt together, therefore these variables are not independent. Source–sink relationships that are not revealed when a single OTU is evaluated can be distinguished when a consortium of OTUs is analyzed simultaneously in an integrated fashion and hence the rationale of using community-based analysis for microbial source tracking (see Sect. 11.1).

Many applications of MST are closely tied to TMDL assessment, which is currently based on FIB concentrations (Santo Domingo et al. 2007), and it is often desirable to evaluate correlations between MST and FIB concentration data. Correlations between multivariate community profiles with FIB concentrations can only be done through multivariate statistics such as direct gradient analysis (see Sect. 11.3.2). Furthermore, when discoveries of OTUs that are indicative of sources are desired, multivariate ordination techniques are more efficient compared to manually counting OTUs that are shared among samples from the same source.

11.3.2 Selection of Multivariate Techniques and Results Interpretation

Common multivariate techniques for the examination of microbial community structure include cluster analysis, principle components analysis (PCA), correspondence analysis (CA), and nonmetric multidimensional scaling (NMDS). These techniques belong to a group called indirect gradient analysis, which aims to reveal community similarities among sites or samples through grouping or ordering the sites or samples into either dendrograms or on a two (2D) or three-dimensional (3D) plot. Direct gradient analysis such as canonical correspondence analysis (CCA), on the contrary, aims to correlate the overall multivariate community profile with environmental variables or FIB concentrations. More details on each individual technique can be found in this review (Ramette 2007) and the references therein. However, methods differ in when and how they should be used, and proper selection of the methods is a very important first step in data analysis. Selection of the multivariate methods must be based on data type (binary, compositional or abundance data), analysis objective, and strengths and limitations of the various multivariate methods (Ramette 2007). Standard statistical software such as R (R Core Development Team 2008) and SAS can be programmed to run multivariate analysis. Specialized multivariate software packages are also available: CANOCO (Microcomputer Power, Ithaca, NY), PC-ORD (MjM Software Design), and Primer (Primer-E Ltd., Plymouth Marine Laboratory, UK).

Microbial ecologists are likely most familiar with cluster analysis, which is historically the basis for constructing phylogenetic trees that reveal similarities between sequences (e.g., OTUs in clone libraries). When discovering similarities between sites or samples is the goal, cluster analysis essentially groups sites or samples according to a similarity coefficient based on OTU data, and its interpretation is mostly intuitive: samples or sites grouped in the same cluster are similar to each other (Ramette 2007). However, because cluster analysis forces the formation of clusters, this method is most appropriate when groupings (i.e., discontinuous changes) of sites or samples are expected, such as when samples are from different known or suspected sources (e.g., animals, sewage, etc.) (Legendre and Legendre 1998). Cluster analysis is not appropriate when changes in communities are either continuous or gradual and discrete groupings are not expected, such as when samples are from upstream to downstream sites.

PCA is a frequently used multivariate technique partially due to its elegant mathematical algorithms (ter Braak 1995); however, it may also be the least appropriate method for analyzing microbial community profiles. A basic assumption for PCA is that OTUs respond to environmental conditions (i.e., environmental gradients) in a linear fashion, which is rarely true because most species have an optimal environmental condition or ecological niche and their response curves to environmental conditions are more similar to unimodal models (ter Braak 1986). However, when the environmental gradient is very short, the unimodal response curve may appear linear, and PCA may be appropriate to use. For example, most bacteria prefer a neutral pH and therefore exhibit a unimodal response to pH; yet, the response may be considered linear if the environmental pH conditions present a very short gradient ranging only from 5.5 to 6.0. Still, the absence of many OTUs in some sites or samples is a clear sign that the linear approximation is not valid and PCA is not an appropriate method. Improper usage of PCA may cause an artifact called the "horseshoe effect" where sites or samples are positioned on the 2D PCA plot resembling the shape of a horseshoe; these positions do not represent either similarity or dissimilarity between sites or samples (Palmer 2006).

CA assumes unimodal species response curves which are more appropriate for analyzing ecological data such as microbial community profiles (ter Braak 1986). CA is also considered a flexible method in that it can accommodate a dataset even when the underlying gradient is short, and thus linear, as long as the composition data (i.e., relative abundance in percentages) instead of absolute abundance data are analyzed (ter Braak and Smilauer 2002). CA results are generally displayed in a 2D plot called a "joint plot" where both sites and OTUs can be displayed as points on the plot, or a sample scatter plot where only sites are displayed. Community similarity between sites is indicated by close proximity of the site positions on the plot. The strong association of OTUs to certain sites (or sources) is implied by close proximity of the OTUs to the sites (or sources), and this can be used to reveal indicative OTUs for developing source-specific qPCR assays. Similar to the "horseshoe effect" in PCA, CA sometimes suffers an artifact called the "arch effect" where positioning of sites along the secondary axis could be arbitrary such that it resembles an arch. Removing the arch effect is achieved by a process called detrending and hence the term detrended correspondence analysis (DCA). Note that the axes on CA plots are meaningful, as they represent latent variables or gradients such as the distance to a point source (e.g., a storm drain discharge). This is useful for discovering trends and potential microbial contamination sources.

NMDS positions sites, or samples, into a 2D (or 3D) plot in a way that the ranks of dissimilarity between these sites are preserved as best as possible, much like positioning cities on a 2D map where the relative, or ranks of, distances between those cities along the Earth's spherical surface are preserved (Clarke and Warwick 2001). Therefore, sites in closer proximity to each other on a NMDS plot have more similar community profiles than those that are further apart on the plot. However, the distances between sites on the NMDS plot do not reflect the original dissimilarity in community profiles between those sites because only the ranks of the dissimilarity are preserved. NMDS has gained popularity because it does not either assume linear or unimodal species response curves or produce "horseshoe" or "arch" artifacts, and its interpretation is intuitive. Drawbacks of NMDS include its great sensitivity to dissimilarity measures, i.e., distance metrics, which must be specified *a priori* by the user. NMDS also cannot simultaneously display sites and OTUs on one plot; thus, associations between sites and OTUs may not be revealed (Palmer 2006).Therefore, if identifying site- or source-specific OTUs is the objective of the analysis, NMDS would not be the method of choice. NMDS is most useful for quickly assessing relative (dis)similarity among sites, or samples.

While indirect gradient analysis is generally considered exploratory, direct gradient analysis offers a means of specific hypothesis testing (ter Braak and Prentice 1988). A popular direct gradient analysis technique is CCA, which is an extension of CA, therefore CCA shares advantages and disadvantages of CA. CCA has been used to test whether community changes are influenced by environmental variations such as inundation (Cao et al. 2006). In the case of MST, for instance, CCA could be useful for testing whether community profiles correlate with FIB concentrations or if community profiles correlate with days after a sewage spill. An extension of CCA is partial canonical correspondence analysis (pCCA), where influences from a covariable can be excluded before evaluating effects on community profiles from another environmental variable. For example, pCCA was successfully used to identify the correlation between denitrifying community changes with heavy-metal contamination after adjusting for influences of dissolved carbon (Cao et al. 2008). pCCA is potentially useful for MST when the effects of geographic location need to be accounted for before correlating changes in microbial community with the magnitude of a microbial pollution source, for instance, the volume of a sewage spill.

11.4 Two Case Studies

11.4.1 A Case Study Using TRFLP

This case study examined bacterial communities using TRFLP during dry weather flow in the Arroyo Burro watershed (Santa Barbara, CA) where elevated FIB concentrations and human-specific *Bacteroides* markers were previously reported (Sercu et al. 2009). A laboratory spike-in experiment for validating the TRFLP technique, and a field study for investigating pollution sources in the Arroyo Burro watershed, were conducted. For the spike-in experiment, fecal samples were collected from suspected fecal sources such as dog, cat, and human (e.g., septage).

Dog feces were acquired from three healthy individuals of different breeds, each from a separate household. Cat feces were acquired from three healthy individuals of mixed breeds from two households. Septic solids, representing the composite material from several residential tanks, were obtained from a local pumping company (MarBorg, Santa Barbara). Relatively unimpacted creek water from a reference site in the watershed was collected in order to create spiked samples using the above fecal sources at various doses.

For the field study, nine sites in the lower Arroyo Burro watershed were selected. The nine sites (site 9 to 0 from upstream to downstream) included a storm drain (site 9) discharging into the Arroyo Burro creek (site 8 to 3) that flowed through the Arroyo Burro lagoon (site 2), and then into the ocean at Hendry's Beach, CA (site 1). Water samples were collected from these sites on 3 consecutive days (August 2005) as described previously (Sercu et al. 2009). No rain occurred at least 48 h prior to or during sampling, and the creek flow rate was 0.013 $m^3\ s^{-1}$. Four sewage influent samples were also collected from the El Estero Wastewater Treatment Plant (Santa Barbara, CA) during the period October 2004–2005. Microbial communities were analyzed by TRFLP using universal primers targeting the domain *Bacteria*. Relative abundance TRFLP data were aligned and analyzed using DCA as before (Cao et al. 2006).

Community (dis)similarity among the spiked samples and the fecal sources reflected differences among the fecal sources and the spiking dose (Fig. 11.6). Communities from water samples spiked with low doses of septic solids (0.01 and 0.1%) were more similar to the reference water sample than to the septic solids, while communities from those spiked with higher doses (1 and 5%) were more similar to the septic solids. Water samples that contained spikes from cat and dog

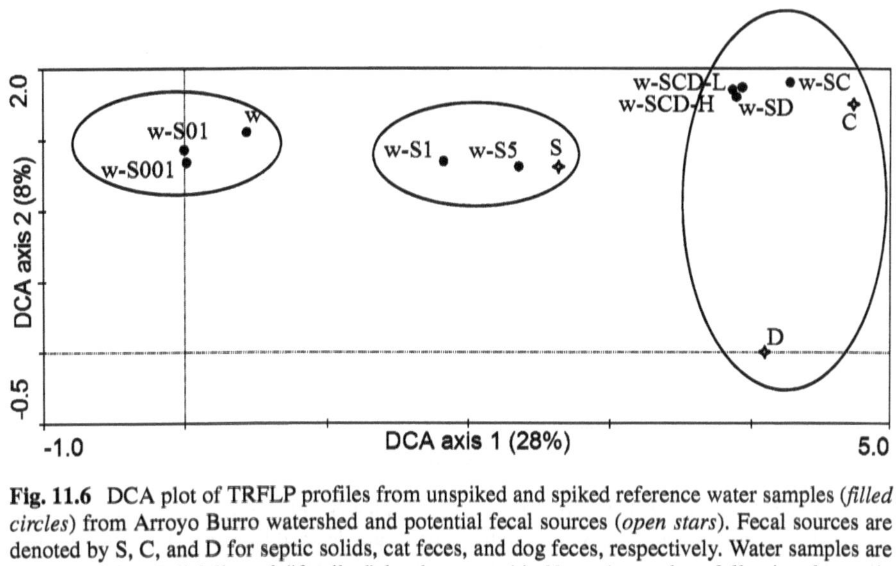

Fig. 11.6 DCA plot of TRFLP profiles from unspiked and spiked reference water samples (*filled circles*) from Arroyo Burro watershed and potential fecal sources (*open stars*). Fecal sources are denoted by S, C, and D for septic solids, cat feces, and dog feces, respectively. Water samples are denoted with a "w" followed (if spiked) by the source(s). *Numeric numbers* following the septic solids ("S") source indicate dose of spiking (e.g., 001, 01, 1, and 5 indicate 0.01, 0.1, 1, and 5%). "L" and "H" indicate a low and a high dose of spiking

feces grouped with cat and dog feces but not with the septic solids and the reference water. The results indicate that TRFLP can identify pollution sources, but relative contributions from different sources affect its sensitivity for MST.

For the field study, bacterial communities showed temporal stability during the 3 sampling days and clear spatial source tracking related to hydrological connectivity (Fig. 11.7). Communities differed as the water flowed downstream from the storm drain (site 9) to site 8, shortly downstream of the drain discharge, to the further downstream creek and lagoon sites (site 7 to 2), and then to the ocean (site 1). The major DCA gradient (DCA axis 1) explained 20% of the total variation in TRFLP profiles and coincided with the creek flow direction. While sewage samples grouped together by themselves and separated from the water samples along the secondary DCA axis, storm drain samples grouped with sewage samples when profiles from dog and cat feces were also included in the DCA (data not shown). These results are consistent with frequent detection of human-specific *Bacteroides* markers at the drain and the creek site following drain discharge (Sercu et al. 2009), and further implicated the storm drain as a source of human fecal pollution. The field study results also demonstrate the ability of community analysis to track sources of pollution on a spatial scale, in addition to its ability to accomplish source identification.

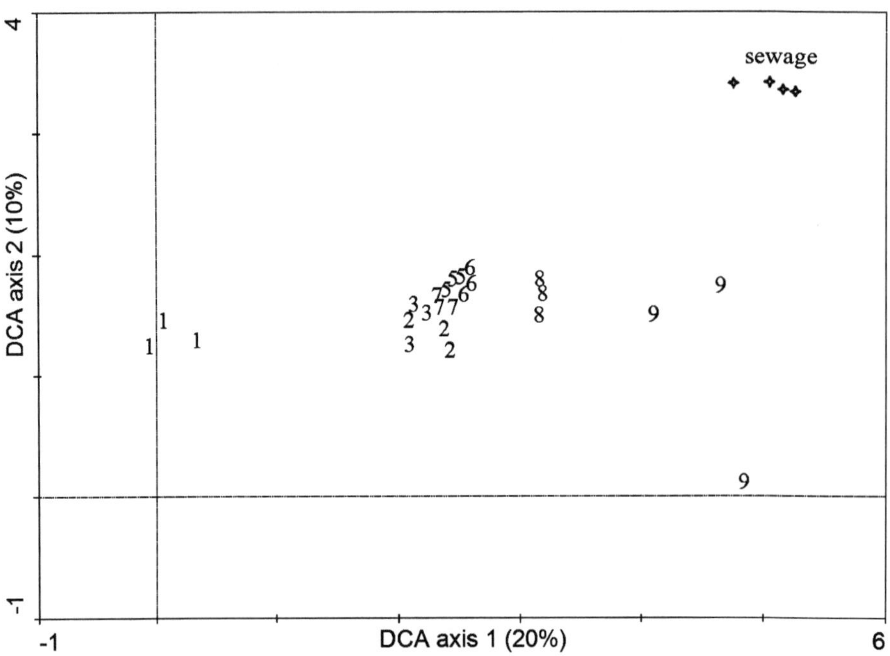

Fig. 11.7 DCA plot of TRFLP profiles from water samples (*numbers*) from Arroyo Burro watershed and sewage samples (*open stars*). Water samples are denoted by increasing numbers from downstream to upstream (e.g., "1" for ocean, "2" for the Arroyo Buroo Lagoon, and "9" for the most upstream urban drain)

11.4.2 A Case Study Using PhyloChip

This case study examined bacterial communities during dry weather flow in the lower Mission Creek and Laguna watersheds (Santa Barbara, CA) where elevated FIB concentrations and human-specific *Bacteroides* markers were previously reported (Sercu et al. 2009). Communities from creek (including storm drains), lagoon, and ocean sites, along with three fecal samples of human origin, were analyzed by the G2 PhyloChip (Wu et al. 2010). Mission Creek and Laguna Channel flow through an urbanized area of downtown Santa Barbara and discharge at a popular bathing beach. As described previously (Sercu et al. 2009), water column samples from 3 consecutive days were collected during the dry season (June 2005) from nine locations within the Mission Creek and Laguna watersheds in Santa Barbara, California. No rain occurred at least 48 h prior to or during the sampling. The creek flow rate in Mission Creek averaged 0.016 $m^3 s^{-1}$. Both watersheds discharged into the same lagoon and then flowed from the lagoon into the ocean.

NMDS with the Jaccard distance measure was used to visualize the dissimilarity between the bacterial communities in the samples (Fig. 11.8). Ocean bacterial communities were different from creek, lagoon, and fecal communities. For the most part, lagoon, and creek communities were different from each other, except for two samples. Fecal samples grouped separately from creek, lagoon, and ocean samples, illustrating the presence of distinct bacteria taxa. These bacteria, which

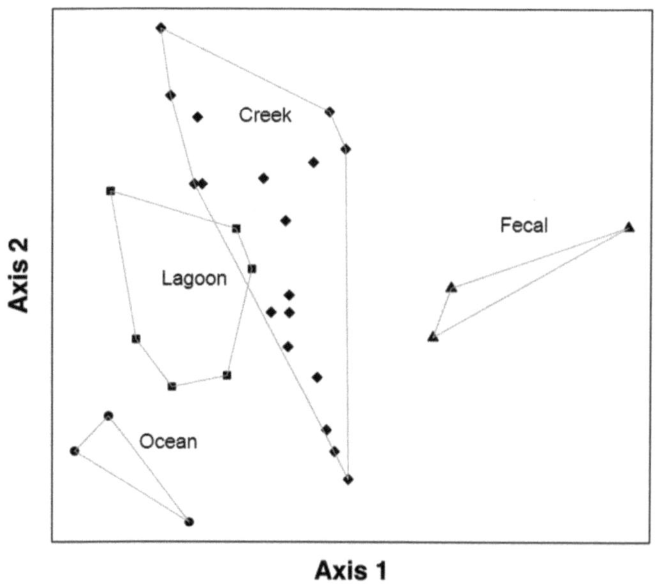

Fig. 11.8 Nonmetric multidimensional scaling (NMDS) plot of Mission Creek watershed samples, using the Jaccard distance measure. Stress = 8.14. Fecal samples grouped separately from ocean, lagoon, and creek samples

were unique to fecal samples, could potentially be used as signature taxa for indicating fecal communities.

PhyloChip data resulted in a comprehensive description of community composition from each of the samples. Comparative analysis of relative richness, which is the number of OTUs detected for each class normalized to total number of OTUs on the PhyloChip, showed that four bacterial classes exhibited the greatest variation across the sample types. Relative richness of *Bacteroidetes*, *Bacilli,* and *Clostridia* was higher in the fecal samples than the other sample types, while *Alphaproteobacteria* relative richness was lower in the fecal samples (Fig. 11.9). The sum of the relative richness of *Bacteroidetes*, *Bacilli,* and *Clostridia* was divided by the relative richness of *Alphaproteobacteria* to obtain a ratio (BBC:A). Thus, the BBC:A ratio incorporated microorganisms prevalent in fecal samples as well as those that were found in "pristine" environments. The BBC:A for fecal, creek, lagoon, and ocean samples were 4.06, 0.92, 0.69, and 0.65, respectively. The BBC:A for fecal samples was ~6-fold higher than that of ocean samples. High concentrations of FIB and human-specific *Bacteroides* markers were also detected on all 3 days at the site with the highest BBC:A (Wu et al. 2010).With further development and validation, this ratio could potentially be a useful tool for identifying fecal pollution. Future research is moving toward identifying signature communities for MST.

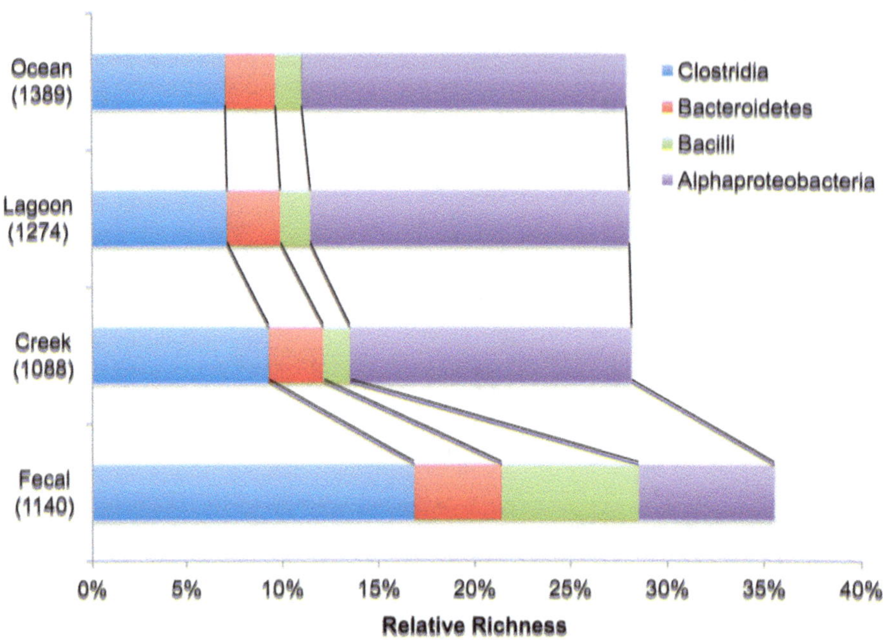

Fig. 11.9 The relative richness of four bacterial classes for each of the sample types. The *number in parentheses* represent the total richness (number of OTUs) detected. Fecal samples have higher relative richness of *Clostridia*, *Bacteroidetes*, *Bacilli*, and lower *Alphaproteobacteria* than the other three sample types

11.5 Relationship of Community Analysis to Multiple Indicator Approaches

As no single indicator can correctly identify microbial contamination sources 100% of the time, multiple indicators are often measured from the same sample and used in a tiered fashion. When measured in this manner, multiple indicators operate as a selected subset of the microbial community. Therefore, methods used to analyze and interpret community analysis results can be applied to analyzing multiple indicator datasets as well. For example, multivariate statistics are rarely used in multiple indicator MST studies, but provide a means to simultaneously view all indicator results, and could greatly aid in discovering patterns and trends. A DCA joint plot where sample IDs and the multiple indicators are simultaneously displayed illustrates this concept, using data from the case study in Sect. 11.4.1 (Fig. 11.10). The close proximity of culturable *E. coli* and *Enterococcus* and of human-specific *Bacteroides* markers to sites 9 and 8 (drain and nearby creek site, respectively), and the proximity of salinity to site 1 (ocean) indicate relatively higher levels of these indicators at the corresponding sites, while total coliform, dissolved oxygen, and pH do not indicate a strong pattern or trend (Fig. 11.10).

Similarly, methods commonly used for multiple indicators such as ratio and predictive modeling (Blanch et al. 2006) can also be applied to community analysis. In addition to using the overall community data, one can choose to reduce the data to a few OTU groups and investigate the ratios between groups as a source identification tool (e.g., as per the case study in Sect. 11.4.2), or to focus on several specific

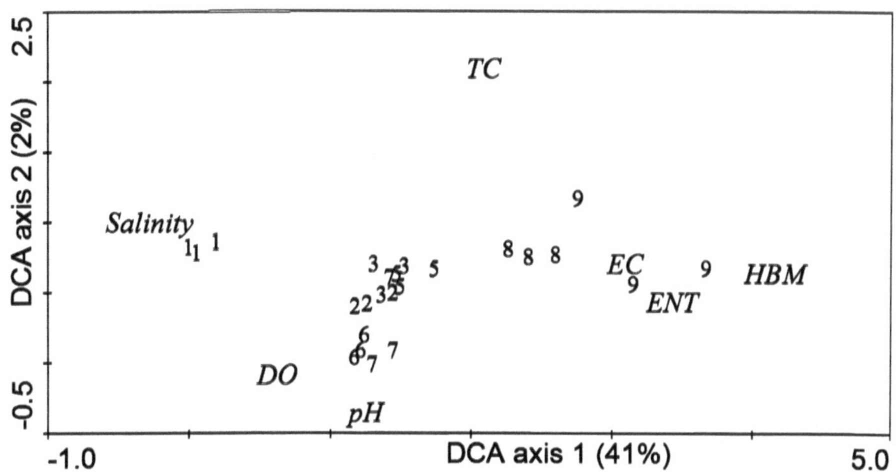

Fig. 11.10 DCA joint plot to illustrate multivariate analysis on multiple indicator data, displaying both samples (*numeric numbers*) and indicators (*italic characters*). Samples are labeled as in Fig. 11.7. Indicators are salinity, DO (dissolved oxygen), pH, TC (total *Coliform*), EC (*E. coli*), ENT (*Enterococcus*), and HBM (human *Bacteroides* marker). All indicator values are normalized to the mean across all samples before DCA is performed

OTUs that may be indicative of specific sources (Bernhard and Field 2000a). Furthermore, common OTUs (or cosmopolitan OTUs, if any) can be removed from the overall community data, and the community profiles can be focused into a more selected dataset where the most predictive OTUs, which can be considered multiple indicators, are measured simultaneously in just one assay.

11.6 Summary and Future Directions of Community Analysis for MST

11.6.1 Advantages

Ideally, MST methodology should include assays targeting specific pathogens that have been identified via epidemiology studies as public health risks in the context of recreational water use (Field and Samadpour 2007). However, to date, epidemiology studies include a very limited number of pathogen measurements. There are two reasons for such limitation. First, prior knowledge about relevant pathogens in a particular water body is often lacking. Second, the cost to perform a complete survey of all possible pathogenic indicators via many single measurements is prohibitive. Nevertheless, sole reliance on specific pathogens could be inadequate for MST, since detection of pathogens would depend both on the presence of fecal material and on the health status of surrounding human or animal populations. Community analyses provide a cost-effective alternative that offers many advantages, most notably: (1) comprehensiveness and relevancy, and (2) data density.

Comprehensiveness and relevancy refer to the inclusion of pathogens, fecal indicators, and other organisms when DNA or other biomarkers are fully extracted and analyzed from a water sample (Fig. 11.3). If a single marker is labile and its environmental fate ill-defined, simultaneous reliance on many singly or interactively relevant markers from a community can enable waste detection even in the absence of the yet-to-be-discovered single, robust marker. Furthermore, when identification of fecal source(s) is based on the entire microbial community, by default it is also based on tracing pathogens within that community (Fig. 11.3). Although resolution for particular pathogens differs with variations of the community analysis techniques (Sect. 11.2.1), the public health relevance of data acquired from community analysis is higher for MST when compared to data from a few individual markers whose transport characteristics and fates are unlikely to mimic fates exhibited by a majority of fecal pathogens.

Data density refers to the inherently multivariate nature of community analysis data, which are extremely versatile in how and what information may be extracted. In an MST study, the multivariate community analysis data also represent multiple lines of evidence, which, as in a trial by jury, increase the certainty of a water-quality diagnosis. While performing many different types of individual assays can also provide the needed multiple lines of evidence, e.g., by various chemical and biochemical host-specific markers, each requires separate procedures and expertise. An additional advantage provided by community analysis includes the capability of using community similarity between sites to conduct spatial source tracking (see the case study in Sect. 11.4.1). This advantage exists because community similarity analysis naturally combines data from source types and loading (Sercu et al. 2009), both of which are often needed for locating the source of contamination in MST studies. Finally, by using community analysis approaches more broadly in MST, more insights across numerous studies and geographical locations can be obtained to define additional individual markers of waste or to define the suite of individual markers within the overall community that best resolves sources of concern.

11.6.2 Critical Issues

Despite its advantages, there are technological, logistical, and implementation-related issues regarding usage of community analysis for MST.

Technologically, the sensitivity and resolution of certain community analysis methods may prevent source identification if the sources contributing to the receiving waters are very diluted or are of complex, mixed origins. Also, more biomass (or DNA) is often needed for community analysis than a single qPCR assay. Temporal variation in source concentrations, a concern for MST studies in general, can add to the issue. It is, therefore, important to examine the characteristics and complexity of a watershed (or an MST study system) before: (1) selecting a community analysis technique, because the different techniques vary in their sensitivity and resolution (Sect. 11.2) as well as cost and feasibility and (2) formulating a sampling design that may capture temporal and spatial variation in source contributions. Multiple community analysis techniques may also be used in one MST study to obtain more sensitivity and resolution in a cost-effective manner. For example, less expensive, high-throughput TRFLP may be applied to screen an entire watershed for "hot zones," where more expensive cloning and sequencing or PhyloChip analysis can be applied to obtain higher resolution source identification. While efficiency of concentration (such as water sample filtration) and DNA extraction methods are important to the reliability of all molecular techniques in general, such concerns and PCR bias may be less for community composition studies performed comparatively among samples or sites. However, such technical issues may hinder quantitative interpretation of community analysis results.

Fate (including die-off, persistence, and growth) and transport of microorganisms contributed from various fecal sources in the environment would also affect application of community analysis for MST. The microbial communities contained in fecal sources such as sewage undergo alterations when migrating through soil or groundwater. Wastewater changes biochemically when passed through reactive porous media (An et al. 2004) such as soil, so that nutrient (Hua et al. 2004; Stogbauer et al. 2004) and microbial (Hua et al. 2003) concentrations may change in predictable ways (Brinkmann et al. 2004). However, how microbial communities in fecal sources change during their mixing and migration in the environment is yet unknown. This is consistent with the state of knowledge for other, host-specific, individual markers (Field and Samadpour 2007) and warrants further research.

Logistically, as most methods need specialized equipment and expertise, community analyses are often more expensive than a PCR assay of a single indicator and are performed in (sometimes very specialized) research laboratories or commercial service laboratories. Routine use of such methods may be limited by the availability of expertise and cost, although high demand of applying such analysis (from MST or other fields such as biomedical research) will drive the availability up, and the cost down.

A common issue related to implementation in the MST field is the lack of standard protocols for performing MST studies. While lack of standardized sample processing and laboratory protocols is common to most MST methods, for community analysis-based methods there are even more needs for standard protocols in data processing, analysis, and interpretation, particularly because of data complexity. Research aimed at developing standardized MST protocols for sampling design, sample processing (including DNA extraction), and data analysis are needed. Lastly, and of largely a practical concern, is the potential difficulty for water-quality managers in communicating microbial community data to the public. There may be multiple dimensions to this issue, including the fact that the nonscientific community is generally unaware that microbial communities exist in nature, and thus could become unnecessarily alarmed by data that reveals the richness of microbial taxa in water even in the absence of fecal contamination. As with the public consumption of voluminous "personal genomic" data (from genetic testing for susceptibility to disease) for which full interpretation is lacking (Wright and Kroese 2010), there is a possibility for public misunderstanding and data misuse. The evolution of microbial community analysis in microbiological water-quality from a research tool into a monitoring tool will require consideration of this and other issues described in this section.

11.6.3 Future Directions

Future directions for community analysis may include: (1) incorporation of these methods into epidemiology studies, (2) assisting in research on indicator persistence and survival, and (3) development of new indicators, customized community analysis, and automation. While advanced molecular technologies did not exist for early epidemiology studies, modern epidemiology studies often archive genetic samples for future analysis. These archived samples can be analyzed by community analysis such that comprehensive water-quality data can be correlated to human health data that are already collected. This would provide a means to discover pathogens or nonpathogenic indicators that are highly predictive of health risk.

Comprehensive community analysis is also used to study the succession of microbial communities along pollution gradients of sewage discharge (Zhang et al. 2009) or after sewage spills (Dubinsky et al. 2009), which can provide insight into indicator survival and persistence after a sewage source is introduced into receiving waters. Similar studies can be conducted on other fecal sources in other environments to provide similar information that are greatly needed for revealing age and contributions of various fecal sources in MST (Blanch et al. 2006).

Although limitations such as read length and complexities of data analysis still exist, the next-generation sequencing technologies and bioinformatics are advancing very rapidly to provide truly high-efficiency and low-cost sequencing in the near future (Shendure and Ji 2008). Sequencing the whole microbial community associated with each source covering diverse geographic locations and individuals would soon be feasible, which will lead to identification of more and better source-specific indicators for development of qPCR-based methods and specialized MST microarrays. Because of the versatility of the community analysis techniques, "customized" community analysis can be designed to provide higher sensitivity and resolution for MST. For example, high-throughput techniques such as TRFLP can be paired with more specific and variable marker genes such as functional genes to offer higher sensitivity by tuning down the background that is generated when universal 16S rRNA primers are used. While the PhyloChip is one form of custom microarray, other microarrays can also be designed (Shiu and Borevitz 2008) to specifically target pathogens and pathogenic genes. Source identification microarrays (i.e., "MST on a chip") can be designed to include thousands of source-specific assays such that each sector on the microarray represents a particular pollution source (human, sewage, gulls, etc.). The "MST on a chip," combined with automated data analysis software, may give a probability estimate of contributions from each source and enable fast diagnosis in a watershed. Ultimately, a high-throughput, close to real-time (within 6–12 h) pipeline (array or qPCR) for processing water samples and obtaining results may be developed for real-time source tracking.

Acknowledgments The authors acknowledge the City of Santa and the Switzer Foundation for support, and the NSF-funded Santa Barbara Long Term Ecological Research project (NSF OCE 9982105 and OCE 0620276) for assistance including stream flow data in Santa Barbara, and the work of Laurie C. Van De Werfhorst and Bram Sercu in sampling, and sample and data processing for the AB and MC case studies herein. Other attributions for the Arroyo Burro and Mission Creek fecal source and sample acquisition plus analysis are as per Sercu et al. (2009). Part of this work was performed at Lawrence Berkeley National Laboratory under Department of Energy contract number DE-AC02-05CH11231.

References

An PL, Hua JM, Franz M et al. (2004) Changes of chemical and biological parameters in soil caused by trickling sewage. *Acta Hydrochim Hydrobiol* **32**: 296–303.

Andersen GL, He Z, DeSantis TZ et al. (2010) The use of microarrays in microbial ecology. In *Environmental Molecular Microbiology*. Liu W-T, and Jansson JK (eds). Caister Academic Press, Norfolk, UK, pp. 87–109.

Bernhard AE, and Field KG (2000a) Identification of nonpoint sources of fecal pollution in coastal waters by using host-specific 16S ribosomal DNA genetic markers from fecal anaerobes. *Appl Environ Microbiol* **66**: 1587–1594.

Bernhard AE, and Field KG (2000b) A PCR assay to discriminate human and ruminant feces on the basis of host differences in *Bacteroides-Prevotella* genes encoding 16S rRNA. *Appl Environ Microbiol* **66**: 4571–4574.

Blanch AR, Belanche-Munoz L, Bonjoch X et al. (2006) Integrated analysis of established and novel microbial and chemical methods for microbial source tracking. *Appl Environ Microbiol* **72**: 5915–5926.

Bodrossy L, and Sessitsch A (2004) Oligonucleotide microarrays in microbial diagnostics. *Current Opinion in Microbiology* **7**: 245–254.

Boehm AB, Fuhrman JA, Mrse RD et al. (2003) Tiered approach for identification of a human fecal pollution source at a recreational beach: Case study at Avalon Bay, Catalina Island, California. *Environ Sci Technol* **37**: 673–680.

Braker G, Ayala-del-Rio HL, Devol AH et al. (2001) Community structure of denitrifiers, *bacteria*, and *archaea* along redox gradients in pacific northwest marine sediments by terminal restriction fragment length polymorphism analysis of amplified nitrite reductase (*nir*S) and 16S rRNA genes. *Appl Environ Microbiol* **67**: 1893–1901.

Breitbart M, Hewson I, Felts B et al. (2003) Metagenomic analyses of an uncultured viral community from human feces. *J Bacteriol* **185**: 6220–6223.

Brinkmann T, Abbt-Braun G, Karle E et al. (2004) Transformation of wastewater-derived dissolved organic matter below leaky sewers - Fate of amino acids and carbohydrates. *Acta Hydrochim Hydrobiol* **32**: 316–327.

Brodie EL, DeSantis TZ, Parker JPM et al. (2007) Urban aerosols harbor diverse and dynamic bacterial populations. *Proceedings of the National Academy of Sciences of the United States of America* **104**: 299–304.

Brodie EL, DeSantis TZ, Joyner DC et al. (2006) Application of a high-density oligonucleotide microarray approach to study bacterial population dynamics during uranium reduction and reoxidation. *Appl Environ Microbiol* **72**: 6288–6298.

Cabelli VJ, Dufour AP, McCabe LJ et al. (1982) Swimming-associated gastroenteritis and water quality. *Am J Epidemiol* **115:** 606–616.

Cao Y, Green PG, and Holden PA (2008) Microbial community composition and denitrifying enzyme activities in salt marsh sediments. *Appl Environ Microbiol* **74**: 7585–7595.

Cao Y, Cherr GN, Córdova-Kreylos AL et al. (2006) Relationships between sediment microbial communities and pollutants in two California salt marshes. *Microb Ecol* **52**: 619–633.

Chivian D, Brodie EL, Alm EJ et al. (2008) Environmental genomics reveals a single-species ecosystem deep within earth. *Science* **322**: 275–278.

Clarke KR, and Warwick RM (2001) *Change in marine communities: An approach to statistical analysis and interpretation, 2nd edition*. PRIMER-E, Plymouth.

Clement BG, Kehl LE, DeBord KL et al. (1998) Terminal restriction fragment patterns (TRFPs), a rapid, PCR-based method for the comparison of complex bacterial communities. *J Microbiol Methods* **31**: 135–142.

Córdova-Kreylos AL, Cao Y, Green PG et al. (2006) Diversity, composition and geographical distribution of microbial communities in California salt marsh sediments. *Appl Environ Microbiol* **72**: 3357–3366.

Cruz-Martinez K, Suttle KB, Brodie EL et al. (2009) Despite strong seasonal responses, soil microbial consortia are more resilient to long-term changes in rainfall than overlying grassland. *ISME J* **3**: 738–744.

DeAngelis KM, Brodie EL, DeSantis TZ et al. (2009) Selective progressive response of soil microbial community to wild oat roots. *ISME J* **3**: 168–178.

DeSantis TZ, Brodie EL, Moberg JP et al. (2007) High-density universal 16S rRNA microarray analysis reveals broader diversity than typical clone library when sampling the environment. *Microb Ecol* **53**: 371–383.

DeSantis TZ, Hugenholtz P, Larsen N et al. (2006) Greengenes, a chimera-checked 16S rRNA gene database and workbench compatible with ARB. *Appl Environ Microbiol* **72**: 5069–5072.

Dick LK, Stelzer EA, Bertke EE et al. (2010) Relative decay of *Bacteroidales* microbial source tracking markers and cultivated *Escherichia coli* in freshwater microcosms. *Appl Environ Microbiol* **76**: 3255–3262.

Dick LK, Bernhard AE, Brodeur TJ et al. (2005) Host distribution of uncultivated fecal *Bacteroidales* bacteria reveal genetic markers for fecal source identification. *Appl Environ Microbiol* **71**: 3184–3191.

Dorigo U, Volatier L, and Humbert JF (2005) Molecular approaches to the assessment of biodiversity in aquatic microbial communities. *Water Res* **39**: 2207–2218.

Dubinsky E, Wu C, Hulls J et al. (2009) A complete microbial community approach to tracking fecal pollution in coastal waters. In *State of the Estuary Conference*. Oakland, CA.

Dunbar J, Ticknor LO, and Kuske CR (2001) Phylogenetic specificity and reproducibility and new method for analysis of terminal restriction fragment profiles of 16S rRNA genes from bacterial communities. *Appl Environ Microbiol* **67**: 190–197.

Eaton AD, Clesceri LS, Greenberg AE et al. (1998) *Standard Methods for The Examination of Water and Wastewater*. American Public Health Association, Washington, DC.

Eckburg PB, Bik EM, Bernstein CN et al. (2010) Diversity of the human intestinal microbial flora. *Science* **308**: 1635–1638.

Ercolini D (2004) PCR-DGGE fingerprinting: Novel strategies for detection of microbes in food. *J Microbiol Methods* **56**: 297–314.

Esseili MA, Kassen II, and Sigler V (2008) Optimization of DGGE community fingerprinting for characterizing *Escherichia coli* communities associated with fecal pollution. *Water Res* **42**: 4467–4476.

Farnleitner AH, Kreuzinger N, Kavka GG et al. (2000) Simultaneous detection and differentiation of *Escherichia coli* populations from environmental freshwaters by means of sequence variations in a fragment of the beta-D-glucuronidase gene. *Appl Environ Microbiol* **66**: 1340–1346.

Field KG, and Samadpour M (2007) Fecal source tracking, the indicator paradigm, and managing water quality. *Water Res* **41**: 3517–3538.

Field KG, Bernhard AE, and Brodeur TJ (2003a) Molecular approaches to microbiological monitoring: Fecal source detection. *Environmental Monitoring and Assessment* **81**: 313–326.

Field KG, Chern EC, Dick LK et al. (2003b) A comparative study of culture-independent, library-independent genotypic methods of fecal source tracking. *J Water Health* **1**: 181–194.

Flanagan JL, Brodie EL, Weng L et al. (2007) Loss of bacterial diversity during antibiotic treatment of intubated patients colonized with *Pseudomonas aeruginosa*. *J Clin Microbiol* **45**: 1954–1962.

Fogarty LR, and Voytek MA (2005) Comparison of *Bacteroides-Prevotella* 16S rRNA genetic markers for fecal samples from different animal species. *Appl Environ Microbiol* **71**: 5999–6007.

Frostegard A, Tunlid A, and Baath E (1993) Phospholipid fatty acid composition, biomass, and activity of microbial communities from two soil types experimentally exposed to different heavy metals. *Appl Environ Microbiol* **59**: 3605–3617.

Hazen TC (1988) Fecal coliforms as indicators in tropical waters–a review. *Toxicity Assessment* **3**: 461–477.

Hua JM, An PL, Winter J et al. (2003) Elimination of COD, microorganisms and pharmaceuticals from sewage by trickling through sandy soil below leaking sewers. *Water Res* **37**: 4395–4404.

Hua JM, An PL, Winter J et al. (2004) Elimination of organic compounds of sewage or sewage fractions in sandy soil below leaking sewers. *Acta Hydrochim Hydrobiol* **32**: 287–295.

Hwang CC, Wu WM, Gentry TJ et al. (2009) Bacterial community succession during in situ uranium bioremediation: spatial similarities along controlled flow paths. *ISME J* **3**: 47–64.

Ibekwe AM, Bold RM, Lyon SR et al. (2008) Microbial community composition in middle Santa Ana River watershed impacted by non-point source pollutants. In *the 108th general meeting for American Society of Microbiology*. Boston, MA.

Ishii S, Hansen DL, Hicks RE et al. (2007) Beach sand and sediments are temporal sinks and sources of *Escherichia coli* in Lake Superior. *Environ Sci Technol* **41**: 2203–2209.

Izbicki JA, Swarzenski PW, Reich CD et al. (2009) Sources of fecal indicator bacteria in urban streams and ocean beaches, Santa Barbara, California. *Annals of Environmental Sciences* **3**: 139–178.

Jeter SN, McDermott CM, Bower PA et al. (2009) *Bacteroidales* diversity in Ring-Billed Gulls (*Laurus delawarensis*) residing at Lake Michigan beaches. *Appl Environ Microbiol* **75**: 1525–1533.

Kaur A, Chaudhary A, Kaur A et al. (2005) Phospholipid fatty acid - A bioindicator of environment monitoring and assessment in soil ecosystem. *Curr Sci* **89**: 1103–1112.

Kent AD, Smith DJ, Benson BJ et al. (2003) Web-based phylogenetic assignment tool for analysis of terminal restriction fragment length polymorphism profiles of microbial communities. *Appl Environ Microbiol* **69**: 6768–6776.

La Duc MT, Osman S, Vaishampayan P et al. (2009) Comprehensive census of bacteria in clean rooms by using DNA microarray and cloning methods. *Appl Environ Microbiol* **75**: 6559–6567.

Lamendella R, Santo Domingo JW, Oerther DB et al. (2007) Assessment of fecal pollution sources in a small northern-plains watershed using PCR and phylogenetic analyses of Bacteroidetes 16S rRNA gene. *FEMS Microbiol Ecol* **59**: 651–660.

LaMontagne MG, Schimel JP, and Holden PA (2003a) Comparison of subsurface and surface soil bacterial communities in California grassland as assessed by terminal restriction fragment length polymorphisms of PCR-amplified 16S rRNA genes. *Microb Ecol* **46**: 216–227.

LaMontagne MG, Griffith JF, and Holden PA (2003b) Comparative analysis of animal fecal bacterial communities using terminal restriction fragment length polymorphisms of bacterial 16S rDNA PCR-amplified from fecal community DNA. In *American Society for Microbiology General Meeting*. Washington, DC.

LaMontagne MG, Leifer I, Bergmann S et al. (2004) Bacterial diversity in marine hydrocarbon seep sediments. *Environ Microbiol* **6**: 799–808.

Leadbetter ER (1997) Prokaryotic diversity: form, ecophysiology, and habitat. In *Manual of Environmental Microbiology*. Hurst CJ, Knudsen GR, McInerney MJ et al. (eds). American Society for Microbiology, Washington, D.C., pp. 14–24.

Legendre P, and Legendre L (1998) *Numerical Ecology*. Elsevier Science BV, Amsterdam.

Ley RE, Hamady M, Lozupone C et al. (2008) Evolution of mammals and their gut microbes. *Science* **32**: 1647–1651.

Li F, Hullar MAJ, and Lampe JW (2007) Optimization of terminal restriction fragment polymorphism (TRFLP) analysis of human gut microbiota. *J Microbiol Methods* **68**: 303–311.

Liu W-T, and Jansson JK (eds) (2010) *Environmental Molecular Microbiology*. Caister Academic Press, Norfolk, U.K.

Liu W-T, Marsh TL, Cheng H et al. (1997) Characterization of microbial diversity by determining terminal restriction fragment length polymorphisms of genes encoding 16S rRNA. *Appl Environ Microbiol* **63**: 4516–4522.

Loffhagen N, Härtig C, and Babel W (2004) *Pseudomonas putida* NCTC 10936 balances membrane fluidity in response to physical and chemical stress by changing the saturation degree and the trans/cis ratio of fatty acids. *Biosci Biotechnol Biochem* **68**: 317–323.

Lu JR, Santo Domingo JW, Lamendella R et al. (2008) Phylogenetic diversity and molecular detection of bacteria in gull feces. *Appl Environ Microbiol* **74**: 3969–3976.

Macalady JL, Mack EE, Nelson DC et al. (2000) Sediment microbial community structure and mercury methylation in mercury-polluted Clear Lake, California. *Appl Environ Microbiol* **66**: 1479–1488.

Maidak BL, Cole JR, Lilburn TG et al. (2001) The RDP-II (Ribosomal Database Project). *Nucleic Acids Res* **29**: 173–174.

Muyzer G, and Smalla K (1998) Application of denaturing gradient gel electrophoresis (DGGE) and temperature gradient gel electrophoresis (TGGE) in microbial ecology. *Antonie van Leeuwenhoek International Journal of General & Molecular Microbiology* **73**: 127–141.

National Research Council (1993) *Managing Wastewater in Coastal Urban Areas.* National Academy Press, Washington, D.C.

Nocker A, Burr M, and Camper AK (2007) Genotypic microbial community profiling: A critical technical review. *Microb Ecol* **54**: 276–289.

Osborn AM, and Smith CJ (2005) *Molecular Microbial Ecology.* Taylor & Francis, New York, NY, USA.

Palmer MW (2006). Ordination methods for ecologists: The ordination webpage. http://ordination.okstate.edu/. Accessed May 11, 2011.

Pennanen T, Frostegard A, Fritze H et al. (1996) Phospholipid fatty acid composition and heavy metal tolerance of soil microbial communities along two heavy metal-polluted gradients in coniferous forests. *Appl Environ Microbiol* **62**: 420–428.

R Core Development Team (2008) *R: A language and environment for statistical* computing. R Foundation for Statistical Computing, Vienna, Austria. ISBN 3-900051-07-0, URL http://www.R-project.org.

Ramette A (2007) Multivariate analyses in microbial ecology. *FEMS Microbiol Ecol* **62**: 142–160.

Rastogi G, Osman S, Vaishampayan PA et al. (2010) Microbial diversity in uranium mining-impacted soils as revealed by high-density 16S microarray and clone library. *Microb Ecol* **59**: 94–108.

Sagaram US, DeAngelis KM, Trivedi P et al. (2009) Bacterial diversity analysis of Huanglongbing pathogen-infected citrus, using PhyloChip arrays and 16S rRNA gene clone library sequencing. *Appl Environ Microbiol* **75**: 1566–1574.

Santo Domingo JW, Bambic DG, Edge TA et al. (2007) Quo vadis source tracking? Towards a strategic framework for environmental monitoring of fecal pollution. *Water Res* **41**: 3539–3552.

Schütte UME, Abdo Z, Bent SJ et al. (2008) Advances in the use of terminal restriction fragment length polymorphism (T-RFLP) analysis of 16S rRNA genes to characterize microbial communities. *Appl Microbiol Biotechnol* **80**: 365–380.

Sercu B, Van De Werfhorst LC, Murray J et al. (2009) Storm drains are sources of human fecal pollution during dry weather in three urban southern California watersheds. *Environ Sci Technol* **43**: 293–298.

Seurinck S, Defoirdt T, Verstraete W et al. (2005) Detection and quantification of the human-specific HF183 *Bacteroides* 16S rRNA genetic marker with real-time PCR for assessment of human faecal pollution in freshwater. *Environ Microbiol* **7**: 249–259.

Shendure J, and Ji HL (2008) Next-generation DNA sequencing. *Nature Biotechnology* **26**: 1135–1145.

Shiu SH, and Borevitz JO (2008) The next generation of microarray research: Applications in evolutionary and ecological genomics. *Heredity* **100**: 141–149.

Shyu C, Soule T, Bent SJ et al. (2007) MiCA: A web-based tool for the analysis of microbial communities based on terminal-restriction fragment length polymorphism of 16S and 18S rRNA genes. *Microb Ecol* **53**: 562–570.

Sigler V, and Pasutti L (2006) Evaluation of denaturing gradient gel electrophoresis to differentiate *Escherichia coli* populations in secondary environments. *Environ Microbiol* **8**: 1703–1711.

Simpson JM, Santo Domingo JW, and Reasoner DJ (2004) Assessment of equine fecal contamination: the search for alternative bacterial source-tracking targets. *FEMS Microbiol Ecol* **47**: 65.

Stogbauer A, Berner Z, and Stuben D (2004) Redox control on the isotopic composition of dissolved sulfate in percolating sewage - An experimental study. *Acta Hydrochim Hydrobiol* **32**: 304–315.

Sunagawa S, DeSantis TZ, Piceno YM et al. (2009) Bacterial diversity and White Plague Disease-associated community changes in the Caribbean coral *Montastraea faveolata. ISME J* **3**: 512–521.

ter Braak CJF (1986) Canonical correspondence analysis: A new eigenvector technique for multivariate direct gradient analysis. *Ecology* **67**: 1167–1179.

ter Braak CJF (1995) Ordination. In *Data analysis in community and landscape ecology*. Jongman RHG, ter Braak CJF, and van Tongeren OFR (eds). Cambridge University Press, New York, pp. 91–173.

ter Braak CJF, and Prentice IC (1988) A theory of gradient analysis. *Advances in Ecological Research* **18:** 271–317.

ter Braak CJF, and Smilauer P (2002) *CANOCO reference manual and CanoDraw for windows user's guide: Software for canonical community ordination (version 4.5).* Microcomputer Power, Ithaca, NY, USA.

Thies JE (2007) Soil microbial community analysis using terminal restriction fragment length polymorphisms. *Soil Sci Soc Am J* **71**: 579–591.

Tsiamis G, Katsaveli K, Ntougias S et al. (2008) Prokaryotic community profiles at different operational stages of a Greek solar saltern. *Res Microbiol* **159**: 609–627.

U.S. Environmental Protection Agency (2005) Microbial Source Tracking Guide Document. In. Cincinnati, OH: Office of Research and Development, p. 150.

Vogel JR, Stoeckel DM, Lamendella R et al. (2007) Identifying fecal sources in a selected catchment reach using multiple source-tracking tools. *J Environ Qual* **36**: 718–729.

Wade TJ, Calderon RL, Sams E et al. (2006) Rapidly measured indicators of recreational water quality are predictive of swimming-associated gastrointestinal illness. *Environ Health Perspect* **114**: 24–28.

Weidhaas JL, Macbeth TW, Olsen RL et al. (2010) Identification of a *Brevibacterium* marker gene specific to poultry litter and development of a quantitative PCR assay. *J Appl Microbiol* **109**: 334–347.

White DC (1994) Is there anything else you need to understand about the microbiota that cannot be derived from analysis of nucleic acids? *Microb Ecol* **28**: 163–166.

Wright CF, and Kroese M (2010) Evaluation of genetic tests for susceptibility to common complex diseases: why, when and how? *Hum Genet* **127**: 125–134.

Wrighton KC, Agbo P, Warnecke F et al. (2008) A novel ecological role of the *Firmicutes* identified in thermophilic microbial fuel cells. *ISME J* **2**: 1146–1156.

Wu CH, Sercu B, Van De Werfhorst LC et al. (2010) Characterization of coastal urban watershed bacterial communities leads to alternative community-based indicators. *PLoS One* 5(6): e11285. doi: 10.1371/journal.pone.0011285.

Yergeau E, Schoondermark-Stolk SA, Brodie EL et al. (2009) Environmental microarray analyses of Antarctic soil microbial communities. *ISME J* **3**: 340–351.

Zelles L (1999) Fatty acid patterns of phospholipids and lipopolysaccharides in the characterisation of microbial communities in soil: A review. *Biol Fertil Soils* **29**: 111–129.

Zhang R, Lau SCK, Ki J-S et al. (2009) Response of bacterioplankton community structures to hydrological conditions and anthropogenic pollution in contrasting subtropical environments. *FEMS Microbiol Ecol* **69**: 449–460.

Zoetendal EG, Collier CT, Koike S et al. (2004) Molecular ecological analysis of the gastrointestinal microbiota: A review. *J Nutr* **134**: 465–472.

Chapter 12
Public Perception of and Public Participation in Microbial Source Tracking

Susan Allender-Hagedorn

Abstract Microbial source tracking (MST) is used to determine the source, extent, and content of water pollution; results from MST studies can be used to ameliorate the sources of pollution. If the general public is involved in such studies, MST can be an extremely valuable tool. ("The public" can include a local government official, a parent concerned about child safety, a congressman deciding about funding, and/or a water-resource manager; i.e., "the public" includes everyone with a stake in clean water.) But MST comes with a price tag, and national and international agencies that fund a great deal of MST work can be swayed by public opinion. Local nongovernmental organizations (NGOs) can help or hinder MST testing and efforts to apply the results to improve polluted waters. If the public is not involved in an informed dissemination and/or application of the results, it will be difficult for MST results to lead to improvements to the affected waters. Without a clear perception of MST by this public, cooperation in making the necessary improvements and acceptance of MST can be hindered.

Keywords Stakeholders • Remediation • Human Genome Project • C.P. Snow • Two cultures • Science literacy • Microbial source tracking • Water pollution • Public perception of science • Public opinion • Non-government organizations • TMDLs

12.1 Introduction

Microbial source tracking (MST) to determine the cause(s) of water pollution can be extremely valuable. Knowing the source, extent, and content of water pollution can lead to environmental improvement on a local to a global scale. As one recent article states:

S. Allender-Hagedorn (✉)
Department of English, 207 Shanks Hall, Virginia Tech, Blacksburg, VA 24061-0112, USA
e-mail: hagedors@vt.edu

C. Hagedorn et al. (eds.), *Microbial Source Tracking: Methods, Applications, and Case Studies*, DOI 10.1007/978-1-4419-9386-1_12,

> At present, faecal pollution in water is one of the main causes of health problems in the world, and is associated with several thousands of deaths per day, being a main vehicle of pathogen transmission (Belanche-Muñoz and Blanch 2008).

However, if not-for-profit nongovernmental organizations (NGOs), university outreach entities, and/or local/national governmental agencies are not involved in an informed dissemination and/or application of the results, it will be very difficult, if not impossible, for those MST results to lead to improvements to the environment. Without a clear perception of MST by the public, cooperation in and approval (and funding) of MST studies can be hindered.

> The public should be viewed as a "partner," and a level of trust needs to be created. Developing this style will be [a] major challenge for business leaders as well as university scientists and government regulators (Hardy 1990).

First, "the public" needs to be identified in the context of this chapter. "The public" can include a local government official facing a redistricting or rezoning referendum, a parent concerned about the safety of his/her children swimming at a beach or the drinking water available in her house, a congressman deciding about funding for the Environmental Protection Agency (EPA), or a water-resource manager deciding how to best juggle limited funds and the need to meet US federal TMDL requirements. In other words, "the public" includes everyone with a stake in clean water. There is a growing recognition of the importance of stakeholders, especially those who make public policy and/or environmental regulations but who are not directly involved in the science because

> policy making is better served when informed by recent scientific information … There are also a variety of emerging issues, such as the discovery of pharmaceuticals and personal care products in wastewater and in drinking water streams, that speak to the need for continuing dialog (Schaefer and Bielak 2005).

Clear public perceptions go beyond the traditional studies of science literacy, involving an understanding of the scientific process itself as well as a comprehension of scientific discussions of risks and costs vs. benefits. Public perceptions of personal risk can impact trust in a nation's regulatory processes which can ultimately determine acceptance or rejection of new technologies like MST (Lemaux 1999).

Public opinion will have an impact on the acceptance and application of MST results, mainly through influence on the funding of MST methodology development, targeted field testing, and application of results to counter the pollution identified in the testing. Public opinion can also influence the success of MST-based evidence in the legal system. For an example of the recognition of the importance of public concern, in USA, the Virginia Department of Environmental Quality (DEQ) requires that all Total Maximum Daily Load (TMDL) projects include public meetings to acquaint the public with the goals of a project and with details on how the project will be conducted. The objectives of this chapter are (1) to demonstrate that if the public is involved and understands the efforts of MST researchers, then the scientific research can be better utilized (and funded) and may lead to more substantial environmental improvement, (2) to present briefly several illustrative case studies of successful public participation, and (3) to present a bibliography for further readings on the topic of public perceptions and public involvement relevant to MST.

> There are two truths in this world: one of the laboratory, and other of the media. What people perceive as the truth is truer in a democracy than some grubby little experiment in a laboratory notebook (Koshland 1990).

12.2 Public Perception and Participation

We all have a stake in the effort to have clean drinking and recreational waters, as well as environmental waters that support ecosystem health. Specific stakeholders for an MST effort can include the residents of the local area being examined, local environmental watchdog organizations, and the regional, state, and federal government agencies that sponsor and pay for MST (as well as those agencies whose activities are impacted by MST study results). MST-based scientific achievements can have a big influence on public policy, but only insofar as scientists who practice MST and regulators who hope to benefit from its results learn to negotiate the science-public boundaries and deal with the different stakeholders involved. A priority in such efforts is determining what communication devices are the most effective in which circumstances.

Most scientists would acknowledge that science is at least partially a social enterprise:

> I have never met a pure scientific realist who views social context as entirely irrelevant, or only as an enemy to be expunged by the twin lights of universal reason and incontrovertible observation (Gould 2000).

12.2.1 C. P. Snow and the Science/Public Divide

Over 50 years have passed since C. P. Snow published his work *The Two Cultures and the Scientific Revolution*, where he detailed how the breakdown in communication between scientists and the rest of humanity hinders progress in solving the world's problems. "Alas, Snow's vision [of cooperation between science and the public] has gone unrealized" (Krauss 2009).

Many scientists cite a low level of science literacy on the part of the public as the problem – nonscientist stakeholders are simply not knowledgeable enough to effectively participate (see Sect. 12.2.2 below). On the contrary, many of those stakeholders claim that scientific information is at worst released to the public in a form that is incomprehensible and inaccessible, and at best, in an impractical manner. Both stances have some truth. According to Frank Oppenheimer,

> for many people science is incomprehensible and technology [is] frightening. They perceive these as separate worlds that are harsh, fantastic and hostile to humanity (Oppenheimer 1968).

And

> there is a significant need for a science-based approach to interpret water-monitoring data and to facilitate the rapid transfer of information to water resource managers and the general public (Kaurish and Younos 2007).

However, cooperation between public stakeholders and scientists can shorten the time, cost, and effort in applying scientific research to environmental problems.

> There is growing consensus that to ensure science better informs the decision-making process, researchers and policy/program managers need to understand and respect each other's way of working, culture and operational timelines (Schaefer and Bielak 2005).

The communication should not be a one-way street – public stakeholders can provide valuable information to scientists: "the public has useful knowledge and concerns that need to be acknowledged" (Beecher et al. 2005).

12.2.2 Science Literacy and Public Understanding

There is a strong need for a nonscientific public to understand science to make rational decisions about many aspects of their lives, including public policy:

> If an industrialized nation is democratic, its citizens must ultimately make the crucial choices about the uses of science and technology. A scientifically illiterate citizenry is thus a prescription for disaster (Hobson 2008).

According to Jon Miller (quoted in *Science Daily*),

> no major industrial nation in the world today has a sufficient number of scientifically literate adults. We should take no pride in a finding that 70 percent of Americans cannot read and understand the science section of the *New York Times* (Science Daily 2007).

Miller defined science literacy in a presentation at the 2007 American Association for the Advancement of Science annual meeting as:

> the level of understanding of science and technology needed to function in a modern industrial society. This … does not imply an ideal level of understanding, but rather a minimal threshold level (Miller 2008).

But a deeper understanding of science does not ensure a blanket approval of a particular application of a science. According to George Gaskell, London School of Economics, when interviewed by the on-line *RTDinfo, the Magazine on European Research*:

> Another claim that is often made is that [members of] the public are ignorant of the underlying scientific principles and that if they knew more about science, they would be more supportive. However, in reality the correlation between scientific knowledge and support for science is very low. Without wishing to downplay the value of education in science, the emphasis on knowledge in the context of support for science misses the point, For the public, science is a means to certain ends and one does not need to understand the details of the means (scientific concepts and methods) in order to have a view on the desirability of the ends (the outcomes for society). So what the public is concerned about are questions such as "is this new development going to be good for me, my family and society or not?" "Is there a slippery slope here, if we have X today, where will science be in ten years time?" (Gaskel 2005).

12.2.3 Public Participation

The level of stakeholders' knowledge of MST and involvement in policy and actions resulting from MST research is crucial to the incorporation of scientific principles into decisions made concerning the achievement of clean water. But increasing stakeholders' understanding can be difficult.

> Although there is consensus that strengthening science-policy linkages is desirable, there is little readily available guidance that provides tangible direction for effectively incorporating scientific information into policy, nor for evaluating the effectiveness of such initiatives (Schaefer and Bielak 2005).

On the one hand, scientists need to acknowledge the public's knowledge of local situations and stake in the support (including political and fiscal support) of research and later application of that research to solve problems. On the other hand, stakeholders need to be scientifically literate enough to balance the risks and benefits of research.

> What the public perceives as a risk, why they perceive it that way, and how they will subsequently behave are thus vital questions for policy makers … (Leiserowitz n.d.).

12.3 Funding Scientific Research

One of the most significant arenas where MST science and stakeholders now interact is in the area of funding, from funding of methods development, to testing of those methods, to application of the methods in the field to determine the sources and content of pollution, to application of those results to finally ameliorate the pollution.

First, it is no revelation to anyone that science, and for the focus of this book, environmental science, is expensive (see Chap. 16). For MST, funding entities have to consider costs against potential benefits of the development of testing methodologies, targeted testing to determine source and extent of water pollution, and finally, application of testing results to remediate surface waters and reduce pollution loadings.

12.3.1 Impact of Funding on Research Emphasis

Scientific research and development (R&D) is funded from two main sources: corporations and governmental agencies – both come with obligations that can be problematic. Both sources can have an effect on the direction or emphasis of the research itself. If a particular direction of research cannot find funding, that research will not be pursued.

Corporate research is profit-oriented. It has to be as follows – corporations do not exist to fund projects that do not hold a promise of future returns. So projects

Table 12.1 Total university research and development funding[a,b]

Funding source[b]	2005	2006	2007	% change	% of total
Federal government	93,734	97,701	98,331	0.6%	26.7%
Industry	207,841	227,276	245,027	7.8%	66.6%
Universities/colleges	8,575	9,282	9,866	6.3%	2.7%
Nonprofits	9,905	10,542	11,647	10.5%	3.2%
Nonfederal government	2,950	3,071	3,226	5.1%	0.9%

[a]American Association for the Advancement of Science (2008)

[b]Expenditures in millions of dollars

without a short-term return are rarely funded. Consider so-called "orphan drugs," drugs that treat comparatively rare diseases with low number of patients and with minor expectations for financial returns. Funding has been so hard to find for orphan drugs that the US government created the Office of Orphan Drug Development to provide incentives for related research, including some reduced requirements for prerelease testing (Phillips n.d.)

In USA, nearly two thirds of R&D funding for drug development comes from corporate sources (see Table 12.1); however, most MST funding comes from government agencies that are concerned with water quality.

Corporate costs rise when incomplete data is released, a failure of complete disclosure of scientific results. For example, consider the fraud suit (2004) against Glaxo SmithKiline for failure to disclose study results; "GSK failed to disclose that in at least four of these studies the drug had performed no better than the placebo used as a control" (Jasanoff 2006).

> It is in these situations - when science moves from contexts of production to contexts of public use and application - that pressures for disclosure most often arise. Communication among and certification by peers and coworkers are no longer sufficient to guarantee the quality and relevance of the science in question (Jasanoff 2006).

Government agencies historically have funded a great deal of research and development (R&D) in USA, and that is often with funds channeled through institutions of higher education. But in restricted economic times, even that source is diminishing:

> NSF's Survey of Research and Development Expenditures at Universities and Colleges showed a decrease in federal funding as a share of the academic R&D total, from 64 percent in fiscal year 2005 to 60 percent in fiscal year 2008 (National Science Foundation 2009).

Other governmental entitites, state and local, have made up some of this loss of university research funding: "Funding from all sources outside the federal government grew 8.3% in 2008" (Nagel 2009).

Government-sponsored research can also be result-oriented. However, many US governmental agencies concentrate on funding more basic research where the result is new knowledge or new or improved methodologies that might not have immediate commercial applications (Robson 1993).

When government-sponsored research funding is channeled through universities, costs can rise even further when institutional costs of maintaining research facilities, assurance of accountability, etc. are added in. "For every dollar attracted from external sponsors of university research, institutions must provide at least 40 cents to support the institutional costs of this research" (Association of Universities and Colleges in Canada 2009). These "overhead" costs, @ 40% or higher in USA, are lower in some countries, such as Canada (which is trying to rectify the situation), and are higher in many European countries (Association of Universities and Colleges in Canada 2009).

According to the Organization for Economic Cooperation and Development (OECD), within its 30 member countries, research funding is not much different, with @2/3 science and technology R&D funded by industry, but with 20% by universities, and 10% by government (Jasanoff 2006; Organization for Economic Cooperation 2009).

12.3.2 Funding the Development of Testing Methodologies

Before MST can be widely applied, appropriate methods to track and identify sources of pollution need to be developed, be tested in the laboratory, be applied in the field, and be refined. Some newer methodologies are much more expensive than others, and in most cases desired results require a combination of methodologies. Belanche-Muñoz and Blanch (2008) compared relative costs of different methods (Table 12.2):

It is obvious that some methodologies are more expensive to develop and conduct than others.

Much basic research to develop MST has been national-government-funded or with national funds channeled through university research laboratories. The increasing cost of research can serve to direct a researcher's funding quest to large governmental and/or corporate organizations that potentially have the required funds. Total research funding in most developed countries ranges between 1.5 and 3% of a country's gross domestic product (GDP) or the total value of the goods and services produced by the country's economy; few exceed 4% (Organization for Economic Cooperation and Development 2009). For the US National Science Foundation (NSF), in 2000, the percent of successful grant proposals was approximately 30%. Since then, governmental appropriations for the agency have barely risen enough to account for inflation, and the percent of successful proposals has dropped to 20–25%, reflecting the rising costs of individual projects, stagnant appropriations from Congress, and greater numbers of submissions (Powers 2008).

Table 12.2 Comparative costs of MST methodologies[a]

Methodologies	Comparative costs[b]
Standardized bacterial aerobic analyses	1 (least expensive)
Standardized bacterial anaerobic analyses	2
Enumeration of bacteriophages under aerobic conditions	2
Enumeration of bacteriophages under anaerobic conditions	3
Molecular techniques	5
Phene-plate phenotyping	6

[a]Adapted from Belanche-Muñoz and Blanch (2008)

[b]According to this cost assignment, a solution can be subjectively described as very feasible (overall [combined method] cost in the range 1–5), feasible (5–10), unfeasible (10–15), and very unfeasible (>15)

At present, MST is largely used in basic development (and comparison) of methodologies and in targeted testing. In USA, the TMDL program has been the regulatory "driver" for most MST research.

> All states have been required by the United States Environmental Protection Agency USEPA) to establish Total Maximum Daily Loads (TMDLs) of pollutants in surface water-impaired areas. The TMDL is the largest loading of the pollutant that will allow water quality goals to be achieved (Kern 2002).

See Chap. 14 (this book) for a complete discussion of TMDLs and watershed modeling.

12.3.3 Funding Targeted Testing

Currently, in the US, the methodologies that are still in development are moving out of the laboratory into the field. Federal and local government-funded projects, such as TMDL studies, are achieving results, which in turn inform refinement of the methods. Even before appropriate methodology for a study can be identified, other questions need answers (Table 12.3).

But even after situational-appropriate organism and methodology choices are made, many questions still remain. Virtually, all of the field-based MST studies that have been conducted recently have raised as many questions as they have answered. For example, what role does geographic scale have on source tracking? Is a methodology more effective in only some circumstances? What are the appropriate size and representativeness for databases, and are library-dependent MST methods cost-effective and interpretable in a watershed-wide context? What is the best way

Table 12.3 First steps in microbial source tracking[a]

Steps	Questions to answer
Defining the problem	What knowledge is needed from source tracking?
	With an impaired body of water, is it necessary to differentiate only between human and "all other" sources? (For example, is human fecal contamination, perhaps from faulty septic systems, the main concern?) Is there a need to differentiate between human, livestock, and "everything else?"
	Is a main concern mostly about runoff from cattle feedlots?
	Is it necessary to identify fecal contamination from migratory or coastal birds? Wildlife in general or particular animals such as deer, raccoons, etc.? Pets such as cats and dogs?
Choosing the MST microorganism	Will the study deal with fresh or salt water? (In general, most source tracking studies use one particular microorganism to indicate the source of fecal pollution. *E. coli* and *Enterococcus* are widely used, but others are being evaluated.)
Choosing the MST method	Will the study use a library-dependant method or a library-independent method? (Library-dependent methods can be less expensive but require more time to establish a baseline data base. Library-independent methods tend to be more expensive to develop and apply.)

[a]Allender-Hagedorn (2004)

to determine the number and size of water samples and frequency of sampling? These questions are addressed in Chap. 3, dealing with library-dependent methods.

12.3.4 Funding Pollution Remediation

Once the levels and sources of water pollution have been identified, then procedures to remediate the pollution need to be chosen and put into action, but acting upon the results of MST studies and addressing the generally deteriorating condition of water and wastewater infrastructure will also be very expensive:

> According to recent estimates, the level of investment that will be required over the next 20 years to repair, replace, or upgrade aging facilities; accommodate the nation's growing population; and meet new quality standards will be very large, up to $1 trillion. Moreover, following the terrorist attacks of September 11, 2001, both drinking water and wastewater utilities may have to make additional investments to increase the security of their operations (U.S. General Accounting Office 2002).

Local entities also have to consider these costs:

> According to the Environmental Protection Agency (EPA) and water utility industry groups, communities will need an estimated $300 billion to $1 trillion over the next 20 years to repair, replace, or upgrade aging drinking water and wastewater facilities; accommodate a growing population; and meet new water quality standards (U.S. General Accounting Office 2002).

Many factors that cause increasing costs include rising labor and material costs, the complexity of research questions, the development of technology to address those questions, and the changing nature of research itself (National Academy of Engineering 1992)

12.3.5 Funding and Effective Application of MST Research

Using MST to identify sources of pollution does not by itself remediate the pollution. Currently, the US TMDL program ends only with documents containing the data resulting from MST studies, which are submitted to EPA. The ultimate goal of the TMDL program is to apply the data in the documents to remedy the pollution, but currently the program is at the information-gathering stage. Since MST is still in development and testing, there are not many studies that reflect the entire spectrum of the process, from identification of a water pollution problem, through targeted MST field testing, to remediation of a problem using the MST data collected. (However, several are discussed briefly in the Case Studies section of this chapter, below, and in Chaps. 14 through 22, this book.) But this is an important part of the process where science passes from a scientific arena to a public one.

> [There are] important differences between regulatory science and research science. Regulatory science, which provides the basis for policy, routinely operates with different goals and priorities and under different institutional and temporal constraints from 'science' done in academic settings without implications for policy (Jasanoff 1995).

Below is a study of how a "big science" project was influenced by public perceptions, followed by examples of international, federal, and local recognition of the importance of public input on the conduct of science.

12.4 Effects and Recognition of Public Perception/ Participation Issues

12.4.1 The Human Genome Project (HGP): An Example of Big Science Public Rhetoric

The Human Genome Project (HGP) was the name of a "big science" program to map and decode the entire complement (genome) of human genetic material. This costly program was projected to revolutionize the conduct of biological sciences in the twenty-first century. The HGP is an example of how public involvement and language led to support and massive funding distributed to many different institutions. The public was involved from the beginning in the discussions of the formation of the project.

> When a research initiative like the Human Genome Project relies on language, in addition to science, to puzzle out and present problems, we need to understand how ideas, issues, and questions are subject to the shaping power of the very people who present them (Ribidous 1997).

Science is expensive – as previously discussed, funding for science can determine which initiatives go forward and which are abandoned or fail.

> Recent headlines show that as conservative thinkers and moralists become more influential, policies based on these cultural influences can proscribe some research avenues, such as stem cell sources (Allender-Hagedorn and Ruggiero 2009).

When the HGP was presented to the US Congress for funding, it had stiff competition from other big-ticket, big-science ventures popular with the public: the NASA space station and the superconducting supercollider. Of these three, only the HGP achieved status in 1990 as a federally funded program. The project's success was in a great part due to the way the project was presented to the members of Congress (the "public" that voted on the appropriations).

> The public (including its representatives, members of Congress) generally present a low level of comprehension of basic scientific facts even at the same time those facts are all being taught at the elementary through high school level (Allender-Hagedorn 2001).

Successful argumentation/appeals for funding utilized commonly understood and for the most part nonscientific references to potential health benefits and to the metaphors of "mapping" the "book of life," as well as appeals to the widely held US ethic of exploration of an "unknown territory." The need to involve the public at large in science projects was again recognized in the *Nature* article announcing the initial sequencing of the human genome:

> But the science is only part of the challenge. We must also involve society at large in the work ahead (International Human Genome Sequencing Consortium 2001).

12.4.2 Science Citizen Juries

Documents from the EU, which has a tradition of introducing small groups of the nonscientific public to study the effects of an intense and in-depth introduction to certain scientific controversies in clear and understandable language, clearly illustrate a recognition of the importance of public involvement. Results from these European science citizen juries or science shops illustrate that people do lose their fear of the science. But they also lose fear of the authority of science. In general, they do not blindly accept the proclamations of benefit from science products but ask harder (and more scientifically grounded) questions than before and are more apt to demand a say in the setting of science policy (Evans and Durant 1995).

> Determination of public science policy should involve recognition of the social impact of science, not just reliance only on public knowledge of a list of scientific facts (Hagedorn and Allender-Hagedorn 1997).

Australia is also actively utilizing citizen juries so that

> decision-makers who watch a citizens jury project in action or listen to a jury's recommendations are able to learn what an informed public wants, and why. This information can be an invaluable resource for elected officials and other decision-makers at the local, state, and national levels (URP Toolbox n.d.)

Advantages of citizen juries include involvement of citizens in a "high-quality dialog" about an important issue and they also create informed citizen input to government and regulatory officials. The jury process is an effective way to involve citizens in developing a thoughtful, well-informed solution to a public problem or issue (Jefferson Center 2002).

12.4.3 Ontario's Water-Quality Monitoring Programs

Ontario, Canada's Ministry of the Environment clearly illustrates a concern over public involvement in water-quality issues. The ministry conducts a massive water-quality monitoring program, covering groundwaters, rivers and streams, and inland lakes. Data recovered are used to measure progress in restoring and maintaining Ontario clean waters and to detect emerging issues that affect that water. Some of the issues identified include toxic substances, elevated nutrients, acid deposition, and effects of climate changes. While results are published in peer-reviewed scientific articles and technical reports,

> the information presented in technical publications is not easily accessed by decision and policy makers and interested members of the public. Plain-language reports (e.g., *Guide to Eating Ontario Sport Fish* and the *Minister's Annual Report on Drinking Water*) are produced by the Ministry to communicate monitoring results to nontechnical audiences (Todd 2008).

Following are a few case studies of MST taken to the remediation level and that included some level of public input.

12.5 Case Studies

12.5.1 Millwood, Virginia: Successful Environmental Group Cooperation and Local Government Sponsorship to Apply MST Data to Remediate Pollution

The village of Millwood, in Clarke County of northwest Virginia, was an unsewered community of approximately 130 people in 82 homes on either lots or small farms.

Millwood is a predominantly minority community located close to more affluent communities served by wastewater treatment facilities. A stream, Spout Run, passes through the center of the village. Most of the homes were over 40 years old, all obtained water from individual wells, and a few homes did not have indoor plumbing. Septic tanks and drainfields (frequently of unknown location, age, or operational status), outhouses, or cesspools provided waste disposal. A 1999 survey conducted by the Virginia Department of Health classified 39 homes (48%) as having unsatisfactory waste treatment systems, and 33 of these 39 homes were located on property that bordered Spout Run.

Ironically, Millwood also is the site of Carter Hall, a colonial plantation that is now the location of the headquarters of Project Hope. In the eighteenth century, Millwood was the slave community that supported the plantation.

Spout Run stream samples for a MST study were collected in 1999 and 2000, as were drinking water samples (well water) from Millwood homes where the owners elected to participate in a voluntary water-testing program. The study identified fecal bacteria of human origin as one source of the pollution in Spout Run (as the stream passed through Millwood) and in some of the drinking water samples from residences. MST results showed that a substantial portion of the fecal pollution in the streams and wells was human in origin (Graves 2000).

Stakeholders have been involved in the project in Clarke County from the start, including local government officials, The Friends of the Shenandoah River, and the Millwood Community Association. This scientific study directly led to the successful application for a $1.76 million Community Development Block Grant from the US Department of Housing and Community Development. The capacity of the wastewater treatment plant in a nearby town was increased, and a sewer line with pumping stations was extended to Millwood and became operational in 2003–2004. A total of 59 homes, five businesses, three churches, and two community centers were placed on the sewer system, and a community drinking water system that serves most of the residences was also established.

Follow-up studies after remediation efforts showed zero well samples positive for fecal coliforms, and the percentage of stream samples that exceeded regulatory thresholds dropped below 10%. MST results showed that none of the remaining stream pollution was human in origin (Graves et al. 2002; Graves 2000).

12.5.2 Virginia Beaches: Combining Environmental Activism and MST Data to Remediate Pollution

In 2004 a library-dependent MST method, antibiotic resistance analysis, was used to classify isolates of *Enterococcus* from humans as well as dogs, birds, and other wildlife into source categories. A chemical method to differentiate between human and nonhuman sources of pollution, fluorometry (which detects detergent optical brighteners in sewers and drainfields) was added as well (Hartel et al. 2007). When the 2004 MST results found human sources of pollution at several beaches, local officials (Hampton, Newport News, and Hampton Roads Sanitation District) used the results to identify probable pollution sources and took steps to remediate the problems (Dickerson et al. 2007).

The area was directly affected by the September 2003 Hurricane Isabel, but sampling events in 2005 and 2006 MST illustrated success, i.e., reduction in the level of human source pollution and improvement in water-quality conditions at beaches that had conducted posthurricane restoration projects.

> This demonstrated the success of using MST to identify sources of fecal pollution in 2004, performing remediation to remove the origins of the pollution in 2005, and then following-up with MST in 2006 to prove that the sources found in 2004 and 2005 were no longer present in 2006. This is the first report where MST results indicated pollution from a particular source was present (human-origin sewage), the origin of the pollution was then located, steps were taken to eliminate the pollution, and subsequent MST results indicated the success of those remediation efforts (Dickerson et al. 2006, 2007).

However, one beach, Fairview, continued to show frequent beach advisories, in part because of the earlier restoration efforts which were destroyed by tropical storm Ernesto in September, 2006. With support of and prompting by local environmental groups, the beach was closely studied. A near-by nature preserve was eliminated as a potential source, as were holiday boating events on the river (Dickerson 2006).

However, one particular drainpipe, identified early by local environmental groups, continued to flow in spite of a drought in the summer of 2007, and MST confirmed that the drainpipe was a source of high numbers of enterococci from human sources by using the *esp* gene of *Ent. faecium.* Work by the Virginia Department of Health and local officials and environmental groups located homes on old septic systems that were then moved to community sewer systems, and advisories have been reduced.

> While progress is slowly being made at Fairview, the drainpipe seems to be the only real issue left, and digging up the parking lot that the drainpipe is under to follow the pipes connections, wherever they might lead, appears to be the final solution that is needed at Fairview Beach (Dickerson et al. 2007).

12.5.3 Successful Water-Quality Citizen Jury in Australia

One example of a successful citizen jury in Ipswich, Queensland, Australia involved the water quality issues of the Bremer River catchment area. The area saw a growing population and a deteriorating ecosystem. In order to receive quick feedback and inform the public about related issues and to disseminate basic scientific information, 24 local citizens (nonscientists) were chosen to participate. Any management plan developed would need community and industry acceptance, and local government needed to know what measures would meet community needs at an acceptable level of funding. Key questions included what further resources should/could be used in the area, what were acceptable uses of these resources, and how much were citizens willing to fund the processes. After the citizen's jury was concluded,

> "Jurors' comments indicated a higher awareness of the need for community environmental education. They did not come up with 'yes/no' answers, but showed they understood the wider implications." (URP Toolbox n.d.).

12.6 Conclusions

According to George Gaskel,

> Science is not an island. The surveys show that science and technology does not attract the interest of a narrow group of people who are particularly trained in [a] single issue. Rather, those who are most engaged with science and technology tend to be 'generalists'—people with interests in a wider range of public affairs, including politics ... Science is not insulated from the way ... problems are managed ... public opinion on science and technology is likely to derive in part from views about the credibility of wider political and societal institutions (Gaskel 2005).

MST encompasses a scientific set of tools that can identify sources, extent, and content of microbial water pollution; these results can be used to then ameliorate the pollution. All members of "the public" are stakeholders in the need for clean water for both health and recreational use. This nonscientifically trained "public" has members ranging from individual citizens to local, national, and international environmental groups. It can also include local, federal, and international governmental and regulatory entities. But if all are not considered and (where applicable) included in the planning and conduct of MST studies and the application of the results of those studies, there is a great chance that the achievement of clean water can be drastically affected.

The public can affect MST in many ways, e.g., from influencing funding sources, identifying potential study sites, disseminating understandable information to localities under study, and providing cooperation on the application of remediation efforts. The public may even take part in such efforts. Without an informed public, it will be more difficult for MST to lead to improvements and avoid degradation of environmental waters. Without a clear perception of MST by the public, cooperation in and approval of MST can be hindered.

References

Allender-Hagedorn S (2001) Arguing the genome: A topology of the argumentation behind the construction of the Human Genome Project. Dissertation, Virginia Tech

Allender-Hagedorn S (2004) Evaluating known source tracking libraries: Artificial clustering. Environ Detection News 2(1):1–3

Allender-Hagedorn S, Ruggiero CW (2009) Connecting popular culture and science: The case of biotechnology. In: Hayhoe GF, Grady HM (eds) Connecting people with technology: Issues in Professional Communication. Baywood, Amityville

American Association for the Advancement of Science (2008) Science and Policy: R&D Budget and Policy Program: Guide to R&D funding data – Total US $&D (1953). http://www.aaas.org/spp/rd/guitotal.html. Accessed 6 Mar 2010

Association of Universities and Colleges of Canada (2009) Building a competitive advantage for Canada. http://www.aucc.ca/_pdf/english/reports/2009/prebudget_11_18_e.pdf. Accessed 11 Mar 2010

Beecher N, Harrison E, Goldstein N et al (2005) Risk perception, risk communication, and stakeholder involvement for biosolids management and research. J Environ Qual 34:122–128

Belanche-Muñoz L, Blanch AR (2008) Machine learning methods for microbial source tracking. Environ Modelling Software 23:741–750

Dickerson JW, Hagedorn C, Hassell A. (2006) Pathogen research symposium: Pathways and monitoring in natural and engineered systems. Symposium sponsored by Virginia Water Resources Research Center, Blacksburg, VA, VWRRC Special Report SR32–2006

Dickerson JW, Hagedorn C, Hassall A (2007) Remediation of human-origin pollution at two public beaches in Virginia using multiple source tracking methods. Water Res. 41:3758–3770

Evans G, Durant J (1995) The relationship between knowledge and attitudes in the public understanding of science in Britain. Pub Understand Sci 4:57–74

Gaskel G (2005) Interview: Public opinion in the science equation. RTDinfo 51: http://ec.europa.eu/research/rtdinfo/special_euro/01/article_3148_en.html. Accessed 24 May 2010

Gould SJ (2000) Deconstructing the 'science wars' by reconstructing an old mold. Science magazine, 287 (14 Jan 2000): 253–261

Graves AK (2000) Determining sources of fecal pollution in water for a rural Virginia community. Thesis, Virginia Tech

Graves AL, Hagedorn C, Teetor A et al. (2002) Determining sources of fecal pollution in water for a rural Virginia watershed. J Environ Qual 31:1300–1308

Hagedorn C, Allender-Hagedorn S (1997) Issues in agricultural and environmental biotechnology: Identifying and comparing biotechnology issues from public opinion surveys, the popular press and technical/regulatory sources. Pub Understand Sci 6:233–245

Hardy DR (1990) Towards a consensus. In: MacKenzie DR, Henry SC (eds) Biological monitoring of genetically engineered plants and microbes. Proceedings of the Kiawah Island Conference, Kiawah Island, 27–30 Nov 1991

Hartel PG, Hagedorn C, McDonald, JL et al (2007) Exposing water samples to ultraviolet light improves fluorometry for detecting human fecal contamination. Water Research 41:3629–3642

Hobson A (2008) The surprising effectiveness of college scientific literacy courses. Physics Teacher 46:404–406

International Human Genome Sequencing Consortium (2001) Initial sequencing and analysis of the human genome. Nature 409(6822):915

Jasanoff S (1995) Procedural choices in regulatory "Science." Tech in Soc 17(3):279–293

Jasanoff S (2006) Sequestered science: The consequences of undisclosed knowledge article: Transparency in public science: Purposes, reasons, limits. Law & Contemp Probs 69(3):21–46

Jefferson Center for New Democratic Processes (2002) The citizen jury process, http://www.jefferson-center.org/citizens_jury.html. Accessed 23 May 2010

Kaurish FW, Younos T (2007) Developing a standardized water quality index for evaluating surface water quality. J Am Water Res Assoc 43(2):533–545

Kern J (2002) Application of source tracking results to performing TMDLs. Environ Detection News 1(1):1–3

Koshland Jr. DE (1990) Two plus two equals five. Science 247:1381

Krauss LM (2009) An update on C. P. Snow's "Two Cultures." Sci Am 31 Sep 2009, http://www.scientificamerican.com/article.cfm?id=an-update-on-cp-snows-two-cultures. Accessed 5 Mar 2010

Leiserowitz A (nd) International public opinion, perception, and understanding of global climate change. http://environment.yale.edu/uploads/IntlPublicOpinion.pdf. Accessed 4 Mar 2010

Lemaux PG (1999) The interplay of public perception issues and federal regulatory policy in agricultural biotechnology: A U.S. perspective. http://www.cdesign.com.au/proceedings_abts1999/papers/P_G_Lemaux2.pdf. Accessed 4 Mar 2010

Miller JD (2008) The impact of college science courses for non-science majors on adult science literacy. Paper presented to symposium on the critical role of college science courses for non-majors, annual meeting AAAS, San Francisco, 18 Feb 2007. In Hobson A (2008) The surprising effectiveness of college scientific literacy courses. Physics Teacher 46:404–406

Nagel D (2009) NSF: Federal role in academic R&D funding has diminished. http://campustechnology.com/articles/2009/10/06/nsf-federal-role-in-academic-r-d-funding-has-diminished.aspx. Accessed 11 Mar 2010

National Academy of Engineering, Institute of Medicine (1992) Responsible science: vol I: Panel on scientific responsibility and the conduct of research. http://www.nap.edu/openbook.php?record_id=1864&page=67. Accessed 6 Mar 2010

National Science Foundation (2009) Federal government's share of university R&D funding drops to 60 percent: Press release 09–182. http://www.nsf.gov/news/news_summ.jsp?org=NSF&cntn_id=115710&preview=false. Accessed 10 Apr 2010

Oppenheimer F (1968) Rationale for a science museum, http://www.exploratorium.edu/frank/rationale/index.html. Accessed 5 Mar 2010

Organization for Economic Cooperation and Development, Directorate for Science, Technology and Industry (2009) Main science and technology indicators (MSTI): 2009/2 edition. http://www.oecd.org/document/26/0,3343,en_2649_34451_1901082_1_1_1_1,00.html. Accessed 5 Mar 2010

Phillips T (nd) About.com guide: What is orphan drug status. http://biotech.about.com/od/faq/f/orphandrugs.html. Accessed 3 Mar 2010

Powers S (2008) National trends in science and engineering funding and research priorities: Updates from the ASEE Engineering Research Council. http://www.asee.org/conferences/erc/2008/Presentations.cfm. Accessed 7 Mar 2010

Ribidous CA (1997) "The Human Genome Project: Novel approaches, probable reasoning, and the advancement of science." In: Advances in the history of rhetoric: The centrality of rhetoric, a collection of selected papers presented at ASHR conferences in 1997 vol2, Am Soc Hist Rhetoric, Fort Worth

Robson MT (1993) Federal funding and the level of private expenditure on basic research. South Econ J 60(1):63–71

Schaefer KA, Bielak AT (2005) CCME: Linking water science to policy: Workshop series final report: Overview and lessons learned. http://www.ccme.ca/assets/pdf/water_wkshp_smryrpt_2004_e.pdf. Accessed Mar 3 2010

Science Daily (2007) Scientific literacy: How do Americans stack up? http://www.sciencedaily.com/releases/2007/02/070218134322.html. Accessed 13 Mar 2010

Todd A (2008) Strategies for reporting results from Ontario's water quality monitoring programs. In: Canadian Council of Ministers of the Environment, CCME National science and technology workshop on water quality monitoring PN 1419, 5–6 Feb 2008, Fredericton. http://www.ccme.ca/assets/pdf/wqm_workshop_2008_02_e_pn1419.pdf. Accessed 10 Apr 2010

URP Toolbox. (nd) Welcome to the citizen science toolbox. https://app.secure.griffith.edu.au/03/toolbox/index.php. Accessed 20 May 2010

US General Accounting Office (2002) Water infrastructure: Information on financing, capital planning, and privatization (GAO-02–764). http://www.gao.gov/new.items/d02764.pdf. Accessed 5 Mar 2010

Further Reading[1]

Bauer M, Durant J, Evan G (1994) European public perceptions of science. Int J Public Opinion Res 6(2):163–186

Collini S (1993) Introduction. In: Snow CP (ed) The two cultures (Canto). Cambridge University Press, Cambridge

Dickson D (1988) The new politics of science (with a new preface). University Chicago Press, Chicago

Environmental Protection Agency (2007) Report of the experts scientific workshop on critical research needs for the development of new or revised recreational water quality criteria Report: EPA-823-R-07. Workshop at Airlie Center, Warrenton, 26–30 Mar 2007

Hayhoe GF, Grady HM (eds) (2009) Connecting people with technology: Issues in professional communication. Baywood Publ Co, Amityville

Jensen ET, McLellan SL (2005) Beach closings: Science versus public perception. ActionBioscience.org. http://www.actionbioscience.org/environment/jensen_mclellan.html Accessed 10 Mar 2010

Lakes, rivers, streams and ponds partnership (2007) What's our water worth? The economic impact of potential decline in New Hampshire water quality: The link between visitor perceptions, usage and spending, Phase IV report, May 2007, http://www.nhrivers.org/documents/Econ%20Study%20Brochure.pdf Accessed 9 Mar 2010

Nelkin D. (1987). Selling science: How the press covers science and technology. W. H. Freeman, Cranbury

Thomas L (1977) The medusa and the snail. Penguin Books, NY

Declaration on science and the use of scientific knowledge. (1999) Proceedings of the science for the twenty-first century, Budapest, Hungary 26 June-1 July, http://www.unesco.org/science/wcs/eng/declaration_e.html Accessed 11 Mar 2010

Wynne B (1995) Public understanding of science. In: Jasanoff S et al. (eds) Handbook of science and technology studies. Sage Publ, Thousand Oaks

[1] Below is a short list for further reading on public perceptions of science in general and on perceptions of environmental science in particular.

Chapter 13
Use of Microbial Source Tracking in the Legal Arena: Benefits and Challenges

Christopher M. Teaf, Michele M. Garber, and Valerie J. Harwood

Abstract Public health risks attributable to microbial pathogens are of serious concern and their evaluation is necessary to provide assurances of safety for food, drinking water supplies, recreational surface waters, beneficial water reuse (e.g., irrigation of areas accessible to the public), health care, and other applications. Litigation or other legal processes that may arise from individual infection/illness claims, from claims of damage to a public resource, or from criminal cases, often are practically limited by the ability to identify an unambiguous source of a putative infectious agent or pollution source with a high degree of certainty. Applied in rigorous fashion, microbial source tracking (MST) has the potential to assist in identification of likely bacterial or viral agents with both accuracy and precision. Satisfaction of the relevant contemporary scientific criteria for demonstration of a persuasive linkage is not necessarily sufficient to satisfy applicable legal criteria. A working understanding of acceptable requirements for both the technical and legal audiences is useful to scientists who seek to apply the principles and practices of MST within the legal process.

Keywords Risk • Public health • Legal • Pathogens • Safety • Litigation

13.1 Why Do We Care? Public Health and Legal Issues of Microbial Pathogens

Human fecal pollution alone may spread dangerous bacterial and viral pathogens, such as those causing shigellosis, hepatitis, and rotavirus infections, while other human pathogens, such as *Cryptosporidium parvum*, *Giardia lamblia*, *Salmonella* spp., and *E. coli 0157:H7* all have been associated with animal fecal pollution as

V.J. Harwood (✉)
Department of Integrative Biology, University of South Florida, 4202 E. Fowler Avenue, Tampa, FL 33620, USA
e-mail: vharwood@usf.edu

C. Hagedorn et al. (eds.), *Microbial Source Tracking: Methods, Applications, and Case Studies*, DOI 10.1007/978-1-4419-9386-1_13,

well as human sources (USEPA 2005; SWCSMH 2006; Edge and Schaefer 2006). Thus, to the extent possible, identification of pollution source(s) in a specific case is essential to prevention, interdiction, or remediation of the impact and ultimate attribution of responsibility from a legal or regulatory perspective. Identifying the sources of water pollution is crucial for assessing public health risks, deciding what management strategies to use in each individual situation, and applying such information in the legal arena (Kon et al. 2009).

The health risks associated with microbial contamination of various media can range from self-limiting, minor gastrointestinal distress to catastrophic illness and death in individuals or populations. The spectrum and severity of disease typically is related to the intensity of the contamination, the route, frequency/duration of exposure, single or multiple microbial species present, and the speed with which the event is recognized, understood, and corrected. This is particularly applicable in situations involving fecal contamination of water bodies or the environment.

Edge and Schaefer (2006) defined *microbial source tracking* (MST) as an approach "based on comparing the similarity of organisms collected from nearby pollution sources in order to make inferences about the likely source of fecal contamination." Hagedorn et al. (2009) similarly defined it as "a method used to determine the source of *fecal bacteria* and establish whether fecal bacteria are being introduced into water bodies through human, wildlife, agricultural or pet wastes." Viruses and protozoa may also be targets of MST methodologies (Blanch et al. 2006; Griffith et al. 2003; Jellison et al. 2009; McQuaig et al. 2009). MST techniques can be applied in the area of microbial forensics that "builds on traditional microbiology and epidemiology within a legal framework" (McEwen et al. 2006). In order to perform a thorough microbial forensic analysis, one must take into account sample handling, collection preservation, method selection, casework analysis, interpretation of results, validation, and quality assurance (Budowle et al. 2005; McEwen et al. 2006). This technical discipline developed in response to the need to examine biological agents used for criminal purposes so that results could be presented as evidence in legal proceedings (Salyers 2004; Harmon 2005; Pattnaik and Jana 2005). Microbial forensics, as distinguished from more conventional forensics typically involving only one species (humans), is much more complex due to the large number of potential bacterial and viral species implicated, and the vast microbial dynamics involved (McEwen et al. 2006). The relationship and rapid development of these related disciplines have expanded the tool chest for environmental or public health managers and decision-makers, including the legal system culminating in the courts.

Notwithstanding the technical complexity and solid biological foundation of the MST and microbial forensics field, its relative novelty is illustrated by the fact that microbial forensics evidence was first introduced in a criminal trial in 1998 (Koblenz and Tucker 2010). In that instance, a judge concluded that the evidence satisfied necessary scientific and evidentiary criteria, as defined by what is known colloquially as the "Daubert Decision" (Daubert 1993; see Sect. 13.4). More recently, techniques of microbial forensics and source tracking have been employed in the scientific and legal investigations related to a bioterrorism event that occurred in 2001 (McEwen et al. 2006; Koblenz and Tucker 2010), in which letters laced with *Bacillus anthracis* spores caused lethal cases of anthrax in several individuals in the USA. Events similar to this clearly can cause great disruption, create fear, and affect economic well-being (Budowle et al. 2005). Fecal contamination to water systems also can involve significant economic losses and serious disturbances of natural and ecological environments, as well as protracted legal proceedings.

Significant gaps still remain in both scientific understanding and operational capability, but microbial forensics can play a major role in the attribution and deterrence of biological warfare and terrorism (Koblenz and Tucker 2010). However, because the object of microbial forensics is not assessment of a man-made device but rather living microorganisms that are self-replicating, evolving, and can usually be obtained from multiple sources, the attribution of a biological attack is extremely challenging (Koblenz and Tucker 2010), as is the presentation of scientific evidence in court.

The prevailing sentiment, with appropriate limitations, is that having a consistent standard of proof modeled after the Daubert test will permit the microbial forensic community to provide decision-makers from the regulatory or judicial community with the most consistent and reproducible degree of technical information with which to conduct their deliberations (Pattnaik and Jana 2005; Koblenz and Tucker 2010).

In those instances, the available techniques, and the practical ability to address issues of environmental, public health, agricultural, veterinary, or criminal significance, often hinge on the reliable identification and possible quantification of the microbial source of pollution and/or pathogens. This is particularly true when the microbial contamination originates from nonpoint sources (e.g., agricultural or urban runoff, wildlife) as opposed to point sources (e.g., effluents from wastewater treatment plants, industrial sources). As discussed elsewhere in this volume, a variety of MST tools and approaches are available, all of which are influenced by the complexity of environmental samples and a host of variables affecting microbial survival and growth (Simpson et al. 2002; Stoeckel 2005; Stoeckel and Harwood 2007; USEPA 2005). The American Academy of Microbiology (AAM 2003) has addressed the scientific underpinnings of microbial forensics, and both the US Environmental Protection Agency and Environment Canada more recently have released detailed volumes from a number of consensus workshop meetings on MST regarding the methods, strengths, and limitations inherent to environmental microbiological investigations (USEPA 2005; Edge and Schaefer 2006). A variety of chemical, culture-based, isolate-by-isolate (library-dependent), sample-specific, and host-specific (library-independent) approaches, may be applicable alone or in combination with one another in particular circumstances (Harwood et al. 2003; Ritter et al. 2003; Indest et al. 2005; Seurinck et al. 2005; Stoeckel and Harwood 2007; Duran et al. 2009), as discussed earlier in this volume.

One of the major assumptions of library-dependent MST (Chap. 3) is that the clonal composition of populations is stable over time and in the environment. However, the intestinal flora of humans and other mammals is an assemblage of populations with continuous turnover of individual strains, which can cause the application of MST to be all the more difficult (Seurinck et al. 2005). The impacts of temporal and geographical variability and the stability of the used markers should be considered when applying MST (Hartel et al. 2002; Seurinck et al. 2005). However, library-independent methods (Chap. 4) generally rely on fewer assumptions than library-dependent methods and may, therefore, prove to be more useful from the standpoint of legal challenges.

Kuntz et al. (2003) developed a targeted sampling protocol to utilize in the process of watershed characterization prior to implementing MST. Using this approach, they found it was much easier and less expensive to identify the sources of fecal contamination. Stapleton et al. (2007) set out to evaluate the utility of MST and concluded that, at the time of their investigation, the techniques did not provide sufficient quantitative source apportionment for their particular study catchment area and, thus, did not meet the standards required to be a useful forensic tool. Nevertheless, those authors did acknowledge that continued research and method development could increase the success of MST in future studies, evidence from which may be utilized in the presenting of forensic evidence in the courtroom. Numerous other authors have concluded in a variety of cases for a particular watershed or geographic area that MST did, in fact, produce useful and applicable results (Graves et al. 2007; Reischer et al. 2008; Hundesa et al. 2010; Weidhaas et al. 2010). Reischer et al. (2008) emphasized the importance of including the hydrological catchment dynamics when applying the quantitative MST methods.

13.2 Why Does It Matter Where They Came From?

Whether the issue is the source of an infectious outbreak, or the source of mail items contaminated with a bioterrorism agent, or a potential source of environmental contamination, or standard of proof in a court of law, it is essential to identify as accurately as possible the origin of the microbial agent. This permits decisions to be made with regard to actions dedicated to correcting the problem, or to allocate legal, remedial, and fiscal responsibility. Simply put, it is difficult or impossible to solve a problem if you do not know where or how that problem started and how it has evolved over time. Furthermore, it is not possible to assign responsibility for damage to the environment or health unless the dominant contamination source is known. As with any environmental forensic tool, MST must be able to demonstrate that it can be used to accurately represent and reproduce the true source(s) of environmental contamination.

Unger (2008) noted that municipalities and other watershed managers require robust science-based tools to more effectively and efficiently deal with the threat to human health from outbreaks of waterborne disease. Beach closures, boil-water advisories, and shellfish harvest closures cost Canadian and American tax payers millions of dollars yearly. "With MST tools to more clearly identify the source of pollution, water quality decision makers can make more strategic investments to better target reduction in contamination in a cost-effective manner" (Unger 2008).

MST methods, along with indicator organism enumeration and hydrological surveys of a region, have been shown to provide valuable information about the sources of fecal microbial contaminants, which can then aid in the management and remediation of impaired water bodies (Griffith et al. 2003; Bitton 2005; Brownell et al. 2007; Korajkic et al. 2009). Demonstration of such patterns is an essential component of establishing the foundation to support expansion of more precise and probative forensic methods, and construction of reproducible predictive models (Blanch et al. 2006). Although standard fecal indicator organisms (e.g., *E. coli*, coliforms) have received extensive technical and legal attention due to the large database of routine water-quality investigations, the extreme genetic heterogeneity and broad host range of those taxonomic groups provide the impetus to develop other bacterial probes (Lasalde et al. 2005). For example, in Korea, Lee et al. (2009) demonstrated the promising use of bacteriophages, which were originally proposed back in the early 1980s (Osawa et al. 1981; Tartera et al. 1989; Hsu et al. 1995; Puig et al. 1999) as the targets of MST. This study and others cited throughout this volume also illustrate that different MST techniques are being applied all over the world.

It is critical to improve on the existing phenotypic and genotypic MST techniques to develop technologies that are reliable, cost-effective, practical, and easily applicable to water environments (Duran et al. 2006, 2009). This will depend highly on increasing our fundamental understanding of complex interactions between microorganisms and their primary hosts, including their behavior once they have reached the water environment (Duran et al. 2009). Reliability assessment of the chosen protocol must be performed prior to and during any attempts to apply MST protocols to a given environmental setting (Stoeckel et al. 2004; Stoeckel and Harwood 2007). Identifying the sources of water pollution is crucial for assessing public health risks, deciding what management strategies to use in each individual situation, and applying such information in the legal arena (Kon et al. 2009).

13.3 Benefit of MST Information to Demonstration of Cause and Effect Relationships

Despite often being considered a "novel" technology, the use of MST is rapidly becoming widespread as more researchers are becoming aware of its considerable potential (Hagedorn et al. 2009). One useful application of MST methods has been the assessment and evaluation of agricultural impacts related to livestock. Beef (Hagedorn et al. 1999; Griffith et al. 2003; Hundesa et al. 2010), poultry (Scott et al. 2003; Buchan et al. 2001; Weidhaas et al. 2010), and wildlife (Hartel et al. 2003; Indest et al. 2005; Scott et al. 2004; Jiang et al. 2007) have been investigated, and methods have been developed to provide reliable predictive information regarding microbial source contributions. State and local officials often use MST results to recommend best management practices (BMPs) for agricultural animals to lower indicator organism populations and meet total maximum daily load (TMDL) goals (Stoeckel et al. 2004; PBS&J 2008). Efforts are also being made in the research arena to adapt new and technically/legally defensible methods for assessing and managing risk posed by microbial pollution (Indest et al. 2005).

It is important to apply the correct MST methodology to each unique situation. As demonstrated by Moore et al. (2005), the library-dependent MST techniques of ribotyping and antibiotic resistance analysis were not well suited for determination of the source(s) of fecal pollution in large urban watersheds. At the time of their investigation, it was believed that these techniques were more applicable to water bodies that were impacted by a very limited number of potential sources in a limited geographical area and over a short time period (Moore et al. 2005).

Advances in MST continue to occur. Kon et al. (2009) concluded that environmentally adapted strains (EAS) of *E. coli* should be included as one of the potential sources in MST studies where that indicator organism is the target, acknowledging that indicator organisms released from original sources can survive, persist, and reproduce in the environment. In addition, MST using fecal *Bacteroidales* has been combined with modeling fate and transport techniques in San Pablo Bay to monitor fecal pollution (Wuertz et al. 2009).

MST can be applied to source water protection, watershed management, bathing beach monitoring and closure, and shellfish bed monitoring, as well as quantitative regulatory TMDL development (Long and Plummer 2008; Stapleton et al. 2007). The TMDL program in the USA was, and remains, a driving force behind the development of methodologies to reliably distinguish between bacteria from human and animal origins. An ideal method for TMDL assessment would also allow discrimination among animal sources and proportional allocation of contaminant loading, since the process requires that the various source load allocations be estimated to derive allowable pollution loads to a watershed, and to exert controls on particular potential or known sources (Bitton 2005; Hagedorn et al. 2009). Whether or not a TMDL program is successful in improving water quality depends largely on the positive identification of microbiological pollution sources and the relative or quantitative loading that is attributable to those sources (Duran et al. 2009 and Chap. 14).

13.4 Legal Applications of MST Information in a Regulatory and Litigation Context

As presented previously, MST and microbial forensics are well-recognized tools in the regulatory community for investigation and management of watersheds and potential pollutant impacts. Those methods play a significant role in the process for developing assessments and recommendations involving TMDLs in many states in the USA and are also used similarly in the UK (Stapleton et al. 2007) and Canada (Edge and Schaefer 2006). It is inevitable that some of the source allocation assessments and implementation plans for remediation that are developed under TMDL programs will enter the legal system, as strategies such as improving wastewater infrastructure and altering animal waste disposal practices can be very expensive and may not be undertaken willingly in some cases. MST evidence was used in a federal district court proceeding in the case of State of Oklahoma et al. vs. Tyson Foods et al. in 2008. In that case, testimony was presented concerning quantitative PCR assay for the 16S rRNA gene for a *Brevibacterium sp.* (Weidhaas et al. 2010) that was used as a marker for poultry litter contamination in the Illinois River watershed. The preliminary injunction, which sought immediate cessation of the practice of land application of poultry litter in the Illinois River watershed, was denied, and the trial for permanent injunction began in the Fall of 2009 without a jury (judge only). No final decision in the case was available as of April 2011.

While applicable legal rules often will differ among jurisdictions, for example in state versus federal court venues, the common thread is that they are designed to ensure the reliability of any testimony for use by the trier of fact (i.e., the judge and/or jury). Legal admissibility of scientific evidence is a fundamental and critical consideration, so an understanding of the constraints that influence admissibility is important for scientists who may be called upon to provide expert testimony in these types of cases (Harmon 2005).

A number of important judicial decisions over the last 20 years have influenced the introduction and application of scientific testimony in court. In many states, the controlling decision since the 1920s had been "*Frye*" (Frye v. United States, 54 App. D.C. 46, 293 F. 1013 1923), which held simply that the methods used by an expert witness must be generally accepted within the scientific community. Frye has been criticized for the notion that it allowed new, but otherwise legitimate, scientific conclusions to be excluded simply because they lacked what was judged to be "technical consensus." It was primarily applied in criminal cases (Bernstein 2002). The countervailing concern about vacating or abandoning Frye was that mere "relevance" of a scientific concept was not judged to be an adequate barrier to "junk science," often defined as "manufacturing uncertainty" (Michaels and Monforton 2005). In 1993, the US Supreme Court addressed some of these issues in its ruling in Daubert vs. Merrell Dow Pharmaceuticals, which appointed judges as the courthouse scientific "gatekeepers." A subsequent pair of high-court decisions (*General Electric* vs. *Joiner* and *Kumho Tire Co.* vs. *Carmichael*) clarified or solidified certain elements of the *Daubert* decision concerning the admissibility of expert testimony.

Following the *Daubert* ruling, Rule 702 of the Federal Rules of Evidence was amended in 2000 to include consideration of the factors shown in Table 13.1 (NRC 2009). That rule, from Article VII (Opinions and Expert Testimony) from Federal Rules of Evidence (2006), states as follows:

> Rule 702. Testimony by Experts
>
> If scientific, technical or other specialized knowledge will assist the trier of fact to understand the evidence or to determine a fact in issue, a witness qualified as an expert by knowledge, skill, experience, training or education, may testify thereto in the form of an opinion or otherwise, if (1) the testimony is based upon sufficient facts or data, (2) the testimony is the product of reliable principles and methods, and (3) the witness has applied the principles and methods reliably to the facts of the case.

Approximately half of the states in the USA have adopted Daubert, while at least as of 2008 the remainder, including some very populous states, continued to be

Table 13.1 Federal Rule 702 and important evaluation factors arising from *Daubert*

Whether experts are proposing to testify about matters growing naturally and directly out of research they have conducted independent of the litigation, or whether they have developed their opinions expressly for purposes of testifying
Whether the expert has unjustifiably extrapolated from an accepted premise to an unfounded conclusion
Whether the expert has adequately accounted for obvious alternative explanations
Whether the expert is being as careful as he/she would be in his/her regular professional work outside his/her paid litigation consulting
Whether the field of expertise claimed by the expert is known to reach reliable results for the type of opinion the expert would give

Source: NRC (2009)

classified as "Frye" states (e.g., FL, CA, IL, PA,), or elect to use modifications to these rulings and Federal Rule 702 (Sharp 2009; Calhoun 2008). Increasing reliance on principles laid out in *Daubert* and its related decisions, as well as increasing use of specific hearings to explore the opinions of experts in the context of *Daubert*, has had the effect of requiring judges to become de facto amateur scientists themselves to discharge their "gatekeeping" obligations (NRC 2009; Tellus Institute 2003).

Concerning legal and regulatory application of MST, as Blanch et al. (2006) observed, "determining the source of fecal contamination in aquatic environments is essential for estimating the health risks associated with pollution, facilitating measures to remediate polluted waterways, and resolving legal responsibility for remediation." As with other scientific disciplines, it is essential for experts to understand and use explicitly defensible and reproducible methods, but the legal fabric of how the data and opinions will be evaluated by the court will vary. Owing to the acknowledged applicability of the Daubert Test, or similar criteria, great emphasis is placed on the methods used in microbial forensics (McEwen et al. 2006), which may serve as a model for MST evidence in the courts. As detailed in NRC 2009, with "more and better educational programs, accredited laboratories, certified forensic practitioners, sound operational principles and procedures, and serious research to establish the limits and measures of performance in each discipline, forensic science experts will be better able to analyze evidence and coherently report their findings in the courts."

The benefits of the use of validated MST techniques clearly relate to the need for unambiguous and reproducible methods by which to identify sources of environmental bacterial populations and to assess their relative importance. This very rapidly developing field has reached a stage in which it is ripe for use in the regulatory and legal sphere. The challenges facing the field of MST are related to standardization processes for methods, and ongoing improvement of techniques to identify and segregate sources with increasing specificity and precision. Furthermore, error rates of the methodologies must be defined, as this is an important component of legal admissibility (Mahle 1999). The fact that processes and techniques are not yet universal, or that they are subject to interpretation, does not render them useless. Rather, it emphasizes the need for the MST community to continue the process of its technical consensus building so that these valuable techniques effectively can be used in the complex world of the legal process as well as they are being used in the regulatory world.

13.5 Summary and Recommendations

MST and microbial forensics represent relatively recent specializations in the use of genetic and biochemical techniques for assessing the relative or specific importance of bacterial and viral sources in samples from environmental, food safety, or criminal matters. As with many nascent disciplines, the science in this instance is developing more rapidly and completely in terms of fundamental knowledge and regulatory applications than has been the case in the legal arena and in the courtroom. There is a tremendous utility to these rapidly emerging and powerful methods that have great promise in the fields of environmental assessment, informed pollution management/regulation, food and product safety, criminology, and medicine. Certainly, prudent verification and validation are essential, and these processes continue to evolve, but care should be taken to heed the admonition that it is unwise and potentially dangerous, paraphrasing Voltaire, to inadvertently permit "a goal of perfection to be the enemy of excellence."

References

AAM (2003) Microbial Forensics: A Scientific Assessment. American Academy of Microbiology/ American Society of Microbiology, Washington, DC.

Bernstein, D. (2002) Disinterested in Daubert: State courts lag behind in opposing "junk" science. Washington Legal Foundation, Washington, DC.

Bitton G (2005) Microbial indicators of fecal contamination: Application to microbial source tracking. Department of Environmental Engineering Sciences University of Florida, Gainesville, Florida.

Blanch A, Belanche-Munoz L, Bonjoch X et al. (2006) Integrated analysis of established and novel microbial and chemical methods for microbial source tracking. J Environ Microbiol 72(9):5915–5926.

Brownell MB, Harwood VJ, Kurz RC et al. (2007) Confirmation of putative stormwater impact on water quality at a Florida beach by microbial source tracking methods and structure of indicator organism populations. Water Res 41:3747–3757.

Buchan A, Alber M, Hodson R (2001) Strain-specific differentiation of environmental *Escherichia coli* isolates via denaturing gradient gel electrophoresis (DGGE) analysis of the 16S-23S intergenic spacer region. FEMS Microbiol Ecology 35:313–321.

Budowle B, Schutzer S, Ascher M, et al. (2005) Toward a system of microbial forensics: from sample collection to interpretation of evidence. Appl Environ Microbiol 71(5):2209–2213.

Calhoun, M C. (2008) Scientific evidence in court: *Daubert* or *Frye*, 15 years later. Washington Legal Foundation, Washington, DC.

Daubert v. Merrell Dow Pharmaceuticals, Inc. (1993) 509 US 579.

Duran M, Haznedaroglu B, Zitomer D (2006) Microbial source tracking using host specific FAME profiles of fecal coliforms. Water Res 40:67–74.

Duran M, Yurtsever D, Dunaev T (2009) Choice of indicator organism and library size considerations for phenotypic microbial source tracking of FAME profiling. Water Sci Tech 60(10):2659–2668.

Edge T, Schaefer K (eds.) (2006) Microbial source tracking in aquatic ecosystems: The state of the science and an assessment of needs. National Water Research Institute, Burlington, Ontario. NWRI Scientific Assessment Report Series No. 7 and Linking Water Science to Policy Workshop Series. 23 p.

Federal Rules of Evidence (2006) http://www.uscourts.gov/rules/Evidence_Rules_2007.

Frye v. United States (1923) 54 App. D. C. 46, 293 F. 1013.

Graves AK, Hagedorn C, Brooks A et al. (2007) Microbial source tracking in a rural watershed dominated by cattle. Water Res 41:3729–3739.

Griffith J, Weisberg S, McGee C (2003) Evaluation of microbial source tracking methods using mixed fecal sources in aqueous test samples. J Water Health 1(4):141–151.

Hagedorn C, Robinson SL, Filtz JR et al. (1999) Determining sources of fecal pollution in a rural Virginia watershed with antibiotic resistance patterns in fecal streptococci. Appl Environ Microbiol 65:5522–5531.

Hagedorn C, Benham B, Zeckoski S (2009) Microbial source tracking and the TMDL (total maximum daily loads) process. Virginia Cooperative Extension Pub 442–554.

Harmon, R. (2005) Admissibility standards for scientific evidence. In: Breeze, R, Budowle, B and Shutzer, S, editors, Microbial Forensics. Elsevier Academic, Amsterdam, The Netherlands.

Hartel P, Summer J, Hill J et al. (2002) Geographic variability of *Escherichia coli* ribotypes from animals in Idaho and Georgia. J Environ Qual 31:1273–1278.

Hartel P, Summer J, Segars W (2003) Deer diet affects ribotype diversity of *Escherichia coli* for bacterial source tracking. Water Res 37:3263–3268.

Harwood V, Wiggins B, Hagedorn C et al. (2003) Phenotypic library-based microbial source tracking methods: Efficiency in the California collaborative study. J Water Health 1(4):153–166.

Hsu F, Shieh Y, van Duin J et al. (1995) Genotyping male-specific coliphages by hybridization with oligonucleotide probes. Appl Environ Microbiol 61:3960–3966.

Hundesa A, Bofill-Mas S, Maluquer de Motes C et al. (2010) Development of a quantitative PCR assay for the quantitation of bovine polyomavirus as a microbial source tracking tool. J Virol Met 163(2): 385–389.

Indest K, Betts K, Furey J (2005) Application of oligonucleotide microarrays for bacterial source tracking of environmental *Enterococcus sp.* Isolates. Int J Environ Res Public Health 2(1):175–185.

Jellison KL, Lynch AE, Ziemann JM (2009) Source tracking identifies deer and geese as vectors of human-infectious *Cryptosporidium* genotypes in an urban/suburban watershed. Environ Sci Technol. 43(12):4267–72.

Jiang S, Chu W, Olson B et al. (2007) Microbial source tracking in a small southern California urban watershed indicates wild animals and growth as the source of fecal bacteria. Environ Biotech 76:927–934.

Koblenz G, Tucker J (2010) Tracing an attack: The promise and pitfalls of microbial forensics. Survival 52(1):159–186.

Kon T, Weir S, Howell E et al. (2009) Repetitive element (REP)-polymerase chain reaction (PCR) analysis of *Escherichia coli* isolates from recreational waters of southeastern Lake Huron. Can J Microbiol 55:269–276.

Korajkic A, Badgley B, Brownell B et al. (2009) Application of microbial source tracking methods in a Gulf of Mexico field setting. J App Microbiol 107:1518–1527.

Kuntz R, Hartel P, Godfrey D et al. (2003) Targeted sampling protocol as prelude to bacterial source tracking with *Enterococcus faecalis*. J Environ Qual 32:2311–2318.

Lasalde C, Rodriguez R, Toranzos G (2005) Statistical analyses: Possible reasons for unreliability of source tracking efforts. Appl Environ Microbiol 71(8):4690–4695.

Lee J, Lim M, Kim S et al. (2009) Molecular characterization of bacteriophages for microbial source tracking in Korea. Appl Environ Microbiol 75(22):7101–7114.

Long S, Plummer J (2008) Using microbial source tracking in watershed management: Is high quality source water sustainable? AWWA Sustainable Water Sources Conference, Reno, NV, February 10–13.

Mahle S (1999) The impact of Daubert v. Merrell Dow Pharmaceuticals, Inc., on expert testimony with applications to securities litigation. The Florida Bar Journal, Volume LXXIII, No. 3. 36 p.

McEwen S, Wilson T, Ashford D et al. (2006) Microbial forensics for natural and intentional incidents of infectious disease involving animals. Rev Sci Tech Off Int Epiz 25(1):329–339.

McQuaig S, Scott T, Lukasik J et al. (2009) Quantification of human polyomaviruses JC Virus and BK Virus by TaqMan quantitative PCR and comparison to other water quality indicators in water and fecal samples. Appl Environ Microbiol 75:3379–88.

Michaels D and Monforton C. (2005) Manufacturing uncertainty: Contested science and the protection of the public's health and the environment. Am J Public Health 95(S1): 39–48.

Moore D, Harwood V, Ferguson D et al. (2005) Evaluation of antibiotic resistance analysis and ribotyping for identification of faecal pollution sources in an urban watershed. J Appl Microbiol 99:618–628.

National Research Council (NRC; 2009) Strengthening Forensic Science in the United States. National Research Council or the National Academies, Washington, DC.

Osawa S, Furuse K, Watanabe I (1981) Distribution of ribonucleic acid coliphages in animals. Appl Environ Microbiol 41:164–168.

Pattnaik P and Jana A (2005) Microbial forensics: Applications in bioterrorism. Environ Forensics 6:197–204.

PBS&J (2008) Fecal BMAP Implementation: Source Identification, Hillsborough River Watershed, Final Summary Report, Prepared for the Florida Department of Environmental Protection, Tallahassee, FL http://publicfiles.dep.state.fl.us/DEAR/BMAP/Tampa/MST%20Report/Hillsborough%20Source%20Identification%20Final%20Report.pdf.

Puig A, Queralt N, Jofre J et al. (1999) Diversity of *Bacteriodes fragilis* strains in their capacity to recover phages from human and animal wastes and from fecally polluted wastewater. Appl Environ Microbiol 65:1772–1776.

Reischer G., Haider J, Sommer R et al. (2008) Quantitative microbial faecal source tracking with sampling guided by hydrological catchment dynamics. Environ Microbiol 19(10): 2598–2608.

Ritter K, Carruthers E, Carson C et al. (2003) Assessment of statistical methods used in library-based approaches to microbial source tracking. J Water Health 1(4):209–223.

Salyers A (2004) Microbes in court: The emerging field of microbial forensics. Available at http://www.actionbioscience.org/newfrontiers/salyersarticle.html; Accessed March 21, 2010.

Scott T, Parveen S, Portier K et al. (2003) Geographical variation in ribotype profiles of *Escherichia coli* isolates from humans, swine, poultry, beef, and dairy cattle in Florida. Appl Environ Microbiol 69(2):1089–1092.

Scott T, Caren J, Nelson GR et al. (2004) Tracking sources of fecal pollution in a South Carolina watershed by ribotyping *Escherichia coli*: A case study. Environ Forensics 5:15–19.

Seurinck S, Verstraete W, Siciliano S (2005) Microbial source tracking for identification of fecal pollution. Rev Environ Sci Biotechnol 4:19–37.

Sharp E (2009) Scientific evidence and the courts. In: Leestma J (ed) Forensic Neuropathology, 2nd edn. Taylor & Francis, Florida.

Simpson J, Santo Domingo J, Reasoner D (2002) Microbial source tracking: State of the science. Environ Sci Tech 36(24):5279–5288.

Soil & Water Conservation Society of Metro Halifax (SWCSMH; 2006) Bacterial Source Tracking (BST) - A Review. http://www.chebucto.ns.ca/ccn/info/Science/SWCS/H-2/bst.html; accessed October 2009.

Stapleton C, Wyer M, Kay D et al. (2007) Microbial source tracking: a forensic technique for microbial source identification. J Environ Monit 9:427–439.

Stoeckel D, Mathes M, Hyer K et al. (2004) Comparison of seven protocols to identify fecal contamination sources using *Escherichia coli*. Environ Sci Technol 38:6109–6117.

Stoeckel D (2005) Selection and application of microbial source tracking tools for water-quality investigations: U.S. Geological Survey Techniques and Methods Book 2, Chapter A3, 43 p.

Stoeckel D and Harwood V. (2007) Performance, design, and analysis in microbial source tracking studies. Appl Environ Microbiol 73(8):2405–2415.

Tartera C, Lucena F, Jofre J (1989) Human origin of *Bacteroides fragilis* bacteriophages present in the environment. Appl Environ Microbiol 55:2696–2701.

Tellus Institute (2003) Daubert: The Most Influential Supreme Court Decision You've Never Heard Of. Project on Scientific Knowledge and Public Policy (SKAPP). Tellus Institute, Boston, MA.

Unger S (2008) http://ec.gc.ca/scitech/default.asp?lang=En&n=4B40916E-1&xsl=privateArticles2,viewfull&po=B2DDBB17.

United States Environmental Protection Agency (USEPA; 2005) Microbial Source Tracking Guide Document. United States Environmental Protection Agency, Office of Research and Development, Cincinnati, OH.

Weidhaas J, Macbeth T, Olsen R et al. (2010) Identification of a *Brevibacterium* marker gene specific to poultry litter and development of a quantitative PCR assay. J Appl Microbiol 109(1):334–347.

Wuertz S, Bombardelli F, Sirikanchana K et al. (2009) Quantitative pathogen detection & microbial source tracking combined with modeling the fate and transport of *Bacteroidales* in San Pablo Bay. Report submitted to The NOAA/UNH Cooperative Institute for Coastal and Estuarine Environmental Technology (CICEET).

Chapter 14
Applications of Microbial Source Tracking in the TMDL Process

Brian Benham, Leigh-Anne Krometis, Gene Yagow, Karen Kline, and Theo Dillaha

Abstract The US Environmental Protection Agency's Total Maximum Daily Load (TMDL) program is frequently cited as a primary driver in the development of microbial source tracking (MST) techniques. As MST techniques continue to mature, it is prudent to identify those areas where further MST-related research is most likely to contribute to the efficient development and implementation of bacterial TMDLs. The objectives of this chapter are to review the basic phases in the TMDL process, to describe current applications of MST within these stages, to identify research needed to increase MST application, and to discuss opportunities for the expanded use of MST data within the TMDL process.

Keywords TMDL • Modeling • Pollutant loading • Pathogen impairment • Watershed

The US Environmental Protection Agency's Total Maximum Daily Load (TMDL) program is frequently cited as a primary driver in the development of microbial source tracking (MST) techniques (Field and Samadpour 2007; Harwood 2007; Santo Domingo et al. 2007; Scott et al. 2002; Simpson et al. 2002); however, most published literature consists of descriptions of watershed-scale case studies (e.g., Hagedorn et al. 1999; Wiggins et al. 1999) with no attempt at a broader examination of the application of MST analyses to the TMDL process on a national scale. As MST techniques continue to mature, it is prudent to identify those areas where further research is most likely to contribute to the efficient development and implementation of TMDL restoration plans. The objectives of this chapter are to review the basic phases in the TMDL process, to describe current applications of MST within these stages, to identify research needed to increase MST application, and to discuss opportunities for the expanded use of MST data within the TMDL process.

B. Benham (✉)
Biological Systems Engineering, Virginia Tech, Blacksburg, VA, USA
e-mail: benham@vt.edu

C. Hagedorn et al. (eds.), *Microbial Source Tracking: Methods, Applications, and Case Studies*, DOI 10.1007/978-1-4419-9386-1_14,

14.1 Overview of the TMDL Program

The TMDL Program was promulgated by the US Congress in sections 303(d) and 305(b) of the 1972 Clean Water Act (CWA); however, the US Environmental Protection Agency (USEPA) did not issue TMDL-related guidance until the early 1990s. In accordance with the CWA, all US water bodies (streams, reservoirs, lakes, and estuaries) must be evaluated in the context of applicable water-quality standards, which include water-quality criteria designed to protect the water's designated uses (e.g., drinking water, swimming, fishing, etc.). Water bodies that do not meet water-quality standards are classified as "impaired." The USEPA, or its equivalent state or tribal agency, is responsible for developing TMDLs for impaired water bodies. A TMDL is defined as the total pollutant load a water body can receive and still meet applicable water-quality standards. It is a quantitative representation of all the contributions of a particular pollutant (e.g., bacteria, sediment) to a water body and is mathematically defined as:

$$\text{TMDL} = \sum \text{WLA} + \sum \text{LA} + \text{MOS},$$

where WLA = waste load allocations (point source loadings such as sewage treatment plant discharges), LA = load allocations (nonpoint source (NPS) loadings such as stormwater runoff), and MOS = a margin of safety. The term "TMDL" is also used to describe the allocation strategy required to reduce existing pollutant loadings to the required mathematical TMDL load. Since each TMDL addresses a specific pollutant (e.g., bacteria, sediment), water bodies with multiple impairments require multiple TMDLs.

In the broadest sense, the TMDL process consists of three basic phases (Fig. 14.1): (1) impairment designation – identification of the type, severity, and extent of the water-quality impairment, (2) TMDL development, which involves (a) determination of the existing and potential future pollutant loads in the watershed, (b) linking

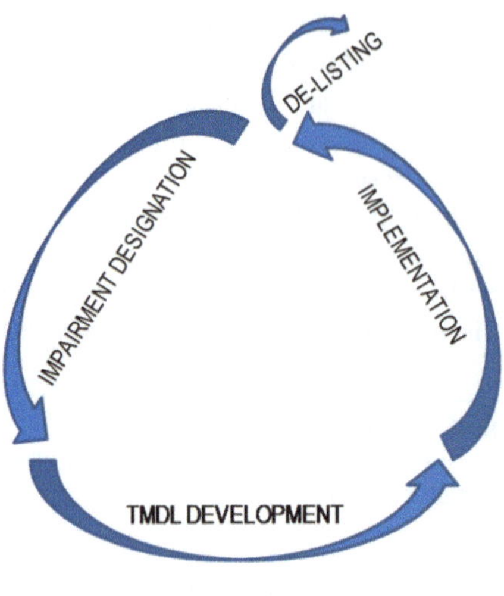

Fig. 14.1 Phases of the total maximum daily load process

those loads to the water quality in the waterbody, and (c) defining the pollutant load reductions required to achieve applicable water-quality criteria, and (3) implementation, which requires development of a watershed-specific plan identifying pollution control measures required to achieve water-quality standards, implementing the control measures, monitoring water quality to document progress toward attainment of water-quality standards, and finally delisting (removal of the water body from the state's impaired waters list) after attainment of water-quality standards.

Forty percent of waters assessed in the US to date have been designated as impaired due to the violation of one or more water-quality criteria. Roughly 40,000 TMDLs have been completed to address the 76,000 impairments identified to date (USEPA 2009a). Pathogens are the leading group classification of impairments and responsible for 14% of identified impairments nationwide (USEPA 2009a). Causes of pathogen group impairment include excessive concentrations of fecal coliform, *E. coli*, *Enterococcus*, and other fecal indicator bacteria (FIB) in waters. These impairments are of particular concern to public health, as microbial contamination of surface waters, particularly by NPSs such as stormwater runoff, has been repeatedly linked to the elevated incidence of waterborne disease (Rose et al. 2000; Gaffield et al. 2003; Arnone and Walling 2007; Hrudey and Hrudey 2007).

In subsequent sections of this chapter, each of the basic phases of the TMDL process (Fig. 14.1) is discussed in greater detail. Current use of MST data (if applicable) within each phase is reviewed along with related gaps in scientific knowledge that currently limit further application. Opportunities are also identified for future MST use within each phase and potential use in other water-quality restoration and protection efforts.

14.2 Impairment Designation

Waters in USA are currently designated as having a pathogen impairment and require subsequent TMDL development if FIB concentrations in monitoring samples exceed state-specific water-quality criteria for the designated water use. In general, pathogen water-quality criteria for most uses are based upon 1986 USEPA recommendations for recreational water quality (USEPA 1986, 2003) and make no distinction between fecal loadings originating from anthropogenic and natural (e.g., wildlife) sources. There is concern that the application of such general criteria renders the TMDL process inefficient, with a high probability of misclassification of actual water-quality threats and a misplaced use of scarce resources (Keller and Cavallaro 2008), since anthropogenic sources are presumed to present greater risk to human health than nonanthropogenic sources of FIB. One of the most common uses of MST during water-quality impairment assessments is to identify sources of FIB impairments (e.g., human, livestock, wildlife) for the purposes of prioritizing TMDL development. Where impairments are demonstrated as primarily from nonanthropogenic bacteria sources such as wildlife, which are not practical to control and are thought to pose less health risk, TMDL development and implementation may not be warranted. In these cases, a Use Attainability Analysis

(UAA) might be more appropriate. A UAA is a structured scientific assessment of the physical, chemical, biological, and economic factors that affect the attainment of a particular designated use by a given water body. If a UAA shows that attaining a designated use is not feasible, the state, after considering public opinion, may choose to modify the waterbody's designated use and to set less restrictive water-quality criteria and a more realistic water-quality goal (USEPA 1991). In reality, changing water-quality standards based on UAA is difficult and rarely attempted.

Widespread application of MST to identify and prioritize impairments will require further research and information about the relative health risks from human and animal fecal contamination, correlations between MST results and FIB water-quality criteria, and regulatory tools to facilitate impairment prioritization based on health risk. Water-quality researchers and managers have long assumed that sewage presents a greater threat to the public health than animal feces, as it is more likely to contain microorganisms preadapted to infect humans (Field and Samadpour 2007). A recent epidemiological study of swimmers in Mission Bay, California supports this assumption, as it indicated that waters exceeding FIB criteria did not necessarily always pose a serious health risk (Stewart et al. 2007). Although the waters of Mission Bay were consistently in violation of California's enterococci criteria, no elevated incidence of illness was observed amongst swimmers, perhaps because MST indicated that over 95% of enterococci were identified as nonhuman in origin (~60% primarily avian). These results suggest that the bacterial impairment of Mission Bay should be a lesser priority in the state TMDL program, with resources more effectively applied to other watersheds with water-quality impairments that pose a more substantial health risk to humans.

Before TMDL prioritization strategies can be refined, further clarification is required on the relative risks of fecal contamination from different species contributing to waterborne disease. The incidence of zoonoses (transmission of disease to humans from nonhuman animals) is increasing rapidly (Taylor et al. 2001; Fayer et al. 2004); however, the evolution and transmission of zoonotic waterborne diseases are poorly understood (Bolin et al. 2004; Field and Samadpour 2007; USEPA 2009b). Further research comparing health risks between exposures to primarily human vs. primarily nonhuman contamination is necessary before MST analyses can be used to assist in determining the relative public health risks and, therefore, priorities of identified impairments. Regardless, it is important to note that because current FIB-based water-quality criteria are based on the concentration and persistence of FIB rather than the origin of the FIB, there is no provision to list or delist based on source identification, even when the impairment is primarily attributed to "natural" sources such as wildlife (Field and Samadpour 2007). Assuming that future research identifies significantly different levels of risk associated with different sources of fecal contamination, existing regulatory programs will have to be altered to allow for prioritizing TMDL development based on such risk-based information.

Accurate interpretation of MST results in relation to existing FIB-based water-quality criteria may be complicated, since MST results are not presently standardized. While current bacterial water-quality criteria and water-quality monitoring programs quantify only FIB, MST results or targets include many additional types of microorganisms (e.g., anaerobic *Bacteroides*, coliphage viruses). In addition, FIB are most often quantified via culture-based methods, while MST efforts may target specific genes via molecular methods such as quantitative polymerase chain reaction (qPCR). The relationships between culturable levels of FIB and results from molecular methods are unclear, as molecular analyses detect not only culturable but also viable but not culturable (VBNC) and lysed cells (Noble and Weisberg 2005; Wade et al. 2006; Harwood 2007). Since it is highly unlikely that new water-quality criteria specific to each of the numerous potential MST targets will be developed for the purpose of impairment classification, correlations between these targets and FIB is necessary to maximize the use of MST data in the impairment assessment phase of the TMDL process.

14.3 TMDL Development

Developing a TMDL to address a FIB impairment typically begins with source characterization and then quantifies FIB loadings from each identified significant source. Next, during the linkage analysis, FIB loads are linked with in-stream water-quality concentrations to determine compliance with or violation of applicable water-quality criteria. Finally, an allocation analysis determines the reductions in FIB loadings required from each source to meet the applicable water-quality criteria.

14.3.1 Source Characterization

Potential sources considered during bacterial impairment TMDL development will vary from one watershed to another and will depend on factors such as land use and population characteristics. Historically, water bodies have been monitored for culturable fecal coliform as a measure of bacterial water quality. In recent years, there has been a movement toward *E. coli*- and enterococci-based water-quality criteria. A recent review of pathogen TMDL reports revealed that the following sources are nearly universal: failing or inadequate septic systems or other home-based wastewater treatment systems, wildlife, livestock, stormwater runoff, and permitted facilities (Table 14.1).

As MST data are not yet sufficiently quantitative to provide accurate and defensible estimates of the relative loadings of fecal contamination from potential upland sources, TMDL developers currently draw upon a variety of other potential information sources to determine pollutant loadings in an impaired watershed, including hydrology, land use and cover, human demographics, agricultural production, and wildlife habitat (Table 14.2). It is worth noting that the information sources listed in Table 14.2 are independent of monitored water-quality data. Spatial data in GIS format may be used to estimate the density and distribution of wildlife or livestock, while federal, state, and local government personnel generally provide information on the human population demographics, permitted discharges (e.g., wastewater,

Table 14.1 Common bacteria sources considered in TMDL studies (from Benham et al. 2006)

Source category	Location or pathway considered	Typical sources
Livestock	Deposition on land, deposition in streams, manure storage and application	Cattle (dairy and beef) Horses Swine Sheep and goats Poultry
Wildlife	Deposition on land, deposition in streams, residential storm runoff	Deer Ducks Geese Beaver Muskrats
Residential	Failing or inadequate septic systems, straight pipes, stormwater runoff, leaking sewer systems, illicit sewer connections	Humans Pets Stormwater runoff
Permitted	Permitted discharges, recorded sewer overflows	National Pollutant Discharge Elimination System (NPDES) permits
	Land application	Municipal biosolids

Table 14.2 Common information sources used to quantify bacteria sources (from Benham et al. 2006)

Information category	Uses	Information sources
Land use and cover	Animal numbers and distribution (agricultural and wildlife); General source characterization (urban, rural, agricultural, etc.)	National Land Cover Dataset (NLCD); USGS Digital Raster Graphic (DRG) or Digital Ortho Quarter Quads (DOQQ); state agencies or organizations
Political boundaries, roads, sewered areas	Human demographics, presentations	US Census Topologically Integrated Geographic Encoding and Referencing (TIGER) data; state agencies or localities
Permitted facilities and activities	Waste load allocation, human sources	State environmental agencies; NPDES permits and others
Livestock	Livestock types, numbers, waste handling, agronomic practices	National Agricultural Statistics Service (NASS); Cooperative Extension and Natural Resource Conservation Service (NRCS) personnel; state agricultural agencies; local citizens
Humans	Potential sources, magnitude of sources	US Census data; county sanitarians
Wildlife	Species present, population numbers	State and federal wildlife agencies; National Trappers Association members; local citizens

confined animal feedlot operations) (Benham et al. 2005a, b), and the extent of sewered areas (Parajuli et al. 2005). Local input from watershed stakeholders, obtained through stakeholder meetings, postal surveys, phone conversations, or other personal contact, is often a critical means of refining these source loading estimates.

14.3.2 Linkage Analysis

The approaches used in pathogen TMDL development to link FIB loadings and water-quality concentrations fall into two categories: nonmodeling approaches that focus on monitoring and/or empirical data, and modeling approaches that simulate FIB loads and concentrations within the impaired water body.

14.3.2.1 Nonmodeling Approaches

Nonmodeling approaches that have been used to develop bacterial impairment TMDLs include mass balance/conservation of mass, statistical analyses, and simple empirical calculations.

Mass Balance Approach

Jarrell (1998) discusses developing TMDLs using a "conservative" mass balance approach that targets a single value of the offending pollutant. This "single-value" is a function of the applicable water-quality criteria and some physical condition of the impaired water body (e.g., a particular design flow or storage volume). Developers of a pathogen TMDL in Mississippi used this approach and concluded that it was best suited to situations where data limitations preclude the use of a more sophisticated method, such as a water-quality simulation model (MDEQ 2006). Stiles (2002) concluded that a single-value mass balance approach has limited utility when dealing with impairments primarily attributable to NPS pollution because it may not take into account water-quality violations associated with larger storm events where NPS loadings dominate. As FIB impairments are frequently associated with stormwater (Gaffield et al. 2003), the mass balance approach is less useful for the development of pathogen TMDLs when compared to other methods.

Statistical Approach

The load-duration curve (LDC) method is frequently used for pathogen TMDL development. Similar to a flow-duration curve, a LDC illustrates the percentage of time that a particular pollutant load is exceeded. A LDC combines observed flow and pollutant concentration data into loads, which can then be compared against an allowable load curve that is formed from a flow frequency curve and an applicable water-quality FIB concentration criterion. The LDC compares observed loads with allowable loads across the range of flows at a specific monitoring location to determine required pollutant reductions (Stiles 2002). The flow conditions under which observed loads exceed allowable loads indicate whether potential pollutant sources are primarily point source or nonpoint source in nature. Required pollutant load reductions are calculated as simple percentages, but not in terms of loads. The LDC method inherently accounts for pollutant fate and transport processes within the watershed because of the reliance on observed data. Further, using the LDC approach can potentially overcome problems of communicating complex watershed phenomena to lay persons because of the simplicity of the calculations and can assist in quantifying the variability and uncertainty associated with the TMDL (Bonta and Cleland 2003). However, depending on the range and complexity of the bacteria sources, the LDC approach may not be useful in determining intrawatershed pollutant contributions. Understanding such spatial variability may have to be accomplished through supplemental sampling or through subsequent simulation modeling that links land uses and pollutant sources with hydrologic response and pollutant loads in the impaired water body (Benham et al. 2006).

Simple Empirical Approaches

Simple empirical approaches rely on existing monitored water-quality data to describe the relationship between pollutant sources and water-quality targets. One example of a simple empirical approach is the percent reduction (PR) method (USEPA 2008), which is often used when flow data is unavailable. This method involves comparing monitoring data to water-quality criteria and includes equations for each criterion. The PR required to meet the criterion is based on the equation:

$$PR = [(Conc_{exist} - Cr) / Conc_{exist}] \times 100,$$

where PR = percent reduction; $Conc_{exist}$ = existing concentration (e.g., maximum, long-term average); and Cr = applicable water-quality criterion.

The Watershed Treatment Model (Caraco 2002) is another simple approach used to link pollutant loads to water quality that was developed by the Center for Watershed Protection (www.cwp.org). The Watershed Treatment Model (WTM) is a straightforward spreadsheet model that uses the Simple Method (USEPA 2008) to estimate pollutant loadings in a watershed. The WTM calculates pollutant loads using areas and typical unit-area loads from various land uses to estimate the annual FIB load in watersheds and resulting in-stream concentrations. It does not address FIB die-off or regrowth.

14.3.2.2 Modeling Approaches

Computer-based water-quality simulation models are used extensively to develop TMDLs for pathogen impairments. The models typically estimate watershed-scale FIB loads over a range of flow conditions and evaluate the effectiveness of proposed control measures in reducing source loads to meet water-quality criteria (DePinto et al. 2004). Borah and Bera (2003), Chapra (2003), and Neilson et al. (2003) published comparisons of the approach, assumptions, and capabilities of various existing water-quality simulation models. While numerous models are available for TMDL development (USEPA 2008), the two most commonly used models, the Hydrological Simulation Program-FORTRAN (HSPF) and the Soil and Water Assessment Tool (SWAT), are discussed here.

Hydrological Simulation Program-FORTRAN

HSPF is an USEPA supported model that is frequently used to develop pathogen TMDLs (USEPA 2001). HSPF is a watershed-scale, process-oriented, lumped-parameter, continuous simulation hydrology and water-quality constituent (including FIB) model. HSPF represents spatial variability in a limited way by dividing the watershed into subwatersheds and land-use segments and simulates hydrology and water-quality processes occurring on pervious areas, impervious areas, and in reaches and reservoirs. While HSPF does not contain modules that were developed specifically for simulating microbial transport, existing modules (PQUAL) have been adapted for this purpose. When using HSPF to develop pathogen TMDLs, FIB loads to the land surface are estimated external to HSPF and then input as monthly loads. Direct FIB loads (e.g., loads deposited directly into a water body by livestock, wildlife, and permitted or nonpermitted point source discharges) are input as time series data. Groundwater and interflow FIB concentrations are input using monthly model parameters.

TMDLs developed using HSPF typically simulate FIB as free-phase (also described variously as planktonic, "dissolved," or water-column) constituents (Ferguson et al. 2003; Jamieson et al. 2004; Pachepsky et al. 2006). Although the model has the capability to discriminate between free-phase and particle-associated microbes, the quantitative data necessary to appropriately parameterize these two phases are generally unavailable. Bacteria die-off on the land surface is indirectly represented via a limit on surface accumulation. A user-defined parameter sets the rate of runoff needed to wash off 90% of the accumulated bacterial load on the land surface. This limit can be specified on a monthly basis and can vary by land use. In-stream die-off is modeled using a temperature-corrected first-order decay function (Chick's law).

TMDLs developed using HSPF typically lump FIB from all sources and use identical die-off and transport parameters. While HSPF theoretically permits simulation of FIB from different sources (e.g., humans, cattle, waterfowl) separately, there is extremely limited data available to quantitatively describe differences in fate and transport between source-specific FIB (e.g., Anderson et al. 2005), making this approach impractical at present. Die-off rates for different sources of fecal indicators is an issue frequently identified as requiring further investigation (Scott et al. 2002; Simpson et al. 2002; Field and Samadpour 2007; Muniesa et al. 2009).

Best management practices (BMPs) for pathogen impairments can be represented in HSPF either through calculated reduction in source inputs or through the application of reduction factors. Source reductions may be used to simulate improvements in wastewater treatment and reduced FIB densities in animal manure treated in a holding system. Land-based BMPs, however, mostly rely on the use of reduction factors. Although BMP performance may vary dramatically across watersheds or between individual storm events (Clary et al. 2008; Krometis et al. 2009), most water-quality models, including HSPF, can only currently represent performance through an average pollutant reduction factor. Reduction factors are multiplied by the source loading from the area impacted by the BMP. For example, a reduction factor could be applied to FIB loads from a pervious area with a BMP, such as pasture with improved grazing. More complex BMPs or systems of BMPs can be simulated with HSPF, but doing so is complicated and requires specific expertise (Donigian et al. 1991).

Soil and Water Assessment Tool

SWAT was developed and is supported by the USDA-ARS and is conceptually similar to HSPF. It is a watershed-scale, process-oriented, lumped-parameter, continuous simulation model developed to predict the impact of land-use and land-management practices on water quantity and quality (Arnold and Fohrer 2005). Runoff, leaching, and pollutant loadings are estimated by hydrologic response units (HRU), which represent a unique combination of soil, land use, and topography in each subwatershed. Pollutant loads and flow are routed through the stream network to obtain total water and pollutant yields for a given watershed. FIB inputs are specified by HRU.

SWAT has a module developed specifically to simulate bacterial transport and includes the following: variable NPS loadings of manure and other bacterial sources to the land surface, point source loadings to streams and reservoirs, die-off and wash-off of bacteria applied to foliage, partitioning and differential die-off/regrowth of bacteria in solution and adsorbed to soil, percolation of solution-phase bacteria past the surface soil layer, movement of solution-phase bacteria from the land surface to the stream network with surface runoff, transport of bacteria adsorbed on eroded sediment to streams, and the transport and die-off/regrowth of bacteria in streams. In addition, SWAT has modules to simulate trapping of bacteria by BMPs. Bacteria-related BMPs that can be simulated by SWAT include composting, rotational grazing, sewage treatment, edge-of-field filters, catchment ponds, and manure incorporation.

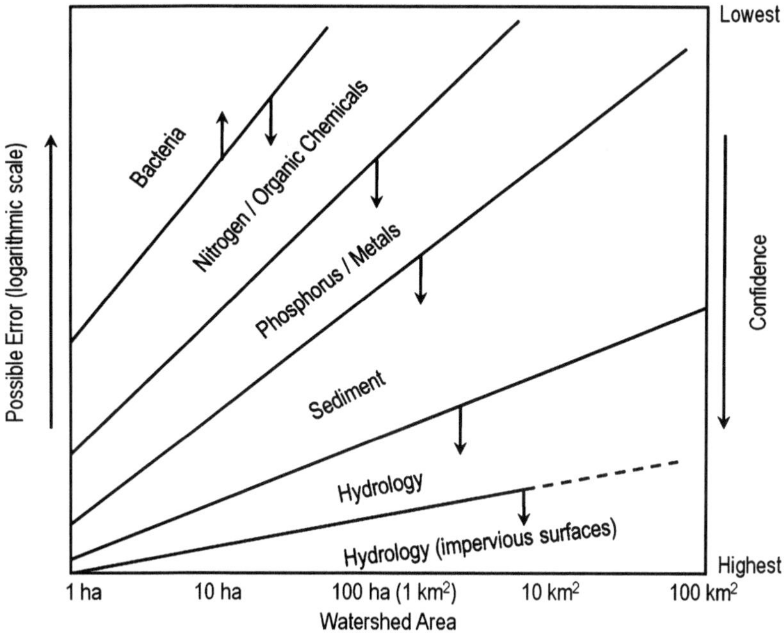

Fig. 14.2 Reliability and possible errors for nonpoint pollution models (after Novotny and Chesters 1981)

14.3.2.3 Uncertainty in Water-Quality Modeling

Water-quality simulation models are frequently used in TMDL development because they predict quantitative values of FIB loadings and resultant concentrations; however, the uncertainty associated with model outputs can be significant (Fig. 14.2). Far less confidence and greater potential error are expected in FIB model outputs when compared to simulations of hydrology and other water-quality components such as sediment and nutrients (Novotny 2003). The low confidence in FIB model outputs is due to the relatively limited current scientific knowledge of microbial fate and transport behavior in the environment, which has often been cited as a primary research need for improved water-quality management (Ferguson et al. 2003; Jamieson et al. 2004; Pachepsky et al. 2006; Arnone and Walling 2007).

In order to account for the limited scientific understanding of bacteria fate and transport in receiving waters, computer simulation models like HSPF and SWAT are calibrated using existing water-quality data prior to TMDL development. Model parameters are adjusted during calibration to improve the degree of agreement between model predictions and observed monitoring data during the calibration period. Calibration, thus, seeks to "ground-truth" modeling with observed data to increase confidence in model predictions and to assess model uncertainty. Unfortunately, historical water-quality data in watersheds with impairments are generally very limited, fail to adequately represent needed hydrologic conditions, and are often of limited value for calibrating models such as HSPF and SWAT.

There is also a concern that the role and power of models are often overstated by users and developers (Muñoz-Carpena et al. 2006). Model developers need to better clarify the proper use and limitations of their models to prevent misuse of models. Further, models may be better suited for assessing relative, rather than absolute, changes in pollutant loads. Benham et al. (2008) suggest that the more detailed the watershed characterization that is performed during the TMDL study, the more likely it is that TMDL implementation will result in water-quality improvement. In conclusion, water-quality simulation models are an important tool in the development of pathogen TMDLs, particularly in watersheds with a variety of spatially distributed sources and a need for the evaluation and identification of spatially specific source load reductions.

14.4 Allocation Analysis

The allocation analysis determines the FIB load reductions from each source required to meet water-quality criteria and the resulting TMDL. Because of the simplified nature of nonmodeling approaches, load reductions developed using nonmodeling approaches can typically only be expressed as PRs in WLAs and LAs, and specific FIB source or source category load reductions loads cannot be quantified. When using water-quality simulation models to link pollutant sources to water quality, required load reductions from each pollutant source and land use can be estimated.

14.4.1 Current Use of MST Data in the TMDL Development Phase

Data from watershed MST studies are currently most commonly incorporated into the TMDL process during the TMDL development phase. To date, the primary role of MST in pathogen TMDL development has been during source assessment to supplement the identification and characterization of FIB sources developed through a watershed inventory. Where multiple sites have been monitored within a watershed over time, MST can be used to identify spatial and temporal trends that may link with specific sources or source-specific characteristics that influence bacterial fate and transport.

The TMDL development protocol developed by the Florida Department of Environmental Protection (FDEP 2006) includes a suite of MST methods (e.g., antibiotic resistance analysis, carbon source utilization, F^{+} coliphage serotyping, ribotyping) to aid in the bacterial source identification process. The Florida protocol recommends that simple chemical (e.g., presence of detergents) or biological tracers be used if FIB sources cannot be identified with existing data or through simple field inspections. If these simple tracers provide insufficient information to identify fecal sources, the protocol recommends the use of MST for source identification.

The Texas Commission on Environmental Quality (TCEQ) and Texas State Soil and Water Conservation Board (TSSWCB) developed a recommended three-tiered protocol for developing pathogen TMDLs and TMDL Implementation Plans (IPs) (Jones et al. 2007). Tier 1 involves development of a GIS inventory for the watershed and the use of the Load Duration Curve (LDC) method to develop the TMDL. If agency personnel and the watershed stakeholders determine that this approach is not sufficiently comprehensive to adequately define the TMDL, then the group may choose to proceed to Tier 2. Tier 2 includes a targeted monitoring plan to fill gaps in previously collected data, the use of qualitative library-independent MST methods (e.g., Bacteriodetes detection via PCR) to determine pathogen sources, and a nonmodeling linkage analysis such as LDC or mass balance calculations. For the most complex watersheds, the Tier 2 approach may be inadequate, and agency personnel and stakeholders may choose to proceed to Tier 3, which involves more extensive targeted monitoring and a quantitative library-dependent MST analysis (e.g., ribotyping, antibiotic resistance analysis).

Virginia and several other states routinely use MST data to corroborate water-quality model results. Model output is compared to ranges of reported MST source distribution data to adjust models to ensure that the range of observed MST data is being reflected in model output (VADEQ 2006). A few states have used MST with nonmodeling linkage analyses to partition loads and to estimate reductions from source categories for some TMDLs (NHDES 2004; VADEQ 2005, 2007). Table 14.3 provides examples of MST data use in the development of pathogen TMDLs. The MST techniques and associated targets, as well as the extent of resultant data use in TMDL development, differ widely between states. Also, it is worth noting that the targets of the MST methods often differ from the TMDL targets (e.g., *Enterococcus* vs. *E. coli*). Since it is broadly assumed that generation and fate and transport characteristics of microorganisms can vary by species or type, there is some question as to the appropriateness of this link, i.e., whether enterococci distributions can be used in attaining a quantitative *E. coli* water-quality criterion.

14.5 Limitations and Opportunities for MST in the TMDL Development Phase

Limitations and opportunities for MST application exist primarily in the linkage analysis portion of TMDL development, especially in regard to modeling applications, and allocation analysis. Recent literature reviews (Pachepsky et al. 2006; Arnone and Walling 2007) identified a fundamental lack of understanding of bacterial fate and transport behavior as potentially seriously compromising the accuracy of quantitative water-quality modeling.

MST data generally do not reflect the climatic, seasonal, or spatial variability, which typically influences FIB loading in a watershed. This limited representation is

Table 14.3 Current examples of the application of MST data to TMDL development

State	MST method	MST target	TMDL load calculation method	TMDL target	TMDL stage	References
Maryland	ARA[a]	Enterococci	LDC[b]	*E. coli*	Source assessment	MDE (2007)
Michigan	PCR[c]/qPCR[d]	Enterococci/ Bacteroidetes	LDC	*E. coli*	Source assessment	Michigan DEQ (2008)
Missouri	RepPCR[e]	*E. coli*	SWAT[f]	Fecal coliform	Model calibration	Baffaut and Benson (2003)
New Hampshire	Ribotyping	*E. coli*	Mass balance	Fecal coliform	Load partitioning	NHDES (2004)
			Percent reduction	*E. coli*	Source assessment	NHDES (2006)
New Jersey	Genotyping/ARA	Coliphage/*E. coli*	WTM[g]	*E. coli*	Source assessment	NJDEP (2007)
New York	PFGE[h]	*E. coli*	WTM	fecal coliform	Source assessment	Battelle (2006)
North Dakota	Ribotyping	*E. coli*	LDC	Fecal coliform	Source assessment	NDDH (2005)
Oregon	Ribotyping/PCR	*E. coli*/enterococci/ Bacteroidetes	LDC	*E. coli*	Source assessment	ODEQ (2007)
South Dakota	PFGE	*E. coli*	HSPF[i], LDC	Fecal coliform	Source assessment	SDDENR (2009)
Texas	Ribotyping	*E. coli*	SWAT, LDC	*E. coli*	Source assessment	TCEQ (2007)
Virginia	PFGE	*E. coli*	HSPF	Fecal coliform	Source assessment	NVRC (2002)
	Ribotyping	*E. coli*	HSPF	Fecal coliform	Model calibration	Hyer and Moyer (2003, 2004)
	ARA	*E. coli*	LDC	*E. coli*	Load partitioning	VADEQ (2005)
			Mass balance	Fecal coliform	Load partitioning	VADEQ (2007)
			HSPF	*E. coli*	Source assessment	VADEQ (2008)

[a]Antibiotic resistance analysis
[b]Load duration curve
[c]Polymerase chain reaction
[d]Quantitative PCR
[e]Repetitive extragenic palindromic PCR
[f]Soil water assessment tool
[g]Watershed treatment model
[h]Pulse field gel electrophoresis
[i]Hydrological Simulation Program-FORTRAN

Table 14.4 ARA results for *E. coli* for North Buffalo Creek, Aycock sample site (after MapTech-HDR 2004)

Sample date	*E. coli* (cfu/100 ml)	Fecal coliform (cfu/100 ml)	Isolates	Bacteria source (%) Human	Domestic	Wildlife
8/23/04	230	160	26	0	4	96
8/30/04	6,000	6,900	48	6	56	38
9/14/04	800	570	48	2	0	98
9/27/04	6,900	6,300	48	17	4	79
10/5/04	760	470	48	27	56	17
10/26/04	700	410	48	0	60	40
11/8/04	800	540	48	69	12	19
11/22/04	231	320	48	15	4	81
12/6/04	173	200	48	2	0	98
12/28/04	134	210	48	2	6	92
1/18/05	226	310	48	46	4	50
1/25/05	72	80	38	8	37	55

typically due to sampling programs, which are typically limited both spatially and temporally. As MST data are often reported in terms of an average percent distribution across a few broadly defined source categories (e.g., human, domestic or livestock, and wildlife), no information regarding temporally or spatial data variability is provided. Table 14.4, illustrates MST data collected at a single location over a 5 month period. In this example, the data reported in the TMDL showed the human signal ranging from 0 to 69%, domestic from 0 to 60%, and wildlife from 17 to 98%.

While an MST estimated distribution of sources can be used to inform simulated fate and transport mechanisms, it should not be used as a substitute for a watershed-scale source characterization. In one TMDL study where an attempt was made to use MST data quantitatively to match simulated average source-specific FIB loads with an MST source distribution, the adjustment produced a highly improbable population for one source (pets, cats specifically) (Moyer and Hyer 2003). Keeling et al. (2005) suggest that MST data can be used qualitatively to ensure that the identified FIB sources are represented in the comprehensive watershed-scale FIB source characterization and are included in the simulation and that the simulated and MST data are in general agreement. Although they recognize the merits of qualitative uses of MST data, Keeling et al. (2005) and others (Field and Samadpour 2007; Santo Domingo et al. 2007) have concluded that using MST data quantitatively (as specific numeric targets) for source loading characterization, model calibration/verification, or load partitioning is not appropriate.

14.6 Implementation

The ultimate objective underlying TMDL development is the attainment of ambient water-quality standards through the control of both point and NPS pollution. While the US CWA has no specific provisions for TMDL implementation planning or execution, TMDL implementation is recognized as essential to achieving program goals and is now a statutory requirement in some states. Support for TMDL implementation is evident in federal guidance (USEPA 1991, 2000; Myers and Wayland 2002). The 1991 "Guidance for Water Quality-Based Decisions: The TMDL Process" (USEPA 1991) recommends the following minimum elements for a TMDL Implementation Plan (IP):

- A description of the implementation actions and management measures
- A timeline for implementing these measures
- Legal or regulatory controls
- Designated time period for attainment of water-quality goals
- Monitoring plans and milestones for attaining these goals

While the most common state recommendations and requirements for implementation plans closely track these minimum elements, states exercise significant flexibility in how they approach their TMDL implementation responsibilities. In lieu of federal requirements for implementation, more than a dozen states have passed laws and/or established regulatory guidance related to TMDL implementation planning. Implementation plans are consequently being implemented across the country using a variety of approaches, with varying levels of detail, stakeholder participation, and success. An examination of TMDL implementation case studies conducted by Benham et al. (2008) indicated that there is no one-size-fits-all approach. Every watershed is unique – not just in terms of environmental features but also in its regulatory landscape, socioeconomic factors, and many other location-specific characteristics, problems, risks, and resources.

The minimum elements listed above describe standard implementation whereby a plan is developed and then followed until water-quality standards are met. By contrast, adaptive implementation is frequently used as a flexible means of responding to the dynamic complexities of most watersheds (Freedman et al. 2004; Shabman et al. 2007). Adaptive implementation can facilitate effective planning and execution if watershed managers have the ability and willingness to adjust the implementation actions based on water-quality response. An adaptive approach to TMDL implementation allows for adjustments to evolving realities on the ground. Adaptive implementation, using interim goals and milestones, helps to reduce uncertainty by utilizing targeted monitoring, ongoing research, and experimental data obtained as the plan is implemented (Shabman et al. 2007).

Application of MST data for TMDL implementation purposes has rarely been cited in the literature; however, this is not surprising, as implementation is just beginning nationwide, and information on the related activities is sparse (Benham et al. 2008). One example of the use of MST analyses within the implementation process is in New Hampshire, where ribotyping (Table 14.3) was used to identify primary bacterial sources in impaired water bodies and to guide restoration efforts once the TMDL was developed (NHDES 2006). Many of the states that performed some type of MST analysis for source characterization also mention use of MST data during TMDL implementation (NDDH 2005; TDEC 2005; Jones et al. 2007), though it is unclear whether these recommendations were implemented.

Periodic MST sampling following implementation of corrective/restorative actions (e.g., the installation of BMPs) would likely be of particular use in adaptive management scenarios to determine whether pollutant load signatures from targeted sources were decreasing. However, it is worth noting that, with the current state of MST science, a single MST sample is very limited in the information it provides. While a series of monthly FIB concentrations can be used to show overall trends in bacteria reductions in response to various treatments, source distributions between two MST samples, one before treatment and one after, can only show relative differences in percent distribution and cannot tell whether any given source has increased or decreased, except in relationship to other sources.

Although MST data cannot be used to quantify load reductions from specific sources, MST may be useful in indicating qualitative changes in specific source categories. For example, if implementation efforts target failing septic systems and illicit sewer discharges and if temporal MST data indicate that bacteria from human feces are declining, this would suggest that restoration is moving in the correct direction. By contrast, MST data could also be used to demonstrate the failure of initial attempts to reduce human sources of contamination and thereby suggest the need to identify new strategies to address these sources.

As in all phases of the TMDL process, integration of MST sampling in the implementation phase will require a reanalysis of current FIB standards in relation to MST data. In particular, before MST data can be used to facilitate water body delisting following successful TMDL implementation, MST targets must not only be correlated with existing water-quality criteria, but the absence of these targets (i.e., failure to detect) must also be strongly correlated with an absence (or acceptable level) of human health risk.

14.7 Conclusion: Future Research Needs

Although the USEPA TMDL program is frequently cited as a primary driver of MST method development, at present there is only very limited use of MST in TMDLs. As discussed previously, MST is currently most often used in parallel with modeling efforts to increase stakeholder confidence and to verify model results. Application of MST to additional phases of the TMDL process may offer substantial opportunities to render the program more successful and efficient in achieving water-quality goals. Recommended research activities to improve the utility of MST in TMDL and other watershed management activities are listed in Table 14.5.

Limitations to the use of MST exist in all stages of the TMDL process. Underlying all of these limitations is the uncertainty with which a derived MST source distribution, obtained from a limited sample data set, can adequately and quantitatively represent the true distribution in time and space of various source populations. At a minimum, sampling theory should be applied to determine minimum sample sizes for estimating source distribution with a desired level of confidence (Kern et al. 2002). Sampling in MST studies is not typically used to discern

Table 14.5 Recommended research activities to improve the utility of MST in TMDL and other watershed management activities

TMDL phase	Primary research needs
Impairment designation	Adoption of standard analytical methods
	Reduction in MST uncertainty
	Correlation of MST targets and indicator organisms/human health risk
	Improved understanding of the relative risks of human and nonhuman fecal contamination to human health
	Development of regulatory tools to prioritize impairments based on health risk
TMDL development	Improved "quantification" of MST targets/reductions in MST uncertainty
	Confirmation of the validity of basic model processes and strategies
	Comparison of similar generation, transport, and fate of MST and model targets
	Integration of differential survival etc. into model frameworks
	Standardized method for incorporating "indeterminate" sources
	Verification of the accuracy of downstream model predictions
Implementation	Development of BMP designs to target identified problem sources
	Improved "quantification" of MST targets/reductions in MST uncertainty
	Development of efficient monitoring strategies
	Improved understanding of the relative risks of human and nonhuman fecal contamination to human health
Delisting	Confirmation of the absence of MST targets as equivalent to the absence of health risk
	Development of regulatory tools to prioritize impairments based on health risk

between spatial and temporal variations, and so statistically based sampling protocols have not yet been developed. However, it is reasonable to expect that the number and location of sampling sites in a watershed will influence the ability to detect trends and patterns in FIB source distribution (Simpson et al. 2002). Additional factors that influence the variability between samples include temperature and other seasonal changes, chemical changes, flow dynamics, and storm characteristics, especially differences between baseflow and stormwater conditions.

The US TMDL Program will only be successful in achieving improvements in the quality of the nation's surface waters if the primary sources of contamination responsible for impairments are accurately identified, quantified, and remediated. Although MST analyses offer substantial potential to address these issues, it is important to emphasize that the TMDL process is inherently quantitative and directly tied to FIB-specific criteria. Future MST research and development should focus on describing the contamination of surface waters by different microbial sources and on the quantification and variability of loadings and risk across space and time to maximize MST application and usefulness to the current TMDL framework.

References

Anderson KL, Whitlock JE, Harwood VJ (2005) Persistence and differential survival of fecal indicator bacteria in subtropical waters amd sediments. Appl Environ Microbiol 71(6): 3041–3048

Arnold JG, Fohrer N (2005) SWAT2000: Current capabilities and research opportunities in applied watershed modeling. Hydrol Process 19(3):563–572

Arnone RD, Walling JP (2007) Waterborne pathogens in urban watersheds. J Water Health 5(1):149–162

Baffaut C, Benson VW (2003) A bacteria TMDL for Shoal Creek using SWAT modeling and DNA source tracking. In: Total Maximum Daily Load (TMDL) Environmental Regulations-II: Proceedings of the November 8-12, 2003 Conference, (Albuquerque, New Mexico, USA). pp. 35-40. St. Joseph, MI: ASAE

Battelle (2006) Final Report for Peconic Bay Pathogens TMDL. Prepared for U.S. Environmental Protection Agency Oceans and Coastal Protection Division. Available at: http://www.peconic-estuary.org/Pathogen_TMDL.pdf. Accessed 19 May 2010

Benham B, Zeckoski R, Yagow G (2008) Lessons learned from TMDL implementation case studies. Water Pract 2(1):1–13

Benham BL, Baffaut C, Zeckoski RW et al (2006) Modeling bacteria fate and transport in watersheds to support TMDLs. Trans ASABE 49(4):987–1002

Benham B, Brannan K, Yagow G et al (2005a) Development of bacteria and benthic total maximum daily loads: a case study, Linville Creek, Virginia. J Environ Qual 34:1860–72

Benham B, Yagow G, Barham B et al (2005b) Total Maximum Daily Load Development: Mill Creek bacteria (*E. coli*) impairment, Page County, Virginia. Richmond, VA: Virginia Department of Environmental Quality. Available at: http://www.deq.virginia.gov/tmdl/apptmdls/shenrvr/millpage.pdfwww.deq.virginia.gov/tmdl/apptmdls/shenrvr/millpage.pdf. Accessed 17 December 2009

Bolin C, Brown C, Rose JB (2004) Emerging zoonotic diseases and water. In: Waterborne zoonoses: Identification, causes, and control, J. A. Cortuvo et al., eds., IWA Publishing, London, 19–26

Bonta JV, Cleland B (2003) Incorporating natural variability, uncertainty, and risk into water quality evaluations using duration curves. J Am Water Res Assoc 39:1481–96

Borah DK, Bera M (2003) Watershed-scale hydrologic and nonpoint-source pollution models: Review of mathematical bases. Trans ASABE 46:1553–66

Caraco D (2002) The Watershed Treatment Model Version 3.1. Center for Watershed Protection, Ellicott City, MD

Chapra SC (2003) Engineering water quality models and TMDLs. J Water Res Plan Manage 129(4):247–56

Clary J, Jones JE, Urbonas BR et al (2008) Can Stormwater BMPs Remove Bacteria? New findings from the International Stormwater BMP Database. Stormwater. 9. Available: http://www.stormh2o.com/may-2008/bacterial-research-bmps.aspx. Accessed 17 December 2009

DePinto JV, Freedman PL, Dilks DM et al (2004) Models quantify the total maximum daily load process. J Environ Eng 130:703–13

Donigian AS, Huber WC, Barnwell TO (1991) Modeling of nonpoint-source water quality in urban and non-urban areas. EP1.23:600/3-91/039. Athens, GA: USEPA Environmental Research Laboratory, Office of Research and Development

Fayer R, Dubey JP, Lindsay DS (2004) Zoonotic protozoa: from land to sea. Trends Parasitol; 20(11):531–536

Ferguson C, de Roda Husman AM, Altavilla N et al (2003) Fate and transport of surface water pathogens in watersheds. Crit Rev Environ Sci Tech 33(3):299–361

Field KG, Samadpour M (2007) Fecal source tracking, the indicator paradigm, and managing water quality. Water Res 41:3517–3538

FDEP (Florida Department of Environmental Protection) (2006) TMDL Protocol. Task Assignment 003.03/05–003

Freedman PL, Nemura AD, Dilks DW (2004) Viewing total maximum daily loads as a process, not a singular value: adaptive watershed management. J Environ Eng 130:695–702

Gaffield SJ, Goo RL, Richards LA et al (2003) Public health effects of inadequately managed stormwater runoff. Am J Public Health 93(9):1527–1533

Hagedorn C, Robinson SL, Filtz JR et al (1999) Determining sources of fecal pollution in a rural Virginia watershed with antibiotic resistance patterns in fecal streptococci. Appl Environ Microbiol 65(12):5522–5531

Harwood VJ (2007) Assumptions and limitations associated with microbial source tracking methods. In: Microbial Source Tracking, J. W. Santo Domingo and M. J. Sadowsky, eds., ASM Press, Washington, DC

Hrudey SE, Hrudey EJ (2007) Published case studies of waterborne disease outbreaks - evidence of a recurrent threat. Water Environ Res 79(3):233–245

Hyer KE, Moyer DL (2003) Patterns and Sources of Fecal Coliform Bacteria in Three Streams in Virginia, 1999-2000. USGS Water-Resources Investigations Report 034115. Available at: http://pubs.usgs.gov/wri/wri034115/wrir03-4115.pdf. Accessed 3 September 2009

Hyer KE, Moyer DL (2004) Enhancing fecal coliform total maximum daily load models through bacterial source tracking. J Am Water Res Assoc 40:1511–1526

Jamieson R, Gordon R, Joy D et al (2004) Assessing microbial pollution of rural surface waters: A review of current watershed scale modeling approaches. Agric Water Manage 70:1–17

Jarrell WM (1998) Getting Started With TMDLs. Oregon Institute of Science and Technology, Portland, OR

Jones CA, Wagner K, Di Giovanni G, et al (2007) Bacteria Total Maximum Daily Load Task Force Final Report. Texas Water Resource Institute, Texas. TR–341

Keeling WG, Hagedorn C, Wiggins BA et al (2005) Bacterial Source Tracking: Concept and Application to TMDL. In: Total Maximum Daily Load: Approaches & Challenges, T. Younos, ed., Pennwell Corp., Tulsa, OK, 207–237

Keller AA, Cavallaro L (2008) Assessing the US Clean Water Act 303(d) listing process for determining impairment of a waterbody. J Environ Manage 86:699–711

Kern J, Petrauskas B, McClellan P et al (2002) Bacterial Source Tracking: A Tool for Total Maximum Daily Load Development. In: Advances in Water Monitoring Research, T. Younos, ed., Water Resources Publications, LLC, Highlands Ranch, CO, 125–142

Krometis LH, Drummey PN, Characklis GW et al (2009) Impacts of microbial partitioning on wet detention pond effectiveness. J Environ Eng 135(9):758–767

MapTech-HDR Team (2004) Pathogen source assessment for TMDL development and implementation in North Carolina Piedmont and Coastal Plain Watersheds. Prepared for North Carolina Department of Environment and Natural Resources Division of Water Quality. Available at: http://h2o.enr.state.nc.us/tmdl/documents/Piedmont-Coastal-BSTReportMay05.pdf. Accessed 08 September 2009

MDE (Maryland Department of the Environmental) (2007) Total Maximum Daily Loads of Fecal Bacteria for the Non-Tidal Gwynns Falls Basin in Baltimore City and Baltimore County, Maryland. USEPA Approval Date: December 4, 2007

MDEQ (Mississippi Department of Environmental Quality) (2006) Fecal Coliform TMDL for Cypress Creek, Tombigbee River Basin, Oktibbeha County, Mississippi. Final Report. [cited 2007 Aug 9]. Available from: URL: http://www.deq.state.ms.us/MDEQ.nsf/pdf/TWB_TombigbeeRBCypressCreekFecalColiformJul06/$File/TombigbeeRBCypressCreekFecalColiformJul06.pdf?OpenElement. Accessed 17 December 2009

Michigan DEQ (Michigan Department of Environmental Quality) (2008) Detroit River & Ecorse River *E. coli* Monitoring to Support TMDL Development Final Report. Prepared for U.S.

Environmental Protection Agency Region V and Michigan Department of Environmental Quality Water Bureau. Task Order No. 2006–39
Moyer DL, Hyer KE (2003) Use of the Hydrological Simulation Program–FORTRAN and Bacterial Source Tracking for Development of the Fecal Coliform Total Maximum Daily Load (TMDL) for Accotink Creek, Fairfax County, Virginia. Water-Resources Investigations Report 03-4160. Richmond, VA: U.S. Geological Survey
Muniesa M, Payan A, Moce-Llivinia L et al (2009) Differential persistence of F-specific RNA phage subgroups hinders their use as single tracers for faecal source tracking in surface water. Water Res 43:1559–1564
Muñoz-Carpena R, Vellidis G, Shirmohammadi A et al (2006) Evaluation of Modeling Tools for TMDL Development and Implementation. Trans ASABE 49:961–65
Myers CF and Wayland RH (2002) Memorandum: Supplemental Guidelines for the Award of Section 319 Nonpoint Source Grants to States and Territories in FY 2002 and Subsequent Years. Washington DC: US EPA, Office of Wetlands, Oceans, and Watersheds. 2001 September Memorandum. Available from: URL: http://www.epa.gov/nps/Section319/fy2002.html. Accessed 17 December 2009
NDDH (North Dakota Department of Health) (2005) Cannonball River, North Dakota Bacteria Total Maximum Daily Load. http://www.ndhealth.gov/WQ/SW/Z2_TMDL/TMDLs_Completed/Cannonball_Bacteria_TMDL_20050623.pdf. Accessed 17 December 2009
Neilson BT, Hornsburg JS, Stevens DK, et al (2003) EPRI's Watershed Analysis Risk Management Framework (WARMF) vs. USEPA's Better Assessment Science Integrating Point and Nonpoint Sources (BASINS). In ASAE Proc. Total Maximum Daily Load (TMDL) Environ. Regulations II; Albuerque (NM), USA. St. Joseph, MI. p. 460–70
NHDES (New Hampshire Department of Environmental Services) (2004) TMDL Study for Bacteria in Hampton/Seabrook Harbor. USEPA Approval Date: May 20, 2004
NHDES (New Hampshire Department of Environmental Services) (2006) TMDL Study for Bacteria in Sand Dam Village Pond Town Beach, Troy, New Hampshire. USEPA Approval Date: November 8, 2006
NJDEP (New Jersey Department of Environmental Protection) (2007) Total Maximum Daily Loads for Pathogens to Address 17 Lakes in the Lower Delaware Water Region. USEPA Approval Date: September 28, 2007
Noble RT and Weisberg SB (2005) A review of technologies for rapid microbial detection in recreational waters. J Water Health 3(4):381–392
Novotny V (2003) Water Quality: Diffuse Pollution and Watershed Management. Wiley, New York, NY
Novotny V, Chesters G (1981) Handbook of Nonpoint Pollution: Sources and Management. Van Nostrand Reinhold. New York, NY
NRVC (Northern Virginia Regional Commission) (2002) Fecal Coliform TMDL Development for Four Mile Run, Virginia. USEPA Approval Date: May 31, 2002. http://www.deq.virginia.gov/tmdl/apptmdls/potrvr/fourmlrn1.pdf. Accessed 12 April 2011
ODEQ (Oregon Department of Environmental Quality) (2007) Bear Creek Watershed TMDL. USEPA Approval Date: October 2, 2007. http://www.deq.state.or.us/wq/tmdls/docs/roguebasin/middlerogue/bearcreek/tmdlchp1sec345.pdf. Accessed 17 December 2009
Pachepsky YA, Sadeghi AM, Bradford SA et al (2006) Transport and fate of manure-borne pathogens: modeling perspective. Agricultural Water Management 86:81–92
Parajuli P, Mankin KR, Barnes PL (2005) Calibration and validation of SWAT/Microbial submodel for fecal coliform bacteria prediction on a grazed watershed. ASAE Paper No. 052126. St. Joseph, MI: ASAE
Rose JB, Daeschmer D, Easterling DR et al (2000). Climate and waterborne disease outbreaks. J Am Waterworks Assoc 92(9):77–87
Santo Domingo JW, Bambic DG, Edge TA et al (2007) Quo vadis source tracking? Towards a strategic framework for environmental monitoring of fecal pollution. Water Res 41:3539–3552
Scott TM, Rose JB, Jenkins TM et al (2002) Microbial source tracking: current methodology and future directions. Appl Environ Microbiol 68(12):5796–5803

Shabman L, Reckhow K, Beck MB et al (2007) Draft Adaptive Implementation of Water Quality Improvement Plans: Opportunities and Challenges. Durham, North Carolina, Nicholas School of the Environment and Earth Sciences, Duke University: 92 pp

Simpson JM, Santo Domingo JW, Reasoner DJ (2002) Microbial source tracking: state of the science. Environ Sci Tech 36(24):5279–5288

SDDENR (South Dakota Department of Environment and Natural Resources) (2009) Fecal Coliform Bacteria Total Maximum Daily Load (TMDL) for Beaver Creek, Fall River County, South Dakota. South Dakota Department of Environment and Natural Resources, South Dakota

Stewart JR, Santo Domingo JW, WadeTJ (2007) Fecal pollution, public health, and microbial source tracking. In: Microbial Source Tracking, J. W. Santo Domingo and M. J. Sadowsky, eds., ASM Press, Washington, DC

Stiles TC (2002) Incorporating Hydrology in Determining TMDL Endpoints and Allocations. In Proc. National TMDL Science and Policy 2002 Water Environment Federation Specialty Conference, Water Environment Federation, Alexandria, VA, 2002: 13 pp

Taylor LH, Latham SM, Wollhouse MEJ (2001) Risk factors for human disease emergence. Philosophical Transactions of the Royal Society of London: Series B, Biological Sciences; 356, 983–989

TCEQ (Texas Commission on Environmental Quality) (2007) Bacteria Total Maximum Daily Load Task Force Report. http://twri.tamu.edu/docs/bacteria-tmdl/FinalReport.pdf. Accessed 12 April 2011

TDEC (Tennessee Department of Environment and Conservation) (2005) TMDL for Pathogens in the Ocoee River Watershed (HUC 06020003) Polk County, Tennessee. USEPA Approval Date: October 17, 2005. http://www.state.tn.us/environment/wpc/tmdl/approvedtmdl/OcoeePathF1.pdf. Accessed 17 December 2009

USEPA (United States Environmental Protection Agency) (1986) Ambient Water Quality for Bacteria - 1986. Office of Water, Washington DC, available: http://www.epa.gov/waterscience/beaches/files/1986crit.pdf. Accessed 17 December 2009

USEPA (United States Environmental Protection Agency) (1991) Guidance for water quality-based decisions: The TMDL process. Technical Document EPA 440/4-91-001. Washington, DC: US Environmental Protection Agency, Office of Water (WH–553)

USEPA (United States Environmental Protection Agency) (2000) Revisions to the Water Quality Planning and Management Regulation and Revisions to the National Pollutant Discharge Elimination System Program in Support of Revisions to the Water Quality Planning and Management Regulation; Final Rules, 40 CRF Part 9 et al. (July 13, 2000). Available from: URL: http://www.epa.gov/owow/tmdl/finalrule/finalrule.pdf. Accessed 17 December 2009

USEPA (United States Environmental Protection Agency) (2001) Protocol for Developing Pathogen TMDLs. EPA 841-R-00-002. Office of Water (4503F). United States Environmental Protection Agency, Washington, DC. 132 pp

USEPA (United States Environmental Protection Agency) (2003) Bacterial Water Quality Standards for Recreational Waters: Status Report (EPA-823-R-03-008). Office of Water, Washington, DC

USEPA (United States Environmental Protection Agency) (2008) Handbook for Developing Watershed Plans to Restore and Protect Our Waters. EPA 841-B-08-002. United States Environmental Protection Agency, Washington, DC. 400 pp

USEPA (United States Environmental Protection Agency) (2009a) Watershed Assessment, Tracking and Environmental Results. http://iaspub.epa.gov/waters10/attains_nation_cy.control?p_report_type=T. Accessed 17 December 2009

USEPA (United States Environmental Protection Agency) (2009b) Review of zoonotic pathogens in ambient waters. In: EPA-822-R-09-002, Office of Water, Washington, DC

VADEQ (Virginia Department of Environmental Quality) (2005) Bacteria TMDL for Great Run, Fauquier County, Virginia. USEPA Approval Date: March 10, 2005. http://www.deq.virginia.gov/tmdl/apptmdls/rapprvr/greatfc.pdf. Accessed 17 December 2009

VADEQ (Virginia Department of Environmental Quality) (2006) Bacteria TMDL for Pigg River, Snow Creek, Story Creek, and Old Womans Creek. USEPA Approval Date: September 11, 2006. http://www.deq.virginia.gov/tmdl/apptmdls/roankrvr/piggec.pdf. Accessed 17 December 2009

VADEQ (Virginia Department of Environmental Quality) (2007) Rappahannock River: Carter Creek TMDL Report for Shellfish Areas Listed Due to Bacterial Contamination. USEPA Approval Date: September 20, 2007. http://www.deq.virginia.gov/tmdl/apptmdls/shellfish/cartersf.pdf. Accessed 17 December 2009

VADEQ (Virginia Department of Environmental Quality) (2008) Bacteria TMDL Development for Hays Creek, Moffatts Creek, Walker Creek, and Otts Creek in Augusta County and Rockbridge County, Virginia. VT-BSE Document No. 2008-0004. USEPA Approval Date: May 19, 2008. http://www.deq.virginia.gov/tmdl/apptmdls/jamesrvr/haysec.pdf. Accessed 17 December 2009

Wade TJ, Calderon RL, Sams R et al (2006) Rapidly measured indicators of recreational water quality are predictive of swimming-associated gastrointestinal illness. Environ Health Persp 114(1):24–28

Wiggins BA, Andrews RW, Conway RA et al (1999) Use of antibiotic resistance analysis to identify nonpoint sources of fecal pollution. Appl Environ Microbiol 65:3483–3486

Chapter 15
Relating MST Results to Fecal Indicator Bacteria, Pathogens, and Standards

Julie Kinzelman, David Kay, and Kathy Pond

Abstract The regulation of coastal, inland, and flowing waters through the assessment of fecal indicator bacteria (FIB) alone, while protective of public health, provides little information regarding contamination sources. The most effective protection of public health comes through the identification and mitigation or removal of the pollution source. Methods capable of assigning relative contributions to contamination from a variety of sources would be most beneficial to practitioners, as water-body health frequently suffers due to multiple inputs rather than a single, readily identifiable hazard (i.e., point source). The contribution of various potential sources to nonpoint source pollution, particularly in a comingled environment, is difficult to assess. Microbial source tracking (MST) offers the potential to distinguish between the multiple sources of catchment (watershed) fecal indicator sources to inform remediation strategies. Under ideal circumstances, FIB concentrations would relate to pathogen and MST marker survival, persistence, and ecology in aquatic habitats targeting source identification, attenuation, and thereby reduction of human health risk.

Keywords Microbial source tracking • Fecal indicator bacteria • Pathogens • Water quality • Standards • Sanitary surveys

15.1 Relating Microbial Source Tracking Results to Fecal Indicator Bacteria

Regulation of microbiological contamination of coastal waters has, until very recently, tended to focus on anthropogenic point source discharges that are mainly derived from human sewage systems. Control strategies to achieve water-quality targets in bathing

J. Kinzelman (✉)
City of Racine Health Department, Racine, WI 53403, USA
e-mail: julie.kinzelman@cityofracine.org

C. Hagedorn et al. (eds.), *Microbial Source Tracking: Methods, Applications, and Case Studies*, DOI 10.1007/978-1-4419-9386-1_15,

and shellfish waters have, in many cases, depended on costly engineering interventions to enhance sewage treatment using storm flow retention tanks and ultraviolet (UV) light for disinfection of treated effluents (Neto et al. 2006; Tessele et al. 2005; Zanetti et al. 2006). Many of these "technical fixes" have proven ineffective in achieving microbiological standards in receiving waters because the nonhuman microbial loading remains unaltered. The nonhuman component alone is often sufficient to cause "noncompliance" (EU terminology) or "impairment" (US terminology) of water quality (Kay et al. 2006a, b, 2010). In response, both the EU and US administrations have enacted catchment-based legislation that is designed to address water-quality deterioration through integrated management of diffuse and point sources of pollution. In the USA, this is encompassed in the Federal Water Pollution Control Act (CWA, USEPA 2005a) and the total maximum daily loads (TMDLs) concept (Kay et al. 2006b; USEPA 2005a). Section 303(d) of the CWA requires that States identify "impaired" water bodies that do not meet defined water-quality standards. The TMDL process investigates these water-quality problems, determines proportional load allocations among various contributing sources of contamination, and designs actions in consultation with stakeholders to effect remediation (see Chap. 14). The EU Water Framework Directive (WFD) (Anon 2000) parallels this process (Kay et al. 2006a) and EU member states are required to manage both point and diffuse sources of pollution to achieve "good" ecological status and "good" water quality by 2015. The chemical and microbiological status required of "good" water quality is specified in daughter Directives such as the Bathing Water Directive (Anon 2006). For a coastal beach to be classified as "good" it must have a 95th percentile intestinal enterococci concentration of less than 200 cfu/100 mL. Member states are required to achieve this standard through implementation of the Water Framework Directive's provisions (Anon 2000). This requires designated agencies in member states to identify pressures and impacts on water bodies (Article 5) and then to design a "program of measures" to achieve "good" water quality and "ecological status" within all EU drainage basins (Article 11). Microbial source tracking (MST) offers the potential to distinguish between the multiple sources of catchment fecal indicator sources to inform remediation strategies.

A series of key questions and observations relevant to the assessment of the operational utility of MST were set out in an excellent review by Santo Domingo et al. (2007). They note the requirement for MST to provide a reliable indication of individual fecal indicator bacteria (FIB) sources within catchment systems. This is essential for the implementation of targeted best management practices (BMPs) to attenuate the principal fluxes causing water-quality impairment. They make three key points of relevance to this chapter, as follows:

1. Perhaps the most critical issue in MST is the lack of performance standards to evaluate the accuracy of any of the existing and emerging methods (as originally noted by Stoeckel and Harwood 2007).
2. To fully validate the potential of MST, long-term, large-scale field studies need to be conducted with the methods that meet standardized performance criteria.

3. Ultimately, quantitative assays will be needed for the TMDL process to establish fecal allocations and to predict the levels of reduction that can be achieved by targeting particular sources. Such assays are also needed to further evaluate the efficacy of management practices at temporal and spatial scales.

Thus, to be of relevance to the catchment science community, which includes the regulators charged with achieving water-quality objectives based on FIB levels, MST must provide quantitative information on the relative importance of FIB fluxes from the different "tracked" contributor species, say ruminant and human. This implies the following:

(a) The MST markers correlate closely with the FIB regulatory parameters.
(b) The MST markers have similar environmental fate and transport characteristics compared to the regulatory FIBs.

Furthermore, the practice of taking a sample from a water body for MST analysis and then using that sample to indicate the percentage contributions of, for example, human and ruminant FIB implies the following:

(c) The water body can be reliably characterized by the MST sample.

As noted by Santo Domingo et al. (2007), objective evaluation of MST against these criteria needs large-scale, long-term investigations (see Sect. 15.2 above), which would be best conducted at field sites where FIB compliance is the principal management objective. Such catchment-scale investigations are sparse but are increasingly represented in the literature (Carroll et al. 2009; Fremaux et al. 2009; Reischer et al. 2008; Shanks et al. 2006; Wapnick et al. 2009).

15.1.1 Relationship of FIB Density to MST Results

The key operational question in assessing the relationship between FIB and MST parameters is "can quantification of an MST parameter or parameters predict, or index, the concentration of FIBs?" This question addresses criterion C above. In theory, if a dominant contributor of fecal contamination is human-based, one may anticipate a correlation to increased FIB but an evidentiary approach must be employed when evaluating MST as an operational tool. Clearly, parametric statistical correlation/regression would be required to quantify the variance explained in the FIB parameters by the MST results. Nonparametric (ranking and or categorical) analysis (Fremaux et al. 2009) might suggest an association and its significance, but they would not provide the policy maker with a clear definition of the variance explained in the key regulatory parameters of interest (i.e., the predictive, and hence operational, utility of the MST result). The numbers of studies that have reported testing a parametric relationship at the catchment scale is not extensive. However, Reischer et al. (2008) provide a credible assessment of the explained variance in *E. coli* provided by ruminant-specific marker (BacR) assessed by qPCR in *Bacteriodales* (Reischer et al. 2006, 2007). The study was conducted on high-flow

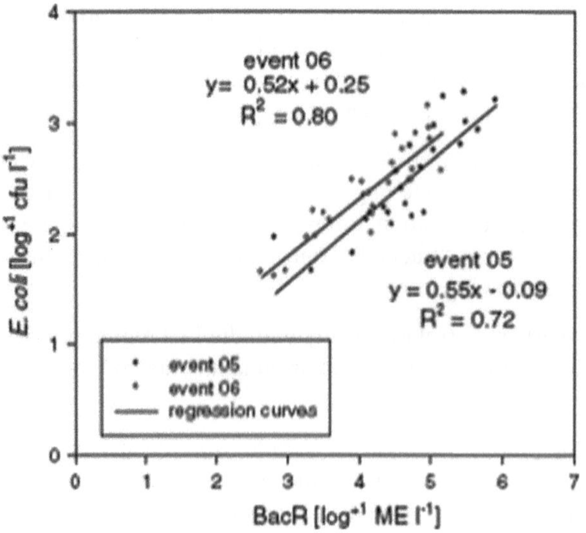

Fig. 15.1 Regression analysis of BacR and *E. coli* data from Events 05 and 06. BacR, ruminant-specific marker; marker equivalents (ME) (source: Reischer et al. 2008: Fig. 4)

event samples in a karstic spring. Both events demonstrated strong correlations between the *E. coli* and BacR marker with explained variances of 80 and 72% for the two datasets, respectively (Reischer et al. 2006, 2007) (Fig. 15.1).

These data are encouraging and suggest a strong relationship between ruminant "marker equivalents" in the spring water, which are reasonably predictive of *E. coli* concentrations during elevations in flow. Nonparametric rank correlations were also reported by the same authors for event and base flow conditions, which suggested higher correlation coefficients between ruminant marker and FIBs than for human marker (see Table 2 in Reischer et al. 2008). It would have been useful to have analysis of the correlation between total *Bacteriodales* marker equivalents (ME), but this is not reported.

A study at marine bathing waters was conducted by regulators and sewage utilities in the UK who sought to address criteria A to C above (Stapleton et al. 2009). Temporal sequences (54 hourly samples) of MST markers, FIBs, and modeled upstream CSO discharges were reported for the principal stream input to the bathing waters (Fig. 15.2). The data suggested good correspondence between periods of CSO discharge and elevations in the human proportion of the MST signal in the stream water, which provides intuitive evidence that the MST marker information reflects actual contributions to the stream FIB loading. However, very different attenuation patterns were observed in FIBs and MST signal parameters, as the local sewage effluent passed through a tertiary treatment plant employing activated sludge and UV disinfection (non-CSO, Fig. 15.3). In addition, two adjacent marine bathing sites were sampled hourly for the same 54-h period to quantify the correlation between FIBs and between the MST markers at the two sites. FIBs (i.e., presumptive

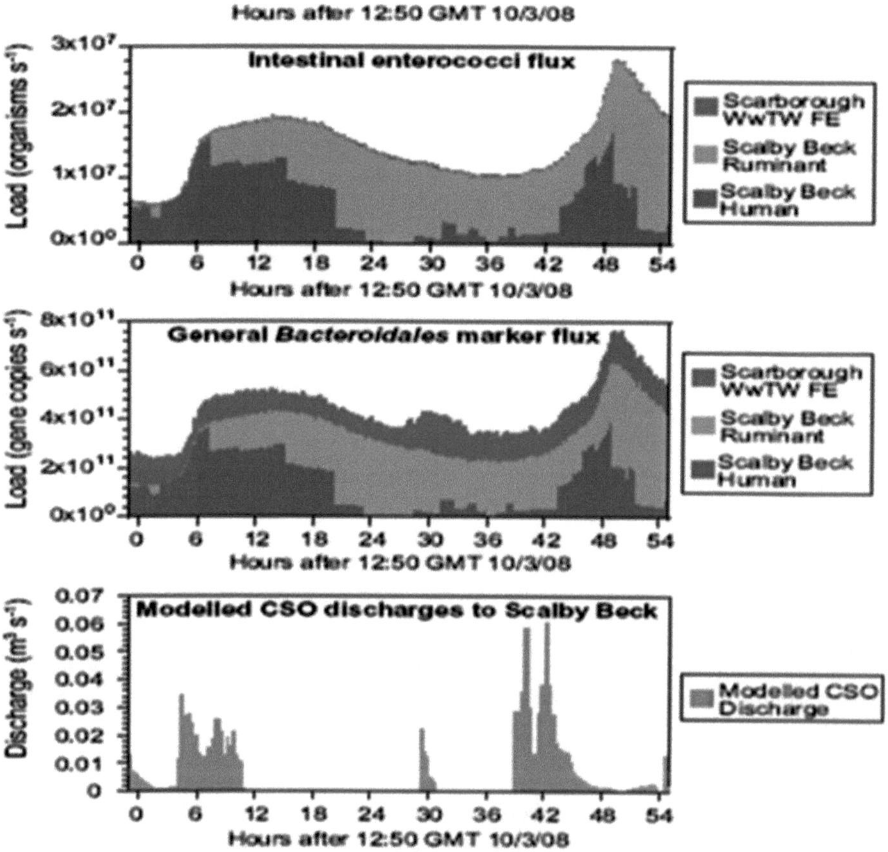

Fig. 15.2 Delivery of fecal coliforms, confirmed intestinal enterococci (organisms sL1) and general Bacteroidales (gene copies sL1) from Scalby Beck and Scarborough WwTW UV disinfected final effluent based on geometric mean concentrations. The Scalby Beck component has been separated into human and ruminant based on the relative ratio of the two specific markers in each sample applied to the period centered on each sample. The WwTW final effluent can be considered a human source. The *lower plot* shows modeled discharge from CSO contributing to Scalby Beck during the study (source: Stapleton et al. 2009: Fig. 3)

total coliform, fecal coliform, and intestinal enterococci) were enumerated by plate counts using standard UK methods (EA 2000), MST markers were host-specific (ruminant and human), and general *Bacteriodales* markers were enumerated by qPCR analysis (Stapleton et al. 2009). The FIBs showed significant parametric correlations ($p<0.05$) between the adjacent sites, but the MST markers showed no significant correlations between the two sites. A possible conclusion arising from this observation was that while either adjacent site could be sampled to characterize the "water body" FIB concentration, an MST sample from one site could not characterize the "water body" in terms of its source proportions of FIBs. These two studies illustrate the partial and, to some extent contradictory, nature of the evidence

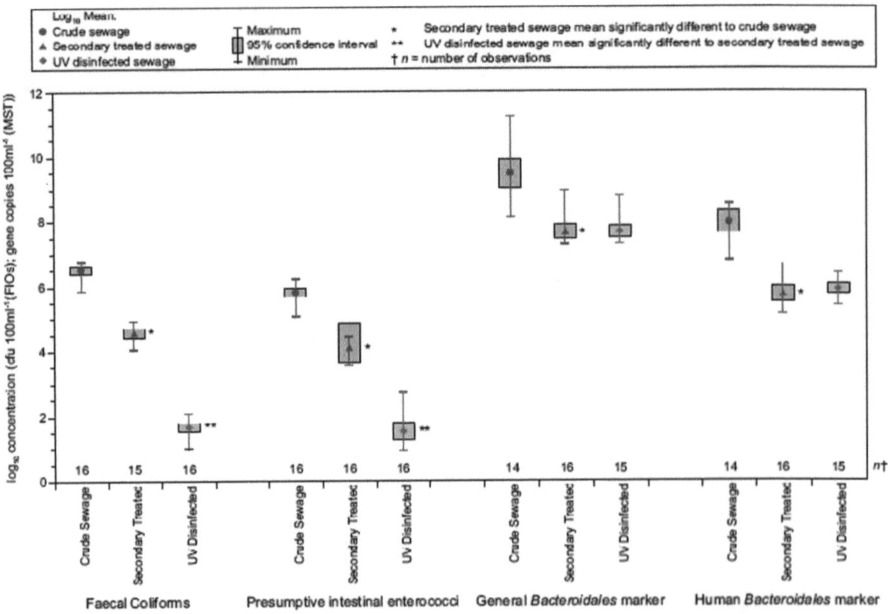

Fig. 15.3 Mean, range and 95% confidence intervals of the mean for log10 transformed fecal indicator organism (FIB) concentrations (cfu 100 mL L1) and *Bacteroidales* marker concentrations (gene copies 100 mL L1) in sewage effluent samples (10/3/08–12/3/08) (source: Stapleton et al. 2009: Fig. 4)

based on available correlations between MST parameters and FIBs. Furthermore, they suggest some pressing questions if the criteria suggested by Santo Domingo et al. (2007) are to be addressed.

There are clearly information gaps and associated hypotheses that require well-designed catchment-scale experiments if MST is to achieve sustainable acceptance by the regulatory and operational communities. Several catchment-scale investigations have reported apparent associations between elevated FIB concentrations and human (Stapleton et al. 2009) and/or ruminant markers (Shanks et al. 2006) in different hydrological components of the catchment. These observations may be entirely appropriate and reflect real catchment-specific differences in contamination source, which is useful to management efforts. However, the key question remaining is whether the implied percentage of human and ruminant contributions, as calculated from the human and ruminant *Bacteriodales* marker proportions, can be used to infer the human and ruminant FIB contribution at a specific point in the catchment or within specific receiving waters. There is very little peer-reviewed catchment-scale information available to conclusively determine the strength of the evidence base for this central policy question. Differential attenuation through common sewage treatment systems of FIBs (enumerated by standard culture methods) and MST markers (qPCR determined) (Fig. 15.3 above) may explain poor correlations reported in receiving waters. However, this discrepancy casts doubt on the utility of commonly deployed qPCR-based MST approaches as indicators of relative human and ruminant contributions of "culturable" (i.e., compliance-relevant) FIBs, where the latter are attenuated during sewage treatment without affecting the MST signal. In addition to the correlation of culturable FIB to MST signal, one must also consider pathogens (viruses, protozoa, and resistant bacteria) that may pass through the treatment process. Little is known about the relationship between the pathogen signal and the MST signal and how this relates to culturable FIB.

It may be possible to envisage a complete paradigm shift in regulation toward the use of MST markers, or their host species (e.g., *Bacteriodales*), enumerated by qPCR, as the principal measure of compliance. This scenario could be potentially complemented by direct pathogen analyses, which are increasingly feasible through molecular methods (Field and Samaspour 2007). However, this change would be imprudent until international standards organizations had validated the precision and reproducibility of the methods for wide-scale application at multiple laboratories, producing statistically comparable results across continental areas that apply uniform regulation, such as the EU and USA. There would also be a need to build a credible evidence base of epidemiological information in which MST markers and/or enumerated pathogens are shown to consistently predict health outcomes before it would be sensible to replace existing regulatory parameters for which such information has been formally evaluated by World Health Organization (WHO) and others (WHO 2003). Construction and consideration of this evidence base for existing regulation has been necessarily meticulous and protracted but must be completed before the regulatory community could consider replacing FIBs as the principal compliance measures for environmental waters.

This caution should not be taken to detract from the considerable potential utility of MST markers as exploratory tools for catchment management. Figures 15.1 and 15.2 present a small sample of the peer-reviewed literature evidence base that this tool offers useful qualitative insight into FIB source at the catchment scale. However, extreme caution is required in the interpretation of single (or a few) MST results, which are often presented without hydrological context. The advice to the UK regulators after the Stapleton et al. (2009) study was that several tens of MST results should be obtained and statistically analyzed before inference concerning proportional source contributions to FIB levels that cause impairment of specific receiving waters could be made with any degree of confidence. MST and sampling are discussed in Chap. 16, and many case studies using a wide variety of MST methods are described in Chaps. 3, 6, 18, and 21.

15.2 Relating MST to Pathogens

15.2.1 Do MST Results Correlate to Pathogen Presence?

The relationship of pathogen occurrence to fecal indicator organisms is poorly understood. Fecal pollution of surface waters for permitting purposes and TMDLs is evaluated using FIB (*E. coli* or enterococci) (Duris et al. 2009), but high FIB levels in recreational and other waters may not correlate with recent fecal pollution because of survival and regrowth in the environment (Carrillo et al. 1985; Desmarais et al. 2002; Shibata et al. 2008). Water managers would benefit from more specific knowledge regarding the presence, abundance, and diversity of pathogens, and the sources of fecal contamination. Location and routes of nonpoint source pollution can be extremely difficult to ascertain, and MST can aid in the identification of the source of microbial pollution at the site of contamination. In addition, MST may also be used to identify agents and sources of agents used in the intentional use of microbial agents for bioterrorism (Chap. 23) or the accidental release of microorganisms or toxins of public health importance into water systems. Although there have been some associations between high levels of indicator bacteria and disease outbreaks (Chao et al. 2003; Chou et al. 2004; Strauss et al. 1995), there is little or no prediction of specific sources of contamination or correlation with human pathogens when using these indicators.

New molecular-based techniques have shown that combined use of conventional and alternative indicators for fecal pollution increases both the detection sensitivity and specificity of fecal pollution and associated pathogens (Savitcheva and Okabe 2006). Methods for fecal source identification can be divided into culture-dependent (Chap. 3) and culture-independent (Chap. 4) methods. Some methods require a library, a set of bacterial isolates or patterns from fecal samples of known origin, beta-tested using the MST method to validate the library's capability for source discrimination, i.e., jackknife analysis (Stoeckel and Harwood 2007). Most library methods are culture-based and require growing environmental isolates from water samples (for a review see: Pond et al. 2004; Stoeckel and Harwood 2007). Source identification occurs by a comparison between test patterns from the library and the environmental isolates. Library-dependent methods often rely on molecular methods to generate characteristic profiles, or patterns, from the DNA of bacterial isolates. Such methods include REP-PCR, rapid amplified polymorphic DNA, restriction fragment length polymorphism, ribotyping, and pulsed-field gel electrophoresis (PFGE). Phenotypic methods for pattern generation may also be used, such as antibiotic resistance analysis (ARA) or carbon source utilization. Culture-dependent, library-independent methods are based on growing source-specific viruses or bacteria, such as culturing enterococci and testing for the *Enterococcus faecium esp* gene (Scott et al. 2005). Alternative fecal indicators such as fecal anaerobes (genera *Bacteroides* and *Bifidobacterium*, spore-forming *Clostridium perfringens*), viruses (*B. fragilis* phage, coliphages (FRNA phage)), and fecal organic compounds (coprostanol) together with conventional fecal markers are very useful to identify the source of fecal pollution (as reviewed by Savitcheva and Okabe 2006). Savitcheva and Okabe (2006) summarized the characteristics of conventional and alternative fecal indicators in terms of prediction of fecal pollution and associated pathogens (Table 15.1).

Probably the most frequently used methodology in the last few years employs the polymerase chain reaction (PCR) to amplify specific nucleic-acid sequences to detectable levels (Stewart et al. 2008). Application of molecular-based methods has provoked interest in directly monitoring for pathogens in surface waters (Scott et al.

Table 15.1 Characteristics of conventional and alternative fecal indicators to predict recent fecal pollution events and associated pathogens [adapted from Savitcheva and Okabe (2006)]

Conventional or alternative fecal indicator	Ability to predict pathogens and/or recent pollution events
Fecal indicator (*E. coli*, fecal coliforms, total coliforms, enterococci)	Correlation with pathogens or waterborne disease, particularly for point source pollution
Bacteroides spp. and *Bacteroidales*	Insufficient data on survival and correlation with pathogens
Bifidobacterium spp	Insufficient data on survival and correlation with pathogens
Clostridium perfringens	Good correlation with some pathogens in some studies (Davies et al. 1995; Morinigo et al. 1990)
F-specific RNA coliphage	Valuable model for viral contamination and inactivation due to their physical similarities (Kott et al. 1974; Sinton et al. 2002).
B. fragilis phage	Related to the level of human enteric viral pollution (Dutka et al. 1987; Tartera and Jofre 1987)
Fecal sterols	No data available on correlations with the presence of pathogens and public health risk

2002, Simpson et al. 2002; US EPA 2008). This approach allows researchers to rapidly and specifically target microbes of public health concern, including those that were previously unexamined because they were difficult or impossible to culture. Application of PCR-based technology has successfully detected the presence of *Salmonella* spp., (Fukushima et al. 2002; Horman et al. 2004); *Campylobacter* spp. (Lubeck et al. 2003), legionellae (Wellinghausen et al. 2001), *Vibrio vulnificus* (Panicker et al. 2004), different pathogenic strains of *E. coli* (Franck et al. 1998; Johnson et al. 2001), protozoan parasites and enteroviruses (Fout et al. 2003; Haramoto et al. 2004; Horman et al. 2004; Johnson et al. 1995; Guy et al. 2003), FIB (Haugland et al. 2005, Griffith and Weisberg 2006), bacterial pathogens (Walters et al. 2007; Hsu et al. 2007), and viral pathogens (Gregory et al. 2006; He and Jiang 2005). As with new indicators, nucleic-acid-based detection of specific pathogens will need to go through testing to determine what level of detection is associated with unacceptable human health risk (Stewart et al. 2008, Sect. 15.1).

Multiplex PCR has been recognized as a rapid and highly sensitive tool for the simultaneous detection of many organisms in a single PCR test (Franck et al. 1998; Fukushima et al. 2002). Singh et al. (2001) report on the use of multiplex PCR to distinguish *V. cholerae* O1, O139, non-O1, and non-O1 39 strains. Applications of 5′-nuclease PCR for quantitative detection of *Listeria monocytogenes* and *Campylobacter jejuni* have been reported by Novga et al. (2000). Ibekwe et al. (2002) describe the same technique for the identification and quantification of *E. coli* O157:H7. Molecular characterization of *Cryptosporidium* spp by Xiao et al. (2001) using RFLP identified and differentiated *Cryptosporidium* spp and *C. parvum* strains in water samples to evaluate sources of fecal contamination. Amplified fragment length polymorphism (AFLP) and enterobacterial repetitive intergenic consensus (ERIC) PCR have been used to identify the differences in pathogenicity between host-specific *E. coli* strains (Guan et al. 2002; Leung et al. 2004; Parveen et al. 1999).

Although PCR-based methods have been widely applied to direct pathogen detection, interpreting the results from a PCR assay can be problematic. A positive result indicates the presence of a pathogen (or pathogen DNA), but a negative result can be difficult to interpret without knowing the precise detection limit for the assay (Loge et al. 2002). Detection limits depend on a number of factors such as the volume of water filtered, the presence of inhibitory factors in the PCR reaction (Loge et al. 2002) and other factors. To increase the possibility of detection of pathogens, preculturing (enrichment) can be carried out, but this does not allow quantitative measurement of pathogens, only presence or absence (Savitcheva and Okabe 2006).

Other methods such as quantitative PCR have been successfully used for quantification of *Salmonella* spp. in environmental water samples (Fey et al. 2004); whole cell in situ hybridization (FISH) with fluorescently labeled oligonucleotide probes has been applied for rapid detection, identification, and enumeration of *E. coli* cells in municipal wastewaters (Stender et al. 2001). Genetic characterization of *Salmonella enterica* strains (Alverez et al. 2003) and detection of various pathogens including *E. coli* O157:H7 (Vora et al. 2004), *Listeria* (Volkhov et al. 2002), and human group A rotaviruses (Chizhikov et al. 2002) have been undertaken using microarray analyses. Further refinement of techniques, including maximizing the efficiency of nucleic-acid extraction and PCR amplification, might improve detection limits for pathogens in water. Knowledge of spatial and temporal variation, reproducibility, reliability of results, and stability of genetic markers are also essential for direct detection of pathogens and MST.

Maintenance of the microbiological quality and safety of water systems used for drinking, recreation, and in the harvesting of seafood is imperative, as contamination of these systems can result in risks to human health as well as significant economic losses due to closures of beaches and shellfish harvesting areas. Waters contaminated with human feces are generally regarded as a relatively great risk to human health, as they carry human-specific enteric pathogens, including *Salmonella enterica* serovar *typhi*, *Shigella* spp., hepatitis A virus, and Norwalk-group viruses (Hoebe et al. 2004). Animals can also serve as reservoirs for a variety of enteric pathogens (e.g., various serotypes of *Salmonella* spp., pathogenic *E. coli*, and *Cryptosporidium* spp.) (Scott et al. 2002). MST of pathogens is useful not only in the water industry but also to investigate foodborne pathogens (Suresh and Vega 2007, see Chap. 26). Gram-negative (bacteria), foodborne pathogens such as *Salmonella enterica*, *Campylobacter* spp., *Shigella* spp., *E. coli*, and *Yersinia enterocolitica* are a cause of morbidity and mortality throughout the world (Foley et al. 2009). Foley et al. (2009) reviewed a number of methods that have been used for molecular subtyping of foodborne pathogens as part of epidemiologic studies. While these methods demonstrated utility for this purpose, being highly discriminatory, the same level of success may not be noted when used as a source tracking tool. Each method has its advantages and disadvantages, i.e., speed versus reproducibility or cost versus discriminatory ability. PFGE appears to be the most suitable for molecular subtyping.

Recent literature indicates that multilocus sequence typing (MLST), multiple locus VNTR analysis (MLVA), and single-nucleotide polymorphism analysis were deemed to perform as well or better than PFGE for subtyping and that results can be obtained in a shorter time (Foley et al. 2009). The disadvantage of newer methods is the cost and specialized equipment required. The choice of the most appropriate method will depend on the requirements. If speed is required, for example, to investigate a limited disease outbreak a PCR-based method may be most appropriate for characterizing the isolates. If conducting an investigation with a wide geographical spread, involving multiple entities, a more robust method such as PFGE may be better to allow sharing of typing results obtained across multiple laboratories. The results need to be evaluated in the context of epidemiological data to be effective at understanding the distribution of foodborne pathogens and associated diseases (Foley et al. 2009). As discussed by Foley (2009), it may be that more than one method is needed in some situations. In these instances, a series of molecular and nonmolecular source attribution techniques may be appropriate.

15.3 Relating Microbial Source Tracking to Standards

15.3.1 Regulatory Approaches and Water-Quality Standards

Regulations and agency approaches (Chap. 13) to water-quality monitoring are currently based primarily on FIB standards. The results of these monitoring assessments are frequently used as the sole or principal element for evidence-based mitigation of water pollution, although recent research indicates that the presence, source, and public health significance may not be reliably assigned using fecal indicator organism assessments alone (Santo Domingo et al. 2007; Haack et al. 2009; Stapleton et al. 2009). While testing water quality against approved standards may provide a framework for gauging water-body health from a human health perspective, it only provides a fraction of the information necessary to initiate remediation when those same standards are not met.

The EU Bathing Water Directive (2006/7/EC), governing the quality of bathing and recreational waters, was revised and ratified in March 2006 (Anon 2006). The directive recommends phylogenetically defined microbial indicators (enterococci and *E. coli* alone rather than total and fecal coliforms) and sets tighter standards than the previous EU Bathing Water Directive (76/160/EEC). These new water-quality standards must be in full force by 2015. Implementation of these revised standards may have serious ramifications for member countries, i.e., potential increases in noncompliance and a downgrading of beach status if no mitigation occurs (Gawler et al. 2007). In the Great Lakes region of USA, the Great Lakes Regional Collaboration (GLRC) calls for a 90–95% reduction in bacterial, algal, and chemical contamination at all coastal bathing water through the identification of diffuse (nonpoint source) pollution sources, public education, and remediation of all potential diffuse pollution sources through source attribution, estimation of relative contribution (based on historical data and sanitary inspection), and mitigation of these sources (GLRC 2005). The GLRC (2005) further recommends that to achieve this goal, the USEPA should build upon existing approaches to water-quality monitoring programs, employing the latest technology for microbial source identification, standardized sanitary surveys, and a holistic watershed assessment.

In Chap. 7 of the *Water Quality Handbook* (USEPA 2007), a water-quality approach to pollution control is presented that is based on both monitoring and identification of impairments (point and nonpoint sources). This approach emphasizes the importance of the overall condition of the water body as well as providing a mechanism by which all pollution entering a water body may be categorized and accounted for. This is accomplished through eight stages; each stage represents a major Clean Water Act program with specific regulatory requirements (Chap. 14) and guidance (USEPA 2007):

- The first stage is to assign a designated use (i.e., primary or secondary contact).
- The second stage is to determine if water-quality standards are being met and detect pollution trends through the use of historic data and new data collected from ongoing monitoring programs (Francy et al. 2005).
- The third stage is to rank water bodies according to the severity of the pollution (i.e., risk to human health accessed via water-quality standards), the designated use, and socioeconomic considerations (recreational, economic, and aesthetic importance). The use of available resources to develop mitigation strategies is also formulated.
- In the fourth stage, the appropriateness of the water-quality standards for specific waters is evaluated with the opportunity to revise or reaffirm the water-quality standards.
- In the fifth stage, mathematical models and/or monitoring data are used to develop an integrated pollution reduction strategy while recognizing that water bodies may often be impacted by multiple pollution sources (TMDLs, Chap. 14).
- In the sixth stage, BMPs for nonpoint source pollution are implemented as the primary means of achieving compliance with water-quality standards.
- In the seventh stage, dischargers are monitored to determine whether or not they meet permit conditions and to ensure that expected water-quality improvements are achieved.
- In stage eight, the effectiveness of the control measures is assessed and a determination of whether water-quality standards have been attained or require revision is conducted. Attainment of water-quality standards may result in beneficial uses such as increased recreational opportunities that contribute to an improved economic condition.

Therefore, while routine monitoring via FIB is a necessary stage in ascertaining water-body health, especially from a human exposure perspective (i.e., setting recreational water-quality guidelines and use designations; as per stages 1–4 above), this approach is limited. Failure to meet FIB-based water-quality standards results in impairments of beneficial uses but does not provide adequate information to develop a pollution reduction strategy and implement BMPs (stages 5–8 above). For example, beaches in Avalon Bay (Catalina Island, California, USA) experienced frequent closures due to exceedences of the enterococci single sample water-quality standard (104 cfu/100 mL) in the absence of an obvious pollutant source (Boehm et al. 2003). As part of an investigative process, Boehm et al. (2003) proposed a multipronged approach utilizing intensive sampling, comprehensive sanitary surveys, and testing for human-specific microbial marker genes. Although a definitive relationship between FIB and the presence of human-associated marker genes was not established, the study outcome was able to rule in (leaking gray water pipe, dry weather runoff from storm drains, and shorebirds) or rule out (pollution from outside the bay) potential point and nonpoint sources of pollution (Boehm et al. 2003). Advances in MST, when appropriately applied in controlled studies, may remove a substantial impediment to fecal source identification (Stoeckel 2005).

A multipronged approach may also prove beneficial when attempting to allocate nonpoint source contributions in the development of TMDLs and control of contaminant loads. Molecular MST techniques, in particular, may have the ability to strengthen these approaches by discriminating against closely related environmental strains of the same organism, i.e., enterococci. This level of discrimination would prove particularly useful in indexing the relative contribution of FIB by source during periods of environmental loading (Siegrist et al. 2007; Giebel et al. 2008; Stapleton et al. 2009). Therefore, monitoring programs, inclusive of source attribution techniques such as MST, have the potential to: enhance new monitoring data, ascertain the severity of the pollution (human versus nonhuman sources), provide supplementary data on the efficacy of current standards, assist in the development of TMDLs and BMPs by elucidating pollutant sources, and be utilized to gauge water-quality improvements (Gawler et al. 2007).

15.3.2 Recommendations for MST

In the USA, the development and application of MST methods have evolved largely from a federal mandate that all impaired waters calculate TMDLs for fecal contaminants (Edge and Schaefer 2005). The US Federal Water Pollution Control Act (Clean Water Act; CWA) calls for the restoration of water quality to a sufficient standard that promotes the propagation of fish, shellfish, and wildlife and provides a safe water supply and recreational outlet (CWA Section 130.3). To accomplish this task, the CWA call for the control of nonpoint source pollution and the elimination of pollutants. The "Clean Coastal Environment and Public Health Act of 2009" would modify provisions of the CWA related to beach monitoring through the revision of water-quality criteria or standards. Monitoring protocols would be specifically amended to require methods most likely to detect contamination that carries pathogens. Under the proposed language, state and public environmental agencies would expand water-quality programs from solely public notification and monitoring performance criteria to include "source tracking, sanitary surveys, and prevention efforts to address identified sources of contamination by pathogens and pathogen indicators in coastal recreation waters adjacent to beaches or similar points of access that are used by the public."

Voted on by the US House of Representatives in July of 2009 and pending in the US Senate, the Clean Coastal Environment and Public Health Act of 2009, if passed as proposed, would encourage and allow federal dollars to be utilized for source attribution approaches such as MST. Previously left to the discretion and expense of local authorities and beach managers, the recommendation for expanded monitoring programs to include source tracking would widely encourage the use of more discriminatory methods for determining pollutant sources for the protection of public health. Steering researchers and regulatory personnel toward source tracking will require guidance so that they have a clear understanding of what these microbial/molecular methods can and cannot do and how they may best be incorporated into existing assessment programs.

Risk-based initiatives in the EU have also generated an interest in the use of MST methods (Edge and Schaefer 2005). The EU Bathing Water Directive requires the establishment of management measures with respect to bathing waters (Anon 2006). These management measures include, among others, the following: monitoring bathing water quality, establishing a bathing water profile, identifying and assessing causes of pollution that may impact recreational value and impair bather health, and taking action to prevent pollution. The development of a bathing water profile for a single beach or group of beaches requires a catchment assessment to identify potential pollution sources. The characterization of contamination within a catchment may be accomplished through the use of source attribution techniques. The segregation of pollution source is important in ascertaining health risk (USEPA 2009). Current WHO recommendations for classifying recreational waters is that health risk can be interpreted based on a combination of fecal indicator organism density and evidence of a human versus nonhuman source (WHO 2004). The ability to attribute pollution to a specific source will also aid in determining the relative contribution from water bodies impacted by multiple sources, a common occurrence, and the advancement of site specific best management/mitigation measures.

15.3.3 Sanitary Surveys

Sanitary surveys, or sanitary inspections, provide a framework for a holistic approach to monitoring water bodies; they generally include site evaluations and spatial/temporal assessments of FIB but may also include source attribution techniques. A sanitary survey is a guided data collection tool capable of providing a comprehensive and accurate review of all probable and potential sources of pollution capable of impairing a water body for its designated use and adversely impacting human health. Sanitary surveys help state and local beach program managers and public health officials identify sources of beach water pollution, assess the magnitude of pollution, and identify priority locations for water testing. This "roadmap"

Table 15.2 Suggested decision tree approach leading to the effective use of MST protocols, study design, data collection, and interpretation [adapted/expanded from Stoeckel (2005)]

MST decision tree				
Define problem	Review historic monitoring data	Conduct sanitary surveys	Expand routine monitoring (spatial & temporal)	Relate FIB to events or conditions
Formulate objectives	Level of discrimination needed	Confirmation of empirical evidence	Coarse (human versus nonhuman)	Fine (to species or host)
	Source apportionment	Relative contribution	Presence/absence of suspected source	
	Degree of certainty	Presence/absence of fecal material	Correlation to FIB concentration	Correlation to human health risk
Choose appropriate MST tool	Phenotypic	Genotypic	Molecular	Human-specific
Define target area	Receiving water	Sediments	Infrastructure	Watershed
Create a sampling plan	Daily	Seasonal	Event-based (weather dependent)	Epidemiological studies
Collect data				
Interpret data	Relationship to FIB (spatial and temporal)	Relationship to observational data	Relationship to disease outbreaks	Relationship to regulatory actions

FIB fecal indicator bacteria; *MST* microbial source tracking

approach allows practitioners to develop appropriate site-specific characterization tools (Table 15.2) (Stoeckel 2005; Pillai and Vega 2007). A sanitary survey should identify all pollutant sources, including human fecal contamination resulting from municipal infrastructure (Marsalek and Rochfort 2003).

Sanitary surveys have been commonly used for assessing potential pollution sources to shellfish harvesting waters (Suoninen 1998). While recommended as part of a beach classification scheme 10 years ago (WHO 1999) the use of sanitary surveys for assessing recreational waters did not garner attention from global legislative bodies until such time as water-quality criteria revisions occurred. The EU Bathing Water Directive (2006/7/EC), Guidelines for Canadian Recreational Water Quality (3rd Edition), and US Clean Coastal Environment and Public Health Act of 2009 (and amendment to the CWA) all contain elements recommending the use of environmental assessments in ascertaining pollutant sources. Public health burden resulting from disease outbreaks and economic concerns over the loss of utility associated with beach closures has brought the need to identify and mitigate pollution sources from the realm of academia, beach managers, and monitoring authorities to the political arena. This is well illustrated within the USA by the GLRC process (GLRC 2005).

The GLRC strategy addresses many of the global concerns regarding the protection and preservation of our surface waters, toxic pollutants, loss of habitat, aquatic invasive species, sustainable development, and coastal health. Each issue area within the GLRC strategy contains a set of goals and recommended actions to achieve those goals. One of the key recommendations of the coastal health chapter was to develop a standardized sanitary tool for the identification of contamination sources. In 2005, the US EPA reported that 95% of beach advisories posted within the Great Lakes were attributed to an unknown source (Kovatch 2006). The development of a sanitary tool for the Great Lakes, based on a WHO model (Bartram and Rees 2000), would provide the necessary tool for beach managers and other monitoring authorizes to identify and subsequently mitigate these contamination sources. The beach sanitary tool was drafted in 2006 and piloted in 2007 (forms and guidance document available online: http://www.epa.gov/waterscience/beaches/sanitarysurvey/). Sixty-one pilot study participants were surveyed regarding the success of the draft beach sanitary tool at identifying localized sources of pollution. Respondants indicated that they collectively achieved a 60% improvement in their ability to attribute water-quality advisories to an origin of pollution or environmentally induced transport mechanism, i.e., transport of fecal material to surface waters via precipitation events or wave action (Rockwell and Wirick 2008). Following suit, the Canadian government has also recommended the use of environmental health and safety surveys in their proposed federal water-quality guideline revisions (Health Canada 2009).

As previously stated, sources of fecal contamination to recreational waters are often unknown (Francy et al. 2005; Rockwell 2006). Determining the spatial distribution of, and identifying factors influencing fluxes in concentration of, FIB may provide insight into sources of fecal contamination (Francy et al. 2005). The choice of which method(s) to employ when one determines fecal pollution sources must be based on specific site characteristics. The results of sanitary surveys, as well as budgetary and time constraints, may drive what tools are used in the investigative process. In some situations, spatial and temporal assessments of FIB density may suffice. In other instances, FIB surveys may need to be augmented with source attribution studies to provide the level of discrimination necessary for effective, targeted mitigation. In some situations, the knowledge of whether the pollution is derived from human versus animal fecal sources, or between domestic animal and wildlife sources, may be sufficient. In other situations, it may be necessary to identify the species of domesticated animal or even the specific herd or flock that is the major contributor of fecal pollution, both of which require more precise MST methods (USEPA 2005b).

15.3.4 MST as a Supplementary Tool to Spatial and Temporal Assessments of FIB

Stapleton et al. (2009) demonstrated correlation of FIB concentrations at adjacent sampling points, indicating that grab samples may provide reasonable water-body characterization in terms of FIB. Although this may be satisfactory from the perspective of binary good/poor assessment programs, technological advances have created the expectation of definitive microbial contamination identification (extent and source) (Pillai and Vega 2007). While definitive source identification may not be feasible, necessary, or cost-effective in every instance, MST techniques may be successfully employed as part of a multitiered investigative approach (Plummer and Long 2007).

The revised *Guidelines for Canadian Recreational Water Quality* (3rd Edition) call for a "multi-barrier" approach to monitoring bathing water (Edge and Schaefer 2005; Health Canada 2009). Within a risk-based water-quality management framework, all potential sources of contamination and physical hazards that could affect the recreational area are identified. Risk to bather health is determined through FIB assessments and pollutant source identification. Mitigation, preemptive whenever possible, is the ultimate goal.

Advancements in risk assessment and MST have made it possible to use these tools in conjunction with spatial and temporal assessments to provide additional evidence regarding the origin of the contamination. A Toronto, Ontario (Canada) study was able to successfully employ two library-dependent MST methods to identify the primary fecal load at two public bathing beaches (Edge et al. 2007). In the Canadian study, avian fecal material was found to be a greater contributor (30–60%) of *E. coli* density in beach sands and surface water at both study beaches (versus pets, sewage, and unknown sources). Results from concurrent ARA and DNA fingerprinting of *E. coli* isolates were consistent with the observation of seagulls, Canada geese, and their feces on the beaches, and absence of pet waste and sewage outfalls (Edge et al. 2007).

In another recreational water-quality study, ARA and DNA fingerprinting techniques were used to inform spatial distribution studies conducted as part of beach sanitary surveys in Racine, Wisconsin (USA) (Kinzelman et al. 2004; Kinzelman and McLellan 2009). Racine's bathing beaches represent a typical urban environment where multiple fecal sources are likely to have an adverse impact on water quality. Sanitary surveys identified avian populations (gulls and Canada geese), domestic animals, storm water, urban runoff, sanitary sewer overflows from neighboring communities, boater waste, and beach sands as potential contributors of fecal contamination. In order to rule in or rule out probable sources, aquatic and terrestrial spatial distribution studies were conducted over a period of 4 years (2001–2005). *E. coli* isolates from surface water, groundwater, and storm water were subjected to ARA and DNA fingerprinting techniques to confirm empirical evidence. MST results corroborated spatial distribution study results. Deposition of gull feces and sanitary sewer infiltration into municipal storm water infrastructure were identified as the primary sources of contamination. Additional molecular techniques (rep-PCR and cellular fatty acid analysis) revealed minimal replication of avian *E. coli* isolates in beach sand (Kinzelman et al. 2004; Kinzelman and McLellan 2009). The unified results of this multiyear study lead to the crafting of source-specific mitigation measures, reducing swimming bans from 66% of the bathing season in 2000 to 5% in 2005. These results have been sustainable, as water-quality advisories have been maintained at five percent or less of the bathing season through the 2009 season.

While Edge et al. (2007) were able to translate the qualitative MST results generated in their study into a percent contribution from avian sources in conjunction with spatial and temporal assessments of FIB, the ability of library-dependent MST methods to provide definitive, quantitative apportionment data is still limited and needs to be utilized within the proper context (Chap. 3) (USEPA 2005b; Edge et al. 2007; Stapleton et al. 2009). Practitioners must take into account naturalized and persistent indicator organisms (Chap. 17) and other confounding factors such as FIB strains that colonize multiple host species when utilizing these methods. A "toolbox" approach is encouraged and conclusions regarding definitive identification of pollutant sources should not be made based on the results of a single MST method or a limited dataset (Field and Scott 2008). Using multiple chemical, viral, or bacterial MST methods in conjunction with sanitary surveys and spatial and temporal assessments of FIB may provide the best chance of success (Field and Scott 2008). Determining which methods, singly or in tandem, are most appropriate to the task at hand at the outset will increase reliability while maintaining costs at a minimum.

15.4 Conclusions

MST can be an effective tool for water-quality management if used judiciously and with a clear understanding of the benefits and limitations of the specific methodology (ies) employed. While it may be difficult to directly relate the results of MST studies to FIB concentrations, or to fully characterize a catchment or water body based solely upon the results of molecular, microbial, or chemical source attribution techniques, these caveats should not detract from the potential utility of these methods as exploratory tools for water-quality management. Maintenance of the microbiological quality and safety of water systems used for drinking, recreation, and in the harvesting of seafood is imperative, as contamination of these systems can result in risks to human health as well as significant economic losses due to closures of beaches and shellfish harvesting areas. Revised global water-quality criteria call for identification and mitigation of pollutant sources for the protection of public health. As an evolving science, MST methods may one day be able to provide accurate prediction of pathogen presence in aquatic environments. Until such time, MST, as one of a suite of methods including sanitary surveys and spatial/temporal distribution studies, has proven its utility to practitioners in identifying and remediating sources of contamination impacting surface waters.

References

Alverez J, Porwollik S, Laconcha I (2003). Detection of a *Salmonella enterica* serovar California strain spreading in Spanish feed mills and genetic characterization with DNA microarrays. Appl. Environ. Microbiol. 69: 7531–7534.

Anon (2000). Council of the European Communities. Directive 2000/60/EC of the European Parliament and of the Council of 23 October 2000 establishing a framework for community action in the field of water policy. Official Journal of the European Union L327, 1–72.

Anon (2006). Directive 2006/7/EC of The European Parliament and of the Council of 15th February 2006 concerning the management of bathing water quality and repealing Directive 76/160/EEC. Official Journal of the European Union L64 (4.3.2006), 37–51.

Bartram J, Rees, G (Eds) (2000). Sanitary inspection and microbiological water quality, p. 113–167. *In* Monitoring Bathing Waters, a Practical Guide to the Design and Implementation of Assessments and Monitoring Programmes. Published on behalf of the USEPA, WHO, and Commission of European Communities. E & FN Spon Publishing, London, UK and New York, NY.

Boehm AB, Fuhrman JA, Mrše RD et al (2003). Tiered approach for identification of a human fecal pollution source at a recreational beach: Case study at Avalon Bay, Catalina Island, California. Environ. Sci. Technol. 37(4): 673–680. doi: 10.1021/es025934x.

Carrillo M, Estrada E, Hazen T (1985). Survival and enumeration of the fecal indicators *Bifidobacterium adolescentis* and *Escherichia coli* in a tropical rain forest watershed. Appl. Environ. Microbiol. 50(2): 468–476.

Carroll SP, Dawes L, Hargreaves M et al (2009). Faecal pollution source identification in an urbanising catchment using antibiotic resistance profiling, discriminant analysis and partial least squares regression. Water Res. 43(5): 1237–1246.

Chao KK, Chao CC, Chao WL (2003). Suitability of the traditional microbial indicators and their enumerating methods in the assessment of fecal pollution of subtropical freshwater environments. J. Microbiol. Immunol. Infect. 36(4): 288–93.

Chizhikov V, Wagner M, Ivshina A et al (2002). Detection and genotyping of human group A rotaviruses by oglionucleotide microarray hybridization. J. Clin. Microbiol. 40: 2398–2047.

Chou CC, Lin YC, Su JJ (2004). Microbial indicators for differentiation of human- and pig-sourced fecal pollution. J. Environ. Sci. Health A 39: 1415–1421.

Davies C, Long J, Donald M et al (1995). Survival of fecal microorganisms in marine and freshwater sediments. Appl. Environ. Microbiol. 61: 1888–1896.

Desmarais TR, Solo-Gabriele HM, Palmer CJ (2002). Influence of soil on fecal indicator organisms in a tidally influenced subtropical environment. Appl. Environ. Microbiol. 68(3): 1165–1172.

Dutka B, Shaarawi A, Martins M (1987). North and South American studies on the potential of coliphage as a water quality indicator. Wat. Res. 21: 1127–1134.

Duris JW, Haack SK, Fogarty LR (2009). Gene and antigen markers of shiga-toxin producing *E. coli* from Michigan and Indiana river water: occurrence and relation to recreational water quality criteria. J. Environ. Qual. 38: 1878–1886. doi:10.2134/jeq2008.0225.

EA (2000). Environment Agency (EA): The microbiology of recreational and environmental waters 2000. S. C. O. Analysts, the Environment Agency, Bristol, United Kingdom.

Edge TA, Hill S, Stinson G, Seto P et al (2007). Tracking sources of microbial pollution in recreational waters: experience from two Toronto beaches. Session 8.3, NOVATECH 2007: 1649–1656.

Edge TA, Schaefer K (2005). Microbial source tracking in aquatic ecosystems: the state of the science and an assessment of needs. NWRI Scientific Assessment Report Series No. 7 and Linking Water Science to Policy Workshop Series Report. March 2005. Environment Canada.

European Union (2006). Directive 2006/7/EC of the European Parliament and of the Council of 15 February 2006.

Fey A, Eichler S, Flavier S et al (2004). Establishment of a real-time PCR-based approach for accurate quantification of bacterial RNA targets in water, using *Salmonella* as a model organism. Appl. Environ. Microbiol. 70: 3618–3623.

Field K, Scott T (2008). Microbial Source Tracking. Michigan State University. Online: http://www.sourcemolecular.com/newsite/pdfs/Microbial%20Source%20Tracking%20White%20Paper.pdf.

Field C, Samaspour M (2007). Fecal source tracking, the indicator paradigm, and managing water quality. Water Res. 41: 3517–3538.

Foley SL, Lynne AM, Nayak R (2009). Molecular typing methodologies for microbial source tracking and epidemiological investigations of Gram-negative bacterial foodborne pathogens. Infect. Genet. Evol. 9: 430–440.

Fout GS, Martinson BC, Moyer MWN et al (2003). A multiplex reverse transcription-PCR method for detection of human enteric viruses in groundwater. Appl. Environ. Microbiol. 69: 3158–3164.

Franck SM, Bosworth BT, Moon HW (1998). Multiplex PCR for enteropathogenic, attaching and effacing, and shiga-toxin-producing *E. coli* strains from calves. J. Clin. Microbiol. 36: 1795–1797.

Francy DS, Struffolino P, Brady AMG et al (2005). A spatial, multivariable approach for identifying proximate sources of *Escherichia coli* to Maumee Bay, Lake Erie, Ohio. USGS Open File Report 2005–1386. http://pubs.usgs.gov/of/2005/1386/pdf/OFR%202005–1386.pdf. Accessed 6 Dec 2009.

Fremaux BJ, Gritzfeld, Boa T, Yost KC (2009). Evaluation of host-specific Bacteroidales 16S rRNA gene markers as a complementary tool for detecting fecal pollution in a prairie watershed. Water Res. 43(19): 4838–4849.

Fukushima M, Kakinuma K, Kawaguchi R (2002). Phylo genetic analysis of *Salmonella*, *Shigella* and *E. coli* strains on the basis of the gyrB gene sequence. J. Clin. Microbiol. 40: 2779–2785.

Gawler AH, Beecher JE, Brandão J et al (2007). Validation of host-specific Bacteriodales 16S rRNA genes as markers to determine the origin of faecal pollution in Atlantic Rim countries of the European Union. Water Res. 41: 3780–3784.

Giebel RA, Fredenberg W, Sandrin TR (2008). Characterization of environmental isolates of *Enterococcus* spp. by matrix-assisted laser desorption/ionization time-of-flight mass spectrometry. Water Res. 42: 931–940.

Great Lakes Regional Collaboration (2005). Great Lakes Regional Collaboration to protect and restore the Great Lakes. http://www.glrc.us, Accessed 7 Dec 2009.

Gregory JB, Litaker RW, Noble RT (2006). Rapid one-step quantitative reverse transcriptase PCR assay with competitive internal positive control for detection of enteroviruses in environmental samples. Appl. Environ. Microbiol. 72(6): 3960–3967.

Griffith J, Weisberg SB (2006). Evaluation of rapid microbiological methods for measuring recreational water quality. Westminster, CA: Southern California Coastal Water Research Project.

Guan S, Xu R, Chen S et al (2002). Development of a procedure for discriminating among *E. coli* isolates from animal and human sources. Appl. Environ. Microbiol. 68: 2690–2698.

Guy RA, Payment P, Krull UJ, Horgen PA (2003). Real-time PCR for quantification of Giardia and Cryptosporidium in environmental water samples and sewage. Appl. Environ. Microbiol. 69(9): 5178–5185.

Haack SK, Duris JW, Fogarty LR et al (2009). Comparing wastewater chemicals, indicator bacteria concentrations, and bacterial pathogen genes as fecal pollution indicators. J. Environ. Qual. 38: 248–258. doi:10.2134/jeq2008.0173.

Haramoto E, Katayama H, Ohgaki S (2004). Detection of noroviruses in tap water in Japan by means of a new method for concentrating enteric viruses in large volumes of freshwater. Appl. Environ. Microbiol. 70: 2154–2160.

Haugland RA, Siefring SC, Wymer LJ et al (2005). Comparison of enterococcus measurements in freshwater at two recreational beaches by quantitative polymerase chain reaction and membrane filter culture analysis. Water Res. 39(4): 559–568.

He JW, Jiang S (2005). Quantification of enterococci and human adenoviruses in environmental samples by real-time PCR. Appl. Environ. Microbiol. 71(5): 2250–2255.

Health Canada (2009). Guidelines for Canadian recreational water quality (DRAFT). Online: http://www.consultations.hc-sc.gc.ca.

Hoebe CJPA, Vennema H, de Roda Husman AM, van Duynhoven YTHP (2004). Norovirus outbreak among primary schoolchildren who had played in a recreational water fountain. J. Infect. Dis. 189(4): 699–705.

Horman A, Rimhanen-Finne R, Manula L et al (2004). *Campylobacter* spp., *Giardia* spp., *Cryptosporidium* spp., noroviruses, and indicator organisms in surface water in southwestern Finland, 2000–2001. Appl. Environ. Microbiol. 70: 87–95.

Hsu WB, Wang JH, Chen PC et al (2007). Detecting low concentrations of *Shigella sonnei* in environmental water samples by PCR. FEMS Microbiol. Lett. 270(2): 291–298.

Ibekwe AM, Watt PM, Grieve CM (2002). Multiplex fluorogenic real-time PCR for detection and quantification of *Escherichia coli* O157:H7 in dairy wastewater wetlands. Appl. Environ. Microbiol. 68: 4853–4862.

Johnson DW, Pieniazek NJ, Griffin DW et al (1995). Development of a PCR protocol for sensitive detection of Cryptosporidium oocysts in water samples. Appl. Environ. Microbiol. 61(11): 3849–3855.

Johnson J, Stell, A (2001). PCR for specific detection of H7 flagellar variant of fliC among extraintestinal pathogenic Escherichia coli. J. Clin. Microbiol. 39: 3712–3717.

Kay D, Lee R, Wyer MD et al (2010). Experience from recreational waters. In: Safe management of shellfish and harvest waters. G. Rees, K. Pond, D. Kay, J. Bartram, J. Santo Domingo (Eds.) London, International Water Association and World Health Organization.

Kay D, McDonald AT, Stapleton CM et al (2006a). The challenges of the water framework directive. Proceedings of the Institution of Civil Engineers, Water Management 159: 58–64.

Kay D, Stapleton CM, Wyer MD et al (2006b). Total maximum daily loads (TMDL). The USEPA approach to managing faecal indicator fluxes to receiving waters: lessons for UK environmental regulation? L. Gairns, C. Crighton and B. E. Jeffrey (Eds.), Agriculture and the Environment VI; Managing Rural Diffuse Pollution. Proceedings of the SAC/SEPA Biennial Conference, Edinburgh, International Water Association, Scottish Agricultural College, Scottish Environmental Protection Agency.

Kinzelman J, McLellan S (2009). Success of science-based best management practices in reducing swimming bans—a case study from Racine, Wisconsin, USA. Aquat. Ecosyst. Health Manage. 12(2): 187–196.

Kinzelman J, McLellan SL, Daniels et al (2004). Non-point source pollution: determination of replication versus persistence of *Escherichia coli* in surface water and sediments with correlation of levels to readily measurable environmental parameters. J. Water Health 2: 103–114.

Kott Y, Roze N, Sperber S et al (1974). "Bacteriophages as viral pollution indicators," Wat. Res. 8: 165–171.

Kovatch C (2006). Beach Sanitary Survey. Presented at the US EPA National Beaches Conference, Niagara Falls, NY, 11 October 2006.

Leung KT, Mackereth R, Tien YC et al (2004). A comparison of AFLP and ERIC-PCR analyses for discriminating *Escherichia coli* from cattle, pig and human sources. FEMS Microbiol. Ecol. 47: 111–119.

Loge FJ, Thompson DE, Call DR (2002). PCR detection of specific pathogens in water: a risk based analysis. Environ Sci. Technol. 36: 2754–2759.

Lubeck PS, Wolffs P, On SLW et al (2003). Toward an international standard for PCR-based detection of food-borne thermotolerant Campylobacters: assay development and analytical validation. Appl. Environ. Microbiol. 69: 5664–5669.

Marsalek J, Rochfort Q (2003). Urban wet-weather flows: sources of fecal contamination impacting on recreational waters and threatening drinking-water sources. J. Toxicol. Environ. Health, Part A, 67: 1–13.

Moriñigo M, Cornax R, Castro D et al (1990). Viability of Salmonella spp. and indicatormicroorganisms in seawater using membrane diffusion chambers. Antonie Leeuwenhoek 57(2): 109–117.

Neto RC, Santos, JU, Franco RMB (2006). Evaluation of activated sludge treatment and the efficiency of the disinfection of Giardia species cysts and Cryptosporidium oocysts by UV at a sludge treatment plant in Campinas, south-east Brazil. Water Sci. Technol. 54(3): 89–94.

Novga HK, Rudi K, Naterstad K et al (2000). Application of 5′-nuclease PCR for quantitative detection of *Listeria monocytogenes* in pure cultures, water, skim milk, and unpasteurized whole milk. Appl. Environ. Microbiol. 66: 4266–4271.

Panicker G, Myers ML, Bej AK (2004). Rapid detection of *Vibrio vulnificus* in shellfish and Gulf of Mexico water by real time PCR. Appl. Environ. Microbiol. 70: 498–507.

Parveen S, Portier KM, Robinson K et al (1999). Discriminant analysis of ribotype profiles of *Escherichia coli* for differentiating human and nonhuman sources of fecal pollution. Appl. Environ. Microbiol. 65: 3142–3147.

Pillai S, Vega E (2007). Molecular detection and characterization tools. In: Microbial source tracking. J.Santo Domingo, M. Sadowsky (Eds.), P 65–91. Washington, DC, American Society for Microbiology Press.

Plummer JD, Long, SC (2007). Monitoring source water for microbial contamination: Evaluation of water quality measures. Water Res. 41(16): 3716–3728.

Pond KR, Rangdale R, Meijer, WG et al (2004). Workshop report: developing pollution source tracking for recreational and shellfish waters. Environmental Forensics, 5: 237–247.

Reischer GH, Haider JM, Sommer R et al (2008). Quantitative microbial faecal source tracking with sampling guided by hydrological catchment dynamics. Environ. Microbiol. 10(10): 2598–2608.

Reischer GH, Kasper DC, Steinborn R et al (2007). A quantitative real-time PCR assay for the highly sensitive and specific detection of human faecal influence in spring water from a large alpine catchment area. Lett. Appl. Microbiol. 44(4): 351–356.

Reischer GH, Kasper DC, Steinborn R et al (2006). Quantitative PCR method for sensitive detection of ruminant fecal pollution in freshwater and evaluation of this method in alpine karstic regions. Appl. Environ. Microbiol. 72(8): 5610–5614.

Rockwell D (2006). US EPA 2006 beach sanitary survey pilot study final report (unpublished).

Rockwell D, Wirick H (2008). An overview of the 2007 pilot sanitary surveys and GLRC coastal health goals. Presented at the Great Lakes Beach Association Annual Conference, Erie, PA, 15–17 September 2008.

Santo Domingo JW, Bambic DG, Edge TA et al (2007). Quo vadis source tracking? Towards a strategic framework for environmental monitoring of fecal pollution. Water Res. 41: 3539–3552.

Savitcheva O, Okabe S (2006). Alternative indicators of fecal pollution: relations with pathogens and conventional indicators, current methodologies for direct pathogen monitoring and future application perspectives. Water Res. 40(13): 2463–2476.

Scott TM, Rose JB, Jenkins TM et al (2002). Microbial source tracking: current methodology and future directions. Appl. Environ. Microbiol. 68(12): 5796–5803.

Scott TM, Jenkins TM, Lukasik J, Rose JB (2005). Potential use of a host associated molecular marker in *Enterococcus faecium* as an index of human fecal pollution. Environ. Sci. Technol. 2005, 39: 283–287.

Shanks OC, Nietch C, Simonich M et al (2006). Basin-wide analysis of the dynamics of fecal contamination and fecal source identification in Tillamook Bay, Oregon. Appl. Environ. Microbiol. 72(8): 5537–5546.

Shibata T, Solo-Gabriele HM, Fleming LE, Elmir S (2008). Monitoring marine recreational water quality using multiple microbial indicators in an urban tropical environment. Water Res. 38(13): 3119–3131.

Siegrist TJ, Anderson PD, Huen WH et al (2007). Discrimination and characterization of environmental strains of *Escherichia coli* by matrix-assisted laser desorption/ionization time-of-flight mass spectrometry (MALDI-TOF-MS). J. Microbiol. Methods 68: 554–562.

Simpson JM, Santo Domingo JW, Reasoner DJ (2002). Microbial source tracking: state of the science. Environ. Sci. Technol. 24: 5279–5288.

Singh DV, Matte MH, Matte GR et al (2001). Molecular analysis of *Vibrio cholerae* O1, O139, non-O1, and non-O139 strains: clonal relationships between clinical and environmental isolates. Appl. Environ. Microbiol. 67: 910–921.

Sinton L, Hall C, Lynch P et al (2002). Sunlight inactivation of fecal indicator bacteria andbacteriophages from waste stabilization pond effluent in fresh and saline waters. Appl. Environ. Microbiol. 68: 3605–3613.

Stapleton CM, Kay D, Wyer M et al (2009). Evaluating the operational utility of a *Bacteriodales* quantitative PCR-based MST approach in determining the source of faecal indicator organisms at a UK bathing water Wat. Res. 43(19): 4888–4899. doi:10.1016/j.watres.2009.09.015.

Stender H, Broomer AJ, Perry-O'Keefe KOH et al (2001). Raid detection, identification, and enumeration of *E. coli* cells in municipal water by chemiluminescent in situ hybridization. Appl. Environ. Microbiol. 67: 142–147.

Stewart JR, Gast RJ, Fujioka RS et al (2008). The coastal environment and human health: microbial indicators, pathogens, sentinels and reservoirs. Environ. Health, 7(Suppl. 2):S3.

Stoeckel D, Harwood V (2007). Performance, design, and analysis in microbial source tracking studies. Appl. Environ. Microbiol. 73: 2405–2415.

Stoeckel DM (2005). Selection and application of microbial source tracking tools for water-quality investigations: U.S. Geological Survey Techniques and Methods Book 2, Chapter A3. http://pubs.usgs.gov/tm/2005/tm2a3/pdf/Book2_Collection%20of%20Environmental%20Data.pdf. Accessed 6 Dec 2009.

Strauss N, Von Schirnding YE, Genthe B et al (1995). The health effects of exposure through swimming in South African marine waters: an overview. Urban Health Newsl. 26: 37–42.
Suoninen WE (1998). Sanitary survey shellfish growing area 46–47 Brigantine to Spray Beach 1992–1997. New Jersey Department of Environmental Protection. Online: http://www.state.nj.us/dep/wms//bmw/Reports/area46–47.pdf.
Suresh SD, Vega E (2007). Molecular detection and characterization tools. *In*: Santo Domingo H and Sadowsky M (Eds) Microbial source tracking, emerging issues in food safety, ASM Press, Washington, DC.
Tartera C, Jofre J (1987). Bacteriophages active against Bacteroides fragilis in sewage-pollutedwaters, Appl. Environ. Microbiol. 53: 1632–1637.
Tessele F, Monteggia LO, Rubio J (2005). Treatment of municipal wastewater UASB reactor effluent by unconventional flotation and UV disinfection. Water Sci. Technol. 52(1–2): 315–322.
United States Environmental Protection Agency (US EPA) (2009). Review of published studies to characterize relative risks from different sources of fecal contamination in recreational water. EPA 822-R-09–001.
US EPA (2008). Literature review of molecular methods for simultaneous detection of pathogens in water. US EPA Environmental Technology Council. EPA/600/R-07/128.
US EPA (2007). Water quality standards handbook: second edition (revised). http://www.epa.gov/waterscience/standards/handbook/. Accessed 6 Dec 2009. EPA-823-B-94–005.
US EPA (2005a). Total maximum daily loads. US EPA, Washington DC. http://oaspub.epa.gov/waters/national_rept.control.
US EPA (2005b). Microbial source tracking guide document. Office of Research and Development, Washington, DC EPA-600/R-05/064. 131 pp.
Volkhov D, Rasooly A, Chumakov K et al (2002). Identification of *Listeria* species by microarray-based assay. J. Clin. Microbiol. 40: 4720–4728.
Vora GJ, Meador CE, Stenger DA et al (2004). Nucleic acid amplification strategies for DNA microarray-based pathogen detection. Appl. Environ. Microbiol. 70: 3047–3054.
Walters SP, Gannon VP, Field KG (2007). Detection of *Bacteroidales* fecal indicators and the zoonotic pathogens *E. coli* 0157:H7, Salmonella, and Campylobacter in river water. Environ. Sci. Technol. 41(6): 1856–1862.
Wapnick CM, Harwood VJ, Singleton T, Morrison G, Staley C, Staley R (2009). Application of the bacteria decision-support tool in the Hillsborough River Watershed. Water Environment Federation TMDL Conference Aug 9–12, 2009, Minneapolis, MN. Online: http://208.88.129.72/Publications/page.aspx?id=5751
Wellinghausen N, Frost C, Maare R (2001). Detection of Legionellae in hospital water samples by quantitative real-time LightCycler PCR. Appl. Environ. Microbiol. 67: 3985–3993.
World Health Organization (WHO) (2004). Cotruvo JA, Dufour A, Rees G, Bartram J, Carr R, Cliver DO, Craun GF, Fayer R, Gannon VPJ, ed. Waterborne zoonoses: identification, causes and control. World Health Organization (WHO). IWA Publishing: London, UK.
WHO (1999). Health based monitoring of recreational waters: the feasibility of a new approach (the Annapolis Protocol). Geneva, WHO. (WHO/SDE/WSH/99.1).
WHO (2003). Guidelines for safe recreational water environments Volume 1: Coastal and fresh-waters. Geneva, Switzerland, World Health Organization.
Xiao L, Singh A, Limor J et al (2001). Molecular characterization of Cryptosporidium oocysts in samples of raw water and wastewater. Appl. Environ. Microbiol. 67: 1097–1101.
Zanetti F, De Luca G, Sacchetti R (2006). Microbe removal in secondary effluent by filtration. Ann. Microbiol. 56(4): 313–317.

Chapter 16
Minimizing Microbial Source Tracking at All Costs

Peter G. Hartel

Abstract Microbial source tracking (MST) is often expensive, time-consuming, and intimidating to those who need it. To make matters worse, some MST researchers think that much MST is unnecessary and a waste of time and money. Simply by combining local knowledge and existing data with a general survey of the waterway, one can identify many sources of fecal bacteria in contaminated waterways without any testing at all. However, this commonsensical approach requires tact and diplomacy, and identifying sources of fecal contamination requires good powers of observation. Because common sense is uncommon, an alternative approach is to add some sort of testing to this general survey. If sources of fecal contamination are still unclear, then targeted sampling, a sampling method based on the children's game of "hot" and "cold" that focuses on hotspots of fecal contamination, can subsequently identify almost all sources of fecal contamination quickly, easily, and inexpensively. Overall, the method works equally well in fresh and marine waters and accounts for both baseflow and stormflow conditions. Although the method has occasional problems with property rights, the approach is well suited for communities that want to comply with state and federal regulations, yet want to minimize MST at all costs.

Keywords *Escherichia coli* • Fecal enterococci • Fresh water • Marine water • Targeted sampling

16.1 Introduction

It may seem ironic that a book on microbial source tracking (MST) has a chapter championing the idea of minimizing MST, yet an often unspoken truth about MST is that it is expensive, time-consuming and that the technology associated with it is

P.G. Hartel (✉)
Department of Crop and Soil Sciences, University of Georgia, Athens, GA, USA
e-mail: pghartel@uga.edu

C. Hagedorn et al. (eds.), *Microbial Source Tracking: Methods, Applications, and Case Studies*, DOI 10.1007/978-1-4419-9386-1_16,

often intimidating to the people who need it the most. Therefore, it is unsurprising that many communities, particularly those that are poor, want to avoid MST. To make matters worse, some MST researchers (the present author included) think that much MST is unnecessary and a waste of time and money.

Even though communities with waters contaminated with fecal bacteria may not be able to afford MST or even understand it, they still want to comply with state and federal regulations because clean water should be everyone's right. Therefore, we have spent over a decade trying to figure out ways to minimize MST. In many cases, we were able to do this by using a simple, commonsensical approach that identified sources of fecal contamination inexpensively and quickly with technology that virtually anyone could do. This chapter outlines how to accomplish this objective.

16.2 The Quest for Local Knowledge

Minimizing MST begins with a quest for local knowledge. If a waterway has a problem with fecal contamination, then there are likely a number of people who have good ideas about what is causing it. Therefore, the quest for local knowledge is simple: talk with as many of the locals as possible and listen to them so that they can give you their ideas as to what they think the sources are. Ideas have ranged from "it's my neighbor's septic tank" to "it's the dirty mops left outside a restaurant." Even if a few of the ideas are nonsensical, it is important to listen to all these ideas carefully.

Over the past decade, experience has taught us that the quest for local knowledge tends to divide people into three groups: a first small group who are knowledgeable about the problem, have good ideas about the potential sources, and are also happy to tell you about it; a second larger group who know little about the problem but are happy to help you however they can (especially when a contaminated waterway is in their backyard); and a third small group who have no interest in volunteering information because they will face either legal and/or financial liabilities, or because you are infringing on their turf.

To find the first small group who are knowledgeable about the problem, have good ideas about the potential sources, and are happy to tell you about it, the best place to begin is often with state and local officials who are responsible for identifying the fecal contamination in the first place. In Georgia (USA), we always begin with one of the two divisions within the Department of Natural Resources: the Coastal Resources Division, which is responsible for coastal waters, or the Environmental Protection Division, which is responsible for all noncoastal waters. The best part of obtaining this local information is that not only are these people knowledgeable but they can also introduce you to other state and local officials who are also knowledgeable and who want to help you. We have also had good success with public health officials as well as officials associated with Regional Development Commissions. The Regional Development Commissions have been such a good

Fig. 16.1 Ongoing sewage spill from a malfunctioning lift station into a creek near Metter, GA. Photo courtesy of Chandra Brown, Ogeechee–Canoochee Riverkeeper (used with permission)

source of information because they know that good economic development is intimately associated with clean water.

Yet another knowledgeable group is riverkeepers and various environmental groups associated with waterways. In the case of riverkeepers, they have been sending us pictures electronically for over a decade on potential sources of fecal contamination in "their" rivers. A typical picture is shown in Fig. 16.1. Because riverkeepers are by nature an outgoing group, they are not shy and reticent about introducing you to other locals who can help you. In the case of environmental groups, many counties have groups who are dedicated to protecting and managing local rivers. For example, the Broad River is a major river near where the author lives, and there is a Broad River Watershed Association (http://www.brwa.org) whose members care for the river.

For the large group who may know little about the problem but are happy to help you, our experiences have been that this group is mostly curious. If the contaminated waterway happens to be in their backyard, then they become more helpful. Neighborhood associations are among the best places to find out about failing septic tanks. These associations are where we get a lot of "my-neighbor's-septic-tank-is-leaking" kind of information. We always talk to these groups because they can also help us avoid problems with trespassing on private property.

Finally, there is one group of people who are not as forthcoming about information because they are invariably concerned about their potential legal and/or financial liabilities if they are found either directly or indirectly responsible for the fecal contamination. However, we always make an attempt to talk to everyone. We do our best to assure to whomever we are talking that we are not lawyers. After all, our solution is to clean up the problem, not to end up in court. Because of its adversarial nature and the easy-to-exploit innate disadvantages of scientific method, our experiences with the court system have been overwhelmingly negative. It is important to avoid going to court at all costs.

The ability to obtain local knowledge requires diplomacy and tact. We always adopt a nonconformational approach. Our job is to listen. Because we are University of Georgia employees, we often wear University of Georgia clothing and we are more likely to start a conversation with "How about them Dawgs?" (The University of Georgia football team's mascot is a bulldog) than about fecal contamination. The ability to obtain local knowledge also requires an ability to talk simply. When we are trying to obtain local knowledge, scientific jargon typically associated with microbiology, biochemistry, and molecular genetics is an impediment to understanding. As a wise professor once told us when talking with the public, "No words more complicated than 'mayonnaise'." We always consider that everyone's time is valuable, and we do our best to ensure that we waste as little of it as possible.

Three groups of people require special attention in obtaining local knowledge. The first is county commissioners. County commissioners are important because the power they wield can help us with the remaining two groups that require special attention: city engineers and journalists. For city engineers, most of our experiences have been positive. However, we have also encountered several instances where city engineers denied that there was any problem of any kind with fecal contamination and the sanitary sewers for which they were responsible. When it became painfully obvious that city or county sanitary sewers were a source of fecal contamination, then their attitude toward solving the problem was, "we're not responsible if anything goes wrong." Our solution to this problem is to always go through the county commissioners whenever we must deal with city engineers. In this manner, city engineers are responding to the county commissioners, not to us.

City commissioners can also help with journalists. Speaking independently to journalists can quickly result in unfavorable publicity for a town or city. For example, county commissioners know that when tourists find out that the beaches they are visiting are contaminated with fecal bacteria, they are going to put their children back in the car, go elsewhere, and probably never come back. For cities dependent on tourism, this understandable human reaction is devastating. Therefore, our solution is similar to that of the engineers: we ensure that all of our findings go through the county commissioners. Therefore, journalists must talk with either a spokesperson for the county commissioners or the commissioners themselves for information.

The number of people we talk to usually depends on the magnitude of the problem. On the one hand, a specific local problem may not warrant talking to more than one or two people. For example, we have worked with a local county agent to resolve a dispute between an upstream property owner whose horses were possibly contaminating water in a small stream that a downstream property owner used for a small fish hatchery. Only the county agent, the upstream property owner, and the downstream property owner were involved. On the other hand, we have had cases where one city was blaming another city for fecal contamination problems in a large creek. In this case, it was important to talk with a large number of people. In some sense, we all live downstream.

16.3 The Quest for Existing Data

Once a reasonable amount of local knowledge has been obtained, it is important to substantiate as much of this knowledge as possible with any ongoing or past records of fecal contamination. It is unfortunate, but waters have been declared as impaired for fecal contamination based on a single data point. This is obvious nonsense. In some instances, the fecal contamination was temporary: a failing septic tank that was replaced, a wastewater treatment plant where a piece of equipment failed, or a broken pipe that was fixed. Alternatively, some sites have extensive data, but the data are not easily accessible because they are not electronically accessible. All of these sources must be checked. In this manner, it is possible to determine how egregious the problem is and its persistence. We have found this information invaluable in limiting the geographic area where the fecal contamination was likely originating.

Much of the information that we obtain is in the form of fecal bacterial counts. Therefore, our quest for data immediately brings up a number of serious limitations with these data. The first of these limitations is that it is not necessarily obvious that the problem is bacterial in nature; it is entirely possible that the problem is caused by viruses or protozoa. Furthermore, the scientific literature is replete with data showing that there is not necessarily a correlation between high counts of fecal bacteria and high counts of viruses or protozoa. In fact, it is well known that high counts of viruses can exist when counts of fecal bacteria are low, and vice versa. These alternative possibilities must be taken into consideration.

In addition to the prospects of a nonbacterial problem, there are other well-known limitations with fecal bacterial data themselves. First, many states have wisely rejected a fecal coliform standard for a standard based on *Escherichia coli*. Even so, many states persist in relying on counts of fecal coliforms instead of *E. coli*. In our freshwater testing, we do not use fecal coliforms, but *E. coli*. However, if we select *E. coli* as a stand-in for fecal coliforms, then there is an ancillary problem in determining the appropriate number of *E. coli* per 100 mL as the threshold of concern. In these instances, our solution has been to select a standard based on neighboring states that do use *E. coli* as a standard (e.g., the standard of 126 *E. coli*/100 mL for Tennessee and Florida serves as a standard for Georgia, a state that continues to use fecal coliforms).

Second, sampling times may be decided by a calendar date than what the weather is doing. Knowing the weather at the time of sampling is important because fecal bacteria typically increase 10- to 100-fold during stormy conditions (e.g., Solo-Gabriele et al. 2000). Therefore, it is important to combine weather data with any fecal bacterial counts. For tidal areas, which typically use fecal enterococci as indicator bacteria, tidal forcing conditions may exist. Tidal forcing are conditions where feces deposited above the normal but below the maximal, high tidemark and are brought into the water by ebbing spring tides (Boehm and Weisberg 2005). In this case, it is important to add tidal data to the picture.

Third, there are many conditions where there is differential survival and potentially even regrowth of fecal bacteria. Even though fecal indicator bacteria should not persist in the environment (Clesceri et al. 1998), yet there are a myriad of conditions where this is not the case. For example, we have observed instances where fecal enterococci regrew after desiccation in sediments (Hartel et al. 2005). It is important to consider these three limitations when dealing with fecal indicator bacteria.

16.4 Conducting a General Survey Without Testing

Once all the local knowledge and existing data have been obtained, it is time to conduct a general survey of the contaminated waterway by physically walking, kayaking, or boating it. Although this sounds easy in theory, we have found that it is rarely practiced. The typical situation is for a community to sample from a convenient bridge or road and only guess at the potential sources of fecal contamination upstream or downstream. In fairness to these communities, they are often only copying the same sort of limited sampling that they observed being done by state and federal agencies.

Generally, water depth and weather determine how a stream is surveyed. For a typical urban stream during baseflow conditions, walking can be easy, but we have had many instances of where we have had to bushwhack through the underbrush. Whenever possible, we always go with a local, usually a city employee, often a wastewater treatment plant employee. Although we do our best to obtain permission whenever a waterway goes through a person's property, sometimes this permission is difficult to obtain because the property owner is not present. In this case, having a city employee gives the survey an "official" air, and in some cases, the city employee knows the property owners. If the stream segment is short enough, two persons are all that are needed for the walk; if the stream segment is longer, then a third person will drive to drop and pick people up.

When we walk a stream, we always walk upstream to avoid disturbing the sediment (which may also obscure the view too). Also, because large numbers of fecal bacteria are often in the sediment (and not in the water column), an easy way to obtain a false picture of fecal contamination is to disturb the sediment. Generally, a walk will go from one major crossover point (bridge or nearby road) to another, as long as the distance is reasonable. Generally, we travel light and carry water, a small first-aid kit, insect repellent, a GPS device, and a cell phone. If there is no cell phone coverage (increasingly unlikely), then we use walkie-talkies. In an age of smart phones, a cell phone often contains a GPS application. Because of possibility of fecal contamination, each person is protected from the water, typically with boots and disposable gloves. However, if waterways are initially determined to be too contaminated, then we do not allow anyone in the water. For example, portions

Fig. 16.2 Gwyneth Moody (*below*) and Sarah Hemmings (*above*) preparing to conduct a general survey of Potato Creek, near Meansville, GA. In addition to a personal flotation device, each kayaker wears disposable gloves for protection

of the Mayagüez River in Puerto Rico had counts of *E. coli* in excess of 240,000 *E. coli*/100 mL (Hartel et al. 2007), and the river was only sampled from the bank. In the southeastern USA, snakes are often present (and occasionally alligators), and we do our best not to disturb them.

If a stream is too deep for walking, then we resort mostly to kayaks because they are highly maneuverable and sufficiently lightweight to portage around the inevitable logs that lie across the stream or creek (Fig. 16.2). We prefer inflatable kayaks designed for biological sampling (e.g., Stearns kayaks) because of their ease of transport and their stability. The only real change for kayaking as opposed to walking a stream is that we paddle downstream, not upstream, for relative comfort. Again, we take care not to disturb the sediment. The greatest risk for kayakers is always strainers, trees or debris in the stream that block the flow of water downstream. A person held in place by the current in front of a strainer is in a potentially life-threatening situation. For this reason, kayakers always travel a creek or river in pairs. In addition, one person always travels with an extra paddle. For salt marshes, we use sea kayaks because it is easier to maneuver with a rudder. Salt marshes also require us to observe tides, and we ensure that we kayak principally during high tides to avoid getting stuck.

In the general survey, baseflow conditions are always conducted first. In this manner, we know the likely dangerous areas when the general survey is done during stormflow conditions. A walk can also identify sources of fecal contamination that are likely to occur during stormflow conditions. For example, many dog owners locate kennels near streams. When there is sufficient overland flow, feces are often washed into the stream. Generally, walking a creek during stormflow conditions is normally done from the banks, rarely in the water. In the case of kayaking, sampling during a stormflow event is rare because of the danger, and the creek is usually sampled from the bank. Large rivers and lakes are typically surveyed from an appropriately sized boat. Tidal areas are normally sampled at high tide to avoid getting stuck, and we often resort to boats with engines to avoid strong currents.

For communities that want to minimize the expenses of MST, the general survey can be done without any kind of testing. The major reason that some MST researchers think that MST is a waste of time and money is that many sources of fecal contamination are obvious. Here are two examples where the sources were obvious, one for septic tanks and one for agricultural animals.

In Georgia, leaking septic tanks are common. The problem usually begins with the home owner, who says, "What, I have a septic tank?" Even for home owners who are aware that they have a septic tank, they often do not inspect the septic tank or know that it has a finite life span. Therefore, the general survey is always looking for failing septic tanks. On the coast, Georgia state law requires that septic tanks be at least 8 m from a water source to function properly. Walker et al. (2003) mapped all septic systems on the coastline of McIntosh County in Georgia. Of the 1,056 septic systems, 100 were within 1 m of coastal waters or marsh, and an additional 11 were between 1 and 8 m from a body of water. The combination of high coastal water tables and sandy soil in the area suggested that 63% of the septic systems were highly susceptible to failure. In fact, 5% of the 1,056 septic systems were visibly failing. There is no need for MST when a person sees a visibly failing septic tank (Fig. 16.3).

Fig. 16.3 Tidal water contaminated by septic tank effluent near Brunswick, GA. Tidal water in Georgia is typically dark-colored because of presence of organic matter, but effluent has turned this water gray. From this site, the water flows directly out into the salt marsh. Photo courtesy of James Holland, Altamaha Riverkeeper (used with permission)

Fig. 16.4 Cattle in an unnamed stream near Eatonton, GA

For agricultural animals, heat dissipation is a function of radiation, wind speed, air temperature, and humidity. If cattle are not fenced from a waterway, then they will spend more and more time in the waterway to cool their body temperature as they become increasingly heat-stressed. In the southeastern USA, where high temperatures and high humidity are common, cattle will spend on average over 5 h/day in streams whenever the temperature humidity index is sufficiently high (Franklin et al. 2009). There is no need for MST when a person sees cattle in a stream (Fig. 16.4).

Regardless of whether there is testing or not, the general survey requires good powers of observation, a sort of "fecal contamination sight," an ability to know where to look for fecal contamination. Here are two examples.

A small, short (<2 km) urban stream near Griffin, Georgia, had persistent fecal contamination for years. Many people had sampled the creek over those years and had observed nothing out of the ordinary. In our group, four people had walked the stream previously and had seen nothing out of the ordinary. The fifth person walking the stream, a person who was known to be especially observant, found a cast iron household sewer line hidden from normal view under a bridge (Fig. 16.5). Although the pipe was well constructed, it was not sufficiently well constructed to withstand the occasional stream flood stage, and the pipe was broken. The line was still in use and small amounts of sewage were flowing out. The person who saw the pipe had "fecal contamination sight." There is no need for MST when a person sees a broken sewer pipe with toilet paper flowing out.

In a second example, the City of Lawrenceville, Georgia, had a persistent fecal contamination problem with a small creek in a residential neighborhood. High counts of fecal bacteria were observed downstream of the neighborhood, but all the homes were on city sewerage, and no sewer lines were located near the stream.

Fig. 16.5 Broken active household sewer line over a small tributary to Potato Creek in Griffin, GA. Note the debris on the broken sewer pipe from a previous flood

A general survey was conducted. At the beginning of the neighborhood, the first person walked upstream, saw a tire swing over the creek, and kept on walking; the second person saw the tire swing and stopped. Closer examination of the soil showed a number of dog tracks. This person looked in the water and saw dog feces. In hot, humid Georgia, the second person had discovered the location where the local kids had brought their dogs to go swimming. There is no need for MST when a person sees dog feces in the water.

Combining local knowledge and existing data with a general survey and no testing is inexpensive and – as long as the individuals surveying the waterway have "fecal contamination sight" – can identify many obvious sources of fecal contamination quickly and easily. However, the method does have a serious limitation: the results are based on common sense, and common sense is insufficient for regulatory agencies that require specific data (e.g., bacterial counts) to delist a contaminated waterway. In these cases, the general survey must be conducted with testing.

16.5 Conducting a General Survey with Testing

If testing is necessary in a general survey, it can be of any type: bacterial, protozoan, or viral. For the purposes of this chapter, only bacterial sampling is considered. In any case, the principles for conducting a general survey with testing are the same as the general survey without testing. As long as water is running, every pipe, every tributary, and anything else that looks interesting are sampled. We use Whirl-Pak

Fig. 16.6 Jennifer McDonald sampling a small tidal waterway on St. Simons Island, GA, with a retractable sampling pole. The *inset* shows a close-up of the sampling pole and how alligator clips hold the Whirl-Pak bag. The alligator clips are secured with hose clamps on the ends of a U-shaped arrangement of plastic tubing and common copper plumbing parts

bags to collect our water samples because the bags are inexpensive and the insides are sterile. To make sampling easier, especially for out-of-reach pipes, as well as avoiding poison ivy, snakes, and brambles, we designed an alligator clip setup attached to a retractable pole (Fig. 16.6). The Whirl-Pak bag is attached to the clips, the top torn off (and pocketed), the sides opened, and the sample obtained. The pole works equally well from a kayak. In addition to the two-person sampling team, a third person is always added to the survey, and this person becomes a "rover" to sample each tributary further upstream.

We have tried a variety of systems to identify fecal bacteria, and we prefer the IDEXX Colilert or Enterolert system to estimate *E. coli* and fecal enterococci in fresh and marine waters, respectively. Not only is it possible to train microbiologically inexperienced people to process their own water samples quickly with this system but we can also process our own samples in a car enroute to the microbiology laboratory (Fig. 16.7). Processed samples are placed in a portable 37°C incubator that runs off the car lighter. The only problem we have with this system is that it is necessary to switch technicians frequently to avoid car sickness.

Depending on the time, duplicate water samples are usually obtained. Typically, only one sample is tested because the sampling will be repeated in time, in which case, the second sample is used only if the first sample is ruined (e.g., leaks). Here are two examples of a general survey with testing, the first survey easy and the second survey more difficult.

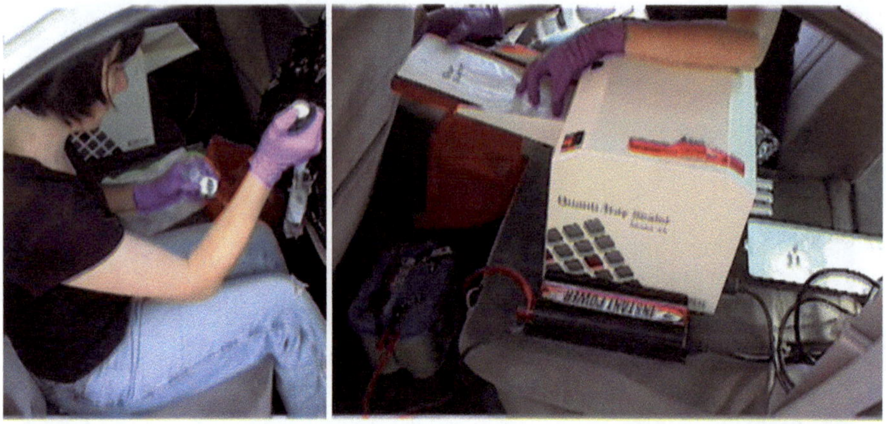

Fig. 16.7 Karen Rodgers processing a water sample in a car enroute to the microbiology laboratory (*left*). The water sample is sealed in an IDEXX Quanti-tray (*right*) using a car battery (barely visible on the car floor) connected to an inverter for electrical power

Stekoa Creek is a major tributary to the most famous part of the Chatooga River in Georgia (the part on which the movie "Deliverance" was filmed and now a major white-water rafting attraction), and this creek has a long history of contaminating the river with fecal bacteria. The marshal of the City of Clayton was concerned the effect of cattle farming on Scott Creek, which flows into Stekoa Creek in the City of Clayton. Local knowledge was obtained from the marshal and local wastewater treatment personnel. Scott Creek had not been tested previously, so there were no existing data. The general survey was done in one afternoon, sampling Scott Creek above and below each of three cattle farms, and Stekoa Creek just above its confluence with Scott Creek. Sampling was done from a law enforcement vehicle, and as expected with an accompanying armed law enforcement officer, there were no private property issues. We observed cattle urinating and defecating into Scott Creek as the cattle went from one pasture to another. The water samples were processed at a local wastewater treatment plant and the results were available the next day (Table 16.1). The results suggest that the cattle are contaminating Scott Creek and that the fecal contamination increases as the number of cattle farms increases. In this case, the general survey with testing confirmed the common sense observation that cattle were a source of fecal contamination to Scott Creek. The solution is for the cattle growers to consider fencing and elevated crossings to keep the cattle out of the creek.

In the second survey, St. Andrews Park on Jekyll Island (one of Georgia's barrier islands), had chronic problems with high counts of fecal bacteria. Wildlife was suspected. In this case, the general survey needed to be done under both baseflow and stormflow conditions because the effect of wildlife feces would become most obvious when it was raining. Since the terms "baseflow" and "stormflow" are inappropriate for tidal waters, we use the terms "calm" and "stormy" conditions here. Also, since these were marine waters, the waters were tested for fecal enterococci,

Table 16.1 Most probable number of *Escherichia coli* estimated by the IDEXX Colilert system for various sites on Scott Creek in Clayton, GA

Number	Site	*E. coli*/100 mL
1	Scott Creek – Headwater	41
2	Scott Creek – Above first cattle farm	146
3	Scott Creek – Below first cattle farm	512
4	Scott Creek – Above second cattle farm	677
5	Scott Creek – Below second cattle farm	1,414
6	Scott Creek – Above third cattle farm	2,224
7	Scott Creek – Below third cattle farm	3,282
8	Stekoa Creek before confluence with Scott Creek	865

Site #7 is located above the confluence of Scott Creek with Stekoa Creek

not *E. coli*. The threshold for fecal enterococci in Georgia with a single grab sample is 104 fecal enterococci/100 mL. When the general survey was conducted during calm weather conditions, all St. Andrews Park beach sites (Sites 22 through 45) and two thirds of the lower reach of Beach Creek (Sites 7 through 21) contained ≤20 fecal enterococci/100 mL (Fig. 16.8a). The remaining one third of the sites in the upper reach of Beach (Sites 1 through 6) had counts between 52 and 228 fecal enterococci/100 mL, generally decreasing from the upper to the lower reach of the creek. Therefore, three sites (Sites 1, 2, and 3) exceeded the State allowable number and were of concern. When the general survey was conducted during stormy conditions and an ebbing spring tide, a total of 35 water samples were collected. Of 15 water samples from St. Andrews Park (Sites 10 through 24), 11 (73%) exceeded the allowable maximum number of 104 fecal enterococci/100 mL, with numbers of fecal enterococci generally decreasing from north to south, away from Beach Creek (Fig. 16.8b). All the numbers of fecal enterococci in the nine Beach Creek sites (Sites 1 through 9) also exceeded the maximum, ranging from 171 to 428 fecal enterococci/100 mL, and generally decreased from the upper to the lower reach of the creek. With the exception of Sites 34 and 35 just north of Beach Creek, all the remaining nine sites defining the "box" around the park (Sites 25 to 35) had 30 or fewer fecal enterococci/100 mL.

Beach Creek was the logical source of fecal contamination to St. Andrews Park. No fecal contamination was observed in waters outside the park during calm or stormy conditions; therefore, sources of fecal contamination must be near or inside the park. During calm conditions, no sources of fecal contamination were observed in the water either near or inside the park except for the extreme upper reach of Beach Creek. However, during stormy conditions with an ebbing spring tide, large numbers of fecal enterococci were observed in the water coming from Beach Creek into St. Andrews Park. Numbers of fecal enterococci decreased north and south of the creek, and most of St. Andrews Park was in violation of the State standard (>104 fecal enterococci/100 mL). Runoff and tidal forcing likely caused this pulse of fecal contamination from Beach Creek.

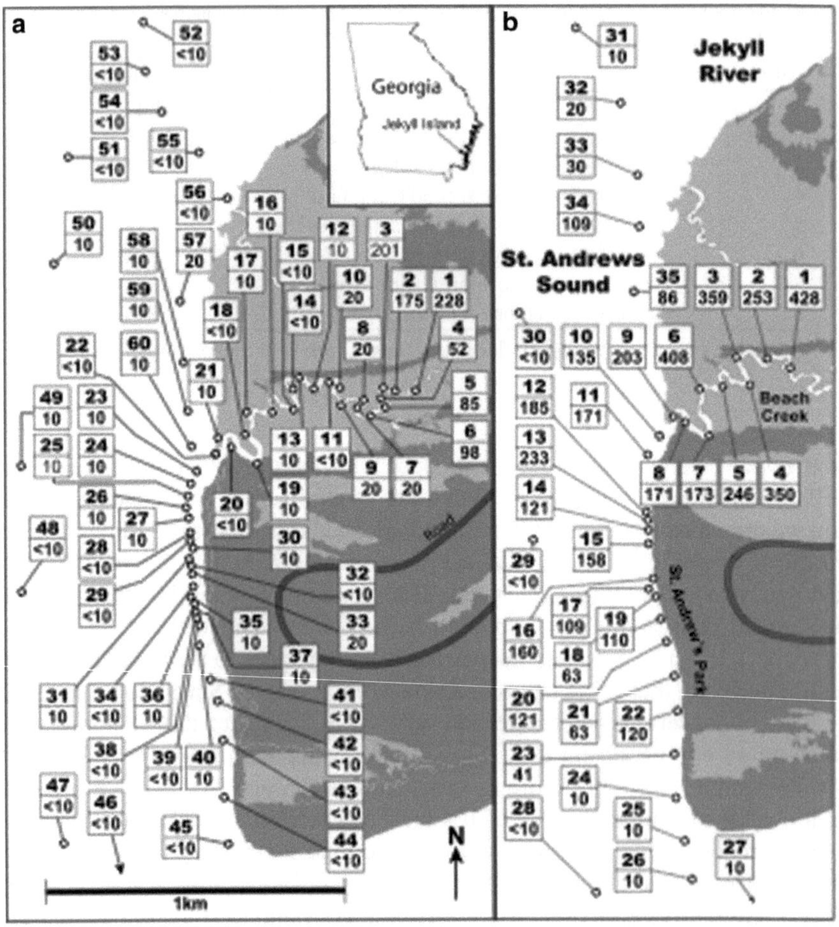

Fig. 16.8 Location of sampling sites at St. Andrews Park on Jekyll Island during (**a**) calm and (**b**) stormy conditions. Each location shows the site number (*boldface top number*) and the number of fecal enterococci/100 mL (*bottom number*). *Darker gray areas* define land and *lighter gray areas* define marsh. The scale is 1:11,000. The figure is adapted from McDonald et al. 2006 (used with permission)

Beach Creek is located in a marsh. Because there is no human habitation anywhere near the marsh, the only possible source of fecal contamination was wildlife. To make matters worse, much of the wildlife is birds, many of which are federally protected. Therefore, the problem of fecal contamination at St. Andrews Park will never be solved. What is needed is to treat marshes more like a wildlife refuge where the beauty of the marsh is emphasized, high numbers of fecal enterococci are accepted, and human activities such as swimming are discouraged, particularly during runoff or when tidal forcing conditions occur.

16.6 Targeted Sampling

In some cases, the sources of fecal contamination are still unclear after a general survey, either with or without sampling. In these cases, targeted sampling is necessary. Targeted sampling is like the children's game of "hot" and "cold," where a waterway is resampled over ever-decreasing distances to determine the exact area where the numbers of fecal bacteria are highest ("hotspots"). In each sampling, the samples are all collected in 1 day because nearly 70% of water-quality exceedences are single-day events (Leecaster and Weisberg 2001), and this kind of sampling reduces bacterial changes with time (Jenkins et al. 2003). The targeted sampling is repeated at the hotspots as necessary to limit the potential source(s) of fecal contamination to as small a geographic area as possible. Limiting the samples to a small geographic area reduces bacterial changes with geography (Hartel et al. 2002) and animal diet (Hartel et al. 2003). The resamplings ensure a site will have persistent fecal contamination and avoid problems where – unbeknownst to the sampler – a site is only temporarily "hot" (e.g., an animal defecating into the waterway). Here is an example of targeted sampling.

The City of Thomaston, Georgia, had chronic problems with fecal contamination in Potato Creek. The first sampling of Potato Creek showed a suspicious tributary with a count of 520 *E. coli*/100 mL and, higher up on the tributary, a count of 1,086 *E. coli*/100 mL from a "rover" sample (Fig. 16.9). Under the auspices of targeted sampling, the tributary represents a hotspot and a potential source of fecal contamination. Therefore, the second targeted sampling focused on this small, short tributary. Counts at Site 1 were high (314 *E. coli*/100 mL) and much lower for Site 2 (107 *E. coli*/100 mL). Only one house was between Site 1 and Site 2: the local dogcatcher. This person used a hose to spray all the dog feces from 20 to 30 outside kennels into the tributary. Counts of *E. coli* in the sediment below Site 1 were 160,582 *E. coli*/100 mL. Changing the practices of the dogcatcher solved the fecal contamination to this reach.

16.7 Considering Cost, Experience, and Other MST Methods

There are three more issues to be considered in the sampling protocols proposed here: cost, experience, and other MST methods. First, outside of the cost for microbial analyses, the protocols still require considerable time on the part of the investigators and their staff. Targeted sampling can involve multiple trips if multiple samplings are required. If waterways are remote from laboratories, then overnight stays in hotels may be required. Therefore, even though this chapter is about minimizing MST, there are still appreciable costs associated with for conducting the protocols. Even so, the protocols should still be relatively less expensive than many other MST approaches.

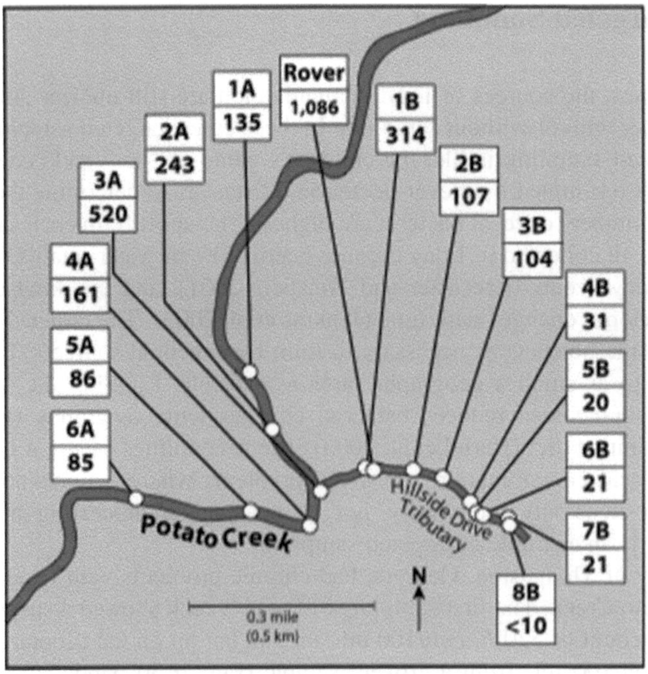

Fig. 16.9 Location of sites during a targeted sampling of Potato Creek near Thomaston, GA. Each location shows the site number (*top boldface number*) and the number of *Escherichia coli*/100 mL (*bottom number*). Top numbers with the designations "A" or "Rover" represent the first sampling; top numbers with the designation "B" represents the second targeted sampling. The creek runs from northeast to southwest (data from Hartel et al. 2008)

Second, the author of this chapter has spent many hours wading in waterways identifying sources of fecal contamination missed by less experienced workers. Yet, for some cultures, the idea of a senior researcher wading in a badly contaminated stream and identifying sources of fecal contamination is an anathema. The human tendency is to want to work in beautiful, clean, air-conditioned laboratories with glamorous methods (e.g., molecular genetics) than in the field where it is often hot, dirty, and uncomfortable. Yet, our experience with the protocols in this chapter is that we consistently obtain our best results with senior researchers because they have better people skills and better powers of observation than junior workers. Experience matters. Senior researchers need to be in the field.

Third, it is unfair to suggest that the sampling protocols in this chapter are never done with other methods of MST. MST methods need to be selected on a "toolbox" approach, where the method is selected after considering each method's cost, reproducibility, discriminatory power, ease of interpretation, and ease of performance (USEPA 2005). For example, a targeted sampling on the tidal Sapelo River in Georgia was combined with the expensive genotypic method, ribotyping (Kuntz et al. 2003). In some cases, other MST methods might even initially supplant the microbial sampling mentioned in this chapter. For example, with recent improvements in fluorometry (Cao et al. 2009; Hartel et al. 2008), this method may well become the method of choice in a targeted sampling for specifically identifying human fecal contamination from broken sewer pipes. Microbial analyses would later simply confirm fluorometric measurements.

16.8 Conclusions

A flow diagram of the chapter for minimizing MST is shown in Fig. 16.10. In many cases, marrying local knowledge, existing data, and conducting a general survey without testing will identify many sources of fecal contamination in a contaminated waterway. This approach will often minimize the need for MST. This solution is best for almost all communities, particularly those that are poor. In the cases where

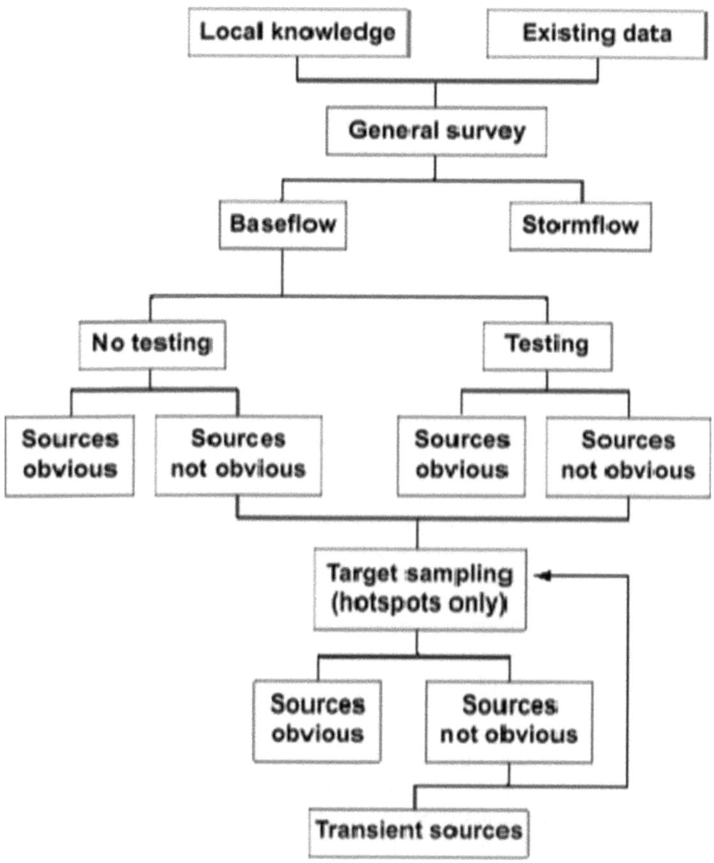

Fig. 16.10 Flow diagram for minimizing MST. The flow diagram for stormflow conditions is identical to that for baseflow conditions

some sort of testing must be done, then a general survey with testing offers a reasonable alternative for supplying the necessary scientific data to determine the sources of fecal contamination or get the waterways delisted. If the general survey with or without testing does not satisfactorily identify the sources, then targeted sampling is possible. Targeted sampling uses the child's game of "hot" and "cold" to narrow the hotspots of fecal contamination, and it uses technology with which most communities are familiar. In this manner, it is possible to identify almost all sources of fecal contamination inexpensively, quickly, and easily.

References

Boehm AB, Weisberg SB (2005) Tidal forcing of enterococci at marine recreational beaches at fortnightly and semi-diurnal frequencies. Environ Sci Technol 39:5575–5583

Cao Y, Griffith JF, Weisberg SB (2009) Evaluation of optical brightener photodecay characteristics for detection of human fecal contamination. Water Res 43:2273–2279 doi:10.1016/j.watres.2009.02.020.

Clesceri LS, Greenberg AE, Eaton, AD (1998) Standard methods for the examination of water and wastewater, 20th ed. American Public Health Association, American Water Works Association, and Water Environment Federation, Washington DC

Franklin DH, Cabrera ML, Byers HL et al (2009) Impact of water troughs on cattle use of riparian zones in the Georgia Piedmont in the United States. J Anim Sci 87:2151–2159 doi:10.2527/jas.2008-1522

Hartel PG, Summer JD, Hill JL, et al (2002) Geographic variability of *Escherichia coli* ribotypes from animals in Idaho and Georgia. J Environ Qual 31:1273–1278

Hartel PG, Summer JD, Segars WI (2003) Deer diet affects ribotype diversity of *Escherichia coli* for bacterial source tracking. Water Res 37:3263–3268

Hartel PG et al (2005) Survival and regrowth of fecal enterococci in desiccated and rewetted sediments. In Hatcher KJ (ed) Proceedings of the 2005 Georgia Water Resources Conference, University of Georgia, Athens

Hartel PG, McDonald, JL, Gentit LC et al (2007) Improving fluorometry as a source tracking method to detect human fecal contamination. Estuaries Coasts 30:551–561

Hartel PG, Rodgers K, Moody GL et al (2008) Combining targeted sampling and fluorometry to identify human fecal contamination in a freshwater creek J Water Health 6:105–116 doi:10.2166/wh.2007.004

Jenkins MB, Hartel PG, Olexa TJ et al (2003) Putative temporal variability of *Escherichia coli* ribotypes from yearling steers. J Environ Qual 32:305–309

Kuntz RL, Hartel PG, Godfrey DG et al (2003) Targeted sampling protocol as prelude to bacterial source tracking with *Enterococcus faecalis*. J Environ Qual 32:2311–2318

Leecaster MK, and Weisberg SB (2001) Effect of sampling frequency on shoreline microbiology assessments. Mar Pollut Bull 42:1150–1154

McDonald JL, Hartel PG, Gentit LC et al (2006) Identifying sources of fecal contamination inexpensively with targeted sampling and bacterial source tracking. J Environ Qual 35:889–897 doi: 10.2134/jeq2005.0328

Solo-Gabriele HM, Wolfert MA, Desmarais TR et al (2000) Sources of *Escherichia coli* in a coastal subtropical environment. Appl Environ Microbiol 66:230–237

US Environmental Protection Agency (USEPA) (2005) Microbial source tracking guide document. EPA/600-R-05-064. Office of Research and Development, Cincinnati, OH

Walker RL, Cotton C, Payne K (2003) A GIS inventory of on-site septic systems adjacent to the coastal waters of McIntosh County, Georgia. Bull 27:1–44. http://www.marsci.uga.edu/gaseagrant/pdf/septic%20text.pdf. University of Georgia Marine Extension Service, Athens.

Chapter 17
Environmental Persistence and Naturalization of Fecal Indicator Organisms

Donna Ferguson and Caterina Signoretto

Abstract Fecal indicator bacteria, including total and fecal coliforms, *Escherichia coli*, and enterococci have been used to indicate the presence of fecal contamination in water used for drinking, shellfishing, and recreation. The assumptions behind the concept of using fecal indicator bacteria (total coliforms, fecal coliforms, *E. coli*, and enterococci) are that these organisms are normal inhabitants of the gastrointestinal tracts of humans and animals and are incapable of long-term survival or replication outside of their hosts. These assumptions must be valid if FIB levels in water are used to ensure the microbial safety in water and to identify "fecally contaminated" areas, which is essential to microbial source tracking (MST) studies. However, after utilization of total coliforms in the 1920s and fecal coliforms in the 1980s, studies showed that these groups of indicator bacteria failed to meet these assumptions because they include strains that are naturally occurring in nonfecal sources (i.e., soil, vegetation, and water) and are capable of persisting and regrowing in the environment. In the 1980s, *E. coli* and enterococci were recommended for use as FIB because they were believed to be more specific to fecal waste. Recent studies have shown that these bacteria also failed to meet the criteria for FIB.

The continuing practice of implementing fecal indicator organisms without understanding their persistence and survivability in the environment has hindered the ability to determine their significance in water and to accurately assess human health risks. This chapter discusses the natural occurrence of total coliforms, fecal coliforms, *E. coli*, and enterococci and explains key survival mechanisms, such as biofilm development and transitioning to the viable but nonculturable (VBNC) state, which can allow them to persist and regrow in various environmental habitats. Understanding how and where fecal indicators can survive in the environment could be useful to MST studies to account for fecal indicator bacteria from natural sources. The information gained from these studies could be applied to studies evaluating MST methods employing new, alternative indicators such as *Bacteroides* spp.

D. Ferguson (✉)
Southern California Coastal Water Research Project, 3535 Harbor Blvd., Suite 110, Costa Mesa, California 92626, USA
e-mail: donnaf@sccwrp.org

C. Hagedorn et al. (eds.), *Microbial Source Tracking: Methods, Applications, and Case Studies*, DOI 10.1007/978-1-4419-9386-1_17,

Keywords Fecal indicator bacteria • Environmental strains • Biofilm • Viable but nonculturable

17.1 Introduction

Fecal pollution as a source of pathogen transmission is widely recognized as a subject of critical concern for waters used for recreation and shellfish harvesting. Levels of fecal indicator bacteria (i.e., total coliforms, fecal coliforms, *Escherichia coli*, and enterococci) are used to identify waterbodies impacted by fecal contamination and to assess the magnitude of contamination.

The term "fecal indicator bacteria" is a working definition, rather than a taxonomic classification, used to describe bacteria that normally inhabit the gastrointestinal tracts of warm-blooded animals. However, numerous studies have shown that fecal indicator bacteria include strains that may survive in a particular setting for a period of time after introduction (persistence) and, perhaps more important, to grow, replicate, and adapt in nonhost environments (naturalization) including water, soil, and vegetation.

The ability of indicator bacteria to survive, adapt, and regrow in the environment invalidates the criteria for fecal indicator organisms. According to Bonde (1966), ideal indicators (1) should always be present in human and warm-blooded animal feces, (2) must not be able to multiply in aquatic environments, (3) must occur in greater numbers than pathogens, and (4) must be unambiguously identifiable by simple, characteristic, and reliable tests. The FIB most commonly used to assess water quality, including total coliforms, *E. coli*, and *Enterococcus* spp., do not meet these criteria. Each of these groups includes species or strains (subgroups within species) that are native to water, soil, and plants, leading to uncertainty about whether or not these organisms have been externally introduced or have the potential to increase health risks.

In terms of water-quality assessment, one of the most general goals of microbial source tracking (MST) is to identify significant sources of fecal contamination (USEPA 2005).

The emergence of molecular MST methods improves researchers' ability to discern one strain of FIB from another, particularly in extraintestinal environments, which greatly enhances the potential for predictive accuracy in assessing human health risks associated with water quality. Certain MST methods target host-specific species of *Bacteroidales* that are anaerobic and, thus, less likely to survive in natural environments. However, little is known about the prevalence and survival of these bacteria in habitats outside of their hosts.

To improve the effectiveness of MST related to water-quality assessment, a better understanding of persistent and naturalized FIB in aquatic environments is essential. This information could useful for enhancing the utility of current FIB measurements and in evaluating new and alternative fecal indicators used to assess water quality and human health risks.

17.2 Evidence of FIB Persistence

Fecal indicator organisms have been relied upon to assess water quality since the 1920s. Total coliforms were the first fecal indicator bacteria group used. However, as microbiological test methods improved and bacterial species were identified, it became evident that total coliforms were ubiquitous in the environment, and thereby unreliable for indicating fecal contamination. Subsequently, fecal coliforms were found to be more specific indicators until environmental counterparts within this group were also identified. In the 1980s, epidemiological studies conducted by the Environmental Protection Agency (EPA) showed that *E. coli* and *Enterococcus* spp. were better correlated than were coliforms to swimmer illness rates (USEPA 1986). However, numerous studies have demonstrated that certain strains of *E. coli* and enterococci are also capable of surviving in the environment.

Persistence of traditional FIB in extraintestinal environments is well documented and presented in this chapter according to fecal indicator group. Strains within each group identified in natural and clinical settings are described to better understand the significance of their occurrence in water and human health risks to exposed populations.

17.2.1 Total Coliforms

In terms of numbers and variety of genera and species, total coliforms comprise the largest group of assessment related FIB. Prior to 1994, the definition of total coliforms was based on the ability of members of this group to produce acid and gas from lactose. Based on this definition, total coliforms were primarily represented by *E. coli,* followed by *Klebsiella* spp., *Enterobacter* spp., and *Citrobacter* spp. Owing to increased use of enumeration methods based on β-galactosidase (e.g., Colilert®), the definition of this group has recently been expanded. Additional genera within the coliforms possessing this enzyme have been added to the total coliform group. These newly added coliforms occur primarily in environmental habitats, further decreasing the specificity of this group as a source indicator (Table 17.1; Leclerc et al. 2001; Stevens et al. 2003).

It has long been accepted that the total coliform group is the least reliable fecal indicator group for assessing the quality of environmental waters and predicting human health risks. *Klebsiella* spp., *Enterobacter* spp., and *Citrobacter* spp. are natural inhabitants of soil, vegetation, farm produce, insects, and wooden reservoirs (Leclerc et al. 2001). Further, species within all three genera have been shown to persist and grow in drinking water distribution systems, demonstrating that the presence of these organisms in water is not necessarily indicative of human health risks (LeChevallier et al. 1990; Camper 1991; USEPA 2002).

In terms of predicting human health risks, environmental strains of *Klebsiella pneumoniae* further diminish the reliability of total coliforms in that this is one of

Table 17.1 Source origin of selected genera within fecal indicator bacteria groups

Indicator	Genera	Fecal origin	Nonfecal origin
Total coliforms[a]	*Escherichia*[b]	+	+
	Klebsiella[b]	+	+
	Enterobacter[b]	+	+
	Citrobacter[b]	+	+
	Serratia[b]	+	+
	Hafnia[b]	−	+
Additional total coliforms isolated using enzyme-based β-galactosidase[c, d]	*Pantoea*	−	+
	Cedecea	−	+
	Ewingella	−	+
	Moellerella	+	−
	Leclericia	−	+
	Rahnella	−	+
	Yokenella	−	+
Fecal coliforms[a] (thermotolerant coliforms)	*Escherichia*	+	+
Enterococci	*Enterococcus*	+	+

[a] Adapted from Leclerc et al. (2001)
[b] Also a member of the fecal coliform group
[c] Considered to be primarily environmental
[d] Source: Stevens et al. (2003)

the most persistent species found in water. In 1984, Caplenas and Kanarek found that recycled water at paper industries, particularly pulp mills, containing particulate matter and wood fibers provided an excellent growth medium for *K. pneumonia*. Although the relationship between health risk and environmental strains of *K. pneumonia* is unclear, some have been identified as plant pathogens (Lacey and Lukezic 2004), while clinical strains have been reported as opportunistic pathogens in humans, causing respiratory and urinary tract infections (Leclerc et al. 2001). In addition to their ability to multiply in aquatic environments, *Klebsiella* spp. fails to meet Bonde's ideal indicator criteria for pathogen to indicator ratios. For example, in surface waters receiving storm-water runoff and industrial discharges, *Klebsiella* spp. are often present in higher densities than fecal coliforms, suggesting that environmental strains of *Klebsiella* are more abundant than fecal strains of the same genus (Geldreich 1996).

Similarly, environmental strains of *Enterobacter* spp. are also unreliable indicators of fecal contamination and potential health risks. *E. aerogenes* and *E. cloacae* are capable of persisting in the environment and within human intestinal tracts without causing illness. In a 1994 study, *E. cloacae* isolates found in a drinking-water supply system were compared to strains obtained from the system's source water and from several local hospitals. Molecular typing by pulsed-field gel electrophoresis (PFGE) revealed that the strains from the drinking-water system were nearly identical to those found in the source water and hospital samples (Edberg et al. 1994).

In addition to environmental waters, total coliforms have been found to persist in many other extraintestinal habitats. *Serratia* spp. are typical of coliform bacteria commonly found in soil and attached to plants and rarely associated with human infections. Still, information important to MST can be derived from the study of such environmental strains. For example, Ganeshwar et al. (2001) reported endophytically established cells of nitrogen-fixating *S. marcescens* within roots, stems, and leaves of different varieties of rice. When rice seedlings were inoculated with *S. marcescens*, the bacteria were observed colonizing in intercellular spaces, root cortical cells, aerynchyma, and xylem vessels. Hypothetically, increased understanding of such colonization characteristics can be extrapolated to improve species-specific source tracking efforts.

17.2.2 Fecal Coliforms

In order to improve the reliability of bacterial tracking in the face of a preponderance of naturally occurring total coliforms, culture-based assessment methods began to focus on fecal coliforms for measuring water quality. Fecal coliforms, a subgroup of total coliforms, have the ability to tolerate relatively higher environmental temperatures (44.5°C), a characteristic that is common to intestinal coliforms and, thus, thought to make their presence in the environment more indicative of the possible presence of fecal pathogens. Although *E. coli* is the most common "thermotolerant" species to be isolated from human feces, certain strains of *Klebsiella*, *Citrobacter*, and *Enterobacter* also have the ability to grow at 44.5°C. McClellan et al. (2001) repeatedly isolated clonal strains of these genera from beach water, suggesting replication of thermotolerant fecal coliforms in aquatic environments. Consequently, thermotolerance alone did not prove to be an effective identifying characteristic for reducing ambiguity in culture-based methods using fecal indicator organisms.

17.2.3 *Escherichia coli*

Although the primary habitat of *E. coli* is the lower intestine of warm-blooded animals, secondary habitats also include soil, vegetation, and water. Some *E. coli* strains isolated from primary habitats have been found to be genetically distinct from natural strains, suggesting possible strain selection due to environmental pressures (Whitam 1989), which can lead to ambiguity in organism source tracking. For example, Topp et al. (2003) demonstrated that *E. coli* populations in swine manure slurry changed dramatically during incubation in soil, suggesting that specific strain types have a selective advantage; however, further studies are needed to more fully characterize strain origination and adaptive qualities.

As with total coliforms, several recent studies have documented environmental strains of *E. coli*. In 2000, Solo-Gabriele et al. identified *E. coli* growth in river water; in 2001, Rozen and Belkin observed *E. coli* surviving in seawater, and in

2009, Filip and Demnorova reported finding persistent *E. coli* in groundwater. Although not entirely dependent on thermotolerant characteristics, FIB persistence in nonhost environments may be significantly related to thermodynamic adaptability and temporal survival qualities. For example, in 1971, Shaw et al. determined the minimum growth temperature for *E. coli* to be 7.5–7.8°C. Subsequently, *E. coli* have been found to persist in an ice-covered river in Alaska (Davenport et al. 1976), a polar marine environment (Smith et al. 1994), and freshwater sand in the Great Lakes area (Whitman and Nevers 2003; Alm et al. 2006).

Other recent environmental FIB studies have documented *E. coli* persistence in groundwater over time. Bitton et al. (1983) calculated the T_{90} (time required to reduce the bacterial population by one order of magnitude) for *E. coli* in groundwater as 6.2 days. More recently, Banning et al. (2003) showed that biofilm-associated *E. coli* in a groundwater reactor persisted longer than planktonic cells. However, additional studies are needed to more fully understand the relationship between thermotolerance and other survival mechanism that lead to persistence over time.

E. coli and, in most cases, enterococci have been shown to persist in the following: clay loam in Oregon (McCoy and Hagedorn 1979), pristine sediments in Hawaii (Fujioka et al. 1988), sediments in Australian urban estuaries (Ferguson et al. 1996), river sediments in Florida (Solo-Gabriele et al. 2000), and beach sands in southern California (Lee et al. 2006; Yamahara et al. 2007) and Florida (Hartz et al. 2008). In addition, *E. coli* and enterococci have been shown to grow on plants, such as bromeliads (Rivera et al. 1988), *Cladophora*, a common algal species found in fresh and marine waters worldwide (Whitman et al. 2003), and *Sarracenia purpurea*, a carnivorous bog plant (Whitman et al. 2005). Notably, the isolates described in these studies were collected from plants located in pristine sites without overt fecal contamination sources.

In addition to environmental persistence, growth of *E. coli* in soils has been recently confirmed by Ishii et al. (2006) using horizontal, fluorophore-enhanced repetitive extragenic palindromic PCR (HFERP) DNA fingerprinting analysis. In this study, identical *E. coli* strains (similarity values of ≥92%) were continually isolated from soil over time. Even after over wintering, the same genotype subsequently grew during the summer months. Maximum growth of *E. coli* strains occurred at 37°C. At 4, 15, and 25°C, no increase in cell densities was reported; however, cells survived longer at these temperatures than at 30 or 37°C. As a result of these findings, the authors proposed that the term "naturalized" *E. coli* be used to describe isolates with unique HFERP DNA fingerprints not similar to strains from known source DNA libraries, which cluster together at similarity values of ≥92% and have the ability to be repeatedly isolated over time. By contrast, environmentally acquired strains of indicator bacteria are generally shed by warm-blooded hosts through urine and feces, with *E. coli* occurring as transient temporary residents without colonizing the host gut (Hartl and Dykhuizen 1984).

Naturally occurring strains of fecal indicators, including *E. coli*, would be expected to have a greater propensity for survival and growth in environmental habitats than enteric or intestinal strains. Specifically, enteric bacteria must be able to withstand drastic changes in physicochemical parameters including nutrient availability and changing levels of osmolarity to survive in aquatic environments. In terms of persistence related to environmental conditions, intestinal bacteria introduced to water may be initially protected from extreme temperature changes, predators, and UV degradation if they are embedded in fecal waste, organic material, or sediment. Gerba and McLeod (1976) demonstrated that the addition of sediment to seawater enhanced survival of *E. coli* due to higher amounts of organics in sediment. Further, a 1990 study indicated that glycine betaine (GB), a potent osmolyte present in many bacteria, may protect enteric bacteria such as *E. coli* from the effects of high ion concentration (Gauthier and LeRudulier 1990). Notably, that study found that GB uptake by *E. coli* was highest in the presence of marine sediment with high total organic carbon.

17.2.4 Enterococci

In the 1940s, *Enterococcus faecalis,* then known as *Streptococcus faecalis,* was first reported in frozen foods and considered to be a possible threat to public health until environmental strains of the species were identified (Mundt 1982). The subsequent discovery that *E. faecalis* was a ubiquitous presence in frozen foods indicated the improbability that this was due solely to fecal pollution and prompted ecological studies to search for nonfecal origins of enterococci.

Similar to coliforms, the genus *Enterococcus* also includes species considered primarily environmental. The most predominant enterococcal species in the lower intestines of humans and animals, *E. faecalis* and *E. faecium*, may cause illness in humans (Devriese et al. 1987; Willey et al. 1999). These species are also the most predominant enterococcal species found in sewage (Lauková and Juriš 1997; Pinto et al. 1999, Manero et al. 2002) and in marine waters (Svec and Sedlacek 1999; Ferguson et al. 2005). This environmental persistence may be attributed to these organisms possessing physiological traits that enable longer periods of persistence and survival in nonhost environments as compared to the other fecal indicator groups. Notably, these bacteria can tolerate oxidative stress; they grow at 10 and 45°C, in 28% NaCl broth, at high (9.6) and low (4.8) pH, and unlike other indicator bacteria, demonstrate resistance to bile salts, azide, detergents, sodium hypochlorite, heavy metals, ethanol, and prolonged desiccation (Huycke 2002).

Environmental enterococci replication in an oxidation pond environment has been recently reported by Moriarty et al. (2008). Unusually high levels of enterococci in oxidation ponds with typical levels of fecal coliforms prompted investigations to determine whether the proliferation of enterococci was due to breakthrough from the treatment process or wildfowl occupying the ponds. Antibiotic resistance analysis (ARA) profiles of high similarity were repeatedly identified over a 4-month period. Ultimately, PFGE typing confirmed that the enterococcal strains with dominant ARA profiles were clonal, suggesting regrowth of enterococci.

Current literature on the association of enterococci with plants is sparse; however, earlier studies clearly document relationships between enterococci and natural sources (Mundt 1961). In addition to natural environmental persistence, enterococci have been found to be capable of limited growth on plants. Mundt suggests that enterococci are temporary residents on plants that may be disseminated by insects and wind. The author also speculates that gravity and rain may transfer enterococci from plants to the ground. Higher densities of these bacteria have been found on plants located in a dense forest than in open areas exposed to sunlight (Mundt 1963). Also, high densities of enterococci and fecal coliforms have been found on aquatic vegetation such as sea wrack, i.e., eel grass and seaweed (Valiela et al. 1991; Anderson et al. 1997). Still, additional information characterizing survival mechanisms and strain attributes is needed to distinguish natural sources from contamination indicators.

Another example of environmental enterococcal persistence can be found in the yellow-pigmented *E. casseliflavus*, *E. mundtii*, and *E. sulfureous*, as well as in nonpigmented *E. faecalis* and *E. faecium,* thought to be part of the microflora of grasses (Ulrich and Müller 1998; Ott et al. 2001). In 2001, Ott et al. found that nearly 70% of 750 enterococcal strains isolated from plant material formed a homogeneous 16S rDNA genotype of an unknown species. Approximately 8% were identified as *E. faecalis*, 8% as *E. mundtii*, 6% as *E. casseliflavus*, and 5% as *E. faecium*. Yellow pigmented enterococci are most often found in soil, plants, and insects (Taylor et al. 1971). More recently, *E. casseliflavus* has been identified as the most common enterococcal species present in urban runoff (Ferguson et al. 2005; Moore et al. 2008), suggesting that vegetation may be a dominant source to surface waters. Regardless of source, pigmentation of bacterial colonies due to carotenoid production can be viewed as a survival mechanism that protects membrane-bound functions from photodynamic damage and contributes to an organism's ability to persist outside of its host environment. Notably, this phenotypic characteristic is more common to environmental species among certain bacteria, yeasts, and fungi than to clinical species in the same genus.

17.3 Fecal Indicator Bacteria (FIB) Survival

Although more comprehensive ecological investigation is needed, survival studies of FIB have greatly contributed to understanding the physicochemical parameters and mechanisms that influence bacterial persistence in the environment. Still, in spite of mounting evidence of environmental FIB persistence, conflicting reports related to die-off rates for FIB in aquatic environments have led to skepticism about the ability of indicator bacteria to adapt and survive outside of humans and warm-blooded animals.

Survival rates determined experimentally vary depending on the bacterial species or strains studied. A number of previous survival studies involving indicator bacteria have assessed the survivability of the indicator group without identifying the species composition, which complicates comparisons between studies or within individual studies conducted over time. In 2002, Sinton et al. found enterococci inactivation in waste stabilization pond (WSP) effluent due to sunlight to be similar to inactivation of fecal coliforms during months of limited sunlight. In that study, enterococci were found to be more rapidly inactivated in the summer than fecal coliforms in winter. Because enterococcal cells were damaged by photooxidation in WSPs, the authors suggested that enterococci may be unsuitable as an indicator group for WSP effluent discharges to natural waters. However, species composition of enterococci and fecal coliforms during winter and summer was not characterized, and it is uncertain whether the results were influenced by or due to seasonal shifts in species composition. Identifying shifts in FIB population over time is critical for the improvement of MST approaches (Anderson et al. 2005).

Differences in survival rates from previously described studies may also be attributed a number of other factors such as (1) the type of innoculum used i.e., laboratory cultures vs. fecal waste, or planktonic vs. cell aggregates, (2) matrix, i.e., filtered or autoclaved water and/or sediment, (3) type of sediment, i.e., coarse grained sands vs. finer sediments, (4) salinity levels, (5) exposure to sunlight and predators, (6) nutrient availability, and (7) type of mesocosm used, i.e., field vs. laboratory and representativeness of mesocosms to environmental conditions. A number of survival studies on FIB were based on introduction of planktonic cells to water, which may have resulted in overestimation of die-off rates. For example, dialysis bags have been frequently used to measure growth of planktonic indicators exposed to water, but survivors attached to or growing on the surfaces of these bags are not typically included in the assessments. Notably, in aquatic environments, bacteria rarely exist as planktonic or free-living cells. More commonly, they are found attached to particles associated with plants and soil, which protects the organisms, to some extent, from UV inactivation, dehydration, predators, and disinfection agents.

Finally, conflicting results among FIB survival studies may be partly due to differences related to cultivation and enumeration of bacteria in the laboratory. Cultures are typically grown at 35°C to reflect human body temperature; yet, temperatures higher than 20°C are often lethal for bacteria adapted to aquatic environments (Roszak and Colwell 1987), leading to potential overestimation of decay rates. Also, bacterial cells can clump together and yet form only a single colony on an agar plate, leading to lower counts of survivors.

17.3.1 Biofilm Formation

In aquatic environments, bacteria are capable of colonizing almost any wetted solid surface (Olson et al. 1991), partially due to adaptive survival mechanisms such as biofilm formation. Biofilm formation is a complex process that leads to development of a protected community. Simply put, biofilm is the slimy material on aquatic surfaces, including rocks, sediment, algae, zooplankton, vegetation, boat bottoms, pier planks, and debris.

FIB, including *Enterococcus* spp., *E. coli*, and *Klebsiella* spp., are known biofilm producers (Camper et al. 1991; Fass et al. 1996; Pratt and Kolter 1998; Tendolkar et al. 2004; Szabo et al. 2006). Bacteria embedded within biofilm are protected from harmful environmental conditions such as desiccation and UV, as well as chemicals such as antimicrobials and disinfectants. In drinking-water distribution systems, *Klebsiella* spp. were found as the predominant species in biofilm from pipe sediments and tubercles in slow-flow sections, dead ends and on walls or sediments in storage tanks. Similarly, many nososcomial (hospital acquired) enterococcal infections occur due to the ability of these organisms to survive for prolonged periods in biofilms associated with hospital environments (Tendolkar et al. 2004). Nosocomial strains of enterococci are presumed to possess virulence traits that distinguish them from strains from the gastrointestinal tracts (Gilmore et al. 2002).

In addition to supporting extraintestinal persistence of FIB, there is significant growth potential for bacteria within biofilms, depending on nutrient availability and temperature. In general, bacterial growth is optimal at warmer temperatures while survival may be longer under colder temperatures.

Bacteria associated with biofilms detach when biofilms mature or are disrupted due to physical forces, such as shearing. Bacterial cells released from the outer layer of mature biofilm can migrate to other locations where they can continue to form biofilms (Dunne 2002). These bacteria can be further transported by increased water flow velocities related to rain events, runoff, and industrial discharges; this increased velocity may shear biofilm on surfaces of storm channels, river bottoms, rocks, sediments, and vegetation. In recent environmental studies, mechanical processes such as shaking and sonication have been used to detach and enumerate bacteria from sediments (Ferguson et al. 2005; Boehm et al. 2009). However, the efficiency of these techniques in terms of detachment efficiency or representativeness to natural scouring processes, such as turbulent water flows or wave action, is yet to be determined. Regardless of detachment mode, biofilm fragments containing bacteria released into the water column and transported to alternate sites may explain some of the variability seen with fecal indicator counts in aquatic environments, particularly when there is no evidence of fecal input.

17.3.2 Viable but Nonculturable (VBNC) State

Biofilm communities also allow bacteria to transition to the VBNC state during unfavorable environmental conditions. Since the 1970s, a discrepancy between cultural vs. microscopic counts has been observed for environmental samples, with microorganisms being detected in higher numbers microscopically than with culture-based methods. It has been suggested that the stressing conditions present in natural environments cause bacterial damage and, thus, are responsible for the loss of culturability (Buck 1979). The idea that damaged bacteria are difficult to grow or incapable of growing on selective media supports this observation. For example, in the 1980s, the pioneering work of Rita Colwell and others showed that bacteria, including those with human health implications, can persist over time in natural environments but are no longer culturable. In this persistent nonculturable state, organisms maintain their pathogenic traits and capacity for regrowth in certain instances when suitable environmental conditions are restored (Kogure et al. 1979; Xu et al. 1982; Roszak and Colwell 1987).

The VBNC hypothesis postulates that human pathogens, which find suitable environmental conditions in terms of nutrient availability and physicochemical conditions within the human body (e.g., in the human gut), when dispersed in natural environments (e.g., fresh or marine water) are faced with low temperatures, nutrient depletion, high or low salinity, and solar radiation inducing bacterial stress, leading to cell division inhibition and death. By convention, viability had formerly been strictly associated with culturability; thus, nonculturable forms of a "normally culturable" enteric microorganism (e.g., *E. coli*) were considered dead (Barer and Harwood 1999). However, according to the VBNC hypothesis, bacteria are capable of surviving in viable form after cessation of culturability. Evidence for bacterial cell viability in the VBNC state has been established through studies examining amino-acid uptake and protein synthesis (Lleò et al. 1998), membrane integrity (Roth et al. 1997; Breeuwer and Abee 2000), and specific gene expression with a resulting characteristic proteome (Lleò et al. 2000; Heim et al. 2002).

Further, specific cell-wall modifications related to entry into the VBNC state have been described in *E. coli* and enterococci (Signoretto et al. 2000, 2002). The observed increase in the peptidoglycan cross-linking for both *E. coli* and *Enterococcus faecalis* has been explained in terms of reinforcement of the cell wall to better tolerate an external stressing environment. Speculatively, the changed peptidoglycan may be reminiscent of that of the spore wall formed by spore-forming bacteria (Lleò et al. 2006). VBNC bacteria have been shown to maintain adhesion capability to human epithelial cells grown in vitro (Pruzzo et al. 2002, 2003). In addition, resuscitation from the VBNC state (i.e., restoration of cell division) has been obtained upon reversion to suitable environmental conditions in both in vitro and in vivo experiments (Oliver and Bockian 1995; Colwell et al. 1996; Lleò et al. 2001). Both adhesion capacity and resuscitation are considered sine qua non events prior to sustaining a new infection process. A notable example may be the ingestion by humans of water contaminated by VBNC microorganisms. These organisms are capable first of adhering to epithelial cells of the gut and then resuscitating, growing, and multiplying, and finally sustaining a new infection process (Signoretto and Canepari 2008).

In addition to adhesion and resuscitation issues, another important consideration in any general application of molecular methods consists in the establishment of a working definition of the time VBNC bacteria remain viable in natural environments (i.e., from a medical point of view, the time they maintain the ability to sustain a new infection process). To do this, reliable markers of cell viability need to be identified for use as targets for detection, i.e., the search for the so-called "point of no return" beyond which bacteria should be considered dead and, thus, no longer capable of sustaining a human infection (Lleò et al. 2006; Signoretto and Canepari 2008).

A list of microorganisms capable of causing human infection and for which entry into the VBNC state has been demonstrated is now available (Oliver 2005, 2009) and is shown in Table 17.2. It includes both gram-negative and gram-positive species. Practically, all current bacterial fecal indicators are capable of entering the

Table 17.2 Bacteria described to enter the VBNC state[a]

Gram-negative	Gram-positive
Aeromonas salmonicida, A. hydrophila	*Bacillus cereus*
Agrobacterium tumefaciens	*Enterococcus casseliflavus, E. faecalis,*
Alcaligenes eutrophus	*E. faecium, E. gallinarum, E. hirae*
Aquaspirillum sp.	*Lactobacillus plantarum, L. lactis*
Burkholderia cepacia, B. pseudomallei	*Micrococcus flavus, M. luteus, M. varians*
Campylobacter coli, C. jejuni, C. lari	*Rhodococcus rhodochrous*
Citrobacter freundii	*Staphylococcus aureus*
Cytophaga allerginae	*Mycobacterium tubercolosis, M. smegmatis*
Enterobacter aerogenes, E. cloacae	
E. agglomerans	
Escherichia coli (including EHEC)	
Erwinia amylovora	
Francisella tularensis	
Helicobacter pylori	
Klebsiella aerogenes, K. pneumoniae	
K. planticola	
Legionella pneumophila	
Listeria monocytogenes	
Pasteurella piscida	
Pseudomonas aeruginosa, P. fluorescens	
P. putida, P. syringae	
Ralstonia solanacearum	
Rhizobium leguminosarum, R. meliloti	
Sinorhizobium meliloti	
Salmonella enteritidis, S. typhi	
S. typhimurium	
Serratia marcescens	
Shigella dysenteriae, S. flexneri, S. sonnei	
Tenacibaculum sp.	
Vibrio anguillarum, V. campbellii,	
V. cholerae, V. fischeri, V. harveyi, V. mimicus,	
V. natriegens, V. parahaemolyiticus,	
V. proteolityica, V. shiloi, V. vulnificus	
Xantomonas campestris, X. axonopodis pv. *citri*	

[a]Adapted from Oliver (2009)

VBNC state. Survival time in this state has been determined for several microbial species in oligotrophic media, ranging from a few weeks for *E. coli* (Boaretti et al. 2003) to at least 3 months for enterococci (Lleò et al. 2001) to several years for *Vibrio* (Roszak and Colwell 1987). Notably, *Vibrio* persistence in a nonculturable form in marine water has been correlated with the existence of an environmental reservoir during an interepidemic period (Colwell 1996).

To test the possibility that VBNC microorganisms may constitute an important reservoir in natural waters, an 18-month survey of the presence of enterococci by comparing culture vs. nonculture methods was conducted in both lake and sea water (Signoretto et al. 2004). In lake water, enterococci were sporadically detected by the culture-based methods only in late spring and early summer. However, when enterococci detection was conducted using a molecular method (qPCR) on the same samples, 59% of the samples proved highly positive, indicating the presence of enterococci all year round. In the course of the same survey, the role of zooplankton as a reservoir of adherent enterococci was evaluated. Notably, during summer and early fall, when the highest numbers of zooplankton were detected, a very substantial number of nonculturable enterococci adherent to zooplankton were counted and no enterococci were detected free in the water column. During the cold months (December to April), when the zooplankton count fell below 500 organism/m^3, enterococci were detected only in water.

Very similar results were obtained in seawater, but due to the fact that zooplankton is detected all year round, enterococci are mostly found adherent to these organisms (Signoretto et al. 2004). These results should be taken into consideration when evaluating the microbiological quality of water, in that enterococci, if adherent to zooplankton, are not uniformly dispersed in the water column but may be concentrated and moved with tides and currents. In addition, laboratory observations indicate that when bound to zooplankton, culturable enterococci enter into the VBNC state more rapidly. This further supports suggestions that, in natural aquatic reservoirs, these microorganisms exist primarily in a VBNC state, possibly organized in a biofilm structure (Signoretto et al. 2005).

17.4 Implications for Microbial Source Tracking

Measuring FIB levels in water can useful for quantifying contributions from overt fecal sources and providing supplementary information in conjunction with host-specific MST methods. However, understanding the occurrence of naturally occurring strains is critical to evaluating the significance of indicator organisms in the environment. For example, sporadic increases in fecal indicator levels in water during the absence of fecal contamination can be a common occurrence. When this problem becomes chronic, MST studies may be initiated to identify possible sources of bacterial pollution. MST investigations have generally focused on identifying fecal sources such as wastewater treatment facilities, sewer systems, livestock, and wild animals and birds. Mitigation efforts directed to fecal sources may not result in reductions of indicator levels needed to meet the designated use of waterbodies. Alternative engineering fixes may be tried with little or no improvement. If exceedances in fecal indicator levels continue to occur despite remediation of obvious fecal sources, then persistence and regrowth related to nonfecal sources should be investigated.

Aquatic biofilms may be important persistent sources of indicator bacteria to source waters. Environmental niches for biofilms in watersheds include sites with high nutrient loading from sewage effluent and runoff associated with storm drains, irrigation, animal feedlots, or agricultural areas. Biofilm formation in areas with reduced water flow rates, such as standing water, ponds, or estuaries should also be investigated. The relationship between changes in water flow, currents, and wave action and elevated levels of indicator bacteria due to biofilm sloughing should be assessed as well. These investigations may be helpful in determining attainable baseline levels of naturally occurring indicator bacteria after obvious fecal sources have been addressed. Certainly, investigating nonfecal sources at the front end of MST studies should be useful for delineating indicator contributions from fecal vs. natural sources.

17.5 Future Studies

The success of MST studies is dependent on the ability of investigators to discern fecal vs. natural populations of microorganisms. Understanding where and how FIB persist in aquatic environments is essential achieving this objective.

The culture-based FIB methods currently used to evaluate the microbiological quality of environmental samples are widely regarded as unreliable due to researchers' inability to distinguish natural vs. fecally-derived indicator bacteria. Also, these methods are not reliable for measuring cells in the VBNC state (Lleò et al. 2006), further emphasizing the need for new culture-independent methods for evaluating the microbiological quality of waters.

Molecular MST methods such as quantitative polymerase chain reaction (qPCR) and PFGE, improve researchers' ability to more fully characterize environmental persistence and naturalization of indicator organisms in aquatic environments, including clonal adaptation and viable but nonculturable (VBNC) subpopulations.

With the emergence of molecular MST methods, the list of measurable microorganisms capable of causing negative human health impacts has increased considerably. Subsequently, the need for additional research related to understanding FIB persistence and survival mechanisms in aquatic environments continues to grow. Molecular methods provide the opportunity to target recently proposed indicators such as *Bacteriodales* spp. and *Bacteroides thetaiotamicron.* These bacteria should be more specific than traditional indicators because as obligate anaerobes, they should not be capable of regrowing in the environment to the extent of coliforms and enterococci. However, *Bacteroides* spp. are capable of producing and living in intestinal biofilm (Costerton et al. 1987; McFarlane and McFarlane 2006) and have also been shown to grow in an aerobic sludge blanket reactor (Tay et al. 2002). Since deeper layers within biofilm structures can be anaerobic, the capability of new indicator bacteria to survive and replicate in natural biofilms should also be investigated.

To date, there are no EPA guidelines for "tracking" indicator organisms from natural sources, including sediment and plants. Although Costerton et al. (1987) proposed that more than 99% of bacteria in the natural world exist in biofilms, little is known about the significance or fate and transport of biofilm bacteria related to environmental water quality. Studies are needed to develop standardized methods for enumerating and characterizing microorganisms from natural materials including biofilms.

References

Alm E, Burke WJ, Hagan E (2006) Persistence and potential growth of the fecal indicator bacteria, *Escherichia coli*, in shoreline sand at Lake Huron. J Great Lakes Res 32:401–405
Anderson SA, Turner SJ, Lewis GD (1997) Enterococci in the New Zealand environment: implications for water quality monitoring. Water Sci Technol 35:325–331
Anderson KL, Whitlock J, Harwood VJ (2005) Persistence and differential survival of fecal indicator bacteria in subtropical waters and sediments. Appl Environ Microbiol 71:3041–3048
Barer MR, Harwood CR (1999) Bacterial viability and culturability. Adv Microb Physiol 41:93–137
Banning N, Toze S, Mee BJ (2003) Persistence of biofilm-associated *Escherichia coli* and *Pseudomonas aeruginosa* in groundwater and treated effluent in a laboratory model system. Microbiology 149:47–55
Bitton G, Farrah SR, Ruskin RH, Butner J and Chou YJ (1983) Survival of pathogenic and indicator organisms in ground water. Ground Water 21:405–410
Boaretti M, Lleò MM, Bonato B, Signoretto C, Canepari P (2003) Involvement of *rpoS* in the survival of *Escherichia coli* in the viable but non-culturable state. Environ Microbiol 5:986–96
Boehm A, Griffith J, McGee C, Edge TA, Solo-Gabriele HM, Whitman R, Cao Y, Getrich M, Jay JA, Ferguson D, Goodwin KD, Lee CM, Madison M, Weisberg S (2009) Faecal indicator bacteria enumeration in beach sand: a comparison study of extraction methods in medium to coarse sands. J Appl Microbiol 107:1740–1750
Bonde JG (1966) Bacteriological Methods for Estimation of Water Pollution. Health Lab Sci 3:24
Breeuwer P, Abee T (2000) Assessment of viability of microorganism employing fluorescence techniques. Intern J Food Microbiol 55:193–200
Buck JD (1979) The plate count in aquatic microbiology. In (eds) Costerton, JW, Colwell RR. Native aquatic bacteria: enumeration, activity and ecology. Amer. Soc. Test. Mat: 19–28
Camper AK, McFeters GA, Charaklis WG, Jones WL (1991) Growth kinetics of coliform bacteria under conditions relevant to drinking water distribution systems. Appl Environ Microbiol 7:2233
Caplenas NR, Kanarek MS (1984) Thermotolerant non-fecal source *Klebsiella pneumoniae*: validity of the fecal coliform test in recreational waters. Am J Public Health 74:1273–1275
Colwell RR (1996) Global climate and infectious diseases: the cholera paradigm. Science 274:2025–2031
Colwell RR, Braiyton P, Herrington PD, Tall B, Huq A, Levine MM (1996) Viable but nonculturable *Vibrio cholerae* O1 revert to a cultivable state in the human intestine. World J Microbiol Biotech 12:28–31
Costerton JW, Cheng KJ, Geesey GG, Ladd TI, Nickel JC, Dasgupta M, Marrie TJ (1987) Bacterial biofilms in nature and disease. Ann Rev Microbiol 41:435–464
Davenport CV, Sparrow EB, Gordon RC (1976) Fecal indicator persistence under natural conditions in an ice-covered river. Appl Environ Microbiol 32:527–536
Devriese LA, Van De Kerckhove A, Kilpper-bälz AR, Schleifer KH (1987) Characterization and identification of *Enterococcus* species isolated from the intestines of animals. Int J Syst Bacteriol 37:257–259
Dunne WM Jr (2002) Bacterial adhesion: seen any good biofilms lately? Clin Microbiol Rev 15:155–166
Edberg SC, Patterson JE, Smith DB (1994) Differentiation of distribution systems, source water, and clinical coliforms by DNA analysis. J Clin Microbiol 32:139–142

Fass S, Dincher ML, Reasoner DJ, Gatel D, Block JC (1996) Fate of *Escherichia coli* experimentally injected in a drinking water distribution pilot system. Water Res 30:2215–2221

Ferguson CM, Coote BG, Ashbolt NJ, Stevenson IM (1996) Relationships between indicators, pathogens and water quality in an estuarine water system. Water Res 30:2045–2054

Ferguson DM, Moore DF, Getrich MA, Zhowandai MH (2005) Enumeration and speciation of enterococci found in marine and intertidal sediments and coastal water in southern California. J Appl Microbiol 99:598–608

Filip Z, K. Demnorova (2009) Survival in groundwater and FT-IR characterization of some pathogenic and indicator bacteria. In: Jones, AAA, Vardania, TG, C Hakopian (eds) Threats to Global Water Security, Springer, Dordrecht p 117–122

Fujioka RS, Tenno K and Kansako S (1988) Naturally-occurring fecal coliforms and fecal streptococci in Hawaii's freshwater streams. Toxicol Assess 3:613–630

Ganeshwar P, EK James, N Mathan, PM Reddy, B Reinhold-Hurek, JK Ladha (2001) Endophytic colonization of rice by a diazotrophic strain of *Serratia marcescens*. J Bacteriol 183: 2634–2645

Gauthier MJ, LeRudulier D (1990) Survival in seawater of *E. coli* seawater cells grown in sediments containing glycine betine. Appl Environ Microbiol 56:2915–2918

Geldreich, EE (1996) Microbial quality of water supply in distribution systems. Lewis Publishers, Boca Raton

Gerba CP, JS McLeod (1976) Effect of sediments on the survival of *Escherichia coli* in marine waters. Appl Eviron Microbiol 32:114–120

Gilmore, MS, Coburn PS, Nallapareddy SR, Murray BE (2002) Enterococcal virulence. In Gilmore MS, Clewell DB, Courvalin P, Dunny GM, Murray BE, Rice LB (eds) The enterococci: pathogenesis, molecular biology and antibiotic resistance, ASM Press, Washington, pp 301–354

Hartl DL, DE Dykhuizen (1984) The population genetics of *Escherichia coli*. Annu Rev Genet18:31–68

Hartz A, M Cuvelier, K Nowosielski (2008) Survival potential of *Escherichia coli* and enterococci in subtropical beach sand: implications for water quality managers. J Environ Qual 37:898–905

Heim S, MM Lleò, B Bonato, CA Guzman, P Canepari (2002) The viable but nonculturable state and starvation are different stress responses of *Enterococcus faecalis* as determined by proteome analysis. J Bacteriol 184:6739–7645

Huycke, MM (2002) Physiology of enterococci. In Gilmore MS, Clewell DB, Courvalin P, Dunny GM, Murray BE and Rice LB (eds) The enterococci: pathogenesis, molecular biology and antibiotic resistance, ASM Press, Washington, pp 133–176

Ishii S, WB Ksoll, RE Hicks, ML Sadowski (2006) Presence and growth of naturalized *Escherichia coli* in temperate soils from Lake Superior watersheds. Appl Environ Microbiol 72:612–621

Kogure S, U Simidu, N Taga (1979) A tentative direct microscopic method for counting living bacteria. Can J Microbiol 25:415–420

Lacey GH, FL Lukezic (2004) Pathogenic Prokaryotes In: RN Trigiano, MT Windham, AS Windham (eds) Plant pathology: concepts and laboratory exercises, CRC Press, Boca Raton.

Lauková A, Juriš, P (1997) Distribution and characterization of *Enterococcus* species in municipal sewages. Microbios 89:73–80

LeChevallier MW, Lowry CD, Lee RG (1990) Disinfection of biofilms in a model distribution system. J Am Water Works Assoc 82:87–99

Leclerc H, DAA Mossel, SC Edberg, CB Struijk (2001) Advances in the bacteriology of the coliform group: their suitability as markers of microbial water safety. Annu Rev Microbiol 55:201–234

Lee CM, TY Lin, C-C Lin, GA Kohbodi, A Bhatt, R Lee, JA Jay (2006) Persistence of fecal indicator bacteria in Santa Monica Bay beach sediments. Water Res 40:2593–2602

Lleò MM, B Bonato, MC Tafi, C Signoretto, M Boaretti, P Canepari (2001) Resuscitation rate in different enterococcal species in the viable but nonculturable state. J Appl Microbiol 91:1095–1102

Lleò MM, S Pierobon, MC Tafi, C Signoretto, P Canepari (2000) mRNA detection by RT-PCR for monitoring viability over time in an *Enterococcus faecalis* viable but nonculturable population maintained in a laboratory microcosm. Appl Environ Microbiol 66:4564–4567

Lleò MM, C Signoretto, P Canepari (2006) Gram-positive bacteria in the marine environment. In Belkin S, Colwell RR (eds) Ocean and Health: pathogens in the marine environment. Springer, New York, p 307–330

Lleò MM, MC Tafi, P Canepari (1998) Nonculturable *Enterococcus faecalis* cells are metabolically active and capable of resuming active growth. Syst Appl Microbiol 21:333–339

Manero A, S Vilanova, M Cerdá-Cuéllar, AR Blanch (2002) Characterization of sewage waters by biochemical fingerprinting of enterococci. Water Res 36:2831–2835

McClellan S, AD Daniels, AK Salmore (2001) Clonal populations of thermotolerant *Enterobacteriaceae* and their potential interference with fecal *Escherichia coli* counts. Appl Environ Microbiol 67:4934–4938

McCoy EL, C Hagedorn (1979) Quantitatively tracing bacterial transport in saturated soil systems. Water Air Soil Pollut 11:467–479

McFarlane S, McFarlane GT (2006) Composition and metabolic activities of bacterial biofilms colonizing food residues in the human gut. Appl Environ Microbiol 72:6204–6211

Moore DF, JA Guzman, C McGee (2008) Species distribution and antimicrobial resistance of enterococci isolated from surface and ocean water. J Appl Microbiol 105:1017–1025

Moriarty E, F Nourozi, B Robson, D Wood, B Gilpin (2008) Evidence of growth of enterococci in municipal oxidation ponds obtained using antibiotic resistance analysis. Appl Environ Microbiol 74:7204–7210

Mundt JO (1961) Occurrence of enterococci: bud, blossom, and soil studies. Appl Environ Microbiol 9:541–544

Mundt JO (1963) Occurrence of enterococci on plants in a wild environment. Appl Environ Microbiol 15:1303–1308

Mundt JO (1982) The ecology of the streptococci. Microbiol Ecol 8:355–369

Oliver JD, R Bockian (1995) In vivo resuscitation, and virulence towards mice, of viable but nonculturable cells of *Vibrio vulnificus*. Appl Environ Microbiol 61:2620–2623

Oliver JD (2005) The viable but nonculturable state in bacteria. J Microbiol. 43:93–100

Oliver JD (2009) Recent finding on the viable but nonculturable state in pathogenic bacteria. FEMS Microbiol Rev. 1-11 DOI:10.1111/j.1574-6976.2009.00200.x.

Olson BH, R McCleary, J Meeker (1991) Background and models for bacterial biofilm formation and function in water distribution systems. In Hurst (ed) Modeling the environmental fate of microorganisms. ASM, Washington. pp 255–282

Ott E.-M, Müller T, Müller M, Fran CMAP, Ulrich A, Gabel M, Seyfarth W (2001) Population dynamics and antagonistic potential of enterococci colonizing the phyllosphere of grasses. J Appl Microbiol 91:54–66

Pinto B, Pierotti R, Canale G, Reali D (1999) Characterization of 'faecal streptococci' as indicators of faecal pollution and distribution in the environment. Lett Appl Microbiol 29:258–263

Pratt LA, R Kolter (1998) Genetic analysis of *Escherichia coli* biofilm formation: roles of flagella, motility, chemotaxis and type I pili. Mol Microbiol 30:285–293

Pruzzo C, R Tarsi, MM Lleò, C Signoretto, M Zampini, RR Colwell, P Canepari (2002) In vitro adhesion to human cells by viable but nonculturable *Enterococcus faecalis*. Curr Microbiol 45:105–110

Pruzzo C, R Tarsi, MM Lleò, C Signoretto, M Zampini, L Pane, RR Colwell, P Canepari (2003) Persistence of adhesive properties of *Vibrio cholerae* after long term exposure to sea water. Environ Microbiol 5:650–658

Rivera SC, TC Hazen, GA Toranzos (1988) Isolation of fecal coliforms from pristine sites in a tropical rain forest. Appl Environ Microbiol 54:513–517

Roszak DB, RR Colwell (1987) Survival strategies of bacteria in the natural environment. Microbiol Rev 51:365–379

Roth BL, M Poot, ST Yue, PJ Millard (1997) Bacterial viability and antibiotic susceptibility testing with SYTOX green nucleic acid stain. Appl Environ Microbiol 63:2421–243

Rozen Y, Belkin S (2001) Survival of enteric bacteria in seawater. FEMS Microbiol Rev 25:513–529

Shaw MK, AG Marr, JL Ingraham (1971) Determination of minimal temperature growth of *Escherichia coli*. J Bacteriol 105:683–684

Signoretto C, G Burlacchini, MM Lleò, C Pruzzo, M Zampini, L Pane, G Franzini, P Canepari (2004) Adhesion of *Enterococcus faecalis* in the nonculturable state to plankton is the main mechanism responsible for persistence of this bacterium in both lake and seawater. Appl Environ Microbiol 70:6892–689

Signoretto C, G Burlacchini, C Pruzzo, P Canepari (2005) Persistence of *Enterococcus faecalis* in aquatic environments with copepods. Appl Environ Microbiol 71:2756–2761

Signoretto C, P Canepari (2008) Towards more accurate detection of pathogenic Gram-positive bacteria in waters. Curr Opin Biotech 19:248–253

Signoretto C, MM Lleò, P Canepari (2002) Modification of the peptidoglycan of *Escherichia coli* in viable but nonculturable state. Curr Microbiol 44:125–131

Signoretto C, MM Lleò, MC Tafi, P Canepari (2000) Cell wall chemical composition of *Enterococcus faecalis* in the viable but nonculturable state. Appl Environ Microbiol 66:1953–1959

Sinton LW, CH Hall, PA Lynch, RJ Davies-Colley (2002) Sunlight inactivation of fecal indicator bacteria and bacteriophages from waste stabilization pond effluent in fresh and saline waters. Appl Environ Microbiol 68:1122–1131

Smith JJ, JP Howington, GA McFeters (1994) Survival, physiological response, and recovery of enteric bacteria exposed to a polar marine environment. Appl Environ Microbiol 60:2977–2984

Solo-Gabriele, HM, Wolfert, MA, Desmarais, TR, CJ Palmer (2000) Sources of *Escherichia coli* in a coastal subtropical environment. Appl Environ Microbiol 66:230–237

Stevens M, N Ashbolt, D Cunliffe (2003) Recommendations to change the use of coliforms as microbial indicators of drinking water quality NHMRC (National Health and Medical Research Council) Accessed Dec 20 2010 http://www.nhmrc.gov.au

Svec P, I. Sedlacek (1999) Occurrence of *Enterococcus* spp. in waters. Folia Microbiol 44:3–10

Szabo JG, EW Rice, L Bishop (2006) Persistence of *Klebsiella pneumoniae* on simulated biofilm in a model drinking water system. Environ Sci Technol 40:4996–5002

Tay STL, S Ivapow, S Yi, W-Q Zhang, J-H Tay (2002) Presence of anaerobic bacteroides in aerobically grown microbial granules. Microbiol Ecol 44:278–285

Taylor RF, M Ikawa, W Chesbro (1971) Carotenoid in yellow-pigmented enterococci. J Bacteriol 105:676–678

Tendolkar PM, AS Baghdayan, MS Gilmore and N Shankar (2004) Enterococcal surface protein, Esp, enhances biofilm formation by *Enterococcus faecalis*. Infect Immun 72:6032–6039

Topp E, Welsh M, Tien YC, Dang A, Lazarovits G, Zhu H (2003) Strain-dependent variability in growth and survival of *Escherichia coli* in agricultural soil. FEMS Microbiol Ecol 44:303–308

Ulrich A, Müller T (1998) Heterogeneity of plant associated streptococci as characterized by phenotypic features and restriction analysis of PCR amplified 16S rDNA. J Appl Microbiol 84:292–303

USEPA (1986) Ambient water quality criteria for bacteria. EPA 440/5-84-002, Washington, DC

USEPA (2002) Health risks from microbial growth and biofilms in drinking water distribution systems. Distribution system white paper, Office of Ground Water and Drinking Water Standards and Risk Management Division, Washington, DC

USEPA (2005) Microbial Source Tracking Guide Document. Office of Research and Development, Washington, DC EPA-600/R-05/064. 131 pp.

Valiela I, Alver M, LaMontagne M (1991) Fecal coliform loadings and stocks in Buttermilk Bay, Massachusetts, USA, and management implications. Environ Manage 15:659–674

Whitam TS (1989) Clonal dynamics of *Escherichia coli* in its natural habitat. Antonie van Leeuwenhoek 55:23–32

Whitman RL, Nevers MB (2003) Foreshore sand as a source of *Escherichia coli* in nearshore water of a Lake Michigan beach. Appl Environ Microbiol 69:555–5562

Whitman RL, Shively DA, Pawlik H, Nevers MB, Byappanahalli MN (2003) Occurrence of *Escherichia coli* and enterococci in *Cladophora* (Chlorophyta) in nearshore water and beach sand of Lake Michigan. Appl Environ Microbiol 69:4714–4719

Whitman RL, Byers SE, Shively DA, Ferguson DM, Byappanahalli M (2005) Occurrence and growth characteristics of *Escherichia coli* and enterococci within the accumulated fluid of the northern pitcher plant (*Sarracenia purpurea* L.) Can J Microbiol 51:1027–1037

Willey BM, Jones RN, McGeer A, Witte W, French G, Roberts RB, Jenkins SG, Nadler H (1999) Practical approach to the identification of clinically relevant *Enterococcus* species. Diagn Microbiol Infect Dis 34:165–171

Yamahara KM, B Layton, AE Santoro, AB Boehm (2007) Beach sands along the California coast are diffuse sources of fecal bacteria to coastal waters. Environ Sci Technol 41:4515–4521

Xu HS, Roberts NC, Singleton FL, Attwill RW, Grimes DJ, Colwell RR (1982) Microbiol Ecol 8:313–323

Chapter 18
Agricultural and Rural Watersheds

Andreas H. Farnleitner, Georg H. Reischer, Hermann Stadler, Denny Kollanur, Regina Sommer, Wolfgang Zerobin, Günter Blöschl, Karina M. Barrella, Joy A. Truesdale, Elizabeth A. Casarez, and George D. Di Giovanni

Abstract Identifying all relevant human and animal fecal sources is a basic requirement for target-oriented water resource management in agricultural and rural watersheds (ARW). As outlined, microbial source tracking (MST) is most suitably applied in concert with other methods within a broader conceptual framework of fecal pollution analysis. Two case studies – covering surface and karstic groundwater resources within ARW – are presented with the following features in common: public importance, problem formulation based on catchment-based pollution source profiling or modeling, and integrated use of several methods and parameters for fecal source characterization and identification at the water resource level. Possibilities and limitations of currently available MST tools, as well as fundamental requirements for their successful application and combination with other methods, are discussed. The use of multiple tools helps overcome specific limitations of individual methods, increases the robustness of the study, improves confidence in the results, or can help identify issues for further investigation.

Keywords Water quality • Fecal pollution • Microbial source tracking • Human vs. animal pollution ground and surface water • Library dependent and independent methods

Andreas H. Farnleitner (✉)
Institute of Chemical Engineering, Research Area Applied Biochemistry and Gene Technology, Research Group Environmental Microbiology and Molecular Ecology, Vienna University of Technology, Gumpendorferstraße 1a, 166/5-2, A-1060 Vienna, Austria
and
InterUniversitary Cooperation Centre for Water and Health (ICC Water & Health), Vienna University of Technology, Gumpendorferstraße 1a, 166/5-2, A-1060 Vienna, Austria
e-mail: a.farnleitner@aon.at

C. Hagedorn et al. (eds.), *Microbial Source Tracking: Methods, Applications, and Case Studies*, DOI 10.1007/978-1-4419-9386-1_18,

18.1 Introduction

Impairment of microbiological water quality is a critical issue, since it can cause severe outbreaks or contribute to the background rate of endemic diseases (Fewtrell and Bartram 2002). In this respect, fecal pollution of water resources in agricultural and rural watersheds (ARW) is considered of paramount importance as fecal material can contain intestinal pathogens in significant numbers. ARW often accommodate a very complex array of potential fecal emission sources covering humans and animals. In these watersheds, animal fecal sources are not limited to livestock, and wildlife impacts may be significant.

Fecal hazards from human pollution sources are often considered of highest priority due to the potential presence of human pathogens and the feasibility of control measures. However, it has to be emphasized that a broad range of zoonotic pathogens can also be present in animal feces, and the risk from such fecal sources must not be underestimated (Cotruvo et al. 2004; Soller et al. 2010). Hazard and risk analysis of fecal pollution must, therefore, include human and animal sources to support sustainable water resources management in ARWs.

The position of microbial source tracking (MST) for fecal pollution analysis can most suitably be characterized by presenting it in concert with other available methods. In this respect, MST methods can be integrated as a "variable set of team players" into a broader conceptual framework contributing to fecal hazard characterization (Fig. 18.1). The actual numbers and types of applied MST

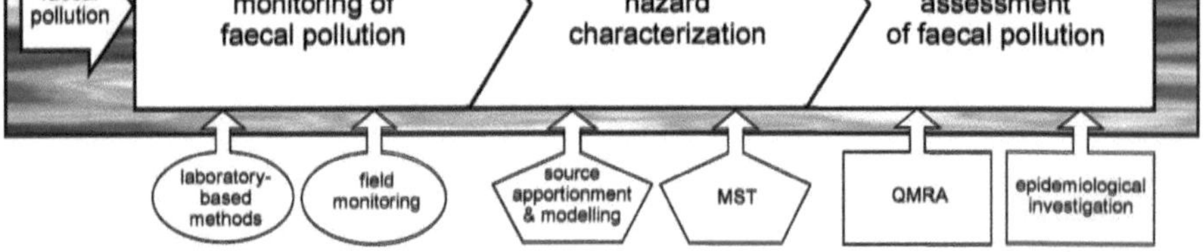

Fig. 18.1 A framework for fecal pollution analysis of water resources (modified after Farnleitner et al. 2008) depicting the position of MST methods. Three interacting levels characterize the backbone of the concept with relevance to the following issues: (1) is there any problem with fecal pollution? (2) If yes, who is responsible for it? (3) What is the actual health risk in relation to the fecal source(s) contributing to the observed pollution? Note that various methods are available at each level. The suggested framework was also referred as "*bottom-up* approach" because starting at the most general level (i.e., level of general fecal pollution monitoring) and becoming more specific as proceeding to the *right* end of the diagram

methods/parameters depend on the availability of reliable methods, the habitat characteristics (e.g., complexity of fecal sources) and the specific issues being addressed (e.g., human vs. animal, differentiation among subgroups of animal sources). Finally, in case MST is used as a basis for further health risk assessment sensitivity levels for the targeted source groups must match the QMRA requirements. Besides MST several other approaches have been applied to characterize fecal pollution levels in ARW. Inspection and documentation of potential fecal pollution sources in the catchment is a useful starting point for any further investigations (Medema et al. 2003). Sanitary surveys of watersheds may use methods ranging from qualitative to quantitative approaches including the application of GIS-based techniques (Grayson et al. 2008). The ultimate goal of fecal hazard characterization is to link potential fecal sources with its actual influence on observed water-quality levels. A great number of modeling approaches on microbial load (i.e., source apportionment) or transport predictions have been applied to tackle this issue (Chaps. 9 and 15).

The integrated use of complementary methods, including those from the MST toolbox, offers the greatest opportunity for successful and accurate watershed fecal pollution analysis. Most importantly, selection of methods and study design must take the characteristics and use of the considered ARW into account. The use of multiple tools helps overcome specific limitations of individual methods, increases the robustness of the study, and improves confidence in the results. It should be highlighted that ARW are often comprised of an unmatched range of water resource types (streams, rivers, dams, lakes, groundwaters, etc.). Furthermore, targeted water qualities may be largely determined by the kind of usage anticipated. For example, a given water body may simply offer space for recreation and sport; may provide water for irrigating crops and public parks, or constitute an essential raw-water resource for drinking water production. The goal of this chapter is to present ARW case studies that exemplify typical microbial sources tracking issues, with a discussion of current technical possibilities and limitations.

18.2 Case Studies

The presented two case studies cover different types of ARW, including a karst groundwater system (Case Study 1), and surface water systems (Case Study 2) with various human and animal fecal inputs. Both studies, although differing somewhat in study design, have several distinct features in common: (1) importance for recreation and/or public water supply, (2) problem formulation based on standard fecal indicator enumeration (i.e., *E. coli* levels), (3) integrated use of several methods and parameters for fecal source characterization, and (4) study results used to guide further target-oriented water management activities. The differentiation and quantification of human vs. animal fecal impacts were the central issue of both investigations.

18.2.1 Case Study 1: MST at Alpine Karstic Spring Water Resources, Austria

18.2.1.1 Background

Alpine Karst Springs and Their Aquifers as Water Resources

As much as 20–25% of the global population depends largely or entirely on groundwater from karst areas (Ford and Williams 2007) with a significant part originating from mountain locations. Approximately 50% of the Austrian water supply is from alpine karstic sources (Fig. 18.2). Mountain karstic spring water from aquifers with prevailing matrix-flow conditions (i.e., water percolating through small fissures within the rock matrix) offer excellent raw-water quality (Farnleitner et al. 2005; Wilhartitz et al. 2009). However, precipitation events can lead to a transient and quick decrease of microbiological water quality due to the input of fecal associated microbes via rapid surface runoff through large and temporarily activated karst conduits (Ford and Williams 2007). Microbiological water safety management is, thus, most suitably based on a multiple barrier system including (1) protection or minimization measures against fecal pollution sources in the catchment, (2) spring abstraction management using spring water only when raw-water quality criteria are below set thresholds, and finally (3) sufficient water treatment to ensure drinking-water quality in agreement with risk-based microbial water-quality targets (WHO 2004; Farnleitner et al. 2008). Within this context, methods for reliable fecal source identification of spring water contaminants are essential analytical elements supporting target-oriented catchment management, providing the basis for evaluating the effectiveness of best environmental practices and guiding microbial hazard and risk assessments (Farnleitner et al. 2008).

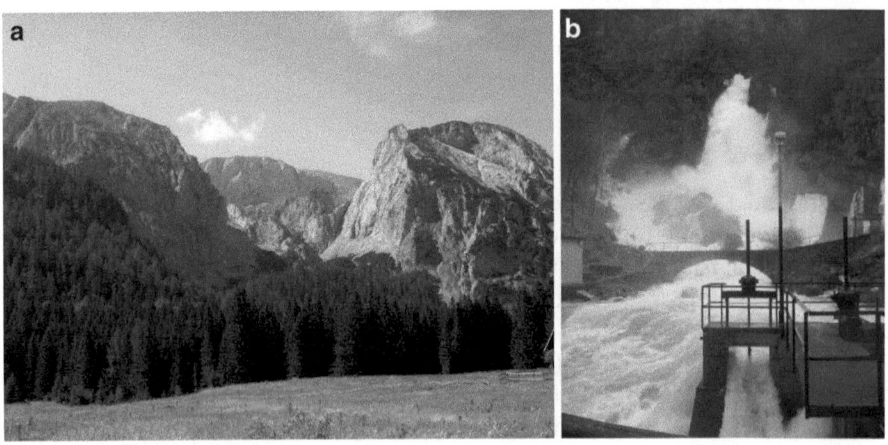

Fig. 18.2 (**a**) A typical alpine mountainous spring catchment in the Austrian calcareous alps, including summer pastures and large areas of wood at the lower to middle altitude regions and stunted tree forests (Krummholz) and alpine grassland at the peak level regions. (**b**) Example of a mountainous karstic spring with water issuing out of the rock at increased flow level

Selected Strategy on MST

Detection of fecal pollution in Austrian alpine spring water resources is based on the cultivation of standard fecal indicator bacteria (SFIB). The general applicability of SFIB – such as *Escherichia coli* – is currently hotly debated (Ishii and Sadowsky 2008). However, a recent evaluation of SFIB for alpine mountainous water resource monitoring revealed *E. coli* to be an excellent indicator (Farnleitner et al. 2010). Furthermore, *E. coli* proved to indicate total fecal pollution levels, being sensitive for human, livestock, and wildlife fecal excreta and sewage (Farnleitner et al. 2010). In addition, high-resolution determination of *E. coli* in the spring water is possible by using cultivation-based field analysis and autosampling (see below) (Stadler et al. 2010; Stadler et al. 2008).

To determine the origin of fecal pollution levels in spring water as determined by *E. coli* enumeration, a recently proposed concept (Reischer et al. 2011) combining information from the catchment and the respective spring water quality is presented (Fig. 18.3). Within this concept, quantitative assessment of fecal pollution sources in the respective catchment – referred to as pollution source profiling (PSP) – is performed as a first step. Data from PSP facilitate the formulation of a working hypothesis on the significance of the potential fecal pollution sources for spring water contamination. It is important to note that the working hypothesis is based on best available estimates of "fecal contamination potentials" by considering the amount of environmentally available fecal material from various sources and the estimated levels of *E. coli* in the catchment. Subsequent study design is directed toward rigorous testing of the working hypothesis by careful selection of available MST tools, as well as parameters for comprehensive hydrological, physicochemical, and microbiological water-quality characterization of the spring. Study design also encompasses a sampling strategy taking the specific hydrogeological situation, hydrological catchment dynamics and characteristics of selected parameters into account. Finally, hypothesis testing is accomplished by statistical analysis of the developed multiparametric dataset.

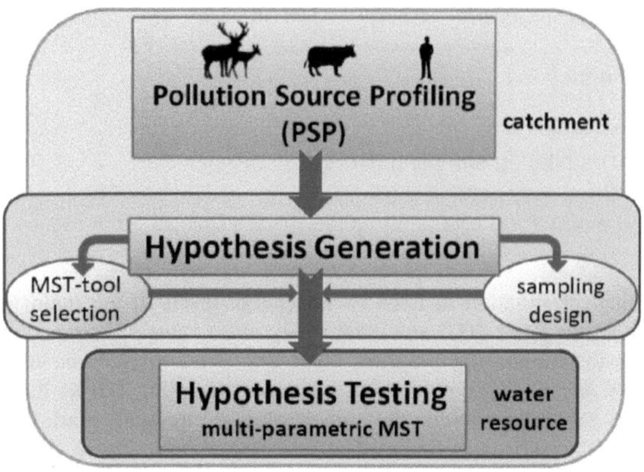

Fig. 18.3 Strategy for conducting a hypothesis-driven fecal hazard characterisation. The strategy makes use of pollution source profiling (PSP) at the catchment scale, hypothesis generation based on PSP and hypothesis testing at the water resource. MST tools and sampling design are selected accordingly to the given habitat characteristic and question being addressed

18.2.1.2 Characterization of the Selected LKAS6 and its Catchment

The limestone karst aquifer spring number 6 (LKAS6) is located in the Northern Calcareous Alps and drains a Triassic limestone aquifer between 820 and 1,828 m above sea level. Based on morphological criteria, the size of the catchment area was estimated at 4 km², where 1.2 km² is included on the plateau of the karst massif with a mean altitude of about 1,750 m. The mean discharge (calculated from daily readings) during the observation period (2007–2008) was 244.4 L s^{-1}, ranging from 117.7 to 1149.5 L s^{-1} (Fig. 18.4 for discharge characteristics). The resulting discharge coefficient of nearly 1:10 is typical for limestone karst aquifers. The course of the runoff is characterized by snow-melting processes, starting usually in April, and storm events during summertime. Although snow-melting processes are the

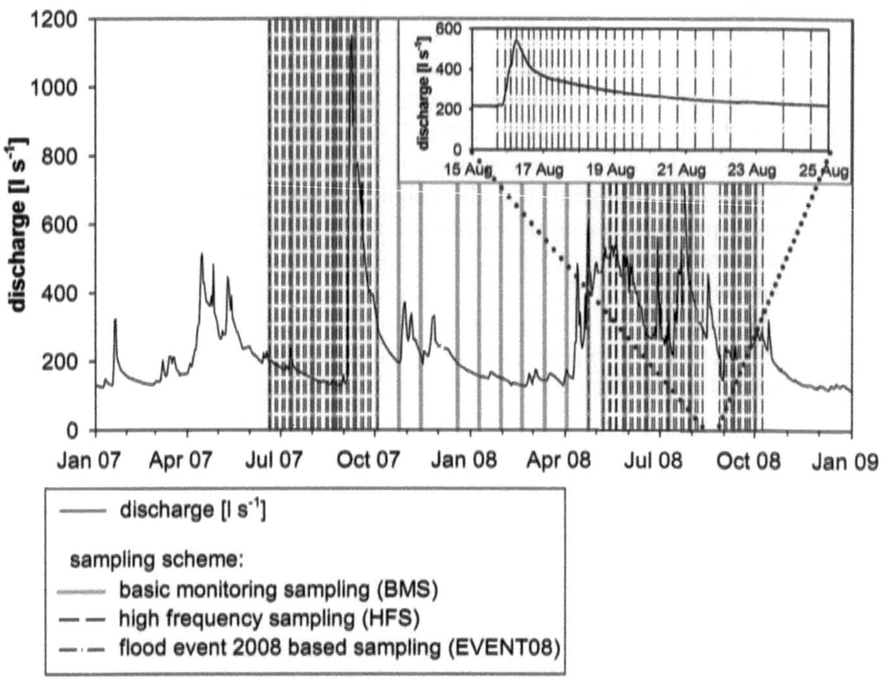

Fig. 18.4 Hydrological situation in LKAS6. Discharge levels in the main graph showing the discharge throughout the years 2007 and 2008 (daily mean values). *Vertical lines* mark the sampling dates during the monitoring program, *light gray lines* mark basic monitoring sampling (BMS) dates, *dark dashed lines* mark high-frequency sampling (HFS) dates during summer months. *Dark dash-dotted lines* within the high-resolution time scale mark the flood event 2008 based sampling (EVENT08) dates (zoomed-in box, *upper right* corner, discharge levels in the zoomed-in box are quarter-hourly measured values). Gaps in the discharge data were due to malfunctions of the instruments

driving forces of mountainous karst springs, single events can reach high discharge peaks, as shown in August 2007 following more than 100 mm of precipitation (Fig. 18.4). The catchment area of the spring LKAS6 shows different land-use patterns covering forests, mountainous grasslands (partially used as seasonal pastures for cattle), and stunted tree forest (Krummholz) areas. Several mountain lodges supporting touristic activities also exist. The data on the LKAS 6 PSP and MST were published elsewhere (Kollanur et al. submitted); however, the herein given work is extended to fit the aim of the book chapter.

18.2.1.3 Methods

Pollution Source Profiling (PSP)

Visual inspection tours were organized to identify fecal pollution sources in the LKAS6 catchment. Human sources (tourists, hikers, alpinists), wildlife populations (red deer, chamois, roe deer), and cattle (kept on pastures during summer months) were identified as potentially important pollution sources. Detailed information for further quantitative characterization of identified fecal sources was obtained from local authorities, official records, and scientific literature (Table 18.1). Produced total fecal matter per source group were estimated by (1) *individuals per source group × ratio of individuals contributing to defecation × amount of defecation per individual*. Since a predominant part of human individuals uses sanitary facilities available in the catchment area (e.g., toilets at mountain lodges), the actual environmentally available fecal material in the LKAS6 catchment was corrected by (2) *amount of total or percentages of produced fecal material per source group × ratio of respective environmental availability*. Finally, the amount of produced *E. coli* per source group was estimated by (3) *amount of environmentally available fecal material per source group × source group specific average E. coli concentrations in excreta or sewage* (Table 18.1 for details).

Hydrological Measurements and Nested Sampling Design

Hydrological characterization including data on discharge, temperature, electrical conductivity, spectral absorption coefficient at 254 nm (SAC254), and turbidity were measured by infield online sensors directly installed at the outlet of the spring as previously described (Farnleitner et al. 2005). Three sampling tiers were performed to cover total system dynamics (Fig. 18.4). Basic monitoring sampling (BMS) was performed on a 3- to 4-weeks interval from June 2007 to November 2008. Additionally, high-frequency sampling (HFS) during summer time (i.e., period of increased hydrological- and fecal pollution dynamics) was performed twice a week. Finally, one flood event (EVENT08) was analysed in detail during late summer 2008. When sampling times overlapped between tiers, the respective

Table 18.1 Pollution source profiling (PSP): estimated environmentally available fecal material and *E. coli* numbers in the catchment of LKAS6

Source		Average produced total fecal material				Average environmentally available fecal material			Average environmentally available *E. coli*	
		Individuals (units/day)[a]	Contributing proportion[b]	Fecal amount per individual (kg wet weight/day)	Total fecal material (kg wet weight/day)	Assumed location of defecation	Environmental availability ratio	Available fecal material (kg wet weight/day)	Daily produced *E. coli* (CFU/d)[c]	Contribution to total *E. coli* numbers
Human	Tourist	45.2[d]	30%[e]	0.15[f]	2.0	100% indoor	1%[g]	0.02	1.99×10^{9}	<0.1%
	Hiker +alpinist	22.6[h]	50%[e]	0.15[f]	1.7	10% outdoor	100%	0.17	1.66×10^{10}	<0.1%
						90% indoor	1%[g]	0.02	1.57×10^{9}	<0.1%
Wildlife	Red deer	90[i]	100%	1.13[j]	102	100% outdoor	100%	102	3.37×10^{12}	6.5%
	Chamois	200[i]	100%	1.13[j]	226	100% outdoor	100%	226	1.49×10^{13}	28.7%
	Roe deer	100[i]	100%	1.13[j]	113	100% outdoor	100%	113	3.74×10^{12}	7.2%
Livestock	Cattle	136[k]	100%	23.6[f]	3,207	100% outdoor	100%	3,207	2.99×10^{13}	57.6%

d day; *CFU* colony-forming units

[a]Average daily human or animal abundance in the investigated catchment of LKAS6 during June to September in 2007 and 2008

[b]Assumed percentage of human or animals defecating in the investigated catchment area

[c]Calculations were based on source group specific *E. coli* concentrations in fecal excreta (tourist = hiker = alpinist, 9.77×10^{7} CFU per g wet weight; red deer = roe deer, 3.31×10^{7} CFU per g wet weight; chamois, 6.61×10^{7} CFU per g wet weight; cattle, 9.33×10^{6} CFU per g wet weight) as given in Farnleitner et al. (2010)

[d]Data recovered from local authorities (personal communication)

[e]Estimated value, taking expected length of stay in the catchment into consideration

[f]According to literature data from Geldreich (1978)

[g]Percentages taking into account (1) the assumed leakage rate during sewage disposal by sewer lines or septic tank storage, or (2) the expected disinfection efficiencies for *E. coli* when sewage is treated with chlorine lime before disposing it into the environment

[h]Assuming that two thirds of total visitors are tourists and one third are hikers and alpinists

[i]Data recovered from local authorities (personal communication)

[j]Owing to lack of data, values were related to average fecal amounts excreted by sheep accordingly to literature data from (Geldreich 1978)

[k]Data on calves and adult numbers (mixed population) recovered from local authorities and converted to adult units by dividing total numbers by 0.7

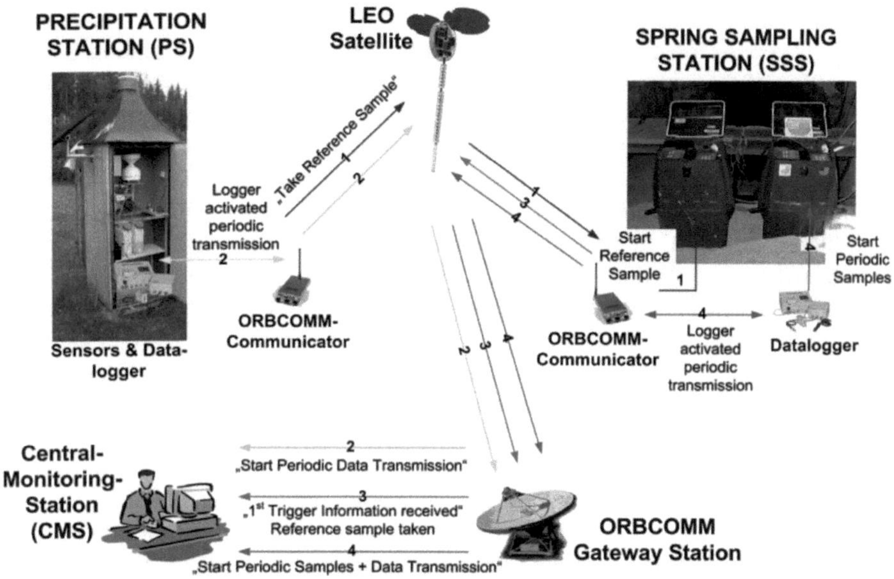

Fig. 18.5 Stream of data and information of an event-triggered LEO satellite hydrology network. The figure shows the data streams (numbers) between precipitation station (PS) – spring sampling station (SSS) and central monitoring station (CMS). Data-stream 1: when precipitation threshold is reached at PS, SSS gets command to take reference sample via LEO-satellite. Data stream 2: simultaneously PS starts periodic transmission of precipitation data to CMS. Data stream 3: in confirmation to data stream 2, SSS sends the first dataset from the spring and is waiting whether increase of discharge exceeds the threshold. Data stream 4: threshold at SSS has exceeded, periodic sampling and data transmission at SSS is starting. All data reaching CMS via LEO-satellites can be observed using a web-interface. A detailed description can be found in Stadler et al. 2008

samples were included in all sample datasets. Manually collected water samples were recovered for all sampling dates in clean and autoclaved sampling bottles (volume 4.2 L, Nalgene Europe Ltd., Hereford, UK), stored in dark cooling boxes at 4°C during transport, and processed within 6 h after collection. Additionally, high-resolution sampling (HRS) during EVENT08 was performed for *E. coli* supported by automated microbiological sampling procedures on 1- to 3-hourly sampling intervals with field analysis using the Colilert system (IDEXX, The Netherlands) and a Low-Earth-Orbit (LEO) satellite-based event monitoring approach (Fig. 18.5 for details).

Determination of Microbiological and MST Parameters

Enumeration of SFIB (*E. coli*, enterococci, presumptive *Clostridium perfringens*) and heterotrophic plate count bacteria at 22°C (HPC22) for BMS, HFS, and EVENT08 was performed as described in the respective ISO standard methods

(ISO 2000a, b, 2002) ISO-HPC. SFIB analysis was based on 1 L volumes of spring water. Aerobic spore-forming bacteria (Aerob) were enumerated by pasteurization of the water sample at 60°C for 15 min, membrane filtration and incubation on yeast extract agar at 22°C for 7 days. For molecular biological MST analysis, a known volume of spring water (usually 4.2 L) was filtered through polycarbonate membrane filters (Isopore™, 45 mm diameter, 0.2 μm pore size, Millipore Corp., Bedford, USA). Immediately after filtration, filters were frozen and stored at −80°C until nucleic-acid extraction. Nucleic-acid extraction was performed as described by (Griffiths et al. 2000), with a DNA precipitation using isopropanol instead of the polyethylene glycol. Recovered DNA was redissolved in 50 μL of sterile double-distilled water and stored at −80°C until further analysis. All extracted sample DNAs were checked for amplifiable bacterial DNA and PCR inhibition by applying a general 16S rRNA gene PCR assay (Winter et al. 2007). MST was based on human (BacH) and ruminant (BacR) specific qPCR assays, targeting host-specific *Bacteroidetes* populations, which were recently developed for the nearby LKAS 2 catchment (Reischer et al. 2006, 2007). In the course of the development of the BacH/R approaches, sensitivity and specificity levels were rigorously determined at animal and human pollution sources at the considered alpine karst catchments (Reischer et al. 2006, 2007). The qPCR assays were performed on an iCycler iQ Real-Time Detection System (Biorad, Hercules, USA) as previously described (Reischer et al. 2008). All sample DNAs were measured in at least duplicate reaction of fourfold DNA dilution steps to rule out the presence of PCR inhibitory substances in the extracts. A total of six tenfold serial dilutions of plasmid standards (10^0–10^5 gene copies) in duplicate, as well as no-template, no amplification, and blank DNA extraction controls, were included for each qPCR run.

For each sample dilution yielding two positive replicates the mean marker copy number per reaction was calculated. The dilution yielding the highest mean value (after taking into account the sample dilution) was used for calculation of the marker concentration of the respective sample. Marker concentrations were expressed as marker equivalents (ME) per liter (Reischer et al. 2008). The filtration volume (usually 4.2 L), the use of 2.5 μL of undiluted DNA extract in qPCR, and the minimal detectable marker concentration per reaction defined the detection threshold as previously established (Reischer et al. 2008). Since the BacH marker occurs in significantly higher concentrations in human fecal material as compared to the BacR marker in ruminant excreta (Reischer et al. 2006, 2007), BacH concentrations were corrected by a factor of 0.1, allowing a best-estimate direct comparisons on the extent of fecal influence from both sources (i.e., BacH corr.; Fig. 18.8).

Data Processing and Statistical Analysis

Logarithmic transformations ($\log^{+1}$) were performed by calculating the $\log_{10}$ after addition of 1 to a given value. Statistical analyses were performed with the

Statistical Package for the Social Sciences, version 14.0 (SPSS Inc., Chicago, IL). For multiple testing, Bonferroni correction was applied. Spearman rank correlation analysis included all samples regardless whether the quantitative MST markers were detectable or not (negative qPCR results were set to the detection threshold of 5 ME L^{-1}). In order to assess the effect of including samples below the detection threshold in the calculations, correlation analysis were repeated only with samples containing >5 ME L^{-1} of both BacR and BacH. No significant change in the results for this reduced dataset was observed.

18.2.1.4 Results

PSP and Formulation of the Hypothesis

Ruminant animals were considered the main fecal emitters in the investigated LKAS6 catchment, producing an estimated amount of 3.6 tons of fecal matter per day (Table 18.1). As much as >99% of the intestinal *E. coli* populations daily deposited in the LKAS6 environment could be allocated to wildlife (42.4%) or livestock ruminants (57.6%) as calculated for the considered sources in the PSP. In contrast to ruminant animal sources, the potential of human fecal pollution sources was considered negligible (Table 18.1). Hence, it was hypothesized that fecal contaminants in the spring water originate from ruminant excreta and would result in a tight statistical relationship between levels of *E. coli* and the ruminant-specific fecal marker.

Challenging the Hypothesis

As expected, all microbiological and most of the physicochemical parameter values showed strong variations in the LKAS6 as illustrated for EVENT08 (Table 18.2). Irrespective of the chosen sampling tier investigated, ruminant BacR marker levels were found to be several orders of magnitude higher than the human BacH marker. BacH levels were frequently close to or below the detection threshold (middle panel, Fig. 18.8).

Comparative correlation analysis between and within sampling tiers further supported the dominance of ruminant fecal pollution of LKAS6 (Table 18.3). Among all investigated parameters, BacR revealed the highest association with *E. coli* and enterococci (ENT), the two SFIB successfully used to indicate total fecal pollution in mountainous water resources (Farnleitner et al. 2010). It is interesting to note that correlation coefficients between BacR and SFIB increased with the rising hydrological dynamics covered in the investigated sampling tiers. For example, correlation coefficients between BacR and ENT increased from 0.59 to 0.80 and 0.96 for the BMS, the HFS and the EVENT08, respectively (Table 18.3). This effect can be explained by differential persistence between applied markers and indicators becoming less influential with decreasing average residence times

Table 18.2 Medians and ranges of parameters determined during flood event 2008 in LKAS6 ($n=29$)

	BacR[a] ME l^{-1}	BacH[a] ME l^{-1}	EC[a] CFU l^{-1}	ENT[a] CFU l^{-1}	pCP[a, b] CFU l^{-1}
Median	4.34	≤0.83[c]	2.56	1.88	0.00
Range	2.31–6.59	≤0.83[c]–3.66	1.08–4.00	0.30–3.81	0.00–1.79

	ASF[a] CFU l^{-1}	HPC22[a] CFU l^{-1}	Dis l s^{-1}	Turb NTU	SAC_{254} m^{-1}	Cond μS cm^{-1}
Median	2.98	2.48	295	0.14	1.28	272
Range	1.96–3.99	0.85–4.27	219–537	0.09–2.32	0.52–8.09	243–273

BacR ruminant-specific marker; *BacH* human-specific marker; *EC Escherichia coli*; *ENT* enterococci; *pCP* presumptive *Clostridium perfringens* spores; *HPC22* heterotrophic plate counts at 22°C; *ASF* aerobic spore-formers; *Dis* discharge; *Turb* turbidity; SAC_{254} spectral absorption coefficient at 254 nm; *Cond* conductivity; *ME* marker equivalents; *CFU* colony-forming units; *NTU* nephelometric turbidity unit; *n* number of samples

[a]Data "$\log_{10}(x+1)$" transformed, with x for value of given variable

[b]Note: only 11 of 29 samples were positive (37.9%)

[c]$\log_{10}(x+1)$ of the "*threshold of detection*"

Table 18.3 LKAS6, correlation analysis of data collected during basic monitoring sampling (BMS; $n=23$), high-frequency sampling (HFS; $n=75$–76) and flood event 2008 based sampling (EVENT08; $n=29$)

Parameter	Spearman correlation coefficient ρ									
	BacR	BacH	Dis	SAC_{254}	Turb	Cond	HPC22	ASF	pCP	ENT
EC	0.59 0.70[a] / 0.95[a]	0.04. 0.06 / 0.22	0.26 0.02 / 0.87[a]	0.50 0.05 / 0.94[a]	−0.13. 0.25 / 0.77[a]	−0.31 0.22 / −0.86[a]	0.34 0.48[a] / 0.91[a]	0.22 0.27 / 0.81[a]	0.08 0.16 / 0.50	87[a] 0.79[a] / 0.95[a]
ENT	0.59 0.80[a] / 0.96[a]	0.05 0.12 / 0.27	0.31 0.14 / 0.94[a]	0.42 0.12 / 0.98[a]	−0.05 −0.07 / 0.85[a]	−0.28 0.08 / −0.94[a]	0.42 0.59[a] / 0.97[a]	0.35 0.38 / 0.84[a]	0.03 0.31 / 0.55	
pCP	−0.29 0.24 / 0.46	−0.18 0.18 / 0.72[a]	−0.32 0.16 / 0.62[a]	0.19 0.21 / 0.60[a]	−0.03 0.12 / 0.64[a]	0.11 −0.14 / −0.64[a]	0.02 0.27 / 0.61[a]	−0.06 0.27 / 0.59[a]		
ASF	0.44 0.55[a] / 0.77[a]	0.24 0.16 / 0.31	0.71[a] 0.65[a] / 0.87[a]	0.42 0.29 / 0.86[a]	0.00 0.10 / 0.79[a]	−0.61 −0.51[a] / −0.87[a]	0.48 0.61[a] / 0.91[a]			
HPC22	0.51 0.65[a] / 0.92[a]	−0.02 0.17 / 0.31	0.45 0.32 / 0.97[a]	0.48 0.25 / 0.99[a]	0.19 −0.08 / 0.87[a]	−0.35 −0.17 / −0.97[a]				
Cond	−0.52 −0.09 / −0.88[a]	−0.34 −0.03 / −0.37	−0.79[a] −0.52[a] / −0.99[a]	−0.76[a] −0.33 / 0.98[a]	−0.41 −0.43[a] / −0.92[a]					
Turb	0.07 −0.11 / 0.78[a]	0.16 0.06 / 0.47	0.23 0.16 / 0.94[a]	0.32 0.51[a] / 0.90[a]						
SAC_{254}	0.54 0.13 / 0.94[a]	0.22 0.06 / 0.32	0.53 0.34 / 0.98[a]							
Dis	0.69[a] 0.31 / 0.88[a]	0.33 0.03 / 0.36								
BacH	0.51 0.21 / 0.24									

basic monitoring sampling(BMS)
―――――――――――――――――――
high frequency sampling (HFS) / event 2008 sampling(EVENT08)

Values above the line are coefficients for the basic monitoring sampling (BMS) dataset; values below the line are coefficients for the high-frequency sampling (HFS; left) and flood event 2008 based sampling (EVENT08; right) datasets, respectively

[a]Correlation significant on the <0.05 level (Bonferroni corrected for multiple testing); for abbreviations see Table 17.1

of the water masses (i.e., less time available for differential die-off or degradation effects). In contrast to BacR, BacH did not show any relevant correlations with *E. coli* or ENT.

The analysis of the EVENT08 in "high-resolution mode" by the LEO-satellite-based system (Stadler et al. 2008) allowed a detailed view of the transfer kinetics of fecal pollution during summer flood events at the LKAS6. The sudden increase in discharge, due to strong precipitation events in the catchment, was followed by an excessive increase in the SAC254 and the *E. coli* concentrations, after a certain lag-time needed for the mass transfer from the catchment to the outlet of the spring (Fig. 18.6). Within a few hours, *E. coli* concentrations increased by more than three orders of magnitude followed by an exponential decline during the following week. The *E. coli* graph was most remarkably reflected by the determined BacR values. Furthermore, regression analysis was able to predict 90% of the *E. coli* variations using BacR values (Fig. 18.7), providing strong support that *E. coli* was predominantly derived from ruminant fecal pollution sources. BacH did not show a clear trend during the investigated event with most values being below the level of detection (Fig. 18.6).

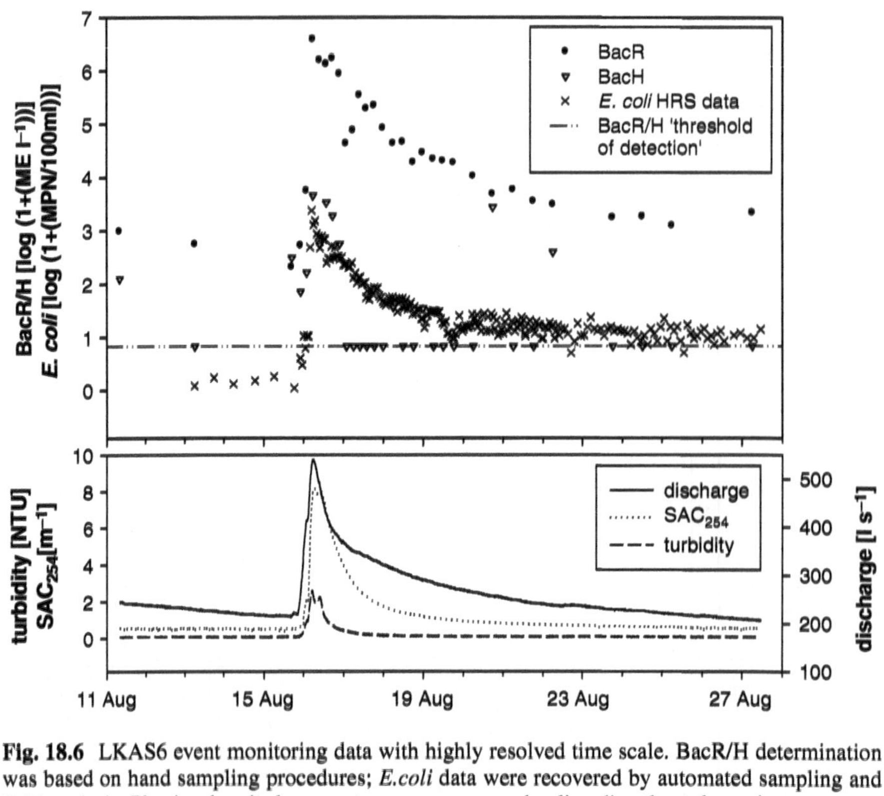

Fig. 18.6 LKAS6 event monitoring data with highly resolved time scale. BacR/H determination was based on hand sampling procedures; *E.coli* data were recovered by automated sampling and field analysis. Physicochemical parameters were measured online directly at the spring outcrops

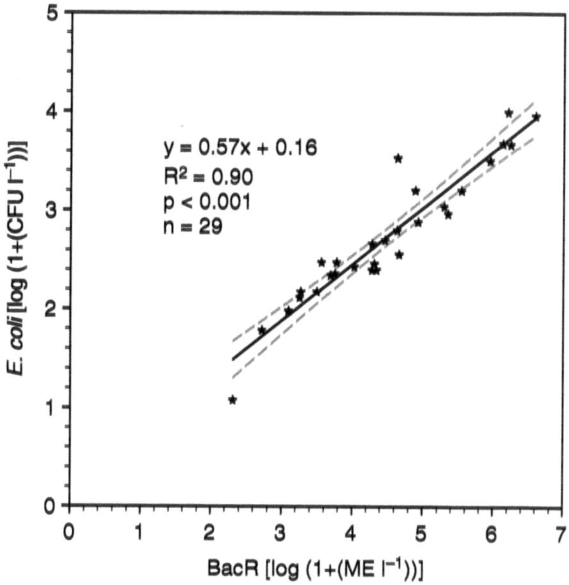

Fig. 18.7 Predictability of *E.coli* CFU concentration by BacR values (linear regression analysis) during the flood event 2008 at LKAS6. *BacR* ruminant-specific marker; *ME* marker equivalents; *CFU* colony-forming units; *solid line* regression curve; *dashed gray lines* 95% confidence intervals; R^2 coefficient of determination for linear regression; data is given after $\log_{10}$-transformation after adding 1 to a given value

18.2.1.5 Discussion: Case Study 1

According to the PSP, we hypothesized that ruminant animals (wildlife and cattle) were the main fecal polluters of the spring within the range of considered sources, being responsible for the crucial part of the daily *E. coli* loading in the catchment. Combining microbiological, MST, and hydrological data from both the catchment and the spring water provided strong evidence in support of our hypothesis. It should be highlighted that the BacR marker proved to be equally sensitive for the detection of all considered ruminant sources in the LKAS6/LKAS8 area including chamois, red deer, roe deer, and cattle. For these populations the 5–95% percentile concentrations of the BacR marker ($n=61$) ranged from $\log_{10}$ 7.4 to 9.4 ME per gram, respectively, of wet feces (Reischer et al. 2011). Fecal pollution from human sources obviously did not play a significant role in spring water contamination in the studied area, despite the presence of significant tourist activities. This is likely due to the state-of-the-art sewage disposal facilities present at restaurants and lodges in the catchment area. Strikingly, in the absence of such measures to reduce the environmental availability of sewage (Table 18.1), PSP would have estimated an increase in the emission potential of human sources by at least two orders of magnitude.

The findings of this study are very similar to investigations performed at other nearby mountainous karstic spring water locations (Reischer et al. 2008; Reischer et al. 2011) where comparable conditions in regard to hydrogeology and land-use pattern exist (Fig. 18.8). Future water safety and health risk management activities will, thus, concentrate on ruminant sources (compare also to the framework on fecal pollution analysis, Fig. 18.1). Research at representative catchments of the

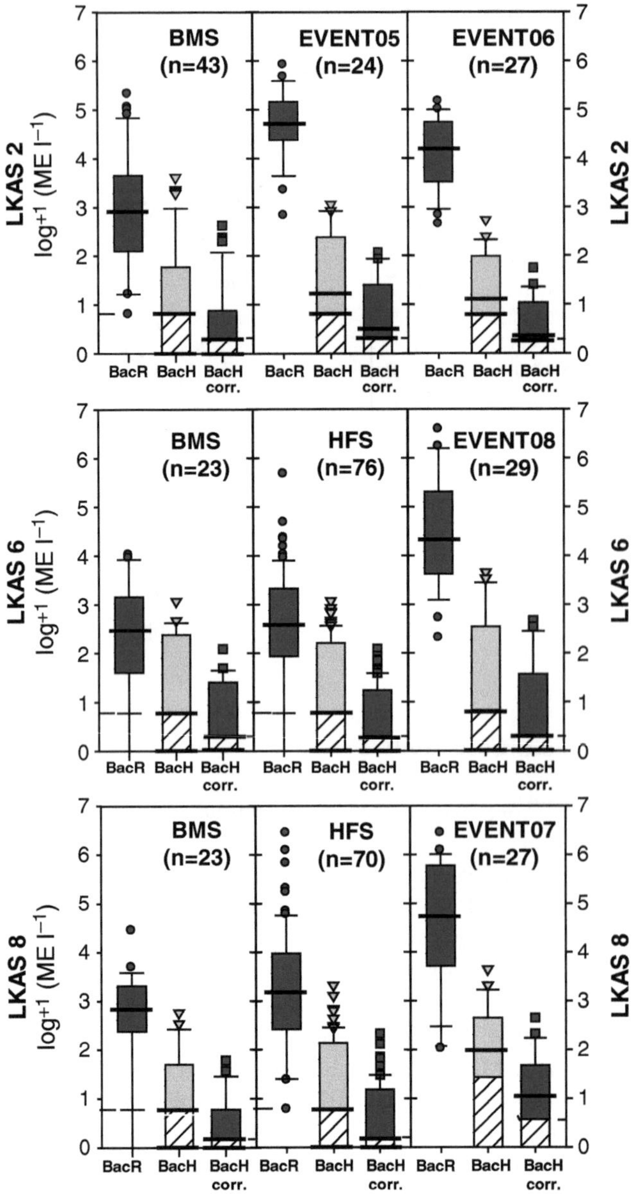

Fig. 18.8 Abundances of BacR and BacH marker in LKAS2, LKAS6, and LKAS8 during BMS, HFS and flood event based sampling (EVENT05, EVENT06, EVENT07, EVENT08). *Box plots* show the distribution of BacR and BacH marker values (*bold whiskers*, 10th and 90th percentile; boxes, 25th and 75th percentile; *bold lines* within the box, median). BacR, ruminant-specific marker (*gray boxes* and *dot symbols*); BacH, human-specific marker (*white boxes* and *triangle symbols*); BacH corr., human-specific marker after correction for different abundance ratio between ruminant and human feces (*gray boxes* and *square symbols*), *ME* marker equivalent; *n* number of samples; data is given after $\log^{+1}$ transformation; *gray-shaded boxes* represent "threshold of detection"; *dash-dot-dot lines* (undiluted) and *short dashed lines* (16-fold diluted, only in flood event in LKAS8) represent "threshold of detection" level for the respective dilutions

Northern Calcareous Alps is already underway to evaluate the actual potential of ruminants as pathogen sources, to establish knowledge on livestock vs. wildlife populations as overlapping zoonotic reservoirs, and to select appropriate indicator pathogens for microbial risk assessment. Recently it was estimated that infection risk associated with human and cattle faecal pollution can reach similar levels (Soller et al. 2010). It is not possible so far to differentiate livestock vs. wildlife ruminant fecal contaminants in the considered spring habitats, although specific information concerning these source groups would be of great practical relevance. Such research issues might be addressed in the future, assuming that reliable deer and cattle-specific MST markers are developed. Besides ruminants, ground-dwelling animals, such as mice or marmot populations, are also expected to represent a significant stock of biomass at the considered mountainous areas (Farnleitner, Ryzinska-Paier 2010). Knowledge on ground-dwelling mammals and their quantitative contribution to fecal pollution is currently extremely limited and warrants further studies. However, the high predictability for total levels of fecal pollution by the BacR marker during the EVENT08 (Fig. 18.7) does not suggest a substantial bias on the outcome of the performed investigation by nonruminant fecal sources. In addition, lower statistical associations of aerobic spore formers (ASF), an indicator of soil influence (Fujioka 2001), with *E. coli* or ENT as compared to BacR suggest fecal input by surface runoff but not soil erosion as the main input mechanism into spring aquifers (Table 18.3). However, increased correlations between values of ASF and *E. coli* or ENT during the EVENT08 were not unexpected, as increased soil erosion processes also happen during such strong thunderstorm events. In conclusion, the integrated use of differing methods for fecal pollution detection and characterization at Case Study 1 revealed a consensus picture on relevant contamination sources, demonstrating the reliability of the results. For a detailed discussion on the reliability of the low BacH results (based on statistical considerations) the reader is referred to a very recent publication from the LKAS 8 catchment (Reischer et al. 2011).

18.2.2 Case Study 2: Lake Granbury and Buck Creek, Texas, USA

Concern for human health is a driving force for regulations requiring that water quality in lakes and streams be suitable for fishing, swimming, and wading, as well as for a healthy aquatic ecosystem. While zoonotic diseases are possible, identifying and eliminating human sources of fecal pollution (sewage, leaking septic systems, and waste from recreationists) must be the highest priority to reduce risks for human infection. MST methods that can detect human fecal pollution and distinguish it from nonhuman animal sources are valuable tools in the source tracking toolbox. However, detection and quantification of human pollution signatures can be complex as highlighted in this case study of two different ARW in Texas, USA. Water-quality monitoring in both watersheds have indicated *E. coli* bacterial levels for routine ambient water samples occasionally over the regulatory limit of 394 colony forming units (CFU)/100 mL. Sanitary surveys found that Lake Granbury is most likely impacted by domestic sewage, while the most probable sources of bacterial pollution in Buck Creek is wildlife and livestock. However, MST results revealed an intriguing and much more complex situation.

18.2.2.1 Background

Lake Granbury, a 48-km long lake created in 1969 by the damming of the Brazos River, is located approximately 48 km southwest of Fort Worth, Texas and is nestled by the town of Granbury, population 5,000. Lake Granbury is a vital water resource for the region, providing drinking water for approximately 150,000 residents. Potential contamination sources have been identified through a sanitary survey of the watershed conducted by Brazos River Authority (BRA). The BRA has determined that monitoring sites consistently high for *E. coli* levels are located in man-made coves (shallow, dead-end inlets off the main lake) with high-density housing (Fig. 18.9) that rely upon septic systems (BrazosRiverAuthority 2006). Most of these septic systems were installed in the 1960s and 1970s prior to current septic tank regulations. The housing developments are densely populated and right on the lake front (some homes as close as 9 m). The soils are not well suited for septic systems, and a historical account from a former septic installer stated that it was not uncommon to run lateral lines in the lake bed and to use 189 L drums as tanks. Both are practices that are now considered illegal. In addition, BRA Rangers have issued citations to people who have bypassed their

Fig. 18.9 A view of a man-made cove on Lake Granbury showing the dense housing development

septic systems, running their household waste directly into the lake. Agricultural runoff may also affect water quality, since the man-made coves may be impacted by runoff from nearby agricultural fields (crop and rangeland). Therefore, Lake Granbury pollution sources are likely domestic sewage, agricultural runoff, wildlife, and possibly pet waste. Indeed, fecal pollution modeling of Lake Granbury performed by consultants indicated that 99% of the *E. coli* in the Port Ridglea East cove water was derived from leaking septic systems (BrazosRiverAuthority 2008).

Buck Creek, part of the Red River Basin, is in the panhandle region of Texas within a predominantly rural and agricultural landscape. Small streams within this region are typically characterized by widely varying flows and high levels of dissolved salts, generally originating from saltwater seeps and springs. Land use in this watershed is predominantly row crops and grasslands (TexasA&MAgriLifeResearch 2006). However, data obtained from water quality monitoring indicate that bacterial levels are sometimes elevated in the creek, particularly after a rainfall. Although these data points provide an indicator of a potential water-quality problem, the data do not provide conclusive evidence of persistent impairment; rather, it suggests a temporal recurring phenomenon. During periods of rainfall, bacteria (*Escherichia coli* specifically) originating from birds, mammals, livestock, inadequately treated sewage, and/or failing septic systems may be washed into streams, potentially impacting recreational use of this waterbody. In contrast to Lake Granbury, the remote stretches of Buck Creek were assumed to be mostly impacted by wildlife and to a lesser extent by livestock, with virtually no human pollution impacts (Fig. 18.10).

Fig. 18.10 A view of the Buck Creek watershed (*left*) near sampling Site 10C (*right*). This region of the watershed area is rural and lacks human development

Because of concerns about the elevated bacterial levels in Lake Granbury and in Buck Creek, separate bacterial source tracking studies were undertaken to identify the likely human and animal sources of fecal pollution impacting these watersheds, although the primary sources of fecal pollution seemed to be obvious. The aims of the studies included aiding the development of watershed protection plans and providing scientific evidence for informed water treatment infrastructure decision making. The overarching goal of the projects was to help protect surface water resources and reduce the risks to public health.

18.2.2.2 Methods

Several MST tools were employed to identify the likely sources of fecal pollution impacting Lake Granbury and Buck Creek, including: *E. coli* enterobacterial repetitive intergenic consensus sequence PCR (ERIC-PCR) and RiboPrinting using a self-validated state-wide library of genetic fingerprints, library-independent *Bacteriodales* PCR, and *Methanobrevibacter smithii* and human polyomavirus PCR for the detection of human fecal pollution.

Water Sample Collection and Processing

Lake Granbury water samples were collected monthly for 6 months, while Buck Creek water samples were collected between 2007 and 2009, mostly representing routine, low-flow conditions. Water grab samples (100 mL) were collected from selected sites for *E. coli* detection using USEPA Method 1603 with modified mTEC medium (USEPA 2006), and for *Bacteroidales* analysis. Water samples for human polyomavirus and *Methanobrevibacter smithii* detection were also collected from Lake Granbury. *E. coli* colonies from modified mTEC plates were isolated, purified, confirmed (using nutrient agar with 4-methylumbelliferyl-β-D-glucuronide (NA-MUG) medium), and archived. Water samples for *Bacteroidales* analysis were filtered, placed in lysis buffer, and frozen. All samples were sent to the Texas AgriLife Research Center at El Paso for source tracking analysis.

Known Source Fecal Samples

Known source fecal samples collected from wildlife, domestic septage/sewage, pets, and livestock from the study areas were used to evaluate the distribution of *Bacteroidales* host-specific markers in the watersheds. In addition, *E. coli* isolated from each fecal sample were screened by DNA fingerprinting to remove identical isolates (clones) and for the selection of isolates to be included in the Texas *E. coli* MST library.

For Lake Granbury, a total of 94 different human and animal fecal samples were analyzed for the presence of *Bacteroidales* markers: 28 samples from livestock,

36 samples from wildlife, 16 samples from domestic human sewage, and 14 samples from pets. After removal of clones, a total of 80 *E. coli* isolates were obtained from 59 of the Lake Granbury human and animal fecal samples: 21 isolates from 17 human sewage samples, 8 isolates from 7 livestock samples, 48 isolates from 33 wildlife samples, and 3 isolates from 2 pet samples.

For Buck Creek, a total of 93 different animal fecal samples were analyzed for the presence of *Bacteroidales* markers: 42 samples from livestock and 51 samples from wildlife. After removal of clones, a total of 31 *E. coli* isolates were obtained from 28 of the Buck Creek animal fecal samples: 10 isolates from 8 livestock samples and 21 isolates from 20 wildlife samples.

ERIC-PCR and RiboPrinting of *E. coli*

E. coli isolates from water and source samples were DNA-fingerprinted using a repetitive sequence polymerase chain reaction (rep-PCR) method known as ERIC-PCR (Versalovic et al. 1994). For source samples, ERIC-PCR was used to identify unique *E. coli* isolates from each sample to maximize the diversity of isolates added to the local library and to eliminate further analysis of identical isolates (clones). At least one *E. coli* isolate from each known fecal source sample was included in the local library, even if it was identical to a previously isolated *E. coli*. Following ERIC-PCR analysis, *E. coli* water isolates and selected source isolates were RiboPrinted using the automated DuPont Qualicon RiboPrinter and the restriction enzyme *Hin*dIII. The RiboPrinter system uses standardized reagents and a robotic workstation, providing a high level of reproducibility.

Jackknife analysis is commonly used for the evaluation of MST library accuracy. Jackknife analysis involves pulling each library isolate one-at-a-time from the library and treating each as an unknown to determine the percentage of isolates correctly identified to the true host source. This is referred to as the rate of correct classification (RCC) (USEPA 2005). One of our previous studies found that a two-method composite dataset that combines the ERIC-PCR and RiboPrinting fingerprints for each isolate (ERIC-RP) gave high rates of correct classification and was able to identify water isolates similar to a four-method composite dataset (Casarez et al. 2007). We also found that the ERIC-RP composite method provided better results than any of the four MST methods used alone.

Analysis of composite ERIC-RP DNA fingerprints was performed using Applied Maths BioNumerics software. Genetic fingerprints of *E. coli* from ambient water samples were compared to fingerprints of known source *E. coli* isolates in the Texas *E. coli* MST library. The ERIC-RP composite patterns of the water isolates were compared to the library using a best match approach and an 80% similarity cutoff as previously described (Casarez et al. 2007). If a water isolate was not at least 80% similar to a library isolate, it was considered unidentified. Although one-to-one matching is performed, identification is to the host source class and not to the individual animal represented by the matching library isolate. Host sources were divided into three groups: (1) human, (2) domestic animals (including livestock and pets), and (3) wildlife (including deer and feral hogs).

Table 18.4 Texas *E. coli* MST library (ver. 1–10) composition and rates of correct classification

Source class (number of isolates/fecal samples)	Library composition and expected random rate of correct classification (%)	Calculated rate of correct classification (%)	Left unidentified (unique patterns) (%)
Human (375/323)	32	90	21
Domestic animals (387/349)	33	80	23
Wildlife (428/391)	36	86	20

Since ERIC-PCR and RiboPrinting are library-dependent source tracking methods, a diverse and robust library of *E. coli* isolate fingerprint patterns from known fecal sources is required for accurate identification. To date, over 5,000 isolates from over 2,000 individual fecal source samples collected during five MST studies throughout the state of Texas have been screened to build a statewide library. For each study, known source samples were collected from the local watershed based on a sanitary survey of likely pollution sources. Three to five *E. coli* isolates from each sample were screened by ERIC-PCR to exclude identical (clonal) isolates from the same sample and to select isolates to maximize the diversity of the library. Local watershed libraries were then analyzed by jackknife analysis to select only those isolates correctly identified to their source class using a seven-way split of source classes (human, pet, cattle, other livestock nonavian, other livestock avian, wildlife nonavian, and wildlife avian) or that were unique and did not match another isolate (unidentified). Source isolates designated as unidentified in a jackknife analysis have unique fingerprints and may still be important for the identification of water isolates. The resulting self-validated isolates from all the studies were combined to form the Texas *E. coli* MST library. The current Texas *E. coli* MST library (ver. 1–10) consists of 1,190 *E. coli* isolates from 1,063 different human and animal fecal source samples. Composition and rates of correct classification for the Texas *E. coli* MST library (ver. 1–10) used in this study are included in Table 18.4. Jackknife analysis revealed an 85% average RCC using a three-way split of source classes.

Since building the state library is a dynamic process, at the time the original analyses for Lake Granbury and Buck Creek were done, the library contained different isolates. However, reanalysis of the Lake Granbury and Buck Creek water isolates using the current Texas *E. coli* MST library (ver. 1–10) provided similar results, with fewer unidentified isolates, and these current results are presented here.

Bacteroidales PCR and Quantitative PCR (qPCR)

Library-independent source tracking methods have been developed as alternatives to the library-dependent methods and may prove to be a more rapid and cost-effective approach for assessment of fecal pollution in source water. The *Bacteroidales* PCR method is a culture-independent molecular method that targets genetic markers of *Bacteroides* and *Prevotella* spp. fecal bacteria that are associated with humans, ruminants (including cattle and deer), and pigs (Bernhard and Field 2000b; Dick et al. 2005), including feral hogs (Lamendella et al. 2009). There is also a general *Bacteroidales* marker (GenBac) that can be used as a general indicator of fecal pollution (Bernhard and Field 2000a). The *Bacteroidales* PCR method has high specificity and moderate sensitivity (Field et al. 2003). For this method, 100 mL water grab samples were concentrated by filtration, DNA extracted from the concentrate and purified, and aliquots of the purified DNA analyzed by PCR (Besner et al. 2010). For qPCR, PCR was performed as described previously (Besner et al. 2010), with the exception that SYBR Green PCR buffer (Applied Biosystems; Foster City, California) was used with a RotorGene 6000HRM real-time PCR thermal cycler (Qiagen, Valencia, California) to generate GenBac marker standard curves. Results were expressed as either the qualitative presence/absence of the host-specific genetic markers or semiquantitative marker abundance as determined by qPCR.

In theory, the GenBac marker detects the majority of the *Bacteroidales* in the samples, including those detected with the host-specific markers. Quantitative PCR GenBac marker quantitation standard curves were developed using 10^0, 10^{-1}, 10^{-2}, and 10^{-3} dilutions of each water sample DNA. Since the actual copy number of GenBac target sequences in each sample was unknown, arbitrary values of 10,000; 1,000; 100; and 10 were assigned to the dilutions, respectively. All GenBac standard curves had R^2 values of ≥ 0.9. The hog, human, and ruminant host-specific markers were quantified using the GenBac standard curve for each water sample. This attempted to make the marker quantitation data for different water samples comparable by accounting for sample-to-sample variation in *Bacteroidales* DNA concentration and any effects of PCR inhibitors on quantitation. This approach makes it possible to compare the relative abundance of each marker between stations or at the same station over time. However, it is not appropriate to directly compare the abundance of one marker to another (e.g., hog vs. human), since that would require DNA extraction controls and marker-specific quantitation standards that were not employed in the current studies.

Methanobrevibacter smithii and Human Polyomavirus

To further detect the presence of human fecal pollution in the Lake Granbury watershed, two other library-independent PCR methods for *Methanobrevibacter smithii* (Ufnar et al. 2006) and human polyomavirus (McQuaig et al. 2009) were used as previously described.

18.2.2.3 Results and Discussion

E. coli and *Bacteroidales* MST results suggested that the Lake Granbury Port Ridglea East site was impacted primarily by animal-derived (wildlife) fecal pollution (Figs. 18.11 and 18.12, respectively). These findings were surprising

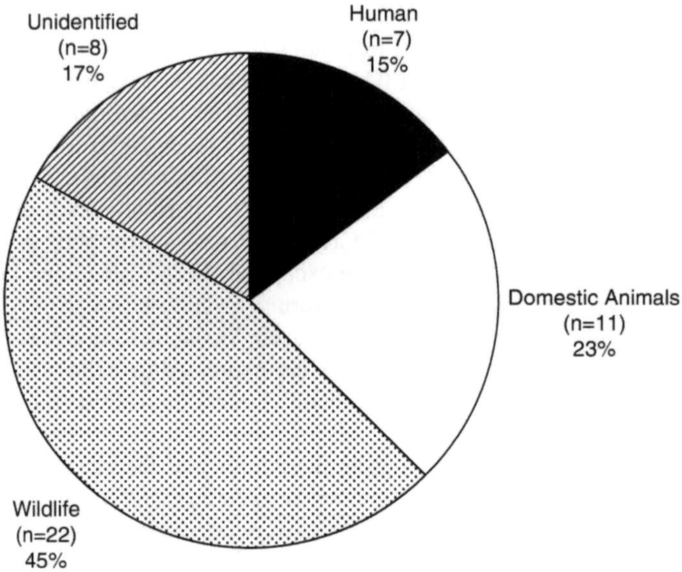

Fig. 18.11 *E. coli* source identification for the Lake Granbury Port Ridglea East site. The number of water isolates identified in each source category is included in parentheses. The *E. coli* long-term geometric mean at this site is high (120 MPN/100 mL)

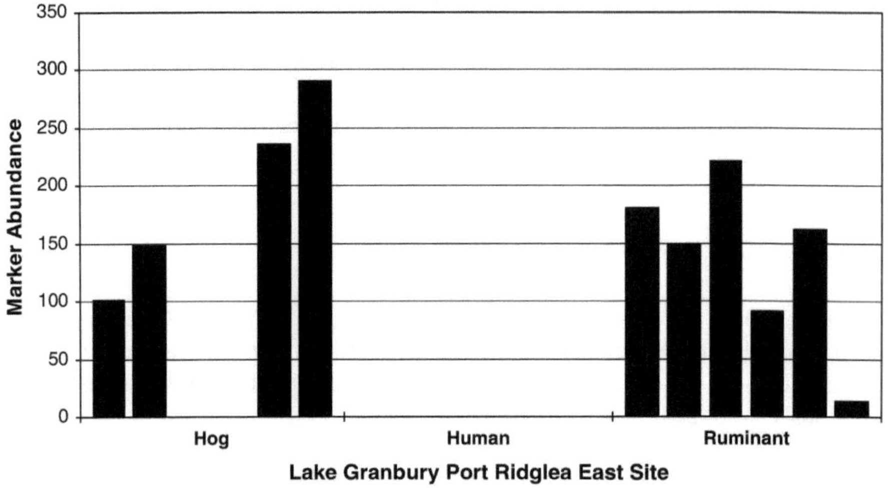

Fig. 18.12 *Bacteroidales* hog, human, and ruminant marker abundance for the Lake Granbury Port Ridglea East site. Six sets of monthly samples were collected and results are reported for each sample. Results are reported as relative GenBac marker arbitrary units per sample. All of the samples tested negative for the human marker

because it was assumed that the site was highly impacted by human fecal pollution from leaking septic systems. Forty five percent of the *E. coli* isolates were identified as originating from wildlife sources, while only 15% were identified originating from human sources. Further, none of the six monthly water samples were positive for the *Bacteroidales* human marker, while all were positive for the ruminant marker. The ruminant marker not only detects cattle and deer fecal pollution but may also cross-react with some nonruminant wildlife sources (Lamendella et al. 2007; Vogel et al. 2007). In most cases, the occurrence of the *Bacteroidales* markers in the 94 Lake Granbury fecal source samples was as expected. The exception was the human marker, which in addition to being detected in the human sewage/septage samples, was also occasionally detected in some animal feces. The human marker was detected in 3 of 8 coyote, 1 of 11 raccoon, 1 of 2 deer, and 5 of 7 rabbit fecal samples. Interestingly, 4 of the rabbit samples testing positive for the human marker were from pet rabbits in close contact with humans.

As a follow-up, more intensive sampling was performed at this site. Two sets of samples were collected approximately 2 weeks apart from ten different locations within the Port Ridglea East cove for *Bacteroidales* analysis and *E. coli* enumeration. In addition, *Methanobrevibacter smithii* and human polyomavirus PCR was performed for the detection of human source pollution. The water samples were collected under base-flow water conditions (no rainfall events for several weeks prior to sampling), so runoff was not a factor. *Bacteroidales* PCR results again revealed the presence of animal fecal pollution and the absence of human source pollution, despite some of the samples having *E. coli* levels up to 2,400 CFU/100 mL. The *Bacteroidales* ruminant marker was detected in 17 of the 20 (85%) follow-up samples, the hog marker was detected in five (25%) of the samples (presumably from feral hogs in the watershed), while all samples tested negative for the human marker. In addition, only one of the follow-up water samples (and its field duplicate) tested positive for human polyomavirus, and none tested positive for human *M. smithii*.

The pollution source modeling of the Port Ridglea East cove performed by the consultants only accounted for surface runoff and not subsurface flow. One possible explanation of the animal fecal pollution signature observed at the site is subsurface flow of wildlife-derived pollution from adjacent undeveloped land. Still, the lack of a significant human pollution signature was puzzling. While several of the residents in the Port Ridglea East cove have complained of backing up septic systems during periods of high lake water levels, no empirical data exists to quantify the amount of septage leaking into the cove. We therefore recommended that the cove be further investigated to determine the source of animal fecal pollution and determine if there are actually significant numbers of leaking septic systems. However, since sanitary surveys and computer modeling using the spatially explicit load enrichment calculation tool (SELECT) (Teague et al. 2009) predicted that 99% of the *E. coli* present in the cove was from failing septic systems (BrazosRiverAuthority 2008), stakeholders decided to move ahead with plans to construct a sanitary sewer system. Since human fecal pollution may contain a variety of human pathogens, this is a prudent course of action with respect to protecting human health. Nevertheless, it will be interesting to see if the *E. coli* levels in the cove show a significant decrease after the improvements.

As expected, wildlife was found to be a significant pollution source for the Buck Creek Site 10C, with 60% of the *E. coli* water isolates indentified as originating from wildlife sources (Fig. 18.13). Further, 90 and 60% of the water samples tested positive for the *Bacteroidales* ruminant and hog (from feral hogs in the watershed) markers, respectively (Fig. 18.14). Surprisingly, 19% of the *E. coli* water isolates were identified as originating from humans. *Bacteroidales* PCR results also indicated the presence of human fecal pollution, with the human marker detected in 60% (12 of 20) of the water samples collected over a 2-year period (Fig. 18.14). The occurrence of human *E. coli* and frequent *Bacteroidales* human marker was particularly unexpected, as this is one of the remote stretches of Buck Creek. There are no readily identifiable sources of human fecal pollution near Station BC10C. One possible explanation is that the abundant populations of coyotes, raccoons, and other wildlife near Station BC10C may be harboring *E. coli* and *Bacteroidales* typically found in humans. Similar to the Lake Granbury study, the human marker was found infrequently in wildlife feces (3 of 51 samples), with one badger and two porcupine samples testing positive. Although results indicate that Buck Creek water quality at Site 10C is impacted by pollution sources, the site generally had low *E. coli* levels with a geometric mean of only 18.9 CFU/100 mL over the 2 year monitoring period.

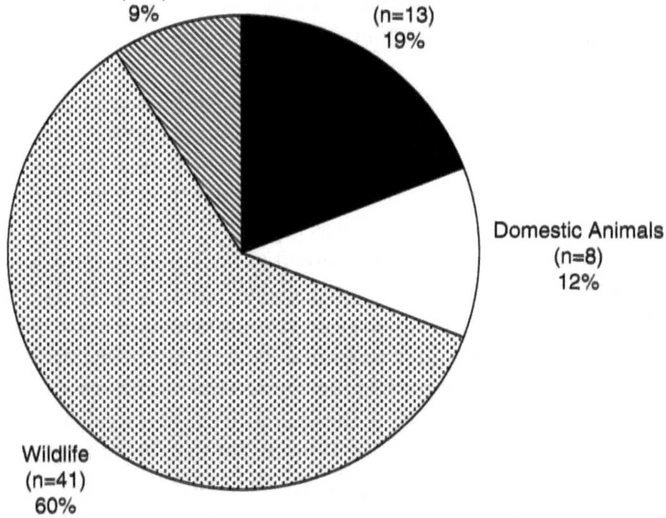

Fig. 18.13 *E. coli* source identification for the Buck Creek Site 10C. The number of water isolates identified in each source category is included in parentheses

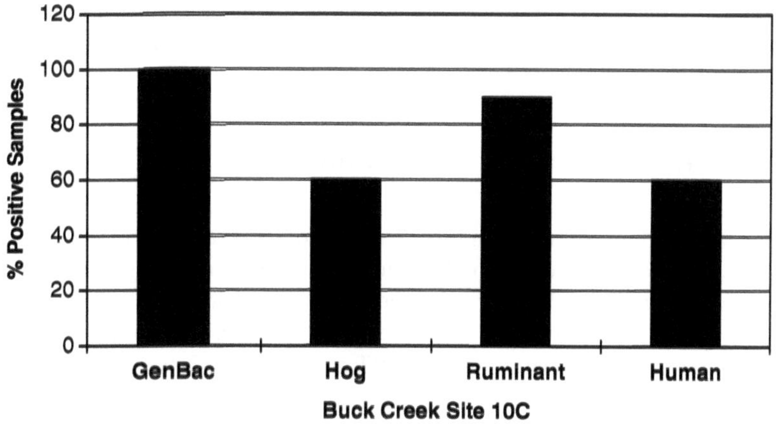

Fig. 18.14 Percentage of Buck Creek Site 10C water samples positive for the presence of the *Bacteroidales* GenBac, hog, human, and ruminant markers. Twenty water samples were collected over a 2-year period of time

Therefore, the site is of interest more from an academic standpoint rather than a watershed management perspective. The Buck Creek case also exemplifies some of the challenges in studying ARW that have low ambient levels of fecal pollution and abundant wildlife populations.

18.2.2.4 Conclusions: Case Study 2

As illustrated, source tracking can sometimes identify unexpected pollution sources, despite seemingly obvious pollution sources. It was assumed that Lake Granbury Port Ridglea East was overwhelmingly impacted by human pollution from leaking septic systems. However, based on the use of several source tracking tools we could find only a faint human pollution signature, while we consistently found evidence of wildlife-derived pollution. By contrast, it was assumed that the majority of fecal pollution at the Buck Creek Site 10C was due to wildlife. While MST data did indeed identify wildlife as the leading pollution source, we also unexpectedly found a consistent human pollution signature that could not readily be explained. Results from these studies provide examples of the dangers in making assumptions of pollution sources. They also stress the importance of well-rounded study designs to properly interpret source tracking results. Study design considerations include in-depth sanitary surveys, hydrogeology, careful field observation, and sufficient analysis of watershed fecal sources for the occurrence and distribution of *Bacteroidales* markers. Currently, the Texas *E. coli* MST library is being refined and challenged with known source isolates from different Texas watersheds to better understand the strengths and limitations of this approach. Additional fecal samples, particularly from wildlife populations, are being analyzed for the presence of *Bacteroidales* markers to further assess marker specificity and distribution.

18.3 Final Conclusion of Case Studies

The two case studies described in this chapter provide examples of different ARW and approaches to fecal pollution characterization based on integrated approaches using hydrological, modeling, physicochemical, microbiological, and MST tools. Careful consideration of methodical issues is of particular importance to successfully manage the unique challenges for MST and the identification of relevant fecal pollution sources in ARW.

18.3.1 Problem Identification, Specification, and Formulation

Basic problem identification in regard to microbiological water quality issues in ARW is typically based on SFIB monitoring. Data on the occurrence of *E. coli* or enterococci in ARW are frequently available from routinely performed monitoring programs. Without doubt, water quality testing based on the application of SFIB has been contributing to a fundamental improvement in safety management of water resources. However, given their reported potential indication bias (Chaps. 3 and 4), the significance of naturalized nonintestinal sub-populations of SFIB should also be considered. It is important to note that both presented case studies verified the main source of *E. coli* being of fecal origin. This was demonstrated by applying a multiparametric correlation analysis approach (Case Study 1) and library based typing procedures (Case Study 2). However, nonapplicability of SFIB for particular questions and/or habitat types may require alternative general fecal pollution indicators (e.g., Chaps. 5–8). For example, fecal hazard and risk assessment in porous groundwater resources would benefit from the additional application of virus and/or bacteriophage monitoring in order to take the lower mobility of bacterial indicators in these media into account.

For further problem specification, both case studies performed a sanitary survey on potential pollution sources within the ARW. The PSP described in Case Study 1 provides an example of how to gather quantitative data on potential fecal sources. Similarly, Case Study 2 included the evaluation of potential pollution sources, historical fecal indicator bacteria data, and land use patterns. In both case studies, the gathered information was then used to formulate the hypothesis or model the inputs of suspected sources and help guide selection of MST tools. It has to be emphasized, that models, irrespective of complexity, only provide estimates of fecal pollution sources based on a variety of assumptions. Adequate field data must be obtained to properly identify or verify suspected pollution sources, even for watersheds where pollution sources appear obvious.

18.3.2 Water Resource Characterization and Sampling Design

Most water resources in ARW are very dynamic systems in space and time. In this respect, different hydrological situations (e.g., base-flow vs. flood-events scenarios) may have profound effects on fecal pollution dynamics or sources involved. Thus, basic hydrological characterization, especially when dealing with highly dynamic systems, is an important prerequisite for proper study design. Ideally, sampling regimes should cover the spectrum of the hydrological dynamics or events within a watershed. Case Study 1 impressively demonstrates possible effects on fecal pollution levels and extent of microbial fluxes that hydrological events can cause (e.g., event analysis, Fig. 18.6). Furthermore, Case Study 1 also demonstrates how basic water resource characterization and sampling design, covering hydrological dynamics, can be realized. In this respect, analytic measurements and sampling is not necessarily linked to manual procedures but is increasingly based on online determination and autosampling (Fig. 18.5) (Stadler et al. 2008, 2010).

18.3.3 Required Performance Characteristics of MST Tools

The identification of human fecal pollution sources may be sufficient for some ARW studies, while other studies may demand the differentiation and quantification of impacts from humans, livestock, and wildlife. Respective of the specific problem formulation, "semiquantitative" and quantitative MST requires information on specific MST populations or markers with respect to (1) occurrence in targeted host groups (sensitivity), (2) occurrence in nontargeted host groups (specificity), (3) respective abundance in fecal pollution sources, (4) persistence and mobility in the water resource, and (5) detection limits and performance characteristics of the respectively applied methods. Proper study design should, thus, include MST verification on all these mentioned methodical levels at the considered pollution sources and water resources studied. It has to be highlighted that required minimum methodical performance levels are governed by the specific situation at the ARW (e.g., complexity and ratio of contributing fecal pollution source groups, extent of fecal pollution) and the question being asked (human vs. animal impacts, differentiation among animal sources, etc.). As mentioned in the introduction, in case MST is also used as a basis for further health risk assessment activities sensitivity levels for the targeted source groups must match the QMRA requirements (i.e. not to overlook important source groups as potentially masked by less important but dominating ones, Soller et al. 2010).

Consequently, MST results have to be checked for their plausibility and likelihood of false-negative (detection limit too low, fast preferential die-off, etc.) or false-positive detection rates (e.g., specificity too low in regard to nontargeted population/target levels). For a detailed discussion on the reliability of BacR versus BacH results (concerning Case Study 1) the reader is referred to recent work by

Reischer et al. (2011). Similar considerations, although on a qualitative basis, are applicable to Case Study 2 for library-based source tracking of *E. coli*. Such considerations on methodical performance criteria, especially under complex situations, may be greatly facilitated by using model-based verification and simulation approaches (Chaps. 4 and 9).

18.3.4 Integrated Approaches for Fecal Hazard Characterization

There is no single, perfect, MST tool available in the real world. As a result, integrated approaches including the wise use of multiple source tracking tools and other supplementary information and methods should be applied whenever possible. In some cases (such as Case Study 1), this approach may provide corroborating field data to support model-based hypotheses. Yet, in other watersheds (such as Case Study 2), multiple MST methods may reliably identify unexpected pollution sources. Although the greatest confidence in study results can be achieved through the use of integrated approaches with multiple tools, there is the inherent risk of disagreement between methods. The MST practitioner should be prepared for this scenario. It is critical to explain the strengths, limitations, and uncertainties of each method to project stakeholders, preferably at the time the study is designed. This can be a challenging task, as existing information on specificity, occurrence, and distribution of markers in animal populations and persistence of MST targets in the environment is still limited.

In conclusion, the presented case studies illustrate that MST has much to contribute in fecal pollution analysis in ARW (Fig. 18.1). However, MST is a young and developing field of applied science, and while it shows great promise for water resource management, there is still much to learn. In this respect, the authors do not give any recommendation for specific MST tools to be applied. Rather, examples of issues to carefully consider for a sound MST study design are provided (Sects. 18.3.1–18.3.4) including (1) a comprehensive problem formulation step, (2) a basic catchment and water resource characterization, (3) and a proper selection of sampling schemes and parameters. Finally, the methodical performance requirements for specific MST tools will largely depend on the problems to be addressed. In this respect, it is very likely that new MST tools that have increased application potential to ARW will be developed in the future. Nonetheless, proper verification of the performance characteristics will always be a key step in generating reliable MST information for water resource management.

Acknowledgment The Austrian part of the work was supported by the Vienna Water Works and the Austrian Science Fund (FWF) translational research project No. L414-B03 and DK plus W 1219-N22 (Vienna Doctoral Programme on Water Resource Systems) granted to A.H.F. The US Environmental Protection Agency Clean Water Act 319(h) program provided funding to G.D.D. for the Lake Granbury study through the Brazos River Authority and Texas Commission on Environmental Quality (Project 582-6-77030), and through the Texas Soil and Water Conservation Board for the Buck Creek study (Project 06–11).

References

Bernhard, A. E. and K. G. Field (2000a). Identification of nonpoint sources of fecal pollution in coastal waters by using host-specific 16S ribosomal DNA genetic markers from fecal anaerobes. Appl Environ Microbiol 66(4): 1587–1594.

Bernhard, A. E. and K. G. Field (2000b). A PCR assay to discriminate human and ruminant feces on the basis of host differences in Bacteroides-Prevotella genes encoding 16S rRNA. Appl Environ Microbiol 66(10): 4571–4574.

Besner, M. C., R. Broseus, et al. (2010). Pressure monitoring and characterization of external sources of contamination at the site of the Payment drinking water epidemiological studies. Environ Sci Technol 44(1): 269–277.

BrazosRiverAuthority (2006). Lake Granbury Watershed Protection Plan In: http://www.brazos.org/gbWPP.asp Retrieved January 21 2010.

BrazosRiverAuthority (2008). Lake Granbury Water Quality Modeling In: http://www.brazos.org/gbWPP/12-3-2008_LG_Modeling.pdf Retrieved January 21 2010.

Casarez, E. A., S. D. Pillai, et al. (2007). Direct comparison of four bacterial source tracking methods and a novel use of composite data sets. J Appl Microbiol 103(2): 350–364.

Cotruvo, J. A., A. Dufour, et al., (eds) (2004). Waterborn Zoonoses. IWA Publishing, London.

Dick, L. K., A. E. Bernhard, et al. (2005). Host distributions of uncultivated fecal Bacteroidales bacteria reveal genetic markers for fecal source identification. Appl Environ Microbiol 71(6): 3184–3191.

Farnleitner, A. H., H. Stadler, et al. (2008). Methods and strategies for alpine karstic water resource management: opening pollution microbiology's "black box". World Water Conference and Exhibition, Vienna, IWA.

Farnleitner, A. H., G. Ryzinska-Paier, et al. (2010). Escherichia coli and enterococci are sensitive and reliable indicators for human, livestock, and wild life faecal pollution in alpine mountainous water resources. J Appl Microbiol. 109(5): 1599–1608.

Farnleitner, A. H., I. Wilhartitz, et al. (2005). Bacterial dynamics in spring water of two contrasting alpine karst aquifers indicate autochthounous microbial endokarst communities. Environmental Microbiology 7: 1248–1259.

Fewtrell, L. and J. Bartram, (eds) (2002). Water Quality: Guidlines, Standards and Health. IWA Publishing, Padstone.

Field, K. G., E. C. Chern, et al. (2003). A comparative study of culture-independent, library-independent genotypic methods of fecal source tracking. J Water Health 1(4): 181–94.

Ford, D. C. and P. Williams (2007). Karst hydrogeology and geomorphology. Wiley, New York.

Fujioka, R. S. (2001). Monitoring coastal marine waters for spore-forming bacteria of faecal and soil origin to determine point from non-point source pollution. Water Sci Technol 44: 181–188.

Geldreich, E. E. (1978). Bacterial populations and indicator concepts in feces, sewage, stormwater and solid wastes. In: G. Berg, (ed) Indicators of viruses in water and food Ann Arbor, MI: Ann Arbor Science Publishers, Inc. pp. 51–97.

Grayson, R., P. Kay, et al. (2008). The use of GIS and multi-criteria evaluation (MCE) to identify agricultural land management practices which cause surface water pollution in drinking water supply catchments. Water Sci Technol 59(9): 1797–802.

Griffiths, R. I., A. S. Whiteley, et al. (2000). Rapid method for coextraction of DNA and RNA from natural environments for analysis of ribosomal DNA- and rRNA-based microbial community composition. Appl Environ Microbiol 66(12): 5488–91.

Ishii, S. and M. J. Sadowsky (2008). *Escherichia coli* in the environment: implications for water quality and human health. Microbes and Environments 23: 101–108.

ISO (2000). Water Quality – Detection and Enumeration of *Escherichia coli* and Coliform Bacteria – Part 1: Membrane Filtration Method *(ISO 9308-1: 2000)*. In: International Organization of Standardization Geneva, Switzerland.

ISO (2000). Water Quality – Detection and Enumeration of Intestinal Enterococci – Part 2: Membrane Filtration Method (ISO 7899-2: 2000). International Organization of Standardization, Geneva.

ISO (2002). Water Quality – Detection and Enumeration of *Clostridium perfringens* – Part 2: Method by Membrane filtration (ISO/CD 6461-2). International Organization of Standardization, Geneva.

Kollanur, D., Reischer, G.H., Sommer, R., Wehrspaun, C., Stadler, H, Mach, R.L., Zerobin. W. and A. H. Farnleitner (submitted) Quantitative Assessment of Faecal Pollution Sources in Alpine Spring Catchments as a Basis for Microbial Hazard – and Risk Assessment. *Water Science and Technology.*

Lamendella, R., J. W. Domingo, et al. (2007). Assessment of fecal pollution sources in a small northern-plains watershed using PCR and phylogenetic analyses of Bacteroidetes 16S rRNA gene. FEMS Microbiol Ecol 59(3): 651–60.

Lamendella, R., J. W. Santo Domingo, et al. (2009). Evaluation of swine-specific PCR assays used for fecal source tracking and analysis of molecular diversity of Bacteriodales-swine specific populations. Appl Environ Microbiol 75(18): 5507–5513.

McQuaig, S. M., T. M. Scott, et al. (2009). Quantification of human polyomaviruses JC Virus and BK Virus by TaqMan quantitative PCR and comparison to other water quality indicators in water and fecal samples. Appl Environ Microbiol 75(11): 3379–88.

Medema, G. J., S. Shaw, et al. (2003). Catchment characterisation and source water quality In: A. Dufour, M. Snozzi, W. Koster, et al., (ed) Assessing microbial safety of drinking water, IWA Publishing, London.

Reischer, G. H., J. M. Haider, et al. (2008). Quantitative microbial faecal source tracking with sampling guided by hydrological catchment dynamics. Environ Microbiol 10: 2598–2608.

Reischer, G. H., D. C. Kasper, et al. (2007). A quantitative real-time PCR assay for the highly sensitive and specific detection of human faecal influence in spring water from a large alpine catchment. Lett Appl Microbiol 44: 351–356.

Reischer, G. H., D. C. Kasper, et al. (2006). Quantitative PCR method for sensitive detection of ruminant faecal pollution in freshwater and evaluation of this method in alpine karstic regions. Appl Environ Microbiol 72: 5610–5614.

Reischer, G. H., D. Kollanur, et al. (2011). A hypothesis-driven approach for the identification of fecal pollution sources in water resources. Environ Sci Technol 45(9): 4038–4045.

Soller, J. A., M. E. Schoen, et al. (2010). Estimated human health risks from exposure to recreational waters impacted by human and non-human sources of faecal contamination. Water Research 44(16): 4674–4691.

Stadler, H., P. Skritek, et al. (2008). Microbiological monitoring and automated event sampling at karst springs using LEO-satellites. Water Sci Technol 58(4): 899–909.

Stadler, H., Klock, E., Skritek, P., Mach, R.L., Zerobin, W. and Farnleitner, A.H. (2010). The spectral absorbance coefficient at 254nm as a near real time early warning proxy for detecting faecal pollution events at alpine karst water resources. Water Sci Technol 62(8): 1898–1906.

Teague, A., R. Karthikeyan, et al. (2009). Spatially explicit load enrichment calculation tool to identify potential E. coli sources in watersheds. Transactions of the ASABE 52(4): 1109–1120.

TexasA&MAgriLifeResearch (2006). Watershed Protection Plan Development for Buck Creek, TSSWCB Project # 06-11, Quality Assurance Project Plan. In: http://www.tsswcb.state.tx.us/files/docs/nps-319/projects/06-11_QA-BUCKCRKWPP-09-24-08.pdf. Retrieved January 21 2010.

Ufnar, J. A., S. Y. Wang, et al. (2006). Detection of the nifH gene of Methanobrevibacter smithii: a potential tool to identify sewage pollution in recreational waters. J Appl Microbiol 101(1): 44–52.

USEPA (2005). Microbial Source Tracking Guide Document. Office of Research and Development, Cincinnati.

USEPA (2006). Method 1603: *Escherichia coli* (*E. coli*) in water by membrane filtration using modified membrane-thermotolerant *Escherichia coli* agar (Modified mTEC). Office of Research and Development, Government Printing Office, Washington.

Versalovic, J., M. Schneider, et al. (1994). Genomic fingerprinting of bacteria using repetitive sequence-based polymerase chain reaction. Methods Mol Cell Biol 5: 25–40.
Vogel, J. R., D. M. Stoeckel, et al. (2007). Identifying fecal sources in a selected catchment reach using multiple source-tracking tools. J Environ Qual 36(3): 718–729.
WHO (2004). Guidlines for drinking-water quality. World Health Organisation, Geneva.
Wilhartitz, I., A. K. T. Kirschner, et al. (2009). Prokaryotic production in karstic alpine spring aquifers and their ecological implications. FEMS Microbiol Ecol 68(3): 287–299.
Winter, C., T. Hein, et al. (2007). Longitudinal Changes In The Bacterial Community Composition Of The Danube River: A Whole River Approach. Appl Environ Microbiol 73: 421–431.

Chapter 19
Case Studies of Urban and Suburban Watersheds

Cheryl W. Propst, Valerie J. Harwood, and Gerold Morrison

Abstract Water quality in urban and suburban watersheds is impacted by many sources, ranging from the mixtures contained in stormwater runoff, to human sewage from failed wastewater handling systems, to domestic animals and urban wild animals such as birds (e.g., gulls, geese), opossums, raccoons, and deer. Many urban and suburban water bodies are chronically contaminated with fecal indicator bacteria, and municipalities frequently make choices about the level of resources devoted to determining microbial sources in their various "impaired" (sub-standard) watersheds based on financial constraints. This chapter discusses such considerations and focuses on work done in Florida, although the issues and solutions can be applied to any US or international water body. We outline a weight-of-evidence approach for investigating the sources of microbial contamination in surface waters that emphasizes stakeholder cooperation and engagement, and is economical and relatively easy to implement. The approach includes a strategy for ranking water bodies according to probable pollution sources and human use that is modified from the World Health Organization's (WHO) Annapolis Protocol. The protocol facilitates decision-making about which areas will be investigated by the more expensive microbial source tracking (MST) methods such as PCR for host-associated microbes, and which will receive more limited attention and resources. Case studies are presented, and the important role of stakeholder cooperation in the process is discussed.

Keywords Microbial source tracking • Urban watersheds • Weight of evidence approach • Stakeholders

---

V.J. Harwood (✉)
Department of Integrative Biology, University of South Florida, 4202 E. Fowler Avenue, Tampa, FL 33620, USA
e-mail: vharwood@usf.edu

C. Hagedorn et al. (eds.), *Microbial Source Tracking: Methods, Applications, and Case Studies*, DOI 10.1007/978-1-4419-9386-1_19,

19.1 Introduction

Across the USA, elevated concentrations of fecal indicator bacteria (FIB) such as fecal coliforms, *Escherichia coli*, and enterococci cause water bodies to be classified as "impaired," meaning that the water quality is below the standard set for its designated uses. Remediation of each impaired water body is required under the US Federal Clean Water Act (1972), and this is achieved by assessment and implementation of total maximum daily loads (TMDL) (see Chap. 14). As FIB levels provide no indication of contamination sources (e.g., human, livestock, wild animals) driving the impairments, the TMDL assessments and associated management actions are frequently inappropriate and ultimately unsuccessful (Wapnick et al. 2009). In addition, as fecal contamination originating from humans and certain animal groups (e.g., cattle, poultry) is known to have a high risk of carrying human pathogens, while others (e.g., pets, wildlife) have either lower or unknown risk (Dorevitch et al. 2010), the determination of contributing sources is needed to accurately assess the human health risk associated with the impairments (Wapnick et al. 2009). In this chapter, we use our experience in urban and suburban watersheds in Florida to demonstrate a practical approach to assess pollution sources in recreational waters and to prioritize further remediation action.

19.1.1 Impaired Florida Waters

In Florida alone, nearly six hundred water bodies have been verified as impaired due to "pathogens," which are not measured but rather are represented by surrogate FIB. This scenario is common across the USA, as most states have TMDL programs and have determined that many surface waters are impaired. The fecal coliform group is the FIB used to determine impairment status in Florida and in many US states. Stakeholders throughout Florida are actively involved in the development and implementation of TMDLs through basin management action plans (BMAPs). As a result, the state needed to identify an integrated and inexpensive methodology to quickly and effectively identify and prioritize water bodies and sources of fecal contamination for management action.

19.1.2 The Role of MST

Decades of frustrating experience engendered by reliance on FIB alone to assess and manage water quality speak strongly to the need for the source identification strategies provided by microbial source tracking (MST). In general, the first question to be asked in urban and suburban watersheds is, "Is it human?" The public and water-quality managers understand that contamination of water by sewage poses a public health threat, and a definitive diagnosis of human sewage contamination carries with it high motivational levels for remediation of pollution sources by the responsible authorities. Thus, most recent field studies involving MST in urban/suburban settings have included one or more tests for human-associated markers (e.g., McQuaig et al. 2006; Noble et al. 2006; Korajkic et al. 2009, 2011; Wyer et al. 2010). Other frequently implicated (but rarely proven) sources in such watersheds include pets (e.g., dogs and cats), small wild mammals such as raccoons, larger animals such as goats, deer, and horses, and birds. Fewer well-vetted MST methods are available for the various nonhuman sources compared to human sources (see Chaps. 4 and 8), although library-dependent methods (Chap. 3) may be useful in certain cases and new MST methods are frequently published.

The analytical methods used for MST are more expensive, more technically demanding, and less widely available than FIB measurements; therefore, their use must be judiciously planned and highly focused. Library-independent MST methods have largely superseded library-dependent methods (see Chaps. 3 and 4), and most of them require end-point (presence–absence) PCR at minimum. Even more recently, quantitative PCR (qPCR) methods, which require more expensive equipment and operator training, are being developed for MST (e.g., McQuaig et al. 2009; Shanks et al. 2009). The level of training for personnel, equipment requirements, and expense of these methods, which cost several to tens of times as much as FIB testing, primarily limit them to university and federal agency laboratories for the time being. Fortunately, such methods tend to decrease greatly in cost as they are standardized and as more facilities use one particular method or group of methods. It can be anticipated that MST testing will be less expensive in the future as methods are commercialized and multiplexed, and as more laboratories become proficient in their use.

19.1.3 The Weight-of-Evidence Approach

The weight-of-evidence approach to identifying microbial sources and prioritizing water bodies for remediation described in this chapter is derived from our previous work (Wapnick et al. 2009). It provides an economical and easy-to-execute process for distinguishing human from nonhuman contamination and targeting the application of the comparatively expensive MST methods. This approach allows MST efforts to be extremely focused and used on an as-needed basis (i.e., in those areas where further identification and confirmation of sources through the use of host-specific microbes is deemed warranted). The approach begins with a detailed review of existing data (usually FIB concentrations) to guide field reconnaissance and sampling efforts, which is critical to ensuring the most efficient and effective sampling strategy (also see Chap. 16). Local knowledge of the watershed (catchment) and its subbasins combined with active involvement of local stakeholders in all aspects of the process is particularly important. This integrative effort facilitates coordination and leveraging of available resources and also builds consensus for addressing the most probable contamination sources. While it is a key element of the weight-of-evidence approach, the involvement of stakeholders in the process requires extensive coordination and ultimately relies upon full collaboration by parties ranging from landowners to regulatory agencies. For example, in a recent Florida study, six agencies were involved (the City of Jacksonville Environmental Quality Division [COJ EQD], the City of Jacksonville Public Works Department [COJ PWD], the Duval County Health Department [DCHD], the Florida Department of Environmental Protection [FDEP], the Florida Department of Transportation [FDOT], and a local utility [JEA]), and numerous homeowners were contacted for information about or access to their property. A point person whose designated responsibility includes coordination of stakeholders is a necessity in this approach.

Many of the impaired basins (subwatersheds or catchments) throughout Florida are large and complex, including hundreds of miles of sewer lines, hundreds of septic systems and stormwater outfalls, over one thousand stormwater inlets, a variety of land uses (e.g., residential, agricultural, commercial, forested), and a mixture of fresh and tidal waters. One of the most significant benefits of the weight-of-evidence approach outlined here is the ability to avoid implication of entire watersheds for wastewater- and stormwater-based management actions (e.g., large-scale septic tank phaseout areas and infrastructure retrofits), when problems are actually localized on a much smaller scale. Instead, this process provides information that can be used to address specific contamination sources and isolated deficiencies within wastewater and stormwater handling systems at significant savings of time and money to the local stakeholders. As some of these impairments apply to recreational beaches, the identification and remediation of sources contributing to elevated FIB levels will result in fewer beach closures, ultimately having a positive effect on the local economy and community.

In addition to the economic benefits, the weight-of-evidence approach documents historic and new findings on microbial water quality, which are all too often measured and archived but not analyzed for informative trends. This approach also documents current and suggested management actions, facilitating the evaluation of "sufficiency of effort" to determine how well the projects designed to restore water quality are actually addressing the source(s) of impairment.

19.2 The Mechanics of the Weight-of-Evidence Approach

Our weight-of-evidence approach to fecal source assessment in environmental waters is based on the "Annapolis protocol," which was developed by the World Health Organization (WHO) following an international workshop that was held in Annapolis, Maryland, to develop improved strategies for managing the microbial quality of recreational waters (WHO 1999, 2003). The protocol combines quantitative FIB concentration data (provided by water-quality monitoring programs) with site-specific evaluations of the potential contaminant contributions and health risks posed by local FIB sources (provided by "sanitary inspections") to prioritize remediation efforts directed toward recreational waters based on their estimated suitability for whole-body contact.

The Annapolis protocol provides a technically sound conceptual approach for managing human health risks for recreational water users (NRC 2004; USEPA 2007). With modifications, it is applicable to assess the relative human health risk posed by recreation in Florida waters. For example, the Annapolis protocol bases water-quality assessment on enterococci levels, while the State of Florida uses fecal coliforms as its primary monitoring tool for ambient water-quality assessment. Similar modifications can be made to apply the tool in other states as well. Furthermore, the Annapolis protocol uses the 95th percentile FIB value observed at a monitoring site to determine FIB exceedances, while monitoring sites in Florida are generally evaluated based on the percentage of samples that exceed the State's 400 CFU/100 mL fecal coliform criterion (Chap. 62-303, Florida Administrative Code). Optional regulatory criteria within the Florida Administrative Code include a geomean value of 200 CFU/100 mL based on a minimum of ten samples taken over a 30-day period (a minimum sampling criterion that is not often met), or a single sample maximum of 800 CFU/100 mL.

In addition, the "sanitary inspection" categories outlined in the Annapolis protocol, while conceptually useful, de-emphasize management responses to situations in which nonhuman fecal sources may pose health risks in recreational waters. Adjustments were made to these categories in our protocol to address situations in which high-risk, nonhuman fecal sources (e.g., livestock) are also an issue of concern for resource managers.

19.2.1 Categorization of Sites by Microbial Water-Quality Assessment (MWQA)

In order to address these issues, a modified version of the WHO's (2003) microbial water-quality assessment (MWQA) matrix was developed (PBS&J et al. 2008) and then applied to six impaired segments of the Hillsborough River basin (PBS&J 2008b) (Fig. 19.1). The site assessment approach developed for the Hillsborough River involved the following steps:

1. Microbial water-quality conditions within each impaired river segment were categorized based on fecal coliform concentrations gleaned from historical monitoring data.
2. Each sampling station within a segment was assigned to a MWQA category based on the monitoring data. The MWQA categories were symbolized as letter grades (A through E) reflecting how frequently the State's fecal coliform criterion of 400 CFU/100 mL was exceeded at a given site. Because long-term monitoring data indicated that sites with higher frequencies of criterion exceedances

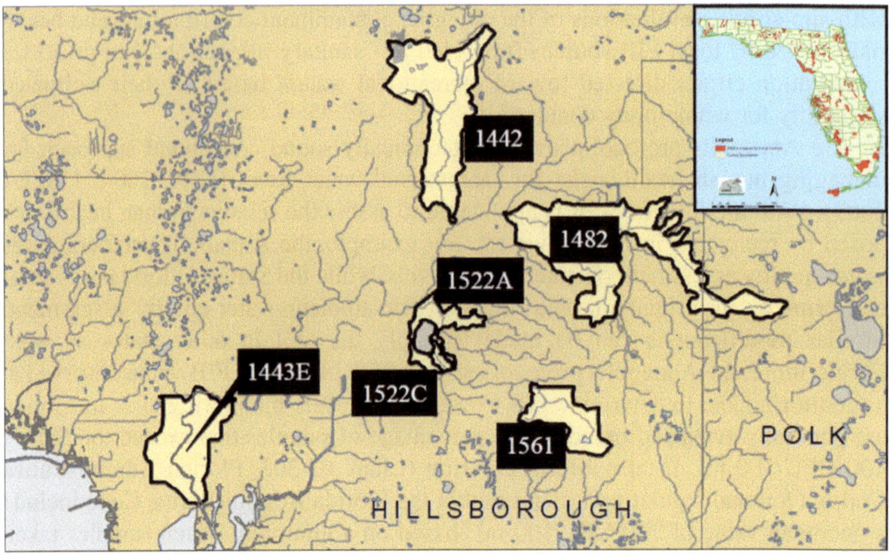

Fig. 19.1 Map of the six impaired Hillsborough County WBIDs discussed in the text and detailed in Tables 19.2 and 19.3. The *inset* is a map of Florida that shows impaired waters due to pathogen exceedances as darkened areas

also tended to exhibit higher overall concentrations of fecal coliforms and enterococci (PBS&J et al. 2008), MWQA categories A through E also represented progressively higher indicator organism concentrations and increasing levels of potential human health risk.

3. For sites at which significantly more than 10% of the samples exceeded the State's 400 CFU/100 mL fecal coliform criterion, contaminant source surveys (CSS), which can be likened to a "sanitary survey," identified the types of probable sources that could contribute to the elevated bacterial concentrations occurring at the site and characterized their potential human health risks.
4. Following the concepts outlined in the Annapolis protocol, each surveyed site was placed in a CSS category (ranging from "very low" to "very high" levels of potential risk; see Sect. 19.2.2) reflecting the types of probable bacterial sources found in the vicinity of the site and the estimated likelihood that those sources pose human health risks. Using a "phased monitoring" concept recommended by the NRC (2004), the intensity of CSS investigation that a site received was based on its MWQA classification. That is, sites that exhibited more frequent (and higher magnitude) exceedances of the State's 400 CFU/100 mL fecal coliform criterion (e.g., sites in MWQA categories C, D, or E) were subject to more intensive CSS investigations than sites exhibiting less frequent (and lower magnitude) exceedances.
5. Once the MWQA and CSS analyses were completed, each site received a two-part classification (Table 19.1), based on the MWQA and CSS categories into which it had been placed.

Table 19.1 A modified classification matrix adapted from the WHO (2003) Annapolis protocol approach, using a combination of fecal coliform measurements (represented by the MWQA group) and contaminant source survey (CSS) information to rank recreational sites based on the apparent likelihood of human health risk

		MWQA group (based on binomial assessment of frequency of 400 CFU/100 mL fecal coliform exceedances)					Exceptional circumstances (e.g., sewer line break)[a]
		A (≤10%)	B (>10–30%)	C (>30–50%)	D (>50–75%)	E (>75%)	
CSS assessment category (likelihood of fecal contamination posing human health risks)	Very low	A1	B1	C1[b]	D1[b]	E1[b]	Immediate Action
	Low	A2[c]	B2	C2	D2[b]	E2[b]	
	Moderate	A3[c]	B3	C3	D3	E3	
	High	A4[c]	B4[c]	C4	D4	E4	
	Very high	A5[c]	B5[c]	C5[c]	D5	E5	
	Exceptional circumstances (e.g., sewer line break)[a]	Immediate action					

See text for explanation of CSS assessment categories

Source: PBS&J et al. (2008)

Modified from WHO (2003)

[a]As explained by WHO (2003), exceptional circumstances involve acute situations known to be associated with higher public health risks, such as sewer-line breaks and other SSOs that contaminate surface waters, which require immediate remedial action

[b]In the authors' interpretation, these outcomes imply that the CSS may be providing an overly optimistic rating of water quality, or the fecal coliform sources in the area may be relatively low-risk or primarily environmental (e.g., wildlife, sediments, soils, vegetation), and the cause(s) of the discrepancy should be verified

[c]In the authors' interpretation, these outcomes imply that the fecal coliform indicator may be providing an overly optimistic MWQA rating, or the CSS may be providing an overly negative assessment, and the cause(s) of the discrepancy should be verified

19.2.2 The Contaminant Source Survey (CSS)

The "sanitary inspection" component of the Annapolis protocol – which is used to rank sites based on the types and magnitudes of potential fecal contaminant sources observed in their hydrological catchments (watersheds) – was revised to incorporate nonhuman fecal sources and renamed the CSS component in the modified matrix.

The CSS categories developed for ranking hydrologic subunits (designated in Florida as "water body identification numbers" or WBIDs) have been defined as follows (PBS&J et al. 2008) to provide a qualitative assessment of the likelihood that fecal contamination posing human health risks would be encountered by recreational users at a given site:

1. *Very low*: No visual evidence of potential sources of human pathogens; natural environment; no or minimal anthropogenic land uses; wildlife present (any density).
2. *Low*: Low-density agricultural and residential sources, including pets, livestock (without direct access to surface waters), or poultry; residences on septic systems.
3. *Moderate*: Urban stormwater sources (including pet waste) present; well-functioning wastewater infrastructure (both sewer and septic); episodic/low volume sanitary sewer overflows (SSOs) reaching surface waters; moderate-density livestock with little direct access to surface waters; Class A residual and/or septage spreading areas may be present.
4. *High*: Major stormwater outfalls present; history of failing wastewater infrastructure (central sewer or on-site systems); episodic or chronic/high volume SSOs reaching surface waters; concentrated livestock or poultry without direct access to surface waters; residual/septage spreading (Class B) may be present.
5. *Very high*: Current failing wastewater infrastructure; chronic/high volume SSOs reaching surface waters; concentrated livestock with direct access to surface waters; evidence of direct sewage inputs (e.g., confirmed illicit discharges).

19.2.3 Implementation of the Weight-of-Evidence Approach

In order to implement this approach to address water-quality impairments within watersheds, three additional steps are needed (HRBWG 2009; Morrison et al. 2010):

1. Prioritize sites for follow-up management actions, based on a weight-of-evidence assessment of their MWQA and CSS classifications and other available information.
2. Implement management actions at priority sites to address the specific FIB and potential pathogen sources that are present there.
3. Track progress in addressing potential pathogen sources and improving water quality.

An example of the site prioritization step, taken from a recent project conducted in a group of hydrologic subunits in the Hillsborough River watershed (Fig. 19.1), is shown in Table 19.2. An overview of the types of water-quality management actions that project participants have committed to implement in each watershed is provided in Table 19.3. Watershed managers are also monitoring FIB levels to determine if water quality is improving and impairments are being reduced.

Table 19.2 Priority ranking, most probable source categories, and recommended management actions for a representative subset of Hillsborough River sampling locations

Sub-basin	WBID	Monitoring location	Classification matrix outcome	Fecal source(s) of concern	Most probable source categories and recommended management actions
Lower Hills-borough River	1443E	DHR4A	D5	Human	Septic systems and possibly sanitary sewer; confirm and address specific source locations and provide public outreach explaining potential presence of health risk at site
		HR1 REF	B3	Human	Homeless camp, sanitary sewer and possibly stormwater; confirm and address specific source locations
Blackwater Creek	1482	BW3	D4	Human	Public or privately owned wastewater facilities; confirm and address specific source locations; provide public outreach explaining potential presence of health risk at site
		BW2	D4	Human	Unknown human source(s); confirm and address specific source location(s); provide public outreach explaining potential presence of health risk at site
				Ruminant	Cattle; install fencing around watercourse
		DBW4	A3	Human	Sanitary sewer; ensure that upstream WWTP (City of Plant City) maintains adequate disinfection of treated effluent
		BW5A	A3	Human	Septic; confirm and address specific source locations

(continued)

Table 19.2 (continued)

Sub-basin	WBID	Monitoring location	Classification matrix outcome	Fecal source(s) of concern	Most probable source categories and recommended management actions
Baker Creek	1522C	BK2	D3	Human	Septic; confirm and address specific source locations; provide public outreach explaining potential presence of health risk at site
				Ruminant	Cattle and/or deer; install fencing around watercourse if cattle are confirmed
Flint Creek	1522A	FL2	C2	Human	Likely septic; identify and address specific source location(s)
		FL1	C2	Human	Possibly upstream septic and livestock sources; identify and address specific source location(s)
Spartman Branch	1561	DSB3	C2	Ruminant	Wildlife sources suspected; no management actions suggested
New River	1442	NR2	C3	Human	Septic; confirm and address specific source locations
				Ruminant	Cattle; install fencing around watercourse upstream if confirmed
		NR1	C3	Ruminant	Cattle and/or wildlife; install fencing around watercourse where cattle present
		NR3	C3	Ruminant	Cattle; install fencing around watercourse upstream if confirmed

Source: PBS&J (2008b)
WBID (water body identification number)

Observations of water-quality conditions and trends provided by the monitoring program will be used to support an adaptive management process, to assess project effectiveness, and to identify possible needs for new management strategies or actions (HRBWG 2009; Morrison et al. 2010).

The weight-of-evidence approach described in this chapter has currently been implemented (in part or in its entirety) in over 20 basins throughout Florida, including tributaries of the lower St. Johns River (Jacksonville), the lower Hillsborough River (Tampa), in upper Peace Creek (Polk County, Florida) and in Orange Creek (Gainesville) (Fig. 19.1). In addition, a portion of this approach was carried out as

Table 19.3 Overview of water-quality improvement projects to be conducted in impaired subunits of the Hillsborough River watershed to implement the weight-of-evidence approach

Subunit	Project category	Number of projects
Blackwater Creek (WBID 1482)	Basic stormwater management program	3
	Education and outreach efforts	5
	Agricultural BMPs	2
	Regulations, ordinances, and guidelines	6
	Special studies, planning, monitoring and assessment	12
	Restoration and water-quality improvement	4
	Wastewater infrastructure management	13
New River (WBID 1442)	Basic stormwater management program	2
	Education and outreach efforts	4
	Agricultural BMPs	1
	Regulations, ordinances and guidelines	2
	Special studies, planning, monitoring, and assessment	9
	Wastewater infrastructure management	7
Spartman Branch (WBID 1561), Baker Creek (WBID 1522C), and Flint Creek (WBID 1522A)	Basic stormwater management program	2
	Education and outreach efforts	10
	Agricultural BMPs	1
	Regulations, ordinances, and guidelines	1
	Special studies, planning, monitoring, and assessment	5
	Wastewater infrastructure management	13
Lower Hillsborough River (WBID 1443E)	Basic stormwater management program	15
	Education and outreach efforts	10
	Regulations, ordinances, and guidelines	1
	Special studies, planning, monitoring, and assessment	7
	Restoration and water-quality improvement	3
	Wastewater infrastructure management	12

Source: HRBWG (2009)

part of a training workshop for the US Environmental Protection Agency (EPA) Region 4 staff in Gainesville, Georgia.

19.3 Case Studies

A critical component of any fecal contamination source identification effort is an intensive field investigation of the impaired basin. While it is important that this investigation be led by a single entity that is responsible for the recording and synthesis of all findings, participation by a team of representatives with local field knowledge of the area is imperative. Information gained through this process is ultimately used to verify initial hypotheses and identify additional potential sources of fecal contamination, to better understand system hydrology, and to develop a sampling implementation plan used to proceed with highly targeted MST testing in areas where it is deemed necessary. Results from this exercise often reveal information about contributing branches and flow patterns, the potential for local flooding, existing private and/or public sewer infrastructure problems, sanitary nuisances from septic systems, and the potential for animal (e.g., livestock, pets, wildlife) contributions. Specific examples of source identification findings from previous field investigations are summarized in Table 19.4, some of which are described in more detail below. These examples demonstrate how intensive field investigations can result in the identification of specific sources of contamination and, in other cases, be used to target and prioritize the use of MST, resulting in considerable time and cost savings.

19.3.1 The Importance of System Hydrology

The Peace Creek Drainage Canal, located in Polk County, Florida, was constructed in the 1920s to help drain the surrounding landscape for agricultural use, which remains a significant land use in this area (PBS&J 2008a). Agriculture is dominated by beef cattle grazing and orange groves. During the intensive field investigation of this basin on March 12, 2008, several drainage ditches originating from the surrounding residential neighborhoods and pasturelands were traced to discharge points in the main channel of the basin. A flooded pastureland populated with cattle had a direct connection to one of the identified drainage ditches. Just downstream, a stormwater culvert that was presumably too high to allow water to flow through was discovered and may have contributed to the flooding upstream. Over 4 in. of rain were recorded in the area the week prior to the investigation. These observations demonstrated the likelihood for cattle to be a contributing source of fecal contamination in this area of the watershed, especially following rainfall events.

19.3.2 Private Sewer Infrastructure

In many instances, contamination sources may not be directly observed during the field investigation; however, information often disclosed by local participants enables specific contributions to be targeted. During the September 8, 2008 reconnaissance efforts in the Big Fishweir Creek basin, located in Jacksonville, Florida, a private lift station was identified as having a history of chronic overflows, presumably associated with a grease trap (PBS&J 2010a). The location of the lift station and associated pipes in a heavily vegetated area and under the restaurant, which was directly adjacent to the main channel of the creek, likely allowed for several SSOs to go unnoticed. The SSOs had the potential to flow directly into surface waters. Awareness of the issue, however, prompted the addition of a sampling station at this location. Within the following 8-month sampling program with monthly sampling, two SSOs were observed and reported, one of which directly

Table 19.4 Examples of source identification findings resulting from intensive field investigations

Potential source	General location	Tributary	Description of observations or reporting	Supporting evidence	Result
Private sewer infrastructure	Jacksonville, FL	Big Fishweir Creek	Chronic SSOs	Witnessed two SSOs during sampling	Lift station currently being monitored by the City of Jacksonville and the FDBPR
Private Sewer Infrastructure	Jacksonville, FL	Christopher Branch	Unlocked, poorly maintained lift station	History of chronic SSOs	Reported to the City of Jacksonville; all repairs completed
Public sewer infrastructure	Jacksonville, FL	Long Branch	Sewer odor prompted sample collection	Hole in gravity main identified	Repairs completed
Public sewer infrastructure	Jacksonville, FL	McCoy Creek	Evidence of unreported manhole SSO after significant rain event	Evidence of an additional unreported manhole SSO at same location	Suggested investigation into potential infiltration/inflow into local sewer system
Public sewer infrastructure	Jacksonville, FL	Craig Creek	Evidence of unreported manhole SSO	Elevated FIB downstream	Suggested investigation into potential infiltration/inflow into local sewer system
Public sewer infrastructure	Gainesville, GA	Flat Creek	Leaking sewer pipe in concrete culvert		Leak confirmed and repair completed
OSTDS	Polk County, Florida	Peace Creek Drainage Canal	Sanitary nuisance at gas station		Reported to Polk County Health Department
OSTDS	Jacksonville, FL	Miramar Creek	Groundwater seeps, pipes in creek bank, and nearby old, unmounded OSTDS	Detection of elevated FIBs and human-specific MST markers	Investigation of nearby OSTDS and PICs (results pending)
Cattle	Hillsborough County, FL	Blackwater Creek	Cattle feces along creek bank and streambed	Ruminant-specific MST marker detected	
Pets	Jacksonville, FL	Miller Creek	Cats, cat feces odor, dog feces, remnants of rabbit cage	Detection of elevated FIBs and no human-specific MST markers	Suggested enforcement of responsible pet practices

impacted surface waters. Although sampling at this location was not necessary for source identification, it allowed the impacts of the SSOs at this location to be documented. This private lift station is currently being monitored by the City of Jacksonville and the Florida Department of Business and Professional Regulation (FDBPR).

19.3.3 Public Sewer Infrastructure

Intensive field investigations often lead to the detection of specific sources. For example, on September 29, 2008, a sewage odor was observed near the headwaters of a stream branch contributing to the Long Branch basin in Jacksonville, Florida (PBS&J 2010b). This observation prompted the collection of a water sample from a nearby stormwater drop inlet that demonstrated elevated fecal coliform levels (6,500 CFU/100 mL). Follow-up investigations by the local utility revealed the presence of a small hole in the top of a 6-in. cast-iron gravity main that crosses through a drop inlet slightly upstream. Although under normal flow conditions leakage would not occur, sewage would have spilled directly into the stormwater system and downstream surface waters under higher flow. Permanent repairs were made on November 5, 2008; four subsequent monthly samples collected downstream from this site revealed greatly reduced fecal coliform levels (<150 CFU/100 mL).

Similarly, an intensive field investigation training exercise performed for the EPA Region IV in September 2008 revealed the potential for a leaking sewer pipe passing through a stormwater culvert and into Flat Creek in Gainesville, Georgia. Within five days the City of Gainesville confirmed and repaired the leak.

19.3.4 On-site Treatment and Disposal System (OSTDS)

Potential septic tank-related sources may be evident in a variety of ways. On occasion, "smoking guns" are apparent in the field. For example, during the March 2008 intensive field investigation of the Peace Creek Drainage Canal, a septic tank overflow associated with a local gas station was observed and reported to the Polk County Health Department (PBS&J 2008a). Although surface waters were not impacted at this time, continued overflow or the addition of precipitation would likely have caused this overflow to reach the stormwater system.

By contrast, the potential for on-site treatment and disposal system (OSTDS) contributions in the Miramar Creek basin in Jacksonville, Florida was less obvious. In this instance, orange colored water, indicative of iron oxides that are locally associated with groundwater, were observed seeping through the banks of the creek (PBS&J 2010d). In addition, several pipes were observed draining water directly into the creek though samples of flowing water did not reveal any contamination at the time. Local knowledge and infrastructure data revealed the presence of old, unmounded OSTDS in the area. As a result, a sampling station and human-specific MST marker tests were utilized in this segment of the creek to further identify the potential for a septic tank-related source. Chronic and extremely elevated FIB in the water (reaching 80,000 CFU/100 mL) and sediments in this area, together with the detection of the human-associated *esp* gene of *Enterococcus faecium* (Scott et al. 2005) on three separate occasions confirmed the presence of a human source of contamination. These results prompted the DCHD to perform local septic tank inspections and for the City of Jacksonville to investigate the area for potential illicit connections (PICs); results of both investigations are pending.

19.3.5 Animal Sources

The large variety of potential animal contributors (e.g., pets, wildlife/birds, and sometimes livestock) in urban and suburban watersheds complicates the use of MST to identify sources. Careful observations regarding the possibility for animal sources are therefore extremely important. The appearance of several cats, a cat feces odor, and four piles of dog feces found near the upstream segment of Miller Creek in Jacksonville, Florida during the field reconnaissance effort on October 1, 2008 suggested that pets were a likely source of fecal contamination in this area of the basin (PBS&J 2010c). Follow-up field efforts also revealed that remnants of a pet rabbit cage, including hay and rabbit feces, had been discarded along the bank of the creek. Subsequent samples of local surface waters revealed elevated FIB concentrations and the lack of human-specific MST markers. These results suggested that pet waste was a dominant source of contamination in this portion of Miller Creek.

The potential for livestock-related sources of contamination may also become obvious during field investigations. For example, reconnaissance of the surface waters that flow through a cattle ranch in Hillsborough County, Florida, revealed the presence of cattle feces along the banks and streambed of Blackwater Creek (PBS&J 2008b). Follow-up surface water samples, including the use of a *Bacteroidales* ruminant-specific MST marker (see Chap. 4), in this area confirmed the presence of a ruminant (most likely cattle) source.

19.3.6 Details of a Beach Study

Determination of FIB sources is particularly crucial at bathing beaches due to the nature of exposure (i.e., swimmers are fully immersed in water and very young children play in the swash zones where bacterial concentrations are usually greatest). MST was employed at a popular beach in Tampa Bay, FL to determine whether human sewage was contributing to frequent exceedances of regulatory standards at the beach (Korajkic et al. 2011). Four sites approximately 100–200 yards apart from each other were sampled and tested for FIB (fecal coliforms, *E. coli*, and enterococci) and human-associated MST markers (*esp* gene of *E. faecium* and human polyomaviruses). Particularly high FIB levels (~log 3 CFU/100 mL for each FIB) were noted at Site 1 (adjacent to portable restrooms) and Site 4 (stormwater drain outfall). The human markers were also detected at these sites.

Field reconnaissance was simple at Site 1, as the portable restrooms were located at the edge of the sea wall. Observation of cleaning practices revealed that after the facilities' contents were collected by a service for disposal, they were hosed down, and contaminated water ran directly into the bathing area. Reconnaissance at Site 4 proved more difficult; initial attempts to identify a cross-connection or other contamination source from the storm drain showed no evidence of a human source. An employee of a local agency, the Hillsborough County Environmental Protection Commission, further investigated the storm drain and a nearby ditch and discovered a leaking sewer connection that was delivering sewage into the ditch. Wave and tidal action carried the contamination plume to the base of the stormwater pipe, creating the false impression that the pipe was the contamination source. After the sewer pipe was repaired and the portable restrooms were moved away from the shoreline, FIB concentrations and the frequency of MST marker detection decreased significantly at the beach (Korajkic et al. 2011).

19.4 Lessons Learned

The importance of the involvement of stakeholders for the success of MST studies cannot be underestimated. State and local agency personnel, as well as citizens, are a wealth of information about historical levels of pollution, potential pollution sources, and recent changes to water-body use or hydrology. Logistical issues such as access to private property can be circumvented with the early involvement of property owners, who may become defensive if they feel that fingers are being pointed at activities occurring on their property, but who also tend to cooperate if they are informed initially of an even-handed, scientific approach to identifying pollution sources. Agency personnel become active participants in field reconnaissance, and in many cases make extraordinary efforts to help with the study, as in the beach example above (Sect. 19.3.6). Often, certain agencies and/or their personnel have a greater stake in the outcome and are willing to put more effort into a study than others, but the exemplary help provided by any one agency can be used as an example and encouragement for others that may not be as motivated. Although the involvement of members of several agencies can be a logistical challenge, it creates a synergistic effect that has proven more than worthwhile in the studies outlined above.

Identification of pollution sources cannot improve water quality unless necessary remediation is applied to remove or diminish inputs. These efforts absolutely require the cooperation of local and state agencies. Their involvement from the study's inception increases their understanding and acceptance of study results, which in turn translates into a greater effort to implement recommendations generated from a given study. Remediation of wastewater and stormwater infrastructure requires considerable effort, coordination, and funding. Few programs are in place in the USA to provide such funding, but they are needed to act upon the findings of MST studies and related efforts to maintain or improve environmental water quality.

References

(1972) Federal Water Pollution Control Act (Clean Water Act) Amendments 33 USC 1311–1313 and USC 1315–1317; PL 92–500, United States

Dorevitch S, Ashbolt NJ, Ferguson CM, Fujioka R, McGee CD, Soller JA, Whitman RL (2010) Meeting report: knowledge and gaps in developing microbial criteria for inland recreational waters. Environmental Health Perspectives 118:871–876

Hillsborough River Basin Working Group (HRBWG). 2009. Hillsborough River basin management action plan. Prepared for the Florida Department of Environmental Protection, Tallahassee, FL. 219 pp.

Korajkic A, Badgley BD, Brownell MJ, Harwood VJ (2009) Application of microbial source tracking methods in a Gulf of Mexico field setting. Journal of Applied Microbiology 107:1518–1527

Korajkic A, Brownell MJ, Harwood VJ (2011) Investigation of human sewage pollution and pathogen analysis at Florida Gulf Coast beaches. Journal of Applied Microbiology 110(1):174–183

McQuaig S, Scott T, Lukasik J, Paul J, Harwood V (2009) Quantification of human polyomaviruses JC Virus and BK Virus by TaqMan quantitative PCR and comparison to other water quality indicators in water and fecal samples. Applied and Environmental Microbiology 75:3379–3388

McQuaig SM, Scott TM, Harwood VJ, Farrah SR, Lukasik JO (2006) Detection of human-derived fecal pollution in environmental waters by use of a PCR-based human polyomavirus assay. Applied and Environmental Microbiology 72:7567–7574

Morrison G, Swanson HN, Harwood VJ, Wapnick CM, Hansen T, Greening HS (2010) Using a 'decision matrix' approach to develop a fecal coliform BMAP for impaired waters in the Hillsborough River watershed. pp. 401–419. In S.T. Cooper (ed.), Proceedings, Tampa Bay Area Scientific Information Symposium, BASIS 5: 20-23 October 2009. St. Petersburg, FL. 538 pp.

Noble RT, Griffith JF, Blackwood AD, Fuhrman JA, Gregory JB, Hernandez X, Liang X, Bera AA, Schiff K (2006) Multitiered approach using quantitative PCR to track sources of fecal pollution affecting Santa Monica Bay, California. Applied and Environmental Microbiology 72:1604–1612

NRC (2004) Indicators for waterborne pathogens, Washington, DC

PBS&J (2008a) Fecal BMAP Development and Implementation: Identification of Probable Sources, Peace Creek Watershed Draft Final Summary Report, Prepared for the Florida Department of Environmental Protection, Tallahassee, FL

PBS&J (2008b) Fecal BMAP Implementation: Source Identification, Hillsborough River Watershed, Final Summary Report, Prepared for the Florida Department of Environmental Protection, Tallahassee, FL

PBS&J (2010a) Big Fishweir Creek Technical Report (WBID 2280), Prepared for the Florida Department of Environmental Protection, Tallahassee, FL

PBS&J (2010b) Long Branch Technical Report (WBID 2233), Prepared for the Department of Environmental Protection, Tallahassee, FL

PBS&J (2010c) Miller Creek Technical Report (WBID 2287), Prepared for the Department of Environmental Protection, Tallahassee, FL

PBS&J (2010d) Miramar Creek Technical Report (WBID 2304), Prepared for the Department of Environmental Protection, Tallahassee, FL

PBS&J, TCC, USF (2008) Development of a decision-support tool to support the implementation of fecal coliform BMAPs in the Hillsborough River Watershed, Prepared for the Department of Environmental Protection, Tallahassee, Florida

Scott TM, Jenkins TM, Lukasik J, Rose JB (2005) Potential use of a host associated molecular marker in *Enterococcus faecium* as an index of human fecal pollution. *Environmental Science and Technology* 39:283–287

Shanks OC, Kelty CA, Sivaganesan M, Varma M, Haugland RA (2009) Quantitative PCR for genetic markers of human fecal pollution. Applied and Environmental Microbiology 75(17):5507–5513

USEPA (2007) Report of the experts scientific workshop on critical research needs for the development of new or revised recreational water criteria, Tallahassee, FL

Wapnick CM, Harwood VJ, Singleton TL, Morrison G, Staley C, Staley ZR (2009) Application of the Bacteria Decision-Support Tool in the Hillsborough River Watershed Proceedings of the Water Environment Federation, Minneapolis, MN

WHO (1999) Health-based monitoring of recreational waters: the feasibility of a new approach (the "Annapolis Protocol"). In. WHO, Geneva, Switzerland, p 50

WHO (2003) Guidelines for safe recreational water environments, WHO, Geneva, Switzerland

Wyer MD, Kay D, Watkins J, Davies C, Kay C, Thomas R, Porter J, Stapleton CM, Moore H (2010) Evaluating short-term changes in recreational water quality during a hydrograph event using a combination of microbial tracers, environmental microbiology, microbial source tracking and hydrological techniques: A case study in Southwest Wales, UK. Water Research 44(16):4783–4795

Chapter 20
Beaches and Coastal Environments

Helena M. Solo-Gabriele, Alexandria B. Boehm, Troy M. Scott, and Christopher D. Sinigalliano

Abstract This chapter summarizes the rationale for using microbial source tracking (MST) methods at beach sites and coastal water bodies (Sect. 20.1), as MST methods are especially useful for evaluating waters impacted by nonpoint sources of pollution. This chapter also describes the most common traditional and alternative MST markers used at beach sites (Sect. 20.2). Two case studies are presented (Sect. 20.3) that describe the use of both biological/chemical MST methods and physical MST methods for identifying sources of microbes at two marine beach sites in USA, one located on the west coast (California) and the other located on the east coast (Florida). The chapter closes with discussion and recommendations concerning the utility and application of MST tools at beach sites impacted by nonpoint-source pollution (Sect. 20.4). Although this chapter focuses on marine beaches, an incredible wealth of MST data has been gathered at freshwater beaches (Byappanahalli et al. 2006; Harwood et al. 2005; Jenkins et al. 2005; Scott et al. 2002; Stapleton et al. 2009; Whitman and Nevers 2003; Whitman et al. 2004), and a comprehensive review of beach studies merits the inclusion of MST work within freshwater systems. The use of MST in freshwater systems is further discussed in Chaps. 18, 19, and 21.

Keywords Indicator microbe • Marine beaches • Coastal environment • Beach management

H.M. Solo-Gabriele (✉)
Department of Civil, Architectural, and Environmental Engineering, University of Miami, Coral Gables, FL, USA
and
University of Miami, Center for Oceans and Human Health, Key Biscayne, FL, USA
e-mail: hmsolo@miami.edu

C. Hagedorn et al. (eds.), *Microbial Source Tracking: Methods, Applications, and Case Studies*, DOI 10.1007/978-1-4419-9386-1_20,

20.1 Introduction: The Need for MST at Nonpoint-Source Beaches and Coastal Water Bodies

Beaches and coastal waters generate billions of dollars annually in tourism revenues in the USA and worldwide. Beach closures due to microbial contamination can have significant economic impacts, and new strategies are continuously being implemented in attempts to predict and prevent infectious disease outbreaks resulting from exposure to contaminated recreational water. Establishing the safety of recreational waters is currently based upon measurements of "indicator" microbes, which are generally bacteria but may also be viruses or protozoa. These microbes are used as "indicators" of the possible presence of sanitary sewage and subsequent risks to human health. Indicator microbes are natural inhabitants of the gastrointestinal tract of humans and are present in large numbers in fecal releases, especially releases from humans and warm-blooded animals (Maier et al. 2008). Indicator microbes are not necessarily pathogenic but are used as surrogates for the presence of pathogenic microbes. For marine waters, enterococci are the indicators recommended by the US Environmental Protection Agency (US EPA). Two criteria have been identified by the US EPA for regulatory purposes: One based upon geometric mean density and the other based upon single sample maximums (US EPA 1986). For marine waters, the acceptable geometric mean level (35 "colony forming units," or "CFU," enterococci per 100 mL of water) is independent of the intended use of that water, whereas the single sample maximums are regulated depending upon the intended use of the recreational area (designated beach area = 104 CFU/100 mL, moderate full body contact = 158 CFU/100 mL, lightly used full body contact = 276 CFU/100 mL, and infrequently used full body contact = 501 CFU/100 mL). In most cases, the 104 CFU/100 mL standard is used.

The support for the current US EPA indicator microbe criteria comes from previously conducted epidemiologic studies that established the relationship between indicator microbes and human health (US EPA 1983). One characteristic of the majority of the sites in these prior epidemiologic studies is that they were impacted by "point" sources of sewage. Point sources of sewage originate from wastewater collection and treatment systems. However, often in USA and worldwide, untreated wastewater is unintentionally released through leaking sewer pipes or other infrastructure failures, such as faulty lift stations. In many cases, treated wastewater is only partially disinfected prior to being released into a coastal water body (most commonly through ocean outfalls, i.e., pipes that extend out into the coastal water body). In such cases, the released wastewater contains high concentrations of microbial indicators and pathogens, as it originates from a large human population.

Therefore, prior epidemiologic studies conducted at sites impacted by point sources of pollution have established relationships between indicator bacterial levels in water and human disease (Cabelli et al. 1979, 1982), given the high probability of finding both fecal indicator bacteria (FIB) and human pathogens in waters impacted by point sources of pollution. In these cases, the levels of fecal bacteria "indicate" a risk to public health.

In developing countries, point sources of wastewater remain extremely problematic. However, in the developed world, wastewater treatment and effluent disposal systems continue to be upgraded by communities, thereby diminishing the dominance of point sources of pollution at beaches and coastal water bodies. Nevertheless, beaches and coastal water bodies worldwide are still impacted by fecal pollution, even though microbial contributions to many of the recreational water bodies are no longer dominated by sewage from ocean outfalls and compromised sewage pipes. Rather, the sources of microbial pollution at many sites, particularly in developed countries, are predominantly "nonpoint" sources, which include contributions from human bathing activities (as opposed to collective fecal contributions from a large community), from animal sources (such as birds, dogs, livestock, and wild animals), from urban and stormwater runoff, and from natural sources (e.g., from persistence and regrowth of indicators in the environment) (Bernhard and Field 2000a; Wright et al. 2009; Desmarais et al. 2002). Thus, at nonpoint-source-impacted beaches, fecal indicator microbes can originate from many different sources.

The statistical and predictive relationships between indicator microbes and pathogens originating from nonpoint sources of fecal pollution are different (less direct) than those between indicators and pathogens present in point sources of pollution. This situation complicates interpretations of indicator microbe data at beaches impacted by nonpoint source pollution, particularly with respect to understanding relationships to public health (WERF 2009). MST markers are well suited for identifying sources of indicator bacteria. As such, MST represents a valuable tool for regulators who wish to identify the source of an indicator microbe signal for purposes of interpreting the public health implications and ultimately remediating the source altogether.

20.2 Indicators and MST Markers

The following section provides background information concerning some of the specific physical, chemical, and biological MST methods that have been used to evaluate the sources of microbial pollution in recreational waters. These methods are split into those that are used currently or have been used for regulatory purposes in the past (i.e., traditional microbial indicators of recreational water quality) and alternative methods (including alternative microbial indicators and physical- and chemical-based methods).

20.2.1 Traditional Indicators of Fecal Pollution

Among the first indicator microbes to be used for water-quality assessment were the coliform bacteria (US EPA 1976). Total coliforms and fecal (thermotolerant) coliforms are still used as indicators by some states for monitoring offshore coastal waters, although newer regulatory guidelines (based upon using *Escherichia coli* for freshwater beaches and enterococci for both fresh and marine waters) have been issued by the US EPA (1986) with funding and subsequent implementation through the Beach Act of 2000 (Public Law 2000).

20.2.2 *Total Coliforms*

Coliform bacteria are Gram-negative, facultative anaerobic rods that ferment lactose to produce acid and gas in 24–48 h at 37°C. While they are found in the intestinal tracts of all warm-blooded animals and birds, they are also ubiquitous in the environment. Therefore, their utility as fecal indicators in environmental waters is limited. The coliforms have utility in drinking-water quality monitoring and are currently used for regulating drinking-water quality (US EPA 1989).

20.2.3 *Fecal Coliforms*

The fecal coliforms, or thermotolerant coliforms, are a subset of the coliforms that are capable of growing at elevated temperatures (44.5°C). The group consists primarily of *E. coli* and *Klebsiella* spp., with the former usually accounting for approximately 90% of the total fecal coliform population in fecal sources (APHA 1992). Fecal coliforms have been used for many years as fecal indicators; however, convincing epidemiological evidence that supports their use as predictors of human health risk is lacking (Cabelli et al. 1975, 1979, 1982; Wade et al. 2003). One drawback is the natural association of *Klebsiella* spp. with plant materials, and an ongoing problem has been the wastewater effluents of paper and pulp mills. These effluents contain very high levels of fecal coliforms, but are often devoid of *E. coli* (Gauthier and Archibald 2001). Fecal coliforms were the predominant indicator used to monitor recreational waters prior to 1986, at which time the US EPA recommended enterococci for monitoring marine waters and enterococci or *E. coli* for monitoring freshwaters.

20.2.4 *Escherichia coli*

E. coli is considered a more specific indicator of fecal pollution than fecal coliforms. They are present in high numbers in the intestines of birds and warm-blooded animals and have been shown in epidemiological studies of freshwater beaches to positively correlate with gastrointestinal (GI) illness. A recent meta-analysis conducted by Wade et al. (2003) further substantiated their utility as predictors of risk of GI illness in fresh water. Because of this correlation, *E. coli* is currently a US EPA recommended standard indicator for freshwater beaches. While it was once assumed that *E. coli* were not easily adapted to survival outside of their warm-blooded hosts, a significant amount of research has recently shown that reservoirs of these organisms can accumulate and persist in the environment in both temperate and subtropical climates (Carillo et al. 1985; Desmarais et al. 2002; Solo-Gabriele et al. 2000; Whitman and Nevers 2003; Mika et al. 2009).

20.2.5 Enterococci

The enterococci are a group of bacteria consisting of at least 23 species belonging to the genus *Enterococcus* (Tyrrell et al. 2002). They are Gram-positive, aerotolerant anaerobes and are present in high numbers in the intestines of both birds and mammals. The enterococci are currently the EPA-recommended indicator standard for marine waters. Their use has been supported by multiple epidemiological studies in waters impacted by point sources of pollution, in which their presence is correlated with GI illness (Cabelli et al. 1975, 1979, 1982). Not all enterococci have a fecal link, and several species are associated with plants, although methods have been developed that differentiate enterococci from human vs. nonhuman sources (Scott et al. 2005). Furthermore, many of the *Enterococcus* species have different survival and persistence characteristics, which can confound interpretations of elevated levels (Leclerc et al. 1996; Rozen and Belkin 2001).

20.3 Alternative Indicators and Source Tracking Methods for Recreational Waters

Many alternative microbial indicators have been proposed in the research literature. Five of the more common alternative microbial indicators (which are discussed below) include *Bacteroidales*, *Clostridium perfringens*, coliphage, *Staphylococcus aureus*, and *Bifidobacterium* spp. Additional microbial-based methods include the use of *Catellicoccus* spp. for identifying contributions from seagulls, the use of host-specific DNA markers in enterococci and the direct detection of pathogens. In addition to microbes, source tracking methods can include chemical markers as well as interpretations of microbial distributions, which is typically based upon an understanding of the water-flow characteristics and release/survival of the microbes in the environment.

20.3.1 Bacteroidales

Bacteroidales (Chap. 4) are obligately anaerobic Gram-negative rods that are abundant in the intestines of warm-blooded animals. In humans, levels of total *Bacteroidales* in the large intestine are 2–3 orders of magnitude greater than that of *E. coli* (Huijsdens et al. 2002). These organisms show promise as alternative fecal indicators due to their direct link to fecal material and their poor persistence in the environment once released, due to their requirement for anaerobic conditions (Scott et al. 2002; Fiksdal et al. 1985). Furthermore, several species or phylotypes are highly host-associated, which renders them useful as MST markers for identifying specific sources of fecal pollution if detected (Bernhard and Field 2000a, b; Kreader 1995; Kildare et al. 2007; Dick et al. 2005; Shanks et al. 2009, 2010). Drawbacks include their more limited environmental persistence, which may make them useful for indicating only recent fecal events. Therefore, they may fail to predict the presence of more persistent pathogens; however, molecular detection has been shown to be possible for days to weeks depending on environmental conditions and predation (Avelar et al. 1998; Betancourt and Fujioka 2006; Field et al. 2003; Kreader 1995; Seurinck et al. 2005; Walters et al. 2009). Quantitative molecular detection of these organisms may show promise as a recent epidemiological study by Wade et al. (2006) showed a moderate correlation between GI illness rates and quantitative polymerase chain reaction (qPCR) quantification of the total *Bacteroidales* marker at Great Lakes beaches. More research is needed to continue to correlate their presence with pathogens and to more clearly determine environmental survival and persistence characteristics (Dick et al. 2010).

20.3.2 *Clostridium perfringens*

C. perfringens is a Gram-positive, endospore-forming bacterium that has been used as a fecal indicator. Currently, the state of Hawaii utilizes this organism when elevated levels of enterococci are detected in an effort to further evaluate the public health relevance of the enterococci signal (Fujioka 2001). The persistence and resistance to environmental stress of this organism is both an asset and a liability. Because clostridia form endospores, they can persist in the environment for indefinite periods and can persist in sediments. The utility of *C. perfringens* as an indicator of diffuse or remote fecal pollution has been substantiated (Payment and Franco 1993; Sorensen et al. 1989), but more epidemiologic studies are needed to determine whether *C. perfringens* levels in recreational waters are related to public health effects.

20.3.3 Coliphages

Coliphages (Chap. 6) are viruses that infect *E. coli* and can be separated into two general groups based upon the mechanisms used to infect their host. Somatic coliphages infect their host through the bacterial cell wall, and F-specific coliphages (also termed "male-specific" or F+) infect through the F-pili of "male" host strains (Long et al. 2005). They are a diverse group that is not phylogenetically coherent; in fact, their genome may be RNA or DNA. F-specific RNA coliphages have been used for many years as an alternative indicator of fecal pollution (Dutka et al. 1987; Hurst et al. 2002; Kott et al. 1974). Although they are present in relatively high concentrations compared to human enteric viruses in sewage influent (usually 1–2 orders of magnitude greater), they are rarely isolated from fecal samples (Long et al. 2005). Once they are released into the environment, it is highly unlikely that they will find acceptable conditions in which to infect a host and multiply (Cornax et al. 1991; Havelaar and Pot-Hogeboom 1988). Because they possess morphological and survival characteristics similar to human enteric viruses, there is an inherent advantage to using them as indicators of viral presence, persistence, and transport (Borrego et al. 1987, 1990; Chung and Sobsey 1993; Havelaar et al. 1993; Boehm et al. 2009a). Recent epidemiological studies have suggested a relationship between the presence of F-specific RNA phage and GI illness at beaches impacted by both point and nonpoint source pollution (Colford et al. 2007; Abdelzaher et al. 2011 in press).

20.3.4 *Staphylococcus aureus*

S. aureus does not have a fecal link; however, it is present on the skin of many humans, and its presence in water is correlated with increased bather density (Fujioka et al. 1999; Charoenca and Fujioka 1993; Plano et al. 2011). *S. aureus* is also considered an opportunistic pathogen and most often is associated with skin infections, although sepsis, meningitis, and other diseases are also possible. One advantage of this organism is its moderate salt tolerance (Bruins et al. 2007), which may make it a reliable indicator in marine systems.

20.3.5 *Bifidobacteria*

Bifidobacteria are Gram-positive, obligate anaerobes that are present in the GI tract of warm-blooded mammals. Certain species may be useful indicators for distinguishing sources of fecal pollution (Bonjoch et al. 2004; Matsuki et al. 2002, 2004; Nebra et al. 2003), although some studies suggest otherwise (Lamendella et al. 2008). Bifidobacteria survive poorly once released into the environment, and their lack of persistence may make them useful only for recent pollution events (Rhodes and Kator 1999). Little is known about their geographic distribution and their correlation to the presence of pathogens in the environment. Nevertheless, new advances in molecular biology have allowed the development of more sensitive methods for their detection, and future research should be directed at assessing the value of this indicator in protecting public health (Plummer and Long 2007).

20.3.6 *Catellicoccus*

C. marimammalium is a recently identified Gram-positive, catalase-negative bacterium that is distantly related to *Enterococcus* and *Vagococcus*. Its growth requirements are complex, and its cultivation is dependent on the use of nutritionally enriched media.

Of particular interest is the fact that this organism cannot be cultivated in broth culture (Lawson et al. 2006). The fastidious nature of this organism may have implications in its survival and persistence in the environment, which should be investigated. Nevertheless, *C. marimammalium* has been shown to be specific to gull fecal samples (with some cross-reactivity to pelican, Lu et al. 2008 and porpoise and seal fecal samples, Lawson et al. 2006) when tested on DNA extracts from different bird and animal species. This marker has been used by researchers in coastal areas to assess the relative impact of fecal pollution from gulls As of the time of this writing, the performance of this Catellicoccus marker was currently being evaluated in a large-scale multi-laboratory MST methods validation study. This Source Identification Pilot Project (SIPP) Study is being conducted by the State of California to assess the performance of a variety of MST markers in various Federal, State, local, and academic labs to detect differing ratios and host sources of fecal contamination in blinded water samples when water-quality parameters exceed regulatory limits (Sinigalliano et al. 2010). As of the time of this writing, the performance of this Catellicoccus marker was currently being evaluated in a large-scale multi-laboratory MST methods validation study. This Source Identification Pilot Project (SIPP) Study is being conducted by the State of California to assess the performance of a variety of MST markers in various Federal, State, local, and academic labs to detect differing ratios and host sources of fecal contamination in blinded water samples.

20.3.7 Enterococci Species Distribution

The enterococci are a subgroup of the fecal streptococci and are characterized by their ability to grow in 6.5% sodium chloride, at low and elevated temperatures (10 and 45°C), and at elevated pH (9.5) (APHA 1992; Facklam et al. 1999). The genus *Enterococcus* includes at least 23 distinct species (Tyrrell et al. 2002), and several studies have reported on the variable distribution of species of enterococci in different animal hosts (Pourcher et al. 1991; Wheeler et al. 2002; Layton et al. 2010). Accurate speciation is difficult, and many biochemical tests currently used for identification of enterococci often produce ambiguous results (Harwood et al. 2004). Virulence genes in *Enterococcus faecium* have also been investigated for use as indicators of human fecal pollution (Scott et al. 2005). Numerous water-quality studies conducted worldwide have validated this approach, but cross-reactivity with other sources may be an issue (Abdelzaher et al. 2010; Ahmed et al. 2008, 2009).

20.3.8 Direct Measurements of Pathogens

The development of analytical methods such as qPCR has progressed to the point where studies can employ direct measures of pathogens rather than relying solely on indicator measurements. One drawback of direct pathogen monitoring is that disease can be caused by any one of many pathogens, and these targets are often present sporadically and at very low levels in recreational waters, requiring the analysis of large volumes of water. Sample concentration methods have been improved with the goal of capturing multiple types of microbes (Hill et al. 2005, 2007; Paul et al. 1991, 1996; Abdelzaher et al. 2008, 2009; Leskinen et al. 2010). Concentrates are then subjected to analysis for multiple targets, including pathogenic bacteria, viruses, and protozoa. For example, Abdelzaher et al. (2010) detected *Giardia* at a recreational beach site using conventional large-volume particle filters (225 L) and also utilizing a dual-filtration technology that requires smaller (5 L) volumes. However, the process of large-volume filtration (5 L is still large in comparison to 100 mL) is still very laborious and expensive and is currently not sensitive enough to replace the need for indicator microbe monitoring. Furthermore, the choice of which pathogens to measure is critical to success, i.e., one must choose pathogens that are prevalent in the contributing sources, which can be very difficult to determine. The ideal scenario would be to eventually supplement measurements of indicator microbe concentrations with direct measures of a cluster of pathogenic targets that are relevant to the pollution sources affecting the study area, in essence employing a "toolbox" approach of multiple target assays that collectively would enhance confidence in microbial water-quality assessment of the relative risk of exposure to recreational waters.

20.3.9 Chemical Source Tracking Technologies

Chemicals associated with sewage have been applied as MST tools (Chap. 8). These include caffeine, fecal sterols, and optical brighteners. A vast majority of adults consume caffeine daily and approximately 3% of ingested caffeine is excreted in urine (Tang-Liu et al. 1983). The presence of caffeine in environmental waters has been linked to wastewater contamination at locations around the globe (Peeler et al. 2006; Weigel et al. 2004; Chen et al. 2002; Knee et al. 2010). Fecal steroids are lipids excreted in animal feces. The source tracking literature makes reference to fecal sterols (such as cholesterol), which are a subset of fecal steroids. The types and relative quantity of various fecal steroids in the environment, often in sediments, have been used as fecal source tracking markers (Leeming et al. 1996; Noblet et al. 2004). Optical brighteners are detergent additives; their presence in the environment has been used for rapid identification of sewage inputs (Dickerson et al. 2007; Hartel et al. 2007, 2008; Cao et al. 2009). Other chemicals that have potential as chemical markers include personal-care products and pharmaceuticals (Kolpin et al. 2002). At the present time, analytical complexities associated with their quantification along with concerns of environmental accumulation and persistence confounds their utility for routine MST, although further refinements of these techniques may prove to be useful.

20.3.10 Physical Methods for Source Tracking

Physical methods are generally based upon investigation of the fate and transport of microbial contaminants in the environment. Physical methods typically require that the source be quantified (CFU per event or CFU per unit time), and this can be accomplished through measurements of the microbe as it is released or transported from its source (Boehm et al. 2005, 2009a). Once the contaminant(s) in the source is quantified, the transport and fate principles are invoked (by considering mass balance and flow characteristics) to evaluate the dilution, dieoff, and potential regrowth of the microbe as it moves through the environment (Nevers and Boehm 2010). Transport is highly dependent upon water currents, which will spread the contaminant from its source to other areas, with concentrations highest at the source and lower in downstream areas that are subject to greater dilution. In addition to currents, hydraulic and density gradients play an important role in dictating transport. Hydraulic gradients describe water elevation changes and since water flows from high elevations (or head) to low elevations, contaminants will be carried down-gradient in the direction of decreasing water head. Of note, the direction of decreasing head can change given changes in water levels induced by tide and by changes in groundwater elevations induced by rainwater infiltration (De Sieyes et al. 2008). Physical methods of source tracking would take into consideration changes in hydraulic gradients. Moreover, water of different densities (these differences can be induced by differences in salinity and temperature) will also impact the movement of microbial pollution, as mixing of water is limited across density gradients.

In addition to the physical movement of the water, the fate of the microbes also plays a significant role in dictating microbe levels in the environment. Enteric microbes are known to generally decay or die-off: (1) under UV light (Boehm et al. 2002a), (2) through predation by other organisms (Avelar et al. 1998; Betancourt and Fujioka 2006; Field et al. 2003), and (3) when nutrients are lacking. In some cases fecal bacteria (*E. coli* and enterococci) have been shown to multiply under simulated field conditions (Desmarais et al. 2002; Yamahara et al. 2009), resulting in the addition of microbes to the system as time progresses. Given the complexities of the physical processes impacting microbe levels, models are in many cases invoked in an effort to evaluate the relationships between environmental parameters and microbe levels at recreational beach sites. Models can range from simple regression models to more complex hydrodynamic fate and transport models (Francy et al. 2003; Olyphant and Whitman 2004; Nevers and Whitman 2005, 2008; Nevers et al. 2007).

20.4 Case Studies

In this section, we outline various studies that took place between the late 1990s through 2010. These studies aimed to identify the source(s) of microbial contamination at two beach sites, one located in California and the other located in Florida. Some of the studies at each of the sites used biological and chemical tools from the MST toolbox. Some investigated the possibility that potential sources contributed to contamination at beaches by using models of microbial fate and transport. These two types of studies will be referred to as "biological and chemical source tracking studies" and "physical source tracking studies."

20.4.1 Case Study #1, Huntington State Beach, CA, USA

Huntington State Beach, CA receives approximately four million visitors per year (Given et al. 2006). The beach is relatively wide (~200 m). The water temperature varies from 10°C in the winter to 20°C in the late summer. This famous section of Californian beach received national attention in the summer of 1999 when the entire shoreline was closed due to high levels of enterococci bacteria. Presently, this stretch of coastline still has elevated levels of bacterial pollution during the summer, and the source(s) remain largely unknown.

There are a number of potential sources of indicator microbes to the Huntington Beach shoreline (Fig. 20.1, Table 20.1) during the summer (dry) season. These include an ocean outfall 7.5 km from the shore, a thermal outfall from a nearby power plant, urban runoff from Talbert Marsh, urban runoff from the Santa Ana River, bird feces from the beach face, polluted groundwater discharge, and beach sands and nearshore sediments. Urban runoff consists of water from lawn watering, car washing, broken sewer infrastructure, and other dry-weather sources of freshwater. The climate in this region is Mediterranean; there is generally no rainfall between April and October; therefore, storm runoff was not identified as a potential source of pollution during this time of year.

Certain characteristics of the microbial pollution problem at Huntington Beach provide clues as to the source. The pollution is modulated by sunlight and tides. The concentration of bacteria (both *E. coli* and enterococci) is much higher at night than during the day, owing to the deleterious effects of solar radiation on the

Fig. 20.1 Map of Huntington Beach, CA (33°38′N, 117°59′W)

Table 20.1 Potential source of indicator bacteria evaluated through research studies conducted at Huntington State beach, California and at Hobie Cat beach, Florida

	Huntington State beach	Hobie Cat beach
Inputs through surface drainage	Talbert Marsh Santa Ana River	Miami river and other canals Localized stormwater from natural drainage ditches
Groundwater	Leaking sewage infrastructure	Bathroom septic tank system
Wastewater outfall	Offshore discharge of primary and secondary treated sewage with and without disinfection	Offshore discharge of primary and secondary treated sewage with disinfection
Animal sources	Bird feces Human bather shedding	Bird feces Dog feces Human bather shedding
Sediments and sands	Beach sands along Huntington State beach, and sediments from the Santa Ana River	Beach sands, including subtidal, intertidal, and supratidal sediments
Other sources	Thermal outfall	

indicator bacteria (Boehm et al. 2002a). Bacterial concentrations are higher during spring tides, when the tide range is largest (Boehm et al. 2002a). In addition, greater bacterial concentrations are observed during low tides compared to high tides (Boehm and Weisberg 2005). The concentration of bacteria is highest at a location removed from the tidal wetland outlets to the south (see star in Fig. 20.1) (Kim et al. 2004).

20.4.1.1 Physical Source Tracking Studies

Here, we review the studies that analyzed the potential for sources in Table 20.1 to contribute bacteria to the surf zone.

The Talbert Marsh. The potential for Talbert Marsh to be a source of bacteria to the surf zone of Huntington Beach was investigated by Grant et al. (2001). The marsh is a tidally influenced wetland that receives some urban runoff from the city of Huntington Beach. It is estimated that approximately 700 m^3/day freshwater is discharged through the mouth of the marsh, and all of this discharge is confined to the ebb tides when water flows out of the marsh to the coastal ocean. Although the urban runoff entering the channels inland of the marsh contained elevated indicator bacteria levels (on the order of 1,000 MPN enterococci/100 mL), most of the enterococci dies off in the channels before reaching the marsh. Interestingly, the study found that the marsh itself was a net source of enterococci, since the observed concentration increased in water that passed through the marsh on both flood and ebb tides. Likely sources within the marsh include bird feces and marsh sediments.

During ebb tides, the discharge of water from the marsh with mean enterococci concentrations of 70 MPN/100 mL was observed, with much higher single-sample concentrations possible. A dye study was conducted to show that discharge from the marsh was entrained in the surf zone and transported along the shore with the littoral current. The study concluded that the marsh was a source of enterococci bacteria and that ebb flow could enter the surf zone and be transported toward Huntington State Beach, where it impacts water quality. However, the researchers cautioned that the marsh could not be the only source of bacterial pollution to Huntington Beach because a marsh source could not explain the highest concentrations of enterococci at the section of beach north of the marsh (see star in Fig. 20.1). If the marsh were the main source, the enterococci concentrations should be highest near the mouth of the marsh and dissipate to the north. In addition, a follow-up study (Jeong et al. 2008) showed that the marsh could, at times, be a smaller fecal bacteria source than originally observed by Grant et al. (2001). In fact, under some conditions the marsh actually reduced indicator bacteria concentrations in runoff that flowed through it. These additional findings cast further doubt on the hypothesis that the marsh is the main source of indicator bacteria to Huntington State Beach.

The Santa Ana River. The Santa Ana River is similar in some respects to Talbert Marsh. During the dry season, it is a tidal wetland that receives urban runoff from its watershed. Several studies have investigated the hypothesis that the wetland is the primary source of microbial pollution to Huntington State Beach. While the Santa Ana River has been found to be a source of indicator bacteria to the surf zone (Kim et al. 2004; Grant et al. 2005; Santoro and Boehm 2007) and certainly contributes bacteria, particularly total coliform, to the surf zone of Huntington State Beach, it cannot explain the high concentrations of enterococci north of the river for reasons that are similar to the Talbert Marsh ebb discharge (i.e., highest concentrations north of the discharge and lower concentrations at the mouth of the inlet).

The wastewater outfall. At the time of the beach closures (summer 1999), the wastewater outfall was discharging a mixture of primary and secondary treated domestic sewage (234 million gallons per day or 10.2 m^3/s) and was operating under a waiver from the USEPA (Boehm et al. 2002b). The wastewater outfall is located far from the shoreline (7.5 km) and discharges the water at 60 m depth with the treated wastewater captured in subthermocline waters (Boehm et al. 2002b). Careful observations of currents near the outfall indicated that the mean current direction was not onshore. Boehm et al. (2003) investigated the potential for nonlinear, internal tides to transport pollutant-laden subthermocline waters onshore. They found that on at least one occasion the wastewater plume was located close to shore, in cold subthermocline waters, and was potentially transported there by internal tides on the shelf. However, they did not find that the plume was transported past the subthermocline to the surf zone. The Orange County Sanitation District began disinfecting its sewage effluent in 2002 to eliminate the possibility that the wastewater plume could be a source of pollutants to the surf zone. The implementation of disinfection appears to have had no impact on the frequency domain character of pollution at Huntington Beach, suggesting that the wastewater outfall was not a major source of pollution to the beach (Noble et al. 2006).

Groundwater discharge. Submarine groundwater discharge consists of saline and fresh waters that discharge from the beach aquifer to the coastal ocean. The possibility that polluted groundwater was contributing to the bacterial pollution at Huntington Beach was investigated by Komex consultants and by Boehm et al. (2004).

Komex installed wells in the region surrounding the beach to access the surface of the unconfined aquifer. They detected enterococci in 3 of 40 wells at concentrations up to 590 CFU/100 mL. The average gradient of groundwater levels was directed inland; therefore, the authors of the study concluded that groundwater was not a major source of contamination to the surf zone. A study of the public restrooms at Huntington Beach employed lithium chloride and Rhodamine WT as tracers. The tracers were flushed down the toilets and were detected in the surf zone within 48 h, indicating leaking sewer infrastructure of the restrooms as a probable bacterial source (Kim et al. 2004).

Boehm et al. (2004) further examined submarine groundwater discharge at this stretch of beach. They used radium isotopes, which are naturally enriched in coastal groundwater, as tracers of groundwater discharge to the coastal ocean. They found that groundwater discharge was occurring in particular phases of the tide and that the discharge was likely primarily composed of recirculated seawater. *E. coli* and enterococci concentrations in the surf zone were well correlated to radium activity levels, suggesting a link between the processes that bring radium and indicator bacteria into the surf zone. However, only 1 in 26 groundwater samples had concentrations of indicator bacteria higher than the single sample contact standards. It was, therefore, concluded that although there could be a link between groundwater discharge and the fecal pollution, it was probably not the main source of pollution to the surf zone.

20.4.1.2 Biological and Chemical Source Tracking Studies

Four other studies used biological and chemical source tracking tools to elucidate the source of pollution at Huntington State Beach. The tools applied include antibiotic resistance profiling of enterococci, fecal sterols, the human-specific marker in *Bacteroidales*, and speciation of enterococci.

Antibiotic resistance profiling. As explained previously in Chap. 3, antibiotic resistance profiling (also known as antibiotic resistance analysis) is a library-dependent method that uses patterns in antibiotic resistance phenotypes of bacterial species in sources (the library) to characterize bacterial species that are present during a pollution event. Choi et al. (2003) characterized antibiotic resistance of enterococci in bird feces, primary treated sewage, urban runoff, sediment from the Talbert Marsh, and from the surf zone of Huntington Beach. They used seven antibiotics each in four different concentrations to characterize resistance phenotypes: ampicillin, erythromycin, tetracycline, chlortetracycline, oxytetracycline, streptomycin, and salinomycin. Enterococci from bird feces, sewage, runoff, and sediment were used to define the library of sources, and enterococci isolates from the surf zone were characterized using this library. In total, 2,491 isolates were characterized by antibiotic resistance profiles, and the data were analyzed using discriminant analysis.

The antibiotic resistance profiling worked well to discriminate between source groups. The "library" was defined by the sources, and then the authors ran their known-source isolates against their library to check for classification accuracy. They found that 85% of bird fecal isolates, 81% of urban runoff isolates, 73% of sediment isolates, and 64% of sewage isolates were correctly classified.

When surf zone water isolates collected over a year were compared to the library of isolates, it was found that 30, 24, 6, and 40% were most similar to bird feces, Talbert Marsh sediments, urban runoff, and sewage, respectively. The surf-zone isolates that comprised the collection were collected on five distinct sampling dates in winter, spring, and summer. When examined by date of collection, between 15 and 53% of the surf zone isolates were classified as sewage isolates, while between 9 and 67% were classified as bird-feces isolates depending on the particular date of collection.

Human-specific Bacteroidales *marker.* This library-independent MST method was used to assess the prevalence of human fecal contamination in the surf-zone of Huntington Beach by Santoro and Boehm (2007). Approximately 50 samples were screened from Huntington State Beach, approximately 50 from the mouth of the Santa Ana River, and approximately 15 from Newport pier during both August 2005 and July 2006 (total 230). Newport Pier was chosen as a site far removed from the Huntington State Beach site; it also has lower concentrations of indicator bacteria. The PCR primers used were HF183 and Bac708R following Bernhard and Field (2000a).

The frequency of observation of the marker varied between years. In 2005, the marker was present in about 23–29% of the samples screened, depending on the site. In 2006, the marker was present in 40–67% of samples screened, with the highest occurrence at the Newport Pier site. The concentrations of FIB (including *E. coli* and enterococci) did not correlate with the occurrence of the human marker. However, it is interesting to note that overall, concentrations of the indicators were higher in 2006 when the occurrence of the marker was also higher.

This study confirmed the results of Choi et al. (2003), in that human sewage is apparently a contributor to fecal pollution at Huntington State Beach. However, pollution from human sources was also observed at Newport Pier to the south, suggesting that human fecal pollution is more widespread and not confined to Huntington State Beach.

Enterococcus *species distribution.* Ferguson et al. (2005) enumerated and speciated enterococci in surf zone water and beach sands along Huntington State beach, and sediments from the Santa Ana River using biochemical tests. They found low levels of enterococci in the beach sands along this stretch of coast (geometric mean of 6 CFU/10 g wet weight), but quite high levels in the sediments of the Santa Ana River (geometric mean of 5,922 CFU/10 g). They found that the species present in the sediments and beach sands were similar to those in the water, and the majority were *Enterococcus faecalis* (~30–40% in shoreline water samples). Unfortunately, the authors reported that the *Enterococcus* species distribution found in the sediments was similar to that expected in animal feces, so this tool could not help identify the source of contamination along the shoreline. However, their work does indicate that sediments could be contributing enterococci to the river ebb discharge and that beach sands are not a likely source of enterococci to beach waters. As discussed earlier, the Santa Ana River has been ruled out as the main source of enterococci pollution to Huntington State Beach (Kim et al. 2004).

Fecal Steroids. These chemicals were quantified in water samples at the outlet of the Santa Ana River to determine whether FIB present in the ebb flow were from sewage or other sources (Noble et al. 2004). The authors collected sewage samples as well as surf-zone water samples and analyzed for a suite of fecal steroids including coprostanol (COP), epicoprostanol (eCOP), cholesterol (CHOE), cholestanol (CHOA), α-cholestanone (αONE), β-cholestanone (βONE), β-sitosterol (βSIT), stigmasterol (STIG), stigmastanol (STAN), and campesterol (CAM). They used the ratio COP/(COP + CHOA) and the ratio βONE/(βONE + αONE) to discriminate among sources (humans, birds, dogs). The authors interpreted a ratio of greater than 0.7 to be indicative of a sewage source. They also used (COP + βCOP + βONE) and (CHOE + CHOA + βSIT) as markers of sewage and bird fecal steroids, respectively.

The results suggest that the source of these fecal-associated chemicals in the surf zone near the river was not sewage. The authors found that while the sewage samples had COP/(COP + CHOA) and βONE/(βONE + αONE) greater than 0.7, the samples from the river had variable ratios spread across the range of values from 0 to 1. Similarly, they found that concentrations of indicator bacteria correlated well with (CHOE + CHOA + βSIT) (bird steroids), but not with (COP + eCOP + βONE) (sewage steroids). The authors concluded that FIB in river ebb discharge were not from sewage, but could be from birds. They also noted that dogs have similar fecal steroid compositions to birds, so the source could also include canine contributions.

20.4.1.3 Summary

Significant resources were invested in determining the source of indicator bacteria (enterococci and *E. coli*) at Huntington Beach, California, yet the main source of bacterial contamination remains unknown. It is now clear that there are human contributions to the microbial pollution at the beach. Birds and ebb flow from tidal marshes also contribute to the microbial pollution. However, the sources all appear to be nonpoint and diffuse in nature, and the "smoking gun" has not been found. Thus, it has not been possible to remediate the pollution problem at Huntington Beach. The research conducted at this shoreline has provided a wealth of knowledge about transport processes at the land–sea interface as well as the microbiology of the urban coastal ocean. The findings have taught stakeholders and researchers that contamination of urban coastal waters is complex and a difficult problem to solve.

20.4.2 Case Study #2, Hobie Cat Beach, Florida, USA

The site is located on Virginia Key, a small island on the eastern edge of Biscayne Bay that is immediately east of the coast of Miami, Florida, USA (Fig. 20.2). This site was chosen because it has periodically been placed under an advisory due to exceedance of microbial water-quality criteria for enterococci and fecal coliforms (i.e., 4.6 days per year on average from 2002 to 2009, Florida Department of Health 2009). A preliminary review of the Florida Department of Health monitoring data revealed no strong indication of the cause of the elevated indicator microbe levels. Hobie Cat Beach is located within a subtropical region characterized by an average ambient temperature of 24.8°C (27.6°C during the summer months and 22.0°C during the winter) and annual average rainfall of 149 cm (total of 109 cm during the wet season, May through September, and 39 cm during the dry season, October through April). The study site is also the only beach within Miami-Dade County where pets are allowed, including dogs. Dog owners are not required to clean up their dogs' waste, even though the beach is a designated recreational swimming area. In addition to dogs and humans, birds are often observed near the shoreline, particularly during the early morning hours. The abundance of seabirds, and gulls in particular, fluctuates seasonally with the highest populations during the dry winter months. A wastewater treatment plant (133 million gallons per day = 5.8 m^3/s) is located on the side of Virginia Key opposite Hobie Cat Beach. The outfall from

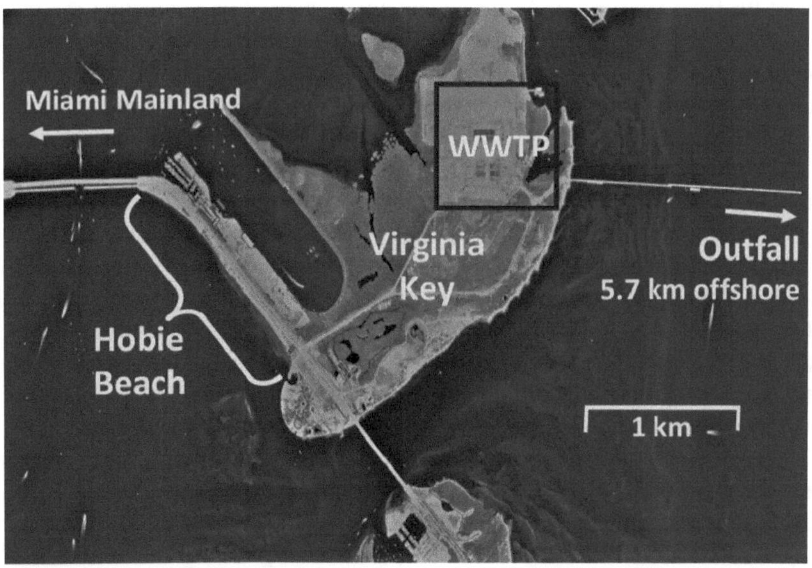

Fig. 20.2 Image of Virginia Key and Hobie Cat Beach, FL (25°44′N, 80°10′W). Background image courtesy of the US Geological Survey

this plant is located approximately 5.7 km further out in the Atlantic Ocean at 28 m depth. Prior to discharge, sewage undergoes secondary treatment (settling and activated sludge process) followed by chlorination.

Hobie Cat Beach is approximately 1.6 km long and the width of the beach is defined by a local-access paved road that runs parallel to the beach at a distance of approximately 13 m from mean high tide. One bathroom facility is located at the northwestern end of the beach. There are no constructed storm drains at the site and runoff from the paved road flows through natural drainage ditches directly to the beach. The beach is characterized by relatively poor water circulation because the bottom slope is shallow and headland features are found at both ends. Movement of water near the shoreline is dominated by tidal action. The average fluctuation in tidal height at the study site is 58 cm. This tidal height results in a 5–12 m horizontal translation of the instantaneous shoreline between high and low tide given the relatively shallow slope (approximately 0.06 m/m) within this area of the beach. The following subsections provide a review of the studies that analyzed some the potential for sources listed in Table 20.1.

20.4.2.1 Physical Source Tracking Studies

Shoreline sands as a source. Shibata et al. (2004) evaluated multiple indicator bacteria (enterococci, *E. coli*, fecal coliform, total coliform, and *C. perfringens*). The study found that the indicator levels did not covary and that measurements taken daily frequently exceeded guideline levels at Hobie Cat Beach for most indicator microbes (except for fecal coliforms). A subsequent study focused on evaluating the location of indicator bacteria inputs to Hobie Cat Beach (Bonilla et al. 2007). Water sampling showed that the concentrations of indicator bacteria were significantly different with distance from the shoreline. The highest concentrations were observed at shoreline points and decreased at offshore points. Furthermore, the highest concentrations of indicator microbe concentrations were observed at high tide, when the swash zone area of the beach was submerged. Beach sands within the swash zone tested positive for all indicator microbes, thereby suggesting that this zone may serve as the source of indicator microbes. Shibata et al. (2004) hypothesized that the dominant sources of indicator microbes to this zone were humans, animals, and possibly the survival and regrowth of indicator microbes. Bonilla et al. (2007) subsequently evaluated the beach sands at Hobie Cat Beach in greater detail and found that bacteria were consistently more concentrated in 100 g samples of nearshore beach sand (2–23 fold in wet sand and 30–460 fold in dry sand) compared to 100 mL samples of water.

These studies collectively show that remote sources, such as the wastewater treatment plant ocean outfall or discharges from canals and the Miami River, are unlikely to impact indicator microbe levels at the study site. The data have consistently shown a decreasing gradient of indicator organism concentrations from nearshore to offshore sites. Concentrations in chest deep waters were generally below detection limits, whereas shallow water near the shoreline typically contained higher concentrations of indicators (Shibata et al. 2004). Thus, the major input of indicator bacteria to this site comes from the intertidal zone (or swash zone) (Wright et al. 2011).

The septic system serving the restrooms at this beach was not implicated as a probable source, as the highest concentrations of indicators were observed during high tide. Septic system sources should be predominant during low tide when the hydraulic gradient is seaward. Moreover, the location of the elevated levels of indicator microbes did not coincide with the location of the septic tank. Since the work of Shibata et al. (2004), the septic system at the beach site has since been connected to a sanitary sewer. During several subsequent studies at this beach, no observable difference has been observed in overall beach water quality or typical indicator microbe levels since this change in wastewater infrastructure at the beach.

Enterococci sources to shoreline sands, humans, and animals. The obvious visible sources of enterococci to Hobie Cat Beach include human bathers, dogs, and birds. In order to evaluate the relative contribution of these sources, a camera was installed to obtain counts of these potential sources over time. Results indicate that the distribution of birds to the beach site is very seasonal, with the highest counts observed during the winter months and very few counts observed during the summer (Wang et al. 2010). The number of humans and dogs that visit the beach showed the opposite trend, with the highest numbers during the summer and lower numbers during the winter. In order to evaluate the significance of each of these sources with respect to enterococci contributions to the water column, the number of enterococci contributed from each of these sources were estimated: human shedding during bathing was estimated to contribute in the order of 10^5 enterococci per human 15-min bathing event (Elmir et al. 2007, 2009), birds were estimated to contribute 10^5 per fecal event on average (5×10^1–5×10^6 depending upon bird species, with seagulls contributing on average 1×10^3 CFU per event), and dogs were estimated to contribute 10^9 enterococci per fecal event (Wright et al. 2009). A comparison of the microbial loads showed that one dog fecal event was equivalent to 7,000 bird fecal events or 3,000 human bathing events. Given the abundance of animals observed on the beach, results of visually observable sources suggest that dogs are the largest contributing animal source.

Direct human and animal contributions vs. indirect contributions from shoreline sands. Zhu et al. (2011) developed a hydrodynamic and fate and transport model to understand the influence of bather shedding, animal fecal sources, and nearshore sand, as well as the impacts of the environmental conditions, on the fate and transport of enterococci at Hobie Cat Beach. The model was based on an existing finite element hydrodynamic and transport model with the addition of a first-order microbe deactivation function due to solar radiation. Results showed that dog fecal events had a major impact (hundreds of CFU/100 mL) on the enterococci concentration within a limited area, while enterococci released from beach sand during high tide caused significantly elevated concentration (tens of CFU/100 mL) along the entire beach shoreline. Bather shedding was found to impose minimal impacts (less than 1 CFU/mL), even during crowded holiday weekends. In addition, weak current velocity near the beach shoreline was found to cause longer dwelling times for the elevated concentrations of enterococci, while solar deactivation was shown to be an effective factor in reducing these microbial concentrations. Stormwater contributions and groundwater were not evaluated in this modeling exercise.

Stormwater as a source. Wright et al. (2011) collected water from natural stormwater drainage ditches during rainfall events at Hobie Cat Beach. Results showed that enterococci levels in the ditches were 12,000 CFU/100 mL on average with beach shoreline water collected at knee depth averaging 200 MPN/100 mL, indicating that runoff concentrations were two orders of magnitude higher than beach water concentrations. Similarly, Shibata et al. (2010) measured enterococci levels in the natural drainage ditches between 25,000 and 120,000 CFU/100 mL. For ankle-deep water, the enterococci concentrations were measured at 7,300 and 8,400 CFU/100 mL and, for knee-deep water, the enterococci levels were 800 and 2,100 CFU/100 mL, suggesting that stormwater runoff is a significant source of enterococci to the shoreline during and shortly after rainfall events.

20.4.2.2 Biological Source Tracking Studies

Enterococcus species distribution. Bonilla et al. (2006) evaluated enterococci species in sand and in nearshore and offshore waters. Results showed that the assemblage of enterococci in offshore waters was different from populations observed nearshore. *E. hirae* was unique to nearshore waters, and four species were unique to offshore waters (*E. raffinosus*, *E. durans*, *E. dispar*, and *E. sulfurous*). This suggests that offshore waters do not account for the enterococci populations routinely detected in nearshore waters. The species compositions in wet sand and nearshore water were notably more similar to each other, contributing to the hypothesis that washout from wet sand into the swash zone during the tidal cycle may be seeding the water column with enterococci. Dry sand was found to contain a greater percentage of *E. faecalis* than wet sand and nearshore waters.

Use of animal-specific source tracking markers. Sinigalliano et al. (2010) and Shibata et al. (2010) evaluated source specific markers, in particular human-specific *Bacteroidales* BacHum-UCD qPCR assay of Kildare et al. 2007; and HF8 qPCR assay of Sinigalliano et al. 2010, dog-specific *Bacteroidales* AOML DogBact qPCR assay (modified from Dick et al. 2005, as described in Sinigalliano et al. 2010), and seagull-specific *Catellicoccus marimammalium* AOML Gull2 qPCR assay (modified from Lu et al. 2008 as described by Sinigalliano et al. 2010). No significant correlations between the concentration of animal or human markers and enterococci levels were observed. Of interest, however, is that the levels of gull marker observed in the water column (Sinigalliano et al. 2010) appear to track the number of seagulls that visit the beach (Wang et al. 2010) on a monthly basis (Fig. 20.3). These data suggest that gulls contribute FIB, including enterococci, predominantly during the winter season with small, almost negligible contributions during the summer.

Of note, during a stormwater sampling event, elevated levels of enterococci were observed in the natural drainage ditches (at 25,000–120,000 CFU/100 mL as mentioned above); however, the levels of human-specific *Bacteroidales* (both BacHum-UCD and HF8) were below detection limits. Low levels of dog-specific

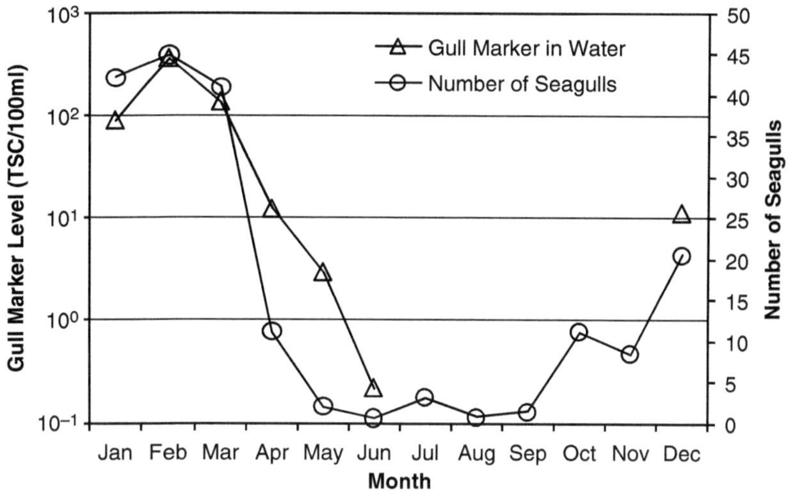

Fig. 20.3 Gull marker levels (target sequence copies per 100 mL plotted on a $\log_{10}$ scale) and number of seagulls (number plotted on an arithmetic scale) observed at Hobie Cat Beach on a monthly basis. No gull marker data were available for July through November

Bacteroidales marker were detected in the drainage ditch water, while the values observed in ankle-deep and knee-deep water were higher. The *Catellicoccus* gull marker, on the contrary, was found at very elevated levels in the drainage ditch water with decreases in ankle and knee deep water, and much lower concentrations in the water column during nonstorm conditions. These results suggest that gull droppings were the most likely source of enterococci during this storm event with negligible contributions from humans and dogs. However, these results do not rule out the possibility that the high levels of enterococci may have also been due to "old" fecal sources containing enterococci, which may persist for long periods of time (i.e., outlive *Bacteroidales*) or from possible regrowth of environmental enterococci (Ferguson et al. 2005). The strict growth requirements of *C. marimammalium* make it unlikely that this organism is persisting or regrowing in the environment; however, more research would be needed to evaluate all of these proposed hypotheses.

Concurrent measurement of pathogens. Abdelzaher et al. (2010) evaluated the presence of and association among indicator microbes, pathogens, and environmental conditions at Hobie Cat Beach. Analyses included FIB (fecal coliform, *E. coli*, enterococci, *C. perfringens*), human-associated MST markers (human polyomaviruses [HPyVs] and *E. faecium esp* gene), and pathogens (*Vibrio vulnificus*, *S. aureus*, enterovirus, norovirus, hepatitis A virus, *Cryptosporidium* spp., and *Giardia* spp.). FIB concentrations in water were below recreational water-quality standards for three of four sampling events, when pathogens and MST markers were also generally undetectable. FIB levels exceeded regulatory guidelines during one event, accompanied by detection of HPyVs and pathogens, including the autochthonous bacterium *V. vulnificus* in sand and water, the allochthonous protozoan

Giardia spp. in water, and *Cryptosporidium* spp. in sand samples. These elevated microbial levels were detected concurrently (during high tide and during low solar insolation levels), indicating a possible relationship between FIB and human-derived microbes and pathogens during specific environmental conditions. Subsequent work by Shah et al. (2011) confirmed the presence of pathogens within beach sand and found that indicator microbes in sand correlated with pathogenic bacteria, yeasts, and helminthes.

During storm events, as mentioned earlier, Shibata et al. (2010) found that FIB are released from natural drainage ditches that are observed at the beach. The large pulses of FIB return to background levels within 6 h after rain events. Abdelzaher et al. (in press) found, however, that FIB are elevated in the water column within 6 h after rainfall conditions, whereas pathogens (*Giardia*, *Cryptosporidium*, and enterovirus) were detected within 24 h after rainfall conditions. Such results suggest that both indicators and pathogens respond to rainfall events at Hobie Cat Beach, but the response of indicator bacteria is much quicker and is associated with direct runoff inputs, whereas the response of pathogens is slower. This discrepancy suggests a slower groundwater associated release of pathogens, perhaps those which are found in deeper sand layers and transported toward the beach due to rainfall infiltration and induced flow of shallow groundwater toward the shoreline.

20.4.2.3 Summary

The identification and quantification of sources of fecal pollution to Hobie Cat Beach, at which pollution is dominated by nonpoint sources, have been challenging as shown by the numerous studies conducted at this location. Initial work has suggested that dogs represent the major direct input of FIB to this beach. Detection of the impacts from this source during routine monitoring would be very intermittent (at best) given that the impacts of direct dog fecal events are very localized and are generally short-lived. Although dogs are the major observable source of fecal inputs to this site, the number of dogs visiting this beach is not sufficient to explain the consistently elevated levels of enterococci. The larger, more sustained source of indicator bacteria to the beach appears to be an indirect source associated with beach sands. Relatively large pulses of indicator bacteria are washed from the beach sands to the beach shore (via natural drainage ditches) during rainfall events, and smaller pulses are believed to be washed in through tidal action. Recent fecal sources do not appear to be the cause of these elevated levels along the shoreline (as the *Bacteroidales*-based markers of human and dog impacts were very low within runoff drainage ditches). The cause of these rainfall and tidal driven pulses may be from gull influences (as the *Catellicoccus* marker was elevated) and the prolonged persistence and possibly regrowth of FIB in the sand. Moreover, evidence exists to suggest that the sands also harbor pathogens and these pathogens (of human or animal origin) can also be released into the water column through possibly tidal action and rain induced groundwater flows toward the shoreline. The ultimate source of pathogens to the sand is not known but can include contributions from direct fecal inputs at the beach site, sorption from intermittently contaminated water, and possibly from contaminated groundwater. A recent epidemiologic study conducted at this site (Fleisher et al. 2010; Sinigalliano et al. 2010) found a relationship between human health effects (skin ailments) and enterococci levels providing support that the enterococci observed at the beach is of human health significance. More research is needed to characterize the source and transport of pathogens at this site.

20.5 Summary, Conclusions, and Recommendations

The case studies presented in this chapter come from two marine beach sites that have been extensively studied. Even with the large amount of data collected at each of these beach sites, the ultimate source(s) of microbial indicators and pathogens have still not been identified with certainty. A common observation between the two case study sites is that solar radiation and tides appear to play important roles impacting the fate (die-off) and transport of the indicator bacteria to the water column. Both studies identified high tides as critical time periods for elevated indicator levels. For Huntington Beach, extensive work was also conducted to evaluate the importance of groundwater inputs, which dominated during certain periods of the tide. For Hobie Cat Beach, groundwater is also suspected given the timing of pathogen occurrence; they seem to appear at the beach shore at a time delay (about 24 h) after rainfall events, as opposed to indicators, which appear to enter the water column within 6 h after the storm. Water circulation patterns were also considered as important factors. Microbe concentrations at both sites were affected by advection and dispersion. With respect to possible offshore point sources, both sites are relatively near wastewater ocean outfalls and possible river influences. For Hobie Cat Beach, the ocean outfall and nearby rivers and channel inputs from across the bay were eliminated as possible sources because of the distribution of bacteria with higher concentrations of indicators detected near the shoreline, while concentrations in samples collected offshore were consistently below detection limits. Further evidence was provided through enterococci speciation studies that showed that the microbial assemblages of nearshore water were similar to the microbial assemblages of sand from the intertidal zone, both of which were different than assemblages offshore. At Huntington Beach, there was evidence of potential impacts from the wastewater treatment plant outfall and from nearby river and marsh inputs. For the ocean outfall at Huntington Beach, transport of indicator bacteria may be occurring through subthermocline currents; however, evidence also suggests that such currents may not be impacting the shoreline as the addition of wastewater disinfection processes did not result in improved waste quality. A nearby river (Santa Ana River) and marsh (Talbert Marsh) were documented to have elevated indicator bacteria levels in their discharges, but these sources could not explain the elevated indicator levels at Huntington Beach in areas which were not impacted by the river or marsh. Overall, comparison of these case study sites indicates that evaluating sources at beach sites impacted by nonpoint sources is very difficult. MST tools are helpful for identifying the sources of pollution; however, one study is generally incapable of quantifying the relative contributions of different sources with certainty. More work is needed to understand the source functions, fate, and transport mechanisms of indicator bacteria at beaches impacted by nonpoint sources of pollution.

The ultimate goal of identifying the sources of indicator microbes to a beach site is to evaluate whether or not the beach is safe from a public-health perspective. Information about sources may provide knowledge on which sorts of pathogens are likely to be present and can provide insight into the potential public health risks of recreational exposure to fecal polluted water. At the present time, MST can serve at least two regulatory purposes. Insight into the sources of contamination obtained from MST can guide remediation efforts, as well as discharge permitting, within a watershed.

In the USA, one of the current shortcomings in the enforcement and application of regulatory water-quality guidelines for recreational water is that current regulations were adopted following studies conducted at beaches impacted by point sources of sanitary sewage. It is becoming increasingly apparent, however, that beaches impacted by "nonpoint" sources may require alternative means of assessing water quality and risk to public health to be effective (Boehm et al. 2009b). This regulatory strategy has been employed because (a) nonpoint sources are difficult to identify, (b) even if identified, the health risk of many pollution sources, such as domestic animal or wild animal feces, and urban stormwater, is so poorly characterized, and (c) of the paucity of epidemiologic studies conducted at nonpoint source beaches. Logic suggests that the relationships between indicators and pathogens are different for point source beaches compared to most nonpoint source beaches (WERF 2009), particularly for environmental reservoir sources of enterococci (i.e., persistent and growing populations). Humans and animals do share pathogens (US EPA 2009a, b), including certain viruses (e.g., hepatitis E, Feagins et al. 2007), bacteria (e.g., *Campylobacter*, *Salmonella*, and pathogenic *E. coli*), and protozoa (e.g., *Giardia*, *Cryptosporidium*, and *Toxoplasma*, Hunter and Thompson 2005); however, the infectivity potential of human sewage is generally estimated to be much greater than that of animal feces, as the number of diseases transmitted among humans is higher than the number of disease transmitted between humans and animals.

Any discussion regarding waterborne enteric microorganisms ultimately raises additional questions concerning disease end points. Gastrointestinal illness is the disease end point for current regulatory guidelines. This type of illness is consistent with a point source dominated by human fecal inputs. Studies show that humans can inadvertently release pathogens of fecal origin during swimming activities (Gerba 2000); however, humans also can transmit skin-related diseases during swimming (Fujioka et al. 1999; Charoenca and Fujioka 1993; Plano et al. 2011). Therefore, microbes found in point sources of sewage would likely originate from humans and would likely cause gastrointestinal disease. Conversely, microbes transmitted by nonpoint sources may originate from humans or other animals and may cause entirely different disease end points ranging from gastrointestinal illness to skin, eye, ear, and respiratory infections. Therefore, the disease end points may also be more variable at nonpoint-source beaches relative to point source beaches.

Because of the complexity of assessing human health risks at beaches influenced by nonpoint source pollution, (i.e., source, fate, and transport of indicator microbes and pathogens), as well as possible disease end points, more studies are needed at such beaches to elucidate the underlying mechanisms and principles that control indicator microbe and pathogen levels. In particular, more data are needed (via epidemiologic studies and quantitative microbial risk analysis) to understand the human health impacts at nonpoint source beaches, including the risks associated with exposure to feces of various animals. MST should be incorporated as part of human health studies and other types of water-quality studies, such as total maximum daily load assessments, at nonpoint source beaches. In this manner, the scientific and regulatory community can gain a better understanding of how microbial sources influence indicator organism and pathogen levels, and how those parameters are related to human health risk.

Acknowledgments ABB acknowledges Dr. Stanley Grant for input on the Huntington Beach case study. ABB acknowledges support from NSF/NIEHS Pacific Research Center for Marine Biomedicine (OCE 0910491). HSG, CS, and TS acknowledge funding support through the National Science Foundation (NSF) and the National Institute of Environmental Health Sciences (NIEHS) Oceans and Human Health Center at the University of Miami Rosenstiel School (NSF 0CE0432368/0911373; NIEHS 1 P50 ES12736) and NSF REU in Oceans and Human Health, and the NSF SGER (NSF SGER 0743987) in Oceans and Human Health. CS acknowledges funding support for development of the canine *Bacteroidales* qPCR assay by the Northern Gulf Institute, a NOAA Cooperative Institute (NOAA's Office of Ocean and Atmospheric Research, US Department of Commerce award NA06OAR4320264). CS also acknowledges logistics and materials support by the Source Molecular Corporation for the development of the gull *Catellicoccus* qPCR assay.

References

American Public Health Association. Standard Methods for the Examination of Water and Wastewater, 18th Edition, Washington: American Public Health Association; 1992., p. 9–66

Abdelzaher AM, Solo-Gabriele HM, Palmer CJ, and Scott TM (2009) Simultaneous concentration of enterococci and coliphage from marine waters using a dual layer filtration system. J. Environ. Qual. 38: 2468–2473

Abdelzaher AM, Solo-Gabriele HM, Wright ME, and Palmer CJ (2008) Sequential concentration of bacteria and viruses from marine waters using a dual membrane system. J. Environ. Qual. 37: 1648–1655

Abdelzaher A, Wright M, Ortega C, Solo-Gabriele H, Miller G, Elmir S, Newman X, Shih P, Bonilla JA, Bonilla TD, Palmer CJ, Scott T, Lukasik J, Harwood VJ, McQuaig S, Sinigalliano C, Gidley M, Plano L, Zhu X, Wang JD, Fleming L (2010) Presence of pathogens and indicator microbes at a non-point source subtropical recreational marine beach. Appl. Environ. Microbiol. 76(3): 724–732

Ahmed W, Powell D, Goonetilleke A, Gardner T (2008) Detection and source identification of faecal pollution in non-sewered catchment by means of host-specific molecular markers. Water Sci. Technol. 58(3): 579–586

Abdelzaher AM, Wright ME, Ortega C, Hasan AR, Shibata T, Solo-Gabriele HM, Kish J, Withum K, He G, Elmir SM, Bonilla JA, Bonilla TD, Palmer CJ, Scott TM, Lukasik J, Harwood VJ, McQuaig S, Sinigalliano CD, Gidley ML, Wanless D, Plano LRW, Garza AC, Zhu X, Stewart JR, Dickerson Jr JW, Yampara-Iquise H, Carson C, Fleisher JM, Fleming LE, (201X) Daily Measures of Microbes and Human Health at a Non-point Source Marine Beach. Journal of Water and Health. In press

Ahmed W, Goonetilleke A, Powell D, Chauhan K, Gardner T (2009) Comparison of molecular markers to detect fresh sewage in environmental waters. Water Res. 43(19): 4908–4917

Avelar KES, Moraes SR, Pinto, LJF, de Souza WDG, Domingues RMCP, de Souza Ferreira MC (1998) Influence of stress conditions on *Bacteroides fragilis* survival and protein profiles. Zentralbl. Bakteriol. 287: 399–409

Bernhard, AE, Field KG (2000a) Identification of nonpoint sources of fecal pollution in coastal waters by using hostspecific 16S ribosomal DNA genetic markers from fecal anaerobes. Appl. Environ. Microbiol. 66: 1587–1594

Bernhard AE, Field KG (2000b) A PCR assay to discriminate human and ruminant feces on the basis of host differences in *Bacteroides prevotella* genes encoding 16S rRNA. Appl. Environ. Microbiol. 66: 4571–4574

Betancourt W Q, Fujioka RS (2006) *Bacteroides spp.* as reliable marker of sewage contamination in Hawaii's environmental waters using molecular techniques. Water Sci. Technol. 54: 101–107

Boehm AB, Fuhrman JA, Mrse RD, Grant SB (2003) Tiered approach for identification of a human fecal pollution source at a recreational beach: case study at Avalon Bay, Catalina Island, California. Environ. Sci. Technol. 37: 673–680

Boehm AB, Grant SB, Kim J-H, Mowbray SL, McGee CD, Clark CD, Foley DM, Wellman DE (2002a) Decadal and shorter period variability of surf zone water quality at Huntington Beach, California. Environ. Sci. Technol. 36: 3885–3892

Boehm AB, Sanders BF, and Winant CD (2002b) Cross-shelf transport at Huntington Beach -Implications for the fate of sewage discharged through an offshore ocean outfall. Environmental Science & Technology. 36: 1899–1906

Boehm AB, Shellenbarger GG, Paytan A (2004) Groundwater Discharge: A potential association with fecal indicator bacteria in the surf zone. Environ. Sci. Technol. 38: 3558–3566

Boehm AB, Weisberg SB (2005) Tidal forcing of enterococci at marine recreational beaches at fortnightly and semidiurnal frequencies. Environ. Sci. Technol. 39: 5575–5583

Boehm AB, Keymer DP, Shellenbarger GG (2005) An analytical model of enterococci inactivation, grazing, and transport in the surf zone of a marine beach. Water Res. 39: 3565–3578

Boehm AB, Yamahara KM, Love DC, Peterson BM, McNeill K, Nelson KL (2009a) Covariation and photoinactivation of traditional and novel indicator organisms and human viruses at a sewage-impacted marine beach. Environ. Sci. Technol., 43: 8046–8052

Boehm AB, Ashbolt NJ, Colford JM Jr, Dunbar LE, Fleming LE, Gold MA, Hansel JA, Hunter PR, Ichida AM, McGee CD, Soller JA, Weisberg SB (2009b) A sea change ahead for recreational water quality criteria. J. Water Health. 7(1): 9–20

Bonilla TD, Nowosielski K, Esiobu N, McCorquodale DS, Rogerson A (2006) Species assemblages of Enterococcus indicate potential sources of fecal bacteria at a south Florida recreational beach Source. Marine Pollution Bulletin. 52(7): 807–810

Bonilla TD, Nowosielski K, Cuvelier M, Hartz A, Green M, Esiobu N, McCorquodale DS, Fleisher JM, Rogerson A (2007) Prevalence and distribution of fecal indicator organisms in South Florida beach sand and preliminary assessment of health effects associated with beach sand exposure. Marine Pollution Bulletin. 54: 1472–1482

Bonjoch X, Balleste E, Blanch AR (2004) Multiplex PCR with 16S rRNA gene-targeted primers of *Bifidobacterium spp.* to identify sources of fecal pollution. Appl. Environ. Microbiol. 70: 3171–3175

Borrego JJ, Morinigo MA, de Vicente A, Cornax R, Romero P (1987) Coliphages as an indicator of faecal pollution in water. Its relationship with indicator and pathogenic microorganisms. Water Res. 21: 1473–1480

Borrego J, Cornax R, Morinigo MA, Martinez-Manzanares E, Romero P (1990) Coliphages as an indicator of faecal pollution in water, Their survival and productive infectivity in natural aquatic environments. Water Res. 24: 111–116

Bruins MJ, Juffer P, Wolfhagen MJ, Ruijs GJ (2007). Salt tolerance of methicillin-resistant and methicillin-susceptible *Staphylococcus aureus*. J. Clin. Microbiol. 45: 682–683

Byappanahalli MN, Whitman RL, Shively DA, Ting WT, Tseng CC, Nevers MB (2006) Seasonal persistence and population characteristics of Escherichia coli and enterococci in deep backshore sand of two freshwater beaches. J. Water Health. 4: 313–20

Cabelli V, Dufour A, Levin M, Habermann P (1975) The impact of pollution on marine bathing beaches: an epidemiological study. In: Middle Atlantic Continental Shelf and the New York Bight: Proceedings of the Symposium, American Society of Limnology and Oceanography, 3–5 November 1975, Waco, TX, USA

Cabelli VJ, Dufour AP, Levin MA, McCabe LJ, Haberman PW (1979) Relationship of microbial indicators to health effects at marine bathing beaches. Am J. Public Health. 69: 690–696

Cabelli VJ, Dufour AP, McCabe L, Levin MA (1982) Swimming-associated gastroenteritis and water quality. Am. J. Epidemiol. 115: 606–616

Cao Y, Griffith JF, Weisberg SB (2009) Evaluation of optical brightener photodecay characteristics for detection of human fecal contamination. Water Res. 43(8): 2273–2279

Carillo M, Estrada E, Hazen TC (1985) Survival and enumeration of the fecal indicator Bifidobacterium adolescentis and E. coli in tropical rain forest watershed. Appl. Environ. Microbiol. 50: 468–476

Charoenca N, Fujioka R (1993) Assessment of staphylococcus bacteria in Hawaii marine recreational waters. Wat. Sci. Tech. 27(3–4): 283–289

Chen Z, Pavelic P, Dillon P, Naidu R (2002) Determination of caffeine as a tracer of sewage effluent in natural waters by on-line solid-phase extraction and liquid chromatography with diode-array detection. Water Res. 36: 4830–4838

Choi S, Chu W, Brown J, Becker S J, Harwood VJ, Jiang SC (2003) Application of enterococci antibiotic resistance patterns for contamination source identification at Huntington Beach, California. Marine Pollution Bulletin. 46: 748–755

Chung H, Sobsey MD (1993) Comparative survival of indicator viruses and enteric viruses in seawater and sediments. Water Sci. Technol. 27: 425–428

Colford JM Jr., Wade TJ, Schiff KC (2007) Water quality indicators and the risk of illness at beaches with nonpoint sources of fecal contamination. Epidemiology. 18(1): 27–35

Cornax R, Morinigo MA, Balebona MC, Castro D, Borrego J (1991) Significance of several bacteriophage groups as indicators of sewage pollution in marine waters. Water Res. 25: 673–678

De Sieyes NR, Yamahara KM, Layton BA, Joyce EH, Boehm AB (2008) Submarine discharge of nutrient-enriched fresh groundwater at Stinson Beach, California is enhanced during neap tides. Limnology and Oceanography. 53: 1434–1445

Desmarais TR, Solo-Gabriele HM, Palmer CJ (2002) Influence of soil on fecal indicator organisms in a tidally influenced subtropical environment. Appl. Environ. Microbiol. 68(3): 1165–1172

Dick LK, Bernhard AE, Brodeur TJ, Santo Domingo JW, Simpson JM, Walters SP, Field KG (2005) Host distributions of uncultivated fecal Bacteroidales bacteria reveal genetic markers for fecal source identification. Appl. Environ. Microbiol. 71: 3184–3191

Dick LK, Stelzer EA, Bertke EE, Fong DL, Stoeckel DM. 2010. Relative decay of Bacteroidales microbial source tracking markers and cultivated Escherichia coli in freshwater microcosms. Appl. Environ. Microbiol. 76(10): 3255–3262

Dickerson JW Jr, Hagedorn C, Hassall A (2007) Detection and remediation of human-origin pollution at two public beaches in Virginia using multiple source tracking methods. Water Res. 41(16): 3758–3770

Dutka BJ, Shaarawi AE, Martins MT (1987) North and south American studies on the potential of coliphage as a water quality indicator. Water Res. 21: 1127–1134

Elmir SM, Wright ME, Solo-Gabriele HM, Abdelzaher A, Fleming LE, Miller G, Rybolowik M, Shih M-T P, Pillai SP, Cooper J A, Quaye EA (2007) Quantitative evaluation of bacteria released by bathers in a marine water. Water Res. 41: 3–10

Elmir SM, Shibata T, Solo-Gabriele HM, Sinigalliano CD, Gidley ML, Miller G, Plano L, Kish J, Withum K, Fleming L (2009) Quantitative evaluation of enterococci and Bacteroidales released by adults and toddlers in marine water. Water Res. 43: 4610–4616

Facklam RR, Sahm DF, Texeira LM (1999) Enterococcus. In *Manual of clinical microbiology*. 7th edition. Edited by: Murray PR, Baron EJ, Pfaller MA, Tenover FC, Yolken RH. ASM press, Washington, DC, p. 297–305

Feagins AR, Opriessnig T, Guenette DK, Halbur PG, Meng X-J (2007) Detection and characterization of infectious Hepatitis E virus from commercial pig livers sold in local grocery stores in the USA. J. Gen. Virol. 88: 912–917

Ferguson DM, Moore DF, Getrich MA, Zhowandai MH (2005) Enumeration and speciation of enterococci found in marine and intertidal sediments and coastal water in southern California. Journal of Applied Microbiology. 99: 598–608

Field KG, Bernhard AE, Brodeur TJ (2003) Molecular approaches to microbiological monitoring: fecal source detection. Environmental Monitoring and Assessment. 81: 313–326

Fiksdal L, Maki JS, LaCroix SJ, Staley JT (1985) Survival and detection of *Bacteroides spp*., prospective indicator bacteria. Appl. Environ. Microbiol. 49: 148–150

Fleisher JM, Fleming LE, Solo-Gabriele HM, Kish JK, Sinigalliano CD, Plano LRW, Elmir SM, Wang JD, Withum K, Shibata T, Gidley ML, Abdelzaher A, He G, Ortega C, Zhu X, Wright M, Hollenbeck J, Backer LC (2010) The BEACHES study: health effects and exposures from non-point source microbial contaminants in subtropical recreational marine waters. International Journal of Epidemiology. 39(5): 1291–1298

Florida Department of Health (2009) Florida Healthy Beaches Program, Florida Dept. of Health. 6 Oct. 2008. http://esetappsdoh.doh.state.fl.us/irm00beachwater/default.aspx

Francy DS, Gifford AM, Darner RA (2003) Escherichia coli at Ohio bathing beaches—Distribution, sources, wastewater indicators, and predictive modeling, Water-Resources Investigations Report 02–4285. U.S. Geological Survey, Columbus, OH, 2003

Fujioka RS, Roll K, Morens D (1999) A Pilot Epidemiological Study of Health Risks Associated with Swimming at Kuhio Beach. Hawaii Water Resources Research Center.

Fujioka RS (2001) Monitoring coastal marine waters for spore-forming bacteria of faecal and soil origin to determine point from non-point source pollution. Water Sci Technol. 44: 181–188

Gauthier F, Archibald F (2001) The ecology of "fecal indicator" bacteria commonly found in pulp and paper mill water systems. Wat Res. 35: 2207–2218

Gerba, CP (2000) Assessment of enteric pathogen shedding by bathers during recreational activities and its impact on water quality. Quant. Microbiol. 2: 55–68

Given S, Pendleton LH, Boehm AB (2006) Regional public health cost estimates of contaminated coastal waters: a case study of gastroenteritis at Southern California beaches. Environ. Sci. Technol. 40: 4851–4858

Grant SB, Sanders BF, Boehm AB, Redman JA, Kim JH, Mrse RD, and others (2001) Generation of enterococci bacteria in a coastal saltwater marsh and its impact on surf zone water quality. Environ. Sci. Technol. 35: 2407–2416

Grant SB, Kim JH, Jones BH, Jenkins SA, Wasyl J, Cudaback C (2005) Surf zone entrainment, along-shore transport, and human health implications of pollution from tidal outlets. Journal of Geophysical Research 110: C10025, doi:10.1029/2004JC002401

Hartel PG, Hagedorn C, McDonald JL, Fisher JA, Saluta MA, Dickerson JW Jr, Gentit LC, Smith SL, Mantripragada NS, Ritter KJ, Belcher CN (2007) Exposing water samples to ultraviolet light improves fluorometry for detecting human fecal contamination. Water Res. 41(16): 3629–3642

Hartel PG, Rodgers K, Moody GL, Hemmings SN, Fisher JA, McDonald JL, Belcher, CN (2008) Combining targeted sampling and fluorometry to identify human fecal contamination in a freshwater creek. J. Water Health. 6(1): 105–116

Harwood V J, Delahoya NC, Ulrich RM, Kramer MF, Whitlock JE, Garey JR, Lim DV (2004) Molecular confirmation of *Enterococcus faecalis* and *E. faecium* from clinical, faecal and environmental sources. Lett. Appl. Microbiol. 38: 476–482

Harwood VJ, Levine AD, Scott TM, Chivukula V, Lukasik J, Farrah SR, Rose JB (2005) Validity of the indicator organism paradigm for pathogen reduction in reclaimed water and public health protection. Appl. Environ. Microbiol. 71(6): 3163–3170

Havelaar, AH, van Olhen M, Drost YC (1993) F-specific RNA coliphages are adequate model organisms for enteric viruses in fresh water. Appl. Environ. Microbiol. 59: 2956–2962

Havelaar AH, Pot-Hogeboom WM (1988) F-specific RNA-bacteriophages as model viruses in water hygiene: ecological aspects. Water Sci. Technol. 20: 399–407

Hill VR, Polaczyk A, Hahn D, Jothikumar N, Cromeans T, Roberts J, Amburgey J (2005) Development of a rapid method for simultaneous recovery of diverse microbes in drinking water by ultrafiltration with sodium polyphosphate and surfactants. Appl. Environ. Microbiol. 71: 6878–6884

Hill VR, Kahler A, Jothikumar N, Johnson T, Hahn D, Cromeans T (2007) Multi-state evaluation of an ultrafiltration-based procedure for simultaneous recovery of enteric microbes in 100-L tap water samples. Appl. Environ. Microbiol. 73: 4218–4225

Huijsdens XW, Linskens RK, Mak M, Meuwissen SG, Vandenbroucke-Grauls CM, Savelkoul PH.(2002) Quantification of bacteria adherent to gastrointestinal mucosa by real-Time PCR. J. Clin. Microbiol. 40(12): 4423–4427

Hunter PR, Thompson RC (2005) The zoonotic transmission of *Giardia* and *Cryptosporidium*. Int J Parasitol. 35(11–12): 1181–1190

Hurst CJ, Crawford RL, Knudsen GR, McInerney MJ, Stetzenbach LD (2002) Manual of Environmental Microbiology, second ed. ASM Press, Washington, DC.

Jenkins T M, Scott TM, Morgan MR, Rose JB (2005) Occurrence of alternative fecal indicators and enteric viruses in Michigan rivers. J. Great Lakes Res. 31: 22–31

Jeong, Y, Sanders B F, McLaughlin K, Grant SB (2008) Treatment of dry weather urban runoff in tidal saltwater marshes: A longitudinal study of the Talbert Marsh in Southern California. Environ. Sci. Technol. 42: 3609–3614

Kildare BJ, Leutenegger CM, McSwain BS, Bambic DG, Rajal VB, Wuertz S (2007) 16S rRNA-based assays for quantitative detection of universal, human-, cow-, and dog-specific fecal Bacteroidales: A Bayesian approach. Water Res. 41: 3701–3715

Kim, JH, Grant SB, McGee C D, Sanders BF, Largier JL (2004) Locating sources of surf zone pollution: A mass budget analysis of fecal indicator bacteria at Huntington Beach, California. Environ. Sci. Technol. 38: 2626–2636

Knee KL, Gossett R, Boehm AB, Paytan A (2010) Caffeine and agricultural pesticide concentrations in surface water and groundwater on the north shore of Kaua'i (Hawai'i, USA). Mar. Poll. Bull. 60(8): 1376–1382

Kolpin DW, Furlong ET, Meyer MT, Thurman EM, Zaugg SD, Barber LB, Buxton HT (2002) Pharmaceuticals, hormones, and other organic wastewater contaminants in U.S. streams, 1999–2000: A National Reconnaissance. Environ. Sci. Technol. 36(6): 1202–1211

Kott Y, Roze N, Sperber S, Betzer N (1974) Bacteriophages as viral pollution indicators. Water Res. 8: 165–171

Kreader CA (1995) Design and evaluation of Bacteroides DNA probes for the specific detection of human fecal pollution. Appl. Environ. Microbiol. 61: 1171–1179

Lamendella R, Santo Domingo JW, Kelty C, Oerther DB (2008) Bifidobacteria in feces and environmental waters. Appl. Environ. Microbiol. 74: 575–584

Lawson PA, Collins MD, Falsen E, Foster G (2006) *Catellicoccus marimammalium* gen. nov., sp. nov., a novel Gram-positive, catalase-negative, coccus-shaped bacterium from porpoise and grey seal. Int. J. Syst. Evol. Microbiol. 56(2): 429–432.

Layton BA, Walters SP, Lam L, Boehm AB (2010) Enterococcus species distribution among human and animal hosts using multiplex PCR. Journal of Applied Microbiology. 109(2): 539–547

Leclerc H, Devriese LA, Mossel DAA (1996) Taxonomical changes in intestinal (faecal) enterococci and streptococci: consequences on their use as indicators of faecal contamination in drinking water. J. Appl. Bacteriol. 81: 459–466

Leeming R, Ball A, Ashbolt N, Nichols P (1996) Using faecal sterols from humans and animals to distinguish faecal pollution in receiving waters. Water Res. 30(12): 2893–2900

Leskinen SD, Brownell M, Lim DV, Harwood VJ (2010) Hollow-fiber ultrafiltration and PCR detection of human-associated genetic markers from various types of surface water in Florida. Appl. Environ. Microbiol. 76(12): 4116–4117

Long SC, El-Khoury SS, Oudejans SJG, Sobsey MD, Vinjé J (2005) Assessment of sources and diversity of male-specific coliphages for source tracking. Environmental Engineering Science 22(3): 367–377 doi:10.1089/ees.2005.22.367

Lu J, Santo Domingo JW, Lamendella R, Edge T, Hill S (2008) Phylogenetic diversity and molecular detection of bacteria in gull feces. Appl. Environ. Microbiol., 74(13): 3969–3976

Maier RM, Pepper IL, Gerba CP (2008) Environmental Microbiology. Second Edition. Academic Press, New York

Matsuki T, Watanabe K, Fujimoto J, Miyamoto Y, Takada T, Matsumoto K, Oyaizu H, Tanaka R (2002) Development of 16S rRNA-gene-targeted group-specific primers for the detection and identification of predominant bacteria in human feces. Appl. Environ. Microbiol. 68: 5445–5451

Matsuki T, Watanabe K, Fujimoto J, Kado Y, Takada T, Matsumoto K, Tanaka R (2004) Quantitative PCR with 16S rRNA-gene-targeted species-specific primers for analysis of human intestinal bifidobacteria. Appl. Environ. Microbiol. 70: 167–173

Mika KB, Imamura G, Chang C, Conway V, Fernandez G, Griffith JF, Kampalath RA, Lee CM, Lin CC, Moreno R, Thompson S, Whitman RL and Jay JA (2009) Pilot- and bench-scale testing of faecal indicator bacteria survival in marine beach sand near point sources. Journal of Applied Microbiology. 107: 72–84

Nebra Y, Bonjoch X, Blanch AR (2003) Use of Bifidobacterium dentium as an indicator of the origin of fecal water pollution. Appl. Environ. Microbiol. 69: 2651–2656

Nevers MB, Whitman RL (2005) Nowcasting modeling of Escherichia coli concentrations at multiple urban beaches of southern Lake Michigan. Water Res. 39: 5250–5650

Nevers MB, Whitman RL, Fric WA, Ge Z (2007) Interaction and influence of two creeks on E. coli concentrations of nearby beaches: Exploration of predictability and mechanisms. J. Environ. Qual. 36, 1338–1345

Nevers MB, Whitman RL (2008) Coastal strategies to predict Escherichia coli concentrations for beaches along a 35 km stretch of southern Lake Michigan. Environ. Sci. Technol. 42(12): 4454–4460

Nevers MN, Boehm AB (2010) Modeling Fate and Transport of Fecal Bacteria in Surface Water. In M. Sadowsky and R. Whitman (eds.) The Fecal Indicator Bacteria, ASM Press, Herndon, VA

Noble R T, Griffith JF, Blackwood AD, Fuhrman JA, Gregory JB, Hernandez X, Liang X, Bera AA, Schiff K (2006) Multitiered approach using quantitative PCR to track sources of fecal pollution affecting Santa Monica Bay, California. Appl. Environ. Microbiol. 72: 1604–1612

Noblet JA, Young DL, Zeng EY, and Ensari S (2004) Use of fecal steroids to infer the sources of fecal indicator bacteria in the lower Santa Ana River Watershed, California: Sewage is an unlikely source. Environ. Sci. Technol. 38: 6002–6008

Olyphant, GA, Whitman, RL (2004) Element of a predictive models for determining beach closures in a real-time basis: The case of 63rd street beach Chicago. Environmental Monitoring and Assessment. 98: 175–190

Paul JH, Jiang SC, Rose JB (1991) Concentration of viruses and dissolved DNA from aquatic environments by vortex flow filtration. Appl. Environ. Microbiol. 57: 2197–2204

Paul JH, Rose JB, Jiang SC, London P, Xhou X, Kellogg C (1996) Coliphage and indigenous phage in Mamala bay, Oahu, Hawaii Appl. Environ. Microbiol. 63: 133–138

Payment P, Franco E (1993) Clostridium perfringens and somatic coliphages as indicators of the efficiency of drinking water treatment for viruses and protozoan cysts. Appl. Environ. Microbiol. 59: 2418–2424

Peeler KA, Opsahl SP, Chanton JP (2006) Tracking anthropogenic inputs using caffeine, indicator bacteria, and nutrients in rural freshwater and urban marine system. Environ. Sci. Technol. 40: 7616–7622

Plano LRW, Garza AC, Shibata T, Elmir SM, Kish J, Sinigalliano CD, Gidley ML, Miller G, Withum K, Fleming LE, and Solo-Gabriele HM (2011) Shedding of Staphylococcus aureus and methicillin-resistant Staphylococcus aureus from adult and pediatric bathers in marine waters. BMC Microbiology. 11(5): 1–10

Plummer JD, Long SC (2007) Monitoring source water for microbial contamination: evaluation of water quality measures. Water Res. 41: 3716–3728

Pourcher AM, Devriese LA, Hernandez JF, Delattre JM (1991) Enumeration by a miniaturized method of Escherichia coli, Streptococcus bovis, and enterococci as indicators of the origin of faecal pollution of waters. J. Appl. Bacteriol. 70(6): 525–530

Public Law, 2000. Beaches Environmental Assessment and Coastal Health Act of 2000. Public Law 106-284-Oct. 10, 2000

Rhodes MW, Kator H (1999) Sorbitol-fermenting bifidobacteria as indicators of diffuse human fecal pollution in estuarine watersheds. J. Appl. Microbiol. 87: 528–535

Rozen Y, Belkin S (2001) Survival of enteric bacteria in seawater. FEMS Microbiology Reviews. 25: 513–529

Santoro AE, Boehm AB (2007) Frequent occurrence of the human-specific Bacteroides fecal marker at an open coast marine beach: Relationship to waves, tides, and traditional indicators. Environmental Microbiology. 9: 2038–2049

Scott TM, Rose JB, Jenkins TM, Farrah SR, Lukasik J (2002) Microbial source tracking: current methodology and future directions. Appl. Environ. Microbiol. 68: 5796–5803

Scott TM, Jenkins TM, Lukasik J, Rose JB (2005) Potential use of a host associated molecular marker in Enterococcus faecium as an index of human pollution. Environ Sci. Technol. 39 (1): 283–287

Seurinck S, Defoirdt T, Verstraete W, Siciliano SD (2005) Detection and quantification of the human-specific HF183 Bacteroides 16S rRNA genetic marker with real-time PCR for assessment of human faecal pollution in freshwater. Environ Microbiol. 7: 249–259

Shah AH, Abdelzaher AM, Phillips M, Hernandez R, Solo-Gabriele HM, Kish J, Scorzetti G, Fell JW, Diaz MR, Scott TM, Lukasik J, Harwood VJ, McQuaig S, Sinigalliano CD, Gidley ML, Wanless D, Agar A, Lui J, Stewart JR, Plano LRW, Fleming LE (2011) Indicator microbes correlate with pathogenic bacteria, yeasts, and helminthes in sand at a subtropical recreational beach site. Journal of Applied Microbiology. 110: 1571–1583.

Shanks OC, White K, Kelty CA, Hayes S, Sivaganesan M, Jenkins M, Varma M, Haugland RA (2010) Performance assessment PCR-based assays targeting bacteroidales genetic markers of bovine fecal pollution. Appl. Environ. Microbiol. 76(5): 1359–1366

Shanks OC, Kelty CA, Sivaganesan M, Varma M, Haugland RA (2009) Quantitative PCR for genetic markers of human fecal pollution. Appl. Environ. Microbiol.75(17): 5507–5513

Shibata T, Solo-Gabriele HM, Fleming L, Elmir S (2004) Monitoring marine recreational water quality using multiple microbial indicators in an urban tropical environment. Water Res. 38: 3119–3131

Shibata T, Solo-Gabriele HM, Sinigalliano CD, Gidley ML, Plano LRW, Fleisher JM, Wang JD, Elmir SM, He G, Wright ME, Abdelzaher AM, Ortega C, Wanless D, Garza AC, Kish J, Scott T, Hollenbeck J, Backer LC, Fleming LE (2010) Evaluation of conventional and alternative monitoring methods for a recreational marine beach with non-point source of fecal contamination. Environmental Science & Technology. 44: 8175–8181

Sinigalliano CD, Fleisher JM, Gidley ML, Solo-Gabriele HM, Shibata TM, Plano LRW, Elmir SM, Wanless D, Bartkowiak J, Boiteau R, Withum K, Abdelzaher A, He G, Ortega C, Zhu X, Wright M, Kish J, Hollenbeck J, Backer LC, Fleming LE (2010) Traditional and molecular analyses for fecal indicator bacteria in non-point source subtropical recreational marine waters. Water Res. 44(13): 3763–3772

Solo-Gabriele H, Wolfert M, Desmarais T, Palmer C (2000) Sources of *E. coli* to a sub-tropical coastal environment. Appl. Environ. Microbiol. 66(1): 230–237

Sorensen DL, Eberl SG, Dicksa RA (1989) *Clostridium perfringens* as a point source indicator in non-point polluted streams. Water Res. 23: 191–197

Stapleton C M, Kay D, Wyer MD, Davies C, Watkins J, Kay C, McDonald AT, Porter J, Gawler A (2009) Evaluating the operational utility of a Bacteroidales quantitative PCR-based MST approach in determining the source of faecal indicator organisms at a UK bathing water. Water Res. 43(19): 4888–4899

Tang-Liu DD, Williams RL, Riegelman S (1983) Disposition of caffeine and its metabolites in man. J. Pharmacol. Exp. Ther. 224: 180–185

Tyrrell GJ, Turnbull L, Teixeira LM, Lefebvre J, Carvalho Mda G, Facklam RR, Lovgren M (2002) *Enterococcus gilvus* sp. nov. and *Enterococcus pallens* sp. nov. isolated from human clinical specimens. J. Clin. Microbiol. 40: 1140–1145

U.S. Environmental Protection Agency (1976) Quality Criteria for Water. Office of Water and Hazardous Materials, Washington D.C. PB-263 943

U.S. Environmental Protection Agency (1983) Health Effects Criteria for Marine Recreational Waters. Health Effects Research Laboratory, Research Triangle Park, NC. EPA-600/1-80-031

U.S. Environmental Protection Agency (1986) Ambient Water Quality Criteria for Bacteria – 1986. Office of Water Regulations and Standards, Washington DC. EPA 440/5-84-002

U.S. Environmental Protection Agency (1989) Drinking Water; National Primary Drinking Water Regulations; Total Coliforms (Inclusing Fecal Coliforms and E. coli); Final Rule. 40 CFR Parts 141 and 142, Federal Register (June 29, 1989), Vol. 54, No. 124: 27544–27568

U.S. Environmental Protection Agency (2009a) Review of Zoonotic Pathogens in Ambient Waters. US Environmental Protection Agency, Office of Water, Washington DC. EPA 822-R-09-002

U.S. Environmental Protection Agency (2009b) Review of Published Studies to Characterize Relative Risks from Different Sources of Fecal Contamination in Recreational Water. US Environmental Protection Agency, Office of Water, Washington DC. EPA 822-R-09-001

Wade TJ, Pai N, Eisenberg JN, Colford JM Jr. (2003) Do U.S. Environmental Protection Agency water quality guidelines for recreational waters prevent gastrointestinal illness? A systematic review and meta-analysis. Environmental Health Perspectives. 111(8): 1102–119

Wade TJ, Calderon RL, Sams E, Beach M, Brenner KP, Williams AH, Dufour AP (2006) Rapidly measured indicators of recreational water quality are predictive of swimming-associated gastrointestinal illness. Environ Health Perspect. 114: 24–28

Wang JD, Solo-Gabriele HM, Abdelzaher AM, Fleming LE (2010) Estimation of enterococci input from bathers and animals on a recreational beach using camera images. Marine Pollution Bulletin. 60(8): 1270–1278

Walters SP, Yamahara KM, Boehm AB. 2009. Persistence of nucleic acid markers of health-relevant organisms in seawater microcosms: Implications for their use in assessing risk in recreational waters. Water Res. 43: 4929–4939

Weigel S, Kallenborn R, Hühnerfuss H (2004) Simultaneous solid-phase extraction of acidic, neutral and basic pharmaceuticals from aqueous samples at ambient (neutral) pH and their determination by gas chromatography-mass spectrometry. J. Chromatogr. A. 1023: 183–195

Water Environment Research Foundation (2009) Report on the Expert Scientific Workshop on Critical Research and Science Needs for the Development of Recreational Water Quality Criteria for Inland Waters, Water Environment Research Foundation, Alexandria, VA

Wheeler A, Hartel PG, Godfrey DG, Hill JL, Segars WI (2002) Potential of Enterococcus faecalis as a human fecal indicator for microbial source tracking. J. Environ. Qual. 31(4): 1286–1293

Whitman RL, Nevers MB (2003) Foreshore sand as a source of Escherichia coli in nearshore water of a Lake Michigan beach. Appl. Environ. Microbiol. 2003 69:5555–5562

Whitman RL, Nevers MB, Korinek GC, Byappanahalli MN (2004) Solar and temporal effects on Escherichia coli concentration at a Lake Michigan swimming beach. Appl. Environ. Microbiol. 70: 4276–4285

Wright ME, Solo-Gabriele HM, Abdelzaher AM, Elmir S, Fleming LE (2011) The inter-tidal zone is the geographic location of elevated concentrations of enterococci. Water Science & Technology. 63(3): 542–549

Wright ME, Solo-Gabriele HM, Elmir S, Fleming LE (2009) Microbial load from animal feces at a recreational beach. Marine Pollution Bulletin. 58: 1649–1656

Yamahara KM, Walters SP, and Boehm AB (2009) Growth of enterococci in unaltered, unseeded beach sands subjected to tidal wetting. Appl. Environ. Microbiol. 75: 1517–1524

Zhu X, Wang JD, Solo-Gabriele HM, Fleming LE, Elmir S (2011) A microbial water quality model for recreational marine beaches, Water Research. 45: 2985–2995

Chapter 21
Source Tracking in Australia and New Zealand: Case Studies

Warish Ahmed, Marek Kirs, and Brent Gilpin

Keywords Fecal source tracking • Sewage pollution • *E. coli* • Enterococci • Genetic markers

This chapter outlines the application of fecal source tracking (FST) methods in waterways in Australia and New Zealand. FST methods used in the case studies include biochemical fingerprinting (BF), antibiotic resistance analysis (ARA), anaerobic bacterial genetic markers, toxin genetic markers, viral markers, fluorescent whitening agents (FWAs), and fecal sterols. These methods were predominantly used to identify human fecal pollution in receiving waters sourced from defective septic systems or discharges from sewage treatment plants (STPs). In some cases, these methods were also used to identify the sources of elevated levels of fecal indicator bacteria in catchment waters. The earlier case studies employed library-dependent FST methods, whereas the recent studies focused on validation and application of library-independent methods. Several case studies reported the presence of human fecal pollution in environmental waters and suggested that genetic markers are appealing because of their high specificity and sensitivity to differentiate and detect human and animal fecal pollution. Few case studies also used a combination of methods and suggested that such an approach can compensate uncertainty when one marker fails to produce satisfactory results. However little is known regarding the persistence of these markers in relation to fecal indicators and pathogens. More research is required regarding the behaviors of these markers in the environments.

W. Ahmed (✉)
CSIRO Land and Water, Ecosciences Precinct,
41 Boggo Road, Brisbane 4102, Australia
e-mail: warish.ahmed@csiro.au

C. Hagedorn et al. (eds.), *Microbial Source Tracking: Methods, Applications, and Case Studies*, DOI 10.1007/978-1-4419-9386-1_21,

21.1 Introduction

Pollution from animal and human waste is one of the major concerns about water bodies that are used for drinking water supply, recreational activities, and harvesting seafood worldwide due to possible exposure of water users to a wide array of pathogenic bacteria, protozoa, and viruses (Fong and Lipp 2005; Hörman et al. 2004). Unsafe drinking water, inadequate sanitation, or insufficient hygiene is to blame for 1.5 million deaths each year, most being deaths of children in the developing world (Prüss-Üstün et al. 2008). While the developing world is hit hardest by the waterborne illnesses, waterborne disease outbreaks also occur in developed countries (Hrudey and Hrudey 2004). Various sources such as agricultural runoff, wild animals, combined sewer overflows, STPs, defective septic systems, and industrial outlets are known to be potential sources of such pollution (Ahmed et al. 2005a, b; McLellan 2004; Parveen et al. 1997). Microbiological quality of water is generally assessed by enumerating indicator bacteria such as *Escherichia coli* and enterococci (USEPA 2000). Indicator bacteria such as these are commonly found in the feces of warm-blooded animals including humans. The presence of these bacteria in water bodies generally indicates pollution and the potential public health risks. However, the presence of fecal indicators does not provide information regarding their sources. The identification of the major polluting source(s) is vitally important to implement appropriate mitigation strategies to minimize pollution and subsequent public health risks (Scott et al. 2002). However, the identification and assigning of indicator bacteria to human and animal sources in environmental waters is difficult due to their cosmopolitan nature, i.e., they are shed in the waste of a wide variety of hosts (Field and Samadpour 2007). In addition, environmental waters are subjected to multiple sources of pollution, making it extremely difficult to implement management plans without knowing the sources. Economically feasible and simple methodologies to estimate water-related health risk and identify pollution sources are needed globally.

Over the last decade, researchers have developed a range of FST tools that can be used to distinguish human fecal pollution from animal. These methods are broadly categorized into library-dependent (i.e., phenotypic and genotypic), library-independent (i.e., molecular markers), and chemical (i.e., sterols, FWAs). The application of some of these methods has been evaluated rigorously. However, other methods require more validation before being used to identify the sources of fecal pollution in environmental waters. This chapter reviews the application of source tracking tools in practical situations in Australia and New Zealand.

21.2 Source Issues in New Zealand

New Zealand is well known for its vast and, seemingly, untouched landscapes. Indeed, the population of New Zealand is relatively low (four million) compared to its size (250,000 km^2). Placed deep in the Pacific, over 2,000 km from the nearest continent, the native fauna of New Zealand is characterized by low terrestrial mammal diversity with only three species of bats considered as true native mammals. While the isolation has limited the mammalian diversity, a number of species have been introduced by both Polynesian and European settlers and are now major components of New Zealand wildlife. These include rabbits, rodents, and possums, which currently outnumber humans fifteen to one (King 2005). Furthermore, farming is a significant contributor to the New Zealand economy (agriculture contributing roughly 5% of the national gross domestic product GDP and most of export), and the livestock numbers are high with sheep (34.1 million), poultry (20 million), and beef and dairy cattle (9.7 million) being the dominant species (Statistics New Zealand 2009). It is estimated (Kirs' unpublished data) that approximately 2.2×10^5 and 4.3×10^4 T of wet weight matter per day is released by cattle and sheep, respectively, in New Zealand, which exceeds the contribution from the human population by 350 fold. Although this untreated waste is dispersed over a large area, livestock usually roam free and often have open access to waterways. Therefore, in the New Zealand context, livestock can be a major contributor to microbial pollution of streams, lakes, and coastal regions and may result in significant human health risks. Municipal sewage treatment facilities have progressively been upgraded in most cities, with high levels of treatment achieved in most areas. There has, however, been an increase in septic systems on the edges of urban centers as an increasing number of people move to "lifestyle" blocks or accept longer commuting distances in exchange for living in less urbanized environments.

21.3 Fecal Source Issues in Australia

Microbial pollution in Australian waters is caused by a combination of point sources (PS) and nonpoint sources (NPS) of pollution. Pollution from PS includes discharge from industries and STPs and defective septic systems. Septic systems are the only waste disposal option in nonsewered catchments throughout Australia. High numbers of septic systems are reported to be failing, are not monitored (Ahmed et al. 2005a), and have the potential to discharge microbial pollutants in surface and ground waters. NPS pollution also poses a significant threat to the catchment and recreational water quality. This is because NPS sources of pollution are comprised of diverse sources of microbial pollutants from undefined sites within a catchment. These include stormwater runoff, urban runoff from agricultural fields, livestock runoff, and defecation of wild animals and pets. With increasing urbanization of the nation's coastal areas, pollution of waterways by NPS is a growing concern.

21.3.1 Fecal Source Tracking Tools

The number of tools available to identify the possible sources of fecal pollution in environmental waters is rapidly increasing. A brief summary of these tools is given below.

21.3.2 Library-Dependent Source Tracking Tools

The majority of the early-developed MST methods are library-dependent, which requires the development of a database or "library" of *E. coli* or enterococci from suspected sources using various genotypic and phenotypic methods. The underlying assumption of these methods is that host specificity of microorganisms is influenced by selective pressure (Wiggins 1996). Phenotypic and genotypic patterns of target strains are then compared to the library to identify their likely sources (Harwood et al. 2000; Wiggins et al. 1996). The commonly used phenotypic library-dependent methods are ARA (Wiggins et al. 1999), carbon source utilization (CSU) (Hagedorn et al. 2003), and BF (Ahmed et al. 2005a, b). The most commonly used genotypic library-dependent methods include pulsed field gel electrophoresis (Moyda et al. 2003), repetitive extragenic palindromic PCR (rep-PCR) (Dombek et al. 2000), and ribotyping (Parveen et al. 1999; Scott et al. 2003). The case studies presented in this chapter include examples of the use of BF and ARA.

21.3.2.1 Biochemical Fingerprinting

This method uses quantitative measurements of the kinetics of several biochemical reactions of bacteria in microtiter plates with dehydrated substrates (Katouli et al. 1997; Möllby et al. 1993). The typing reagents used in this method are specifically chosen for different groups of bacteria to give an optimal discriminatory power and reproducibility (Möllby et al. 1993). For each bacterial isolate, it yielded a biochemical fingerprint made of several quantitative data which were used with the PhPlate software version 4001 (PhPlate system, PhPlate AB, Stockholm) to calculate the level of similarity between the tested isolates. Similarities between the isolates were calculated as correlation coefficients and clustered according to the unweighted-pair group method with arithmetic averages (UPGMA) (Sneath and Sokal 1973). All data handling, including optical readings, calculations of correlation coefficients, diversity indexes, and Sp values, as well as clustering and printing dendrograms, was performed using the PhPlate software.

21.3.2.2 Antibiotic Resistance Analysis

Antibiotics are used to prevent and treat infections in humans and domestic animals, as well as to promote growth in animals. Microorganisms develop resistance to antibiotics to which they are regularly exposed, and intrinsic resistance to certain antibiotics is found in many bacteria. The ARA method is based on the hypothesis that bacteria present in the intestine of different animals, subjected to different types and concentrations of antibiotics, would result in host-specific resistance profiles (Wiggins 1996). ARA fingerprints of unknown environmental isolates can be compared to a reference library of several known host groups. There is currently no standard panel of antibiotics and concentrations used for this method. Antibiotics are basically selected on the basis of their uses in different host groups. This method has shown to be successful in discriminating *E. coli* and enterococci isolated from animal species (Parveen et al. 1997; Wiggins et al. 1996, 1999).

21.3.3 Library-Independent Source Tracking Tools

There are many microorganisms other than fecal coliforms, *E. coli* and enterococci present in feces, which have greater specificity to human and animal hosts. Difficulties in culturing and identifying some of these organisms have limited their useful application to source identification. An alternative approach is to extract total DNA from a water sample and examine the sample using PCR for DNA from source-specific organisms. These methods do not require the development of a library and, therefore, are known as library-independent methods. Library-independent methods could be categorized into three groups:

- Anaerobic bacterial genetic markers: Some members of the *Bacteroides* and *Bifidobacterium* genus appear to be host-associated (Allsop and Stickler 1985; Kreader 1995). These anaerobes constitute a larger portion of gut microflora than do coliforms and enterococci (Sghir et al. 2000) and have limited reproduction capacity in the environment. A research study reported the identification of human- and bovine-specific *Bacteroides-Prevotella* 16S rRNA gene markers using length heterogeneity PCR and terminal restriction fragment length polymorphism (T-RFLP) and concluded that these markers could be used to detect human or bovine origin pollution (Bernhard and Field 2000). Other researchers have also exploited this group of bacteria, developing PCR assays targeted at *Bacteroides* markers specific to human (Layton et al. 2006; Okabe et al. 2007; Reischer et al. 2007), dog (Kildare et al. 2007), ruminant (Reischer et al. 2006), pig, and horse (Dick et al. 2005) for the identification of pollution. Recently, quantitative PCR (qPCR) methods have been developed to quantify human and ruminant specific *Bacteroides* markers in sewage and environmental waters (Layton et al. 2006; Reischer et al. 2007; Seurinck et al. 2005). Other species can also be targeted. For example Devane et al. (2007) isolated from duck feces a novel bacteria (most closely related to members of the *Desulfovibrio* genus) and then designed PCR primers targeting the 16S rRNA of this organism, which could be used as wildfowl-specific marker.
- Viral markers: Viral genetic markers have been used to identify the sources of pollution in environmental waters. The most commonly used viral markers are human adenoviruses (Choi and Jiang 2005; Fong et al. 2005; Noble et al. 2003), human polyomaviruses (McQuaig et al. 2006, 2009), F+ specific RNA phages (Love and Sobsey 2007). Human-, bovine- and porcine-specific adenoviruses and polyomaviruses have been used to identify the sources of pollution in USA and Spain (Choi and Jiang 2005; Fong and Lipp 2005; Hundesa et al. 2006;

Table 21.1 General and host-specific PCR markers used in the case studies in Australia and New Zealand

Assays	Host-marker designation	References
Bacteroides general	General-Bac32	Bernhard and Field (2000)
Bacteroides human	HF183	Bernhard and Field (2000)
Bacteroides ruminant	CF128	Bernhard and Field (2000)
Bacteroides dog	Dog-BacCan	Kildare et al. (2007)
E. faecium human	Human-*esp*	Scott et al. (2005)
JCV–BKV human PCR	Human polyomavirus-JCV and BKV	McQuaig et al. (2006)
M. smithii human PCR	Human-*nifH*	Ufnar et al. (2006)
B. adolesscentis human	*B. adolesscentis*	Matsuki et al. (2004)

Maluquer de Motes et al. 2004; McQuaig et al. 2006; Pina et al. 1998). qPCR method has also been developed for the quantitative detection of these viruses in environmental waters.

- Bacterial toxin genetic markers: PCR assays have been developed to detect host-specific toxin genes in indicator bacteria such as *E. coli* and enterococci. These include the pig-specific ST1b (Khatib et al. 2003) and cattle-specific LTIIa*E. coli* toxin gene (Chern et al. 2004; Khatib et al. 2002), and the enterococcal surface protein (*esp*) gene found in *Enterococcus faecium* (Scott et al. 2005). A recent study reported the development of a qPCR assay for the quantitative detection of *esp* markers in sewage and environmental waters (Ahmed et al. 2008c).

A selection of these methods has been used in some of the case studies described in this chapter. These are detailed in Table 21.1

21.3.4 Chemical Source Tracking Methods

FWAs (also called optical brighteners) are man-made fluorescent organic compounds that absorb ultraviolet light and reemit most of the absorbed energy as blue light. To improve whiteness, FWAs are used in the manufacture of paper and clothing. As FWAs are lost from clothing during wear and washing, most laundry powders contain 0.10–0.15% (w/w) FWA to maintain whiteness. It is estimated that between 20 and 95% of the FWAs bind to the fabric during washing with the remainder being discharged with the washing liquor (Poiger et al. 1998, Stoll and Giger 1998). Most household plumbing mixes effluent from toilets with "grey water" from washing machines. Consequently, FWAs are usually associated with human fecal pollution in both septic tanks and community wastewater systems. FWAs absorb light at 350 nm (the excitation wavelength) and reemit the light as fluorescence at a higher wavelength (emission wavelength) in the range 430 nm (Poiger et al. 1993). The detection of FWAs, therefore, indicates the potential presence of human fecal pollution from a sewage system. In general, levels of FWA greater than 0.2 ppb typically indicate recent or local source of human pollution, while lower levels indicate increasingly dilute or distant sources of human pollution. Levels below 0.1 ppb typically are not associated with local source of human pollution (Devane et al. 2006).

Fecal sterols are a group of C27-, C28-, and C29-cholestane-based sterols found mainly in animal feces. The sterol profile of feces depends on the interaction of three factors. First, the animal's diet determines the relative quantities of sterol precursors (cholesterol, 24-ethylcholesterol, 24-methylcholesterol, and/or stigmasterol) entering the digestive system. Second, animals differ in their endogenous biosynthesis of sterols (for example, human beings on a low cholesterol diet synthesize cholesterol). Third and perhaps most importantly, anaerobic bacteria in the animal gut biohydrogenate sterols to stanols of various isomeric configurations (Devane et al. 2006).

The sterol, cholesterol, can be hydrogenated to one or more of four possible stanols. In humans, cholesterol is preferentially reduced to coprostanol where it constitutes 60% of the total sterols found in human feces (Leeming et al. 1996). By contrast, cholesterol is predominantly reduced to cholestanol in the environment. Similarly, plant-derived 24-ethylcholesterol is reduced to 24-ethylcoprostanol and 24-ethylepicoprostanol in the gut of herbivores, whereas in the environment it is primarily reduced to 24-ethylcholestanol. As a consequence, analysis of the sterol composition of feces can generate a sterol fingerprint, which can be distinctive from one species to another – particularly in the case of differentiating human from animal pollution. In the case studies described in this chapter, ten different ratios are presented (Table 21.2). Ratio 1 (coprostanol/cholestanol) and ratio 2 (24-ethylcoprostanol/24-ethylcholestanol) are typically >0.5 in human and ruminant fecal material (Leeming et al. 1996). These ratios in wildfowl and canine feces may not exceed these thresholds. Ratios 3–6 are all indicative of human feces when thresholds are exceeded and are based around elevated relative levels of coprostanol (Grimault et al. 1990, Reeves and Patton 2001). Ratios 5–8 are indicative of herbivore feces (Leeming et al. 1998, Gilpin et al. 2002). High levels of ratio 8 (24-ethylcholesterol/24-ethylcoprostanol) suggest either plant decay or a diet of

Table 21.2 Fecal sterol ratios and data interpretation

Sterol ratio		Interpretation
Ratio 1	Coprostanol/cholestanol	>0.5 Fecal
Ratio 2	24-Ethylcoprostanol/24-ethylcholestanol	>0.5 Fecal
Ratio 3	Percent coprostanol	>5–6% Human fecal pollution
Ratio 4	Coprostanol/(coprostanol+cholestanol)	>0.7 Human fecal pollution
Ratio 5	Coprostanol/24-ethylcoprostanol	<1.0 Herbivore; ≥1.0 human
Ratio 6	Coprostanol/coprostanol+24-ethylcoprostanol	<30% Herbivore; >75% human
Ratio 7	Percent 24-ethylcoprostanol	>5–6% Herbivore
Ratio 8	24-Ethylcholesterol/24-ethylcoprostanol	<1.0 Herbivore; >4.0 plant decay
Ratio 9	24-Ethylcholestanol/(24-ethylcholestanol+24-ethylcoprostanol+24-ethylepicoprostanol)	>30% Avian
Ratio 10	Cholestanol/(cholestanol+coprostanol+epicoprostanol)	>67% Avian

plant material such as that seen in wildfowl. Ratios 9 and 10 when both exceeding thresholds suggest wildfowl source of sterols (B. Gilpin, unpublished).

Fecal sterols analysis in each of the case studies in this chapter was performed by filtering up to 4 L of river water onto glass fiber filters. Filters were stored frozen until they were analyzed. Solvent extraction was performed prior to hydrolysis, which was followed by back-extraction into hexane. The sterol fraction is eluted into methanol and silylated prior to analysis by GC-MS (Gregor et al. 2002). Each sterol and stanol detected is expressed in parts per trillion (ppt).

21.4 Case Studies

21.4.1 Library-Dependent Methods

21.4.1.1 Case Study 1: Biochemical Fingerprinting Method Identifies Human and Animal Pollution in Eudlo Creek, Southeast Queensland, Australia

Situation: Urban creek water samples were collected from five sites on Eudlo Creek, Southeast Queensland, Australia. The primary aim was to identify human sewage pollution in the creek, which may have entered via defective septic systems. A secondary aim was to identify domestic and wild animal pollution (Ahmed et al. 2005b).

Tools used: BF libraries comprising of 4,057 enterococci and 3,728 *E. coli* isolates from horses, cattle, sheep, pigs, ducks, chickens, deer, kangaroos, dogs and septic tanks were used to identify the sources of unknown environmental *E. coli* and enterococci using cluster analysis.

Results: A total of 248 enterococci biochemical phenotypes (BPTs) were obtained from creek water samples, of which 26 BPTs (10%) were identical to BPTs from septic tanks and 152 BPTs (61%) were identical to various animals (Table 21.3). Of the 282 *E. coli* BPTs from the same water samples, 36 BPTs (13%) were identical to

Table 21.3 Comparison of biochemical phenotypes (BPTs) from environmental water samples from sampling sites (EC1–EC5) on Eudlo Creek, Queensland, Australia with enterococci (Ent) and *E. coli* libraries

	No. of BPTs found		Septic UQ-BPTs		Animal BPTs		Unknown BPTs	
Sampling sites	Ent	*E. coli*	Ent	*E. coli*	Ent	*E. coli*	Ent	*E. coli*
EC1	60	71	9	11	38	37	13	23
EC2	72	84	8	11	42	47	22	26
EC3	71	85	8	8	45	51	18	26
EC4	22	13	1	3	14	5	7	5
EC5	23	29	0	3	13	11	10	15
Total	248	282	26	36	152	151	70	95

BPTs from septic tanks, and 151 BPTs belonged to animals. The sources of the remaining 70 enterococci BPTs and 95 *E. coli* BPTs could not be identified.

Conclusions: The study reports the use of BF method as a potential tool for MST studies. *E. coli* and enterococci libraries were capable of identifying the sources of more than 65% of indicator bacteria in the studied creek. The authors reported the presence of human fecal isolates in the studied creek originating from defective septic tanks as well as animal sourced isolates. However, the percentage of isolates that were identified as animals was higher than humans.

21.4.1.2 Case Study 2: Antibiotic Resistance Analysis for Detecting Pollution from Septic Systems in Surface Waters in Queensland, Australia

Situation: ARA was used to determine the significance of septic systems as a major contributor to fecal pollution in two mixed land-use catchments, Bonogin Valley and Tallebudgera Creek, in the Gold Coast region, Queensland, Australia (Carroll et al. 2005).

Tools used: Antibiotic resistance patterns (ARPs) were established for a library of 717 known source *E. coli* isolates obtained from human, domesticated animals, livestock ,and wild sources. Discriminant analysis (DA) was used to differentiate between the ARPs of isolates from various sources and to classify each isolate from water (unknown source) into a source category.

Results: A total of 256 (from five sites from Bonogin Creek catchment) and 169 (from three sites from Tallebudgera Creek catchment) isolates from water were subjected to ARA analysis. By applying DA to the water isolates, and utilizing the human vs. nonhuman source category, the percentage of human source isolates contained in each water sample was obtained. From the discriminant analysis of samples obtained from Bonogin Creek, 40, 55, 10, 52, and 56% of the isolates from sites BOS1 to BOS5, respectively, were classified as human source (Table 21.4). For Tallebudgera Creek, 24, 37 and 47% of isolates obtained from TA1 to TA3, respectively, were also classified as human source.

Table 21.4 Source identification of unknown environmental *E. coli* isolates from the Bonogin Valley and Tallebudgera Creeks by antibiotic resistance analysis in Queensland, Australia

Sampling sites	No. of unknown isolates	Source identification (%) of unknown source isolates	
		Human	Animals
BOS1	45	40	60
BOS2	48	55	46
BOS3	23	10	90
BOS4	93	52	48
BOS5	46	56	44
TA1	51	24	76
TA2	74	37	63
TA3	43	47	53

Conclusions: The results suggested the presence of human fecal pollution within the investigated catchments originating from septic systems. From the other classified sources, it was evident that in the upper regions of both catchments, the major source of pollution was from animals. The information obtained through this study has been utilized by the local regulatory authority to implement more appropriate management practices to reduce the human fecal pollution of water resources caused by high numbers of failing septic systems.

21.4.1.3 Case Study 3: Biochemical Fingerprinting and Antibiotic Resistance Analysis to Identify Dominant Sources of Pollution in a Coastal Lake, Southeast Queensland, Australia

Situation: Water samples were collected from five sampling sites on the Tooway, a recreational coastal lake, Queensland, Australia to identify the sources of elevated levels of indicator bacteria (Ahmed et al. 2008b).

Tools used: BF and ARA were used to identify the sources of enterococci and *E. coli* in the studied lake. A population similarity (Sp) analysis was used to compare the overall similarity between bacterial populations from suspected sources with those found in the environmental water samples.

Results: Five sampling sites (T1–T5) were chosen at various points along the length of the lake. Water samples ($n=20$) were collected fortnightly on four occasions. The BPTs of enterococci and *E. coli* isolates from each site were compared to the BPTs of the suspected sources and host groups. However, only *E. coli* isolates from water samples were typed by ARPs and were classified according to host source by ARA. BF of enterococci populations showed that isolates from waterfowl were most similar (showed the highest Sp-coefficient) (0.46 ± 0.09) to water samples and showed the next highest similarity to isolates from STPs (0.31 ± 0.06) (Table 21.5). Similar patterns were also observed when *E. coli* were subjected to BF (0.32 ± 0.03; waterfowl, and 0.27 ± 0.09; STP). Both bacterial populations from all sampling sites showed the highest similarities with those of waterfowl. In contrast, bacterial populations from dogs and chickens generally showed low similarities to water samples. High similarity values were also observed for both bacterial populations from STP and water samples with higher values found in sites T2 and T3 located below the submerged sewerage pipes collecting domestic wastewater. When *E. coli* populations from each site were compared to those of the ARPs, the highest similarity (0.27 ± 0.07) was found between STP and water samples followed by waterfowls and water samples (0.16 ± 0.07) (Table 21.6). *E. coli* populations from dogs and chickens generally showed low similarities with those from water samples.

Conclusions: BF identified waterfowl as a major source of contamination. Each method individually also identified the STP as a source of pollution. The author concluded that Sp-analysis is a simple, rapid, and reliable approach and could be used for comparing bacterial populations from known fecal sources with those from water samples to predict the sources of pollution. However, this approach should be limited to small catchments with limited possible sources of pollution.

Table 21.5 Comparison of population similarity (Sp) coefficient based on biochemical fingerprinting of enterococci (Ent) and *E. coli* isolates from sources and water samples collected from sites T1 to T5 on Tooway Lake, Queensland, Australia

	Population similarity (Sp) coefficient to water samples									
	T1		T2		T3		T4		T5	
Sources	Ent (n=116)	*E. coli* (n=85)	Ent (n=97)	*E. coli* (n=92)	Ent (n=97)	*E. coli* (n=87)	Ent (n=98)	*E. coli* (n=83)	Ent (n=100)	*E. coli* (n=88)
STP	0.32	0.22	0.27	0.38	0.35	0.29	0.40	0.31	0.24	0.14
Waterfowl	0.26	0.32	0.48	0.37	0.46	0.31	0.47	0.27	0.46	0.34
Dog	0.09	0.03	0.15	0.39	0.11	0.17	0.13	0.07	0.10	0.13
Chicken	0.07	0.11	0.13	0.09	0.16	0.03	0.06	0.04	0.04	0.02

Table 21.6 Comparison of population similarity (Sp) coefficient based on antibiotic resistance patterns (ARPs) of *E. coli* isolates from sources and water samples collected from sites T1 to T5 on Tooway Lake, Queensland, Australia

	Population similarity (Sp) coefficient to water samples				
Sources	T1 (n=31)	T2 (n=69)	T3 (n=46)	T4 (n=52)	T5 (n=26)
STP	0.19	0.35	0.29	0.34	0.20
Waterfowl	0.11	0.14	0.27	0.09	0.29
Dog	0.02	0.11	0.07	0.04	0.22
Chicken	0.06	0.05	0.06	0.01	0.03

Table 21.7 Quantitative PCR results of the enterococci surface protein (*esp*) marker in environmental waters collected from Ningi Creek, Queensland, Australia

Sampling sites (event 1)	Enterococci	Gene copies/100 mL	Sampling sites (event 2)	Enterococci	Gene copies/100 mL
NC1	4.1×10^3	1.1×10^2	NC1	3.7×10^3	–
NC2	3.2×10^3	–	NC2	1.0×10^2	–
NC3	1.3×10^4	1.6×10^2	NC3	3.9×10^3	–
NC4	1.9×10^4	5.3×10^2	NC4	5.6×10^4	4.3×10^2
NC5	4.3×10^4	–	NC5	3.9×10^4	3.1×10^2
NC6	2.8×10^4	5.2×10^2	NC6	2.1×10^3	–
NC7	3.9×10^3	–	NC7	9.1×10^2	–
NC8	1.4×10^3	–	NC8	9.2×10^2	–

21.4.2 Library-Independent Methods

21.4.2.1 Case Study 4: Quantitative PCR Assay for the Quantitative Detection of Human-Specific Enterococci Surface Protein (*esp*) Marker in Queensland's Environmental Waters

Situation: Quantitative PCR (qPCR) was used to estimate the levels of human-specific *esp* markers in environmental waters in Ningi Creek, Southeast Queensland, Australia. Environmental samples (n=16) were collected after storm events and tested with the qPCR along with the enumeration of enterococci for the quantitative detection of human pollution (Ahmed et al. 2008c).

Tools used: qPCR of sewage associated enterococcal surface protein (*esp*) markers from *E. faucium*.

Results: The specificity of the *esp* marker to distinguish between human and animal pollution was determined by screening a large number of human and animal samples. The *esp* marker was detected in 90.5% of combined sewage and septic tank samples (n=42) and was not detected in any of the fecal samples (n=155) from the nontarget animals tested. The overall specificity of this marker to distinguish between sewage and animal pollution was 1.0 (100%). The concentration of culturable enterococci in water samples collected from the studied creek ranged between 9.1×10^2 and 4.3×10^4 cfu/100 mL (Table 21.7). Of the 16 samples tested, six (38%)

were positive for the *esp* marker, and the concentration ranged between 1.1×10^2 and 5.3×10^2 gene copies/100 mL of water.

Conclusions: The evidence presented in this study demonstrated that the *E. faecium esp* marker appears to be host-specific and promising for human pollution tracking in environmental waters in Southeast Queensland, Australia. The study successfully demonstrated the application of a newly developed qPCR assay to quantify the *esp* marker in environmental waters.

21.4.2.2 Case Study 5: Application of Human-Specific HF183 and HF134 *Bacteroides* Markers for the Detection of Human Pollution in Nonsewered Catchments in Southeast Queensland

Situation: Stormwater samples were collected from the Bergin Creek, Four Mile Creek and River Oaks Drive nonsewered catchments within the Pine Rivers Shire in Southeast Queensland. The primary aim of this study was to assess whether human-specific *Bacteroides* markers (indicative of human pollution) could be detected in stormwater samples potentially contaminated by defective septic systems (Ahmed et al. 2008d).

Tools used: PCR detection of human-specific *Bacteroides* HF183 and HF134 markers.

Results: Prior to field application, the specificity of each marker was tested by screening a large number of samples from nontarget fecal species. The overall specificity of the tested markers to differentiate between human and animals was 1.0 (HF183) and 0.95 (HF134), respectively, as the HF134 marker was detected in a number of dog samples. The number of *E. coli* and enterococci in storm water samples collected from the three catchments is shown in Table 21.8. Of the four samples tested from the Bergin Creek on four occasions, three were positive for both the markers. Of the three samples tested from the Four Mile Creek on three occasions, two were positive for the HF134.

Table 21.8 The number of *E. coli* and enterococci and PCR positive/negative results of human-specific *Bacteroides* markers in environmental water samples collected from three nonsewered catchments

Catchments	Events	*E. coli*	Enterococci	HF183	HF134
Bergin Creek	Event 1	2.6×10^3	2.7×10^3	+	+
	Event 1	3.9×10^3	4.3×10^3	+	+
	Event 2	4.0×10^3	3.1×10^3	+	+
	Event 3	4.1×10^3	3.4×10^3	–	+
Four mile Creek	Event 1	1.4×10^3	1.8×10^3	–	+
	Event 2	9.6×10^3	8.5×10^3	+	+
	Event 3	2.6×10^3	2.5×10^3	–	–
River Oaks	Event 1	2.7×10^3	2.4×10^3	–	+
	Event 2	2.1×10^3	1.8×10^3	–	–
	Event 3	1.6×10^3	1.4×10^3	–	–

Conclusions: The HF183 marker is specific to human sewage and is a reliable marker for detecting human fecal pollution in Southeast Queensland, while the use of HF134 marker alone may not be sufficient enough to provide evidence of human pollution because of its presence in dog feces.

21.4.2.3 Case Study 6: Application of Human Associated JCV and BKV Polyomaviruses for the Detection of Sewage Pollution in a Coastal River in Southeast Queensland, Australia

Situation: Environmental water samples were collected from five locations (MR1 – MR5) in Maroochy River, Sunshine Coast Region, Queensland, Australia. The primary aim of this study was to evaluate the host specificity of a PCR method to detect JCV and BKV polyomaviruses, and a secondary aim was to identify sewage pollution in the studied river (Ahmed et al. 2010).

Tools used: PCR detection of human-specific JCV and BKV polyomaviruses.

Results: The host specificity of the markers was tested by screening wastewater/ samples from nontarget sources such as chickens, dogs, ducks, kangaroos, wild birds, cattle, pigs, and sheep. The overall host specificity of the JCV and BKV PCR assay to differentiate between human and animal wastewater/samples was 0.99. The concentration of *E. coli* in water samples ranged between <1 and 2,906 ± 300 cfu/100 mL of water (Table 21.9). The concentration of enterococci in water samples ranged between <1 and 1,586 ± 180 cfu/100 mL of water. Of the 20 samples tested, five (25%) were positive for JCV and BKV indicating the presence of human wastewater in various sites in the Maroochy River.

Table 21.9 Concentrations of *E. coli* and enterococci along with PCR positive results for sewage-specific JCV and BKV polyomaviruses at sampling sites on Maroochy River, Southeast Queensland, Australia

Sampling sites (occasion)	*E. coli* (cfu/100 mL)	Enterococci (cfu/100 mL)	PCR positive results for HS-PVs
MR1 (1)[a]	$4.8 \times 10^1 \pm 1.1 \times 10^1$	$7.3 \times 10^1 \pm 1.2 \times 10^1$	–
MR2 (1)[a]	$5.2 \times 10^1 \pm 3.0 \times 10^1$	$1.2 \times 10^2 \pm 2.0 \times 10^1$	+
MR3 (1)[a]	$1.5 \times 10^2 \pm 5.5 \times 10^1$	$2.3 \times 10^2 \pm 3.0 \times 10^1$	–
MR4 (1)[a]	$3.6 \times 10^2 \pm 8.0 \times 10^1$	$4.0 \times 10^2 \pm 1.0 \times 10^2$	–
MR5 (1)[a]	$2.9 \times 10^3 \pm 3.0 \times 10^2$	$1.6 \times 10^3 \pm 1.8 \times 10^2$	–
MR1 (2)	$1.7 \times 10^1 \pm 1.0 \times 10^1$	$1.6 \times 10^2 \pm 3.8 \times 10^1$	–
MR2 (2)	$1.3 \times 10^1 \pm 5.0 \times 10^0$	$2.2 \times 10^2 \pm 6.0 \times 10^1$	–
MR3 (2)	$2.1 \times 10^1 \pm 7.0 \times 10^0$	$3.3 \times 10^1 \pm 8.0 \times 10^0$	–
MR4 (2)	$3.3 \times 10^1 \pm 7.0 \times 10^0$	$6.2 \times 10^1 \pm 1.3 \times 10^1$	–
MR5 (2)	$2.1 \times 10^1 \pm 6.0 \times 10^0$	$1.3 \times 10^2 \pm 5.7 \times 10^1$	–
MR1 (3)	$<1.0 \times 10^0$	$6.0 \times 10^1 \pm 2.0 \times 10^1$	+
MR2 (3)	$4.4 \times 10^1 \pm 5.0 \times 10^0$	$2.0 \times 10^1 \pm 5.0 \times 10^0$	–
MR3 (3)	$1.5 \times 10^1 \pm 5.0 \times 10^0$	$3.0 \times 10^1 \pm 1.0 \times 10^1$	+
MR4 (3)	$7.0 \times 10^1 \pm 1.5 \times 10^1$	$6.5 \times 10^1 \pm 4.1 \times 10^1$	+
MR5 (3)	$1.0 \times 10^2 \pm 1.1 \times 10^1$	$< 1.0 \times 10^0$	–

(continued)

Table 21.9 (continued)

Sampling sites (occasion)	*E. coli* (cfu/100 mL)	Enterococci (cfu/100 mL)	PCR positive results for HS-PVs
MR1 (4)	$1.3\times10^1\pm5.0\times10^0$	$4.0\times10^1\pm1.0\times10^1$	–
MR2 (4)	$6.0\times10^0\pm4.0\times10^0$	$1.5\times10^1\pm6.0\times10^0$	–
MR3 (4)	$<1.0\times10^0$	$1.8\times10^1\pm1.1\times10^1$	–
MR4 (4)	$6.7\times10^1\pm1.5\times10^1$	$4.0\times10^1\pm1.4\times10^1$	–
MR5 (4)	$6.0\times10^0\pm2.0\times10^0$	$4.2\times10^1\pm7.0\times10^0$	–

ND not detected
[a]Study area received >100 mm rainfall 2 days prior sampling

Conclusions: JCV and BKV viruses are highly specific to human sewage, and they proved a reliable marker for detecting human fecal pollution in this coastal river in Southeast Queensland, Australia. The presence of JCV and BKV viruses in water samples indicate potential public health risks as the studied river is used for recreational activities including swimming, fishing, and water sports.

21.4.2.4 Case Study 7: Application of a Suite of MST Markers in an Urbanized Waterway, New Zealand

Situation: A MST study was initiated due to the chronic elevated fecal indicator bacteria concentrations in the urban sections of the lower Maitai River, New Zealand.

Tools used: Water samples were collected along the river on three separate dates and tested for a suite of MST markers (human-associated *E. faecium esp* gene, human- and ruminant-associated *Bacteroides* spp. (HF183 and CF128), human-associated *Methanobrevibacter smithii nifH* gene (*nifH*), and human JCV and BKV polyomaviruses by end-point PCR analyses.

Results: All samples collected expressed a signal of ruminant-associated fecal pollution; however, a strong human component was detected in all samples collected at or downstream of stormwater drains entering from the urbanized side of the river. The human-associated fecal pollution detection was supported by a minimum of two different human markers at those sites and was coupled with elevated fecal indicator bacterial concentrations (Table 21.10). Therefore, while the upstream land runoff was probably the major source of ruminant-associated fecal pollution, leaking sewage collection systems or cross-connections that impact the groundwater or stormwater systems was indicated as the major source of human-associated fecal pollution in the river.

Conclusions: While the actual source at the time of this study was not clear, follow-on engineering work commissioned by the regional council located an aged and leaking sewage drainage system in a historic residential area upstream of the stormwater drainage entering the city side of the river. This outcome highlights the utility of using MST markers in association with traditional fecal indicator bacteria tests to assist in confirming and pinpointing a fecal pollution problem area and also reveals the risk of fecal pollution from aging sewage reticulation systems that many

Table 21.10 Fecal indicator bacteria concentrations and presence/absence of MST markers (positive signal/number of samples tested) in water samples collected in the Maitai River (New Zealand). More than one number is given where multiple samples for fecal indicator bacterial concentrations were collected.

	Fecal indicator concentrations		General	Ruminant	Human			
Sites Maitai River	Fecal coliforms (cfu/100 mL)	Enterococci (cfu/100 mL)	Bac32	CF128	HF183	*esp*	*nifH*	JCV-BKV
1	1.4×10^3	2.9×10^2	1/1	1/1	1/1	0/1	0/1	0/1
2	5.0×10^2	2.2×10^2	3/3	3/3	2/3	0/3	0/3	0/3
3	5.7×10^2	2.3×10^2	2/2	2/2	2/2	1/2	1/2	2/2
4	4.0×10^2	4.0×10^2	1/1	0/1	1/1	1/1	1/1	1/1
5A[a]	1.4×10^3	1.6×10^3	2/2	2/2	1/2	1/2	1/2	2/2
5B[a]	$>2.5\times10^4$	$>1.0\times10^4$	1/1	1/1	1/1	1/1	1/1	1/1
6	2.5×10^3	1.7×10^3	1/1	1/1	1/1	1/1	1/1	1/1
7[a]	NA	1.3×10^2	1/1	0/1	1/1	0/1	0/1	0/1
8	4.4×10^2	1.2×10^2	1/1	1/1	1/1	0/1	0/1	1/1
9	1.1×10^2	8.7×10^1	1/1	1/1	1/1	1/1	1/1	1/1
10[a]	9.2×10^1	6.3×10^1	1/1	1/1	1/1	1/1	0/1	1/1

[a]Indicates samples collected in stormwater outfalls

older developed areas may face in the future. Frequent seismic activity can be a contributing factor in New Zealand.

21.4.3 Combinations of Methods

21.4.3.1 Case Study 8: Application of FWAs and Molecular Methods for Determination of Fecal Sources in a Large River in Two New Zealand Rivers

Situation: An investigation of fecal sources was undertaken on two New Zealand rivers. The primary interest for both rivers was whether human sources of fecal pollution were present, and secondarily, whether dairy cows further up the catchment of each river were contributing to fecal pollution of each river.

Tools used: Five samples were analyzed from various sites on river A, and three samples from river B. Samples were analyzed for the presence of FWAs, and with PCR-based assays for *Bacteroides* general, human, and ruminant markers. River B water samples were also tested for the presence of the *B. adolescentis* PCR marker.

Results: Levels of the microbial indicator *E. coli* differed markedly between the two rivers. All sites on river B contained high levels of *E. coli* ($>5.0\times10^3$ *E. coli*/100 mL), while in River A only site 5 contained similar levels.

FWAs were detected in all samples from river B, only from the site 2 samples in river B (Table 21.11). These levels all support a human source of fecal contamination. The general *Bacteroides* marker was detected in all samples. The human *Bacteroides* marker was present in all river B samples, but from river A, only in the site 2 sample.

Table 21.11 *E. coli*, FWAs and molecular markers detected in rivers A and B

Sample	*E. coli*[a]	General	Human	*B. adolescentis*	Ruminant	FWA[b]
River A, site 1	5.2×10^1	+	–	Not tested	–	<0.01
River A, site 2	1.5×10^2	+	+	Not tested	–	0.2
River A, site 3	9.8×10^1	+	–	Not tested	+	<0.01
River A, site 4	6.3×10^1	+	–	Not tested	–	<0.01
River A, site 5	8.7×10^3	+	–	Not tested	+	<0.01
River B, site 1	1.5×10^4	+	+	+	–	0.38
River B, site 2	5.6×10^3	+	+	+	–	0.45
River B, site 3	7.5×10^3	+	+	+	–	0.55

[a]MPN/100 mL
[b]Parts per billion (ppb) equivalent to μg/L

This human source was supported in river B samples by the presence of the *B. adolescentis* marker. The ruminant marker in contrast was not detected in any of the river B samples. It was, however, detected in river A site 3 and 5 samples.

Conclusion: River A: At the time of sampling, *E. coli* levels were very low in samples from four of the sites, with the fifth (site 5) containing very high levels of *E. coli*. This site was negative for human-specific PCR marker, negative for FWAs and positive for the ruminant marker. Subsequent investigation identified that this site was a drain which was being used for illegal dumping of stock effluent. Of the other four sites with low levels of pollution, the general *Bacteroides* marker could be detected in all of these samples, with evidence of human source of pollution at site 2 (human-associated *Bacteroides* and elevated FWAs). Site 2 is impacted directly by a stormwater drain from a residential area, and these results indicate that human sewage is likely to be entering this system. The low levels of *E. coli* in samples from other sites limit the conclusions that can be made, although site 3 was positive for the ruminant marker.

River B: All sites sampled on this river contained significant evidence of the entry of human sewage into this stream and the very high levels of *E. coli* reinforce the likely health risk. Further investigation of site 1 identified evidence of recent construction of a retaining wall, which may have disrupted sewage lines.

21.4.3.2 Case Study 9: Application of FWAs, Fecal Sterols, and Molecular Methods for Determination of Fecal Sources in Two Streams in Auckland, New Zealand

Situation: A series of small streams run between residential properties in the city of Auckland. These discharge into the marine environment. In this study, three sites on stream C and 4 sites on stream D were sampled. For comparison, a duck pond adjacent, but not visibly connected, to stream D was also tested.

Tools used: Fecal sterols, FWAs, and PCR markers for human-associated *Bacteroides* and *B. adolescentis*.

Results: *E. coli* levels were relatively low in stream C, while in stream D levels were tenfold higher. FWAs were detected at low levels in stream C but were at or below the

Table 21.12 Measured levels of microbial indicators, FWAs, and human DNA markers in streams C and D

	Microbial Indicators	Human effluent indicators		
Sample	*E. coli*[a]	FWAs[b]	*Bacteroidetes*	*B. adolescentis*
Stream C, site 1(3)	5.2×10^2	0.03	+	+
Stream C, site 2	5.2×10^2	0.04	+	+
Stream C, site3(1)	3.1×10^2	0.02	+	+
Stream D, site 1	7.8×10^3	<0.01	–	–
Stream D, site 2	4.8×10^3	0.01	–	–
Stream D, site 3	3.6×10^3	<0.01	–	–
Stream D, site 4	3.6×10^3	<0.01	–	–
Duck pond	2.2×10^3	<0.01	–	–

[a]MPN/100 mL
[b]Parts per billion (ppb) equivalent to μg/L

detection limit in stream D. The human-associated *Bacteroides* and *B. adolescentis* markers were both detected in all stream C sites, but not in stream D sites. The duck pond while containing high levels of *E. coli*, contained no detectable FWAs or human-associated *Bacteroidetes* and *B. adolescentis* markers (Table 21.12). Total levels of fecal sterols contrasted between streams C and D. Stream C contained high levels of sterols, with ratios strongly supporting the presence of human fecal material (Table 21.13). The duck pond also contained high levels of sterols but with quite different ratios, clearly not indicative of a human source. The stream D samples contained much lower levels of sterols, particularly of coprostanol and 24-ethylcoprostanol. The low levels of these two sterols makes interpretation of some of the ratios difficult, but on the basis of the ratios of sterols present, a human source is not indicated.

Conclusions: These two streams contrasted markedly in levels of *E. coli* and the presence of human-associated markers. For stream C, the three source specific tools used all indicated the presence of human effluent, although the levels of FWAs are at very low levels. The low level of FWAs may be indicative of the lack of use of washing powders at the time of sampling, and may suggest relatively few contributing households. In light of the source specific indicators, it would seem likely most of the *E. coli* present were of human origin. Stream D samples contained no evidence of human effluent in the samples analyzed. The absence of FWAs, human specific molecular markers, and fecal sterols, all suggested a nonhuman source of the *E. coli* present.

21.4.3.3 Case Study 10 Application of FWAs, Fecal Sterols, and Molecular Methods for Determination of Fecal Sources in a Rural River in the South Island of New Zealand

Situation: This investigation focussed on two sites on a river in the lower South Island of New Zealand. Possible fecal sources included human septic tanks, dairy cows, and wildfowl. An adjacent duck pond was also sampled for comparison.

Table 21.13 Fecal sterol analysis of streams C, D, and a duck pond

	Stream C			Stream D					
Sterol[a]	Site 1	Site 2	Site 3	Site 1	Site 2	Site 3	Site 4	Duck pond	Interpretation
Coprostanol	3,620	2,150	2,410	35	50	46	34	1,110	
24-Ethylcoprostanol	1,460	914	941	52	38	46	46	4,380	
Epicoprostanol	29	52	56	21	33	23	31	630	
Cholesterol	5,950	4,920	5,110	1,310	1,060	1,150	2,500	14,890	
Cholestanol	414	432	467	270	223	213	289	5,640	
24-Methylcholesterol	932	715	669	466	334	391	632	6,170	
24-Ethylepicoprostanol	24	5	5	28	5	5	5	1,000	
Stigmasterol	457	329	322	339	275	289	387	7,640	
24-Ethylcholesterol	2,860	2,120	2,180	1,640	1,000	1,330	2,190	31,920	
24-Ethylcholestanol	391	340	362	207	269	210	554	8,490	
Total sterols	16,137	11,977	12,522	4,368	3,287	3,703	6,668	81,870	
Ratio 1	8.74	4.98	5.16	0.13	0.22	0.22	0.12	0.20	>0.5 Fecal
Ratio 2	3.73	2.69	2.60	0.25	0.14	0.22	0.08	0.52	>0.5 Fecal
Ratio 3	22.4	18.0	19.2	0.8	1.5	1.2	0.5	1.4	>5–6% Human contamination
Ratio 4	0.90	0.83	0.84	0.11	0.18	0.18	0.11	0.16	>0.7 Human contamination
Ratio 5	2.48	2.35	2.56	0.67	1.32	1.00	0.74	0.25	<1.0 Herbivore; ≥1.0 human
Ratio 6	71.26	70.17	71.92	40.23	56.82	50.00	42.50	20.22	<30% Herbivore; >75% human
Ratio 7	9.0	7.6	7.5	1.2	1.2	1.2	0.7	5.3	>5–6% Herbivore
Ratio 8	1.96	2.32	2.32	31.54	26.32	28.91	47.61	7.29	<1.0 Herbivore; >4.0 plant decay
Ratio 9	20.85	27.01	27.68	72.13	86.22	80.46	91.57	61.21	>30% Avian
Ratio 10	10.19	16.40	15.92	82.82	72.88	75.53	81.64	76.42	>67% Avian

[a]Results all parts per trillion

Table 21.14 *E. coli*, molecular markers and fecal sterols detected in water samples from two sites on a South Island River and an adjacent duck pond

Analyte	Site 1	Site 2	Duck pond	Interpretation
E. coli[a]	2.8×10^3	1.0×10^3	5.4×10^3	
PCR markers				
General *Bacteroides*	+[b]	+	+	
Herbivore *Bacteroides*	+	–	–	
Human *Bacteroides*	–	+	–	
Fecal sterols				
Coprostanol	362[c]	4,243	561	
24-Ethylcoprostanol	1,425	2,410	961	
Epicoprostanol	63	295	349	
Cholesterol	1,914	6,672	11,795	
Cholestanol	296	952	5,143	
24-Methylcholesterol	562	2,820	4,422	
24-Ethylepicoprostanol	424	486	246	
Stigmasterol	627	2,680	4,631	
24-Ethylcholesterol	1,712	6,036	14,397	
24-Ethylcholestanol	1,095	1,268	1,106	
Total sterols	8,480	27,862	43,611	
Ratio 1	1.22	4.46	0.11	>0.5 Fecal
Ratio 2	1.30	1.90	0.87	>0.5 Fecal
Ratio 3	4.3	15.2	1.3	>5–6% Human fecal pollution
Ratio 4	0.55	0.82	0.10	>0.7 Human fecal pollution
Ratio 5	0.25	1.76	0.58	<1.0 herbivore; ≥1.0 human
Ratio 6	20.26	63.78	36.86	<30% All herbivore; >75% all human
Ratio 7	16.8	8.6	2.2	>5–6% Herbivore
Ratio 8	1.20	2.50	14.98	<1.0 Herbivore; >4.0 plant decay
Ratio 9	37.19	30.45	47.82	>30% Wildfowl
Ratio 10	41.05	17.34	84.97	>67% Wildfowl

[a]MPN/100 mL
[b]+ equals detection of marker, – equals not detected
[c]Sterol results all parts per trillion

Tools applied: Samples were analyzed for presence of fecal sterols, and for PCR markers specific for *E. coli*, human, ruminant, and general *Bacteroides* markers (Table 21.1).

Results: All three sites contained fairly similar levels of *E. coli* and elevated levels of sterols (Table 21.14).

Fecal sterol ratios 1 and 2 were both elevated above the typical human and herbivore fecal thresholds at both these sites. The sterols at site 1 didn't meet any of the human-associated ratio thresholds (ratios 3–6), while the herbivore indicative ratios 3–5, were satisfied. While ratio 6 was not below 1 in site 1, it is close to this point. In contrast at site 2 the ratios exceeded thresholds for human ratios 3–5. Ratio 10 was significantly less than 67% in both sites 1 and 2, indicating that wildfowl contribution to pollution is not significant. Analysis of the duck pond samples confirmed that the fecal sterol profile of this water was quite different from either of the water samples. These conclusions were also supported by the PCR analysis with the human-associated marker only detected at site 2, while the herbivore specific marker was only detected in site 1.

Conclusions: These two sites provided a strong contrast in terms of the identified source of fecal pollution. The molecular and fecal sterol signatures in site 1 were consistent with an herbivore source of pollution, while site 2 samples produced a profile consistent with a human source. Sterols and molecular markers analyzed from the duck pond confirmed that water containing feces from ducks would not falsely be identified as being either of human or herbivore origin.

21.4.3.4 Case Study 11: A Combination of Source Tracking Methods to Identify Human Sourced Pollution in Stormwater via Defective Septic Systems in Pine Rivers Shire, Queensland, Australia

Situation: Storm water samples were collected from Bergin Creek, Four Mile Creek and River Oaks Drive to determine whether the water was contaminated by human pollution from possible defective septic systems (Ahmed et al. 2007).

Tools used: A battery of methods, (1) library-dependent BF of *E. coli* and enterococci (2) human-specific *Bacteroides* HF183, HF134 and (3) human-specific enterococci surface protein (*esp*) markers, were used to detect human sourced pollution in the nonsewered, residential catchments studied.

Results: In all, 550 *E. coli* and 700 enterococci were isolated and biochemically fingerprinted to compare these fingerprints with existing libraries (Ahmed et al. 2005b). Of the 18 samples tested, 7 samples were also analyzed for the presence of human-specific markers using PCR methods. A total of 305 *E. coli* BPTs and 299 enterococci BPTs were obtained from water samples. The source of 105 *E. coli* BPTs and 93 enterococci BPTs were identified in water samples from River Oaks Drive catchment. Of these, 10 and 9% were identified as human-source *E. coli* and enterococci BPTs, respectively. Similarly, of the 83 *E. coli* BPTs and 93 enterococci BPTs from the Bergin Creek catchment site, 8% *E. coli* BPTs and 9% enterococci BPTs were identified as human-source isolates. The number of *E. coli* and enterococci assigned to human origin in the Four Mile Creek site were 4 and 3% respectively. Of the seven samples tested, both HF134 and *esp* markers were detected in five samples, and the HF183 marker was detected in four samples (Table 21.15). Human fecal pollution was detected in six out of seven water samples by at least one of these markers. The methods were not always in agreement in detecting human fecal pollution in water samples.

Table 21.15 Detection of human fecal pollution using library-dependent and library-independent methods

Catchments	Storm events	*E. coli*	Ent	*Bacteroides* HF183	*Bacteroides* HF134	Enterococci surface protein (*esp*)
BC	1	–	+	Not tested	Not tested	Not tested
	3	+	+	Not tested	Not tested	Not tested
	5	–	–	+	+	+
	5	–	+	+	+	+
	6	–	–	+	+	–
FMC	1	–	–	Not tested	Not tested	Not tested
	2	–	+	Not tested	Not tested	Not tested
	3	–	–	Not tested	Not tested	Not tested
	4	–	–	Not tested	Not tested	Not tested
	4	+	+	Not tested	Not tested	Not tested
	5	–	+	–	+	+
	6	+	+	+	+	+
ROD	1	+	+	Not tested	Not tested	Not tested
	2	+	+	Not tested	Not tested	Not tested
	3	+	–	Not tested	Not tested	Not tested
	4	+	+	Not tested	Not tested	Not tested
	5	+	–	–	–	+
	6	+	–	–	–	–

Ent enterococci

Conclusions: The results suggested that human fecal pollution is present in stormwater from these catchments. The *E. coli* and enterococci libraries used in this study were capable of detecting human fecal pollution. The presence of host-specific markers further confirmed the presence of human fecal pollution. This study demonstrated the value of a combination of methods for source tracking to obtain a better understanding regarding the pollution sources.

21.4.3.5 Case Study 12: Detection of Human and Animal Fecal Pollution in a Coastal Creek Located in Southeast Queensland, Australia Using Multiple Host-Specific PCR Markers

Situations: Environmental samples ($n = 16$) were collected from Ningi Creek urban catchment to identify the sources of fecal pollution using PCR along with the enumeration of *E. coli* and enterococci (Ahmed et al. 2008e).

Tools used: PCR detection of the human-specific HF183, HF134, *esp* markers, cattle-specific markers, and dog-specific markers.

Results: The specificity of these markers were determined by testing 197 samples from sewage/septage, ducks, kangaroos, cattle, horses, dogs, chickens, pigs, pelicans, goats, deer, wild birds, and sheep. The overall specificity of the *Bacteroides* HF183 and HF134 markers to differentiate between sewage/seepage and animal host groups was 1.0 and 0.95, respectively. The *Bacteroides* CF128 markers also showed high specificity (0.93) for ruminant feces, which included cattle.The *Bacteroides* BacCan

Table 21.16 Concentrations of *E. coli* and enterococci and detection of MST markers signifying human or animal pollution in water samples from Ningi Creek, Queensland, Australia

Sampling sites	*E. coli*	Enterococci	HF183	HF134	CF128	BacCan	*esp*
Occasion 1							
NC1	2.1×10^3	4.1×10^3	+	–	–	–	+
NC2	3.6×10^3	3.2×10^3	+	–	+	+	–
NC3	4.9×10^3	1.3×10^3	+	+	+	+	+
NC4	4.1×10^3	1.9×10^3	+	+	–	+	+
NC5	1.2×10^4	4.3×10^4	+	+	+	–	–
NC6	3.9×10^3	2.8×10^4	+	–	+	–	+
NC7	3.1×10^3	3.9×10^3	–	–	+	–	–
NC8	3.4×10^3	1.4×10^3	–	–	+	–	–
Occasion 2							
NC1	3.1×10^3	3.7×10^3	–	–	+	–	–
NC2	9.1×10^2	1.0×10^2	+	–	–	–	–
NC3	4.9×10^4	3.9×10^3	+	+	–	+	–
NC4	4.4×10^4	5.6×10^4	+	+	+	+	+
NC5	4.2×10^4	3.9×10^4	+	+	+	–	+
NC6	1.1×10^3	2.1×10^3	–	–	+	–	–
NC7	1.6×10^3	3.1×10^2	–	–	+	–	–
NC8	2.1×10^3	1.2×10^2	–	–	–	–	–

marker (i.e., dog markers) was detected in samples from sewage/septic, chickens and pigs and the specificity was low (i.e., 65%). The *esp* marker also exhibited high specificity for differentiation between human and animal feces. The concentrations of FIB in the water samples ranged between 9.1×10^2 and 1.2×10^4 cfu/100 mL (for *E. coli*), and 1.2×10^2 and 5.6×10^4 cfu/100 mL (for enterococci) (Table 21.16). At least one host-specific marker was detected in 14 (87%) out of 16 samples. Human-specific *Bacteroides* HF183 and HF134 markers were detected in 9 (56%) and 6 (37%) samples, respectively. This figure for human-specific *esp* marker was also 6 (37%). Cattle-specific marker CF128 was detected in 11 (69%) samples, whereas dog-specific marker BacCan was detected in 5 (31%) samples.

Conclusion: The host-specific PCR markers are reliable tools for detection of fecal pollution from humans and animals. Among all markers, *Bacteroides* HF183 and *esp* performed well in terms of specificity and identifying the sources of human fecal pollution. However, a combination of multiple human-specific markers provides greater reliability regarding the presence/absence of human fecal pollution when one marker is not sufficient to identify human fecal pollution. The CF128 marker also performed well in identifying ruminant fecal pollution.

21.5 Conclusions

This series of case studies conducted in Australia and New Zealand demonstrate the application of FST tools in a range of water systems. The primary question that arises in many situations is whether a water body contains human derived fecal pollution. Human fecal pollution usually is considered to represent the greatest health risk (Field and Samadpour 2007; Leclerc et al. 2002), has the lowest threshold of public acceptability, and once identified often provides an opportunity to rectify the situation. This corrective action may involve considerable expenditure, and therefore, avoidance of false-positive results is very important. Indeed, demonstrating an absence of human fecal pollution allows water managers to prioritize resources to other areas and thereby achieve improvements in water quality.

Library-dependent methods such as BF and ARA, as illustrated in case studies 1 and 2, can be effective in source identification. However, the need to generate a large source library and potential concerns over validity of a library beyond the spatial and temporal constraints in which it was derived from can make library-dependent methods both time-consuming and expensive.

Library-independent source tracking methods, such as source specific PCR marker approaches, are appealing as in theory these markers may be temporally and spatially more stable than libraries. The case studies in this chapter indicate that the tested markers indeed exhibit similar sensitivity and specificity in Australia and New Zealand compared to results obtained overseas. Most of the markers showed higher specificity, although the sensitivity was not always high. For example, the *esp* markers in the case study 4 could not be detected in all wastewater samples collected from septic tanks. Some cross-reactivity has been observed for some markers as in case study 5, where HF134 markers were detected in dog fecal samples. Nonetheless, the application of an array of markers and/or combination of MST techniques can compensate for any uncertainty associated with a single marker. FWAs are useful indicators of human pollution, as are fecal sterols. Increasing use of sterols in water-quality analysis is also improving our understanding of these chemicals in nonhuman sources (Devane et al. 2006).

The cost of assays often limits the willingness of water managers to invest in sufficient replication of analysis to be able to understand variability of results. "Murphy's Law" and the inherent variability of aquatic systems can also create the situation where a river with historically high levels of pollution may have low levels when samples are taken for analysis by FST tools. This is demonstrated in case studies 8 and 9, and as illustrated in case study 9, the generation of meaningful results may still be possible in this situation. However, these must be interpreted with care. Are the sources of pollution at lower *E. coli* levels the same as at higher levels, or do the higher levels of pollution come from a different, intermittent source? When a human source is detected as in case study 9, this may be sufficient evidence for water managers to take action even if water standards are not exceeded. Unless all pathogens have been removed, and only indicators are present, any human pollution is usually unacceptable. Certainly, the avoidance of the use of this water for recreational or aquaculture is preferable. If used for drinking water, a very high level of treatment is required to ensure that any possible viral or protozoan pathogens in particular are inactivated.

While we are beginning to build up knowledge on the degradation, absorption, sedimentation, and transport of these new fecal source indicators (Bae and Wuertz 2009; Okabe and Shimazu 2007; Walters and Field 2006; Walters and Field 2009), our understanding is incomplete and in most cases untested in real-life situations. Preferential absorption or transport may result in some markers behaving differently compared to traditional *E. coli* indicators and also behaving differently compared to pathogens that are of ultimate concern. The adoption of multiple source indicators does to a degree counter this issue. Additional tools do, however, increase both the time and expense in analyzing an area and may not be available in some areas. Cost-effectiveness of the use of multiple tools must also be considered. Devane et al. (2008) explored the development of decision trees to begin addressing these issues. While a work in progress, they do provide guidance, enabling users of these tools to compare various scenarios and identify cost-effective implementation strategies. This is critical as the cost and complexity have been the key factors hampering the implementation of MST technologies in water-management programs (Sagarin et al. 2009).

Collectively, these case studies indicate that current MST technology can successfully be applied for source identification and lead to meaningful and productive management decisions. There is room for significant refinement of these tools, and a continued investment in research to achieve these improvements is required. However, MST technology, even in its current, developing form can and is being used to improve water-quality outcomes.

References

Ahmed W, Neller R, Katouli M (2005a) Evidence of septic system failure determined by a bacterial biochemical fingerprinting method. J Appl Microbiol 98:910–920

Ahmed W, Neller R, Katouli M (2005b) Host-species specific metabolic fingerprint database for enterococci and *Escherichia coli* and its application to identify sources of fecal contamination in surface waters. Appl Environ Microbiol 71:4461–4468

Ahmed W, Stewart J, Gardner T, Powell D, Brooks P, Sullivan, D, Tindale, N (2007) Sourcing fecal pollution: A combination of library-dependent and library independent methods to identify human fecal pollution in non-sewered catchments. Water Res 41:3771–3779

Ahmed W, Stewart J, Powell D, Gardner T (2008a) Evaluation of host-specificity and prevalence of enterococci surface protein (*esp*) marker in sewage and its application for sourcing human faecal pollution. J Environ Qual 37:1583–1588

Ahmed W, Hargreaves M, Goonetilleke A, Katouli, M (2008b) Population similarity analysis of indicator bacteria for source prediction of fecal pollution in a recreational coastal lake. Mar Pollut Bull 56:1469–1475

Ahmed W, Stewart J, Gardner T, D Powell (2008c) A real-time polymerase chain reaction assay for the quantitative detection of the human-specific enterococci surface protein marker in sewage and environmental waters. Environ Microbiol 10:3255–3264

Ahmed W, Stewart J, Powell D, Gardner T (2008d) Evaluation of Bacteroides markers for the detection of human faecal pollution. Lett Appl Microbiol 46:237–242

Ahmed W, Powell D, Goonetilleke A, Gardner T (2008e) Detection and source identification of faecal pollution in non-sewered catchment by means of host-specific molecular markers. Water Sci Technol 58:579–586

Ahmed W, Wan C, Goonetilleke A, Gardner T (2010) Evaluation of human associated JCV and BKV polyomaviruses for the detection of sewage pollution in a coastal river. J Environ Qual 39:1743–1750

Allsop K, Stickler JD (1985) An assessment of *Bacteroides fragilis* group organisms as indicators of human faecal pollution. J Appl Bacteriol 58:95–99

Bae S, Wuertz S (2009) Rapid decay of host-specific fecal *Bacteroidales* cells in seawater as measured by quantitative PCR with propidium monoazide. Water Res 43:4850–4859

Bernhard AE, Field KG (2000) A PCR assay to discriminate human and ruminant feces on the basis of host differences in *Bacteroides-Prevotella* genes encoding 16S rRNA. Appl Environ Microbiol 66:4571–4574

Carroll S, Hargreaves M, Goonetilleke A (2005) Sourcing faecal pollution from onsite wastewater treatment systems in surface waters using antibiotic resistance analysis. J Appl Microbiol 99:471–482

Chern EC, Tsai YL, Olson BH (2004) Occurrence of genes associated with enterotoxigenic and enterohemorrhagic *Escherichia coli* in agricultural waste lagoons. Appl Environ Microbiol 70:356–362

Choi S, Jiang SC (2005) Real-time PCR quantification of human adenoviruses in urban rivers indicates genome prevalence but low infectivity. Appl Environ Microbiol 71:7426–7433

Devane M, Saunders D, Gilpin B (2006) Faecal sterols and fluorescent whiteners as indicators of the source of faecal contamination. Chemistry in New Zealand 70:74–77

Devane ML, Robson B, Nourozi F, Scholes P, Gilpin BJ (2007) A PCR marker for detection in surface waters of faecal pollution derived from ducks. Water Res 41:3553–3560

Devane M, Gilpin B, Robson B, Nourozi F, Wood, D (2008) Decision analysis trees for identification of faecal sources in water. The NZWWA Journal 157:23–31

Dick LK, Bernhard AE, Brodeur TJ, Santo Domingo JW, Simpson JM, Walters SP, Field KG (2005) Host distributions of uncultivated fecal *Bacteroidales* reveal genetic markers for fecal source identification. Appl Environ Microbiol 71:3184–3191

Dombek PE, Johnson LK, Zimmereley ST, Sadowsky MJ (2000) Use of repetitive DNA sequences and the PCR to differentiate *Escherichia coli* isolates from human and animal sources. Appl Environ Microbiol 66:2572–2577

Field KG, Samadpour M (2007) Fecal source tracking the indicator paradigm and managing water quality. Water Res 41:3517–3538

Fong TT, Griffin DW, Lipp EK (2005a) Molecular assays for targeting human and bovine enteric viruses in coastal waters and their application for library-independent source tracking. Appl Environ Microbiol 71:2070–2078

Fong TT, Lipp EK (2005b) Enteric viruses of humans and animals in aquatic environments: health risks detection and potential water quality assessment tools. Microbiol Mol Biol Rev 69:357–371

Gilpin BJ, Gregor JE, Savill MG (2002) Identification of the source of faecal pollution in contaminated rivers. Water Sci Technol 46:9–15

Gregor J, Garrett N, Gilpin B, Randall C, Saunders D (2002) Use of classification and regression tree (CART) analysis with chemical faecal indicators to determine sources of contamination. NZ J Mar Freshwater Res 36:387–398

Grimault JO, Fernandez P, Bayona JM, Albaiges J (1990) Assessment of faecal sterols and ketones as indicators of urban sewage inputs to coastal waters. Environ Sci Technol 24:357–363

Hagedorn C, Crozier JB, Mentz KA, Booth AM, Graves AK, Nelson NJ, Reneau Jr RB (2003) Carbon source utilization profiles as a method to identify sources of fecal pollution in water. J Appl Microbiol 94:792–799

Harwood, VJ, Whitlock J, Withington, V (2000) Classification of antibiotic resistance patterns of indicator bacteria by discriminant analysis: use in predicting the source of fecal contamination in subtropical waters. Appl Environ Microbiol 66:3698–3704

Hrudey SE, Hrudey EJ (2004) Safe drinking water: Lessons from recent outbreaks in affluent nations. IWA Publishing, London

Hundesa A, Maluquer de Motes C, Bofill-Mas S, Albinana-Gimenez N, Girones R (2006) Identification of human and animal adenoviruses and polyomaviruses for determination of sources of fecal contamination in the environment. Appl Environ Microbiol 72:7886–7893

Hörman A, Rimhanen-Finne R, Maunula L, von Bonsdorff C-H, Torvela N, Heikinheimo A, Hanninen M-L (2004) *Campylobacter* spp. *Giardia* spp. *Cryptosporidium* spp. Noroviruses and indicator organisms in surface water in south-western Finland 2000–2001. Appl Environ Microbiol 70:87–95

Katouli M, Lund A, Wallgren P, Kuhn I, Sodermind O, Mollby R (1997) Metabolic fingerprinting and fermentative capacity of the intestinal flora of pigs during pre- and post-weaning periods. J Appl Microbiol 83:147–154

Khatib LA, Tsai YL, Olson BH (2002) A biomarker for the identification of cattle fecal pollution in water using the LTIIa toxin gene from enterotoxigenic *E. coli.* Appl Microbiol Biotechnol 59:97–104

Khatib LA, Tsai YL, Olson BH (2003) A biomarker for the identification of swine fecal pollution in water using the STII toxin gene from enterotoxigenic *E. coli.* Appl Microbiol Biotechnol 63:231–238

Kildare BJ, Leutenegger CM, McSwain BS, Bambic DG, Rajal VB, Wuertz S (2007) 16S r RNA-based assays for quantitative detection of universal human- cow- and dog-specific fecal *Bacteroidales*: a Bayesian approach. Water Res 41:3701–3715

King CM (Ed) (2005) The Handbook of New Zealand Mammals, Second Edition. Oxford University Press, USA

Kreader CA (1995) Design and evaluation of *Bacteroides* DNA probes for the specific detection of human fecal pollution. Appl Environ Microbiol 61:1171–1179

Layton A, McKay L, Williams D, Garrett V, Gentry R, Sayler G (2006) Development of *Bacteroides* 16S rRNA gene TaqMan based real-time PCR assays for estimation of total human and bovine fecal pollution in water. Appl Environ Microbiol 72:4214–4224

Leclerc H, Schwartzbrod L, Dei-Cas E (2002) Microbial agents associated with waterborne diseases. Crit Rev in Microbiol 28:371–409

Leeming R, Ball A, Ashbolt N, Nichols P (1996) Using faecal sterols from humans and animals to distinguish faecal pollution in receiving waters. Water Res 30:2893–2900

Leeming R, Bate N, Hewlett R, Nichols PD (1998) Discriminating faecal pollution: A case study of stormwater entering Port Phillip Bay Australia. Water Sci Technol 38:15–22

Love DC, Sobsey MD (2007) Simple and rapid F+ coliphage culture, latex agglutination and typing assay to detect and source track fecal contamination. Appl Environ Microbiol 73: 4110–4118

Maluquer de Motes C, Clemente-Casares P, Hundesa A, Martin M, Girones R (2004) Detection of bovine and porcine adenoviruses for tracing the source of fecal contamination. Appl Environ Microbiol 70:1448–1454

Matsuki T, Watanabe K, Fujimoto J, Kado Y, Takada T, Matsumoto K, Tanaka R (2004) Quantitative PCR with 16S rRNA-gene-targeted species-specific primers for analysis of human intestinal bifidobacteria. Appl Environ Microbiol 70:167–173

McLellan SL (2004) Genetic diversity of *Escherichia coli* isolated from urban rivers and beach water. Appl Environ Microbiol 70:4658–4665

McQuaig SM, Scott, TM, Harwood VJ, Farrah, SR, Lukasik JO. 2006. Detection of human-derived fecal pollution in environmental waters by use of a PCR-based human polyomavirus assay. Appl Environ Microbiol 72:7567–7574

Mcquaig SM, Scott TM, Harwood VJ, Farrah SR, Lukasik JO. 2009. Quantification of human polyomaviruses JC virus and BK virus bt Taqman Quantittaive PCR and comparison to other water quality indicators in water and fecal samples. Appl Environ Microbiol 75:3379–3388

Möllby R, Kühn I, Katouli, M (1993) Computerized biochemical fingerprinting – a new tool for typing of bacteria. Rev Med Microbiol 4:231–241

Moyda SP, Carson CA, Fuhrmann JJ, Hahm BK, Hartel PG, Kuntz RL, Nakatsu CH, Sadowsky MJ, Samadpour M, Yampara-Iquise, H (2003) Comparing genotypic bacterial source tracking methods that require a host origin database. J Wat and Health 1:167–180

Noble RT, Allen SM, Blackwood AD, Chu W, Jiang SC, Lovelace GL, Sobsey MD, Stewart JR, Wait DA. 2003. Use of viral pathogens and indicators to differentiate between human and non-human fecal contamination in a microbial source tracking comparison studu. J Water Health 1:195–207

Okabe S, Shimazu Y (2007) Persistence of host-specific *Bacteroides*-Prevotella 16S rRNA genetic markers in environmental waters: effects of temperature and salinity. Appl Microbiol Biotechnol 76:935–944

Okabe S, Okayama N, Savichtcheva O, Ito T (2007) Identification and quantification of host-specific *Bacteroides-Prevotella*16S rRNA genetic markers for assessment of fecal pollution in freshwaters. Appl Microbiol Biotechnol 74:890–901

Parveen S, Murphree RL, Edminston L, Kaspar CW, Portier KM, Tamplin ML (1997) Association of multiple antibiotic resistance profiles with point and nonpoint sources of *Escherichia coli* in Apalachicola Bay. Appl Environ Microbiol 63:2607–2612

Parveen S, Portier KM, Robinson, K, Edmiston L, Tamplin ML (1999) Discriminant analysis of ribotype profiles of *Escherichia coli* for differentiating human and nonhuman sources of fecal pollution. Appl Environ Microbiol 65:3142–3147

Pina S, Puig M, Lucena F, Jofre J, Girones R (1998) Viral pollution in the environment and in shellfish: human adenovirus detection by PCR as an index of human viruses. Appl Environ Microbiol 64:3376–3382

Poiger T, Field JA, Field TM, Giger W (1993) Determination of detergent-derived fluorescent whitening agents in sewage sludges by liquid chromatography. Analytical Methods and Instrumentation 1:104–113

Poiger T, Field JA, Field TM, Siegrist H, Giger W (1998) Behaviour of fluorescent whitening agents during sewage treatment. Water Res 32:1939–1947

Prüss-Üstün AR, Bos F, Gore, J Bartram (2008) Safer Water Better Health: Costs Benefits and Sustainability of Interventions to Protect and Promote Health. Geneva: World Health Organization

Reeves AD, Patton D (2001) Measuring change in sterol input to estuarine sediments. Physics and Chemistry of the Earth Part B: Hydrology Oceans and Atmosphere 26:753

Reischer GH, Kasper DC, Steinborn R, Farnleitner AH, Mach RL (2007) A quantitative real-time PCR assay for the highly sensitive and specific detection of human faecal influence in spring water from a large alpine catchment area. Lett Appl Microbiol 44:351–356

Reischer GH, Kasper DC, Steinborn R, Mach RL, Farnleitner AH (2006) Quantitative PCR method for sensitive detection of ruminant fecal pollution in freshwater and evaluation of this method in Alpine Karstic regions. Appl Environ Microbiol 72:5610–5614

Sagarin R, Carlsson J, Duval M, Freshwater W, Godfrey MH, Litaker W, Munoz R, Noble R, Schultz T, Wynne B. (2009) Bringing molecular tools into environmental resource management: untangling the molecules to policy pathway. PloS Biol 7:e69

Scott TM, Jenkins TM, Lukasik J, Rose JB (2005) Potential use of a host associated molecular marker in *Enterococcus faecium* as an index of human pollution, Environ Sci Technol 39:283–287

Scott TM, Rose JB, Jenkins TM, Farrah SR, Lukasik J (2002) Microbial source tracking: current methodology and future directions. Appl Environ Microbiol 68:5796–5803

Scott TM, Parveen S, Portier KM, Rose JB, Tamplin ML, Farrah SA, Koo A, Lukasik J (2003) Geographical variation in ribotype profiles of *Escherichia coli* isolates from humans swine poultry beef and dairy cattle in Florida. Appl Environ Microbiol 69:1089–1092

Seurinck S, Defoirdt T, Verstraete W, Siciliano SD (2005) Detection and quantification of the human-specific HF183 *Bacteroides* 16S rRNA genetic markers with real-time PCR for assessment of human fecal pollution in freshwater. Environ Microbiol 7:249–259

Sghir A, Gramet G, Suau A, Rochet V, Pochart P, Dore J (2000) Quantification of bacterial groups within human faecal flora by oligonucleotide probe hybridization. Appl Environ Microbiol 66:2263–2266

Sneath PHA, RR Sokal (1973) Numerical taxonomy: the principles and practice of numerical classification. W H Freeman, San Francisco, CA

Stoll JMA, Giger W (1998) Mass balance for detergent-derived fluorescent whitening agents in surface waters of Switzerland. Water Res 32:2041–2050

Ufnar JA, Wang SY, Christiansen JM, Yampara-Iquise H, Carson CA, Ellender RD (2006) Detection of the *nifH* gene of *Methanobrevibacter smithii:* a potential tool to identify sewage pollution in recreational waters. J Appl Microbiol 101:44–52

US Environmental Protection Agency (2000) Improved Enumeration Methods for the Recreational Water Quality Indicators: Enterococci and *Escherichia coli* EPA/821/R-97/004. Office of Science and Technology, Washington DC, 55 pp

Walters SP, Field KG (2006) Persistence and growth of fecal *Bacteroidales* assessed by bromodeoxyuridine immunocapture. Applied and Environmental Microbiology 72:4532–4539

Walters SP, Field KG (2009) Survival and persistence of human and ruminant-specific faecal *Bacteroidales* in freshwater microcosms. Environmental Microbiology 11:1410–1421

Wiggins BA (1996) Discriminant analysis of antibiotic resistance patterns in fecal streptococci a method to differentiate human and animals sources of fecal pollution in natural waters. Appl Environ Microbiol 62:3997–4002

Wiggins BA, Andrews RW, Conway RA, Corr CL, Dobratz EJ, Dougherty DP, Eppard JR, Knupp SR, Limjoco MC, Mettenburg JM, Rinehardt JM, Sonsino J, Torrijos RL, Zimmerman ME (1999) Use of antibiotic resistance analysis to identify nonpoint sources of fecal pollution. Appl Environ Microbiol 65:3483–3486

Chapter 22
Microbial Source Tracking in China and Developing Nations

Charles Hagedorn, Joe Eugene Lepo, Kristen Nicole Hellein, Abidemi O. Ajidahun, Liang Xinqiang, and Hua Li

Abstract In less developed countries (LDCs) with poor water quality, sources of pollution are often obvious (especially point sources) and not difficult to find. Under such circumstances, microbial source tracking (MST) is not always necessary. However, there are situations such as identifying nonpoint sources and determining the relative contributions of multiple point and nonpoint sources where MST can be very useful. In this chapter, three water quality improvement case studies from different parts of the world, Malaysia (Asia-pacific), Poland (Eastern Europe), and Colombia (Latin America) are briefly described as successful examples for other LDCs. The actions in these countries resulted in reductions in pollution loads and improvements in water quality that other nations might emulate. While each case study has unique and different political and institutional structures, the three cases illustrate that pollution control is possible – but in all three it was enforcement that proved to be the key. This chapter then describes benefits, issues, and solutions that are common to all countries and presents different scenarios that can result: business as usual, sustainable water use, or water crisis. Six LDCs are then described (China, India, the Philippines, Mexico, Chile, and Nigeria), and a summary of each is presented based on what appears to be the most likely of the scenarios. Throughout the chapter, the role that MST can play to assist researchers and officials in accurately determining the sources of fecal pollution and how such information can be most effectively used in LDCs is described.

Keywords Developing countries and water quality [China, India, Philippines, Mexico, Chile, Nigeria] • Microbial Source Tracking (MST) • Fecal pollution of water • Water and public health • Drinking water contamination

---

C. Hagedorn (✉)
Department of Crop and Soil Environmental Sciences, 401 Price Hall, Virginia Tech, Blacksburg, VA 24061, USA
e-mail: chagedor@vt.edu

C. Hagedorn et al. (eds.), *Microbial Source Tracking: Methods, Applications, and Case Studies*, DOI 10.1007/978-1-4419-9386-1_22,

22.1 Introduction

Estimates from 2006 indicate that nearly 1.5 billion people lack safe drinking water; almost 250 million illness cases reported each year are attributed to waterborne diseases, with about ten million deaths (United Nations Development Programme (UNDP) 2006; World Health Organization (WHO) 2009a–e). One in five people living today does not have access to safe drinking water, and half the world's population does not have adequate sanitation. This is most acute in Asia where the majority of the world's poor people live. Not surprisingly, water- and sanitation-related diseases are widespread and increasing (UNDP 2006; WHO 2009a–e). Diarrhea alone kills more than two million children in developing countries. A 1998 report noted that "at any given time, 50% of the population in developing countries is suffering from water-related diseases caused either by infection, or indirectly by disease-carrying organisms" (Gleick 1998). This situation has not improved since that time and has further deteriorated in many regions of the world (UNDP 2006).

Perhaps the most important reason for developing a worldwide program to monitor and restrict global pollution is the fact that most forms of pollution do not respect national boundaries. The first major international conference on environmental issues was held in Stockholm, Sweden, in 1972 and was sponsored by the United Nations (UN). This meeting was controversial because many developing countries were fearful that a focus on environmental protection was a means for the developed world to keep the undeveloped world in an economically subservient position (Shiklomanov 1997). The most important outcome of the conference was the creation of the United Nations Environmental Program (UNEP). UNEP was designed to be the environmental conscience of the UN, and in an attempt to allay fears of the developing world, it became the first UN agency to be headquartered in an LDC (Nairobi, Kenya). In addition to attempting to find a scientific consensus about major environmental issues, a primary focus for UNEP was the study of ways to encourage sustainable development while increasing standards of living without destroying the environment. At the time of UNEP's creation in 1972, only 11 countries had environmental agencies. Twenty years later, that number had grown to 126, of which 85 (67%) were in LDCs (United Nations Industrial Development Organization (UNIDO) (1996)).

Water quality is closely linked to water use and economic development. In industrialized countries, bacterial contamination of surface water caused serious health problems in major cities throughout the mid 1800s. By 1900, cities in Europe and North America began building sewer networks to route domestic wastes downstream of water intakes. Development of such sewage networks and waste treatment facilities in LDCs has expanded tremendously in the past 2 decades. However, the rapid growth of urban populations (especially in Latin America and Asia) has outpaced the ability of governments to adequately expand sewage and water infrastructures. While waterborne diseases have been mostly eliminated in the developed world, outbreaks of cholera and other similar diseases still occur with an alarming frequency in LDCs (Donoso and Melo 2006). In developed countries, different types of pollution that impact water such as eutrophication, nitrification, and acidification have occurred sequentially, with the result that most developed countries have had time to develop strategies to deal with them. By contrast, however, newly industrialized countries are now facing all these issues simultaneously (Gleick 1998). The World Health Organization (WHO 2010) has developed water quality standards recommended for LDCs in different parts of the world. While many LDC governments have adopted the WHO standards, enforcement from adequate regulatory authorities is frequently sporadic at best or largely nonexistent.

Clearly, the problems associated with water pollution have the capabilities to disrupt life on Earth to a great extent. Governments around the world have passed laws to try to combat water pollution, thus acknowledging the fact that water pollution is a serious issue. But governments alone cannot solve the entire problem, although it is their responsibility to both provide an adequate water and sewer infrastructure and to enforce existing environmental laws. Some developing countries do not have a stable government that can provide either an infrastructure or any kind of environmental protection (WHO 2010; World Water Council 2000). It is ultimately up to citizens to be informed, responsible, and involved when it comes to the problems that nations face with their water (a scenario nearly impossible in poor countries with dysfunctional governments). Also, in most LDCs low-income, minority, and indigenous communities have been historically underrepresented in the regulatory decision-making process. In the twenty-first century, awareness and education will most certainly emerge as the two most important ways to prevent water pollution (Fidelia 2008). If these measures are not taken and water pollution continues at current rates in many countries around the world, people will continue to suffer, and disease outbreaks will continue (United Nations Commission on Sustainable Development 1997; UNDP 2006).

Global environmental collapse is not inevitable. But the developed world must work with the developing world to ensure that new industrialized economies do not add to the world's environmental problems. Politicians must think of sustainable development rather than economic expansion. Conservation strategies have to become more widely accepted, and people must learn that energy use can be dramatically diminished without sacrificing comfort. In short, with the technology that currently exists, it is possible to reverse years of global environmental mistreatment, but accomplishing this will not be easy or inexpensive (World Water Council 2000; Fidelia 2008).

22.2 The Role of MST in Developing Countries

Many different MST techniques for detecting and tracking microbes and chemicals of public health concern are described in detail in this book (Chaps. 3–8, 10, and 11). These techniques will be useful for assessing health risks for humans in recreational waters, drinking water, and harvested fish or shellfish. Such MST technology will also allow the sources of microbes to be identified, assisting regulators in formulating strategies to reduce contamination. The most suitable MST methodologies for developing countries will be rapid, sensitive, accurate, and be as inexpensive as possible. As described in Chap. 16, most MST approaches in developed nations have concentrated on the traditional fecal indicator bacteria (FIB), but this may not be the most useful approach for LDCs. With the wide range of pathogens present in many polluted waters in LDCs, new MST methodologies are needed that target these pathogens directly rather than continued emphasis on the FIB. The largest source categories of fecal pollution in both LDCs and developed nations are municipal effluents, livestock/poultry, and wildlife (Chaps. 15 and 18). There is a role for MST indentifying all three of these sources.

22.2.1 Municipal Effluents

Communities and cities need the capability to track sources of municipal wastewater, combined sewer overflows (CSOs), and stormwater contamination as quickly as possible because of the relatively higher potential for the occurrence of waterborne pathogens of human health concern (Chap. 19). Fecal pollution can come from inadequately treated effluents from sewage treatment plants, sewage treatment plant bypasses, stormwater, and CSOs. Leaking septic tanks and shipboard wastes or "gray water" can be other sources of human fecal contamination found in aquatic ecosystems. One complication for MST is that municipal wastewater may not contain microbial contaminants exclusively of human origin. Municipal wastewater can also contain fecal contamination from food processing activities and from urban runoff sources such as pets and urban wildlife. Fecal contamination occurs frequently in urban waters as a result of discharges of various municipal effluents, among which wet-weather flows, stormwater, and CSOs are particularly important. Both stormwater and CSO discharges can be highly contaminated with fecal bacteria and pathogens and widely distributed throughout urban areas. As such, they need to be addressed in planning the protection of all waters. In many LDCs, direct discharge of urban sewage without treatment is all too common (WHO/UNICEF 2010). Stormwater characterization data indicate concentrations of *Escherichia coli* or fecal coliforms in the range from 10^3 to 10^5 CFUs/100 mL, many-fold higher in raw sewage. Such concentrations may be attenuated prior to discharge into open waters by stormwater management measures or by disinfection. The levels of FIB in CSOs or direct discharges are much higher than in stormwater and can be as high as 10^8 *E. coli* per 100 mL (Shanks et al. 2009). Consequently, the abatement of fecal contamination in receiving waters is among the primary drivers behind costly infrastructure improvement programs (WHO/UNICEF 2010). Abatement options comprise combinations of storage and treatment, in which the treatment process generally includes disinfection, particularly where outfalls or discharges are located upstream of recreational waters. With aging and overloaded sewage infrastructures, many of these same issues apply to cities in developed nations as well (Chap. 19; Shanks et al. 2009).

22.2.2 *Agriculture*

While it is possible to treat livestock/poultry fecal wastes effectively and apply manure to agricultural lands safely, poor farming practices or storms and surface-water runoff can result in fluxes of fecal pollution downstream into aquatic ecosystems. Communities need to be able to track sources of livestock fecal pollution quickly to prevent contamination of source waters used for drinking, irrigation, or recreation (Anderson et al. 2006). Increasingly intensive rearing practices for livestock animals such as cattle, hogs, and poultry present significant animal waste management challenges in both LDCs and developed nations. Management of aquatic ecosystems in agricultural watersheds will need to consider potential livestock fecal pollution sources (e.g., droppings on pastures, manure lagoons) and the timing of events such as manure spreading when investigating potential fecal pollution sources. Livestock numbers, densities, and manure production have increased in almost every nation over the last decades. However, the impact of this trend differs both among various countries and within regions inside each country. Technological and structural changes in the livestock sector and increased demand for livestock products as nations develop are causes of the rapid increase in livestock and poultry numbers. The trend is toward specialized larger farms employing a smaller immediate land base in some cases and the proliferation of many small farms in others (Chap. 18).

22.2.3 *Wildlife*

Wildlife can present an unpredictable and difficult source tracking challenge that is not so amenable to control with familiar waste treatment practices (Noble et al. 2006). Notable are the growing numbers of birds such as shorebirds (gulls) and migratory birds in many countries (Chap. 20). Where aquatic ecosystems occur near large wildlife populations (e.g., bird colonies), consideration needs to be given to monitoring wildlife populations, their fecal droppings, and their seasonal migrations or behavior characteristics that could contribute to fecal contamination. Fecal pollution from wildlife species has been shown to contribute to impairment of recreational waters in many developed countries. For example, fecal droppings from birds along beaches or from birds roosting under bridges can lead to significant increases in waterborne FIB in urban habitats (Chap. 19). In some areas, efforts to enhance biodiversity habitat and establish buffer strips along streams may also facilitate increased loadings of fecal pollution from wildlife (Chap. 14). MST studies need to evaluate wildlife species as possible sources of fecal pollution and to consider the significance of local wildlife populations such as aquatic mammals or birds (e.g., gulls and waterfowl) and the timing of wildlife movements and migrations (Chap. 20; Noble et al. 2006).

22.3 Water Quality Improvement Case Studies: Examples for Developing Countries

In attempting a chapter to address water-quality issues and appropriate uses for MST in LDCs, only certain countries could be included, and many had to be left out. A chapter covering nearly all LDCs would be a book in itself, so the approach in this chapter was to select certain LDCs as examples that were representative of other nations in similar circumstances within the same region. Kathuria (2006) examined three countries in detail from different parts of the world – Malaysia (Asia-Pacific), Poland (Eastern Europe), and Colombia (Latin America) where substantial improvements in water quality have been achieved. The improvements in each case were due to very different solutions but can serve as three distinct and relatively successful options for other LDCs (as described in Sect. 22.5).

22.3.1 Malaysia

Malaysia is the world's leading producer of palm oil, and the scenario in Malaysia was essentially one of controlling this single industry as it rapidly grew and expanded. By mid-1977, 42 rivers in Malaysia were so critically polluted from palm oil mill discharges that freshwater fish could not survive in them. By 1992, improvements had occurred in all rivers, and only 12 were listed as impaired. The solution in Malaysia most closely resembled the approach taken in developed countries, consisting of creating an environmental government agency with both regulatory and enforcement powers that was able to impose penalties or fines, and close down problem mills if pollution abatement programs were ignored or failed to meet standards. In addition to regulation, the government established a strong program emphasizing research and development that resulted in effective and relatively inexpensive mill effluent treatment technologies along with numerous useful by-products produced from the effluent.

The Malaysian experience in effluent control in the palm oil industry demonstrated that a set of well-designed environmental policies, coupled with R&D support, can be very effective in controlling industrial pollution in a developing country. It also demonstrated that pollution reduction and industrial expansion can occur simultaneously, a lesson from an LDC that is worth exporting to developed countries as well. That an industry was economically important was not used as justification for not addressing the pollution problems caused by it. Perhaps the most important lesson from the Malaysian example is that compliance required a regulator to fulfill multiple roles – a credible regulator, a facilitator, and an enforcer. The credibility of developing regulatory standards was established when the industry was included in the process of creating the standards. The facilitator role of the regulator became apparent when the agency allocated some time to the industry to develop and construct treatment facilities and acquire some experience before implementing new regulations. The enforcer role consisted of unannounced visits and sample collection, penalties for defaulters, and the closing of the worst problem mills until pollution abatement was implemented. The success in Malaysia (simplified by being essentially a single industry situation) also demonstrated the importance of an independent judicial system and centralized government that supported the roles and actions of the regulators (Kathuria, 2006).

22.3.2 Poland

Poland represents a much more complicated situation than Malaysia, as Poland was a satellite of the Soviet Union for over 40 years and had little political power to control pollution from sewage produced by different Soviet-operated heavy industries. The Soviets were more concerned with water supply and industrial production and essentially ignored water quality. In 1991, almost 11% of the Poland was considered severely environmentally threatened, and of the 118 rivers monitored in 1990, only 6% of the rivers were rated in the Class I category (i.e., of drinkable quality with the use of disinfectants only). Eighty percent of the total river length in Poland was deemed nonclassifiable (nonusable) according to biological characteristics, and 36% according to chemical/nutrient properties. This increased to over 83% and 40%, respectively, by 1992.

Poland had developed a system of environmental/resource charges and fines as part of an environmental management system as early as in the 1970s, and pollution fees and environmental protection funds were first levied in the early 1980s. However, the fines levied during this period were toothless as Poland was an occupied country, and the Soviets ignored Polish law. However, with the change in the political and economic scenario in 1989, the environmental charges became institutionalized. In the post-1989 era, the fines were reestablished with emphasis on incentive impacts and pollution reduction. The most important lesson from the Polish example is that it did not rely exclusively on charges; rather, it was the combined use of discharge permits based on environmental quality standards with fees and fines, public-funded environmental subsidies, and a widely publicized list of the worst polluters that brought the pollution problem under control. These were complemented by long-term regulatory consistency, gradual tightening of enforcement, and limited administrative discretion to exempt polluters from fines and enforcement. Intensive monitoring in the first few years post-Soviet occupation to identify and publicize the worst polluters, and then dealing with them on a case-by-case basis that involved public assistance funds, has proven over time to be a very effective approach. Although Poland has made good progress in cleaning up pollution hotspots, Polish rivers are still too polluted for industrial or agricultural purposes. Poland can be taken as an example of many emerging eastern-European nations that were dominated by the former Soviet Union (the Czech Republic, Slovenia, *etc.*). Poland is best viewed as a work in progress, but the implemented strategies that are now in place should continue to produce water-quality improvements in the future (Kathuria, 2006).

22.3.3 Colombia

Colombia has abundant renewable water resources with more than 1,000 river systems and 720,000 microwatersheds. Rainfall level exceeds 2,000 mm per year over 88% of the country, with a national average of 3,000 mm per year. Despite this high reserve of water sources, water bodies are highly contaminated due to discharge mainly from industries and cities. As recently as 1997, 95% of municipal wastewater, 70% of industrial wastewaters, and 90% of agro-industrial wastewaters flowed into Colombia's watershed ecosystems completely untreated. The health costs associated with this water pollution have been estimated at around US $17 million per year. Colombia introduced water user charges as early as 1942, but the real introduction of water charges took place only in 1993, with the passage of Law 99. Implementation occurred initially in 1997 in seven regions of the country, with each region allowed to vary pollution charges until the target reduction had been achieved. This action resulted in significant reductions in pollution across the country for all types of discharges, although Colombia is also a work in progress, and much improvement is still needed before rivers can be considered sufficiently clean for other types of uses such as recreation.

Colombia has come as far as it has with a combination of a strong judicial system, powerful regional administrations, and popular and active public support. The judicial system laid the foundation for public support with legal instruments such as the "Citizens' Rights Action" (Acción de Tutela, 1991), the "Compliance Action," and the "People's Action." The latter specifies that anyone who files a People's Action has the right to compensation between 10 and 15% of the total value of the work necessary to correct the environmental damage caused. This provided a strong incentive for citizens to sue noncompliant firms, and it also represented a potentially powerful substitute for ineffective centralized administrative enforcement. The effectiveness of these instruments has been proven in a short time: in 3 years there have been almost 300 "tutelas" related to environmental disputes. Another approach that worked in Colombia was the development of an advanced environmental information system on the internet to inform the public about polluters. Through the Web site, the public can observe industrial discharges into rivers, and this information has served as a strong tool to demand that polluters improve their environmental performance, and it constitutes a significant incentive for environmental investment. Industry support was slow in developing, but acceptance has been improving after numerous meetings in which regulators and international experts presented credible information about abatement costs. Fees for noncompliance were kept high enough to affect managers' financial calculations significantly, making it more profitable to treat than to pay charges, another important message from the Colombian case.

Widespread political support was also essential for successful pollution reduction despite two changes in the national administration and three different environmental ministers since 1997. The support continued because the program's local constituencies remain politically potent, although there are still implementation problems.

Without strong centralized administration, regulation and enforcement is left to the different regional governments, with the expected variation due to political activities of industry, local corruption, the tendency of each region to set its own standards and timetables, etc. Most of the successful pollution reduction occurred during the most severe economic recession Colombia has faced in decades. Although the country appears to have emerged from the recent recession, at present, much infrastructure remains to be built (Kathuria 2006).

22.4 Benefits, Issues, and Solutions Relating to Water Quality (Rosegrant et al. 2002)

22.4.1 Benefits and Issues

Water stimulates annual food production and covers 16 different food-based commodities just in agriculture alone. Water also drives food demand and commodity prices, plus trade at a global level for both irrigated and rainfall-based food production. Improved water quality also results in less morbidity and mortality regarding both endemic and seasonal disease outbreaks via contaminated water (fecal-oral route). Healthier human populations are more robust and productive and are less likely to challenge authorities over poor living conditions resulting from environmental pollution. Issues that impact the benefit of improving water quality include the following: (i) increasing competition for water severely limits irrigation, and constraints on food production leads to conflicts between countries; (ii) if the current slow progress in extending access to safe drinking water continues, water quality will decline, and the amounts of water for environmental uses will be inadequate; and (iii) declines by governments regarding enforcement of water policies and investments could lead to full-blown water crises for many countries.

22.4.2 The Desired Outcome and Three Different Scenarios (Rosegrant et al. 2002)

The three case studies presented in Sect. 3 (Malaysia, Poland, and Colombia) all represent situations of countries moving from Water Crisis (22.4.3.2, below, the unfortunate situation in most LDCs) toward the Sustainable Water-Use scenario (22.4.3.3, below). While all three countries have a long way yet to go, they collectively demonstrate that fundamental changes and improvements in water management and policy can produce a sustainable future for water and food (desired outcome). The scenarios are defined as follows (and will be used to "evaluate" each developing country in Sect. 22.5):

22.4.2.1 Business as Usual Scenario

Assumes continuation of existing policies and trends:

Continued decline in research investments – MST used moderately in some situations
Declining investment in irrigation expansion and reservoir storage
Limited institutional and management reform
Slow water-use efficiency increases
Slow growth in harvested area
Production increases mainly through yield growth
Low priority of rain-fed agriculture
Expansion of groundwater pumping
No increase in environmental flows

22.4.2.2 Water Crisis Scenario

Assumes worsening of existing policies and trends:

Sharp reduction in research investments – little support for MST
Degradation of irrigation infrastructure and management
Reduced water–use efficiency
Lower investment in crop breeding and slower growth in rain-fed crop yields
Increased erosion and sedimentation
Decline in net water storage due to reduced investment and sedimentation
Reduction in environmental flows, less pollution restrictions
Low investment in water supply systems, decline in access to household water services

22.4.2.3 Sustainable Water-Use Scenario

Assumes improvement of existing policies and trends and focus on the environment:

Increased investment in research and higher substantial use and support for MST
Elimination of pollution sources (demonstrated by MST) necessary to achieve success
Growth in water storage and reduced sedimentation
Higher water-use efficiency due to water management reform
Better government regulation and enforcement of regulations
More effective use of rainfall
Increased water prices and higher investment in water supply systems
Sharp increase in environmental flows rather than reducing flows with dams

22.5 Examples of Developing Countries and Likely Scenarios

22.5.1 People's Republic of China

22.5.1.1 Current Status and Perspective

Rapid economic expansion in China may elevate the country to "superpower" status, but the country faces some of the most serious environmental challenges in the world. China's extraordinary economic growth, industrialization, and urbanization, coupled with inadequate investment in basic water supply and treatment infrastructure, have resulted in widespread water pollution (CIA, Central Intelligence Agency 2010f). Water supply is scarce in the populous north, annual flooding endangers lives and land in the south, and growing municipal and industrial pollution jeopardizes fast-developing regions throughout the country. Currently, around 700 million people in China drink water that fails to meet state standards for FIB and that is presumably contaminated from fecal sources (Boxer 1998).

According to data of the WHO/UNICEF Joint Monitoring Program (WHO/UNICEF JMP, 2010), 215 million Chinese did not have access to improved water sources in 2008; water source improvements would include household connections, standpipes, protected wells and springs, boreholes, or rainwater collection. About 600 million did not have access to improved sanitation, which includes connection to public sewer or septic systems or improved latrines. Improvements and modernization in rural areas lag far behind those of urban areas. Although improved water supply and sanitation have dramatically increased over the past 2 decades in parallel with economic growth, these advances have not ensured access to safe water.

China has as much water overall as Canada, but 100 times more people, approximately 1.4 billion. China is the most populous country in the world, and the Woodrow Wilson International Center report (Boxer 1998) that China's per capita water reserves of 2,500 m^3 are one-fourth the global average is sobering. So-called "water pollution accidents" are often triggered by weather: rainfall, floods, droughts, and an overwhelmed infrastructure for wastewater storage and treatment releases industrial wastewater or sewage overflows and reduces the supply of fresh water available to dilute pollutants (Cao and Xu 1989).

22.5.1.2 China's Water-Quality Standards

China's bacteriological drinking water standards include fewer than 100 CFU/ml for total bacteria and fewer than 3 CFU/mL for total coliforms (Chinese National Environmental Protection standards cited in Wu et al. 1999). Although currently 26 chemicals are routinely assessed (including arsenic, chloride, nitrate, and silver) and have regulatory limits, regulations regarding bacteriological indicators are sparse; for instance, no drinking water standards have been implemented for the detection of *E. coli* (Pakistan Council of Research in Water Resources 2002, (World Water Council 2003).

Chao et al. (2003) studied the correlation in subtropical fresh waters of northern and central Taiwan and attempted to correlate traditional culture-based assays for total coliforms, fecal coliforms and enterococci, with culture-based detection of potential pathogens such as *E. coli* and *Salmonella* spp. The authors argued against the utility and validity of reliance on commonly used culture-based FIB for microbial water quality in estimating risk of co-contamination by human pathogens commonly found in feces. However, one might counter that argument by positing that implementation of conventional culture-based fecal indicator standards is much better than none at all.

An examination of China's major rivers reveals that most of them do not meet government standards for primary drinking water supplies, and some rivers are unsuitable even for agricultural purposes due to industrial pollution (Boxer 1998; Ministry of Environmental Protection 2009). The level of water pollution in China is alarming, as such contamination not only degrades the environment but also severely threatens public health (Wu et al. 1998). China's water pollution crisis made international headlines following a 2005 petro-chemical plant explosion which released 100 tons of benzene into the Songhua River (The World Bank 2007a). Drinking water source pollution has also been the result of toxic algal blooms in the Tai Lake in May 2007 (The World Bank 2007a). Such spills may encourage the Chinese government to change its approach to water pollution, specifically as it relates to public access to information, enforcement of pollution laws, and accountability.

The latest World Health Statistics Report (2009) states there is access to improved drinking water sources in 98% of urban areas and 81% of rural areas; however, there is much less access to improved sanitation methods, 74% of urban areas and 59% of rural areas. As of 2005, 364 of 661 cities in China have wastewater treatment plants, resulting in about 45% of the country having the capacity to treat human waster (The World Bank 2007b). The water resources and public health literature in China identify three principal threats to human health from water pollution and degraded water quality: (1) rapid and unregulated expansion of industrial activities, (2) growth of urban and suburban areas without adequate investments in water supply infrastructure, and (3) adoption of green revolution technologies together with a continued reliance on sewage irrigation (Wu et al. 1999). Since 1999, China has made substantial improvements in all of these areas. The deployment of new water-monitoring technologies, of which MST is just one example, along with much more substantial regulatory and enforcement policies, will all be important in reversing the historic degradation of China's water resources. However, much work is needed to bring China's microbial water quality on par with that of developed Western countries (World Water Council 2003).

22.5.1.3 Economic Expansion and Rapid Industrialization

During the past 3 decades, China's economy has changed from a command (state-controlled) economy that was largely closed to international trade to a market-dominated economy that fosters a rapidly growing private sector and has promoted China to a major player in the global economy (CIA 2010f). Over 10 years ago, Wu et al. (1999) noted the environmental impact of this transition from economic production by state-owned enterprises to that of township-village enterprises and other parts of the private sector. More recently, such high economic growth rates have exerted extraordinary pressure on natural resources, particularly water. In China, as in many rapidly developing LDCs, inadequate stormwater collection systems in cities allow groundwater infiltration into the drainage pipes and overflows of untreated wastewater into receiving water bodies (World Water Council 2003).

To mitigate environmental water-quality issues in China, the Asian Development Bank has provided substantial funding to finance integrated water environment management projects to improve water environment, urban flood-control facilities, and urban watercourse ecological systems. Although commendable, recent projects within the Jiaozhou Municipal Government (Qingdao) (Asian Development Bank 2010a) and the northeastern city of Harbin (Asian Development 2010b) are both entirely focused on chemical pollutants and do not addresses microbial or fecal borne pollution. While there are situations where MST can play an important role in China, point sources in both urban areas (sewage discharges) and rural regions (intense livestock farms with direct access to streams and rivers) are obvious. Also, in many Chinese rivers, chemical pollution is so bad that it overshadows microbial pollution and must be dealt with first. Adapting MST methods to track specific pathogens such as *Camplyobacter* and *Vibrio* rather than the FIB may be a more useful approach in most of China.

22.5.1.4 Conclusions

At present, most of China is in the Water Crisis Scenario and its challenge for the future is to avoid moving upwards to the Business as Usual scenario and being satisfied with that. China must take steps to move toward the Sustainable Water-Use scenario. Chinese decision-makers and researchers face daunting challenges of not only how to strengthen research but also how to establish a legislative and regulatory mechanism, as well as a policy framework to guide the costly efforts of water pollution control and remediation (World Water Council 2003). How well this can be accomplished will largely determine China's future in the world.

22.5.2 India

22.5.2.1 Current Status and Perspective on India's Microbiological Water Quality

India is another large country that is both demographically and ecologically diverse. Although generalizations applicable to the entire country are difficult to formulate, the overall situation in India regarding water quality and quantity is one of crisis in the future – unless dramatic changes are made soon. India is on track toward becoming a severely water-stressed nation. There is increasing demand for potable water in the absence of adequate measures to provide clean, safe water to the country's citizens (Veeralapalam et al. 2009). As with other large, rapidly developing countries, problems are largely due to the increase in urban population, depletion of nearby water sources, water pollution, inefficient use of water, inefficient management of water supply systems, and multiple institutional and bureaucratic arrangements (National Institute for Urban Affairs 2005).

India is the second most populous country in the world, after China, with a population projected to reach 1.2 billion in 2010 (CIA Factbook 2010e). Rainfall and the snowmelt of glaciers in the Himalayas that supply the country's major rivers (and subsequently into the groundwater) are the two sources of water in India (FAO 2000). Groundwater is being depleted at an unsustainable rate and is also showing intrusion of seawater (FAO 2000). The rural population in India is largely dependent upon this groundwater, and people in urban areas rely mainly on surface waters (India Water Portal 1991). In a list of 122 countries rated on quality of potable water, India ranks near the bottom at 120 (Bansil 2004). Discharge of untreated or partially treated wastes from industry, domestic sewage and fertilizer, and agricultural pesticide runoff have resulted in most of the country's water resources being polluted, and tap water is not considered to be potable throughout the country (WHO 2007b; CIA Factbook 2010e).

There are two different perspectives on water quality in India. Data from the Indian government present a very optimistic scenario. For example, the national Department of Drinking Water Supply estimates that 94% of rural habitations and 91% of urban households have access to safe drinking water. However, some experts point out that these data are misleading, simply because the coverage refers to installed capacity and not actual supply that is available to consumers. The World Health Organization (2007a) estimates that the urban population with access to improved drinking-water sources was 96% as of 2006. The urban population with access to improved sanitation was 52% in urban areas and 18% in rural areas (World 2007a). Aggregate figures are also misleading since there is considerable spatial and temporal variation in rainfall (CIA Factbook, 2010e). Some areas receive slight rainfall, whereas others experience monsoon conditions that can result in flooding, loss of life, and increased poverty (Veeralapalam et al. 2009). The grim reality appears much different from government reports: the World Bank estimates 21% of communicable diseases in India are water related, and in 1999 diarrhea killed over 700,000 Indians (estimated, DeNormandie and Sunita 2002). High nitrate content in water is another serious concern, with fertilizers being the main sources of nitrate contamination (Suthar et al. 2009).

Perhaps the most widespread water-quality problem in India is microbial contamination resulting in diarrhea, cholera, and waterborne viral hepatitis (DeNormandie and Sunita 2002). Bacteriological levels in Indian waters are supposed to adhere to the following guidelines: <10 CFU/100mL for total coliforms and 0 CFU/100mL of *E. coli* (India Water Portal, 1991). Diseases resulting from contaminated water are of major importance in India, including bacterial diarrhea, hepatitis A and E, and typhoid fever (CIA Factbook 2010e). Cholera, dysentery, hepatitis, and cryptosporidiosis have resulted in disease outbreaks due to drinking contaminated water for the entire South East Asia region (WHO 2007b). Many sources of pollution in India are so apparent (open sewers, cattle in public waters, *etc.*) that there is little role for MST until such obvious sources are addressed first.

22.5.2.2 Conclusions

At this point, it is hard to see any future for India's water-quality situation other than the Water Crisis scenario. In most ways, India's microbiological water-quality conditions suffer the same political and bureaucratic impediments as does that of China. Thus, India's challenge is to avoid being trapped in the Business as Usual scenario and to move toward the Sustainable Water-Use scenario. Like China, how well India meets these challenges will determine quality of life for all of its citizens.

22.5.3 Republic of the Philippines

22.5.3.1 Current Status and Perspective on Water Quality in the Philippines

The Philippines has a population of approximately 98 million people, the 12th most populous country in the world (CIA 2010e; Kingston 2006). The Philippine islands constitute an archipelago in the Pacific Ocean off the coast of Southeast Asia, and the population density is 306.6/km^2, ranking 44th in the world (Republic of the Philippines 2007; United Nations 2009). The Philippines includes some 7,100 islands, but only 2,000 are inhabited, and people are unevenly distributed throughout the islands (Kingston 2006). The Philippines has a tropical marine climate and is hot and humid year round (Philippine Department of Health, PDH 2009). The Philippines is one of the world's most ecologically diverse countries (US Department of State 2009a, b). The Philippines (like Malaysia, Sect. 3.1) can serve as a surrogate for much of SE Asia, especially the tropical island nations, as the Philippines is representative of similar ecological and climatic conditions also found in Laos, Cambodia, Thailand, and Vietnam. The Philippines in general suffers from a poorer economy and a lower standard of living for its people than do most other SE Asian island nations, a circumstance that exacerbates the problems of poor microbiological water quality and the country's inability to ameliorate them (Kingston 2006).

Water is an abundant resource in the Philippines. In addition to large amounts of rainfall, there is substantial water storage capacity; approximately 70% of the land area is in watersheds containing lakes, springs, and streams (Kingston 2006). The agricultural sector uses 86% of nation's water supply, commerce and industry use about 8%, and domestic use amounts to 6%, with numerous agencies handling distribution for public consumption (World Bank 2005). Demand is estimated at one third of the actual supply. Problems, however, lie in quality and distribution. Saltwater intrusion and pollution offset abundant freshwater, and demand is increasing rapidly (FAO 2005). At the national level the PDH is responsible for water-quality monitoring and coordination of the development of drinking water standards, and it has defined microbe-safe water as that free of bacteriological, viral, or other disease-causing organisms. The Philippine National Standards for Drinking Water of 1993 outlines parameters for drinking water quality (Philippine Dept. of Health 2009). There are currently 56 bacteriological, physical, chemical, radiological, and biological parameters to be monitored, including bacteriological standards based on FIB (Kingston 2006). Sanitary inspectors take samples to provincial or regional laboratories, but testing is limited to bacteriological quality and physical parameters, and if the samples pass these tests, a "certificate of potability" is issued.

Critically inadequate management of solid and liquid wastes in the Philippines results in pollution that directly affects water quality and the nation's food resources; this is particularly true in high population urban areas. Nearly all surface waters in metropolitan Manila are considered biologically dead during the dry months. In 2002, around 80% of the urban population had access to sanitary toilets compared to 61% in the rural communities, and overall nearly one quarter of households did not have access to sanitary facilities (WHO 2009c). Approximately 5% of the total population is connected to a sewer network, while the vast majority uses flush toilets connected to septic tanks (often defective and/or poorly maintained). Sludge treatment and disposal facilities are rare, and most effluents are discharged without treatment (World Bank 2005). According to the Asian Development Bank, the Pasig River is one of the world's most polluted rivers and is highly contaminated with human fecal material (Asian Development Bank 2007).

Waterborne diseases remain a severe public health concern in the country. About 4,200 people in the Philippines die each year due to contaminated drinking water (World Bank 2005). The CIA Factbook (2010e) designated the degree of waterborne disease risk to be high, with food or waterborne diseases of particular hazard, mainly bacterial diarrhea, hepatitis A, typhoid fever, and leptospirosis. Diarrhea remains the number one cause of illness in all age groups and is the number three cause of morbidity (Philippine Dept. of Health 2009). These waterborne diseases account for more than 500,000 morbidity and 4,200 mortality cases a year. Investigations of these disease outbreaks (with little need for the sophistication of MST methodologies) have identified contaminated sources of drinking water, improper disposal of human waste, and unsanitary food handling practices as the main causes (Kingston 2006).

There are government initiatives aimed at protecting water sources. The PDH set a goal in 1999 to increase the proportion of households with access to safe water to 91% by 2004. In rural areas, this goal has not yet been met as of 2010, though in urban areas 96% of the population has access to improved drinking water (WHO 2009c). It will be a daunting task for the Philippines to achieve the Millennium Development Goal of increasing formal access to water supply to 90% by 2012 (World Bank 2005). On the plus side, water-quality improvements have begun to make a difference in the health of the citizens. Outbreaks of cholera have averaged less than one per year during the period 2000–2003, compared to 12 outbreaks a year in 1998. The morbidity trends for typhoid and paratyphoid fever have decreased from 33 cases per 100,000 population in 1995 to 17.1 per 100,000 in 2000, while the mortality trend has remained consistently low from 1980 to 2000 (Kingston 2006).

22.5.3.2 Conclusions

The Philippines is currently in the Water Crisis scenario in many rural and most urban regions, and the Business as Usual scenario appears most likely for the future. The prevalence of typhoons that regularly damage urban areas and the lack of economic resources and/or political apathy will exacerbate the potential for a perfect storm: the collapse of the sanitary water infrastructure in the Philippines.

22.5.4 Mexico

22.5.4.1 Current Status and Perspective on Mexico's Microbiological Water Quality

While many Mexican water supply and sanitation service providers rank among the best in Latin America, in general the Mexican water and sanitation sector is characterized by the following issues (1) poor technical and commercial efficiency of service provision; (2) inadequate water service quality; (3) inadequate sanitation service quality, particularly concerning wastewater treatment; and (4) inadequate coverage, in particular in poorer rural areas (CIA Factbook 2010c). During the past decade, the Mexican water and sanitation sector made major strides in service coverage with water supply and sanitation coverage. In urban areas, almost 100% of the population is estimated to have access to improved water supply and 91% to adequate sanitation. In rural areas, the respective shares are 87% for water and 41% for sanitation. Coverage levels are particularly low in the south of the country. Some 17% of Mexicans (over 18 million) have no access to a potable water supply (WHO 2009d).

Quality of service also leaves much to be desired. The 2000 census indicated that 55% of Mexican households with access to piped water received services on an intermittent basis, in particular in smaller municipalities and poor areas. About 36% of wastewater was being treated in 2006, a share that is more than twice as high as the average for Latin America. However, the percentage of Mexican treatment plants that do not comply with standards for effluent discharge is unknown.

In 2006, 63% of the Mexican water was extracted from surface sources, such as rivers or lakes. The remaining 37% came from underground aquifers. Owing to the strong growth of population and internal migration toward arid and semiarid regions, many water resources in North and Central Mexico are now overexploited. According to the National Water Commission (2010), groundwater extraction comprises almost 40% of total groundwater use. In addition, government officials estimate that 52% of surface waters are very polluted, and only 9% of surface waters are in an acceptable condition (WHO 2009d). Despite scarce resources in many Mexican regions, water consumption is at a high level, partly favored by poor payment rates and low tariffs. In 2006, more than three quarters (76.8%) was used for agriculture, while public supply only used up 13.9%, the remainder being used by thermal power station (5.4%) and industry (3.8%). In 2006, some 77.3 billion m^3 of water were consumed in Mexico, of which 10.7 billion m^3 were used for domestic consumption. This means that the average domestic use per capita and day was 270 L. This amount is not sustainable, especially in arid regions of Mexico, and does not reflect the disparity between areas served by an infrastructure and poor regions where no service is available. Like many developing countries, Mexico is one drought and crop failure away from disaster (WHO 2009d).

Mexico ranks 11th in terms of global human population (around 107,450,000 people, CIA Factbook 2010c). Seventy percent of Mexicans reside in the cities, and the population around Mexico City is about 20 million, which ranks it among the most densely populated area in the world. The highest structure in population is observed between the age group of 15–64. Population growth rate was estimated at 1.16% in 2006. Birth rate was 20.69 per 1,000, and the death rate was 4.74 per 1,000. Based on these birth and death rates, Mexico's population will continue to increase. The population has seen a sharp rise in recent years due to migration that has taken place from places such as Guatemala, Peru, Cuba, Panama, and Venezuela (CIA Factbook 2010c).

22.5.4.2 Conclusions

Based on the three scenarios described in Sect. 22.4.2, Mexico will likely trend toward the Water Crisis scenario in many rural and arid regions, while trying to maintain the Business as Usual scenario in most urban areas; an achievement that will be very difficult based on ever-increasing urban populations and civil unrest caused by drug trafficking that has now spread far beyond the border region in both Mexico and USA.

Mexico and Chile (Sect. 22.5.5) were chosen as representative of Latin America, thus omitting Brazil, the world's fifth largest country, both by geographical area and by population, and the largest in South America. Brazil has been described as a heterogeneous blend between Colombia and Chile/Argentina. Chile and Argentina are relatively "clean" countries, which have a better foundation on addressing microbial water quality than does Colombia (which is more like Peru, Ecuador, and Nicaragua). Brazil combines both the best and worst of South America. Any discussion of Latin America also lends itself to briefly include the greater Caribbean Region. Throughout the Caribbean, access to water and sanitation remains insufficient, particularly in rural areas and for the poor. Access also differs substantially among and within countries. According to the World Health Organization/UNICEF Joint Monitoring Programme (2010), in 2004 the share of population that was connected to an improved water source varied from 54% in Haiti to 100% in Uruguay. Some 50 million people, or 9% of the population of Latin America and the Caribbean, did not have access to improved water supply in 2004, and 125 million or 23% did not have access to improved sanitation. Increasing access remains a challenge, especially given the poor finances of wastewater authorities and central and local governments. Only an estimated 15% of the collected wastewater actually enters wastewater treatment plants in the Caribbean region, and these plants often are not functioning properly. Even for those having access to water supply, poor service is often experienced in the form of intermittent supply, low pressure, and poor drinking water quality. However, differences in service quality between countries and cities in Latin America and the Caribbean are vast, and some service providers achieve a quality of service on par with developed countries (WHO 2009d).

22.5.5 Republic of Chile

22.5.5.1 Current Status and Perspective on Chile's Microbiological Water Quality

Chile has an estimated population of 16,601,707 as of July 2009, ranking as the 60th most populous country in the world with a population density of approximately 22 persons/km^2 (CIA 2010b). The largest urban area is the capital city of Santiago. Chile is comprised of three distinct geographic regions with an irregular distribution of rainfall totals: the arid north, the Mediterranean central region, and the cool and damp south (US State Dept. 2009a, b). The central region of Chile is home to roughly two-thirds of the country's population (Speiser 2009) and is the most variable in terms of rainfall totals, being prone to both droughts and flooding (Vargas 2008). Most of the water in Chile comes from its rivers. The Baker and BíoBío rivers are used to generate hydropower. The Maipo provides water to the city of Santiago and is also used for irrigation. The Loa, the longest river in Chile, has suffered pollution from copper mining. The Maule is predominantly used for agriculture. Chile has water in aquifers, but they are used much less than the surface waters in the country. Water is split between industry (25%), domestic use (11%), and agriculture (63%) (CIA 2010b).

Chile is not as dependent as other Latin American countries such as Argentina and Bolivia on glacier water from the Andes (UNEP 2008). Chile is considered a model throughout Latin America as a provider of clean, safe potable water to its citizens. According to the WHO, in 2006, 72% of the population in rural areas and 98% in urban areas of the country had access to improved drinking water sources. Also, 74% of the population in rural areas and 97% of the population in urban areas has access to improved sanitation (WHO 2009b). This is largely due to the free-market model Chile applies to its water services. In 1981, the Water Law created property rights for water, resulting in water rights being independent of the land it is associated with. Water prices fluctuate based on availability, and this seems to encourage water rationing. This also results in improved services and a government water subsidy introduced in 1989 assured that the poorer citizens of Chile would also have clean water made available to them (de la Luz Domper 2009).

The Chilean government has put into place organizations that manage and monitor water services. Both the Ministry of Public Health (Ministry of Health 2009) and the Superintendencia de Servicios Sanitarios (SISS 2007) police water quality in rural and urban areas. There are laws in place to describe what the water-quality standards should be. The Official Chilean Standard (NCh 409/1.Of84) describes the physical, chemical, radioactive, and bacteriological requirements for any water system (Paratori 2009). Chile is not without water problems. Despite its successes in water-quality management, there are still several issues to be considered. Copper mining, hydroelectricity, agriculture, inadequate sewage treatment, and climate change all impact the cleanliness and availability of water. As Bauer (2004) addresses in his book on Chilean water supplies, problems include poor "environmental protection, river basin management, public interest, social equity, coordination of multiple uses, and resolution of water conflicts."

Inadequate sewage treatment is one of the greatest threats to Chile's water supply currently. While all major Chilean cities have municipal sewage systems, it is estimated that only 5–15% of that sewage is currently treated (Speiser 2009). This can be seen, for example, in the open drainage canals that run through Santiago. During time of flooding, the sewage would overflow into the nearby streets and the nearby Maipo River and contaminate the city's drinking water. The resulting diarrheal diseases have been particularly devastating to children. A study was performed in 2005 in which the groundwater in an area of central Chile was tested to determine if it complied with the Chilean Water Standard. One hundred percent of the sites tested exceeded government mandated levels for total coliforms, fecal coliforms, and fecal streptococci (Valenzuela et al 2009).

Despite the water challenges facing Chile, it is a country that seems to have acknowledged several of its shortcomings and is working to remediate them. In 2006, the SISS shut down several companies in the agro-industrial, wine, salmon, and frozen food industries for failing to comply with new, more stringent regulations of underground wastewater discharges (Medalla 2006). As of April 2009, miners were using 11.5 m^3/s of water, which is down from 15 m^3/s of water in 2006 (Vargas and Velez 2009). Copper mining is very susceptible to droughts. This has led miners to consider desalinization as a means to provide water to plants. Also the Ministry of Public Health is currently exploring implementation of tertiary wastewater treatment systems at their sewage plants (Speiser 2009). The National Commission on the Environment (CONAMA) has a plan set in motion to have wastewater treatment programs in place for 95% of raw domestic wastewaters in 2020 (Donoso and Melo 2006). As with other LDCs further along in development, there is more of a role for MST as obvious sources such as open sewers are dealt with, leaving the more difficult nonpoint sources to be addressed.

22.5.5.2 Conclusions

Based on Chile's current water issues, the Business as Usual scenario seems to most accurately fit the country's profile. However, the implementation of new strategies to decrease water usage and increase wastewater sanitization indicates that Chile is moving toward the Sustainable Water-Use scenario. Higher water-use efficiency and better government regulation and enforcement are increasing over time. Chile, like Colombia (Sect. 22.3.3), is often described as an example for water quality and management throughout Latin America. As a result of a devastating earthquake in 2010, much of the water infrastructure in affected areas of Chile will have to be completely rebuilt.

22.5.6 The Federal Republic of Nigeria

22.5.6.1 Current Status and Perspective on Nigeria's Microbiological Water Quality

Water plays an important role in the occurrence and spread of diseases, particularly in LDCs such as Nigeria, a West African nation known for its crude oil. This oil, along with other human activities, has posed immense problems with the availability and quality of potable water. Nigeria has an area of about 923,768 km^2 (356,669 sq mi, 32nd largest in the world) and a population of about 148 million (CIA 2010a). It is considered the most populous country in Africa and the eighth most populous country in the world (Lagos state government 2009). Nigeria has a tropical climate and many water bodies, with two main rivers, the Niger and Benue, that converge to form a "y" shape and end in the Niger Delta located in the southern part of the country (Human and Physical Characteristics of Nigeria 2009). Nigeria has an estimated growth rate of 1.99% (CIA 2010a), and 70 million people lack access to safe, potable water (UNICEF 2009).

The WHO (2009a) estimated the life expectancy of the average Nigerian to be 48 years for men and 49 years for women. The short life expectancy can be attributed to the high morbidity and mortality rates in the country, partly from water-related diseases such as cholera, typhoid fever, dysentery, and schistosomiasis,. The WHO (2009a) reported schistosomiasis and dysentery as prevalent diseases in Nigeria. Dysentery prevalence rate is 386 for every 100,000 people. (WHO 2002–2007a, b). According to the WHO records in 2001, 80% of diarrhea cases worldwide and 119,700 diarrhea cases in Nigeria were attributed to unsafe water or poor hygiene situations (Prüss-Üstün et al. 2008). Similarly, 68,000 schistosomiasis cases in Nigeria and 16.7% mortality in 2002 were related to improper water sanitation and hygiene. Surface waters in Nigeria are generally unsafe for drinking because they are more likely to be contaminated by sewer systems, industrial wastes, oil spills, and pollution from farmland runoff (Isiorho and Oginni 2007). Groundwater sources such as springs and boreholes have become alternate water sources used to circumvent polluted surface waters in many African nations such as Tunisia, Algeria, Libya, Morocco, Namibia, Zimbabwe, and Botswana; as a result, groundwater accounts for over 75% of Africa's source of drinking water (UNICEF 2009).

Like many tropical LDCs, Nigeria is overpopulated, and water is not readily available in many communities. In rural areas, people typically walk to water sources with containers to get water needed for their households. Residents use these same sources for bathing, washing, farming, fishing, cooking, and drinking. These activities pollute surface water and subsequently lead to disease outbreaks such as the regional cholera epidemics of 2001 and 2004 (WHO 2001, 2004). As a result of an unreliable government, many households depend on water from local sources, typically only 5 m (16 ft) deep, which do not meet WHO water standard requirements and hence are a threat to public health (Yusuf 2007). Water pollution in Nigeria is further complicated by the government because of the lack of proper regulation of wastewater disposal, placement of boreholes, oil spill remediation, and coordination of hydrological studies (Isiorho and Oginni 2007). In a 2000 report, more than 90% of Nigeria's oil was from the Niger delta, and oil accounted for 80% of the government's revenue (Johansen 2002). The combination of crude oil extraction in Nigeria, coupled with an ineffective government and lack of water monitoring, has resulted in widespread pollution of surface water sources with oil, especially within the heavily populated Niger delta (Johansen 2002).

A Nigerian Standard for Drinking Water Quality (NSDQW) was established in 2007, with the help of guidelines from institutions such as WHO and International Organization of Nigeria (ISO). Some internationally recognized organizations are attempting to do what the government has not done, provide guidelines and educational resources to assist with promoting the achievement of safe drinking water. One example is the Water Aid organization (Water Aid Nigeria 2008), whose approach involves investing in the provision of water and better sanitation conditions at the local government level.

22.5.6.2 Conclusions

Nigeria is a surrogate for most of the 20–25 countries that make up sub-Saharan Northwest- and Central-Africa (Kenya, Mali, Niger, Congo, Senegal, etc.), with political, social, and economic problems that typify much of Africa. South Africa is more "Western" and a contrast but faces an uncertain future. The National Council of Water Resources (NCWR), during its 16th annual meeting in 2007, initiated the development of national water quality standards for Nigeria as a result of the following reasons: overall poor water quality, lack of acceptable drinking-water quality standards, weakly coordinated water monitoring programs, absence of nationally acceptable and enforceable regulations for safe drinking water, inadequate water-quality data, and weak collaboration among the key agencies (UNICEF 2009). The same reasons can be applied to most LDCs around the world.

Nigeria is an African nation that naturally has an abundant water supply but yet experiences poor water quality because of poverty and most importantly, ineffective governmental policies and monitoring programs (WHO 2009a). These conditions increase the mortality and morbidity rates in the country, thereby crippling the economy of the nation. MST applications under current circumstances in Nigeria might be considered both crucial, and ironically, moot: whereas MST can provide highly useful information concerning the sources of fecal-derived microbial pollution, sadly, in the majority of situations in Nigeria, contamination with human feces is a given (Africa Water 2006). Moreover, at present Nigeria lacks infrastructure and perhaps the political will to implement adequate water monitoring, and in most cases it lacks the capability of applying more advanced technologies such as MST. With continuous water-quality interventions from organizations such as the Water Aid, WHO, and UNICEF, the "giant of Africa" may be able some day to fully harness its potential. For the present, like India, it is hard to see any future for Nigeria's water-quality situation other than continuing in the Water Crisis scenario.

22.6 Summary and Conclusions

22.6.1 Infrastructure Recommendations for LDCs

The World Water Council's global imperative from 1999 (World Water Council 2000) to ensure that at least 95% of human beings have safe water and sanitation by 2020 seems unreachable in 2010. Perhaps the Millennium Development Goal (MDG) of the WHO/UNICEF (2010) Joint Monitoring Programme (JMP) for Water Supply and Sanitation in the United Nations (MDG 7, Target 7c), to: *Halve, by 2015, the proportion of people without sustainable access to safe drinking-water and basic sanitation* is a more realistic goal. Regardless of whether or not such imperatives can be attained, there are many commonalities for any LDC regarding what can be done to improve both water quality and the lives of their citizens. A framework for improvement includes the following:

1. Approving more stringent laws and standards within water-quality sanitation codes, coupled with adequate enforcement and compliance.
2. Developing databases to track and compare water quality for different regions, detect trends, identify high risk areas, and prioritize parameters to be monitored.
3. Increasing the number of accredited laboratories involved in national water-quality monitoring and analysis; this will require additional laboratory infrastructure, trained personnel, and implementation of microbial monitoring (and perhaps MST) methodologies, particularly by local governments.

Incorporating MST technology where appropriate will allow the sources of microbes to be identified, assisting regulators in formulating strategies to reduce contamination. The most suitable MST methodologies for developing countries will be rapid, sensitive, accurate, and as inexpensive as possible (Chap. 2). Most MST approaches in developed nations have concentrated on the traditional FIB, but this will be a much less useful approach for LDCs. With the wide range of pathogens present in many polluted waters in LDCs, new MST methodologies are needed that target these pathogens directly rather than continued emphasis on the FIB.

22.6.2 Public Education and Water Quality

Many reports on global water quality suggest that it is ultimately up to citizens to be informed, responsible, and involved when it comes to the problems that their nations face with water (a scenario nearly impossible in poor countries with dysfunctional governments, such as Haiti and Nigeria). Citizen responsibility is easier said than done. As described in Chap. 12, informing the public and interpreting water quality, public health, and MST science in a manner that the general public can understand and empowering people with knowledge is never easy, especially with governments that show little interest in doing so. In much of Africa, for example, waterborne diseases continue to spread because victims do not understand the value of water treatment and have not made the connection between dirty water and sickness (Africa Water 2006). Uninformed populations do not realize that unsafe water sources result in waterborne diseases, the intake of toxic substances in drinking water, and heavy metal poisoning (UNDP 2006). In Africa, many people continue to drink utility water, though it is typically contaminated by the very pipes it travels through (Africa Water 2006). They also fail to understand the dangers of vended water, which has rarely been given proper treatment (also true for much of the rest of the world).

At least in Africa, one helpful approach has been to teach basic sanitation and hygiene that includes convincing people to use toilets. Such advocacy is necessary to create the sense that using latrines and managing waste is not taboo (it is in many tribal traditions) and is really acceptable to everyone (Lyimo 2009). Many African governments have failed to keep up with raw sewage generated by cultures where defecating in the streets is commonplace (Africa Water 2006). As bad as this may sound, people are frequently left without any other option. In Kenya, for example, the government's sanitation budget is lowest in the Nairobi slums, where the rate of urban growth, and thus the need for sanitation infrastructure has increased most rapidly. People in the slums migrated from rural areas, and living conditions are extremely congested (Africa Water 2006). With no basic provision for water and sanitation provided by the government or any other organization, it is little wonder that waterborne diseases flourish. Using MST to demonstrate human-origin pollution at such a location would be completely unnecessary.

Still, it is critical that citizens of LDCs become familiar with local water resources and learn how to dispose of harmful household wastes (including sewage) so that these wastes do not end up in sewage treatment plants that cannot handle them or in landfills not designed to receive such materials (Fidelia 2008). There are other action options for those who live in LDCs as well. For example, people must determine whether additional nutrients are needed before fertilizers are applied and must look for alternatives where fertilizers might run off into surface waters. Citizens need to preserve existing trees and plant new trees and shrubs to help prevent soil erosion and promote infiltration of water into the soil. These are just a few of the many ways where individuals have the ability to combat water pollution (Fidelia 2008). In the twenty-first century, awareness and education will most certainly emerge as the two most important ways to prevent water pollution (Rosegrant et al. 2002).

22.6.3 Clean Water on a Global Scale

Under current water policies and investments in many parts of the world, global food production will decline, targets for safe water access will not be met, and water quality will also decline. There is a link between water quality and food production, and this link is critical in feeding a growing human population. Increasing complacency by governments and citizens will lead to dramatic worsening of these trends, while improved policies and investments can produce a far more sustainable world (IPPC 2001).

It is critical that politicians think of sustainable development rather than economic expansion, as is already happening in China to a limited degree (World Water Council 2003). Conservation strategies have to become more widely accepted, and people must learn that water use can be dramatically diminished in many cases without sacrificing comfort or food production. These reforms imply fundamental changes in the way nations manage water and will take time and resources to implement and enforce. Time is very important as there is one additional factor that must be mentioned at the close of this chapter – climate change. This is already having an impact on water availability and quality and is frequently more visible and obvious in LDCs. From declining coral reefs throughout the Caribbean, to saltwater intrusions into freshwater aquifers in coastal areas around the globe, to the melting of mountain glaciers worldwide, especially in the Andes and the Himalayas, climate change will have a vast and permanent impact on people everywhere (Mahesha and Nagaraja 1996; Ranjan et al. 2006).

The worst scenario for climate change is in Asia. The Tibetan plateau, with 46,000 glaciers at an average height of 4,300 m above sea level, is the Earth's third largest ice mass after the Arctic and Antarctic, and it is often referred to as the "third pole" (Qui 2008). In the past half-century, 82% of the plateau's glaciers have retreated. In the past decade, 10% of its permafrost has melted. As the changes continue or even accelerate, their effects will resonate far beyond the isolated plateau, greatly reducing the water supply for nearly three billion people and altering atmospheric circulation over half the planet (Qui 2008). The so-called third pole is the water source for all major rivers across Southeast Asia, including the Yangtze, Yellow, Indus, Ganges, Brahmaputra, Salween, and Mekong (CIA 2010f). If glaciers continue to retreat and snowpack shrinks on the plateau, the water supplies of all nations in Southeast Asia and most of India, Pakistan, and China will be in danger of drying up (Cyranoski 2005). In Qumalai County, China, for example, near the headwaters of the Yangtze, wells have recently gone dry, and smaller rivers have vanished completely. If climatologists are correct, these water shortages may be a harbinger of the future along the entire region. The implications of large-scale and permanent water shortages in this highly populated region caused by the melting of the third pole, plus continued saltwater intrusions in coastal areas, are too terrible to contemplate (Qui 2008).

The time to act is now.

References

Africa Water (2006) Report says Africa's water is one of "induced scarcity." March 16, 2006. A newsletter of UN/Water Africa in support of the African Minister's Council on Water

Anderson MA, Whitlock JE, Harwood VJ (2006) Diversity and distribution of *Escherichia coli* genotypes and antibiotic resistance phenotypes in feces of humans, cattle, and horses. Appl Environ Microbiol. 72(11):6914–6922

Asian Development Bank (2010a) PRC: Qingdao water resources and wetland protection project 2010. http://www.adb.org/Documents/Resettlement_Plans/PRC/40017/40017-01-prc-rp.pdf. Accessed 4 June 2010

Asian Development Bank (2010b) People's Republic of China: Harbin water supply project. 2010. http://www.adb.org/documents/reports/validation/prc/33177-prc-pcv.pdf Accessed 4 June 2010

Asian Development Bank; Asia-Pacific Water Forum (2007) Country paper Philippines. Asian water development outlook 2007. p. 4. http://www.adb.org/Documents/Books/AWDO/2007/cr09.pdf. Accessed 4 June 2010

Bansil PC (2004) Water management in India. Pg. 288. Concept Publishing Company, New Dehli, India.

Bauer CJ (2004) Siren song: Chilean water law as a model for international reform. p. 48. Resources for the Future, Washington, DC

Boxer B (1998) China's water problems in the context of U.S.-China relations. In: China environment series 2.Washington, DC: Woodrow Wilson International Center for Scholar's Environmental Change and Security Project, Washington, DC

Cao SR, Xu F (1989) Cases of illness associated with the contamination of water supplies [in Chinese]. Wei Sheng Yen Chiu. J. Hyg. Res. 18:26–27

Chao KK, Chao CC, Chao WL (2003) Suitability of the traditional microbial indicators and their enumeration methods in assessment of fecal pollution of subtropical freshwater environments. J. Microbiol. Immunol. Infect. 36:288–293

Chinese Ministry of Environmental Protection (2009) The state of the environment of China in 2008. http://english.mep.gov.cn/News_service/news_release/200906/t20090618_152932.htm. Accessed 11 August 2010

CIA (Central Intelligence Agency) The world factbook (2010b) South America: Chile. https://www.cia.gov/library/publications/the-world-factbook/geos/ci.html. Accessed 11 August 2010

CIA (Central Intelligence Agency) The world factbook (2010f) East and Southeast Asia: China. https://www.cia.gov/library/publications/the-world-factbook/geos/ch.html. Accessed 16 June 2010

CIA (Central Intelligence Agency) The world factbook (2010a) Africa: Nigeria. https://www.cia.gov/library/publications/the-world-factbook/geos/ni.html. Accessed 11 August 2010

CIA (Central Intelligence Agency) The world factbook (2010c) North America: Mexico. https://www.cia.gov/library/publications/the-world-factbook/geos/ci.html. Accessed 11 August 2010

CIA (Central Intelligence Agency) The world factbook (2010d) East & Southeast Asia: Philippines. https://www.cia.gov/library/publications/the-world-factbook/geos/rp.html. Accessed 11 August 2010
CIA (Central Intelligence Agency) The world factbook (2010e) South Asia: India. https://www.cia.gov/library/publications/the-world-factbook/geos/in.html. Accessed 14 June 2010
Cyranoski D (2005) Climate change: The long-range forecast. Nature. 438:275–276
de la Luz Domper M (2009) Can Uganda learn from Chile's water revolution? http://www.sdnetwork.net/main/page.php?type=article&article_id=1462. Accessed 11 August 2010
DeNormandie J, Sunita J (2002) Combating diarrhoeal disease in India through safe drinking water. http://www.who.int/mediacentre/multimedia/2002/ind_sanitation/en/index.html. Accessed 14 June 2010
Donoso G, Melo O (2006) Water quality management in the Americas; Chapter 5: Water quality management in Chile: Use of economic instruments." pp. 229–251. Springer, Berlin Heidelberg. http://www.springerlink.com/content/n41176g02h408317/. Accessed 11 August 2010
Fidelia NN (2008) Best management practices (BMPs) for surface water pollution in developing countries. World Environment and Water Resources Congress 2008: Ahupua'a. Copyright American Society of Chemical Engineers (ASCE)
Food and Agriculture Organization (FAO) (2000) Country profile: India. http://www.fao.org/nr/water/aquastat/countries/india/index.stm. Accessed 14 June 2010
Food and Agriculture Organization of the United Nations (2005) Report of the regional workshop on salt-affected soils from sea-water intrusion: Strategies for rehabilitation and management. http://www.fao.org/ag/tsunami/docs/salf-affect.PDF. Accessed 11 August 2010
Gleick PH (1998) The World's Water 1998–1999: The Biennial Report on Freshwater Resources. Washington, DC: Island Press
India Water Portal (1991) Indian standard for drinking water – Specification IS10500. http://www.indiawaterportal.org/sites/indiawaterportal.org/files/Indian%20Standard%20for%20Drinking%20Water%20as%20per%20BIS%20specifications_2010.pdf. Accessed 14 June 2010
Isiorho SA, Oginni FA (2007) Waste disposal and 'pure water' quality issues from Lagos State, Nigeria. UNICEF. www.gwpc.org/. Accessed 11 August 2010
IPCC (2001) Climate Change 2001: The Scientific Basis. Contributions of Working Group I to the Third Assessment Report of the Intergovernmental Panel on Climate Change. Houghton JT, Ding Y, Griggs DJ, *et al.* (Eds), Cambridge University Press: Cambridge, UK. ISBN: 0521014956
Johansen BE (Updated 2002) Nigeria: The Ogoni: Oil, blood, and the death of a homeland. http://www.ratical.org/ratville/IPEIE/Ogoni.pdf. Accessed 11 August 2010
Kathuria V (2006) Controlling water pollution in developing and transition countries – Lessons from three successful cases. J. Environ. Management. 78:405–426
Kingston P (2006) Philippines. http://www.wpro.who.int/NR/rdonlyres/368CD64E-E45D-47FD-A5EC-881076EE143F/0/Philippines.pdf. World Health Organization. Accessed 11 August 2010
Lagos State Government, Nigeria (Updated 2009) http://www.lagosstate.gov.ng/index.php?page=home. Accessed 11 August 2010
Lyimo TJ (2009) Microbial and nutrient pollution in the coastal bathing waters of Dar es Salaam. Aquatic Conserv. Mar. Freasw. Ecosyst. 19:S27–S37
Mahesha A, Nagaraja SH (1996) Effect of natural recharge on sea water intrusion in coastal aquifers, Journal of Hydrology. 174:211–220
Medalla E (2006) SISS: Company closures imminent due to water regulations – Chile. http://www.bnamericas.com/news/waterandwaste/SISS:_Company_closures_imminent_due_to_water_regulations. Accessed 11 August 2010
Ministry of Public Health (2010) Santiago, Chile. http://www.minsal.cl/. Accessed 4 June 2010
National Institute for Urban Affairs (2005) Status of water supply, sanitation, and solid waste management in urban areas New Deli, India. http://www.scribd.com/doc/23362613/Status-of-Water-Supply-Sanitation-and-Solid-Waste-Management-in-Urban-Areas-Part-1. Accessed 14 June 2010

National Water Commission: CONAGUA (2010) Statistics on water in Mexico, 2010 edition. http://www.conagua.gob.mx/english07/publications/EAM2010Ingles_Baja.pdf. Accessed 11 August 2010
Nigeria Background Information (2009) The human and physical characteristics of Nigeria. http://www.uni.edu/gai/Nigeria/Background/Standard4.html. Accessed 11 August 2010
Noble R T, Griffith JF, Blackwood AD, Fuhrman JA, Gregory JB, Hernandez X, Liang X, Bera AA, Schiff K (2006) Multitiered approach using quantitative PCR to track sources of fecal pollution affecting Santa Monica Bay, California. Appl Environ Microbiol. 72:1604–12
Pakistan Council of Research in Water Resources (2002) National water quality monitoring programme: Water quality standards. http://www.pcrwr.gov.pk/wq_standards.htm. Accessed 11 August 2010
Paratori C (2009) Persistent organic pollutants. Chile: Current situation. http://www.chem.unep.ch/Pops/POPs_Inc/proceedings/Iguazu/PARATORI.html. Accessed 10 August 2010
Philippine Department of Health (2009) Laboratory for water analysis. http://portal.doh.gov.ph/templates/water. Accessed 11 Aug 2010
Prüss-Üstün A, Bos R, Gore F et al (2008) Safer water, better health: costs, benefits and sustainability of interventions to protect and promote health. World Health Organization, Geneva. http://www.who.int/water_sanitation_health/publications/safer_water/en/. Accessed 11 August 2010
Qui Jane (2008) China: the third pole. Nature. 454:393–396
Ranjan SP, Kazama S, Sawamoto M (2006) Effects of climate and land use changes on groundwater resources in coastal aquifers, Journal of Environmental Management. 80(1):25–35
Republic of the Philippines. National Statistics Office (2007) Census of population – Population and annual growth rates by region, province, and city/municipality: 1995, 2000, and 2007. http://www.census.gov.ph/data/sectordata/2007/municipality.pdf. Accessed 11 August 2010
Rosegrant MW, Cai X, Cline SA (2002) World water and food to 2025: dealing with scarcity. International Food Policy Research Institute. Washington, D.C. ISBN: 0-89629-646-6
Shanks OC, Kelty CA, Sivaganesan M, Varma M, Haugland RA (2009) Quantitative PCR for genetic markers of human fecal pollution. Applied and Environmental Microbiology. 75(17):5507–5513
Shiklomanov IA (1997) Assessment of water resources and water availability in the world. Report prepared for the comprehensive assessment of freshwater resources of the world. United Nations: St. Petersburg
Speiser RM (2009) War over water – Climate change and clean tech opportunities in Chile. http://blog.cleantechies.com/2009/08/10/climate-change-war-over-water-clean-tech-opportunities-chile/. Accessed 11 August 2010
Superintendencia de Servicios Sanitarios (SISS) (2007) Informe de gestión del sector sanitario 2006. http://www.siss.cl/articles-3687_informegestion.pdf. Accessed 11 August 2010
Suthar S, Bishnoi P, Singh S et al (2009) Nitrate contamination in groundwater of some rural areas of Rajasthan, India. J. Haz. Materials. 171:189–199
UN WATER/AFRICA (2009) Africa water vision for 2025: Equitable and sustainable use of water for socioeconomic development. http://www.uneca.org/awich/african%20water%20vision%202025.pdf. Accessed 11 August 2010
UNICEF (2009) Nigeria fact sheet: Water, sanitation and hygiene in Nigeria..http://www.unicef.org/nigeria/ng_media_WASH_fact_sheet_Oct_2009.pdf. Accessed 11 August 2010
United Nations Commission on Sustainable Development (1997) Overall progress achieved since the United Nations Conference on Environment and Development. Report of the Secretary General, UN: New York
United Nations Development Programme (UNDP) Human Development Report (2006) Beyond scarcity: Power, poverty and the global water crisis. 1 UN Plaza, New York, New York, 10017, USA ISBN 0-230-50058-7
United Nations Environment Programme (UNEP) (2008) Glaciers are melting faster than expected. http://www.sciencedaily.com/releases/2008/03/080317154235.htm. Accessed 11 August 2010
United Nations Industrial Development Organization (UNIDO) (1996) Global assessment of the use of freshwater resources for industrial and commercial purposes. Industry, sustainable development, and water programme formulation, technical report. New York: UN

United Nations: Department of Economic and Social Affairs Population Division (2009) World population prospects, table A.1. 2008 revision. United Nations. http://www.un.org/esa/population/publications/wpp2008/wpp2008texttables.pdf. Accessed 11 August 2010
United States Department of State (2009a) Chile. http://www.state.gov/p/wha/ci/ci/. Accessed 11 August 2010
United States Department of State (2009b) Background note: Philippines. http://www.state.gov/r/pa/ei/bgn/2794.htm. Accessed 11 August 2010
Valenzuela M, Lagos B, Claret, M et al (2009) Fecal contamination of groundwater in a small rural dryland watershed in central Chile. Chilean J. Agric. Res. 69:235–243
Vargas C (2008) Chile: Drought raises likelihood of energy rationing. http://ipsnews.net/news.asp?idnews=41223. Accessed 11 August 2010
Vargas M, Velez P (2009) Chile calls on miners to further cut water use. http://in.reuters.com/article/oilRpt/idINN0648821120090406. Accessed 11 August 2010
Veeralapalam, Pradesh A, Teku L (2009) When the rains fall, a briefing on India's water crisis. The Economist. Sept.11, pp 27–30
WaterAid Nigeria: Effective financing of local governments to provide water and sanitation services (2008) Water Aid Report pp 3–5 http://www.wateraid.org/documents/plugin_documents/local_financing__nigeria.pdf Accessed 18 August 2010
WHO/UNICEF JMP (2010) Joint monitoring program for water and sanitation: Estimates for the use of improved sanitation facilities and estimates for the use of improved water facilities. http://www.wssinfo.org/download?id_document=928. Accessed 11 August 2010
World Bank (2005) Philippines: Meeting infrastructure challenges. http://siteresources.worldbank.org/INTEAPINFRASTRUCT/Resources/PHInfra.pdf. Accessed 11 August 2010
World Bank (2007a) Water pollution emergencies in China: Prevention and response. http://siteresources.worldbank.org/INTEAPREGTOPENVIRONMENT/Resources/Water_Pollution_Emergency_Final_EN.pdf. Accessed 16 June 2010
World Bank (2007b) Stepping up: Improving the performance of China's water utilities. http://www/wds.worldbank.org/external/default/WDSContentServer/WDSP/IB/2007/09/25/000310607_20070925111156/Rendered/PDF/409640P0704130Public.pdf. Retrieved Accessed 16 June 2010
World Health Organization (WHO) (2001) Cholera in Nigeria. http://www.who.int/csr/don/2001_12_03/en/index.html & http://www.who.int/csr/don/2001_11_27/en/index.html. Accessed 11 August 2010
World Health Organization (WHO) (2002-2007) WHO country cooperation strategy: Federal Republic of Nigeria. http://www.who.int/countries/nga/about/ccs_strategy02_07.pdf. Accessed 11 August 2010
World Health Organization (WHO) (2004) Cholera in Nigeria. http://www.who.int/csr/don/2004_12_03/en/index.html . Accessed 11 August 2010
World Health Organization (2007a) Country health system profile: India. http://www.searo.who.int/EN/Section313/Section1519_10851.htm. Accessed 14 June 2010
World Health Organization (2007b) Sustainable development and healthy environment: water, sanitation, and health: drinking water quality in south east Asia region. http://www.searo.who.int/EN/Section23/Section1000_15448.htm. Accessed 14 June 2010
World Health Organization (WHO) (2009a) WHO: African region: Nigeria. http://www.who.int/countries/nga/nga/en/. Accessed 28 July 2010
World Health Organization (WHO) (2009b) Chile. http://www.who.int/countries/chl/en/. Accessed 11 August 2010
World Health Organization (WHO) (2009c) Core health indicators: Philippines. http://apps.who.int/whosis/database/core/core_select_process.cfm?countries=phl&indicators=PopAccessImprovedWaterUrban&indicators=PopAccessImprovedWaterRural&indicators=PopAccessImprovedSanitationUrban&indicators=PopAccessImprovedSanitationRural. Accessed 11 August 2010
World Health Organization (WHO) (2009d) Mexico. http://www.who.int/countries/mex/en/. Accessed 11 August 2010
World Health Organization (WHO) (2009) World Health Statistics. ISBN: 97892-4-156381-9. France

World Health Organization (WHO) (2010) Water sanitation and health guidelines. http://www.who.int/water_sanitation_health/dwq/guidelines/en/. Accessed 11 August 2010

World Health Organization/UNICEF Joint monitoring programme for water supply and sanitation (2010) Progress on sanitation and drinking-water. ISBN: 978-92-4-156395-6 http://www.wssinfo.org/home/introduction.html/. Accessed 18 August 2010

World Water Council (2000) A water secure world: Vision for water, life, and the environment. Commission Report for the World Water Council. http://www.worldwatercouncil.org/fileadmin/wwc/Library/Publications_and_reports/Visions/CommissionReport.pdf. Accessed 18 August 2010

World Water Council (2003) Country report of the Peoples Republic of China. Chinese Ministry of Water Resources, Beijing. ISBN: 92-95017-05-6

Wu C, Gottlieb M, Davis D (1998) In: World resources 1998-1999. Washington, DC: World Resources Institute. 1998:120–122

Wu C, Maurer C, Wang Y et al (1999) Water pollution and human health in China. Environ. Health Perspective. 107:251–256

Yusuf, KA (2007) Evaluation of groundwater quality characteristics in Lagos-City. J. Applied Science. 7:1780–1784

Chapter 23
A National Security Perspective of Microbial Source Tracking

Stephaney D. Leskinen and Elizabeth A. Kearns

Abstract Protecting a nations' waters and those that utilize these waters for drinking, recreation, and/or shellfishing is of paramount importance from both a public health and national security perspective. Molecular methods for microbial source tracking (MST) have been developed over the last 2 decades to address needs regarding the source of natural microbial contamination of these critical water resources. MST has gained a foothold in the water-quality arena over the past decade as the limitations of the fecal indicator paradigm pertaining to protecting public health have been recognized. Microbial forensics (MF) has been developed since approximately 1996 but only became a national security priority following the anthrax attacks of 2001. The research efforts in the fields of MF and MST have been developed along parallel lines to address the needs of security and health, respectively. This chapter addresses the similarities and differences between MST and MF and how methodologies developed for MST can be applied to forensic analysis of public health incidents involving possible intentional use of microorganisms. Examples from natural contamination incidents that could provide lessons for investigations of possible intentional releases are highlighted. Challenges facing the use of MST and MF techniques, as well as future research needs, are also discussed.

Keywords Microbial forensics • Water security • Microbial signatures • Bioterrorism • Biocrime

S.D. Leskinen (✉)
Department of Cell Biology, Microbiology and Molecular Biology, University of South Florida, 4202 E. Fowler Avenue, BSF 218, Tampa, FL 33620-5150, USA
e-mail: Leskinen@usf.edu

C. Hagedorn et al. (eds.), *Microbial Source Tracking: Methods, Applications, and Case Studies*, DOI 10.1007/978-1-4419-9386-1_23,

23.1 Current State of Microbial Water Quality Assessments

Clean water is, and will continue to be, an increasingly critical resource as global populations increase, larger numbers of Third-World countries begin to industrialize, and climate change continues to affect the availability of water reserves throughout the world (Chap. 22). The vital nature of water to human health and daily economic activities makes this resource an especially attractive target to terrorist groups. This is in addition to the inherent vulnerability of recreational and potable water supplies to unintentional contamination.

Fecal contamination of waters used for recreation, shellfishing, or as sources for drinking water poses a grave threat to human health due to the increased risk of exposure to human pathogens (Scott et al. 2002; Korajkic et al. 2009). Determining sewage impacts to these waters can prevent illness through posting of advisories and/or environmental remediation. It is currently impractical, from both a cost and manpower perspective, to test directly for all enteric pathogens that may contaminate waters for human use (Field and Samadpour 2007; Korajkic et al. 2009). The indicator paradigm states that certain microorganisms act as sentinels for the possible presence of human pathogens of fecal origin. Bacterial indicator organisms have been utilized for over a century as surrogates for the assessment of water quality and safety (Scott et al. 2002). The US government specifies the collection of water samples for assessing the quality of recreational and shellfishing waters (United States Environmental Protection Agency 1986), as well as drinking waters (United States Congress 1977).

Standard water monitoring technologies for fecal contamination currently involve culturing indicator microorganisms (e.g., *Escherichia coli*; *Enterococcus* spp.) (U.S. Environmental Protection Agency 2000). During the incubation period recreational water users may be exposed to unsafe conditions before beach advisories are issued. There is a need for rapid methods for identification of fecal pollution events to minimize exposure of the public to microbial pathogens for the safety of those using a nation's recreational waters. Various procedures have been recently developed in an attempt to meet this need that would permit "same-day" warning systems. Rapid detection of biological targets in water is needed but is complicated by their low levels in ambient samples, which makes concentration necessary (Hill et al. 2005; Olszewski et al. 2005; Kearns et al. 2008; Leskinen et al. 2009; Leskinen et al. 2010). Coincident with the need for rapid methods is the need for source determination of fecal contaminants.

Microbial source tracking (MST) encompasses a group of methods, initially library-dependent (Chap. 3) but now essentially library-independent (Chap. 4), used to determine possible source(s) of fecal contamination in a water body. Human feces (i.e., in sewage or from septic tanks) is known to contain pathogens that are a serious health risk (Harwood et al. 2005a; Lee et al. 2008). Epidemiological studies conducted over the past several decades have correlated the levels of certain indicator organisms with the risk of gastrointestinal illness in recreational water users (Cabelli 1983; Fleisher et al. 1998). The ubiquitous nature of indicator bacteria and the possible long-term survival of certain indicator organisms in the environment, however, obscure this relationship (Byappanahalli and Fujioka 1998; Anderson et al. 2005).

Indicator bacteria are normal inhabitants of the gastrointestinal tract of warm-blooded and some cold-blooded animals. Furthermore, certain *Escherichia coli* strains are capable of long-term survival and even re-growth in the environment after release in sewage (Anderson et al. 2005; Anderson et al. 2006). These factors make indicator bacteria unsuitable for tracing the source of fecal contamination (Scott et al. 2002). The possibility of human feces contaminating a water body and the opportunity of preventing such an event if the source of the contamination is located and mitigated makes the ability to distinguish human from non-human fecal material exceedingly important for protection of public health. MST methods were developed with the goal of differentiating contamination from various fecal sources.

23.2 Application of Microbial Source Tracking to Microbial Forensics

MST is most commonly associated with water-quality monitoring; however, the basic concepts also apply to national security investigations related to a biological terrorism (BT) or biocrime event. The methods for investigating a suspected BT event are referred to as microbial forensics (MF) and have many similarities to those utilized by MST scientists. MF can be defined as a scientific discipline dedicated to analyzing evidence from a suspected BT incident or biocrime for the purpose of attribution. History is replete with examples from as far back as the Assyrians and Romans using rye ergot or dead carrion, respectively, to poison enemy water supplies (Robertson and Robertson 1995). Regardless of the terminology used, the basic concept for both MST and MF is the same: identify and characterize the microbial contaminant so that the source of the organism can be determined.

As implied by inclusion of "forensics" in MF, application of MST techniques to the investigations of biocrimes and BT events requires a bit more rigor than their application to natural contamination events (Chap. 13). Data gathered in these investigations must be unequivocal enough to serve as evidence in criminal proceedings or to provide unquestionable support for the use of sanctions or military force against a country implicated in perpetrating the event. Hence, MF requires that samples be collected, processed and stored according to strict procedures that prevent compromise of the sample and maintain chain-of-custody (Budowle et al. 2005), a requirement not yet generally employed for natural contamination incidents. MF may also involve concomitant collection and characterization of signatures other than molecular that might be associated with the organism involved in the event to reveal the methods, means, processes, and locations used by the perpetrators to prepare for and conduct the attack (Budowle et al. 2005). Characterization of evidence may include an analysis of physical and chemical properties that can help reveal whether a biological agent is natural or artificial, how it was weaponized, what materials were used to process it, and how recently has been prepared (Velsko 2005). While currently-employed chemical and physical analysis techniques have use in the forensics of microorganisms dispersed through air, it is likely that many of the signatures examined by these methods would be substantially modified or destroyed for organisms distributed *via* water due to the differences between the water and air environments. For example, techniques to determine the appropriate culture medium for microorganisms or the chemicals used to prevent clumping of organisms may be less reliable in or degraded by, respectively, disinfectants present in drinking water that are not found in air. Furthermore, current forensic techniques do not explore the biological aspect of microorganisms.

Clean fresh water is an increasingly critical resource that will become progressively more precious as additional societies industrialize and population pressure increases (Chap. 22). The critical nature of potable water resources increases the vulnerability of water supplies to political and terrorist activities. With this expanded vulnerability, the importance of forensic analysis in management of water resources and protection from contamination increases (White et al. 2003).

Development of a system of methods to detect and track biological contamination events, natural or intentional, overt or covert, will lead to greater safety and national security. All are a public health concern. The two scientific communities of MST and MF investigators are more closely connected than has been previously discussed in regards to the safety of a nation's water resources. The link between MST and MF has been previously discussed from a food safety perspective (Nakatsu et al. 2007). Drinking and recreational waters are considered part of a nation's critical infrastructure and protecting recreational, source, and drinking waters is integral to national security. The established methods utilized by MST scientists would provide significant value to forensic investigations (Chap. 2).

The separate development of the fields of MST and MF is in part due to the governing body to which each has generally been assigned. MF is analogous to microbial source tracking, but has historically been applied to investigations of microorganisms used as bioweapons, and the field has been primarily developed under the auspices of government law enforcement agencies (Budowle et al. 2003). MST has been primarily developed in academic laboratories and only recently has begun transitioning to regulatory agencies for water quality monitoring purposes (Chap. 14).

Whereas the effort to use MST for Total Maximum Daily Load (TMDL) monitoring in the USA has been spearheaded primarily at the state level, the federal government has been leading the effort to implement a standardized national MF system since the anthrax attacks of 2001 (Budowle et al. 2003; Budowle et al. 2008). Law enforcement has had the traditional role and infrastructure for investigating crimes and is now enhancing its capabilities to confront the new challenge of biological weapon usage and BT through new partnerships with the scientific community. The Federal Bureau of Investigation (FBI) initiated the Scientific Working Group on Microbial Genetics and Forensics (SWGMGF) in 2002 to establish a foundation for the MF field (Budowle et al. 2003). The focus of the SWGMGF initially included (1) defining quality assurance (QA) guidelines for laboratories performing MF analyses, (2) establishing criteria for forensic method development and validation for various threat agents for attribution of criminal acts, (3) prioritizing research efforts for pathogens that would be most likely to be used as bioagents, and (4) increasing the microbial population genetic dataset and establishing design criteria for information databases for event interpretation (Budowle et al. 2005).

23.3 Microorganisms of Concern and Methods for Detection

In general, bacteria serve as a good choice for weaponization because they can be mass produced from a single cell without sophisticated instrumentation or highly trained personnel. A number of bacteria (Table 23.1) and/or their products, such as toxins, pose serious public health concerns, as well as threaten food supplies and the environment (Pattanik and Jana 2005) (Table 23.1). While the 'ideal' biothreat agent does not exist, the five basic characteristics of such an agent have been identified (Zilinskas 1986). These include the following traits: (1) high virulence linked to a high level of host specificity, (2) controllability (attacks only those targeted and not those disseminating the agent), (3) resistance to environmental degradation, (4) general lack of available countermeasures, and (5) easily disguised. The potential for use of biological agents as weapons for warfare, terrorism, and crimes has been demonstrated throughout history regardless of the lack of an "ideal" agent (Robertson and Robertson 1995; Carus 2000). The intentional mailings of anthrax spores in 2001 confirmed the use of these biothreat agents in the modern era (Budowle et al. 2005). Although attacks involving biological weapons are rare compared to other types of crimes, the possibility raises concerns about the ability to provide MF data that can be used to identify the source of the microorganisms that are used as weapons.

There are well-established procedures for handling and analyzing pathogenic agents (Fleming and Hunt 2000), but these techniques address epidemiological concerns for tracking the spread of disease. Furthermore, these techniques are generally

Table 23.1 Selected bacteria that pose a serious health concern if used as a biological weapon and that could be used for intentional contamination of water bodies

Microorganism	Environmental sources	Disease	Route(s) of human infection
Bacillus anthracis	Soil	Anthrax	Skin, lungs, gastrointestinal tract
Burkholderia mallei	Horses	Glanders	Skin, mucosal surfaces, contact with infected animal
Clostridium perfringens	Soil	Gas gangrene	Puncture, gastrointestinal tract
E. coli O157:H7	Cattle	Gastrointestinal illness, hemolytic uremic syndrome	Gastrointestinal tract
Francisella tularensis	Perhaps rodents/rabbits[a]	Tularemia	Gastrointestinal tract, lungs, contact with infected reservoir
Vibrio cholerae	Autochthonous in water, sewage	Cholera	Gastrointestinal tract

[a]Reservoir remains unknown

culture-based and time-consuming. The source of the pathogen is not determined by these methodologies, but microbial "signatures" are critical for tracing the microbe's source (Pattanik and Jana 2005). Phenotypic evidence has proved to be unreliable for MF investigations and more emphasis is being given to molecular signatures or molecular markers (i.e., DNA fingerprints based on polymorphism of certain genes), which are more reliable and quantifiable (Pattanik and Jana 2005). This approach is similar to that used in current MST methodologies, making MST researchers a valuable resource with extensive experience in using the analytical tools needed for MF investigations.

MST methods have advanced beyond being comprised mostly of library-dependent methods, such as antibiotic resistance analysis and ribotyping, to use of library-independent methods over the past decade (Scott et al. 2002). These methods are a subset of MST that focus on detection of a specific target gene sequence using PCR. Target genes are specific to, or at the least highly associated with, feces from a particular host organism (Bernhard and Field 2000; United States Environmental Protection Agency 2005; McQuaig et al. 2006; Ufnar et al. 2006). PCR-based methods are being used with increasing frequency in MST studies for source apportionment in TMDL applications (Harwood et al. 2005b; Wapnick et al. 2007; Giacalone et al. 2009).

Methodologies used by both MST and MF include a variety of techniques for determining the source of the pollution or contamination, be it from a natural cause or an intentional release. Examples of shared techniques include the identification of specific genetic markers (e.g. genes for virulence or antibiotic resistance) and alignment of DNA sequence patterns for source matching to libraries (MST terminology) or "curated genes" (MF terminology) (Budowle et al. 2005). Results from an American Academy of Microbiology Colloquium recommended that such databases contain sequence data for a minimum of three to twenty diverse strains, depending on the diversity of the organism and its priority level, as well as one close relative for each included pathogen (Keim 2003). This would improve the relevance and accuracy of sequence data to determine the source of biological agents. Again the similarity between the techniques and the complexity of samples that require a multiple method, or toolbox, approach belies the semantic difference between these fields.

23.4 Potential Sources of Targets of National Security Concern

Release of microbial targets into the nation's waters could result from intentional events or unintentional events that occur either naturally or are anthropogenically mediated. To date no known intentional contamination events have occurred in US critical water resources. However, waterborne disease outbreaks over the past decade have resulted from microbial contamination and provide clues as to what organisms might make good weapons for contaminating critical water resources. Waterborne disease outbreaks in drinking and recreational waters occur every year and are attributed to a variety of microorganisms, any one of which could potentially be used to intentionally contaminate a water resource. The continual occurrence of these outbreaks from natural sources points to the potential for intentional contamination in the future. The outbreaks highlighted below demonstrate the potential intersection of the MST and MF fields during an event investigation, and the role that MST techniques could play in the case of an intentional contamination event (Yoder et al. 2008).

The first recognized outbreak of Legionnaires' disease occurred on 27 July 1976 at the Bellevue Stratford Hotel in Philadelphia, Pennsylvania (Edelstein 2008). Members of the American Legion, a US military veterans association, gathered there for the American Bicentennial celebration. Attendees began falling ill within 2 days of the event's start with as many as 221 people given medical treatment and 34 deaths. Research continued for months with various theories discussed in scientific and mass media. Hypothesized causes of the outbreak ranged from toxic chemicals to terrorism, be it domestic or foreign. The CDC mounted an unprecedented investigation and shifted the focus from outside causes, such as a disease carrier, to the hotel environment itself. The causative agent, a bacterium later named *Legionella pneumophila*, was identified and isolated in January 1977. The bacteria were found in biofilms in the cooling tower of the hotel's air conditioning system and from there spread throughout the building air droplets distributed via the air handling system (Edelstein 2008). Many additional outbreaks associated with cooling towers have been reported since this initial incident, making *Legionella* potential candidate for intentional release.

In May 2000 an investigation of a contaminated water distribution system in the town of Walkerton, Ontario, was performed by the Bruce-Grey-Owen Sound Health Unit, with the assistance of Health Canada and the Ontario Ministry of Health and Long-Term Care. The purpose of the investigation was to determine the scope, likely cause and contributing factors of an *E. coli* O157:H7 outbreak. This investigation represented the first documented outbreak of *E. coli* O157:H7 associated with a treated municipal water supply in Canada and one of the largest waterborne outbreaks in Canada. Investigative techniques included hydrogeological studies, environmental transport assessments, fate and transport studies and genetic characterization of contaminants. The probable factors implicated in the bacterial contamination included the use of cast-iron pipes, the thickness and concentration of biofilm sediment and mineral, the infiltration of contaminated surface runoff due to heavy rain or flooding, and/or flow velocity related to the diameter of the pipe. Genetic fingerprinting using pulsed-field gel electrophoresis showed the strain of bacteria found in victims matched closely to one found on a farm located near a contaminated well, implicating this as the source of the contamination (Danon-Schaffer 2001).

A large outbreak of salmonellosis resulted from contamination of an in-ground storage tank in Alamosa, CO in 2008 (Falco and Williams 2009). Approximately 1,300 people were sickened, and one death was attributed to ingestion of the contaminated water. The outbreak lasted 3 weeks, while the investigation of the incident lasted 18 months. MST analysis of the possible contamination sources consisted of collection of animal fecal material, including that from water fowl and deer near the reservoir. Based on molecular analysis, the conclusion of the Colorado Department of Health and Environment was that animal fecal contamination was the source and the incident was facilitated by the lack of chlorination of the Alamosa water supply. The investigation into the Alamosa outbreak of 2008 confirmed that physical, regulatory, and human infrastructure all played a role in the failure to protect the public's safe drinking water supply.

Contamination events of water supplies are not limited to *E. coli* and *Salmonella.* They have also been documented for biothreat agents, but were not associated with intentional contamination activities. An outbreak of acute melioidosis that caused three deaths was documented in a small coastal community in northwestern Australia in 2000 (Inglis et al. 2000). An investigation was conducted the following year and led to the discovery of *Burkholderia pseudomallei* in the community's water treatment facility. Forensic analyses confirmed that the strain was identical to the clinical isolate from the patient, and the source of the contamination was likely a storage tank and aerator. Outbreaks of tularemia, which is caused by *Francisella tularensis,* have been associated with contaminated food and water (Hoel et al. 1991). Outbreaks of Lassa fever caused by the Lassa virus have been described in Nigeria following ingestion of water contaminated with the fecal material of rats or water used for bathing the bodies of the deceased in funerary rites (Inegbenebor et al. 2010). The disease outbreaks associated with these events are testament to the importance of MST/MF in investigations of an incident to determine if it was intentional or unintentional.

23.5 Current Challenges and Limitations Facing Use of MST in Forensics

One major hurdle to the application of MST to forensic investigations is the establishment of standards for evidence/sample collection, analysis and interpretation of results. The foremost challenge is the collection of specimens at the site of a suspected attack. When a BT event or biocrime is suspected, samples must be properly collected and stored to ensure that reliable analytical data are obtained. Chain of custody must be maintained throughout the collection and analysis process. In addition, since many agents that might be used in an attack are extremely hazardous, first responders and investigators at the attack site face a substantial health risk. Therefore, proper education and training of personnel who will respond to a suspected incident is critical, both to protect them from exposure and to ensure that evidence is not compromised by improper handling (Keim 2003).

Another urgent need is the development of systems for the integration of state-of-the-art analysis, sampling logistics, and sterilization procedures for effective handling of suspect samples. The function of those examining the polluted area will be to obtain samples from victims and/or from the environment for MF or MST analysis, respectively, to identify the contaminating microorganisms. Analyses of samples that could help identify the source of the contamination will also be included. Preservation of evidence is vital to MF investigations as the evidence may need to be used in criminal proceedings or to take action against a specific country or international terrorist group, and may also be important in MST investigations, especially those with regulatory ramifications. Standard protocols should be followed for sampling (American Public Health Association 2005; Budowle et al. 2005). Proper handling (chain of custody) and analyses of the specimens collected by first responders, utility employees or microbiologists pose another challenge after the samples are delivered to the laboratory.

The most formidable challenge to analyzing samples for criminal investigations is validation of the methods utilized by establishing the limits of detection, sensitivity, specificity, and reproducibility/robustness of each technique. This challenge also applies to MST analysis methods applied to regulatory monitoring (Chap. 2). The validation criteria for these methods are largely determined on an adhoc basis by scientists in the field. Standard operating procedures or regulatory-approved criteria for assessing the performance of these methods do not currently exist (Stoeckel and Harwood 2007). A general concern exists regarding newly developed molecular analytical methods in that they may be so narrowly focused for a particular species or strain of microorganism generally considered to be potential bioagents that their widespread use for events that are not biological attacks will be limited (Budowle et al. 2005). The same concern could be expressed for new MST methodology. An example of the need for method validation involves a plague outbreak in India in 1994 that caused widespread panic (Raza et al. 1997). A lack of validated forensic evidence limited the reliability of collected data and prevented definitive conclusions from being drawn as to whether the outbreak was truly a case of bioterrorism (Khushiramani et al. 2004). Further complicating the interpretation of results for these types of incidents is that molecular typing data is not available in many developing countries for outbreak strains not associated with terrorist-related activity. Without these data, using molecular protocols to determine the source of the infection and collection of reliable legal evidence for prosecution of bioterrorism activity may prove exceedingly difficult.

Another hurdle facing the incorporation of MF and MST into event investigations is that aggressive research programs are needed to integrate trace evidence analysis with MF/MST data. This will allow these new disciplines of microbiology and forensic science to flourish for national security and public health needs. Powerful tools for MF and MST will be available to detect and prevent future terrorist acts after methodological procedures are standardized and genetic databases of different potential bioterrorism pathogens are established (Budowle et al. 2005).

23.6 Recommendations for the Future of MST and MF Research

Development of molecular techniques for use in either MST or MF investigations is in its early stages. Current methods have their basis in epidemiology, but their speed needs to be increased as it relates to responding to possible BT events.

Furthermore, overcoming the problems caused by the dilution of targets in water must be addressed, possibly through concentration, to aid in the detection of bioagents. Time and funding are necessary to standardize procedures and define quality assurance guidelines for performing analyses to avoid data variability between laboratories (Pattanik and Jana 2005). For BT/biocrime investigations, a current list of most likely potential pathogens needs to be maintained and updated regularly (Keim 2003), as well as a database of genotypes of these agents. Research efforts should be prioritized toward pathogens and toxins that would most likely be used in intentional contamination events, which are expected to vary depending on regional concerns such as differences in crop irrigation protocols and use of reclaimed water. Criteria for the validation of methods used to characterize various agents should be established in a way that can be used to attribute criminal acts. Researchers seeking to improve MST methodology can contribute to these efforts by focusing on these goals. Investigations of natural contamination events will also benefit, especially as they apply to regulatory applications. The availability of genomic information in a universal database format would enhance molecular methods used for both MF and MST analysis and method validation. The successful utilization of MF or MST in environmental investigations is contingent upon a combined knowledgebase composed of validated methodologies, genomic databases and, increasingly, bioinformatics (Pattanik and Jana 2005; Chap. 13).

Experts have maintained that national MF guidance must rely on at least three major components to be successful: (1) information storehouses, (2) scientific partnerships and (3) the Scientific Working Group on Microbial Genetics and Forensics (SWGMGF) (Budowle et al. 2003). The first component involves establishing centers of knowledge composed of databases of nucleic acid and protein sequences, and standardized methodologies for microbiology, genomics, proteomics, and bioinformatics. The second key involves maintaining partnerships between existing government, academic, and private-sector assets, as well as fostering new scientific associations between such groups. No single laboratory or institution can or will be able to address all MF requirements. These partnerships will include scientists from the Departments of Defense, Energy, and Health and Human Services, the National Science Foundation, the various National Laboratories, state departments of health and environmental resources, and academic laboratories (Budowle et al. 2005). Owing to the similarities between tracking the source of natural contamination and tracking the source of a BT/biocrime, research laboratories that have been working in the field of MST will make ideal partners in this consortium. The final component is a formal group or panel to establish standardized quality assurance guidelines for laboratory analysis of samples. The SWGMGF was established to serve this purpose for MF work and has produced a guidance document (Scientific Working Group on Microbial Genetics and Forensics (SWGMGF) 2003). MST researchers wanting to make contributions to the field of MF should become familiar with these standards, the use of which could also improve the relevancy of MST data used in regulatory monitoring. Additionally, guidance and recommendations produced by the US Environmental Protection Agency (United States Environmental Protection Agency 2005) and the World Health Organization (World Health Organization 2004) provide useful information to guide development of MF molecular protocols.

23.7 Conclusions

Application of MST to investigation of potential cases of bioterrorism or biocrimes represents a relatively new application of techniques that have been used for years to investigate natural contamination of waters. It remains difficult to definitively determine the origin of (a) microorganism(s) used in an intentional attack, as well as for a natural or accidental release. Currently, in both fields there is generally a lack of detailed data about the genetic variability and/or genetic marker density of target microorganisms. Local, regional, national and international diversity of microorganisms and the influence of biotic and abiotic factors in their environment are poorly understood and not well defined. These limitations are common to both MST and MF and reduce confidence in conclusions regarding the source(s) of microbes responsible for both natural and intentional contamination. Further development of molecular techniques and their databases would benefit both MF and MST. Standardized guidelines are needed to establish common and efficient practices that will be accepted as evidence in criminal proceedings or putative actions against perpetrators of terror and will also benefit regulatory compliance efforts to prevent or investigate natural contamination events. A powerful tool to detect and prevent future terrorist acts will become available once reliable procedures are standardized and genetic databases are established for target biological agents. MST scientists can play a substantial role in thwarting the use of biological weapons and preventing public health crises by helping to develop these tools so that they can reliably characterize, detect and determine the source of pathogens involved in contamination events, whether natural, accidental or intentional. The robust development of molecular techniques for both MF and MST will enhance national security and public health beyond the physical borders of a country.

References

American Public Health Association (2005) *Standard methods for the examination of water and wastewater, 20th ed.* Washington, DC: American Public Health Association.

Anderson, K.L., Whitlock, J.E. and Harwood, V.J. (2005) Persistence and differential survival of fecal indicator bacteria in subtropical waters and sediments. *Appl Environ Microbiol* **71**, 3041–3048.

Anderson, M.A., Whitlock, J.E., and Harwood, V.J. (2006) Diversity and Distribution of *Escherichia coli* Genotypes and Antibiotic Resistance Phenotypes in Feces of Humans, Cattle, and Horses. *Appl Environ Microbiol* **72**(11), 6914–6922.

Bernhard, A.E. and Field, K.G. (2000) A PCR Assay To Discriminate Human and Ruminant Feces on the Basis of Host Differences in *Bacteroides-Prevotella* Genes Encoding 16S rRNA. *Appl Environ Microbiol* **66**, 4571–4574.

Budowle, B., Murch, R. and Chakraborty, R. (2005) Microbial forensics: the next forensic challenge. *International Journal of Legal Medicine* **119**, 317–330.

Budowle, B., Schutzer, S.E., Einseln, A., Kelley, L.C., Walsh, A.C., Smith, J.A.L., Marrone, B.L., Robertson, J. and Campos, J. (2003) PUBLIC HEALTH: Building Microbial Forensics as a Response to Bioterrorism. *Science* **301**, 1852–1853.

Budowle, B., Schutzer, S.E., Morse, S.A., Martinez, K.F., Chakraborty, R., Marrone, B.L., Messenger, S.L., Murch, R.S., Jackson, P.J., Williamson, P., Harmon, R. and Velsko, S.P.

(2008) Criteria for Validation of Methods in Microbial Forensics 10.1128/AEM.00966-08. Appl Environ Microbiol **74**, 5599–5607.

Byappanahalli, M.N., Fujioka R.S. (1998) Evidence that tropical soil environment can support the growth of *Eschericia coli. Water Science and Technology* **38**, 171–174.

Cabelli, V.J. (1983) *Health effects criteria for marine recreational waters* No. EPA-600/1-80-031. Research Triangle Park: U.S. Environmental Protection Agency.

Carus, S.W. (2000) *Bioterrorism and Biocrimes: the Illicit Use of Biological Agents Since 1900.* The Netherlands: Fredonia Books.

Danon-Schaffer, M.N. (2001) Walkerton's Contaminated Water Supply System: a Forensic Approach to Identifying the Source. *Environmental Forensics* **2**, 197–200.

Edelstein, P.H. (2008) Legionnaires' Disease: History and Clinical Findings. In *Legionella: Molecular Microbiology* eds. Heuner, K. and Swanson, M. Portland, Oregon: Caister Academic Press.

Falco, R. and Williams, S.I. (2009) Waterborne Salmonella Outbreak in Alamosa, Colorado March and April 2008: Outbreak identification, response and investigation ed. Division, S.D.W.P.W.Q.C.: Colorado Department of Public Health and Environment.

Field, K.G. and Samadpour, M. (2007) Fecal source tracking, the indicator paradigm, and managing water quality. *Water Research* **41**, 3517–3538.

Fleisher, J.M., Kay, D., Wyer, M.D. and Godfree, A.F. (1998) Estimates of the severity of illnesses associated with bathing in marine recreational waters contaminated with domestic sewage. *Int J Epidemiol* **27**, 722–726.

Fleming, D.O. and Hunt, D.A. (2000) *Biological Safety: Principles and Practices, 4th Edition.* Washington, D. C.: ASM Press.

Giacalone, K., Miskewitz, R., Phelps, C., Walsh, S. and Obropta, C. (2009) Fecal Coliform TMDL Implementation Plan and the Use of Microbial Source Tracking- A New Jersey Case Study. In *USDA CSREES National Water Conference*. St. Louis, MO.

Harwood, V.J., Levine, A.D., Scott, T.M., Chivukula, V., Lukasik, J., Farrah, S.R. and Rose, J.B. (2005a) Validity of the Indicator Organism Paradigm for Pathogen Reduction in Reclaimed Water and Public Health Protection. *Appl Environ Microbiol* **71**, 3163–3170.

Harwood, V.J., Shehane, S.D. and Ulrich, R.M. (2005b) *Microbial Source Tracking: Tools for Refining Total Maximum Daily Load Assessments.* Tallhassee, Florida.

Hill, V.R., Polaczyk, A.L., Hahn, D., Narayanan, J., Cromeans, T.L., Roberts, J.M. and Amburgey, J.E. (2005) Development of a rapid method for simultaneous recovery of diverse microbes in drinking water by ultrafiltration with sodium polyphosphate and surfactants. *Appl Environ Microbiol* **71**, 6878–6884.

Hoel, T., Scheel, O., Sandvik, T. and Nordahl, S.H.G. (1991) Water- and airborne *Francisella tularensis* biovar *palaearctica* isolated from human blood. *Infection* **19**, 348–350.

Inegbenebor, U., Okosun, J. and Inegbenebor, J. (2010) Prevention of Lassa fever in Nigeria. *Transactions of the Royal Society of Tropical Medicine and Hygiene* **104**, 51–54.

Inglis, T.J.J., Garrow, S.C., Henderson, M., Clair, A., Sampson, J., O'Reilly, L. and Cameron, B. (2000) *Burkholderia pseudomallei* traced to water treatment plant in Australia. *Emerging Infectious Diseases Journal* **6**, 56–59.

Kearns, E.A., Magana, S. and Lim, D.V. (2008) Automated concentration and recovery of micro-organisms from drinking water using dead-end filtration. *J Appl Microbiol* **105**, 432–442.

Keim, P. (2003) *Microbial forensics: a scientific assessment.* Analysis. Washington, D.C.: American Academy of Microbiology.

Khushiramani R, T.U., Shukla J, and Batra HV. (2004) Characterization of outer membrane proteins of *Yersinia pestis* and *Yersinia pseudotuberculosis* strains isolated from India. *Indian Journal of Experimental Biology* **42**, 508–514.

Korajkic, A., Badgley, B.D., Brownell, M.J. and Harwood, V.J. (2009) Application of microbial source tracking methods in a Gulf of Mexico field setting. *Journal of Applied Microbiology* **107**, 1518–1527.

Lee, D.-Y., Lauder, H., Cruwys, H., Falletta, P. and Beaudette, L.A. (2008) Development and application of an oligonucleotide microarray and real-time quantitative PCR for detection of wastewater bacterial pathogens. *Science of The Total Environment* **398**, 203–211.

Leskinen, S.D., Brownell, M., Lim, D.V. and Harwood, V.J. (2010) Hollow fiber ultrafiltration and PCR detection of human-associated genetic markers from various surface water types in Florida. *Appl Environ Microbiol*, AEM.00025-00010.

Leskinen, S.D., Harwood, V.J. and Lim, D.V. (2009) Rapid dead-end ultrafiltration concentration and biosensor detection of enterococci from beach waters of Southern California. *J Water Health* **7**, 674–684.

McQuaig, S.M., Scott, T.M., Harwood, V.J., Farrah, S.R. and Lukasik, J.O. (2006) Detection of Human-Derived Fecal Pollution in Environmental Waters by Use of a PCR-Based Human Polyomavirus Assay. *Appl Environ Microbiol* **72**, 7567–7574.

Nakatsu, C.H., Pesenti, P.T. and Rhodes, A. (2007) Application of Microbial Source Tracking to Human Health and National Security. In *Emerging Issues in Food Safety: Microbial Source Tracking* eds. Santo Domingo, J.W. and Sadowsky, M.J. pp.211–234. Washington D.C.: ASM Press.

Olszewski, J., Winona, L. and Oshima, K.H. (2005) Comparison of 2 ultrafiltration systems for the concentration of seeded viruses from environmental waters. *Can J Microbiol* **51**, 295–303.

Pattanik, P. and Jana, A.M. (2005) Microbial Forensics: Applications in Bioterrorism. *Environmental Forensics* **6**, 197–204.

Raza, G., Dutt, B. and Singh, S. (1997) Kaleidoscoping public understanding of science on hygiene, health and plague: a survey in the aftermath of a plague epidemic in India. *Public Understanding of Science* **6**, 247–267.

Robertson, A.G. and Robertson, L.J. (1995) From asps to allegations: biological warfare in history. *Military Medicine* **160**, 369–373.

Scientific Working Group on Microbial Genetics and Forensics (SWGMGF) (2003) Quality assurance guidelines for laboratories performing microbial forensic work. *Forensic Science Communications (online:* http://wwwfbigov/hq/lab/fsc/backissu/oct2003/2003_10_guide01htm*)*

Scott, T.M., Rose, J.B., Jenkins, T.M., Farrah, S.R. and Lukasik, J. (2002) Microbial source tracking: current methodology and future directions. *Appl Environ Microbiol* **68**, 5796–5803.

Stoeckel, D.M. and Harwood, V.J. (2007) Performance, Design, and Analysis in Microbial Source Tracking Studies. *Appl Environ Microbiol* **73**, 2405–2415.

U.S. Environmental Protection Agency (2000) *Improved enumeration methods for the recreational water quality indicators: Enterococci and Escherichia coli.* Washington D.C.

Ufnar, J.A., Wang, S.Y., Christiansen, J.M., Yampara-Iquise, H., Carson, C.A. and Ellender, R.D. (2006) Detection of the nifH gene of *Methanobrevibacter smithii*: a potential tool to identify sewage pollution in recreational waters. *Journal of Applied Microbiology* **101**, 44–52.

United States Congress (1977) Clean Water Act, 303(d).

United States Environmental Protection Agency (1986) *Ambient water quality criteria for bacteria* No. EPA 440/5-84-002. Washington, D.C.: U.S. Environmental Protection Agency.

United States Environmental Protection Agency, U. (2005) *Microbial source tracking guide document.* Guide Document No. EPA/600/R-05/064. Cincinnati, OH: National Risk Management Research Laboratory, Office of Research and Development.

Velsko, S.P. (2005) Physical and chemical analytical analysis: a key component of bioforensics. In *AAAS Annual Conference* ed. (AAAS), A.A.f.t.A.o.S. p.10. Washington, D.C.: AAAS.

Wapnick, C.M., Korajkic, A. and Harwood, V.J. (2007) Application of Microbial Source Tracking (MST) Methods in Assessment of the Sources of Fecal Pollution in Tributaries. In *9th Biennial Conference on Stormwater Research and Watershed Management*. Orlando, FL.

White, D.C., Gouffon, J.S., Peacock, A.D., Geyer, R., Bierbacki, A., Davis, G.A., Pryor, M., Tabacco, M.B. and Sublette, K.L. (2003) Forensic Analysis by Comprehensive Rapid Detection of Pathogens and Contamination Concentrated in Biofilms in Drinking Water Systems for Water Resource Protection and Management. *Environmental Forensics* **4**, 63–74.

World Health Organization, W. (2004) Secton 1: Expert concensus. In *Waterborne zoonoses: identification, causes and control* ed. Cotruvo J. A., D.A., Rees G., Bartram J., Carr R., Cliver D.O., Craun G.F., Fayer R., Gannon V.P.J. London, U.K.: IWA Publishing.

Yoder J.S., Hlavsa. M.C., Craun G.F., Hill V., Roberts V., Yu P.A., Hicks L.A., Alexander N.T., Calderon R.L., Roy S.L., Beach M.J. (2008) Surveillance for waterborne disease and outbreaks associated with recreational water use and other aquatic facility-associated health events–United States, 2005-2006. *Morbidity and Mortality Weekly Rreport Surveillance Summary* **57**, 1–29.

Zilinskas, R.A. (1986) Recombinant DNA Research and Biological Warfare. In *The Gene Splicing Wars Reflections on the Recombinant DNA Controversy* eds. Zilinskas, R.A. and Zimmerman, B.K. pp.167–203. New York: Macmillan Publishers.

Chapter 24
Applications of Quantitative Microbial Source Tracking (QMST) and Quantitative Microbial Risk Assessment (QMRA)

Jack F. Schijven and Ana Maria de Roda Husman

Abstract For the estimation of public health risk from exposure to pathogens, quantitative microbial risk assessment (QMRA) can be used. QMRA entails hazard identification, exposure assessment, dose response and risk characterisation. Here, QMRA is focussed on infection risks from exposure to water-borne pathogens due to drinking-water consumption, recreational water activities or eating shellfish. Quantitative microbial source tracking (QMST) is microbial source tracking (MST) using quantitative data on the presence of pathogens and/or indicators in sources, in this case in surface water and faecal sources, such as wastewater and animal manure and has, therefore, a prominent role in hazard identification. Quantitative data requirements for QMRA and QMST are the same. For QMRA, pathogen data in source waters are required. For QMST, pathogen data are preferred over indicator data. Estimating probabilities of specific micro-organisms passing treatments and barriers and modelling of transport between locations are integral part of QMST and QMRA and can be used to identify, select and prioritise between sources of contamination and provide the basis for preventive measures and mitigation strategies. Modelling of transport of pathogens requires data on the fate and behaviour of those pathogens or of representative indicator organisms that have followed the same pathway.

Keywords Hazard identification • Transport model • Quantitative data • QMRA • QMST

J.F. Schijven (✉)
Expert Centre for Methodology and Information Services, National Institute of Public Health and the Environment, P.O. Box 1, 3720 BA Bilthoven, The Netherlands
e-mail: Jack.Schijven@rivm.nl

C. Hagedorn et al. (eds.), *Microbial Source Tracking: Methods, Applications, and Case Studies*, DOI 10.1007/978-1-4419-9386-1_24,

24.1 Introduction

QMST refers to quantitative microbial source tracking using quantitative data on the presence of a particular micro-organism, or any characteristic associated with that micro-organism, in specific sources. Also, the relation of QMST with and significance of QMST for quantitative microbial risk assessment (QMRA) is discussed.

Application of microbial source tracking (MST) to identify, and sometimes quantify, human and animal sources of contamination is rapidly increasing in the fields of water quality and food safety. In the field of water quality, MST has mainly been driven by the need to comply with quality standards for faecal indicator bacteria (Santo Domingo et al. 2007). Culture-independent and library-independent PCR-methods gain popularity in MST, but only a limited number of field applications exist because the main focus is on method development (Santo Domingo et al. 2007). Because MST is mainly aimed at identifying the origin of a contamination, most MST data are qualitative data. QMST uses quantitative data on contamination sources and enables estimation of the attribution of sources.

QMRA for drinking water or recreational water requires quantitative data on the presence of water-borne pathogens in the source water. If QMST data can be associated with the presence of a pathogen (quantitative pathogen source tracking, QPST), it becomes integral part of QMRA, specifically of hazard identification, the first step in QMRA. Most QMST data concern faecal indicator organisms (Field and Samadpour 2007), implying that these data cannot directly be used to estimate risks of infection. Nevertheless, because such QMST data may provide information on the origin of faecal indicator organisms, it may be used to take measures aimed at reducing contamination and associated risks, although the risk reduction may not be estimated. Note that throughout this chapter, the concepts MST, QMST and QPST are used. MST encompasses QMST and QPST, while QMST encompasses QPST.

Goss and Richards (2008) pointed out that evaluation of the health risk from a water resource requires knowledge of the strength (pathogen load) of the hazard and understanding the modification of pathogen numbers (attenuation) together with characteristics of the transport pathways (surface or subsurface). QMST data on pathogens and/or indicator organisms alone may not be sufficient to conclude that a pathogen's origin was identified. In the case where a source of contamination has been identified, inevitably the question arises whether this source was indeed the origin of the contamination.

To answer this question, not only quantitative data on the presence of the pathogen or indicator micro-organisms are needed but also their ability to pass a treatment or barrier and reach a specific location or product (water/food) needs to be explained. This is where models that describe treatment and/or transport processes that determine passage of pathogens and indicator organisms through a treatment or barrier come into play. Such models account for the environmental conditions and the fate and behaviour of the micro-organisms of concern along their travel. Such models are mathematical formulations of the processes that take place by which micro-organisms, on the one hand, are removed from the system or, on the other hand, may increase in numbers. Therefore, these models are a key part of QMST and QMRA. The combination of QMST data and transport modelling enable estimating the likelihood that a specific pathogen came from a particular source.

On the basis of QMST and transport modelling, mitigation strategies can be defined to prevent contamination, to improve or protect source water quality. With the help of QMRA, the effects of such strategies can be evaluated in terms of risk reductions.

In the following sections, the elements of QMST are discussed in more detail. In Sect. 24.2, the general principles of QMRA are described with references to leading articles and reviews. Hazard identification is the first element in QMRA. Because MST has a prominent role in hazard identification for QMRA, Sect. 24.3 deals with hazard identification and MST in more detail. Section 24.4 discusses the requirements for quantitative data for QMRA as well as for QMST. Section 24.5 deals with describing fate and behaviour of pathogens in the environment as well as models that predict their transport.

24.2 Principles of QMRA

24.2.1 Introduction

A general approach to QMRA for drinking water is outlined in the third edition of the guidelines for drinking-water quality of the World Health Organisation (WHO 2008, pp. 127–131), which is based on the ILSI-RSI framework (ILSI 1996). This approach is quoted below and supplemented with other literature. In brief, QMRA encompasses (1) hazard identification, (2) exposure assessment, (3) dose response and (4) risk characterisation. Here, QMST is integral part of hazard identification. Detailed descriptions of the QMRA methodology is given by Benford (2001), Haas et al. (1999), Haas and Eisenberg (2001), and Teunis et al. (1997). Figure 24.1 shows a schematic representation of the relation of the four major steps in QMRA for drinking water and recreational water, including QMST and modelling of transport, treatment and barriers. QMRA follows a tiered approach, implying that the steps in QMRA may be passed through repeatedly. The arrows in Fig. 24.1 with the treatment/barrier models link hazard identification and exposure assessment. Depending on the data and on the problem, treatment or barrier models may be included as part of either hazard identification or exposure assessment.

The WHO guidelines for drinking water (2008) outline a framework for safe drinking water encompassing health-based targets established by a competent health authority; adequate and properly managed systems (adequate infrastructure, proper monitoring and effective planning and management); and a system of independent surveillance. Pathogens of faecal origin are the principal concerns in setting health-based targets for microbial safety.

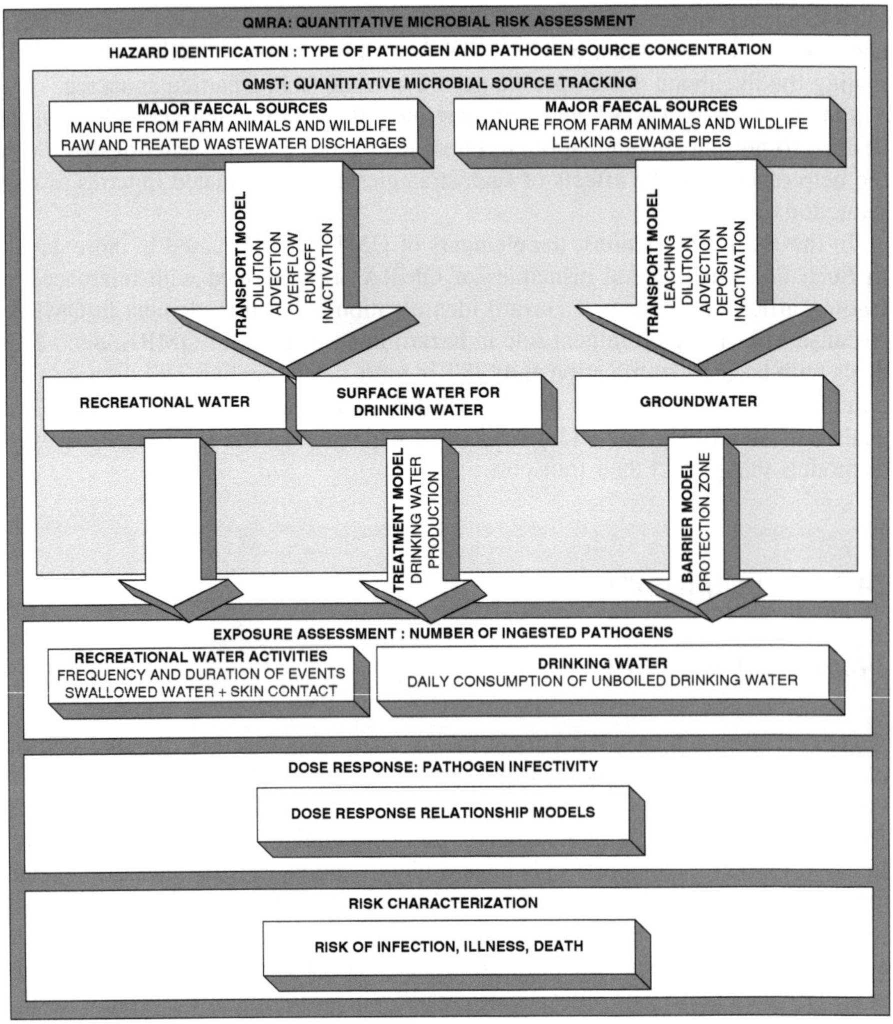

Fig. 24.1 Schematic representation of the relation of the four major steps in QMRA for drinking water and recreational water with QMST and modelling of transport, treatment and barriers

In the WHO guidelines for drinking water (2008), it is explained that some microbial hazards may have clearly defined adverse health effects that may particularly occur in developing countries. If in those cases, effects of measures to reduce exposure to drinking water can readily be monitored, the reduction of the disease risk can be the health-based target. Otherwise, health outcome targets may be the basis for evaluation of results through quantitative risk assessment models, where health outcomes are estimated based on information concerning exposure and dose–response relationships (WHO 2008). The results may be employed directly as a basis for the specification of water-quality targets or provide the basis for development of the other types of health-based targets. Health outcome targets based on information on the impact of tested interventions on the health of real populations are ideal but rarely available. More common are health outcome targets based on defined levels of tolerable risk, either absolute or fractions of total disease burden, preferably based on epidemiological evidence or risk assessment studies (WHO 2008). For example, Dutch legislation for drinking water has implemented that pathogenic micro-organisms in drinking water may not exceed a limit associated with a risk of infection of one per ten thousand persons per year as microbiological health-based target (De Roda Husman and Medema 2005).

24.2.2 Problem Formulation and Hazard Identification

All potential hazards, sources and events that can lead to the presence of these hazards should be identified and documented for each component of the drinking-water system, regardless of whether or not the component is under the direct control of the drinking-water supplier (WHO 2008). Here, the hazard of major concern is the water-borne pathogen, its presence in source waters and where it came from, which is exactly the focus of QPST. The major sources of water-borne pathogens include point sources, such as raw and treated wastewater discharges into surface water and leaking sewage pipes contaminating groundwater, as well as diffuse sources, such as farm animal manure on agricultural land that is contaminating surface water by runoff and groundwater by leaching (Fig. 24.1). Climate and seasonal changes clearly affect the input of faecal sources to surface water and groundwater (Schijven and de Roda Husman 2005). Extreme events such as heavy rainfall lead to increases in runoff, storm water overflow and resuspension of river sediments, which all lead to peak concentrations of pathogens in surface water and represent peak concentrations that determine the risk (Schijven and de Roda Husman 2005; Westrell et al. 2006b). Clearly, MST has a very prominent place in hazard identification (Fig. 24.1) because pathogens or indicator organisms are identified according to specific characteristics that, for example, may distinguish between human and animal origin. This distinction is especially relevant for zoonotic pathogens and also because pathogens from human origin are usually expected to pose a higher risk to public health than pathogens of animal origin because of host specificity. For this reason, Sect. 24.3 is included to elaborate in more detail on hazard identification and MST. For the purpose of QMRA, representative organisms are selected that, if controlled, would ensure control of all pathogens of concern. Typically, this implies inclusion of at least one virus, bacterial pathogen and protozoan parasite (WHO 2008). For example, QMRA for drinking according to Dutch legislation encompasses enteroviruses, *Campylobacter*, *Cryptosporidium* and *Giardia* as the so-called index pathogens (De Roda Husman and Medema 2005).

24.2.3 Exposure Assessment

Exposure assessment involves estimation of the number of pathogenic microbes to which an individual is exposed, principally through ingestion. Exposure assessment is a predictive activity that often involves subjective judgement. It inevitably contains uncertainty and must account for variability of factors such as concentrations of micro-organisms over time, volumes ingested, etc. Exposure can be considered as a single dose of pathogens that a consumer ingests at a certain point of time or the total amount over several exposures (WHO 2008). Commonly, infection risk assessment that is associated with drinking-water consumption is integrated over a year (e.g. De Roda Husman and Medema 2005), whereas a daily infection risk may provide guidance for exercising control over short-term adverse risk fluctuation events and their causes (Signor and Ashbolt 2009). In the case of diving and swimming, a risk assessment per diving (Schijven and de Roda Husman 2006) or swimming event (Schets et al. 2010) is the most obvious basis for communication of risk in this regard.

Here, exposure is determined by the concentration of microbes in drinking water, or recreational water, and the volume of water consumed. In the case of recreational water, skin contact may also be included in the exposure assessment. Data on the volume of swallowed water during swimming and diving are published (Dufour et al. 2006; Schets et al. 2010; Schijven et al. 2006). Schets et al. (2010) also collected data on the frequency and duration of swimming events, including head emersions.

The number of pathogens that are swallowed (dose) during a swimming event can be calculated as follows:

$$D = C \times 1 / R \times V, \tag{24.1}$$

where D is the dose (number of swallowed pathogens) per swimming event; C is the concentration of pathogens in the bathing water (number per litre); R is the recovery efficiency of the enumeration method for the pathogen, for example, if 50% of the pathogens that were present were actually detected and counted, then $R=0.5$ and, consequently, D will be corrected by a factor of 2; V is the volume (litre) of swallowed water per swimming event. Note that for a simplified approach, arithmetic average values are applied (point estimates), but it is more appropriate to sample from the associated distributions (Monte Carlo drawings) to derive a distribution of the dose.

It is rarely possible or appropriate to directly measure pathogens in drinking water on a regular basis. More often, concentrations in source waters are assumed or measured, and estimated reductions – for example, through treatment – are applied to estimate the concentration in the consumed water. Pathogen measurement, when performed, is generally best carried out at the location where the pathogens are at highest concentration (generally source waters). Estimation of their removal by sequential control measures is generally achieved by the use of indicator organisms. E.g. in The Netherlands, bacteriophages, *Escherichia coli* and spores of sulphite-reducing clostridia are used as the reference indicator organisms for enteroviruses, *Campylobacter* and the parasitic protozoa *Cryptosporidium* and *Giardia*, respectively (De Roda Husman and Medema 2005).

The other component of exposure assessment of drinking water, which is common to all pathogens, is the volume of unboiled water consumed by the population, including person-to-person variation in consumption behaviour and especially consumption behaviour of at-risk groups. For microbial hazards, it is important that the unboiled volume of drinking water, both consumed directly and used in food preparation, is used in the risk assessment, as heating will rapidly inactivate pathogens. The daily exposure of a consumer can be assessed by multiplying the concentration of pathogens in drinking water by the volume of drinking water consumed. For the purposes of the WHO Guidelines (2008), unboiled drinking-water consumption is assumed to be 1 L of water per day. However, detailed information on the variability of drinking-water consumption exists and can be used (Mons et al. 2007; Teunis et al. 1997; Westrell et al. 2006a; US-EPA 2006).

The number of pathogens that were swallowed (dose) by drinking unboiled tap water can be calculated as follows:

$$D = C \times 1/R \times Z_1 \times \cdots \times Z_n \times V, \tag{24.2}$$

where D is the dose (number of swallowed pathogens) by drinking-water consumption; C is the concentration (number of pathogen per litre); R is the recovery efficiency of the enumeration method for the pathogen; Z_1 to Z_n are the fractions of pathogen numbers that were able to pass n subsequent treatment steps, for example, if 10% of the pathogens are able to pass a treatment step 1,

$Z_1 = 0.1$, which is the same as one $\log_{10}$ removal; V is the volume (litre) of swallowed unboiled tap water per day. Obviously, in drinking-water production, towards finished water, the probability of not detecting micro-organisms increases considerably. Teunis et al. (2009) presented three simple models allowing statistical analysis of series of counts before and after treatment: distribution of the ratio of concentrations, and distribution of the probability of passage for unpaired and paired water samples. They demonstrated performance of these models for virus removal by several drinking-water treatment processes.

24.2.4 Dose–Response Assessment

The probability of an adverse health effect following exposure to one or more pathogenic organisms is derived from a dose–response model. Available dose–response data have been obtained mainly from studies using healthy adult volunteers. Several subgroups in the population, such as children, the elderly and immunocompromised persons, are more sensitive to infectious disease; currently, however, adequate data are lacking to account for this (WHO 2008). The conceptual basis for the infection model is the observation that exposure to the described dose leads to the probability of infection as a conditional event. For infection to occur, one or more viable pathogens must have been ingested. Furthermore, one or more of these ingested pathogens must have survived in the host's body. An important concept is the single-hit principle (i.e. that even a single organism may be able to cause infection and disease, possibly with a low probability) (WHO 2008).

In general, well-dispersed pathogens in water are considered to be Poisson-distributed, i.e. homogeneously distributed. When the individual probability of any organism to survive and start infection is the same, the dose–response relation simplifies to an exponential function (24.3). If, however, there is heterogeneity in this individual probability, this leads to the beta-Poisson dose–response relation (24.4), where the individual probabilities among pathogens (and hosts) are beta distributed. Variability of infectivity of a pathogen can for part be ascribed to variability in host specificity. At low exposures, such as would typically occur in drinking water, the dose–response model is approximately linear and can be represented simply as the probability of infection resulting from exposure to a single organism.

$$p_{\mathrm{I}} = 1 - \mathrm{e}^{-rD}, \tag{24.3}$$

$$p_{\mathrm{I}} = 1 - \left(1 + \frac{D}{\beta}\right)^{-\alpha}, \tag{24.4}$$

where p_{I} is the fraction of a population who will experience a risk of infection, D is the dose, r is the survival of each of the pathogens in any host, and α and β are the parameters of the Beta distribution describing variability of the pathogen survival in the host (Haas 2002).

Although it is very difficult to obtain dose–response data, meanwhile, a considerable body of publications on dose–response relations has appeared (Bartrand et al. 2008; Haas 2002; Huang and Haas 2009; Medema et al. 1996; Pujol et al. 2009; Teunis and Havelaar 2000; Teunis et al. 1996, 1997, 1999b, 2002a, b, 2005, 2008).

The dose–response relationships that are used generally apply to a group of a certain type of pathogen, whereas the pathogen that was measured in the source water may only be a subpopulation. Nevertheless, it is assumed that the general dose–response relationship applies because the dose–response relationship specific for the pathogen that is actually present is unknown. Therefore, the same assumption may be applied to a pathogen that has specific characteristics as determined by MST methodology. QPST opens the possibility of applying a dose–response relationship specific to that specific pathogen, if such a relationship would be known. For this reason, it is interesting to relate specific characteristics of a pathogen as determined in QPST with its infectivity or pathogenicity.

24.2.5 Risk Characterisation

Risk characterisation brings together the data collected on pathogen exposure, dose–response, severity and disease burden. The probability of infection can be estimated as the product of the exposure by drinking water and the probability that exposure to one organism would result in infection. The probability of infection per day is multiplied by 365 to calculate the probability of infection per year (WHO 2008; De Roda Husman and Medema 2005), or more appropriately as:

$$p_y = 1 - \prod_{i-1}^{365} (1 - p)_i. \tag{24.5}$$

Here, p_y is the probability of infection per person year and p_i that per person day. It is assumed that different exposure events are independent, in that no protective immunity is built up. This simplification is justified for low risks only.

Not all infected individuals will develop clinical illness; asymptomatic infection is common for most pathogens (e.g. Noroviruses: Gallimore et al. 2004; *Campylobacter*: Havelaar et al. 2009; *Cryptosporidium*: Fayer 1997). The percentage of infected persons that will develop clinical illness depends on the pathogen, but also on other factors, such as the immune status of the host. Risk of illness per year is obtained by multiplying the probability of infection by the probability of illness given infection.

The WHO guidelines for drinking water (2008) further describe that to translate the risk of developing a specific illness to disease burden per case, the metric disability adjusted life years (DALY) is used (Havelaar and Melse 2003). This should reflect not only the effects of acute end points (e.g. diarrhoeal illness) but also mortality and the effects of more serious end points (e.g. Guillain–Barré syndrome associated with *Campylobacter*). Disease burden per case varies widely. This considerable difference in disease burden results in far stricter treatment requirements in low-income regions for the same source water quality to obtain the same risk (expressed as DALYs per year).

Only a proportion of the population may be susceptible to some pathogens because immunity developed after an initial episode of infection or illness may provide lifelong protection. Examples include HAV and rotaviruses. It is estimated that in developing countries, all children above the age of 5 years are immune to rotaviruses because of repeated exposure in the first years of life. In developed countries, rotavirus infection is also common in the first years of life, and the illness is diagnosed mainly in young children, but the percentage of young children as part of the total population is lower. This translates to an average of 6% of the population in developed countries being susceptible.

The uncertainty of the risk estimate is the result of the uncertainty and variability of the data collected in the various steps of the risk assessment. Risk assessment models should ideally account for this variability and uncertainty. Theoretical considerations show that risks are directly proportional to the arithmetic mean of the ingested dose. Hence, arithmetic means of variables such as concentration in raw water, removal by treatment and consumption of drinking water are recommended. This recommendation is different from the usual practice among microbiologists and engineers of converting concentrations and treatment effects to log-values and making calculations or specifications on the log-scale. Such calculations result in estimates of the geometric mean rather than the arithmetic mean, and these may significantly underestimate risk. Analysing site-specific data may, therefore, require going back to the raw data rather than relying on reported log-transformed values. Benke and Hamilton (2008) showed with Monte Carlo simulations that the choice for arithmetic mean holds even with very skewed distributions that often occur in environmental samples.

24.3 Hazard Identification and MST

This section highlights the prominent role of MST in hazard identification (Fig. 24.1).

Figure 24.2 gives a schematic representation of QMST and transport modelling as part of hazard identification in QMRA to emphasise identification and enumeration of pathogens and/or indicator organisms in faecal sources as well as in water used for drinking water, recreational purposes and irrigation. Quantitative data on the presence of pathogens and/or indicator organisms in faecal sources as well as in source waters may be correlated. For example, detection of a ruminant-specific and human-specific faecal *Bacteroidetes* marker by qPCR could combined with standard microbiological and online hydrological data in a large karstic spring catchment in Austria (Reischer et al. 2008). Multiparametric analysis of all data allowed linking the ruminant-specific marker to general faecal pollution indicators, especially during flood events.

Typical for QMRA for drinking water is the primary interest in the presence of pathogens in the groundwater that is pumped up at the production well or the surface water prior to the additional treatment steps for drinking-water production. In the case of surface water, often this may include open storage reservoirs, which still may get additional faecal contamination from faecal sources. For a proper

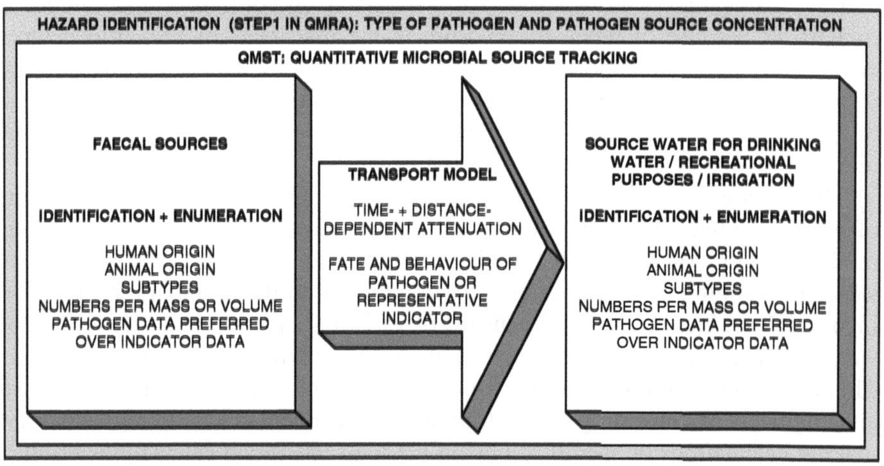

Fig. 24.2 Schematic representation of QMST and transport modelling as part of hazard identification in QMRA

QMRA, these storage reservoirs should be considered the starting point, instead of the intake water (De Roda Husman and Medema 2005). MST studies mainly are focussed on the presence and identification of pathogens and indicator organisms in raw and treated wastewater, surface water (for drinking-water production and/or recreational purposes) and groundwater because MST focuses on the origin of a source of contamination. See for example, the study of Edge et al. (2007) who tracked faecal pollution at two lake beaches in Toronto, Ontario using antibiotic resistance analysis and DNA fingerprinting and identified birds as the dominant sources of *E. coli*. Korajkic et al. (2009) quantified and identified human faecal pollution in a Florida estuary that serves shellfishing and recreational activities. Orosz-Coghlan et al. (2006) identified passerines as the most important faecal source of constructed wetlands in Arizona by means of antibiotic risk profiling and biochemical fingerprinting. Xiao et al. (2000) used a small-subunit rRNA-based PCR-restriction fragment length polymorphism technique to identify species and sources of *Cryptosporidium* oocysts in 29 water samples collected from a stream in New York and found 12 genotypes in 27 positive samples.

Recall that risk assessment requires quantitative data on the presence of pathogens. Field and Samadpour (2007) pointed out that, generally, pathogens may be rare, difficult to culture and patchy in distribution, yet highly infectious even at low doses. Quantitative data on the presence of indicator organisms between origins of faecal contamination and water resources, may be useful in combination with transport modelling to estimate the likelihood from which faecal source a pathogen may have originated, but this requires knowledge of the processes that determine fate and behaviour of the indicator as well of the pathogen during transport from one location to the other and indicator and pathogen may not behave the same (Figs. 24.1 and 24.2). In most cases, quantitative data on processes such as inactivation, growth, deposition are not or only sparsely available. Time and distance are the scales that determine this likelihood. Also recall that this approach provides a basis for formulating mitigation strategies. Often such strategies are aimed at reducing faecal contamination, where strategies aimed at reducing pathogens are more appropriate (Field and Samadpour 2007). Only if pathogen and associated indicator organism have followed the same route as well as have been subject to the same removal processes, such as inactivation and deposition, to the same extent, then reductions of the concentrations of the indicator organisms may be assumed to hold for those of the pathogen as well, and consequently, a relative risk reduction may be approximated.

A number of studies have been undertaken to investigate relationships between indicator micro-organisms and water-borne pathogens in surface waters. Mostly Spearman rank correlations between indicators and pathogens are relatively weak, but significant and primarily positive, seasonally dependent and site specific (e.g. Dechesne and Soyeux 2007; Hörman et al. 2004; Payment et al. 2000; Till et al. 2008; Wilkes et al. 2009). Stapleton et al. (2009) found no significant correlations between index faecal indicator organism and *Bacteriodales* 16S rRNA as determined by qPCR in a variety of streams, effluent and bathing water and concluded that the latter does not reliably characterise faecal contamination.

Because often there is no detectable relationship between faecal indicator bacteria and enteric pathogens in environmental samples, QPST is to be preferred when the goal is to reduce a specific pathogen or health risks (Santo Domingo et al. 2007). Fong et al. (2005) used reverse transcription- and nested-PCR to target human and bovine enteroviruses and human adenoviruses in river water samples. The presence of these viruses could not significantly be related with faecal and total coliform concentrations. However, presence of virus was directly related to dissolved oxygen and stream flow, and inversely related to water temperature, preceding rainfall and chlorophyll-*a* concentrations. Hundesa et al. (2009, 2010) developed qPCR assays for porcine adenoviruses and bovine polyomaviruses, respectively, as QMST tools. Hundesa et al. (2009) showed a difference of 5 log in genome copies of porcine adenoviruses between pooled faecal samples and river water samples. Both studies are examples of the development of quantitative assays for faecal indicators that are highly sensitive and highly specific to animal faecal contamination. Nevertheless, Hundesa et al. (2009, 2010) mentioned that further studies are necessary to collect information on the distribution of porcine adenoviruses and bovine polyomaviruses in various matrices and geographical areas. Although highly specific to porcine and bovine faecal contamination, data on the quantitative relation with zoonotic pathogens is not available.

Choi and Jiang (2005) quantified adenovirus concentrations in two Californian rivers using real-time PCR. They found 10^2–10^4 genomes per litre of water, but were not able to detect adenovirus by plaque assay. They found significant correlations between total and faecal coliforms, enterococci and coliphages, but not with the human adenoviruses as determined by PCR. Their findings indicated that non-infectious virus particles were determined by PCR and they concluded that genome-based detection methods alone are inadequate for direct assessment of human health risk.

Targeting enteric viruses can be potentially be used as an indication of human health risk, but the presence of infectious viral particles must be assessed to use the data in risk-assessment models (De Roda Husman et al. 2009; Santo Domingo et al. 2007).

The approach of QMST as schematised in Fig. 24.2 fits well into policies to manage and control water quality for drinking water production, recreational and agricultural purposes. The European Union Water Framework Directive (EU 2000) and the United States Clean Water Act (USEPA 2002) both prescribe a catchment-wide approach for the management of microbial pollution of river basins that necessitates the consideration of point- and diffuse-pollution sources to effect control of water quality where it is used for ecosystem maintenance, drinking-water production, recreation or fisheries.

Control of the microbial water quality within the EU has been redirected towards the catchment-wide approach through the European bathing water directive 2006/7/EC (EU 2006), which requires that bathing water profiles are established for a single bathing water body or a number of contiguous bathing waters. This entails an inventory of possible sources of contamination. Regulations in USA use the approach of total maximum daily loads (TMDLs), which requires that water bodies not complying with defined water quality standards are assigned an overall maximum load of the pollutant that the water body can tolerate. TMDL investigations then seek to determine pollutant fluxes from all potential sources within the catchment, before recommending remediation measures to reduce inputs to a level that complies with the specified TMDL. By means of QMST, the relative contributions of pollution sources can be determined by enumerating the differently identified types of pathogens and/or indicator organisms (Fig. 24.2). Stapleton et al. (2007), therefore, concludes that apportionment of pollutant fluxes to the various sources within a catchment is key to successful application of water management both in Europe and USA.

Obviously, in the case of a QMRA for recreational water, data on the presence of pathogens in that water that is in use for recreational activities are needed. However, bathing water regulations require monitoring of faecal indicator bacteria, such as *E. coli* and enterococci (EU 2006). Pathogen concentrations are not monitored on a regular basis; moreover, their concentrations may even be below detection limits. Regardless of compliance of recreational water to bathing water regulations, pathogen concentrations may still be high enough to have an undesirable health effect (Graczyk et al. 2007; Schets et al. 2008). The strategy of the EU Bathing Water Directive (2006) to characterise beaches by listing contamination sources to enable taking appropriate measures to improve the bathing water quality is already much better than simply monitoring water for the presence of indicator organisms. Nevertheless, QPST would strengthen this strategy even more.

Groundwater as the resource for drinking-water production is a particular case. Quantitative data on the presence of pathogens in groundwater are much harder to get, if at all possible. In those cases where pathogens were found to be present in the groundwater, commonly the origin of those pathogens remains unknown. Nevertheless, it may be useful to make a profile of potential sources in the vicinity of the well, whereby QPST data in those sources should be collected. The use of groundwater as a source for drinking-water production is often preferred because of its generally good microbial quality in its natural state as compared with for instance fresh surface water. Nevertheless, it may be readily contaminated, and outbreaks of disease from contaminated groundwater sources are reported in countries at all levels of economic development (Howard et al. 2006). The contribution of groundwater to the global and significant incidence of water-borne disease cannot be assessed easily because of many competing transmission routes (Howard et al. 2006). In this regard, viruses are considered to be the most critical pathogens for groundwater contamination, because of their ability to travel through the subsurface and their high infectivity (Schijven and Hassanizadeh 2000).

Human pathogenic viruses such as enterovirus, adenovirus, norovirus, reovirus, rotavirus and hepatitis A viruses have been detected in groundwater with molecular and/or cell culture techniques with prevalence rates varying from 8 to 23% (Fout et al. 2003; Borchardt et al. 2003, 2007). Contamination of drinking water from groundwater with human pathogenic viruses may lead to epidemics that cause deaths and severe illness (Maurer and Stürchler 2000; Parshionikar et al. 2003; Kim et al. 2005; Jean et al. 2006; Gallay et al. 2006). Note that in the case of outbreaks and/or where a high prevalence rate of viruses in groundwater samples was found, it concerned vulnerable geologic settings such as fractured rock or cross-connecting well bores or leaking well cases in sandstone and shale aquifers (Powell et al. 2003; Borchardt et al. 2007) in combination with the presence of significant sources of contamination, such as wastewater treatment facilities, septic tanks, animal manure (Parshionikar et al. 2003; Gallay et al. 2006; Jean et al. 2006). Frost et al. (2002) stated that the occurrence of virus contamination in groundwater may be overestimated because many studies selected high-risk wells for testing.

Often, groundwater contamination occurs as a peak event, hampering detection. Borchardt et al. (2003) found that contamination was transient, since none of the wells in their study was virus positive in two sequential samples. In addition, when molecular detection is used for virus enumeration, it is important to take into consideration that the fraction of infectious virus is time and temperature dependent (de Roda Husman et al. 2009). By using an optimised cell culture-PCR assay for detection of rotavirus strains in naturally contaminated source waters, Rutjes et al. (2009) found that the broad variation observed in the ratios of rotavirus RNA and infectious particles demonstrates the importance of detecting infectious viruses instead of viral RNA for the purposes involving estimations of public health risks. Given the difficulties in monitoring for and interpretation of groundwater contamination, one should better aim at preventing contamination, or include treatment prior to use.

24.4 Quantitative Data Requirements for QMRA and QMST

In order to conduct QMRA for a particular pathogen, one needs to have quantitative data on the presence of the pathogen in the source and of the pathogen or associated indicator organisms before and after treatments or barriers. In the case of drinking water, the sources are groundwater and surface water. Obviously, the latter is also the source when it is in use for recreational purposes. Groundwater and, even more, surface water may be contaminated with water-borne pathogens mainly originating from faecal sources as discharges of raw and treated wastewater from wastewater treatment plants, septic tanks and leaking sewers, but also from animal manure runoff and leaching and from direct input by wildlife. Collecting quantitative data on the presence of pathogens in source water may be limited because concentrations are too low, especially on groundwater. The limitations to measure pathogen concentrations in source water may be even stronger if specific subpopulations of a specific micro-organism, and, therefore, lower numbers, are considered. According to Scott et al. (2005), library-dependent methods based on *E. coli* libraries can result in high source miscalculation rates or the inability to classify many unknown source isolates. For example, the concentrations of pathogenic *E. coli* can be 2–3 orders in magnitude less than those of non-pathogenic *E. coli*.

It is desirable to include variability and uncertainty into QMRA (WHO 2008), which puts particular demands on the type and format of quantitative data.

Table 24.1 Estimation of distribution of concentrations from microbiological data

Type of data	Data format	Concentration distribution	
Counts in samples	Counts ($N \geqq 0$) and sample sizes (V)	Fit negative binomial distribution with parameters r and λ of N and V of all samples	Concentrations gamma-distributed with parameters r and λ
Most probable numbers	± and sample size scheme (Vs)	Concentration estimate assuming Poisson distribution	Fit gamma distribution with parameters r and λ to concentrations
Presence/ absence	± and sample size (V)	Need more than one sample	One concentration estimate assuming Poisson distribution

Table 24.1 gives an overview of the type of data, the format of data and the distribution of concentrations. Not only should data on the presence of pathogens and/or indicator organisms be determined throughout the year to account for seasonal variation, and include peak events, for a correct weighing of the data, but it is also required to use actual counts and sample sizes (Teunis et al. 1999a, 2009). Often, only calculated concentrations are reported, whereby information on the precision of the data is lost. Also, actual investigated sample size is often confused with the size of the sample that was taken. For example, it is reported that 560 pathogens were counted in 10 L of sample, but actually 10 L of sample were collected, but only 56 pathogens were counted in one litre of that sample.

In the case of most probable number methods, it is desirable to use raw data in the form of presence/absence data and the associated scheme of samples sizes. And, obviously, in the case of presence absence data, the presence/absence should be reported together with the actual size of the analysed sample (Table 24.1). In most cases, it is assumed that micro-organisms are randomly distributed in water, which implies that counts of micro-organisms in a water sample are Poisson-distributed. Because of many environmental factors, counts between samples vary in time and space. This variability can be described by a gamma distribution; consequently, the counts between samples follow a negative binomial distribution, which is a mixture of the Poisson and gamma distribution (Haas et al. 1999; Teunis et al. 1999a). Thus, to raw data in the form of counts and actual sample sizes a negative binomial distribution is fitted, giving estimates of the distribution parameters r and λ, and, consequently, micro-organism concentrations are gamma-distributed with parameters r and λ (Table 24.1). In the case of most probable number data, concentrations for each sample are calculated assuming micro-organisms in the sample are Poisson-distributed or by consulting MPN-tables. Then, a gamma distribution can be fitted to these concentrations. MPN concentrations tend to be positively biased as a consequence of the probabilistic basis for calculation the MPN concentration (Gronewold and Wolpert 2008). To account for this bias, adaptive rejective Monte Carlo sampling using the Metropolis-Hasting algorithm can be used (Westrell et al. 2006b), or bootstrapping methods (McBride et al. 2003; Smeets et al. 2008). In the case of presence/absence data, more than just one sample needs to be analysed. These data (presence or absence and sample size) can be used as in an MPN scheme to obtain one concentration estimate.

Note that quantitative data on the presence of micro-organisms are only an estimate of their true value and that this estimate very much depends on the performance characteristics of the enumeration method. Performance characteristics of the enumeration method, which includes all the handling from sampling to counting, are given by detection limit (related to sample size), recovery efficiency (the fraction of micro-organisms present in the sample that were actually detected), sensitivity (the fraction of true positives) and specificity (the fraction of true negatives). In QMRA, it is desired to include information on variability and uncertainty as much as possible. Therefore, it is useful to relate performance characteristics of the enumeration methods to sources of variability such as the matrix (type of water), seasonality and laboratory. This requires data collection on performance characteristics encompassing sources of variability and may lead to distributions for recovery, sensitivity and specificity for which the concentration distribution can be corrected (e.g. Rutjes and de Roda Husman 2004; Schets et al. 2004). For obvious reasons, if using QMST data for QMRA, the same data requirements apply.

QMST data can be direct or indirect. For example, direct QMST data consist of numbers of a specific DNA-sequence per sample volume. Indirect QMST data refer to the fraction of isolates that contained a specific characteristic. Likewise, as for example for recoveries, one would like to know the distribution of the fraction of specific isolates for usage in QMRA. One may assume that the fraction of isolates having a specific characteristic follows a binomial distribution. Currently, there is a lack of performance standards to evaluate the accuracy of MST methods, and data are sparse (Santo Domingo et al. 2007; Stoeckel and Harwood 2007). New molecular-based techniques have shown that combined use of conventional and alternative indicators for faecal pollution increases both the sensitivity and specificity of faecal pollution and associated pathogens (Savitcheva and Okabe 2006).

The validation of MST methods requires analysis of a large number of samples from different geographic locations over time; however, this is commonly not part of MST studies (Santo Domingo et al. 2007). To provide 100% correct classification rates, several approaches are needed simultaneously to fully characterise the array of faecal sources impacting water systems (Santo Domingo et al. 2007). In that regard, library-independent PCR methods have the advantage of being fast, inexpensive, sensitive, quantitative, amenable to automation and able to target multiple microbial populations. Harwood et al. (2009) evaluated library-independent MST methods for human sewage detection by conventional PCR and determined specificities and sensitivities. Scott et al. (2005) presented a method for PCR detection of a putative virulence factor, the enterococcal surface protein *esp* in *Enterococcus faecium*, as an index of human pollution. The gene was detected in 97% of sewage and septic tank samples but was not detected in any livestock waste lagoons or in bird or animal faecal samples. MST methods should be specific and applicable over a broad geographic region. Although PCR assays are theoretically capable of detecting a few gene copies, because of inhibitors and competing targets, detection limits may be higher and difficult to estimate (Santo Domingo et al. 2007). Hundesa et al. (2009, 2010) developed qPCR assays for porcine adenoviruses and bovine polyomaviruses as QMST tools and tested efficiency, sensitivity and specificity of these qPCR assays. They reported detection limits of 1–10 genome copies per test tube.

24.5 Modelling Fate, Behaviour and Transport of Micro-organisms in the Aquatic Environment

In QMRA for drinking water, commonly, pathogen concentrations in the source water are measured, and concentrations of indicator organisms are measured before and after the following drinking-water treatment steps. Indicator organisms usually occur in higher concentrations and are easier to enumerate than pathogens. Removal of those indicator organisms by treatment should be less or equal to that of the pathogens they represent. Concentrations of pathogens in drinking water are below detection limits; therefore, their concentrations are calculated from their concentration in the source water and their removal by treatment (De Roda Husman and Medema 2005).

Models for treatment efficiency used in QMRA mostly are stochastic models (e.g. Neumann et al. 2007; Teunis et al. 1999a, 2009; Smeets et al. 2008), but may also be deterministic process models (e.g. Hijnen et al. 2010; Page et al. 2009), or a combination of those (Schijven et al. 2006). Schijven et al. (2006) presented a model for calculation of groundwater protection zones using distributions for the model parameters describing inactivation and attachment of virus to sand grains.

Stochastic modelling can be applied to compare distributions of concentrations, in this case before and after a treatment, such as applied in drinking-water production or in a wastewater treatment plant, or before and after a barrier, such as the subsurface, or simply between to locations, such as wastewater and surface water. Data of micro-organisms before and after a treatment or barrier or between locations can either be paired or unpaired. If there is no time delay, or a negligible time delay between the concentrations of the treatment, barrier, or locations, concentration data may be considered to be paired. Other examples of paired data are split samples, where different enumeration methods are applied for comparison, or where the half is spiked with a known number of a reference micro-organism for determining recovery efficiency of the method, or also, as in QMST for comparison of the total population and a specific subpopulation of micro-organisms in the same sample. If such relations between samples are not the case, or unknown, one may consider sets of concentration data unpaired. In the case of unpaired data, which can be counts and volumes or MPN data (Table 24.1), it is possible to compare the gamma distributions of both concentration sets. When fitting the data to either the negative binomial distributions (counts and samples sizes) or gamma distributions

(MPN concentrations), maximum likelihood values are obtained as well (McCullagh and Nelder 1989; Teunis et al. 2009). By means of maximum likelihood ratio testing, it can be determined if both distributions are significantly different or not. If not, the concentrations of both data sets can be considered to be part of the same distribution, otherwise not. From both distributions, concentrations can be generated by means of Monte Carlo sampling. Next, ratios of the randomly ordered concentrations of both data sets can be calculated. This is the gamma-ratio model resulting in a distribution of the relative difference between the two concentration sets, representing the fraction of micro-organisms that were able to pass a treatment or barrier or that were able to travel from one location to the other. The gamma-ratio model is not restrictive; it merely calculates ratios of concentrations.

If the concentration sets are paired, again, by means of maximum likelihood ratios, distributions can be compared. The ratio of both concentrations can be constant, meaning that each micro-organism has the same probability of passing a treatment or barrier, or the same probability of getting to the other location. The ratio can also be variable, which can be described by a beta distribution (Teunis et al. 1999a, 2009). A beta distribution for the ratio is in comparison to the gamma-ratio, more restrictive, because this ratio must lie between 0 and 1. The gamma-ratio can be larger than one. In the case of only presence/absence data, one has only one concentration estimate before and after a treatment or barrier, or for each of two locations; therefore, one can only calculate one ratio (point estimate).

Probably more than in drinking-water treatment processes, between a faecal source and source water for drinking water or recreational purposes (Fig. 24.2), a time-delay effect is more prevalent. To account for large time delays, transport modelling comes into play. An example of such transport modelling is given by Schijven et al. (2005) where the probability of infecting dairy cows that were drinking FMD virus contaminated surface water due to illegal discharges of contaminated milk was assessed. Following scenarios where contaminated milk was dumped illegally in ditches, transported to wastewater treatment plants, discharged with the treated wastewater in surface water from which cows drank at kilometres of distance downstream the point of discharge, it could be concluded that high risks of infection were possible via this route.

Using literature data for virus inactivation in groundwater (Pedley et al. 2006) and a low estimate for the attachment rate of viruses to sand grains (Schijven and Hassanizadeh 2000), Schijven et al. (2006) calculated protection zones for shallow unconfined sandy aquifers that would allow protection against virus contamination to the level that the infection risk of one per 10,000 persons per year is not exceeded with a 95% certainty. In those cases, instead of 60 days, 1–2 years of travel time were needed, corresponding to setback distances of about 200–400 m. Wuijts et al. (2008) investigated the presence of viruses and bacteria large volumes samples taken from four possibly vulnerable unconfined sandy aquifers in The Netherlands. Only in a few samples one bacteriophage was detected. By means of the model of Schijven et al. (2006) it could be calculated that a number of potential faecal contamination sources were too far away from the production wells to actually cause contamination.

Reeves et al. (2004) describe a series of field studies aimed at identifying the spatial distribution and flow forcing of faecal indicator bacteria in dry and wet weather runoff from a highly urbanised coastal watershed in southern California. During storms, the load of faecal indicator bacteria in runoff follows a power law that assumes that faecal indicator bacteria in storm runoff originate from the erosion of contaminated sediments in drainage channels or storm sewers. Scaling properties of pollutants in stormwater runoff may harbour information about the origin of non-point source pollution.

The studies of Reeves et al. (2004), Schijven et al. (2005) and Wuijts et al. (2008) are examples in which the likelihood of a pathogen contamination to reach a location was evaluated by means of transport modelling. Following this approach it may also explain the origin of contamination. Even more, when considering more contamination sources, QMST and transport modelling allow assessment of the relative contributions of these sources to the total contamination, which in turn is a sound basis for prioritising sanitation and mitigation measures. Targeted sampling prior to engaging in intensive MST studies may identify hot spots, using local community knowledge of potential sources, and studying hydrological flow data to identify areas that should be scrutinised more rigorously with an MST method (Santo Domingo et al. 2007). Moreover, here, transport modelling may be very helpful too because model predictions may be used to select the potential sources of contamination that may be relevant.

By means of a Bayesian belief network, which is a probabilistic graphical model that represents a set of random variables and their conditional independencies via a directed acyclic graph probabilistic relationships between contaminations at different locations, or even risks and contaminations sources can be investigated by using the network to compute the probabilities of the presence of various contamination sources.

When including QMST data that represent a subpopulation of a pathogen or indicator organism and the total population, it is possible to compare the behaviour of the subpopulation with the total population, which provides insight into the significance of this subpopulation. This might even lead to the further development of a specific indicator that has improved indicator functions in terms of better mimicking pathogen behaviour and ecology. Simpson et al. (2002) argues that commensal bacteria are not a good choice as target organisms because of the high level of genetic diversity. A wide variability in transport ability, more specifically sticking efficiencies determining attachment of *E. coli* strains to sand, has been demonstrated by Foppen et al. (2010). In that regard, Savitcheva and Okabe (2006) pointed out that ecological and survival characteristics of viral, bacterial, and parasitic pathogens vary under environmental conditions, indicating that probably no single indicator organisms can predict the presence of all enteric pathogens for all types of water and different host-associated faecal pollution. Although data about the quantitative presence and behaviour of pathogens is preferred, for practical reasons and lack of data, one will remain to rely on indicators. Therefore, one needs to follow a cautious approach, such as using indicator organisms that survive as good as or better than the pathogens they represent.

24.6 Conclusions

Hazard identification is the starting point for QMRA to estimate public health risk from exposure to water-borne pathogens. It requires collecting quantitative data on the presence and origin of relevant pathogens. QMST using quantitative data on the presence of pathogens and/or indicators in sources has a prominent role in hazard identification, and it is rapidly developing. Nevertheless, most data from MST studies are still rather qualitative, i.e. presence/absence data with focus on characterisation and identification. Collection of quantitative data in the form of counts and actual investigated samples is crucial to QMRA. Quantitative data not only allow QMRA, QMST in combination with transport modelling but also enable estimating probabilities of specific micro-organisms passing treatments and barriers.

Acknowledgements This investigation has been performed by order and for the account of General Directorate for Environmental Protection, directorate of Drinking Water, Water and Agriculture, The Hague, The Netherlands.

References

Bartrand TA, Wier MH, Haas CN (2008) Dose-response models for inhalation of *Bacillus anthracis* spores: Interspecies comparisons. Risk Anal 28(4):1115–1124

Benford D (2001) Principles of risk assessment of food and drinking water related to human health. In ILSI Europe Concise Monograph SERIES. International Life Science Institute, Brussels, pp. 1–43

Benke KK, Hamilton AJ (2008) Quantitative microbial risk assessment: uncertainty and measures of central tendency for skewed distributions. Stoch Environ Res Risk Assess 22:533–539

Borchardt MA, Bertz PD, Spencer SK, Battigelli DA (2003) Incidence of enteric viruses in groundwater from household wells in Wisconsin. Appl Environ Microbiol 69(2):1172–80

Borchardt MA, Bradbury KR, Gotkowitz MB, Cherry JA, Parker BL (2007) Human enteric viruses in groundwater from a confined bedrock aquifer. Environ Sci Technol 41(18):6606–6612

Choi S, Jiang S (2005) Real-time PCR quantification of human adenoviruses in urban rivers indicates genome prevalence and low infectivity. Appl Environ Microbiol 71 (11):7426–7433

De Roda Husman AMR, Lodder WJ, Rutjes SA, Schijven JF, Teunis PFM (2009) Long-term inactivation study of three enteroviruses in artificial surface and groundwaters using PCR and cell culture. Appl Environ Microbiol 75(4):1050–1057

De Roda Husman AMR, Medema GJ (2005) Inspectorate Guideline – Assessment of the microbial safety of drinking water, Inspectorate of the Ministry of Housing, Physical Planning and the Environment, Art. 5318, The Hague

Dechesne M, Soyeux E (2007) Assessment of the source water pathogen contamination. J Wat Health 05.1:39–50

Dufour AP, Evans O, Behymer TD et al. (2006) Water ingestion during swimming activities in a pool: a pilot study. J Wat Health 04(4):425–430

Edge TA, Hill S, Stinson G, Seto P, Marsalek J (2007) Experience with the antibiotic resistance analysis and DNA fingerprinting in tracking faecal pollution at two lake beaches. Wat Sci Technol 56(11):51–58

EU (2000) Commission of European Committees (CEC) (2000) Directive 2000/60/EC. Off J Eur Comm L237:1–72

EU (2006) Directive 2006/7/EC of the European Parliament and of the Council of 15 February 2006 concerning the management of bathing water quality and repealing Directive 76/160/EEC. Official Journal of the European Union L64:37–51

Fayer R (ed.) (1997) Cryptosporidium and Cryptosporidiosis. CRC Press, Boca Raton, 251 pp.

Field KG, Samadpour M (2007) Fecal source tracking, the indicator paradigm, and managing water quality. Wat Res 41:3517–3538

Fong TT, Griffin DW, Lipp EK (2005) Molecular assays for targeting human and bovine enteric viruses in coastal waters and their application for library-independent source tracking. Appl Environ Microbiol 71(4):2070–2078

Foppen JW, Lutterodt G, Röling WF, Uhlenbrook S (2010) Towards understanding inter-strain attachment variations of Escherichia coli during transport in saturated quartz sand. Wat Res. 44(4):1202–1212

Fout GS, Martinson BC, Moyer MW, Dahling DR (2003) A multiplex reverse transcription-PCR method for detection of human enteric viruses in groundwater. Appl Environ Microbiol 69(6):3158–3164

Frost FJ, Kunde TR, Craun GF (2002) Is contaminated groundwater an important cause of viral gastroenteritis in the United States? J Environ Health 65(3):9–14

Gallay A, Valk H de, Cournot M, Ladeuil B, Hemery C, Castor C, Bon F, Mégraud F, Le Cann P, Desenclos JC (2006) A large multi-pathogen waterborne community outbreak linked to faecal contamination of a groundwater system, France, 2000. Clin Microbiol Inf 12(6):561–570

Gallimore CI, Cubitt D, Plessis N du, Gray JJ (2004) Asymptomatic and Symptomatic Excretion of Noroviruses during a Hospital Outbreak of Gastroenteritis. J Clin Microbiol 42(5):2271–2274

Goss M, Richards C (2008) Development of a risk-based index for source water protection planning, which supports the reduction of pathogens from agricultural activity entering water resources. J Environ Manag 87:623–632

Graczyk TK, Sunderland D, Tamang L, Lucy FE, Breysse PN (2007) Bather density and levels of *Cryptosporidium*, *Giardia*, and pathogenic microsporidian spores in recreational bathing water. Parasitol Res 101:1729–1731

Gronewold AD, Wolpert RL (2008) Modeling the relationship between most probable number (MPN) and colony forming unit (cfu) estimates of fecal coliform concentrations. Wat Res 42(13):3327–3334

Haas CN (2002) Conditional dose-response relationships for microorganisms: Development and application. Risk Anal 22(3):455–463

Haas CN, Eisenberg J (2001) Risk Assessment. In Water Quality—Guidelines, Standards and Health: Assessment of Risk and Risk Management for Water-related Infectious Disease (ed. L. Fewtrell & J. Bartram). IWA Publishing, London, (Chapter 8)

Haas CN, Rose JB, Gerba CP (1999) Quantitative Microbial Risk Assessment. John Wiley and Sons, New York

Harwood VJ, Brownell M, Wang S, Lepo J, Ellender RD, Ajidahun A, Hellein KN, Kennedy E, Ye X, Flood C (2009) Validation and field testing of library-independent microbial source tracking methods in the Gulf of Mexico. Wat Res, doi:10.1016/j.watres.2009.06.029

Havelaar AH, Melse JM (2003) Quantifying Public Health Risk in the WHO Guidelines for Drinking-water Quality. RIVM report 734301022/2003.

Havelaar AH, Pelt W van, Ang CW, Wagenaar JA, Putten JPM van, Gross U, Newell DG (2009) Immunity to Campylobacter: its role in risk assessment and epidemiology. Crit Rev Microbiol 35(1):1–22

Hijnen WA, Suylen GM, Bahlman JA, Brouwer-Hanzens A, Medema GJ (2010) GAC adsorption filters as barriers for viruses, bacteria and protozoan (oo)cysts in water treatment. Wat Res 44(4):1224–1234

Hörman A, Rimhanen-Finne R, Maunula L, von Bonsdorff CH, Torvela N, Heikinheimo A, Hänninen ML (2004) *Campylobacter* spp., *Giardia* spp., *Cryptosporidium* spp., Noroviruses and indicator organisms in surface waters in Southwestern Finland, 2000-2001. Appl Environ Microbiol 70(1):87–95

Howard G, Jahne J, Frimmel FH, McChesney D, Reed B, Schijven JF (2006) Human excreta and sanitation: Information needs, in Protecting groundwater for health, Groundwater Monograph, Edited by Schmoll, O., Howard, G., Chilton, J. and Chorus, I. World Health Organization, London, Chapter 10:275–308

Huang Y, Haas CN (2009) Time-dose-response models for microbial risk assessment. Risk Anal 29(5):648–661

Hundesa A, Maluquer de Motes C, Albinana-Gimenez N, Rodriguez-Manzano J, Bofill-M S, Suñen E, Girones R (2009) Development of a qPCR assay for the quantification of porcine adenoviruses as an MST tool for swine fecal contamination in the environment. J Vir Meth 158:130–135

Hundesa A, Bofill-Mas S, Maluquer de Motes C, Rodriguez-Manzano J, Bach A, Casas M, Girones R (2010) Development of a quantitative PCR-assay for the quantification of bovine polyomavirus as a microbial source tracking tool. J Vir Meth 163:385–389

ILSI Risk Science Institute Pathogen Risk Assessment Working Group (1996) A conceptual framework to assess the risks of human disease following exposure to pathogens. Risk Analysis, 16(6):841–848

Jean JS, Guo HR, Chen SH, Liu CC, Chang WT, Yang YJ, Huang MC (2006) The association between rainfall rate and occurrence of an enterovirus epidemic due to a contaminated well. J of Appl Microbiol 101(6):1224–1231

Kim SH, Cheon DS, Kim JH, Lee DH, Jheong WH, Heo, YJ, Chung, HM, Jee, Y and Lee, JS (2005) Outbreaks of gastroenteritis that occurred during school excursions in Korea were associated with several waterborne strains of norovirus. J Clin Microbiol 43(9):4836–4839

Korajkic Am, Badgley BD, Brownell MJ, Harwood VJ (2009) Application of microbial source tracking methods in a Gulf of Mexico setting. J Appl Microbiol 107:1518–1527

Maurer AM, Stürchler D (2000) A waterborne outbreak of small round structured virus, campylobacter and shigella co-infections in La Neuveville, Switzerland, 1998. Epidemiol Inf 125(2):325–332

McBride GB, McWhirter JL, Dalgety MH (2003) Uncertainty in most probable number calculations for microbiological assays. J AOAC Int 86(5):1084–1088

McCullagh P, Nelder JA (1989) Generalized linear models, volume 37 of Monographs on statistics and applied probability. Chapman and Hall, London

Medema GJ, Teunis PF, Havelaar AH, Haas CN (1996) Assessment of the dose-response relationship of *Campylobacter jejuni*. Int J Food Microbiol 30(1–2):101–111

Mons MN, van der Wielen JM, Blokker EJ, Sinclair MI, Hulshof KF, Dangendorf F, Hunter PR, Medema GJ (2007) Estimation of the consumption of cold tap water for microbiological risk assessment: an overview of studies and statistical analysis of data. J Water Health 5.1:151–170

Neumann MB, von Gunten U, Gujer W (2007) Uncertainty in prediction of disinfection performance. Water Res 41(11):2371–2378

Orosz-Coghlan PA, Rusin PA, Karpiscak MM, Gerba CP (2006) Microbial source tracking of *Escherichia coli* in a constructed wetland. Wat Environ Res 78:227–232

Page MA, Shisler JL, Mariñas BJ (2009) Kinetics of adenovirus type 2 inactivation with free chlorine. Water Res 43(11):2916–2926

Parshionikar SU, Willian-True S, Fout GS, Robbins DE, Seys SA, Cassady JD, Harris R (2003) Waterborne outbreak of gastroenteritis associated with a norovirus. Appl Environ Microbiol 69(9):5263–5268

Payment P, Berte A, Prévost M, Ménard B, Barbeau B (2000) Occurrence of pathogenic microorganisms in the Saint Lawrence River (Canada) and comparison of health risks for populations using it as their source of drinking water. Can. J. Microbiol. 46(6):565–576

Pedley S, Yates MV, Schijven JF, West J, Howard G, Barrett M (2006) Pathogens: Health relevance, transport and attenuation, in Protecting groundwater for health, Groundwater Monograph,

Edited by Schmoll, O, Howard, G, Chilton, J and Chorus, I World Health Organization, London, Chapter 3:49–80

Powell KL, Taylor RG, Cronin AA, Barrett MH, Pedley S, Sellwood J, Trowsdale SA, Lerner, DN (2003) Microbial contamination of two urban sandstone aquifers in the UK. Wat Res 37(2):339–352

Pujol JM, Esienberg JE, Haas CN, Koopman JS (2009) The effect of ongoing exposure dynamics in dose-response relationships. Plos Comp Biol 5(6):1–12

Reeves RL, Grant SB, Mrse RD, Copil Oancea CM, Sanders BF, Boehm AB (2004) Scaling and management of fecal indicator bacteria in runoff from a coastal urban watershed in southern California. Environ Sci Technol 38(9):2637–2648

Reischer GH, Haider JM, Sommer R, Stadler H, Keiblinger KM, Hornek R, Zerobin W, Mach RL, Farnleitner AH (2008) Quantitative microbial faecal source tracking with sampling guided by hydrological catchment dynamics. Environ Microbiol 10(10):2598–2608

Rutjes SA, Lodder WJ, van Leeuwen AD, de Roda Husman AM (2009) Detection of infectious rotavirus in naturally contaminated source waters for drinking water production. J Appl Microbiol 107(1):97–105

Rutjes SA, Roda Husman AM de (2004) Procedure for virus detection in water for use in the Dutch drinking-water guidelines RIVM report 330000007 National Institute for Public Health and the Environment, Bilthoven, the Netherlands, in Dutch, http://wwwrivmnl/bibliotheek/rapporten/330000007.html

Santo Domingo JW, Bambic DG, Edge TA, Wuertz S (2007) Quo vadis source tracking? Towards a strategic framework for environmental monitoring of fecal pollution. Wat Res 41:3539–3552

Savitcheva O, Okabe S (2006) Alternative indicators of fecal pollution: Relations with pathogens and conventional indicators, current methodologies for direct monitoring and future application perspectives. Wat Res 40:2463–2476

Schets FM, Medema GJ, Schijven JF (2004) The recovery of the detection method for Cryptosporidium and Giardia in water RIVM report 330000008 National Institute for Public Health and the Environment, Bilthoven, the Netherlands, in Dutch, http://wwwrivmnl/bibliotheek/rapporten/330000008.html

Schets FM, Schijven JF, de Roda Husman AM (2011) Exposure assessment for swimmers in surface waters and swimming pools. Wat Res 45:2392–2400

Schets FM, van Wijnen JH, Schijven JF, Schoon H, de Roda Husman AM (2008) Monitoring of waterborne pathogens in surface waters in Amsterdam, the Netherlands, and the potential health risk associated with exposure to *Cryptosporidium* and Giardia in these waters. Appl Environ Microbiol 74(7):2069–2078

Schijven JF, de Roda Husman AM (2006) A survey of diving behavior and accidental water ingestion among Dutch occupation and sport divers to assess the risk of infection with waterborne pathogenic microorganisms. Environ Health Persp 114(5):712–717

Schijven JF, Mülschlegel JHC, Hassanizadeh SM, Teunis PFM, de Roda Husman, AM (2006) Determination of protection zones for Dutch groundwater wells against virus contamination – Uncertainty and sensitivity analysis. J Wat Health 4(3):297–312

Schijven JF, Rijs G, de Roda Husman AM, Verstappen G (2005) Estimation of the risk of infection of dairy cows by food and mouth disease virus spread by way of surface water. Risk Analysis 25(1):13–21

Schijven JF, de Roda Husman AM (2005) Effect of climate changes on waterborne disease in the Netherlands. Wat Sci Technol 51(5):79–87

Schijven, JF, Hassanizadeh, SM (2000) Removal of viruses by soil passage: overview of modeling, processes and parameters. Crit Rev Environ Sci Tech 30:49–127

Scott TM, Jenkins TM, Lukasik J, Rose JB (2005) Potential use of a host associated molecular marker in *Enterococcus faecium* as an index of human pollution. Env Sci Tech 39:283–287

Signor R S, Ashbolt N J (2009) Comparing probabilistic microbial risk assessments for drinking water against daily rather than annualised infection probability targets. J Wat Health 7(4):535–543

Simpson JM, Santo Domino JW, Reasoner DJ (2002) Microbial source tracking: State of the science. Env Sci Techn 36(24):5279–5288

Smeets PW, Dullemont YJ, Van Gelder PH, Van Dijk JC, Medema GJ (2008) Improved methods for modelling drinking water treatment in quantitative microbial risk assessment; a case study of Campylobacter reduction by filtration and ozonation. J Wat Health 6(3):301–314

Stapleton CM, Kay D, Wyer MD, Davies C, Watkins J, Kay C, McDonald AT, Porter J, Gawler A (2009) Evaluating the presence of a Bacteriodales quantitative PCR-based MST approach in determining the source of fecal indicator organisms at a UK bathing water. Wat Res 43:4888–4899

Stapleton CM, Wyer MD, Kay D, Crowther J, McDonald AT, Walters M, Gawler A, Hindle T (2007) Microbial source tracking: a forensic technique for microbial source identification? J Envi Monit 9:427–439

Stoeckel DM, Harwood VJ (2007) Performance, design, and analysis in microbial source tracking studies. Appl Env Microbiol 73(8):2405–2415

Teunis PFM, Medema GJ, Kruidenier L, Havelaar AH (1997) Assessment of the risk of infection by Cryptosporidium or Giardia in drinking water from a surface water source. Wat Res 31(6):1333–1346

Teunis PF, Rutjes SA, Westrell T, de Roda Husman AM (2009) Characterization of drinking water treatment for virus risk assessment. Wat Res 43(2):395–404

Teunis PFM, Chappell CL, Okhuysen PC (2002a) *Cryptosporidium* dose response studies: variation between isolates Risk Anal 22(1):175–183

Teunis PFM, Chappell CL, Okhuysen PC (2002b) *Cryptosporidium* dose-response studies: variation between hosts. Risk Anal 22(3):475–485

Teunis PFM, Evers EG, Slob W (1999a) Analysis of variable fractions resulting from microbial counts. Quant Microbiol 1:63–88

Teunis PFM, Havelaar AH (2000) The Beta Poisson model is not a single hit model. Risk Anal 20:511–518

Teunis PFM, Heijden OG van der, Giessen JWB van der, Havelaar AH (1996) The dose-response relation in human volunteers for gastro-intestinal pathogens. RIVM Report 284550002, National Institute for Public Health and the Environment, Bilthoven, The Netherlands

Teunis PFM, Nagelkerke NJD, Haas CN (1999b) Dose response models for infectious gastroenteritis. Risk Anal 19(6):1251–1260

Teunis PFM, Ogden ID, Strachan NJC (2008) Hierarchical dose response of *E. coli* O157:H7 from human outbreaks incorporating heterogeneity in exposure. Epid Inf 136:761–770

Teunis PFM, Van den Brandhof W, Nauta M, Wagenaar J, Van den Kerkhof H, Van Pelt W (2005) A reconsideration of the *Campylobacter* dose-response relation. Epid Inf 133(4):583–592

Till D, McBride G, Ball A, Taylor K, Pyle E (2008) Large-scale freshwater microbiological study: rationale and risks. J Wat Health 064:443–460

USEPA (2002) Federal Water Pollution Control Act (FWPCA), as amended through P.L. 107–303, US Environmental Protection Agency, Washington DC, http://www.epa.gov/r5water/cwa.htm

US EPA (2006) Economic Analysis for the Final Ground Water Rule EPA 815-R-06-014

Westrell T, Andersson Y, Stenström TA (2006a) Drinking water consumption patterns in Sweden. J Wat Health 4(4):511–522

Westrell T, Teunis P, van den Berg H et al. (2006b) Short- and long-term variations of norovirus concentrations in the Meuse river during a 2-year study period. Wat Res 40(14):2613–2620

World Health Organization, 3rd ed 2008 *Guidelines for Safe Recreational-Water environments Volume 1: Coastal and Fresh Waters.* WHO, Geneva, Switzerland

Wilkes G, Edge T, Gannon V, Jokinen C, Lyautey E, Medeiros D, Neumann N, Ruecker N, Topp E, Lapen DR (2009) Seasonal relationships among indicator bacteria, pathogenic bacteria, *Cryptosporidium* oocysts, *Giardia* cysts, and hydrological indices for surface waters within an agricultural landscape. Wat Res 43:2209–2223

Wuijts S, Rutjes SA, Van der Aa NGFM, Mendizabal I, De Roda Husman AM (2008) The influence of human and animal sources on groundwater abstraction From field study to protection policy. RIVM report 734301031, National Institute for Public Health and the Environment, Bilthoven, the Netherlands, in Dutch, http://wwwrivmnl/bibliotheek/rapporten/734301031.html

Xiao L, Alderisio K, Limor J, Royer M, Lal AA (2000) Identification of species and sources of Cryptosporidium oocysts in storm waters with a small-subunit rRNA-based diagnostic and genotyping tool. Appl Environ Microbiol 66(12):5492–5498

Chapter 25
Food Safety and Implications for Microbial Source Tracking

Alexandria K. Graves

Abstract As globalization of food marketing and distribution in our present interconnected society increases, the potential for the distribution of contaminated food increases as well, highlighting the need for tools to identify the types and sources of pathogen contamination. The concept of tracing pathogens to their origin using microbiological, genotypic, and phenotypic methods has been termed microbial source tracking (MST). The application of MST in identifying pathogens and the host source of pathogens linked to foodborne outbreaks of human illness can lead to targeted research, improved investigative and inspection practices, and other food system interventions. This chapter reviews microbial food safety concerns in various food industries and examines the applicability of molecular MST tools in food safety studies.

Keywords *Salmonella* • *E. coli* O157:H7 • *Campylobacter* • Antibiotic resistance • Product traceability • Molecular analysis • Pathogen contamination

25.1 Introduction

Annually, foodborne diseases are responsible for an estimated 76 million illnesses: 325,000 hospitalizations and 5,000 deaths in USA (Mead et al. 1999). Over 250 diseases are known to be transmitted through food consumption and the US Centers for Disease Control and Prevention (CDC) publishes annual reports on surveillance data from foodborne disease outbreaks (Bryan 1982; Lynch et al. 2006; CDC 2010). The outbreak reports from the CDC include the reporting state, number of ill persons, and the associated food pathogen. However, globalization

A.K. Graves (✉)
Department of Soil Science, North Carolina State University,
3411E Williams Hall, Raleigh, NC 27695-7619, USA
e-mail: alexandria_graves@ncsu.edu

C. Hagedorn et al. (eds.), *Microbial Source Tracking: Methods, Applications, and Case Studies*, DOI 10.1007/978-1-4419-9386-1_25,

of food marketing and distribution in our present interconnected society increases vulnerability of widely dispersed cases of foodborne disease outbreaks beyond the confines of USA (Tauxe et al. 2009). As international trade increases, the potential for the distribution of contaminated food increases as well. Researchers estimate that over the last 60 years, 30% of emerging infections through all nations included pathogens commonly transmitted through food handling and consumption (Jones et al. 2008).

Hundreds to thousands of food products can be recalled based on the identification of one single contaminated food ingredient. The concept of tracing pathogens to their origin using microbiological, genotypic, and phenotypic methods has been termed microbial source tracking (MST) (Scott et al. 2002). These methods are increasingly being used to identify sources of pathogen contamination in a number of situations that involve polluted waters, disease outbreaks at animal facilities, and food attribution studies (Smith and Perdek 2004). The ability to associate cases of foodborne disease to the food, food ingredient, or other source responsible for the illnesses is known as food attribution (Batz et al. 2005). For the purposes of attributing illnesses to foods, food sources must be grouped into categories, such as poultry, eggs, row crops, tree crops, etc. The classification can also include the level of processing (raw, frozen, etc.) and the origin (domestic or imported) (Greig and Ravel 2009). Food attribution tools can be grouped into two broad categories designated as epidemiological approaches and microbiological approaches. Epidemiological approaches involve the evaluation of case control studies as well as public-health surveillance and foodborne disease outbreak data. Microbiological approaches involve the evaluation of data generated from subtyping pathogens collected from human, animal, and food sources (Batz et al. 2005). MST methods and pathogen subtyping such as those used in Denmark's *Salmonella* Accounts are examples of microbiological approaches to food attribution (Batz et al. 2005).

Food attribution tools have contributed to numerous advancements in identifying hazards and prioritizing food safety interventions (Painter et al. 2009). Nevertheless, limitations within both categories of food attribution tools exist. The Food Attribution Working Group reported the lack of a common food categorization scheme among the CDC, state health departments, Food and Drug Administration (FDA), US Department of Agriculture (USDA) and their constituent agencies as a critical issue, resulting in data that often cannot be directly compared between agencies (Batz et al. 2005). Other limitations include underreporting of illnesses and undercollection of stool specimens from diseased patients for pathogen identification, resulting in an inability to apply MST to identify the contaminated food and pathogen source (Doyle and Erickson 2006).

The application of MST in identifying pathogens and the source of pathogens linked to foodborne outbreaks of human illness can lead to targeted research, improved investigative and inspection practices, and other food system interventions (DeWaal and Barlow 2002; Batz et al. 2005; Sofos 2008; Painter et al. 2009). This chapter reviews microbial food safety concerns in various food industries and examines the applicability of molecular MST tools in food safety studies.

25.2 Food Safety Challenges

There are three broad categories of food safety challenges associated with microbial pathogens. These categories include the following: (1) current pathogens of concern, (2) emerging pathogens of concern, and (3) the environmental connection to pathogens of concern. As pathogens emerge with increased antibiotic resistance (IFT 2006), increased virulence (Fratamico et al. 2005), and resistance to food preservation measures (Samelis and Sofos 2003), the need for more efficient pathogen detection and source identification methods, to enhance pathogen control in food safety, is illuminated. The application of MST within these categories has the potential to provide information on the specific type of pathogen (e.g., *Escherichia coli*) and the specific source (e.g., deer), thereby providing an opportunity to implement-corrective actions within a particular food industry to remediate current contamination and prevent future contamination.

25.2.1 Current Pathogens of Concern

A number of enteric bacteria currently pose a threat to food safety including, but not limited to, *Salmonella*, *E. coli* O157:H7, and *Campylobacter* sp. (Bacon and Sofos 2003; Brandl et al. 2004; Fratamico et al. 2005; Jay et al. 2007). Viruses such as hepatitis A (Wheeler et al. 2005) and norovirus (Widdowson et al. 2005) contribute to numerous foodborne illness outbreaks, with norovirus considered as the most common cause of foodborne illness in USA (Widdowson et al. 2005). Despite decades of control efforts for many of these known pathogens of concern, foodborne illness outbreaks continue (Swaminathan et al. 2006).

25.2.2 Emerging Pathogens of Concern

Pathogen evolution has resulted in the development of adaptation and resistance mechanisms that represent a major concern for food safety (IFT 2006; Samelis and Sofos 2003). Some pathogens have developed measures of protection against food preservation measures such as low pH, extreme temperatures, and chemical additives. After exposure to some of the traditional food preservation measures, pathogens may develop increased virulence (Samelis and Sofos 2003). The potential for coselection of increased virulence is sometimes associated with antibiotic resistance (IFT 2006).

Pathogens resistant to antibiotics used in animal production may pose a concern in human medicine because the mode of action for resistance may be similar to the antibiotic used in human clinical settings (Doyle and Erickson 2006). Five antibiotic-resistant bacterial strains are of particular concern in USA. These bacterial strains include vancomycin-resistant *Enterococcus faecium* and *E. faecalis*, ciprofloxacin-resistant *Campylobacter*, *Salmonella typhimurium* DT104, R-type

ACSSuT penta-resistant (ampicillin, chloramphenicol, streptomycin, sulfamethoxazole, and tetracycline) strains, and *Salmonella* Newport R-type MDR-Amp C strains resistant to nine antimicrobials (ampicillin, chloramphenicol, streptomycin, sulfamethoxazole, tetracycline, amoxicillin-clavulanic acid, cephalothin, cefoxitin, and ceftiofur). The *Salmonella* Newport R-type MDR-Amp C strains also demonstrate lower sensitivity to ceftriaxone (IFT 2006).

While the role of antibiotic use in veterinary and human clinical settings in the development of antibiotic resistance among pathogens is important, the occurrence of pathogens resistant to food preservation methods and antibiotics in the food industry may be influenced by environmental issues such as manure management and water quality.

25.2.3 Environmental Connection to Pathogens of Concern

Food can become contaminated at any point in the chain of food production; however, the opportunity for contamination is highest during three stages: in the field, during processing, and during preparation in the home kitchen or restaurant. Field contamination may come from wild animals that may directly contaminate fruits and vegetables or indirectly contaminate the water used to irrigate the plants. Field contamination may also be associated with field workers with inadequate access to restroom and hand-washing facilities. During processing, likely sources of contamination include contaminated water used for washing, chill tanks, or sprays and shipping ice. In meat and poultry production, pathogens in the feces may be transferred to the hide and carcass of the animal and persist throughout processing (Mor-Mur and Yuste 2010). The restaurant or home kitchen serves as the final stage of potential contamination. During the final stage, an elevated risk for pathogen transmission exists if the food is handled with contaminated hands, prepared with contaminated implements, prepared on contaminated surfaces, and/or undercooked.

While MST efforts have largely focused on detecting and identifying sources of fecal bacteria in waterways, the same tools can be adapted to contamination associated with food. Foodborne disease outbreak investigations involving MST will allow tracking of an outbreak-associated pathogen strain from infected patients back to contaminated food from a restaurant/kitchen/market to the processing facility and to the field or farm of origin.

Environmental cross-contamination of enteric pathogens such as *E. coli* O157:H7, *Salmonella*, and *Shigella* from animal hosts to a number of food products of nonanimal origin has become a common occurrence in recent years (Bowen et al. 2006; CDC 2002, 2006, 2007). These events demonstrate the need for coordinated management efforts in manure disposal, water-quality management, and food safety practices (Crohn and Bianchi 2008).

Inefficient manure management may lead to surface runoff contaminated with pathogens entering waters used to irrigate or wash produce. However, measures intended to minimize contamination via runoff often serves as a wildlife habitat, thus inadvertently increasing the contamination potential. For example, in 2006 an

E. coli O157:H7 outbreak associated with a spinach harvest from California's central coast (Porter and Lister 2006) was determined to be a result of two potential environmental sources: (1) the mixing of ground water with contaminated surface water and (2) contamination by wild feral pigs (CDHS and FDA 2007; Jay et al. 2007). California's central coast's surface waters are heavily impacted by agriculture (Hunt et al. 2006; Los Huertos et al. 2006). In response to the food safety concerns that evolved from the 2006 *E. coli* O157:H7 spinach outbreak, farmers began removing water-quality management practices that attract wildlife (RCDMC 2007). Water-quality management practices that have been targeted for removal include, but are not limited to, irrigation reservoirs, grassed waterways, and buffer strips (RCDMC 2007). While the removal of these practices may reduce the wildlife population in a given area, the potential for an increase in the export of nutrients, chemical fertilizers, pesticides, and pathogens to surface waters is exasperated. As a result, management alternatives that can jointly address food safety and water-quality hazards should be investigated (Crohn and Bianchi 2008).

25.3 Product Traceability

Nationwide foodborne illness outbreaks within the livestock and produce industry have increased interest in traceability in the US food supply chain (Felmer et al. 2006). In 2003, the first case of Bovine Spongiform Ecephalopathy (BSE or "mad cow disease") was discovered in USA. It was believed that without a formal tracking system in place, determining the origin of the diseased cow would take weeks to several months (Clemetson and Simon 2003). However, within 1 week the diseased cow was traced back to Canada because of cooperation among those involved in cattle production (herd owners, livestock dealers, and market personnel) and review of records between the US and Canadian food agencies. However, herd mates were never fully traced (Clemetson and Simon 2003; Felmer et al. 2006).

The 2006 *E. coli* spinach outbreak demonstrates issues with traceability for the produce industry. The contaminated spinach was easily traced back to the food packaging plant; however, MST was necessary to identify the farm of origin (Jay et al. 2007; Porter and Lister 2006). Genetic fingerprints generated from *E. coli* strains isolated from infected patients matched the genetic fingerprints of *E. coli* strains from animals at one of four farms in Monterey and San Benito Counties in California (CDHS and FDA 2007). There are many financial and legal motivations for traceability of animals or produce back to the farms of origin (Golan et al. 2004). The cost associated with foodborne disease outbreaks can reach astronomical proportions. Based on the 1996 annual cost of living, the Economic Research Service estimated that the costs of premature deaths, medical costs, and productivity losses associated with foodborne disease outbreaks ranged from \$6.6 to \$37.1 billion (Buzby and Roberts 1996), although this estimate is conservative in that only seven foodborne pathogens (*Campylobacter jejuni*, *Clostridium perfringens*, *E. coli*

O157:H7, *Listeria monocytogenes*, *Salmonella* sp., *Staphylococcus aureus*, and *Toxoplasma gondii*) are considered (Buzby and Roberts 1996). However, from a food safety standpoint, reliable traceability efforts that include MST could provide critical information about product quality problems that occur throughout the farm-to-table continuum in the food industry.

25.4 Foodborne Illness Outbreaks Associated with Produce

Fruits and vegetables are increasingly recognized as a source of foodborne disease outbreaks worldwide (Lynch et al. 2009). In USA, the proportion of foodborne disease outbreaks associated with produce increased by more than 5% over a 20-year period covering the 1970s to the 1990s (Sivapalasingam et al. 2004).

Recurrent foodborne outbreaks associated with produce often involve specific pathogen–produce combinations. Melons, tomatoes, and bean sprouts have been implicated in *Salmonella* infections (Bowen et al. 2006; CDC 2007; Mohle-Boetani et al. 2009), whereas spinach has been implicated in *E. coli* O157:H7 infections (CDC 2006; CDHS and FDA 2007). These pathogen–produce combinations may provide information in regard to the ecological/environmental aspects of contamination.

Melons grown in Mexico (CDC 2002) and Australia (Munnoch et al. 2009) were both implicated in foodborne disease outbreaks of salmonellosis. Problems associated with the same type of produce from different geographical growing areas suggest that similar growing conditions may have contributed to the contamination. Tomatoes within a particular area of USA have been associated with recurrent outbreaks of salmonellosis due to the same strain of *Salmonella* Newport (Greene et al. 2008). Recurrent problems associated with the same type of produce, the same pathogen, within the same growing region suggest that ecological conditions that favor the survival of a particular pathogen persists over time (Greene et al. 2008).

The ecological/environmental connection between foodborne pathogens and produce is quite complex (Brandl 2006; Tyler and Triplett 2008). *Salmonella* have been shown to persist on the leaves of cilantro (Bradl and Mandrell 2002) and on the surface of tomatoes (Zhuang et al. 1995) for weeks. *Campylobacter* has the ability to persist in plant root zones for weeks (Brandl et al. 2004), and *E. coli* may persist on alfafa and mung seeds indefinitely (Taormina and Beuchat 1999; Jaquette et al. 1996). Human pathogens have also been shown to internalize fruits such as apples (Burnett et al. 2000) and tomatoes (Guo et al. 2001). Alfalfa sprouts from seeds contaminated with *E. coli* O157:H7 and *Salmonella* have demonstrated growth of these bacteria throughout the tissues of the sprouts (Itoh et al. 1998; Charkowski et al. 2002). In addition to the implementation of MST to detect and identify pathogens associated with produce contamination, the survival mechanisms of pathogens outside the animal host, including the ability to survive in or on the surface of produce, must be vigorously investigated in an effort to accurately assess contamination loads and risk to the consumer.

25.5 Foodborne Illness Outbreaks Associated with Meat and Poultry

Foodborne disease outbreaks associated with meat, poultry, and meat products have brought meat safety issues to the forefront of public-health concerns (Mor-Mur and Yuste 2010; Sofos 2008). A number of disease outbreaks caused by *E. coli* have been associated with the consumption of undercooked contaminated ground beef. A notable example, the Jack in the BOX *E. coli* O157:H7 outbreak occurred in several states across the Pacific Northwest, USA and involved undercooked hamburger. Over 700 illnesses and 4 deaths occurred during this 1992–1993 outbreak, and these events led to major changes in the US meat inspection system (Sofos 2008). *E. coli* O157:H7 outbreaks have been reported worldwide and include UK, Japan, and Australia. Food sources involved in the reported outbreaks include, but are not limited to, beef, venison, salami, and turkey (Meng et al. 2007). Meat and meat products have also been implicated in disease outbreaks due to *Campylobacter* infections. However, most *Campylobacter* infections are associated with handling raw or eating undercooked poultry (Altekruse et al. 1997). *C. jejuni* is the most frequently implicated species and has been associated with 53 foodborne outbreaks in USA from 1979 to 1987 (Mor-Mur and Yuste 2010).

Poultry and pork, more often than beef consumption, has been implicated in cases of foodborne disease outbreaks associated with *Salmonella* (Mor-Mur and Yuste 2010). The food sources of salmonellosis have included raw and minced pork, beef, deli meats, chicken, and turkey. Meat has been associated with 17 major outbreaks of salmonellosis from 1984 to 2005, involving over 2,100 cases (D'Aoust and Maurer 2007). In any case of foodborne disease associated with meat and poultry, the application of MST in attributing an outbreak to a specific pathogen, specific contamination source and specific food source will allow regulators to revise or implement quality-control measures to prevent future outbreaks.

25.6 Foodborne Illness Outbreaks Associated with Shellfish

Shellfish include a number of aquatic invertebrates, such as mollusks (oyster, mussel, clam, conch, octopus, etc.), crustaceans (shrimp, crab, and lobster), and echinoderms (sea urchins). Filter-feeding shellfish utilize their gills to generate currents that transport water and particles toward the mouth and digestive tract, resulting in concentration of bacteria and viruses in tissues (Newell and Jordan 1983). Consequently, pathogen levels in shellfish can reach proportions over 100 times greater than pathogen levels found in the surrounding water body (Geary and Davies 2003). This leads to an elevated health risk associated with ingesting raw or undercooked shellfish. For example, in 1924, a ruinous typhoid fever outbreak in Chicago was linked to consumption of raw oysters (Canzonier 1991). As a result of public protest, the National Shellfish Sanitation Program (NSSP) was established, and by 1927 a certification plan set minimum standards that were regulated by state health departments to protect consumers from contaminated shellfish.

A number of studies have correlated the consumption of raw or undercooked shellfish to an increase in the risk of contracting shellfish related foodborne illness (Desenclos et al. 1991; Kohn et al. 1995). Despite improved sanitation and diligent consumer information crusades, foodborne diseases from shellfish consumption continue to rise (Potasman et al. 2002). Cooking shellfish may significantly reduce the risks of contracting a bacterial illness, but viral contamination may persist even after cooking (Kirkland et al. 1996; McDonnell et al. 1997). Viruses or bacteria are typically associated with shellfish contamination, and studies have shown that some viruses such as the noroviruses may survive steaming (Dolin et al. 1972). The noroviruses are believed to be the most common pathogen associated with shellfish-related foodborne illnesses in USA, and disease outbreaks have been reported worldwide (Cheng et al. 2005; Luthi et al. 1996; Morse et al. 1986; Shieh et al. 2003). Hepatitis A virus has also been associated with shellfish-related foodborne disease outbreaks (Chironna et al. 2002; Coelho et al. 2003; Conaty et al. 2000; Formiga-Cruz et al. 2003; Leoni et al. 1998), and in 1988, it was implicated in the largest recorded foodborne disease outbreak linked to shellfish consumption, in which over 292,000 people in China were infected after eating clams (Halliday et al. 1991).

Enteric bacteria such as *Salmonella*, *Shigella*, and *Campylobacter* have also been implicated in shellfish-related foodborne illnesses (Martinez-Urtaza et al. 2003; Reeve et al. 1989; Terajima et al. 2004; Wilson and Moore 1996). Although outbreaks of *Salmonella* gastroenteritis are still attributed to seafood, no cases were associated with oysters between the years 1978 and 1987 (Ahmed 1991), and the association of oyster consumption and salmonellosis is now infrequent (Rippey 1994). However, the naturally occurring bacteria in the marine environment such as the *Vibrio* sp. pose a significant threat. *Vibrio*-related illness increased 40% over the last years, and shellfish-associated *Vibrio* spp. were traced to 75% of all seafood-associated diseases (Jones and Oliver 2009). Thus, the microbiological safety of shellfish can only be ensured when primary human pathogens and commensal shellfish microorganisms pathogenic to humans are eliminated (Feldhusen 2000; Blake et al. 1979). MST has been widely applied to identify sources of pathogens in shellfish waters (Meschke and Boyle 2007) as a management tool and to establish bacterial total maximum daily loads (Stewart et al. 2007; Zhang et al. 2002, 2003), leading to safer waters with lower bacterial counts. However, the reduction or removal of naturally occurring bacteria such as the *Vibrio* spp. in the marine environment may not be a practical objective or achievable with MST or any other technology.

25.7 Microbial Source Tracking Applications Available for Food Safety

The principal supposition for MST applications is that subgroups of bacteria become adapted to a particular host, whether it is the intestinal tract of a dog or the sediments of a river. Presumably, once microorganisms become adapted to a particular host and establish residency, the progeny will be genetically identical. Ultimately, a group of microorganisms within a particular host or environment should possess a similar or indistinguishable genetic fingerprint, which will differ from those microbes adapted to a different host or environment. This same type of rationale is used for MST methodologies that address phenotypic traits within different lineages of bacteria except that in this case the focus is on the expression of traits that may have been acquired from exposure to different host species or environments (Scott et al. 2002).

Multiple molecular MST methods are available to determine the source and distribution of pathogens isolated from sick patients, food, water, and other environmental matrices (Scott et al. 2002). With these procedures, success often depends on the size of the reference genetic fingerprint database that is being used for comparison, the method used for image analysis and band pattern recognition, and the statistical methods used for comparison of environmental isolates with the reference collection (Lu et al. 2004). Both the creation of national databases and additional studies of factors contributing to variable fingerprints within a host will prove valuable to improving the utility of MST methods in the food industry. The following section covers the three molecular-based categories that can be applied to food safety studies: (1) restriction analysis of pathogen DNA, (2) polymerase chain reaction (PCR) amplification of pathogen DNA, and (3) sequencing-based methods.

25.7.1 Restriction Analysis of Pathogen DNA

25.7.1.1 Pulsed-Field Gel Electrophoresis

The pulsed-field gel electrophoresis (PFGE) process involves embedding whole bacterial cells of a known optical density in an agarose matrix prior to cell lysis. Detergents and enzymes are used to lyse the cells; however, the DNA is immobilized in the agarose plugs. Cellular debris is removed from the plugs with a TE buffer wash. The DNA is digested and the plug containing the digested DNA is added to an agarose gel followed by electrophoreses. The polarity of the current is switched at regular intervals, allowing separation of 10–20 large DNA fragments ranging in size from 20 to 80 kb (Lukinmaa et al. 2004; van Belkum et al. 2007). Gels are stained for DNA fragment visualization and image capture (Olive and Bean 1999). *Salmonella, Shigella,* and *E. coli* DNA are commonly digested with XbaI, BlnI, or SpeI (Ribot et al. 2006), while SmaI or KpnI are commonly used to digest *Campylobacter* DNA (Michaud et al. 2001). The DNA from *Salmonella enterica* serovar Branderup (H9812), restricted with Xbal, is used as a "universal" molecular-weight standard for normalizing the PFGE fingerprints in a database for bacterial pathogens such as *Salmonella* and *E. coli* (Hunter et al. 2005). The (H9812) isolate is used by PulseNet and deposited with the American Type Culture Collection (ATCC) under the accession number ATCC BAA-664 (Ribot et al. 2006).

Often considered the "gold standard" of molecular methods for bacterial identification of foodborne pathogens (Foley and Walker 2005; Olive and Bean 1999), criteria have been developed to aid in the determination of genetic relatedness between the bacterial isolates analyzed by PFGE (Barrett et al. 2006; Tenovar et al. 1995). Isolates are grouped into three categories defined as (1) closely related, (2) possibly related, and (3) different/unrelated (Tenovar et al. 1995). Closely related isolates are those that differ by a single genetic event, indicated by 2–3 band differences, whereas isolates defined as possibly related or unrelated differ by two genetic events (4–6 band differences) and ≥3 genetic events (≥7 band differences), respectively. However, these criteria are most appropriate in studies with limited genetic variability among isolates, typically modest regional studies (Olive and Bean 1999), as the criteria inadequately address strain variations related to genetic events such as horizontal gene transfer (Barrett et al. 2006).

Nevertheless, the use of standardized PFGE protocols has led to highly discriminatory and reproducible interlaboratory databases on an international scale as seen in PulseNet-USA and PulseNet-Europe (Lukinmaa et al. 2004; van Belkum et al. 2007; Gerner-Smidt et al. 2006; Gerner-Smidt and Scheutz 2006; Swaminathan et al. 2006). Successful MST involving PFGE in typing *Salmonella* and *E. coli* O157:H7 from contaminated food, water, animals, and from inanimate objects has been widely reported (Aktas et al. 2007; Kam et al. 2007; Nayak et al. 2004; Foley et al. 2006; Nayak and Stewart-King 2008).

When compared to other typing methods, PFGE generally has the greatest discriminatory power. Yet, this method is labor-demanding and often requires 2–4 days to execute the procedure and analyze the results. Moreover, changes in genetic events that do not influence the electrophoretic mobility of the enzyme-digested DNA may not be recognized as a separate genotype (Foley et al. 2007). This shortcoming has led many researchers and public-health laboratories to use multiple restriction enzymes to fully understand the genetic variability among the isolates (Zheng et al. 2007).

25.7.1.2 Plasmid Analysis

Plasmid analysis has been extensively used for the characterization of bacterial foodborne pathogens, such as *Salmonella* (Nayak et al. 2004; Threlfall et al. 1982; 1990), *E. coli* (Domingue et al. 2003; Johnson et al. 2007), *Campylobacter* (Stanley et al. 1994), and *Shigella* (Gebre-Yohannes and Drasar 1991; Liu et al. 1995). Many plasmids are known to confer a selective advantage to their host strain by carrying genes that encode resistance to antibiotics, heavy metals, as well as increase of virulence (Threlfall et al. 1990). Restriction analysis of plasmids minimizes problems associated with conformational changes and migration issues that are associated with standard plasmid analysis (Nauerby et al. 2000). However, a limitation of plasmid analysis stems from the fact that a number of bacteria strains lack plasmids (Liebana et al. 2001). Complicating the issue further, plasmids are transferable between bacterial strains by conjugation, and they can act as mobile elements that are gained or lost under selective pressure (Kumao et al. 2002). As a result, the mobility of plasmids can be detrimental in deciphering the genetic relatedness of bacterial isolates.

25.7.1.3 Restriction Fragment Length Polymorphism Analysis

Researchers have found success in restriction fragment length polymorphism (RFLP) for typing of foodborne pathogens. This method was first used to differentiate *Salmonella* serotypes, such as *S. typhimurium*, *S. dublin*, and *S. enteritidis* (Tompkins et al. 1986). Nayak et al. (2006) have shown that RFLP was suitable for discriminating isolates within the *Campylobacter coli* and *C. jejuni* species. In these investigations, chromosomal DNA from pathogens were either (1) digested using unique restriction enzymes followed by separating DNA fragments using PFGE or (2) digested using unique enzymes followed by DNA fragment transfer to membranes for labeled probe hybridization to repetitive DNA Fragments.

Polymerase chain reaction-restriction fragment length polymorphism (PCR-RFLP) is an alternative to the traditional RFLP, and it involves PCR amplification of specific conserved DNA sequences in the pathogen followed by a digestion of the PCR amplicons with restriction enzymes to generate a DNA banding pattern (Mohran et al. 1996; Nayak et al. 2006). The PCR-RFLP using the *fli*C gene target demonstrated varied results with *Salmonella* spp. This technique effectively discriminated *S. gallinarium* isolates but was incapable of discriminating other *Salmonella* serotypes (Dauga et al. 1998; Kwon et al. 2000). However, PCR-RFLP has been extensively used with success in typing *Campylobacter* isolates, in which the *fla*A gene is amplified and digested with the enzymes Ddel or Alul (Nielsen et al. 2000). The *fla*-RFLP approach provided better discriminatory results than ribotyping but was less discriminatory than PFGE and random amplified polymorphic DNA-PCR (RAPD-PCR) (Nielsen et al. 2000).

25.7.1.4 Ribotyping

Ribotyping involves the cutting of genomic DNA with a frequent cutting restriction enzyme. The restriction fragments are separated by gel electrophoresis, transferred to a membrane, and incubated with a probe specific for a conserved region of the ribosomal (r)RNA genes (Bouchet et al. 2008). Differences in the number of rRNA genes and genetic variability produce distinct restriction fragment band profiles that can be used to discriminate between bacterial strains (Bouchet et al. 2008).

Ribotyping has been used to discriminate between *Salmonella*, *E. coli*, and *Campylobacter* strains. Restriction enzymes such as PvuII, PstI, and SphI have been used to digest *Salmonella* genomic DNA (Bailey et al. 2002; Chisholm et al. 1999; Landeras and Mendoza 1998). Restriction fragments have been hybridized with 16S or 23S rRNA gene-specific probes resulting in successful discrimination (Chowdry et al. 1993; Grimont and Grimont 1986; Threlfall et al. 1998).

E. coli digestions with HindIII enzymes, followed by southern blotting restriction fragments with RNA probes, have provided useful discriminatory results (Picard et al. 1991; Tarkka et al. 1994). *Salmonella, E. coli*, and *Campylobacter* have been typed with automated ribotyping systems such as the RiboPrinter® Microbial Characterization System by Qualicon, Inc. (Wilimington, DE). While automated results are reproducible, the discriminatory ability of ribotyping is limited by the number of rRNA genes within a bacterial species. A lack of rRNA genes reduces the discriminatory ability when compared to other typing methods such as PFGE (Clermont et al. 2001; Hahm et al. 2003; Ge et al. 2006; Nielsen et al. 2000).

25.7.2 Polymerase Chain Reaction Amplification of Pathogen DNA

25.7.2.1 Random Amplified Polymorphic DNA-PCR

The use of generic PCR primers under low-stringency conditions that allow separation of multiple amplicons to generate a DNA fingerprint for pathogen comparison is known as RAPD-PCR analysis (Franklin et al. 1999). This method has been used with some success to identify foodborne pathogens such as *Salmonella* and *E. coli* (Guerra et al. 2000; Kruger et al. 2006). However, the reproducibility of this method is limited when compared to PFGE. Limitations in the reproducibility of RAPD-PCR are linked to the low-stringency conditions including amplification conditions, analysis parameters, and even changes in chemical reagents (Meunier and Grimont 1993; Micheli et al. 1994).

25.7.2.2 Repetitive Element-PCR

Repetitive element-PCR (Rep-PCR) involves the amplification of two repeated DNA sequences distributed throughout the genome of pathogens (Versalovic et al. 1991). Banding pattern diversity is evaluated after gel electrophroesis, and also differences in the number of repetitive elements and their position within the bacterial genome (Versalovic et al. 1991). Multiple repeat sequences such as the enterobacterial repetitive intergenicconsensus (ERIC), repetitive extragenic palindromic (REP), and BOX sequences (Versalovic et al. 1991) have been used to distinguish bacterial strains, but results are usually less discriminatory than results produced by PFGE (Olive and Bean 1999). ERIC sequences are 126-bp conserved motifs found in a number of enteric bacteria associated with foodborne disease outbreaks such as *Salmonella*, whereas REP sequences contain only 38 bp of palindromic DNA sequences (Versalovic et al. 1991). The BOX sequences are inverted repeat elements that are present in a number of bacterial species and include the BOX A1R elements of *E. coli* that have been successfully used to distinguish *E. coli* strains in a number of studies (Dombek et al. 2000; Mohapatra et al. 2007).

The lack of repetitive sequences in a given strain may make the ability to distinguish the strain from others difficult. Nevertheless, the short turnaround time, minimal amount of required DNA, and relatively high discriminatory power make Rep-PCR an attractive tool for foodborne disease outbreak studies (Swaminathan and Barrett 1995; Wise et al. 2009), although issues with reproducibility are concerning and may occur as a result of variances in reagents, sample processing, and method analysis.

25.7.2.3 Variable Number of Tandem Repeat (VNTR) Analysis and Multiple Locus VNTR Analysis

Many bacterial genomes contain regions with directly repeated DNA motifs ranging from a small number of base pairs to over 100 bp that can be used to discriminate bacteria strains (Denoeud and Vergnaud 2004; Lindstedt et al. 2005). Variable number of tandem repeat (VNTR) analysis utilizes differences in the number of repeated DNA motifs at a single specific location (also known as an array) among bacteria strains (Benson 1999; Lindstedt et al. 2005). However, to accomplish maximum discrimination, it is frequently necessary to analyze multiple array regions, and this approach is referred to as multiple locus VNTR analysis (MLVA) (Lindstedt et al. 2005; Noller et al. 2003; Keim et al. 2000). Both VNTR and MLVA have been reported to have similar and sometimes greater discriminating ability than PFGE when subtyping *E. coli* O157:H7, *Salmonella*, and *Shigella* isolates (Keys et al. 2005; Liang et al. 2007).

25.7.3 Sequencing-Based Methods

25.7.3.1 Multilocus Sequence Typing

Microbial genome variability due to mutation or recombination can be used in molecular MST applications to determine the genetic connection among bacterial strains (Foley et al. 2009). In multilocus sequence typing (MLST), sequences of multiple genes are compared for nucleotide base changes as opposed to differences in DNA fragment size to establish genetic associations (Maiden et al. 1998; Spratt 1999). Genes that are required for basic cellular function (housekeeping genes) are commonly sequenced because they exist in all isolates within a species and are not rapidly susceptible to sequence changes attributable to an assortment of selective environmental pressures (Foley et al. 2009). Nevertheless, housekeeping gene sequences have adequate variability to permit strain differentiation (Kotetishvili et al. 2002; Maiden et al. 1998). Similar to PFGE and the PulseNet System, MLST databases have been generated for multi-laboratory use (Enright and Spratt 1999; Swaminathan et al. 2001, 2006).

A number of studies involving *E. coli*, *Campylobacter*, and *Salmonella* have evaluated MLST against more established molecular tools such as the "gold standard" PFGE. Foley et al. (2004) found PFGE to be better than MLST and Rep-PCR when discriminating *E. coli* strains. In studies that involved *Campylobacter* isolates, PFGE emerged as the most discriminating method as well (Levesque et al. 2008; Sails et al. 2003). However, conflicting reports exist with studies that involve *Salmonella* strains. Fakhr et al. (2005) found PFGE to be better than MLST when differentiating serovar Typhimurium isolates. Conversely, Kotetishvili et al. (2002) and Foley et al. (2006) found MLST to provide comparable results and sometimes better results than PFGE when evaluating *Salmonella* strains. The disparity in study results may be attributed to the number and types of genes evaluated in the individual studies (Fakhr et al. 2005; Foley et al. 2006; Kotetishvili et al. 2002).

25.7.3.2 Single-Nucleotide Polymorphism Analysis

Mechanisms such as recombination events, horizontal gene transfer, and nucleotide mutation have all contributed to bacterial evolution, thus allowing strains of bacteria to deviate from one another genetically. Single-nucleotide polymorphism (SNP) analysis involves the evaluation of nucleotide mutations at specific locations on the bacterial genome to discriminate strains of bacteria. Depending on the type and location of the nucleotide changes, the SNPs can elicit a nonsynonymous mutation (encoded amino acid) or a synonymous mutation (amino acid that is not encoded) (Zhang et al. 2006). A very limited number of studies that involve SNP analysis of foodborne pathogens have been conducted to date. Much of the work has concentrated on discovering SNPs that can be used for pathogen identification and differentiation (Zhang et al. 2006; Octavia and Lan 2007). Quinolone resistance (Esaki et al. 2004; Levy et al. 2004) and flagellar antigens (Octavia and Lan 2007) have been evaluated for possible sources of SNPs for *Salmonella*. Standard MLST genes and chromosomal genes have been evaluated for sources of SNPs for *Campylobacter* and *E. coli* isolates, respectively (Price et al. 2006; Zhang et al. 2006). As more informative SNPs are identified (Clawson et al. 2009), SNP analysis will likely be used more frequently as a tool in source tracking foodborne pathogens.

25.8 Future Directions and Research Needs

Given that a plethora of variations in molecular tools used for microbial analysis exist, differences in turnaround time, discriminatory ability, and reproducibility are of paramount concern when selecting a method to use in foodborne disease outbreak investigations. At present, these differences may require the use of a combination of molecular tools to identify and differentiate pathogens and host sources of pathogens. While PFGE has been the molecular tool used most extensively in the detection and identification of pathogens associated with foodborne disease outbreaks, a number of other molecular tools are emerging with great potential. As major advances continue in microbial genome sequencing, the effectiveness of molecular subtyping in source tracking studies will likely be maximized in the future. There is the need for the development of standardized protocols for all molecular tools used in foodborne disease outbreaks studies; standardization will increase reproducibility, thus increasing confidence in the methods. An important consideration for all sequence-based methods under development includes selection of the appropriate type and number of genes adequate to distinguish between closely related isolates for the most accurate identification.

Public-health officials and environmental regulatory agencies worldwide recognize the importance of pinpointing the specific type pathogen, host source of fecal pathogens, and the food source involved in foodborne disease outbreaks. A number of studies have demonstrated the usefulness of MST in response to outbreaks of foodborne disease. Future efforts should also include the employment of MST as a food safety prevention tactic within regulatory compliance monitoring among various food industries. MST results from within a monitoring framework would allow for more accurate quality control, risk assessment, and a reduction or elimination of exposure from identified pathogen sources.

References

Ahmed FE (1991) Microbiological and parasitic exposure. In: Ahmed FE (ed) Seafood safety. National Academy Press, Washington, DC

Aktas Z, Day M, Kayacan C, Diren S, Threlfall E (2007) Molecular characterization of *Salmonella typhimurium* and *Salmonella enteritidis* by plasmid analysis and pulsed-field gel electrophoresis. Int J Antimicrob Agents 30: 541–545

Altekruse SF, Cohen ML, Swerdlow DL (1997) Emerging foodborne diseases. Emerg Infect Dis 3: 285–293

Bacon RT, Sofos JN (2003) Food hazards: biological food; characteristics of biological hazards in foods. In: Schmidt RH, Rodrick G (eds) Food Safety Wiley, New York

Bailey JS, Fedorka-Cray PJ, Stern NJ, Craven SE, Cox NA, Cosby DE (2002) Serotyping and ribotyping of *Salmonella* using restriction enzyme Pvull. J Food Prot 65: 1005–1007

Barrett TJ, Gerner-Smidt P, Swaminathan E (2006) Interpretation of pulsed-field gel electrophoresis patterns in foodborne disease investigations and surveillance. Foodborne Pathog Dis 3: 20–31

Batz MB, Doyle MP, Morris G Jr, Painter J, Singh R, Tauxe RV, Taylor MR, Lo Fo Wong DM, Food Attribution Working Group (2005) Attributing illness to food. Emerg Infect Dis 11: 993–999

Benson G (1999) Tandem repeats finder: a program to analyze DNA sequences. Nucleic Acids Res 27: 573–580

Blake PA, Merson MH, Weaver RE, Hollis DG, Heublein PC (1979) Disease caused by a marine *Vibrio*. Clinical characteristics and epidemiology. N Engl J Med 300: 1–5

Bouchet V, Huot H, Goldstein R (2008) Molecular genetic basis of ribotyping. Clin Microbiol Rev 21: 262–273

Bowen A, Fry A, Richards G, Beuchat L (2006) Infections associated with cantaloupe consumption: a public health concern. Epidemiol Infect 134: 675–685

Brandl MT, Mandrell RE (2002) Fitness of *Salmonella* enteric serovar Thompson in the cilantro phyllosphere. Appli Environ Microbiol 68: 3614–3621

Brandl MT, Haxo AF, Bates AH, Mandrell RE (2004) Comparison of survival of *Campylobacter jejuni* in the phyllosphere with that in the rhizosphere of spinach and radish plants. Appl Environ Microbiol 70: 1182–1189

Brandl MT (2006) Fitness of human enteric pathogens on plants and implications for food safety. Annu Rev Phytopathol 44: 367–392

Bryan FL (1982) Diseases Transmitted by Foods (A classification and summary), 2nd edn, Centers for Disease Control, Georgia

Burnett SL, Chen J, Beuchat LR (2000) Attachment of *Escherichia coli* O157:H7 to the surfaces and internal structures of apples as detected by confocal scanning laser microscopy. Appli Environ Microbiol 66: 4679–4687

Buzby J, Roberts T (1996) ERS updates US foodborne disease costs for seven pathogens. Food Review 19: 20–25

CDC, Centers for Disease Control and Prevention (2002) Multistate outbreaks of *Salmonella* serotype Poona infections associated with eating cantaloupe from Mexico – United States and Canada 2000–2002. MMWR 51: 1044–1047

CDC, Centers for Disease Control and Prevention (2006) Ongoing multistate outbreak of *Escherichia coli* serotype O157:H7 infections associated with consumption of fresh spinach – United States. MMWR 55: 1045–1046

CDC, Centers for Disease Control and Prevention (2007) Multistate outbreaks of *Salmonella* infections associated with raw tomatoes eaten in restaurants – United States 2005–2006. MMWR 56: 909–911

CDC, Centers for Disease Control and Prevention (2010) US foodborne disease outbreaks 1990–2007. [cited 2010 Mar 8] Available from http://www.cdc.gov/outbreaknet/surveillance_data.html

CDHS and FDA, California Department of Health Services and US Food and Drug Administration. (2007) Investigation of an *Escherichia coli* O157:H7 outbreak associated with Dole prepackaged spinach. [cited 2010 Mar 8]. Available from http://www.cdph.ca.gov

Canzonier WJ (1991) Historical perspective on commercial depuration of shellfish. In: Otwell WS, Rodrick GE Martin R (eds) Molluscan Shellfish Depuration. CRC Press, Florida

Charkowski AO, Barak JD, Sarreal CZ, Mandrell RE (2002) Differences in growth of *Salmonella enterica* and *Escherichia coli* O157:H7 on alfalfa sprouts. Appl Environ Microbiol 68: 3114–3120

Cheng PK, Wong DK, Chung TW, Lim WW (2005) Norovirus contamination found in oysters worldwide. J Med Virol 76: 593–597

Chironna M, Germinario C, De Medici D, Fiore A, Di Pasquale S, Quarto M, Barbuti S (2002) Detection of hepatitisA virus in mussels from different sources marketed in Puglia region (South Italy). Int J Food Microbiol 75: 11–18

Chisholm SA, Crichton PB, Knight HI, Old DC (1999) Molecular typing of *Salmonella* serotype Thompson strains isolated from human and animal sources. Epidemiol Infect 122: 33–39

Chowdry N, Threlfall EJ, Rowe B, Stanley J (1993) Genotype analysis of faecal and blood isolates of *Salmonella dublin* from humans in England and Wales. Epidemiol Infect 110: 217–225

Clawson ML, Keen JE, Smith TP, Durso LM, McDaneld TG, Mandrell RE, Davis MA, Bono JL (2009) Phylogenetic classification of *Escherichia coli* O157:H7 strains of human and bovine origin using a novel set of nucleotide polymorphisms. Genome Biol. 10: R56

Clemetson L, Simon B (2003) "US Officials Say Ill Cow Is Linked to Alberta Herd" The New York Times December 28: p 11

Clermont O, Cordevant C, Bonacorsi S, Marecat A, Lange M, Bingen E (2001) Automated ribotyping provides rapid phylogenetic subgroup affiliation of clinical extraintestinal pathogenic *Escherichia coli* strains. J Clin Microbiol 39: 4549–4553

Coelho C, Heinert AP, Simões CM, Barardi CR (2003) Hepatitis A virus detection in oysters (*Crassostrea gigas*) in Santa Catarina State, Brazil, by reverse transcription-polymerase chain reaction. J Food Prot 66: 507–517

Conaty S, Bird P, Bell G, Kraa E, Grohmann G, McAnulty JM (2000) Hepatitis A in New South Wales, Australia from consumption of oysters: the first reported outbreak. Epidemiol Infect 124: 121–130

Crohn D and Bianchi M (2008) Research priorities for coordinated management of food safety and water quality. J Environ Qual 37: 1411–1418
Dauga C, Zabrovskaia A, Grimont PA (1998) Restriction fragment length polymorphism analysis of some flagellin genes of *Salmonella enteric*. J Clin Microbiol 36: 2835–2843
D'Aoust J Y, Maurer J (2007). *Salmonella* species. In: Doyle MP, Beuchat LR (eds) Food microbiology: fundamentals and frontiers, 3rd edn. ASM, Washington, DC
Denoeud F, Vergnaud G (2004) Identification of polymorphic tandem repeats by direct comparison of genome sequence from different bacterial strains: a web based resource. BMC Bioinformatics 5: 4
Desenclos JC, Klontz KC, Wilder MH, Nainan OV, Margolis HS, Gunn RA (1991) A multistate outbreak of hepatitis A caused by the consumption of raw oysters. Am J Public Health 81: 1268–1272
DeWaal CS, Barlow K. (2002) Outbreak Alert! Closing the Gaps in Our Federal Food-Safety Net. Center for Science in the Public Interest, Washington, DC
Dolin R, Blacklow NR, DuPont H, Buscho RF, Wyatt RG, Kasel JA, Hornick R, Chanock RM (1972) Biological properties of Norwalk agent of acute infectious nonbacterial gastroenteritis. Proc Soc Exp Biol Med 140: 578–583
Dombek PE, Johnson LK, Zimmerley SI, Sadowsky MJ (2000) Use of repetitive DNA sequences and the PCR to differentiate *Escherichia coli* isolates from human and animal sources. Appl Environ Microbiol 66: 2572–2577
Domingue G, Willshaw GA, Smith HR, Perry N, Radford D, Cheasty T (2003) DNA-based subtyping of verocytotoxin-producing *Escherichia coli* VIEC 0128ab:H2 strains from human and raw meat sources. Lett Appl Microbiol 37: 433–437
Doyle MP, Erickson MC (2006) Emerging microbiological food safety issues related to meat. Meat Sci. 74: 98–112
Enright MC, Spratt BG (1999) Multilocus sequence typing. Trends Microbiol 7: 482–487
Esaki H, Noda K, Otsuki N, Kojima A, Asai T, Tamura Y, Takahashi T (2004) Rapid detection of quinolone-resistant *Salmonella* by real time SNP genotyping. J Microbiol Methods 58: 131–134
Fakhr MK, Nolan LK, Logue CM (2005) Multilocus sequence typing lacks the discriminatory ability of pulsed-field gel electrophoresis for typing *Salmonella enterica* serovar Typhimurium. J Clin Microbiol 43: 2215–2219
Feldhusen F (2000) The role of seafood in bacterial foodborne diseases. Microb Infect 2: 1651–1660
Felmer R, Chavez R, Catrileo A, Rojas C (2006) Current and emergent technologies for animal identification and their use in animal traceability. Archiv De Medic Veterin 38: 197–206
Foley SL, Simjee S, Meng J, White DG, McDermott PF, Zhao S (2004) Evaluation of molecular typing methods for *Escherichia coli* 0157:H7 isolates from cattle food and humans. J Food Prot 67: 651–657
Foley SL, Walker R (2005) Methods of differentiation among bacterial foodborne pathogens. Iowa State University Ames IA
Foley SL, White DG, McDermott PF, Walker RD, Rhodes B, Fedorka-Cray P, Simjee S, Zhao S (2006) Comparison of subtyping methods for differentiating *Salmonella enterica* serovar Typhimurium isolates obtained from food animal sources. J Clin Microbiol 44: 3569–3577
Foley SL, Zhao S, Walker RD (2007) Comparison of molecular typing methods for the differentiation of *Salmonella* food borne pathogens. Foodborne Pathog Dis 4: 253–276
Foley SL, Lynne AM, Nayak R (2009) Molecular typing methodologies for microbial source tracking and epidemiological investigations of Gram-negative bacterial foodborne pathogens. Infect Genet Evol 9: 430–440
Formiga-Cruz M, Allard AK, Conden-Hansson AC, Henshilwood K, Hernroth BE, Jofre J, Lees DN, Lucena F, Papapetropoulou M, Rangdale RE, Tsibouxi A, Vantarakis A, Girones R (2003) Evaluation of potential indicators of viral contamination in shellfish and their applicability to diverse geographical areas. Appl Environ Microbiol 69: 1556–1563
Franklin RB, Taylor DR, Mills AL (1999) Characterization of microbial communities using randomly amplified polymorphic DNA (RAPD). J Microbiol Methods 35: 225–235

Fratamico PM, Bhunia AK, Smith JL (2005) Foodborne pathogens: microbiology and molecular biology. Caister Academic, United Kingdom

Ge B, Girard W, Zhao S, Friedman S, Gaines SA, Meng J (2006) Genotyping of *Campylobacter* spp from retail meats by pulsed-field gel electrophoresis and ribotyping. J Appl Microbiol 100: 175–184

Geary PM, Davies CM (2003) Bacterial source tracking and shellfish contamination in a coastal catchment. Water Sci Technol 47: 95–100

Gebre-Yohannes A, Drasar BS (1991) Molecular epidemiology of plasmid patterns in *Shigella flexneri* types 1-6. Epidemiol Infect 107: 321–334

Gerner-Smidt P, Hise K, Kincaid J, Hunter S, Rolando S, Hyytia-Irees E, Ribot EM, Swaminathan B (2006) PulseNet USA: A five-year update. Foodborne Pathog Dis 3: 9–19

Gerner-Smidt P, Scheutz F (2006) Standardized pulsed-field gel electrophoresis of Shiga toxin-producing *Escherichia coli*: the PulseNet Europe Feasibility Study. Foodborne Pathog Dis 3: 74–80

Golan E, Krissoff B, Kuchler F, Calvin L, Nelson K, Price G (2004) Traceability in the US food supply: Economic theory and industry studies. Agricultural Economic Report Number 830, Economic Research Service, US Department of Agriculture, Washington, DC

Greene SK, Daly ER, Talbot EA, Demma LJ, Holzbauer S, Patel NJ, Hill TA, Walderhaug MO, Hoekstra RM, Lynch MF, Painter JA (2008) Recurrent multistate outbreak of *Salmonella* Newport associated with tomatoes from contaminated fields, 2005. Epidemiol Infect 136: 157–165

Greig JD, Ravel A (2009) Analysis of foodborne outbreak data reported internationally for source attribution. Int J Food Microbiol 130: 77–87

Grimont F, Grimont PA (1986) Ribosomal ribonucleic acid gene restriction patterns as potential taxonomic tools. Ann Inst Pasteur Microbiol 137B: 165–175

Guerra B, Laconcha I, Soto SM, Gonzalez-Hevia MA, Mendoza MC (2000) Molecular characterisation of emergent multiresistant *Salmonella enterica* serotype [4,5,12:i:-] organisms causing human salmonellosis. FEMS Microbiol Lett 190: 341–347

Guo X, Chen J, Brackett RE, and Beuchat LR (2001) Survival of salmonellae on and in tomato plants from the time of inoculation at flowering and early stages of fruit development through fruit ripening Appl Environ Microbiol 67: 4760–4764

Hahm BK, Maldonado Y, Schreiber E, Bhunia AK, Nakatsu CH (2003) Subtyping of foodborne and environmental isolates of *Escherichia coli* by multiplex-PCR rep-PCR PFGE ribotyping and AFLP. J Microbiol Methods 53: 387–399

Halliday ML, Kang LY, Zhou TK, Hu MD, Pan QC, Fu TY, Huang YS, Hu SL (1991) An epidemic of hepatitis A attributable to the ingestion of raw clams in Shanghai, China. J Infect Dis 164: 852–859

Hunt JW, Anderson BS, Phillips BM, Tjeerdema RS, Richard N, Connor V, Worcester K, Angelo M, Bern A, Fulfrost B, Mulvaney D (2006) Spatial relationships between water quality and pesticide application rates in agricultural watersheds. Environ Monit Assess 12: 245–262

Hunter SB, Vauterin P, Lambert-Fair MA, Van Duyne MS, Kubota K, Graves L, Wrigley D, Barrett I, Ribot E (2005) Establishment of a universal size standard strain for use with the PulseNet standardized pulsed-field gel electrophoresis protocols: converting the national databases to the new size standard. J Clin Microbiol 43: 1045–1050

IFT, Institute of Food Technologists (2006) Antimicrobial Resistance – Implications for the Food System. Comprehensive Reviews in Food Science and Food Safety 5: 71–137

Itoh Y, Sugita-Konishi Y, Kasuga F, Iwaki M, Hara-Kudo Y, Saito N, Noguchi Y, Konuma H, Kumagai S (1998) Enterohemorrhagic *Escherichia coli* O157:H7 present in radish sprouts. Appl Environ Microbiol 64: 1532–1535

Jaquette CB, Beuchat LR, Mahon BE (1996) Efficacy of chlorine and heat treatment in killing *Salmonella* Stanley inoculated onto alfalfa seeds and growth and survival of the pathogen during sprouting and storage. Appli Environ Microbiol 62: 2212–2215

Jay MT, Cooley M, Carychao D, Wiscomb GW, Sweitzer RA, Crawford-Miksza L, Farrar JA, Lau DK, O'Connell J, Millington A, Asmundson RV, Atwill ER, Mandrell RE (2007) *Escherichia coli* O157: H7 in feral swine near spinach fields and cattle central California coast Emerg Infect Dis 13: 1908–1911

Johnson JR, Sannes MR, Croy C, Johnston B, Clabots C, Kuskowski MA, Bender J, Smith KE, Winokur PL, Belongia EA (2007) Antimicrobial drug-resistant *Escherichia coli* from humans and poultry products Minnesota and Wisconsin 2002–2004. Emerg Infect Dis 13: 838–846

Jones KE, Patel NG, Levy MA, Storeygard A, Balk D, Gittleman JL, Daszak P (2008) Global trends in emerging infectious diseases. Nature 451: 990–993

Jones MK, Oliver JD (2009) *Vibrio vulnificus*: disease and pathogenesis. Infect Immun 77: 1723–33

Kam KM, Luey KY, Chiu AW, Law CP, Leung SF (2007) Molecular characterization of *Salmonella enterica* Serotype Typhi isolates by pulsed-field gel electrophoresis in Hong Kong 2000–2004. Foodborne Pathog Dis 4: 41–49

Keim P, Price LB, Klevytska AM, Smith KL, Schupp JM, Okinaka R, Jackson PJ, Hugh-Jones ME (2000) Multiple-locus variable-number tandem repeat analysis reveals genetic relationships within *Bacillus anthracis*. J Bacteriol 182: 2928–2936

Keys C, Kemper S, Keim P (2005) Highly diverse variable number tandem repeat loci in the *E. coli* 0157:H7 and 055:H7 genomes for high-resolution molecular typing. J Appl Microbiol 98: 928–940

Kirkland KB, Meriwether RA, Leiss JK, Mac Kenzie WR (1996) Steaming oysters does not prevent Norwalk-like gastroenteritis. Public Health Rep 111: 527–530

Kohn MA, Farley TA, Ando T, Curtis M, Wilson SA, Jin Q, Monroe SS, Baron RC, McFarland LM, Glass RI (1995) An outbreak of Norwalk virus gastroenteritis associated with eating raw oysters. Implications for maintaining safe oyster beds JAMA 273: 466–471

Kotetishvili M, Stine OC, Kreger A, Morris Jr JG, Sulakvelidze A (2002) Multilocus sequence typing for characterization of clinical and environmental *Salmonella* strains. J Clin Microbiol 40: 1626–1635

Kruger A, Padola NL, Parma AE, Lucchesi PM (2006) Intraserotype diversity among Argentinian verocytotoxigenic *Escherichia coli* detected by random amplified polymorphic DNA analysis. J Med Microbiol 55: 545–549

Kumao T, Ba-Thein W, Hayashi H (2002) Molecular subtyping methods for detection of *Salmonella enterica* serovar Oranienburg outbreaks. J Clin MicrobioI 40: 2057–2061

Kwon HJ, Park KY, Yoo HS, Park JY, Park YH, Kim SJ (2000) Differentiation of *Salmonella enterica* serotype gallinarum biotype pullorum from biotype gallinarum by analysis of phase 1 flagellinC gene fliC. J Microbiol Methods 40: 33–38

Landeras E, Mendoza MC (1998) Evaluation of PCR-based methods and ribotyping performed with a mixture of Pstl and Sphl to differentiate strains of *Salmonella* serotype Enteritidis. J Med Microbiol 47: 427–434

Leoni E, Bevini C, Degli Esposti S, Graziano A (1998) An outbreak of intrafamiliar Hepatitis A associated with clam consumption: epidemic transmission to a school Community. Eur J Epidemiol 14: 187–192

Levesque S, Frost E, Arbeit RD, Michaud S (2008) Multilocus sequence typing of *Campylobacter jejuni* isolates from humans chickens raw milk and environmental water in Quebec Canada. J Clin Microbiol 46: 3404–3411

Levy DD, Sharma B, Cebula TA (2004) Single-nucleotide polymorphism mutation spectra and resistance to quinolones in *Salmonella enterica* serovar Enteritidis with a mutator phenotype. Antimicrob Agents Chemother 48: 2355–2363

Liang SY, Watanabe H, Terajima J, Li CC, Liao JC, Tlung SK, Chiou CS (2007) Multilocus variable-number tandem-repeat analysis for molecular typing of *Shigella sonnei*. J Clin Microbiol 45: 3574–3580

Liebana E, Guns D, Garcia-Migura L, Woodward MJ Clifton-Hadley FA, Davies RH (2001) Molecular typing of *Salmonella* serotypes prevalent in animals in England: assessment of methodology. J Clin Microbiol 39: 3609–3616

Lindstedt BA, Vardund T, Aas L, Kapperud G (2005) Multiple-locus variable number tandem repeats analysis for genetic fingerprinting of pathogenic bacteria. Electrophoresis 26: 2567–2582

Liu PY, Lau YJ, Hu BS, Shyr JM, Shi ZY, Tsai WS (1995) Analysis of clonal relationships among isolates of *Shigella sonnei* by different molecular typing methods. J Clin Microbiol 33: 1779–1783

Los Huertos M, Gentry L, Shennan C (2006) Land use and water quality on California's central coast: Nutrient levels in coastal waterways. Research Brief No 2 UC Santa Cruz Center for Agroecology & Sustainable Food Systems, Univ of California, Santa Cruz CA

Lu L, Hume ME, Sternes KL, Pillai SD (2004) Genetic diversity of *Escherichia coli* isolates in irrigation water and associated sediments: implications for source tracking. Wat Res 38: 3899–3908

Lukinmaa S, Nakari U-M, Eklund M, Siitonen A (2004) Application of molecular genetic methods in diagnostics and epidemiology of food-borne bacterial pathogens. APMIS 11211-12: 908–929

Luthi TM, Wall PG, Evans HS, Adak GK, Caul EO (1996) Outbreaks of foodborne viral gastroenteritis in England and Wales: 1992 to 1994. Commun Dis Rep 6: R131–6

Lynch M, Painter J, Woodruff R, Braden C (2006) Surveillance for foodborne-disease outbreaks – United States 1998–2002. MMWR Surveill Summ 55: 1–42

Lynch MF, Tauxe R, Hedberg CW (2009) The growing burden of foodborne outbreaks due to contaminated fresh produce: risks and opportunities. Epidemiol Infect 137: 307–315

Maiden MC, Bygraves JA, Feil E, Morelli G, Russell JE, Urwin R, Zhang Q, Zhou J, Zurth K, Caugant DA, Feavers IM, Achtman M, Spratt BG (1998) Multilocus sequence typing: a portable approach to the identification of clones within populations of pathogenic microorganisms. Proc Natl Acad Sci USA 95: 3140–3145

Martinez-Urtaza J, Saco M, Hernandez-Cordova G, Lozano A, Garcia-Martin O, Espinosa J (2003) Identification of *Salmonella* serovars isolated from live molluscan shellfish and their significance in the marine environment. J Food Prot 66: 226–232

McDonnell S, Kirkland KB, Hlady WG, Aristeguieta C, Hopkins RS, Monroe SS, Glass RI (1997) Failure of cooking to prevent shellfish-associated viral gastroenteritis. Arch Intern Med 157(1): 111–116

Mead PS, Slutsker L, Dietz V, McCaig LF, Bresee JS, Shapiro C, Griffin PM, Tauxe RV(1999) Food-related illness and death in the United States. Emerg Infect Dis 5: 607–625

Meschke JS, Boyle D (2007) Shellfish and Microbial Source Tracking. In: Santo Domingo JW, Sadowsky MJ (eds) Emerging issues in food safety, microbial source tracking . ASM, Washington, DC

Meng J, Doyle MP, Zhao T, Zhao S (2007) Enterohemorrhagic *Escherichia coli*. In: Doyle MP, Beuchat LR (eds) Food microbiology: fundamentals and frontiers, 3rd edn. ASM, Washington, DC

Meunier JR, Grimont PA (1993) Factors affecting reproducibility of random amplified polymorphic DNA fingerprinting. Res Microbiol 144: 373–379

Michaud S, Menard S, Gaudreau C, Arbeit RD (2001) Comparison of SmaI defined genotypes of *Campylobacter jejuni* examined by KpnI: a population based study. J Med Microbiol 50: 1075–1081

Micheli MR, Bova R, Pascale E, D'Ambrosio E (1994) Reproducible DNA fingerprinting with the random amplified polymorphic DNA RAPD method. Nucleic Acids Res 22: 1921–1922

Mohapatra BR, Broersma K, Nordin R, Mazumder A (2007) Evaluation of repetitive extragenic palindromic-PCR for discrimination of fecal *Escherichia coli* from humans and different domestic- and wild-animals. Microbiol Immunol 51: 733–740

Mohle-Boetani JC, Farrar J, Bradley P, Barak JD, Miller M, Mandrell R, Mead P, Keene WE, Cummings K, Abbott S, Werner SB (2009) *Salmonella* infections associated with mung bean sprouts: epidemiological and environmental investigations. Epidemiol Infect 137: 357–366

Mohran ZS, Guerry P, Lior H, Murphy JR Gendy AM, Mikhail MM, Oyofo BA (1996) Restriction fragment length polymorphism of flagellin genes of *Campylobacter jejuni* and/or *C. coli* isolates from Egypt. J Clin Microbiol 34: 1216–1219

Mor-Mur M, Yuste J (2010) Emerging bacterial pathogens in meat and poultry: An overview. Food Bioprocess Technol 3: 24–35

Morse DL, Guzewich JJ, Hanrahan JP, Stricof R, Shayegani M, Deibel R, Grabau JC, Nowak NA, Herrmann JE, Cukor G (1986) Widespread outbreaks of clam- and oyster-associated gastroenteritis. Role of Norwalk virus. N Engl J Med 314: 678–681

Munnoch SA, Ward K, Sheridan S, Fitzsimmons GJ, Shadbolt CT, Piispanen JP, Wang Q, Ward TJ, Worgan TL, Oxenford C, Musto JA, McAnulty J, Durrheim DN (2009) A multi-state outbreak of *Salmonella* Saintpaul in Australia associated with cantaloupe consumption Epidemiol Infect 137: 367–374

Nauerby B, Pedersen K, Dietz HH, Madsen M (2000) Comparison of Danish isolates of *Salmonella enterica* serovar enteritidis PT9a and PTll from hedgehogs *Erinaceus europaeus* and humans by plasmid profiling and pulsed-field gel electrophoresis. J Clin Microbiol 38: 3631–3635

Nayak R, Stewart-King I (2008) Molecular epidemiological analysis and microbial source tracking of *Salmonella enterica* serovars in a preharvest turkey production environment. Foodborne Pathog Dis 5: 115–126

Nayak R, Stewart T, Nawaz M, Cerniglia C (2006) In vitro antimicrobial susceptibility genetic diversity and prevalence of UDP-glucose 4-epimerase galE gene in *Campylobacter coli* and *Campylobacter jejuni* from Turkey production facilities. Food Microbiol 23: 379–392

Nayak R, Stewart T, Wang RF, Lin J, Cerniglia CE, Kenney PB (2004) Genetic diversity and virulence gene determinants of antibiotic-resistant *Salmonella* isolated from preharvest turkey production sources. Int J Food Microbiol 91: 51–62

Newell RIE, Jordan SJ (1983) Preferential ingestion of organic material by the American oyster *Crassostrea virginica*. Mar Ecol Prog Ser 13: 47–53

Nielsen EM, Engberg J, Fussing V, Petersen L, Brogren CH, On SL (2000) Evaluation of phenotypic and genotypic methods for subtyping *Campylobacter jejuni* isolates from humans poultry and cattle. J Clin Microbiol 38: 3800–3810

Noller AC, McEllistrem MC, Pacheco AG, Boxrud DJ, Harrison LH (2003) Multilocus variable-number tandem repeat analysis distinguishes outbreak and sporadic *Escherichia coli* 0157:H7 isolates. J Clin Microbiol 41: 5389–5397

Octavia S, Lan R (2007) Single nucleotide polymorphism typing and genetic relationships of *Salmonella enterica* serovar Typhi isolates. J Clin Microbiol 45: 3795–3801

Olive DM, Bean P (1999) Principles and applications of methods for DNA-based typing of microbial organisms. J Clin Microbiol 37: 1661–1669

Painter JA, Ayers T, Woodruff R, Blanton E, Perez N, Hoekstra RM, Griffin PM, Braden C. (2009) Recipes for foodborne outbreaks: a scheme for categorizing and grouping implicated foods. Foodborne Pathog Dis 6: 1259–64

Picard B, Picard-Pasquier N, Krishnamoorthy R, Goullet P (1991) Characterization of highly virulent *Escherichia coli* strains by ribosomal DNA restriction fragment length polymorphism. FEMS Microbiol Lett 66: 183–188

Porter DV, Lister SA (2006) Food safety: Federal and state response to the spinach *E. coli* outbreak. Available online. http://wwwnationalaglawcenterorg/assets/crs/RL33722pdf. Accessed 08 Mar 2010

Potasman I, Paz A, Odeh M (2002) Infectious outbreaks associated with bivalve shellfish consumption: a worldwide perspective. Clin Infect Dis 35: 921–928

Price EP, Thiruvenkataswamy V, Mickan L, Unicomb L, Rios RE, Huygens F, Giffard PF (2006) Genotyping of *Campylobacter jejuni* using seven single nucleotide polymorphisms in combination with jlaA short variable region sequencing. J Med Microbiol 8: 1061–1070

RCDMC, Resource Conservation District of Monterey County (2007) A grower survey: Reconciling food safety and environmental protection. Available online. http://wwwrcdmontereyorg/pdf/Food_Safety_Environmental_Protection_2006pdf. Accessed 08 Mar 2010

Reeve G, Martin DL, Pappas J, Thompson RE, Greene KD (1989) An outbreak of shigellosis associated with the consumption of raw oysters. N Eng J Med 321: 224–227

Ribot EM, Fair MA, Gautom R, Cameron DN, Hunter SB, Swaminathan B, Barrett TJ (2006) Standardization of pulsed-field gel electrophoresis protocols for the subtyping of *Escherichia coli* 0157:H7 *Salmonella* and *Shigella* for PulseNet Foodborne Pathog Dis 3: 59–67

Rippey SR (1994) Infectious disease associated with molluscan shellfish consumption. Clin Mirobiol Rev 7: 419–425

Sails AD, Swaminathan B, Fields PI (2003) Utility of multi locus sequence typing as an epidemiological tool for investigation of outbreaks of gastroenteritis caused by *Campylobacter jejuni*. J Clin Microbiol 41: 4733–4739

Samelis J, Sofos JN (2003) Strategies to control stress-adapted pathogens and provide safe foods. In: Yousef AE, Juneja VK (eds) Microbial adaptation to stress and safety of new-generation foods. CRC Press, Florida

Scott TM, Rose JB, Jenkins TM, Farrah SR, Lukasik J (2002) Microbial source tracking: current methodology and future directions. Appl Environ Microbiol 68: 5796–5803

Shieh YC, Baric RS, Woods JW, Calci KR (2003) Molecular surveillance of enterovirus and norwalk-like virus in oysters relocated to a municipal-sewage-impacted gulf estuary. Appl Environ Microbiol 269: 7130–7713

Sivapalasingam S, Friedman CR, Cohen L, Tauxe RV (2004) Fresh produce: a growing cause of outbreaks of foodborne illness in the United States 1973 through 1997. J Food Prot 67: 2342–2353

Smith JE, Perdek JM (2004) Assessment and management of watershed microbial contaminants. Crit Rev Environ Sci and Technol 34: 109–139

Sofos JN (2008) Challenges to meat safety in the 21st century. Meat Sci 78: 3–13

Spratt BG (1999) Multilocus sequence typing: molecular typing of bacterial pathogens in an era of rapid DNA sequencing and the internet. CurR Opin Microbiol 2: 312–316

Stanley J, Jones C, Burnens A, Owen RJ (1994) Distinct genotypes of human and canine isolates of *Campylobacter upsaliensis* determined by 16S rRNA gene typing and plasmid profiling. J Clin Microbiol 32: 1788–1794

Stewart J R, Santo Domingo JW, Wade TJ (2007) Fecal pollution, public health, and microbial source tracking. In: Santo Domingo JW, Sadowsky MJ (eds.) Emerging issues in food safety, microbial source tracking, ASM Washington, DC

Swaminathan B, Barrett TJ (1995) Amplification methods for epidemiologic investigations of infectious disease. J Microbiol Methods 2: 129–139

Swaminathan B Barrett Ij Hunter SB Iauxe RV (2001) PulseNet: the molecular subtyping network for foodborne bacterial disease surveillance United States Emerg Infect Dis 7: 382–389

Swaminathan B, Gerner-Smidt P, Ng LK, Lukinmaa S, Kam KM, Rolando S, Gutierrez EP, Binsztein N (2006) Building PulseNet International: an interconnected system of laboratory networks to facilitate timely public health recognition and response to foodborne disease outbreaks and emerging foodborne diseases. Foodborne Pathog Dis 3: 36–50

Tarkka E, Ahman H, Siitonen A (1994) Ribotyping as an epidemiologic tool for *Escherichia coli* Epidemiol Infect 11: 2263–274

Taormina PJ, Beuchat LR (1999) Behavior of enterohemorrhagic *Escherichia coli* O157:H7 on alfalfa sprouts during the sprouting process as influenced by treatments with various chemicals. J Food Prot 62: 850–856

Tauxe RV, Doyle MP, Kuchenmüller T, Schlundt J, Stein CE (2009) Evolving public health approaches to the global challenge of foodborne infections. Int J Food Microbiol doi:101016/jijfoodmicro200910014

Tenovar FC, Arbeit RD, Goering RV, Mickelsen PA, Murray BE, Persing DH, Swaminathan B (1995) Interpreting chromosomal DNA restriction patterns produced by pulsed-field gel electrophoresis: criteria for bacterial strain typing. J Clin Microbiol 33: 2233–2239

Terajima J, Tamura K, Hirose K, Izumiya H, Miyahara M, Konuma H, Watanabe H (2004) A multi-prefectural outbreak of *Shigella sonnei* infections associated with eating oysters in Japan. Microbiol Immunol 48: 49–52

Threlfall EJ, Bhat MB, Sharma KB (1982) Genetic analysis of plasmids coexisting in *Salmonella* newport isolated from a salmonellosis outbreak in Delhi Indian. J Med Res 76: 358–363

Threlfall EJ, Frost JA, Ward LR, Rowe B (1990) Plasmid profile typing can be used to subdivide phage-type 49 of *Salmonella typhimurium* in outbreak investigations. Epidemiol Infect 104: 243–251

Threlfall EJ, Hampton MD, Ridley AM (1998) Application of molecular methods to the study of infections caused by *Salmonella* spp. In: Woodford N, Johnson AP (eds) Molecular bacteriology: protocols and clinical application. Humana Press, Totowa

Tompkins LS, Troup N, Labigne-Roussel A, Cohen ML (1986) Cloned random chromosomal sequences as probes to identify *Salmonella* species. J Infect Dis 15: 4156–162

Tyler HL, Triplett EW (2008) Plants as a habitat for beneficial and/or human pathogenic bacteria Annual Review of Phytopathology 46: 53–73

van Belkum A, Tassios PT, Dijkshoorn L, Haeggman S, Cookson B, Fry NK, Fussing V, Green J, Feil E, Gerner-Smidt P, Brisse S, Struelens M (2007) Guidelines for the validation and application of typing methods for use in bacterial epidemiology. Clin Microbiol Infect 13 (Suppl 3): 1–46

Versalovic J, Koeuth T, Lupskij R (1991) Distribution of repetitive DNA sequences in eubacteria and application to fingerprinting of bacterial genomes. Nucleic Acids Res 19: 6823–6831

Wheeler C, Vogt TM, Armstrong GL, Vaughan G, Weltman A, Nainan OV, Dato V, Xia G, Waller K, Amon J, Lee TM, Highbaugh-Battle A, Hembree C, Evenson S, Ruta MA, Williams IT, Fiore AE, Bell BP (2005) An outbreak of hepatitis A associated with green onions. N Engl J Med 353: 890–897

Widdowson MA, Sulka A, Bulens SN, Beard RS, Chaves SS, Hammond R, Salehi ED, Swanson E, Totaro J, Woron R, Mead PS, Bresee JS, Monroe SS, Glass RI (2005) Norovirus and foodborne disease United States 1991–2000. Emerg Infect Dis 11: 95–102

Wilson IG, Moore JE (1996) Presence of *Salmonella* spp. and *Campylobacter* spp. in shellfish. Epidemiol Infect 116: 147–153

Wise MG, Siragusa GR, Plumblee J, Healy M, Cray PJ, Seal BS (2009) Predicting *Salmonella enterica* serotypes by repetitive sequence-based PCR. J Microbiol Methods 76: 18–24

Zhang HX, Harrington M, Mauro J, Fillmore LA, Wheeler J (2002) Bacterial source tracking in pathogen TMDL development and implementation. In: Proceedings of the water environment federation, WEFTEC 2002: Session 21 through Session 30. Water Environment Federation, Virginia

Zhang HX, Harrington M, Mauro J, Fillmore LA, Wheeler J (2003) Bacterial source tracking in pathogen TMDL development and implementation part II: challenge and opportunity. In: Proceedings of the water environment federation, WEFTEC 2003: Session 1 through Session 10. Water Environment Federation, Virginia

Zhang W, Qi W, Albert TJ, Motiwala AS, Alland D,Hyytia-Trees EK (2006) Probing genomic diversity and evolution of *Escherichia coli* 0157 by single nucleotide polymorphisms Genome Res 16: 757–767

Zheng J, Keys CE, Zhao S, Meng J, Brown EW (2007) Enhanced subtyping scheme for *Salmonella enteritidis*. Emerg Infect Dis 13: 1932–1935

Zhuang RY, Beuchat LR, Angulo FJ (1995) Fate of *Salmonella* montevideo on and in raw tomatoes as affected by temperature and treatment with chlorine. Appli Environ Microbiol 61: 2127–2131

Chapter 26
Training Future Scientists: Teaching Microbial Source Tracking (MST) to Undergraduates

J. Brooks Crozier and Maria Alvarez

Abstract While major universities have the largest numbers of science majors and are recognized for their productivity in peer-reviewed research, liberal arts institutions and more recently community colleges are increasingly important in providing research training for students. Undergraduate research is an increasingly important aspect of the undergraduate educational experience to learn critical thinking and problem-solving skills and is an important factor for student recruitment and retention. This chapter describes the efforts of Roanoke College (a small 4-year liberal arts college) and El Paso Community College in training undergraduate students in environmental microbiology, both successes and challenges, including research techniques such as microbial source tracking (MST). Examples of MST projects conducted by undergraduate students and simple protocols are offered, as well as resources for educators of undergraduate students.

Keywords Undergraduate Education • Undergraduate Research • STEM • MST • ARA • PCR

26.1 Introduction

Large universities have long been recognized for having large numbers of declared science majors and for placing strong emphasis on peer-reviewed research. However, liberal arts institutions, and more recently community colleges, are increasingly important in the nation's research venue and have proven very effective at preparing undergraduate students for life within and outside of their declared major. At Roanoke College, for example, the students are commonly accepted into a wide variety of

J.B. Crozier (✉)
Department of Biology, Roanoke College, Salem, VA, USA
e-mail: crozier@roanoke.edu

C. Hagedorn et al. (eds.), *Microbial Source Tracking: Methods, Applications, and Case Studies*, DOI 10.1007/978-1-4419-9386-1_26,

health-related and graduate programs in and out of state, occasionally including international programs, and the students at Roanoke College are quite competitive in the pursuit of competitive internship opportunities. Liberal arts institutions are preparing students for their future in that a wide range of courses are required, especially out of the major; students have opportunities for study or courses taught abroad, and smaller colleges have very good student–teacher ratios at approximately 15:1 depending on the school. It is not unusual to have 150–200 students or more attending freshman biology class at large universities, while most science courses at community colleges and liberal arts colleges are capped at 17–30 students.

Community colleges enroll half of all US undergraduates and more than half of all minority and female students, i.e., the groups that have been traditionally underrepresented in Science, Technology, Engineering, and Mathematics (STEM) fields. Because of their open admissions, low tuition, flexible schedules, small class size, and excellent support services for students, community colleges are extremely important in terms of access to higher education for minority and disadvantaged students. Most community colleges are becoming comprehensive institutions offering Associate of Science degrees and preparing students to transfer to the nation's universities. Although community colleges typically are not research institutions, various programs from the National Science Foundation (NSF), the National Institutes of Health (NIH), and the US Department of Energy (DOE) provide funding for community college students and faculty to conduct research.

While a large university may have hundreds of science majors per department, liberal arts institutions and community colleges typically have majors numbering in the 10–50 range for individual departments. Universities also have larger budgets, greater numbers of faculty with exceedingly diverse backgrounds, and reduced teaching loads and stronger emphasis on research productivity. Liberal arts institutions and community colleges are typically considered "teaching institutions," but recently greater pressure has been placed on their faculty to attend professional conferences and publish more frequently while maintaining their teaching responsibilities. Teaching and research budgets at these institutions are smaller, and fewer faculty must cover a more diverse series of courses such as freshman seminar, nonmajors science courses, or related courses, which can drift from the faculty members specific field of training.

An advantage of liberal arts and community colleges is the high level of interaction that students have with faculty. At community colleges, laboratory courses are taught by the faculty teaching the lecture and not by graduate teaching assistants. The mentoring relationships that develop as faculty interact with students in a laboratory setting are so effective that students at smaller colleges have a great opportunity to be well prepared to move into a productive major-related profession and succeed in a graduate program or professional school. Two distinct categories delineate the successes and concerns in microbiology and MST education among undergraduates and can serve as a model on how undergraduate research can be effective and contribute positively to peer-reviewed scientific literature.

They are as follows:

- Student Learning and Exposure to Research
- Peer Support and Networking

26.2 Student Learning and Exposure to Research

Microbiology education and the training required for MST research is very feasible at smaller institutions because of the small faculty–student ratio and the desire (need) of faculty to collaborate with colleagues and publish in the field. Without graduate students, undergraduates are conducting research along with faculty mentors. Many MST protocols are not overly difficult (antibiotic resistance analysis, Biolog, basic PCR procedures), have minimal equipment needs, and teach microbiology in a more inquiry-based manner (see Sect. 26.4 for two examples). Research at undergraduate institutions such as Roanoke College (Fig. 26.1) including MST research is not without its challenges, however. Commitments to college-wide committees, designing and teaching courses, advising students, departmental duties, and the difficulties associated with faculty at small schools obtaining significant funding can add challenges to an active research program.

26.2.1 *The Role of Liberal Arts College in MST Education: Roanoke College*

Helping students develop into mature, responsible, knowledgeable individuals is the goal of any educational institution. Those teaching at undergraduate institutions such as Roanoke College (2,000 students total) with smaller numbers of students have the unique position of being able to closely mentor prospective scientists at an

Fig. 26.1 Front entrance to Roanoke College, a liberal arts college with approximately 2,000 students in southwest Virginia where faculty are expected to remain active in teaching, research, and service to the college

early stage in the student's career. These students take a full course load plus work in research labs and are often their mentor's academic advisee as well. Mentors are able to work closely with students and can help the student develop his or her pursuit of graduate work, professional school, or find employment. Therefore, Student Learning and Exposure to Research is a critical stage in the student's development as a prospective member of the scientific community and can be broken into four subsections: (1) addressing how students learn and their classroom experience, (2) mentoring and the research lab experience, (3) guidance on how to present scientific information, and (4) determining how the research impacts the student's future and the local community.

In the undergraduate liberal arts setting, a wide variety of students take the general microbiology course, and an even wider cross-section of students are science majors. These students often think they know what they want to do as freshmen (many hands go up with the answer of "doctor" in the typical introductory science class), but in reality the students often develop their interests over time in the college setting. Additionally, nonmajors are taught topic-focused science courses at most liberal arts institutions today, allowing instructors to incorporate fairly advanced microbiological information for the students who will not develop careers in science.

26.2.1.1 The Microbiology Classroom

General microbiology at undergraduate institutions has taken on a greater role over the past several years because many graduate and professional programs (i.e., pharmacy, dental) either require or highly recommend general microbiology for admission. Additionally, greater numbers of students today are interested in environmental science, and therefore, most will elect to take microbiology. Typically, the course material, therefore, must encompass a broad set of topics, including both clinical and environmental settings. The labs for a course such as this need to be wide ranging as well and should include methods and concepts that may be needed in a future job, or graduate or professional school.

At Roanoke College, a series of labs that teach basic microbiological knowledge and techniques are employed that focus on both clinical and environmental topics. Students enumerate microorganisms in soil (both bacterial and fungal), use selective media such as Hektoen Enteric agar and m-Enterococcus agar to differentiate among potentially harmful bacteria on food or in water, determine the growth curve of *Escherichia coli*, explore GenBank, the use of BLAST searching and primer design, identify an unknown, and explore the growth of wheat plants in the presence of (potentially) plant-growth-promoting rhizobacteria (PGPRs). While a complex concept at times for the student, the techniques used and purpose of a given lab are intentionally discussed in lab and by the student in the lab report to offer the student the larger outcome of knowing something in that area of microbiology, such as the clinical importance of identifying an unknown or using selective and differential media.

The students at this level design their own experiments that they perform under supervision and then present at the end of the course; the students essentially first learn how the science of microbiology is conducted and then do it for themselves. The lab reports are in the format of Applied and Environmental Microbiology (AEM), thus teaching them the important skill of following a professional scientific format and reporting data (albeit at an introductory level). In this way, lecture material introduces the topic of environmental microbiology, and it is reinforced in the laboratory setting. Independent projects in the lab have included water-quality testing in some form for those interested in this facet of microbiology. Something as simple as membrane filtration with a focus on fecal bacteria in a local watershed makes a fine first-year microbiology project in which students not only report their findings but must also conduct a literature search for the introduction. The student is, at the conclusion of the course, required to present the full project in oral form. In a sense, students learn and conduct a simple version of what professional scientists do regularly. Those students who find that they have a strong interest in water quality after this classroom experience then seek out opportunities to conduct projects in a research setting, an ideal situation to incorporate MST into their research. At large universities with hundreds of students in one class, opportunities for giving oral presentations or having an independent-project aspect to the microbiology curriculum is limited or logistically impossible.

Also at Roanoke College, a series of general curriculum topic-focused courses are taught to nonmajors. A new course has been recently developed, which is essentially soil and environmental microbiology for nonscience majors. While the lecture focuses on soil-organismal biology and chemistry, the lab component is very much a "how does science work" type of curriculum where students collect and analyze data, primarily from an environmental perspective. By the end of the course nonmajors will have learned how scientists observe, make hypotheses, design experiments, collect data, and communicate that information in written form. While the labs require the students to collect many types of data, students are required to collect water and perform membrane filtration to count fecal bacterial colonies and make a statement about the quality of the water. Bringing applied microbiological science to majors and nonmajors alike is not difficult, and greatly enhances these students' understanding of how such science is conducted and what the impact of the results may be. Whether the students end up in a graduate science program or something as different as law school, they have been exposed to how science works and should be able to better interpret the impact of scientific information.

As a resource for undergraduate educators, ASMCUE (American Society for Microbiology Conference for Undergraduate Educators) offers a venue for undergraduate educators to meet and discuss microbiology education. The conference includes everything from new protocols to assessment, and speakers to poster sessions. Additionally, it offers a resource for obtaining peer-reviewed education research in JMBE (the *Journal of Microbiology and Biology Education*). An example of how this venue can help the undergraduate educator is a project assessing the new nonmajors soil microbiology course at Roanoke College (Crozier et al. 2010) in which pre-post testing as well as a Rikert scale was used to assess how well the students learned microbiology concepts. Additionally, ASM's MicrobeLibrary offers peer-reviewed protocols, images, and other resources for undergraduate educators to use in microbiology courses (http://www.microbelibrary.org/index.asp).

26.2.1.2 The Research Laboratory

In today's environment, a greater emphasis has been placed on professional development at smaller teaching-focused institutions that includes research and attendance at national and international professional meetings, plus obtaining extramural funding. Faculty at smaller schools are realizing that collaboration with colleagues at larger institutions is not only desirable but also necessary to successfully compete for grant funding and be competitive when submitting manuscripts. The "grants and publications" expectations for tenure have been slowly increasing. This is not all bad, however, in that the faculty at smaller schools have just as much a responsibility to educate and prepare students for a career or higher education as faculty at universities, and therefore need to be current and productive in their field. The application pools for employment and admittance to graduate or professional school are filled with students from both large and small institutions. At small institutions, there are both factors that enhance the student's education and challenges that are sometimes difficult to overcome.

While undergraduates at large universities may have a chance to get into a research lab in their junior or senior year, the students at Roanoke College have a very good chance to conduct an ongoing research project in a faculty research lab even as a freshman (Fig. 26.2). The students learn quickly how to perform DNA extractions, conduct PCR, and run gels, for example, at a very early stage in their college career. Typically, the students at this level are very capable and learn microbiological and molecular methods quite easily. Performing PCR, gel electrophoresis, DNA extractions, prepping samples for sequencing, membrane filtration, preparing plates and plating cultures are all procedures easily learned and performed by interested and motivated undergraduates. In some cases, theory and design of experiments may be achieved, including primer design and alignment of DNA sequences. All of this competes with a rigorous undergraduate science major course load, but dedicated students typically handle these challenges and enjoy the experience and opportunity. An important aspect of the teaching of undergraduate research students is the "turning" of a student who may be wandering academically. Some students significantly rejuvenate a stagnating education by working in a hands-on problem-solving environment. While students with stellar GPAs will likely perform well in a lab anyway, good students with a slightly lower GPA often have the chance to really see what they can do; these students often realize that graduate school or work in industry is absolutely within reach.

Roanoke College has a program for incoming freshmen called URAP (Undergraduate Research Assistant Program), a competitive program for qualified incoming students. One perk is that the student is paid each semester for working in the faculty mentor's laboratory. Faculty write a proposal for funding to the

Fig. 26.2 Students working on an MST project in Dr. Crozier's lab at Roanoke College. Shown from back to front are Dr. Crozier and students Caitlin O'Callaghan (a Undergraduate Research Assistant Program (URAP) student) and Marina Salama, both of whom started in the lab as freshmen and went on to medical school

College's Faculty Development Committee (FDC). In turn, interested students apply to the program in the summer prior to entry to Roanoke College, and interested students and faculty communicate their mutual research interests, at which point the student may be accepted into the lab. Another program offered at Roanoke College is the Summer Scholars program, a competitive research program in which a student-written proposal is sent to the FDC for review. Successful students are paid for a 10-week period over the summer, and a final presentation is required. Typically, this has been a fantastic program for attracting quality students into a research setting in which a final product such as a poster or oral presentation is required. While currently faculty members are not financially assisted to have a URAP student in the laboratory, the faculty member does receive compensation for working with a Summer Scholar student over the summer, and faculty receive some compensation for having a research student in the lab if they are registered to receive academic credit for that experience.

The relationship between student and research mentor at a small school is one of the most rewarding teaching experiences a faculty member can have, and that is exactly what it is: teaching. Students typically enter a faculty member's lab because they have an interest in enhancing their resume with methods and knowledge they may need for future employment or admission to graduate or professional schools. While at the faculty member's discretion students are often admitted to the lab on a trial basis to determine whether they are genuinely interested in the research being conducted, and if they have the aptitude for such work (unless they have come in through the URAP), one advantage for the small-school faculty member is the opportunity to observe student's habits and demeanor prior to inviting them into the lab, or being asked by the student if they may join the research team. Faculty members at small schools often meet, have as an academic advisee, or have in class many of the prospective majors in the pool of potential research students. Faculty members, therefore, often already know students who show interest and can be told beforehand if there is a potential position in the lab. It is best if faculty have a strategy both for continuation or termination of the student's time in the lab. Working with colleagues is a good way to standardize the "acceptance" and "release" of students from the lab. One option is to have a formal contract with each student to avoid awkward situations. This is a learning experience for the student, not only about the research project but also how to interact with others and "hold" a job.

Faculty at undergraduate institutions should be well read and aware of the level of research that graduate institutions are performing. The research conducted at the undergraduate institution, to be competitive, must be at the level of quality of that being conducted at a university with publication an important outcome. While the experience is really first an educational one for the undergraduate, there is a challenge in attempting to bring the level of research to that expected for funding and publication (Sect. 26.4 offers suggestions for sources of funding). Equipment needs, budgets, and time are all barriers to higher levels of productive output at the undergraduate institution. Faculty members at undergraduate institutions are usually 9-month employees; however, productive faculty must work essentially full-time (12 months) in the research lab to be competitive. Research output can be reduced significantly during the academic semesters but must continue to be productive. While some internal grant funding is likely available, or in some cases grants exclusively for smaller institutions, the high cost of specialty equipment is a barrier, and the time to write the necessary grant proposals is fleeting. The Council on Undergraduate Research (CUR, http://www.cur.org) is an excellent resource for those seeking to apply for funding. Additionally, at the undergraduate institution faculty are training undergraduates, not graduates with already established greater background and understanding. Good students, therefore, significantly aid in the research program by working continually in one lab during the academic year and, if possible, over the summer as a Summer Scholar. Budgets are usually too small for what is required to conduct research at the level needed for peer-reviewed publication, requiring extramural funding and collaboration. However, multiple projects over time, if carefully planned, can result in a publication.

26.2.1.3 Student Presentation

While highly variable, college funds might be available for student travel to conferences, and ASM and other organizations offer travel scholarships. While the costs to attend the ASM General Meeting, for example, are significant, the experience for serious undergraduates is worthwhile. Regional or local meetings are more affordable and likely would offer a meaningful and educational experience.

In some cases, conferences are designed specifically for undergraduates; the MARCUS (Mid-Atlantic Regional Conference for Undergraduate Scholarship), for example, is a format where undergraduates can meet with peers, share ideas, and develop new projects. Because undergraduate research is often the faculty member's project, work presented at such a conference would need to be in theory a side project that is unique enough to stand on its own such that the faculty member can publish it, but as projects continue often work can be presented as it develops. Students could also present at a regional ASM branch meeting as well. Committed students who have helped design experiments and conduct a significant level of the work should be included as authors on publications, giving them a sense of accomplishment as well as very useful entries for their résumé.

As an educational experience, teaching undergraduates how to communicate information is a significant role of the faculty advisor. Students in the research lab are expected to develop posters of the team's research with the goal of attending a conference, or in some cases the work may be written up even if it is not yet at a publishable stage. Teaching students how to use appropriate software to produce a professional looking document teaches organizational skills as well as the skills associated with searching literature, and writing and communicating findings in a logical and meaningful manner. Posters presented at college functions and displayed prominently on campus not only allow students to present their ongoing work but also give the student the sense that their efforts are meaningful and appreciated. Additional venues such as a college or departmental publication or the college or departmental Web page are also possibilities for showcasing work.

In some cases, students receive various levels of academic credit for their work, and so the level of the presentations can vary. The students at Roanoke College can receive academic credit and "honors in the major" or the project might be part of the College's Honors Program. In either case, the work is presented in written and oral form to a committee much as would be found in a graduate program. Faculty members on the committee may offer assistance during the research process and will evaluate both the written work and oral presentation. Committee members' signatures are required for completion of the academic credit. These presentations could then be presented orally at a conference or as a poster.

26.2.1.4 Local Research and Examples

If one is collaborating with a colleague at a large university like Virginia Tech, for example, field aspects of any MST project might be in a locality far from the undergraduate institution, often in a completely different environment. Bringing the research project to the local community is an important and rewarding component for the student. Both student and faculty mentor additionally gain contacts with local and regional personnel. MST research has introduced the students from Roanoke College to internship and job opportunities: Novozymes (an international biotechnology company), the Department of Environmental Quality

(DEQ), and the United States Geological Survey (USGS), for example. One student in particular was selected to work with the USGS locally on a project dealing with the Roanoke River in which he was responsible for monitoring equipment, eventually leading to employment after graduation. The student had a much greater sense of the importance of the MST research he was conducting in the lab and ended up with a good contact that led to his employment.

MST research at Roanoke has evolved right along with the field. In the late 1990s and early part of the twenty-first century, work was focused on ARA and the first glimpses of molecular MST methods. One paper was published on ARA in the Roanoke River (Crozier et al. 2002) where students significantly participated on the project in our local watershed. One of these students then became employed in industry in a field of microbiological research. The students have also been a part of other projects and collaborations in which they were part of work that led to publication. For example, undergraduate students helped with our understanding of the use of Biolog to track *Enterococcus* (Hagedorn et al. 2003) in which the students assisted with collecting isolates and conducted other field and lab work.

More recently, students have worked to help develop and test molecular MST methods. Students have quickly learned how to perform PCR, restriction digests, electrophoresis, design primers, and learned about DNA sequencing, alignments, and BLAST searches on the computer. One project in which students significantly helped dealt with amplifying the ITS region in *Enterococcus* as a MST tool (Dickerson et al. 2007). In brief, PCR was used to amplify DNA from the ITS regions in *Enterococcus*, and restriction digests then generated DNA fingerprints to separate the putative animal source. Students helped with much of the initial phase of this project in running PCR and looking for appropriate restriction enzymes to digest the amplified ITS region.

As the lab has migrated to searching for specific markers and virulence genes in water, a paper on human-source Bacteroidetes in canine fecal matter will soon be submitted. Students have significantly assisted recently in a collaboration with a university lab in the sampling and testing of field samples for the presence of human-source Bacteroidetes and other molecular fecal and pathogen markers (see 26.4.2 for an example protocol). Roanoke College students have been integrally involved in working collaboratively with the university's lab personnel and graduate students in the development of the detection procedures. Not only the students have been able to interact and collaborate with university faculty, staff, and students but the scope of the research has also been local to distant-regional dealing with both fresh- and saltwater environments and multiple indicators. While not all the students who have conducted MST research at Roanoke have ended up in industry or with a water-related job, most have ended up in equally impressive positions, medical and pharmacy school for example. Those that do end up in a water-quality related position from a program such as this potentially assist all who are committed to clean safe water on both local and global scales.

26.2.2 The Role of Community Colleges in MST Education: El Paso Community College

On specific scientific skill subscales measured in PISA 2006, the average score of US students was below the Organization for Economic Cooperation and Development (OECD) average in explaining phenomena scientifically and in using scientific evidence (Baldi et al. 2007).

Numerous science education publications stress the urgency of getting students to understand and appreciate the investigative nature of science and the importance of having more students trained in the scientific method and improving their analytical skills (Alvarez 2005; National Academy of Science 2007). Part of the problem is that we have been teaching our science courses using the "cookbook" approach where students are expected to get expected results. Many students are turned off to science early in their education by this approach. One way to expose the students to the investigative and exciting nature of science is to get them to participate in investigative laboratory projects and scientific research. At El Paso Community College (EPCC) getting students involved in what we call "student-oriented research" has proven to be an effective science learning strategy. EPCC students have participated in a variety of research experiences dealing with issues that affect our USA–Mexico Border Community. With funding from the Minority Biomedical Research Support-Research Initiative for Scientific Enhancement (MBRS-RISE) from NIH, EPCC students have participated in research internships, tutoring/mentoring, supplemental instruction, and travel to present at scientific meetings. There are several student-oriented national meetings including the Annual Biomedical Research Conference for Minority Students (ABRCMS), the annual meetings of the Society for the Advancement of Chicano and Native American Scientists (SACNAS), and the Council for Undergraduate Research. Students that have no funding can apply for travel grants to participate in these meetings. Since these are student-oriented meetings, the opportunity for networking, exposure to role models (mentors), and learning about summer research internships, scholarships, and graduate programs is emphasized. Many EPCC students have been recruited at these meetings to participate in summer internships at prestigious, research-intensive institutions. The ultimate goal is to get these students to transfer to those schools for their baccalaureate and graduate degrees.

EPCC student projects dealing with water quality have been published and presented at various professional meetings including ABRCMS, SACNAS, and ASM national and branch meetings (Bland et al. 2005; Mendoza et al. 2004; Ryou et al. 2005). The accomplishments of EPCC RISE student participants are summarized in (Fig. 26.3).

When funds are not available to get students into paid research internships, it is possible to expose students to the scientific method through investigative learning projects in the classroom. Investigative learning is a new approach that exposes the students to the scientific method by allowing them to ask their own scientific questions, design their own experiments, collect and analyze data, and present their

- The retention of RISE participants was 98%, which exceeds the 64% value for non-RISE Biology and Chemistry students
- The transfer rate of RISE students was 61% (in six years) , which is higher than the 14% value for non-RISE Biology and Chemistry students
- 160 Students participated in research activities
- 89 presentations at 32national and regional scientific meetings
- 12 RISE students received awards
- RISE student cohort received 40 BS, 5 MS, and 2 Ph degrees in an eight-year period.
- 10 Summer research internships at NIH, Baylor, Princeton, the Jackson Laboratories, University of MD, Univ. of Wisconsin and an REU International in Cadiz Spain

Fig. 26.3 Accomplishments of El Paso Community College (EPCC) Research Initiative for Scientific Enhancement (RISE) Students

results to an audience. Having students exposed to investigative learning exercises using projects that have environmental relevance to their local communities is the ideal way to get students excited about science and research. At EPCC, the students enrolled in General Microbiology and Research Techniques in Biology (courses for majors) have conducted research on a variety of projects dealing with water quality (Figs. 26.4 and 26.5). Since the Rio Grande River serves as the natural boundary between USA and Mexico and there are differences in the regulation of waste disposal between the two countries, MST is extremely important in the area. In conjunction with the International Boundary and Water Commission Clean Rivers Program (IBWC-CRP), EPCC students are involved in the collection and analysis of water samples from the Rio Grande River. The students have been exposed to a variety of techniques including isolation, quantification, and identification of fecal coliforms and *E. coli,* as well as F-RNA and somatic coliphage analyses, antibiotic resistance analysis, ribotyping, and detection of optical brighteners. Most of the equipment necessary for these projects has been purchased with funding from NSF Research Instrumentation programs, NIH Supplements to the MBRS-RISE grant and the IBWC-CRP (Fig. 26.6). EPCC RISE students and the students enrolled in the Research Techniques and General Microbiology courses are required to formulate a hypothesis, design and conduct the experiments, collect and analyze the data, and present their research. The students are required to participate in seminars on Friday afternoons. During Friday seminars, the students present their research goals at the beginning of the semester and give a full presentation of their research project at the end of the semester. EPCC students that have transferred to the university are hired as tutors/mentors through the RISE program and also participate in Friday meetings. The peer mentoring that develops as a result of the relaxed and informal interaction during these meetings has been

Fig. 26.4 EPCC Student Violeta Chavez collecting samples at the Rio Grande River. Dr. Violeta Chavez is now a postdoctoral fellow at Columbia University in New York

Fig. 26.5 EPCC students Jamie Ireland and Chris Bland presenting a poster at a national meeting. They graduated with a PhD from Washington University in St. Louis and Baylor College of Medicine, respectively.

extremely beneficial to the RISE students. One hundred and twenty students have participated in independent projects dealing with water quality at EPCC. One hundred percent of the student participants received a score of 90 or above in their project presentations. Over 50% of the students have been transferred to the university and are pursuing a science career.

Participation of community college faculty at national professional association meetings is also extremely important. It is through participation at these meetings that community college faculty can get updated in their fields and establish productive

Fig. 26.6 The Rio Grande River at Fort Hancock, TX showing the presence of chemical contaminants. The river receives partially treated or untreated effluents from Mexico at this site

collaborations with faculty at research-intensive institutions. Many of these national meetings provide workshops on research techniques as well as grant-writing and program funding information. At EPCC, 11 science faculty members have participated in grant-writing training activities through the Extramural Associates Program at NIH (http://www.nichd.nih.gov/about/org/dsp/ea/eap.cfm). All faculty participants have participated in the writing of individual or collaborative grant applications to secure funding for research training.

26.3 Peer Support and Networking

Mentoring undergraduate researchers at a small undergraduate institution is a highly rewarding experience but is a difficult one as well. Time spent preparing for teaching, grading, and working with advisees and time spent on committees unrelated to one's field of expertise, including travel to and attendance at off-campus meetings related to matters unrelated to teaching or research, take its toll. All these detract from the time spent writing research grant proposals or time at the bench in the research lab. Collaboration and networking is critical if one is serious about staying active in MST research. Judicious use of a good grants officer at a small college is useful if available, and collaboration with colleagues at larger institutions on grant proposals is often a must to financially stay active. Getting funding to hire more advanced students to serve as tutors and peer-mentors of the new students in the laboratory is essential and has proven to be a very successful strategy at EPCC. Student-oriented meetings including ABRCMS and SACNAS, and the national ASM meeting, hold special sessions where students have the opportunity to network with students and faculty from other colleges and universities. In addition, professional organizations including ASM and AAAS provide online mentoring. For example, ASM offers three different online tools to find ASM members that have volunteered to act as mentors: (1) The Science Educators Network (microbiologists available to work with, advise, and mentor middle- and high-school students and teachers), (2) International Mentors (ASM members worldwide interested in mentoring undergraduate students), and (3) Minority Mentor Program (ASM members interested in mentoring minority microbiologists at all levels in their careers). The American Association for the Advancement of Science's "MySciNet" is a place for scientists and students from diverse backgrounds to network and build the personal and professional connections needed to succeed in the sciences. Another networking site is JustGarciaHill, a virtual community for minorities in science (http://www.justgarciahill.org/). Additionally, the CUR is an excellent resource for faculty at undergraduate institutions looking for ideas on how to run an undergraduate research laboratory (http://www.cur.org) promoting "high-quality undergraduate student–faculty collaborative research and scholarship."

26.4 Example MST Protocols for Undergraduates

Provided here are (1) an ARA protocol and (2) a protocol for the molecular assessment of the source of fecal contamination from organisms in the Bacteroidetes (Bacteroidales or *Bacteroides–Prevotella* group) ribosomal DNA (Bernhard and Field 2000a, b). While protocols are found throughout this book, the protocols presented here are for undergraduates and teach specific research methodologies. Students can easily transform these skills into real-world MST projects. Not every detail is given here; it is suggested one may gain further information by going to the appropriate chapter in this book or contacting the corresponding author.

26.4.1 Using Antibiotic-Resistance Analysis to Expose Students to Environmental Research

This laboratory exercise conducted at EPCC exposes students to experimental design, environmental sample collection procedures, antibiotic-resistance analysis techniques, data collection and analysis by statistical methods, scientific writing, and presentation to a scientific audience. Cultures of microorganisms isolated from environmental samples should be handled carefully. Students must be trained in aseptic techniques and microbial isolation and culture procedures before conducting these experiments.

Standard microbiology laboratory equipment and materials include inoculating loops, Bunsen burners, disposable plates and gloves, polyethylene bottles. 0.45 μm filters and filtration apparatus, 48-prong replica platers, 96-microwell trays, ultraviolet lamp, biosafety cabinet, and autoclave are required. Trypticase Soy Agar (TSA), m-TEC agar, Nutrient Agar MUG (4-methylumbelliferyl-β-D-glucoronide), and Colilert™ broth are also required. The following antibiotics have been used but can be modified to optimize source identification of isolates: amoxicillin (10, 45, 75, and 85 μg/mL), chloramphenicol (5, 30, 45, and 80 μg/mL), chlortetracycline (10, 45, 60, and 90 μg/mL), erythromycin (5, 35, 60, and 85 μg/mL), neomycin sulfate (25, 55, 70 and 80 μg/mL), oxytetracycline hydrochloride (25, 65, 95, and 110 μg/mL), streptomycin sulfate (10, 30, 75, and 95 μg/mL), tetracycline (5, 25, 45, and 85 μg/mL), and vancomycin (10, 25, 40, and 55 μg/mL). Growth in all four concentrations of vancomycin should be observed and used as a positive control. JMP software from SAS is used for discriminant analysis of data. Additionally, a library of known patterns from known-animal-source fecal matter must be created and maintained for accurate results.

In a classroom–laboratory setting, students will work in groups of four. The instructor can assign the surface water sites to be sampled, or alternatively students can collect their own samples from areas near their homes. Sampling can be conducted over several weeks to determine any temporal variations. Students will survey the area for possible sources of contamination and use the information to develop a hypothesis for the experiment. All groups should use the same microbial library provided by the instructor. Students often conduct these experiments as part of the class or as an independent honors project.

Students in a classroom setting receive a pretest to determine their knowledge of the scientific method, environmental sampling techniques, experimental design, data collection and analysis and a posttest after conducting the experiments and presenting them to the class. Students are required to present their results to the class as a PowerPoint poster presentation.

Protocol. The ARA protocol used at EPCC is based on Hagedorn, personal communication (2001), Parveen et al. (1997), and Graves et al. (2002). While ARA methods are available elsewhere in this book, the basic protocol used by undergraduate students is outlined here. A 250-mL grab sample is obtained from the sampling sites as needed and placed on ice for transport to the laboratory. *E. coli* isolates are obtained from m-TEC agar and confirmed using MUG. Individual isolates are placed into a sterile 96-well plate and a 48-prong replica-plater is used to replicate isolates from the microwells onto individual TSA plates containing the nine different antibiotics at four different concentrations as described above. After 24 h of incubation at 37°C, the antibiotic-resistance patterns (ARPs) are determined by recording of the growth/no growth of isolates as 1's and 0's, respectively. Discriminant analysis is used to find the closest match of ARPs between the known library and unknown samples using the JMP software from SAS. The average rate of correct classification (ARCC) is the number of library isolates assigned to the correct source group when the library is queried using "hold-out" or Jackknife analyses. Students must remove known isolates from the database and run them as unknowns against the database. The degree of similarity of the removed isolate to those remaining in each source group is determined, and the ARCC is calculated. Libraries containing fewer entries cannot capture the genetic diversity of the population and will have lower ARCC values.

26.4.2 DNA Extraction, Amplification, and Visualization to Detect Human-Specific Bacteroidetes in Fecal Matter and Water

This library-independent MST protocol was developed by J. Brooks Crozier and an undergraduate student Ms Katie Scott at Roanoke College from protocols in the literature (Bernhard and Field 2000a, b) and has been adapted for ease of student use. While the costs of implementing such a project is higher than for ARA, this type of work is becoming more accessible as costs of this equipment and the supplies has declined over time. A thermalcycler, gel apparatus, PCR mix components, DNA stain, and a DNA visualization setup are required for this protocol, along with standard lab equipment such as pipettors, tips, microcentrifuge tubes, and agarose. While DNA can be extracted in many ways, a DNA extraction kit is suggested to obtain quality clean DNA and for consistency from sample to sample. Additionally, brand names are given in that it is established that they work; however, multiple producers of these components likely have products that would work equally well. This is an easily performed library-independent molecular biology protocol that students can learn quickly, and the results are positive or negative, meaning that there is little interpretation needed if controls were used.

Protocol. Samples are collected aseptically (fresh fecal matter or water sample); 200 mg or 200 μL solid or concentrated liquid waste from known sources, OR 1 mL influent/effluent placed into 100 mL sterile water for filtration procedure (dilutes highly concentrated samples for filtration), OR 100 mL environmental sample for filtration. Extract DNA using a Qiagen QIAamp® DNA Stool Mini Kit following the "Isolation of DNA from Stool for Pathogen Detection" protocol. For DNA isolation from filters (0.45 μm), the water sample is filtered and then the filter is aseptically placed into a sterile 15-mL Falcon tube. The first buffer (ASL, 1.4 mL) is added to the tube and vortexed to remove bacteria from filter. The sample may be refrigerated as the DNA will be protected at this stage, and is a good time-saver for students with busy schedules. The whole tube is incubated at 70°C for 5 min (6 min if refrigerated). Liquid from this tube is then placed into a fresh sterile 2-mL tube for the InhibitEX tablet step. Again, following Qiagen's DNA extraction protocol, set up PCR reactions in 200-μL PCR tubes using Promega PCR Master Mix (Table 26.1). The name of the PCR mix is "Promega PCR Master Mix."

Table 26.1 Components of PCR mixture

Components	For single tube (μL)	"Master" mix (μL)
Promega PCR Master Mix	12.5	125
BSA[a]	1	10
Forward primer	1	10
Reverse primer	1	10
Nuclease-free water	7	70

The "master" mix (meaning a complete mix except for template DNA) is made to supply all PCR reagents except DNA template for ten reactions. 22.5 μL "master" mix and 2.5 μL template DNA are combined to bring reaction mixes to 25 μL total
[a]BSA is at a final concentration of 0.1 μM in the 25 μL reaction mix

Table 26.2 PCR protocol for the two primer sets

Bac32F/Bac708		HF183/Bac708		
T°C	Time (min)	*T*°C	Time	Cycles
95	5	95	5 min	1
94	1	94	30 s	35
53	1	59	1 min	
72	1.5	72	2 min	
72	5	72	10 min	1

Table 26.3 Oligonucleotide sequences for the two primer sets and expected product size

Primers	Oligonucleotide Sequence (5′–3′)	bp
Bac32F	AAC GCT AGC TAC AGC CTT	700
Bac708	CAA TCG GAG TTC TTC GTG	
HF183	ATC ATG AGT TCA CAT GTC CG	520
Bac708	CAA TCG GAG TTC TTC GTG	

It is recommended that one make a "master mix" to perform PCR, meaning all of the components except template are combined to decrease the chance of single-tube errors. The term "master," therefore, is used in two different contexts in Table 26.1. Set up the thermalcycler to run a PCR protocol as instructed below with primers with the appropriate sequence (Tables 26.2 and 26.3). Use the Bac32 forward primer and Bac708 reverse primer to detect Bacteroidetes bacteria in general (used as a control to see if Bacteroidetes DNA is in the sample). Use the HF183 forward primer and Bac708 reverse primer to detect human-specific Bacteroidetes. Make a 1% agarose gel using cold 1× TAE. Run the gel at 120 V for 45 min in cold 1× TAE (for a 10 cm gel length). Include a DNA ladder to confirm size of any products noted on the gel. Stain the gel for 20 min using 10,000× SYBR green at a concentration of 1 μL to 10 mL 1× TAE and visualize stained DNA under UV light.

A single band at the correct position on the gel is desired for known-source samples. A fuzzy band under 100 bp is primer dimer and should not be interpreted as a positive result. A known-source sample should always be included so that one can see if PCR worked and if the band on the gel is at the correct position.

26.5 Conclusion

Faculty at undergraduate institutions, both liberal arts and community college alike, are teaching a greater number of students by introducing them to research at a stage or to methods of instruction often not practical at larger universities and, therefore, are able help students earlier with critical thinking and problem-solving skills. With exposure to hands-on opportunities at an early stage, these experiences potentially help the wavering or ambitious student alike pursue a profession or program they may not have had the opportunity to pursue if at a larger institution. Research may keep both the high-achieving and the nontraditional students engaged and contribute to higher retention. Students recruited into a lab are more likely to decide to attend a particular institution or remain in a program of study or at the institution as well, and contribute positively to the program. It is likely then that a greater number of students emerging from these programs will successfully compete with those cosubmitting applications to professional or graduate schools, or to the workplace.

References

Alvarez ME (2005) Visualization techniques in the new biology. In: Lynn Arthur Steen (Ed.) Math & Bio 2010. Linking Undergraduate Disciplines 35–43. The Mathematical Association of America, Washington, DC.

Baldi SJ, Skewer YM, Green PJ, et al (2007) Highlights From PISA 2006: Performance of U.S. 15-year-old students in science and mathematics literacy in an international context (NCES 2008), data from the OECD program for international student assessment (PISA), 2006.

Bernhard AE and Field, KG (2000a) Identification of nonpoint sources of fecal pollution in coastal waters by using host-specific 16S ribosomal DNA genetic markers from fecal anaerobes. Appl. Environ. Microbiol. 66:1587–1594.

Bernhard AE and Field, KG (2000b) A PCR assay to discriminate human and ruminant feces on the basis of host differences in *Bacteroides-Prevotella* genes encoding 16S rRNA. Appl. Environ. Microbiol. 66:4571–4574.

Bland CS, Ireland JM, Lozano M, et al (2005) Mycobacterial ecology of the Rio Grande. Appl. Environ. Microbiol. 71:5719–5727.

Crozier JB, Watkinson JI, Filer KL (2010) Student learning in an undergraduate non-majors soil microbiology course in a new general education curriculum. JMBE. 11:74–75.

Crozier JB, Clark B, Weber H (2002) Identifying sources of fecal pollution in the Roanoke River, Roanoke County, Virginia. Virginia Journal of Science. 53:157–165.

Dickerson JW Jr., Crozier JB, Hagedorn C et al (2007) Assessment of the 16S-23S rDNA intergenic spacer region in *Enterococcus* spp. for microbial source tracking. J. Environ. Qual. 36:1661–1669.

Graves AK, Hagedorn C, Teetor A et al (2002) Antibiotic resistance profiles to determine sources of fecal contamination in a rural Virginia watershed. J. Environ. Qual. 31:1300–1308.

Hagedorn C, Crozier JB, Mentz KA, et al (2003) Carbon source utilization profiles as a method to identify sources of fecal pollution in water. Journal of Applied Microbiology. 94(5):792–799.

Mendoza J, Botsford A, Vazquez R. et al (2004) Microbial contamination and chemical toxicity of the Rio Grande River. BMC Microbiology. 4:17

National Academy of Sciences, National Academy of Engineering, Institute of Medicine (2007) Rising above the gathering storm: energizing and employing America for a brighter economic future. Washington, DC: National Academies Press online: http://www.nap.edu/catalog/11463.html#orgs.

Parveen S, Murphree RL, Edminston L et al (1997) Association of multiple antibiotic resistance profiles with point and non-point sources of *Escherichia coli* in Apalachicola Bay. Appl. Environ. Microbiol. 63:2607–2612.

Ryou H, Alum H, Alvarez M, et al (2005) An assessment of water quality and microbial risk in Rio Grande basin in the United States-Mexican border region. Journal of Water and Health. 3.2:209–218.

Index

C. Hagedorn et al. (eds.), *Microbial Source Tracking: Methods, Applications, and Case Studies*, DOI 10.1007/978-1-4419-9386-1,

Q

R

S

MIX
Papier aus verantwortungsvollen Quellen
Paper from responsible sources
FSC® C105338

If you have any concerns about our products,
you can contact us on
ProductSafety@springernature.com

In case Publisher is established outside the EU,
the EU authorized representative is:
Springer Nature Customer Service Center GmbH
Europaplatz 3, 69115 Heidelberg, Germany

Printed by Libri Plureos GmbH
in Hamburg, Germany